Preliminary Edition

INTRODUCTION TO
STATISTICAL
INVESTIGATIONS

Nathan Tintle | **Beth L. Chance** | **George W. Cobb** | **Allan J. Rossman**
Soma Roy | **Todd Swanson** | **Jill VanderStoep**

WILEY

Publisher	Laurie Rosatone
Acquisitions Editor	Joanna Dingle
Associate Editor	Jacqueline Sinacori
Freelance Project Editor	Anne Scanlan-Rohrer
Editorial Program Assistant	Courtney Welsh
Director of Marketing	Jeffrey Rucker
Marketing Assistant	Jesse Adler
Design Director	Harry Nolan
Product Designer	David Dietz
Cover Designer	Wendy Lai
Senior Content Manager	Karoline Luciano
Production Editor	Sandra Dumas
Editorial Operations Manager	Yana Mermel

This book was printed and bound by Bind-Rite Robbinsville. The cover was printed by Bind-Rite Robbinsville.

Founded in 1807, John Wiley & Sons, Inc. has been a valued source of knowledge and understanding for more than 200 years, helping people around the world meet their needs and fulfill their aspirations. Our company is built on a foundation of principles that include responsibility to the communities we serve and where we live and work. In 2008, we launched a Corporate Citizenship Initiative, a global effort to address the environmental, social, economic, and ethical challenges we face in our business. Among the issues we are addressing are carbon impact, paper specifications and procurement, ethical conduct within our business and among our vendors, and community and charitable support. For more information, please visit our website: www.wiley.com/go/citizenship.

This book is printed on acid-free paper. ∞

ISBN 13   978-1118-95667-0

Printed in the United States of America.

About the Authors

Beth L. Chance is Professor of Statistics at California Polytechnic State University. She is co-author with Allan Rossman of the *Workshop Statistics* series and *Investigating Statistical Concepts, Applications, and Methods*. She has published articles on statistics education in *The American Statistician, Journal of Statistics Education*, and the *Statistics Education Research Journal*. She has also collaborated on several chapters and books aimed at enhancing teacher preparation to teach statistics, and has been involved for many years with the Advanced Placement Statistics program. She is a Fellow of the American Statistical Association and received the 2002 Waller Education Award for Excellence and Innovation in Teaching Undergraduate Statistics. The Rossman/Chance collection of online applets for exploring statistical concepts was awarded the 2009 CAUSEweb Resource of the Year Award and a 2011 MERLOT Award for Exemplary Learning Materials.

George W. Cobb is Robert L. Rooke Professor Emeritus of Statistics at Mount Holyoke College and has extensive knowledge of statistics education, expertise in developing imaginative and innovative curricular materials and the honor of having brought the conversation on randomization-based approaches to introductory statistics to the mainstream via his 2005 USCOTS presentation and 2007 paper. He served as the first chair of the Joint Committee on Undergraduate Statistics of the American Mathematical Association and American Statistical Association (1991–98) editing that committee's 1992 report, "Teaching Statistics." He served for three years on the National Research Council's Committee on Applied and Theoretical Statistics, and recently served as vice-president of the American Statistical Association. He a Fellow of the ASA and received the ASA's Founders Award in 2007. He has published/edited a number of books.

Allan J. Rossman is Professor and Chair of the Statistics Department at Cal Poly – San Luis Obispo. He earned a Ph.D. in Statistics from Carnegie Mellon University. He is co-author with Beth Chance of the *Workshop Statistics* series and *Investigating Statistical Concepts, Applications, and Methods*, both of which adopt an active learning approach to learning introductory statistics. He served as Program Chair for the 2007 Joint Statistical Meetings, as President of the International Association for Statistical Education from 2007–2009, and as Chief Reader for the Advanced Placement program in Statistics from 2009 – 2014. He is a Fellow of the American Statistical Association and was received the Mathematical Association of America's Haimo Award for Distinguished College or University Teaching of Mathematics in 2010.

Soma Roy is Assistant Professor of Statistics at California Polytechnic State University. She is an Associate editor for *Journal of Statistics Education* and has presented talks related to the randomization-based curriculum and student learning at national meetings. She writes and reviews assessment tasks for the Illustrative Mathematics Project, an initiative to support adoption of the K-12 core standards for statistics. She co-lead, with her colleagues at Cal Poly, a teacher-preparation workshop for AP Statistics teachers. She also has an active research program in health statistics involving undergraduates.

Todd Swanson is an Associate Professor of Mathematics at Hope College. He is a co-author of *Precalculus: A Study of Functions and their Applications, Understanding our Quantitative World* and *Projects for Precalculus*, which was an INPUT Award winner. He has published articles in *Journal of Statistics Education, Statistics Education Research Journal*, and *Stats: The Magazine for Students of Statistics*. He has presented at numerous national meetings, workshops, and mini-courses about innovative ways to teach mathematics and statistics that focus on guided-discovery methods and projects.

Nathan L. Tintle is an Associate Professor of Statistics at Dordt College. He has led efforts to develop and institutionalize randomization-based curricula at two institutions (Hope College 2005-2011; Dordt 2011-present), and currently leads the curriculum development project. He has been an invited panelist for a number of statistics education sessions at national meetings, was recently elected to the Executive Committee of the Section of Statistical Education of the ASA, recently received the Waller Education Award for teaching and innovation in Introductory Statistics, and serves as a member of a national advisory committee to the ASA President on training the next-generation of statisticians. He has co-authored two articles on student learning using the randomization curriculum, one of which recently won an award for best paper of the year from the Journal of Statistics Education.

Jill L. VanderStoep is an Adjunct Assistant Professor of Mathematics at Hope College. She has participated in efforts to develop and implement randomization-based curricula at Hope College since 2005. She has presented on the curriculum and assessment results at national conferences and has co-led workshops on introducing and implementing the randomization-based curriculum. She has co-authored two peer-reviewed articles looking at student learning differences between randomization-based curriculum and traditional curriculum. She has extensive experience in the evaluation of assessment data to drive curricular reform.

Preface

This book leads students to learn about the process of conducting statistical investigations from data collection, to exploring data, to statistical inference, to drawing appropriate conclusions. We focus on genuine research studies, active learning, and effective use of technology. In particular, we use simulation and randomization tests to introduce students to statistical inference, yielding a strong conceptual foundation that bridges students to theory-based inference approaches, which are presented throughout the book. This approach allows students to see the logic and scope of inference in the first chapter and to cycle through these ideas, too often lost to introductory statistics students, repeatedly throughout the course. Our implementation follows the GAISE[1] recommendations endorsed by the American Statistical Association.

Approach

We adopt several distinctive features:

1. Spiral approach to statistical process. We introduce the six-step process of conducting statistical investigations in the very first section (see Figure 1).

We introduce this process in its entirety beginning in the Preliminaries chapter. Then we present a complete implementation in Chapter 1, involving research questions focused on a process probability. This relatively simple scenario enables us to introduce students to the fundamental concept of statistical significance, along with some issues related to collecting data and drawing conclusions, early in the course.

Our goal is for students to begin to develop an understanding of important and challenging concepts such as p-value from the beginning, and then deepen their understanding as they encounter such ideas repeatedly in new scenarios:
- Single binary variable (inference for a population proportion)
- Single quantitative variable
- Comparing a binary variable between two groups (inference for 2×2 table)
- Comparing a quantitative variable between two groups
- Comparing a categorical variable across multiple groups
- Comparing a quantitative variable across multiple groups
- Association between two quantitative variables

---

http://www.amstat.org/education/gaise/

With each of these scenarios, students reconsider and apply the six-step Statistical Investigation process. Students also revisit, at deeper and deeper levels each time, the core ideas of statistical inference. They learn that the fundamental reasoning process of statistical inference remains the same in all scenarios that they study.

Figure 1. The six-step statistical investigation process

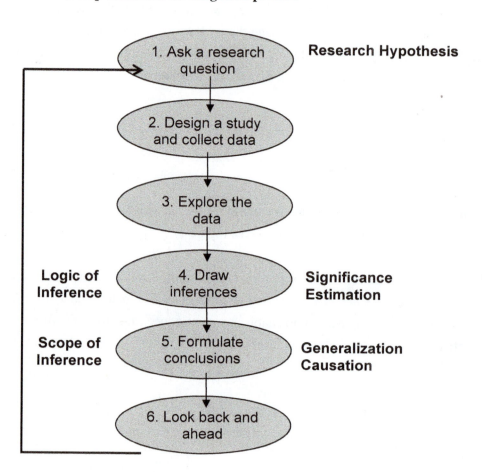

2. Randomization-based introduction to statistical inference. A randomization-based approach to statistical inference is key to its successful early introduction. For every scenario that students encounter in this book, they first learn how to make inferences using simulations of chance models. Then we introduce students to theory-based procedures for statistical inference, often based on the normal distribution and its derivatives, as an alternative approximation to the randomization-based methods. Some of the advantages for starting with the randomization-based approach, as spelled out by Cobb (2007), include:

- More intuitive for students to understand

- Easily generalizable to other situations and statistics
- Takes advantage of modern computing
- Closer to what founders of statistical inference (e.g., R. A. Fisher) envisioned

3. Focus on logic and scope of inference. For virtually every study we present, we ask students to consider two questions related to the *logic* of inference and two questions related to *scope* of inference:

- Are the study's results unlikely to have arisen by chance alone, indicating that the difference between the observed data and the hypothesized model is **statistically significant**?
- How large do you estimate the difference/effect to be, and how **confident** can you be in this estimate?
- To what group can conclusion from the study reasonably be **generalized**?
- Can a **cause/effect** conclusion be legitimately drawn between the variables? (This applies to studies involving at least two variables, starting in Chapter 4.)

The first pair of questions addresses the two key issues of statistical inference: significance and confidence. Answering the second pair depends on examining how subjects were selected for the study and how the groups were formed, involving whether random sampling or random assignment (or both) was used.

4. Integration of exposition, examples, and explorations. Every section includes at least one example that illustrates the ideas and methods presented, and at least one exploration that students work through to learn about and gain experience with applying the topic. We offer much flexibility for instructors to decide on the order in which they will present these components, and what they will ask students to do in class vs. outside of class. *To facilitate this flexibility, examples and explorations within a section are written so that neither depends on the other,* allowing the instructor to present either one first. The only exception is in the Preliminaries, where there are two examples and one exploration, each of which introduces new concepts. We make this exception to encourage instructors to *finish the Preliminaries in no more than 2-3 class periods,* while still introducing the text's flexibility for use with both lecture-based and activity-based class periods.

5. Easy-to-use technology integrated throughout. Rather than ask students to learn to use a statistical software package, we have designed easy-to-use web-applets that enable students to conduct all of the simulations and perform all of the analyses presented in this book. Instructors may also ask students to use a commercial software package, but this is not required.

6. Real data from genuine studies. We utilize real data from genuine research studies throughout the book. These studies are taken from a variety of fields of application and popular culture... Each chapter also includes a detailed investigation and a research article, giving students even more exposure to genuine applications of statistics.

Changes in Content Sequencing

Inference. This book puts inferential statistics at the heart of the curriculum. Thus, the course starts with core concepts of inference immediately in the Preliminaries and Chapter 1 and continues focusing on ideas of inference throughout. We introduce students to fundamental ideas such as statistical significance and p-values in Chapter 1. We engage students with thinking about these crucial, and challenging, issues from the very start of the course, setting the stage for revisiting these core concepts repeatedly in new settings throughout the course. With this spiraling approach we expect students to deepen their understanding of the inferential process each time it is revisited.

Descriptive statistics. We take a case-study approach that focuses on the Statistical Investigation process as a whole. Thus, descriptive statistics are integrated throughout this curriculum. The curriculum cycles through different types of data and numbers of variables in each chapter, so students are introduced to basic descriptive methods as they are necessary for analysis. By the end of the course, the content covered is very similar to a traditional course, but the content has been introduced in context through genuine applications.

Probability. Students see probability concepts in this book, but in a way that differs substantially from how probability is taught in the traditional curriculum. Specifically, we expect students to explore notions of probability through tactile and computer-based simulations. Students use chance models to obtain approximate sampling and randomization distributions of statistics. These concepts are seen throughout the curriculum, and closely tied to specific research studies, instead of covered in only one or two chapters with "probability" In the chapter title. Our approach requires no formal training in formal probability theory or rules. Initially, we choose examples where the simulation procedure is natural and intuitive to students, such as coin flipping. Later we explain how normal-based methods connect to these simulations and randomization tests. At that point, because students already understand the logic of inference, normal-based tests are presented as the long-run behavior of the simulation under certain conditions. With this approach, students can grasp normal-based tests without getting bogged down in the technical cogs of the procedures.

Changes in Pedagogy

Active learning. We have also substantially changed the pedagogical approach from passive (e.g. listening to lectures) to active learning, which engages the full range of students' senses. Each chapter contains a number of explorations for the students to complete, in addition to example-driven exposition of concepts. These materials allow for a variety of instructor-determined approaches to content delivery including approaches where examples/concepts are presented first by the instructor, then explored by the student or vice versa.

Explorations. Student explorations involve a variety of tactile learning experiences like shuffling decks of cards and flipping coins to estimate their own p-values, using computer based simulations, using web-applets, collecting data, running experiments, and (potentially) using computer software to help interpret results. The majority of explorations are flexibly designed to be completed by students working individually, in small or large groups, either inside or outside of class.

Examples and FAQs. Concepts are introduced using compelling examples explained in an easy-to-understand format that limits technical jargon and focuses on conceptual understanding. In addition to this, we have included Frequently Asked Question "dialogues" that help students understand difficult concepts and answer some of their common questions. We have also included Key Idea boxes and Thought Questions to help students understand what they read, identify core concepts, and be engaged readers. Overall, we advocate utilizing a small amount of instructor-led interactive lecture and discussion, but mainly focusing on engaging and strengthening different student learning processes by way of a variety of active, self-discovery learning experiences for students.

Exercises and Investigations. Each chapter contains an extensive set of exercises. Almost all of these are based on real studies and real data. We also include an investigation in each chapter, an in-depth exercise exploring the entire 6-step statistical process so that the single assignment can assess a variety of concepts. Each chapter also challenges students to develop their critical reading skills by including a research article for students to read followed by a series of questions about the article.

Real data. The GAISE recommendations argue that statistics courses should make use of real data. We go a step further and argue that statistics courses should use real data *that matters*. Statistics should be viewed less as a course in which students see "cute" but impractical illustrations of statistics in use, and more about examples where statistics is used to make decisions that have health, monetary, or other implications impacting hundreds, thousands, or millions of people. Our approach has two-fold benefits, first in improving students' statistical literacy and second by helping students to recognize that statistics is the indispensable, inter-disciplinary language of scientific research.

To The Student

We know from decades of experience as teachers of statistics that many students never master the most important but hardest ideas of our subject in their first course. Partly, that's just because the ideas are truly hard. Partly it's because learning the formulas of statistics often gets in the way of learning the ideas of statistics. And partly it's because the hardest and most important ideas are too often saved for the end of the course, when time is running short.

This intro stat book is different. We show you the most important ideas up front, even though we know they are hard. We downplay formulas, especially at the start, in order to put the ideas first. This approach asks more of you up front, but we have become convinced from our own classes that, long run, this approach will pay off for you, the reader. Students leave our classes better prepared to use statistical thinking in their science, social science, and business courses, and in their careers after graduation.

At the same time, we also recognize that this approach may put you in an uncomfortable position. We are asking you at the beginning of the term to start working at understanding ideas that may take several weeks of thinking, effort, and practice to become clear. Many of the most important ideas in all subjects are like that. What we ask of you is continued effort and patience. In return, we offer our understanding that some of the goals we have set for you cannot be achieved in just a week or two or three.

Acknowledgements

This project, which has encompassed many years, has benefitted from input from many individuals and groups.

First, we thank the National Science Foundation for two grants (1140629 and 1323210) that have supported the development and evaluation of this curriculum. We also appreciate funding received to support initial development from the Howard Hughes Medical Institute Undergraduate Science Education Program, the Great Lakes College Association Pathways to Learning Collegium, and the Teagle Foundation.

This book has improved considerably based on very helpful suggestions from class testers who have implemented earlier drafts:

- Karl Beres, Ripon College
- Julie Clark, Hollins University
- Linda Brant Collins, University of Chicago
- Lacey Echols, Butler University
- Julia Guggenheimer, Greenwich Academy
- Chester Ismay, Ripon College
- Gary Kader, Appalachian State University
- Lisa Kay, Eastern Kentucky University
- Dave Klanderman, Trinity Christian College
- John Knoester, Holland Christian High School
- Henry Kramer, Unity Christian High School
- Tom Linton, Central College
- Mark Mills, Central College
- Ron Palcic, Johnson County Community College
- Bob Peterson, Mona Shores High School
- Betsi Roelofs, Holland Christian High School
- Scott Rifkin, San Diego State University
- Michelle Wittler, Ripon College

We also thank our departmental colleagues at Dordt College, Hope College, and Cal Poly who have used earlier versions with their students and provided valuable feedback:

- Airat Bekmetjev, Hope
- Aaron Cinzori, Hope
- Dyana Harrelson, Hope
- Vicki-Lynn Holmes, Hope
- Karen McGaughey, Cal Poly

- Mary Nienhuis, Hope
- Kevin Ross, Cal Poly
- Brian Yurk, Hope College
- Valorie Zonefeld, Dordt

We also benefited greatly from reviewers of the text

Ali Arab, Georgetown University

Audrey Brock, Eastern Kentucky University

Stephanie Casey, Eastern Michigan University

Seo-eun Choi, Arkansas State University

Kathy Chu, Eastern Michigan University

Julie Clark, Hollins University

K. B. Boomer, Bucknell University

Bob Dobrow, Carleton College

Jay Emerson, Yale University

John Emerson, Middlebury College

Michael Ernst, St. Cloud State University

Larry Feldman, Indiana University of Pennsylvania

Brian Garant, Prairie State University

Chi Giang, Westchester Community College

Carla L. Hill, Marist College

Tisha Hooks, Winona State University

Gary Kader, Appalachian State University

Lisa Kay, Eastern Kentucky University

Amy Kimbrough, Virginia Commonwealth University

Tom Linton, Central College

Chris Malone, Winona State University

Mark A. Mills, Central College

Tony Ng, Southern Methodist University

Eric Nordmoe, Kalamazoo College

Ron Palcic, Johnson County Community College

Scott Rikfin, University of California at San Diego

Douglas Ruvolo, University of Akron

Jose Sanqui, Appalachian State University

Sean Simpson, Westchester Community College

Jill Thomley, Appalachian State University

We also received valuable help from students Virginia Burroughs, Andrea Eddy, Kayla Lankheet, Brooke Quisenberry and Kyliee Topliff

Finally, and most importantly, we thank all of our students who have studied with this curriculum as it has evolved. Our students have inspired us to develop curriculum materials and provide a learning environment to support them in the important goal of learning statistical ideas and methods.

CONTENTS

Preliminaries

Introduction to Statistical Investigations

Have you ever heard statements like these?

- "Don't get your child vaccinated. I vaccinated my child and now he is autistic."
- "I will never start jogging because my friend's dad jogged his whole life but he died at age 46 of a heart attack."
- "Teenagers shouldn't be allowed to drive. Everyone knows they are too easily distracted."

The people making these statements are using ***anecdotal evidence*** (personal observations or striking examples) to support broad conclusions. The first person concludes that vaccinations cause autism, based solely on her own child. The second concludes that running is too risky and could cause heart attacks, based entirely on the experience of one acquaintance. The third uses hearsay to draw a broad conclusion.

Scientific conclusions should not be based on anecdotal evidence or single observations. Science requires evidence from a well collected data set. ***Statistics*** is the science of producing useful data to address a research question, analyzing the resulting data, and drawing appropriate conclusions from the data.

For example, suppose you are running for a student government office and have two different campaign slogans in mind. You're curious about whether your fellow students would react more positively to one slogan than the other. Would you ask only for your roommate's opinion, or several of your friends? Or could you conduct a more systematic study? What might that look like? The study of Statistics will help you see how to design and carry out such a study, and you will see how Statistics can also help to answer many important research questions from a wide variety of fields of application.

Section P.1: Introduction to the Six-Step Method

In this section, we will highlight the six steps of a statistical investigation through an example. These six steps are a framework for all of Statistics, and are something you will see throughout this book. We will also use this section introduce and define some key terms.

Example P.1: Organ Donations

Even though organ donations save lives, recruiting organ donors is difficult. Interestingly, surveys show that about 85% of Americans approve of organ donations in principle and many states offer a simple organ donor registration process when people apply for a driver's license. However, only about 38% of licensed drivers in the United States are registered to be organ donors. Some people prefer not to make an active decision about organ donation because the topic can be unpleasant to think about. But perhaps phrasing the question differently could affect people's willingness to become a donor?

Johnson and Goldstein (2003) recruited 161 participants for a study, published in the journal *Science*, to address the question of organ donor recruitment. The participants were asked to imagine they have moved to a new state and are applying for a driver's license. As part of this application, the participants were to decide whether or not to become an organ donor. Participants were presented with one of three different default choices:

- Some of the participants were forced to make a choice of becoming a donor or not, without being given a default option (the "neutral" group).
- Other participants were told that the default option was *not* to be a donor but that they could choose *to* become a donor if they wished (the "opt-in" group).
- The remaining participants were told that the default option was to *be* a donor but that they could choose *not* to become a donor if they wished (the "opt-out" group).

What did the researchers find? Those given the "opt-in" strategy were much less likely to agree to become donors. Consequently, policy makers have argued that we should employ an "opt out" strategy instead. Individuals can still choose not to donate, but would have to more actively do so rather than accept the default. Based on their results, Johnson and Goldstein stated that their data "suggest changes in defaults could increase donations in the United States of additional thousands of donors a year." In fact, as of 2010, 24 European countries had some form of the opt-out system–which some call "presumed consent"–with Spain, Austria, and Belgium yielding high donor rates.

Why were Johnson and Goldstein able to make such a strong recommendation? Because rather than relying on their own opinions or on anecdotal evidence, they conducted a carefully planned study of the issue using sound principles of science and statistics. These principles can be summarized as the six steps of a statistical investigation, which is in line with the scientific method.

Six Steps of a Statistical Investigation

- **Step 1: Ask a research question** that can be addressed by collecting data. These questions often involve comparing groups, asking whether something affects something else, or assessing people's opinions.

- **Step 2: Design a study and collect data**. This step involves selecting the people or objects to be studied and deciding how to gather relevant data on them.
- **Step 3: Explore the data**, looking for patterns related to your research question as well as unexpected outcomes that might point to additional questions to pursue.
- **Step 4: Draw inferences beyond the data** by determining whether any findings in your data reflect a genuine tendency and estimating the size of that tendency.
- **Step 5: Formulate conclusions** that consider the scope of the inference made in Step 4. To what underlying process or larger group can these conclusions be generalized? Is a cause-and-effect conclusion warranted?
- **Step 6: Look back and ahead** to point out limitations of the study and suggest new studies that could be performed to build on the findings of the study.

Let's see how the organ donation study followed these steps.

Step 1: Ask a research question. The general question here is whether a method can be found to increase the likelihood that a person agrees to become an organ donor. This question was then sharpened into a more focused one: Does the default option presented to driver's license applicants influence the likelihood of someone becoming an organ donor?

Step 2: Design a study and collect data. The researchers decided to recruit various participants and ask them to pretend to apply for a new driver's license. The participants did not know in advance that different options were given for the donor question, or even that this issue was the main focus of the study. These researchers recruited participants for their study through various general interest discussion forums on the internet. They offered an incentive of $4.00 for completing an online survey. After the results were collected, the researchers removed data arising from multiple responses from the same IP address, surveys completed in less than five seconds, and respondents whose residential address could not be verified.

Step 3: Explore the data. The results of this study were:
- 44 of the 56 participants in the neutral group agreed to become organ donors,
- 23 of 55 participants in the opt-in group agreed to become organ donors, and
- 41 of 50 participants in the opt-out group agreed to become organ donors.

The proportions who agreed to become organ donors are $44/56 \approx .786$ (or 78.6%) for the neutral group, $23/55 \approx .418$ (or 41.8%) for the opt-in group, and $41/50 = .820$ (or 82.0%) for the opt-out group. The *Science* article displayed a graph of these data similar to Figure P.1.

Figure P.1: Percentages for Organ Donation Study

These results indicate that the neutral version of the question, forcing participants to make a choice, and the opt-out option (default organ donor), produced a higher percentage who agreed to become donors than the opt-in option (default not organ donor).

Step 4: Draw inferences beyond the data. Using methods that you will learn in this course, the researchers analyzed whether the observed differences between the groups was large enough to indicate that the default option had a genuine effect, and then estimated the size of that effect. In particular, this study reported strong evidence that the neutral and opt-out versions generally lead to a higher chance of agreeing to become a donor, as compared to the opt-in version currently used in many states. In fact, they could be quite confident that the neutral version increases the chances that a person agrees to become a donor by between 20 and 54 percentage points, a difference large enough to save thousands of lives per year in the United States.

Step 5: Formulate conclusions. Furthermore, as we will also discuss later in the book, based on the analysis of the data and the design of the study, it is reasonable for these researchers to conclude that the neutral version *causes* an increase in the proportion who will agree to become donors. But because the participants in the study were volunteers recruited from internet discussion forums, generalizing conclusions beyond these participants is only legitimate if they are representative of a larger group of people.

Step 6: Look back and ahead. The organ donation study provides strong evidence that the neutral or opt-out wording could be helpful for improving organ donation proportions. One limitation of the study is that participants were asked to imagine how they would respond, which might not mirror how people would actually respond in such a situation. A new study might look at people's actual responses to questions about organ donation or could monitor donor rates for states that adopt a new policy. Researchers could also examine whether presenting educational material on organ donation might increase people's willingness to donate. Another improvement would be to include participants from wider demographic groups than these volunteers.

Part of looking back also considers how an individual study relates to similar studies that have been conducted previously. Johnson and Goldstein compare their study to two others: one by Gimbel et al. (2003) that found similar results with European countries and one by Caplan (1994) that did not find large differences in the three proportions agreeing to donate.

Figure P.2 displays the six steps of a statistical investigation that we have identified:

Figure P.2: Six Steps of a Statistical Investigation

Four Pillars of Statistical Inference

Notice from Figure P.2 that Step 4 can be considered as the *logic* of statistical inference and Step 5 as the *scope* of statistical inference. Furthermore, each of these two steps involves two components. The following questions comprise the four pillars of statistical inference:

1. **Significance**: How *strong* is the evidence of an effect?

   You will learn how to provide a measure of the strength of the evidence provided by the data. For example, how strong is the evidence that the neutral and opt-out versions increase the chance of agreeing to become an organ donor, as compared to the opt-in version?

2. **Estimation**: What is the *size* of the effect?

   You will learn how to estimate how different two groups are. For example, how much larger (if at all) are the chances someone agrees to donate organs when asked with the neutral version instead of the opt-in version.

3. **Generalization**: How *broadly* do the conclusions apply?

   You will learn to consider what larger group of individuals you believe your conclusions can be applied to. For example, what are the characteristics of the individuals who participated in the organ donation study, and how are they similar or different than typical drivers?

4. **Causation**: Can we say what *caused* the observed difference?

   You will learn whether you can legitimately identify what caused the observed difference. For example, can you conclude that the way the researchers asked the organ donors question was the cause of the observed differences in proportions of donors?

These four concepts are so important that they should be addressed in virtually all statistical studies. Chapters 1-4 of this book will be devoted to introducing and exploring these four pillars of inference. To begin our study of the six steps of statistical investigation, we now introduce some basic terminology that will be used throughout the text.

Basic Terminology

Before we go any further let's introduce you to a few basic terms that we will use throughout the book.
- *Data* can be thought of as the values measured or categories recorded on individual entities of interest.
- These individual entities on which data are recorded are called *observational units*.
- The recorded characteristics of the observational units are the *variables* of interest.

o Some variables are **quantitative**, taking *numerical* values on which ordinary arithmetic operations make sense. For example, height, number of siblings, and age are quantitative variables.

o Other variables are **categorical**, taking *category* designations. For example, eye color, marital status, and whether or not you voted in the last election are categorical variables.

- The **distribution** of a variable describes the pattern of value/category outcomes.

In the organ donation study, the observational units are the participants in the study. The two variables recorded on these participants are the version of the question that the participant received, and whether or not the participant agreed to become an organ donor. Both of these are categorical variables. The graph in Figure P.1 displays the distributions of the donation variable for each default option category.

The observational units in a study are not always people. For example, you might take the Reese's Pieces candies in a small bag as your observational units, on which you could record variables such as the color (a categorical variable) and weight (a quantitative variable) of each individual candy. Or you might take all of the Major League Baseball games being played in one week as your observational units, on which you could record data on variables such as the total number of runs scored, whether the home team wins the game, and the attendance at the game.

Think about it: For each of the three variables just mentioned (about Major League Baseball games), identify the type of variable: categorical or quantitative.

The total number of runs scored and attendance at the game are quantitative variables. Whether or not the home team won the game is a categorical variable.

Think about it: Identify the observational units and variable for a recent study (Ackerman, Griskevicius, and Li, 2011) that investigated this research question: Among heterosexual couples in a committed romantic relationship, are men more likely than women to say "I love you" first?

The observational units in this study are the heterosexual couples, and the variable is whether the man or the woman was the first to say "I love you."

Section P.2: Exploring Data

In this section, you will look more closely at Step 3 of the Six-Step Statistical Investigation Method which concerns exploring data to help answer interesting questions and to help make informed decisions. In Example P.2, you will encounter data arising from a natural "data-generating" process that repeats the same "random event" many, many times, which allows us to see a pattern (distribution) in the resulting data.

Example P.2: Old Faithful

Millions of people from around the world flock to Yellowstone Park in order to watch eruptions of Old Faithful geyser. But, just how faithful is this geyser? How predictable is it? How long does a person usually have to wait between eruptions? Suppose the park ranger gives you a prediction for the next eruption time, and then that eruption occurs five minutes after that predicted time. Would you conclude that predictions by the Park Service are not very accurate? We hope not, because that would be using anecdotal evidence. To investigate these questions about the reliability of Old Faithful, it is much better to collect data.
(A live webcam of Old Faithful and surrounding geysers is available at:
http://www.nps.gov/yell/photosmultimedia/yellowstonelive.htm.)

Researchers collected data on 222 eruptions of Old Faithful taken over a number of days in August 1978 and August 1979. Figure P.3 contains a graph (called a ***dotplot***) displaying the times until the next eruption (in minutes) for these 222 eruptions. Each dot on the dotplot represents a single eruption.

Figure P.3: Times between eruptions of Old Faithful geyser

> **Think about it:** What are the observational units and variable in this study? Is the variable quantitative or categorical?

The observational units are the 222 geyser eruptions, and the variable is the time until the next eruption, which is a quantitative variable. The dotplot displays the ***distribution*** of this variable, which means the values taken by the variable and how many eruptions have those values. The dotplot helps us see the patterns in times until the next eruption.

The most obvious point to be seen from this graph is that even Old Faithful is not perfectly predictable! The time until the next eruption varies from eruption to eruption. In fact, *variability* is the most fundamental property in studying Statistics.

We can view the times until the next eruption as observations from a *process*, an endless series of potential observations from which our data constitute a small "snapshot." Our assumption is that these observations give us a representative view of the long-run behavior of the process. Although we don't know in advance how long it will take for the next eruption, in part because there are many factors that determine when that will be (e.g., temperature, season, pressure), and in part because of unavoidable, natural variation, we may be able to see a predictable pattern overall if we record enough inter-eruption times. Statistics helps us to describe, measure, and, often, explain the pattern of variation in these measurements.

Looking more closely at the dotplot, we can notice several things about the distribution of the time until the next eruption:

- The shortest time until the next eruption was 42 minutes, and the longest time was 95 minutes.
- There appear to be two clusters of times, one cluster between roughly 42 and 63 minutes, another between about 66 and 95 minutes.
- The lower cluster of inter-eruption times is centered at approximately 55 minutes, whereas the upper cluster is centered at approximately 80 minutes. Overall, the distribution of times until the next eruption is centered at approximately 75 minutes.
- In the lower cluster, times until next eruption vary between 42 and 63 minutes, with most of the times between 50 and 60 minutes. In the upper cluster, times vary between 66 and 95 minutes, with most between 75 and 85 minutes.

What are some possible explanations for the variation in the times? One thought is that some of the variability in times until next eruption might be explained by considering the length of the previous eruption. It seems to make sense that after a particularly long eruption, Old Faithful might need more time to build enough pressure to produce another eruption. Similarly, after a shorter eruption, Old Faithful might be ready to erupt again without having to wait very long. Fortunately, the researchers recorded a second variable about each eruption: the duration of the eruption, which is another quantitative variable. For simplicity we can categorize each eruption's duration as short (less than 3.5 minutes) or long (3.5 minutes or longer), a categorical variable. Figure P.4 displays dotplots of the distribution of time until next eruption for short and long eruptions separately.

Figure P.4: Times between eruptions of old faithful geyser, separated by duration of previous eruption (less than 3.5 minutes or at least 3.5 minutes)

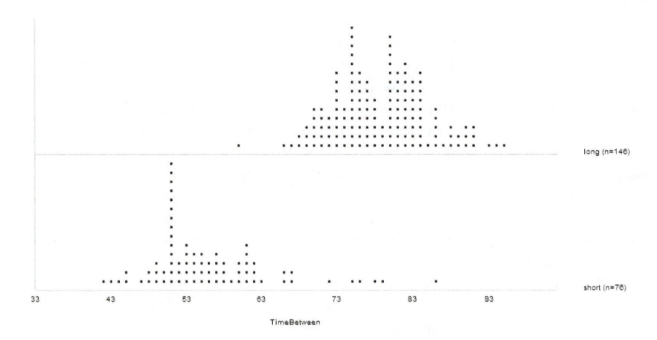

We can make several observations about the distributions of times until next eruption, comparing eruptions with short and long durations, from this dotplot:

- The **shapes** of each individual distribution no longer reveal the two distinct clusters (bi-modality) that was apparent in the original distribution before separating by duration length. Each of these distributions seems to have a single peak.
- The **centers** of these two distributions are quite different: After a short eruption, a typical time until the next eruption is between 50 and 60 minutes. In contrast, after a long eruption, a typical time until the next eruption is between 75 and 85 minutes.
- The **variability** in the times until the next eruption is much smaller for each individual distribution (times tend to fall closer to the mean within each duration type) than the variability for the overall distribution, as we have been able to take into account one source of variability in the data. But of course the times still vary, partly due to other factors that we have not yet accounted for and partly due to *natural variability* inherent in all random processes.

One way to measure the center of a distribution is with the average, also called the mean. One way to measure variability is with the **standard deviation**, which can be roughly interpreted as the typical distance (in absolute value) between the data values and the mean of the distribution. Hence, the more varied or spread out a data set is, the larger its standard deviation will be. (See the Calculation Details Appendix for details about calculating standard deviation. We will also explore the idea of standard deviation more in Chapter 3 and throughout the book.) These values for the time until next eruption, both for the overall distribution and for short and long eruptions separately, are given in Table P.1.

Table P.1: Means and Standard Deviations of Inter-Eruption Times

	Mean	**Standard deviation**
Overall	71.0	12.8
After short duration	56.3	8.5
After long duration	78.7	6.3

Notice that the standard deviations (SD) of time until next eruption are indeed smaller for the separate groups than for the overall dataset, as suggested by examining the variability in the dotplots.

Figure P.5: Times between eruptions of old faithful geyser, separated by duration of previous eruption, with mean and standard deviation shown

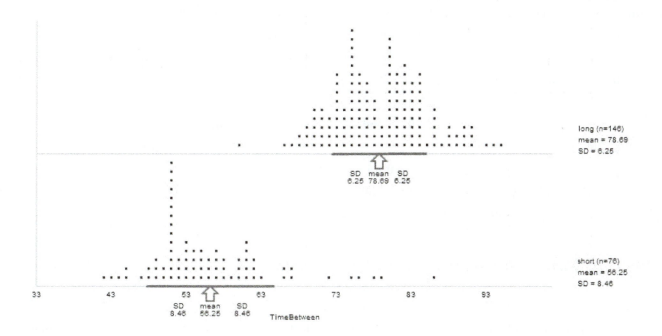

So, what do you learn from this analysis? First, you can better predict when Old Faithful will next erupt if you know how long the previous eruption lasted. Second, with that information in hand, Old Faithful is rather reliable, because it often erupts within six to nine minutes of the time you would predict based on the duration of the previous eruption. So if the park ranger's prediction was only off by five minutes, that's pretty good.

Basic Terminology

From this example you should have learned that a graph such as a dotplot can display the distribution of a quantitative variable. Some aspects to look for in that distribution are:

- *Shape*: Is the distribution symmetric? Mound-shaped? Are there several peaks or clusters?
- *Center*: Where is the distribution centered? What is a typical value?
- *Variability*: How spread out are the data? Are most within a certain range of values?
- Unusual observations: Are there outliers that deviate markedly from the overall pattern of the other data values? If so, identify them to see whether you can explain why those observations are so different. Are there other unusual features in the distribution?

You have also begun to think about ways to measure the center and variability in a distribution. In particular, the standard deviation is a tool we will use quite often as a measure of variability. At this point, we want you to be comfortable visually comparing the variability among distributions and anticipating which variables you might expect to have more variability than others.

> **Think about it:** Suppose that Mary records the ages of people entering a McDonald's fast-food restaurant near the interstate today, while Colleen records the ages of people entering a snack bar on a college campus. Who would you expect to have the *larger* standard deviation of these ages: Mary (McDonald's) or Colleen (campus snack bar)? Explain briefly.

The customers at McDonald's are likely to include people of all ages, from young children to elderly people. But customers at the campus snack bar are most likely to be college-aged students, with some older people who work on campus and perhaps a few younger people. Therefore the ages of the customers at McDonald's will vary more than the ages of those at the campus snack bar. Mary is therefore more likely to have a larger standard deviation of ages than Colleen.

Section P.3: Exploring Random Processes

In this section, you will work through Exploration P.3 to discover for yourself how to explore data from a random process (like rolling dice), in order to use that information to make better decisions. Similar to naturally occurring data-generating processes (such as times between eruptions at Old Faithful), we can often generate sufficient outcomes from a random process to help us learn about long-term patterns. In subsequent chapters, you will analyze both data-generating processes and random processes, often with the goal of seeing how well a random process models or 'behaves like' what you find in data.

Exploration P.3: Cars or Goats

A popular television game show, *Let's Make a Deal*, from the 1960s and 1970s featured a new car hidden behind one of three doors, selected at random. Behind the other two doors were less appealing prizes (e.g., goats!). When a contestant played the game, he or she was asked to pick one of the three doors. If the contestant picked the correct door, he or she won the car!

1. Suppose you are a contestant on this show. Intuitively, what do you think is the probability that you win the car (i.e., that the door you pick has the car hidden behind it)?

2. Give a one-sentence description of what you think *probability* means in this context.

Assuming there is no set pattern to where the game show puts the car initially, this game is an example of a ***random process****:* Although the outcome for an individual game is not known in advance, we expect to see a very predictable pattern in the results if you play this game many, many times. This pattern is called a ***probability distribution***, which is similar to a data distribution like you examined with Old Faithful inter-eruption times. We are interested in features such as how common certain outcomes are – e.g., are you more likely to win this game (select the door with the car) or lose this game?

To investigate what we mean by the term probability, we ask you to play the game many times. We will *simulate* (artificially re-create) playing the game, keeping track of how often you win the car.

3. Use three playing cards with identical backs, but where two of the card faces match and one should differ (e.g., one red and two black). The different card represents the car. Work with a partner (playing the role of game show host), who will shuffle the three cards and then randomly arrange them face down. You pick a card and then reveal whether you have won the car or selected a goat. Play this game a total of 15 times, keeping track of whether you win the car (C) or a goat (G) each time:

Game #	1	2	3	4	5	6	7	8	9	10	11	12	13	14	15
Outcome (car or goat)															

4. In what proportion of these 15 games did you win the car? Is this close to what you expected? Explain.

These fifteen "trials" or "repetitions" mimic the behavior of the game show's random process, where you are introducing randomness into the process by shuffling the cards between games. To get a sense of the long-run behavior of this random process, we want to observe many, many more trials. Because it is not realistic to ask you to perform thousands of repetitions with your partner, we will use a computer to generate a large number of outcomes from this random process.

5. Suppose that you were to play this game 1000 times. In what proportion of those games would you expect to win the car? Explain.

6. Use a website (for example, http://www.grand-illusions.com/simulator/montysim.htm) to simulate playing this game 10 times. Record the proportion of wins in these 10 games. Then simulate another 10 games, and record the overall proportion of wins at this point. Keep doing this in multiples of 10 games until you reach 100 games played. Record the overall proportions of wins after each additional multiple of 10 games in the table below.

Number of games	10	20	30	40	50	60	70	80	90	100
Proportion of wins										

7. What do you notice about how the proportion of wins changes as you play more games? Does this proportion appear to be approaching some common value?

8. Now keep adding sets of 100 more games played until you reach a total of 1000 games played. What percentage of the time did you win a car? Is this close to what you expected in question #5?

Number of games	100	200	300	400	500	600	700	800	900	1000
Proportion of wins										

You should see that the proportion of wins generally gets closer and closer to 1/3 (or 0.3333) as you play more and more games. This is what it means to say that the *probability* of winning is 1/3: If you play the game repeatedly under the same conditions, then after a very large number of games, your proportion of wins should be very close to 1/3. For example, when tossing a coin, it is assumed that the probability of heads is equal to 0.5, meaning that we expect heads to occur about 50% of the time if you toss the coin forever!

Definition: The *probability* of an event is the long-run proportion of times the event would occur if the random process were repeated indefinitely (under identical conditions).

Figure P.6 displays a graph showing how the proportion of wins changed over time for one simulation of 1000 games. Notice that the proportion of wins bounces around a lot at first but then gradually settles down and approaches a long-run value of 1/3.

Figure P.6: Proportion of wins as more and more games are played

Now consider a fun twist that the game show host adds to this game: Before revealing what's behind your door, the host will first reveal what's behind a different door that the host *knows* to be a goat. Then the host asks whether you (the contestant) prefer to stay with (keep) the door you picked originally or switch (change) to the remaining door.

9. Prediction: Do you think the probability of winning is different between the "stay" (keep) and "switch" (change) strategies? If so, what do you think the probability of winning with the switch strategy is?

Whether the "stay" or "switch" strategy is better is a famous mathematical question known as the Monty Hall Problem, named for the host of the game show. What makes this problem famous is that the answer is not intuitive for many people! Fortunately for us, this is not a problem since we can use simulation to easily estimate the probability and help us decide if one strategy is better than the other in the long-run.

10. Investigate the probability of winning with the "switch" strategy by playing with three cards for 15 games. This time your partner should randomly arrange the three cards in his/her hand, making sure that your partner (playing the role of game show host) *knows* where the car is but

you do not. You pick a card. Then your partner reveals one of the cards *known* to be a goat but not the card you chose. Play with the "switch" strategy for a total of 15 games, keeping track of the outcome each time:

Repetition	1	2	3	4	5	6	7	8	9	10	11	12	13	14	15
Outcome (car or goat)															

11. In what proportion of these 15 games did you win the car? Is this more or less than (or the same as) when you stayed with the original door? (question #3)

12. To investigate what would happen in the longer run, again use a website (for example, http://www.grand-illusions.com/simulator/montysim.htm). Notice that you can change from "keep" your original choice to "change" your original choice. Clear any previous work and then simulate playing 1000 games with each strategy, and record the number of times you win/lose with each:

	"Stay" strategy	"Switch" strategy
Wins (cars)		
Losses (goats)		
Total	1000	1000

13. Do you believe that the simulation has been run for enough repetitions to declare one strategy as superior? Which strategy is better? Explain how you can tell.

14. Based on the 1000 simulated repetitions of playing this game, what is your estimate for the probability of winning the game with the "switch" strategy?

15. How could you use simulation to obtain a better estimate of this probability?

16. The probability of winning with the "switch" strategy can be shown mathematically to be 2/3. (One way to see this is to recognize that with the "switch" strategy, you only lose when you had picked the correct door in the first place.) Explain what it means to say that the probability of winning equals 2/3.

Extension

17. Suppose that you watch the game show over many years and find that door #1 hides the car 50% of the time, door #2 has the car 40% of the time, and door #3 has the car 10% of the time. What then is your optimal strategy? In other words, which door should you pick initially, and then should you stay or switch? What is your probability of winning with the optimal strategy? Explain.

Basic Terminology

Through this exploration you should have learned:

- A **random process** is one that can be repeated a very large number of times (in principle, forever) under identical conditions with the following property:
  - Outcomes for any one instance cannot be known in advance, but the proportion of times that particular outcomes occur in the long run can be predicted.
- The **probability** of an outcome refers to the long-run proportion of times that the outcome would occur if the random process were repeated a very large number of times under identical conditions.
- **Simulation** (artificially re-creating a random process) can be used to estimate a probability.
  - Simulations can be conducted with both tactile (by-hand) methods (e.g., cards) and with computers.
  - Using a larger number of repetitions in a simulation generally produces a better estimate of the probability than a smaller number of repetitions.
- Simulation can be used for making good decisions involving random processes.
  - A "good" decision (in this context) means you can accurately predict which strategy would result in a larger probability of winning. This tells you which strategy to use if you do find yourself on this game show, but of course does not guarantee you will win!

Preliminaries Summary

This concludes your study of the preliminary but important ideas necessary to begin studying Statistics. We hope you have learned that:

- Collecting data from carefully designed studies is more dependable than relying on anecdotes for answering questions and making decisions.
- Statistical investigations, which can address interesting and important research questions from a wide variety of fields of application, follow the six steps illustrated in Figure P.2.
- Some data arise from processes that include a mix of systematic elements and natural variation.
- Almost all data display variability. Distributions of quantitative data can be analyzed with dotplots, where we look for shape, center, variability, and unusual observations.
- Standard deviation is a widely used tool for quantifying variability in data.
- Random processes that arise from chance mechanisms display predictable long-run patterns of outcomes. Probability is the language of random processes.
- Beginning in Chapter 1, we will use a known random process to model a data-generating process in order to assess whether the data process appears to behave similarly to the random process.
- Using data to draw conclusions and make decisions requires careful planning in collecting and analyzing the data, paying particular attention to issues of variability and randomness.

Preliminaries Glossary

Preliminaries Exercises

Section P.1

P.1.1* For each of the following research questions identify the observational units and variable(s).
   a)   An article in a 2006 issue of the *Journal of Behavioral Decision Making* reports on a study involving 47 undergraduate students at Harvard. All of the participants were given $50, but some (chosen at random) were told that this was a "tuition rebate," while the others were told that this was "bonus income." After one week, the students were contacted again and asked how much of the $50 they had spent and how much they had saved. Researchers wanted to know whether those receiving the "rebate" would tend to save more money than those receiving the "bonus."
   b)   How much did a typical American consumer spend on Christmas presents in 2012?
   c)   Do college students who pull all-nighters tend to have lower grade point averages than those who do not pull all-nighters?
   d)   Is the residence situation of a college student (on-campus, off-campus with parents, off-campus without parents) related to how much alcohol the student consumes in a typical week?
   e)   Can you predict how far a cat can jump based on factors such as its length?

P.1.2 For each of the following research questions identify the observational units and variable(s).
   a)   Subjects listened to ten seconds of the Jackson 5's song "ABC" and then were how long they thought the song snippet lasted. Do people tend to overestimate the song length?
   b)   Are newborns from couples where both parents smoke less likely to be boys than newborns from couples where neither parent smokes?
   c)   There many different types of diets, but do some work better than others? Is low-fat better than low-carb, or is some combination best? Researchers conducted a study involving three popular diets: Atkins (very low carb), Zone (40:30:30 ratio of carbs, protein, fat), and Ornish (low fat). They randomly assigned overweight women to one of the three diets. The 232 women who volunteered for the program were educated on their assigned diet and were observed periodically as they stayed on the diet for a year. At the end of the year, the researchers calculated the changes in body mass index for each woman and compared the results across the three diets.
   d)   An instructor wants to investigate whether the color of paper (blue or green) on which an exam is printed has an effect on students' exam scores.
   e)   Statistical evidence was used in the murder trial of Kristen Gilbert, a nurse who was accused of killing patients. More than one thousand 8-hour shifts were analyzed. Was the proportion of shifts with a death substantially higher for the shifts that Gilbert worked?

P.1.3* Refer back to Exercise P.1.1. For each of the research questions (a) – (e), identify which variables are quantitative, and which categorical. Also classify categorical variables according to whether they only have two possible outcomes (sometimes called a binary variable), or more than two possible outcomes.

P.1.4 Refer back to Exercise P.1.2. For each of the research questions (a) – (e), identify which variables are quantitative, and which categorical. Also classify categorical variables according to whether they only have two possible outcomes (sometimes called a binary variable), or more than two possible outcomes.

P.1.5* Consider the students in your class as the observational units in a study.
   a)   Identify three quantitative variables that could be recorded on these students.
   b)   Identify three categorical variables that could be recorded on these students.

P.1.6 Consider all of the countries in the world as the observational units in a study.
   a)   Identify three quantitative variables that could be recorded on these countries.
   b)   Identify three categorical variables that could be recorded on these countries.

Expert Performance*

P.1.7 A famous study titled "The Effect of Deliberative Practice in the Acquisition of Expert Performance," published in *Psychological Review* (Ericsson, Krampe, and Tesch-Romer, 1993) led to the now-conventional wisdom that 10,000 hours of deliberate practice are necessary to achieve expert performance in skills such as music and sports. The researchers in this study asked violin students at the Music Academy of West Berlin to keep a diary indicating how they spent their time. The researchers also asked the students' professors to indicate which were the top students with the potential for careers as international soloists, which were good violinists but not among the best, and which were studying to become music teachers. Researchers found that those in the top two groups devoted much more time per week to individual practice than did those in the group studying to become music teachers.
   a)   Identify the observational units in this study.
   b)   Identify the quantitative variable in this study.
   c)   Identify the categorical variable in this study.

Spinning racquets

P.1.8 Tennis players often spin a racquet to decide who serves first. The spun racquet can land with the manufacturer's label facing up or down. A reasonable question to investigate is whether a spun tennis racquet is equally likely to land with the label facing up or down. (If the spun racquet is equally likely to land with the label facing in either direction, we say that they spinning process is fair.) Suppose that you gather data by spinning your tennis racquet 100 times, each time recording whether it lands with the label facing up or down.
   a) What are the observational units?
   b) What is the variable?
   c) Is the variable categorical or quantitative?

Tattoos*

P.1.9 Consider the students in your class as the observational units in a statistical study about tattoos. For each of the following, indicate whether it is a categorical variable, a quantitative variable, or a research question and therefore not a variable at all.
   a) How many tattoos the student has
   b) Whether male students are more likely to have a tattoo than female students
   c) Whether or not the student has a tattoo
   d) Subject that the student is majoring in
   e) Whether students majoring in the Liberal Arts tend to have more tattoos than students majoring in Engineering

Hand washing

P.1.10 In August of 2005, researchers for the American Society for Microbiology and the Soap and Detergent Association monitored the behavior of more than 6300 users of public restrooms. They observed people in public venues such as Turner Field in Atlanta and Grand Central Station in New York City. For each person they kept track of the person's gender and whether or not the person washed his/her hands, along with the person's location.
   a) Identify the observational units in this study.
   b) Identify the three variables that were recorded on each observational units.

Consider the following two graphs produced from the resulting data:

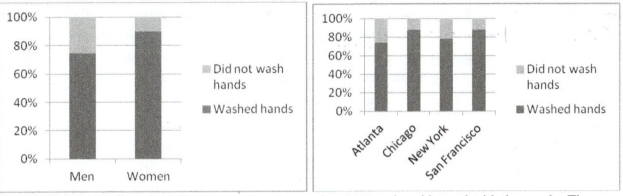

c) For the graph on the left, state a research question that can be addressed with the graph. Then answer the question based on the graph.

d) Repeat part c) for the graph on the right.

Skydiving and Anxiety*

P.1.11 A recent study investigated self-reported anxiety levels immediately before a sky-dive among 11 first-time skydivers and 13 experienced skydivers (at least 30 jumps) recruited from a parachute center in Northern England (Hare et al. 2013). The researchers found that anxiety levels were substantially higher (average anxiety score of 43, higher means more anxiety) among the first-time skydivers as compared to the experienced skydivers (average anxiety score of 27) on a standard test for anxiety (Spielberger State-Trait Anxiety Inventory). Identify the six-steps of a statistical investigation in this study by answering the following questions about this study

a) **Step 1. Ask a research question.** What is the research question being investigated in this study?

b) **Step 2. Design a study and collect data.**

    i) What are the observational units in the study?

    ii) How were the observational units obtained?

    iii) What are the variables in this study? Classify each as categorical or quantitative

c) **Step 3. Explore the data.**

    i) What are the average anxiety scores in the two groups?

    ii) How many individuals are in each group?

d) **Step 4. Draw inferences beyond the data.** What is the researchers' conclusion about the difference in anxiety levels between the two groups?

e) **Step 5. Formulate conclusions.** In order to make their conclusions more relevant, what are the researchers hoping about the 24 skydivers in this study as compared to all skydivers in general?

f) **Step 6. Look back and ahead.** The average age of novice sky divers was 21 and the average age of experienced sky divers was 28, explain how this could impact the conclusions of the researchers. How could this we addressed in a future study?

Skydiving and cortisol

P.1.12 In the same study as described in P.1.11., researchers found little difference in the average amount of cortisol levels in the novice and expert skydivers. The average amount of cortisol was approximately 10 in experienced group and 8 in the inexperienced group. Cortisol is known as the 'stress' hormone and is released in response to fear or stress. Identify the six-steps of a statistical investigation in this study by answering the following questions about this study.

a) **Step 1. Ask a research question.** What is the research question being investigated in this study?

b) **Step 2. Design a study and collect data.**

i) What are the observational units in the study?

ii) How were the observational units obtained?

iii) What are the variables in this study? Classify each as categorical or quantitative.

c) **Step 3. Explore the data.**

i) What are the average cortisol scores in the two groups?

ii) How many individuals are in each group?

d) **Step 4. Draw inferences beyond the data.** What is the researchers' conclusion about the difference in cortisol levels between the two groups?

e) **Step 5. Formulate conclusions.** In order to make their conclusions more relevant, what are the researchers hoping about the 24 skydivers in this study as compared to all skydivers in general?

f) **Step 6. Look back and ahead.** The researchers note that novice skydivers used an automatically deploying parachute, whereas experts deployed their own parachute. Explain the impact this could have on the results of the study and how it could be addressed in future studies.

Your own research study*

P.1.13 Find a research study or article about a study that interests you. After reading the research article, describe how all six steps of the statistical investigation method were applied in the study, as was done in ***Example P.1: Organ Donations***. Some websites that often report interesting studies include *LiveScience* (www.livescience.com), *Science Daily* (www.sciencedaily.com/), *Yahoo News* (http://news.yahoo.com/) and *USA Today* (www.usatoday.com).

Emergency room patients

P.1.14 Suppose that the observational units in a statistical study are the patients arriving at the emergency room of a particular hospital for one month.

   a) Identify two *categorical* variables that could be recorded for these patients.
   b) Identify two *quantitative* variables that could be recorded for these patients.
   c) State two research questions that hospital administrators might want to investigate with these data.

Online purchasing*

P.1.15 Suppose that the observational units in a statistical study are purchases made on a particular website (think of amazon.com) for one month.

   a) Identify two *categorical* variables that could be recorded for these purchases.
   b) Identify two *quantitative* variables that could be recorded for these purchases.
   c) State two research questions that the company executives might want to investigate with these data.

Online dating service

P.1.16 Suppose that the observational units in a statistical study are people who submit information about themselves to an online dating service (think of match.com).

   a) Identify two *categorical* variables that could be recorded for these people.
   b) Identify two *quantitative* variables that could be recorded for these people.
   c) State two research questions that the dating service managers might want to investigate with these data.

About you*

P.1.17 State a research question relevant to your own life that that you could investigate by collecting data. Also identify the observational units and variables for this statistical study.

Section P.2

P.2.1* Suppose that Nellie records the high temperature every day for one year in New York City, while Sandy does the same in San Diego.

   a) Who would you expect to have the larger *mean* of these temperatures? Explain.
   b) Who would you expect to have the larger *standard deviation* of these temperatures? Explain.

P.2.2 Consider three students with the following distributions of 24 quiz scores:

Amanda: 8, 8, 8, 8, 8, 8, 8, 8, 8, 8, 8, 8, 8, 8, 8, 8, 8, 8, 8, 8, 8, 8, 8, 8

Barney: 5, 6, 6, 6, 6, 7, 7, 7, 7, 7, 7, 7, 8, 8, 8, 8, 8, 8, 8, 9, 9, 9, 9, 10

Charlene: 0, 0, 0, 0, 0, 0, 0, 0, 0, 0, 0, 0, 10, 10, 10, 10, 10, 10, 10, 10, 10, 10, 10, 10

Answer the following without bothering to do any calculations. Explain your choices.

a) Which student has the *smallest* standard deviation of quiz scores?

b) Which student has the *largest* standard deviation of quiz scores?

P.2.3* Consider the following four distributions of quiz scores for a class of 10 students:

Arrange these in order from smallest standard deviation to largest standard deviation. Also explain how you make your choices.

Grisham books

P.2.4 The following dotplot displays the distribution of sentence lengths (number of words in a sentence) for 55 sentences selected from John Grishham's novel *The Confession*. Describe what this graph reveals, paying attention to shape, center, variability, and unusual observations. Write as if you are describing this distribution to someone with no knowledge about sentence lengths.

NFL replacement referees*

P.2.5 In the 2012 National Football League (NFL) season, the first three weeks' games were played with replacement referees because of a labor dispute between the NFL and its regular referees. Many fans and players were concerned with the quality of the replacement referees' performance. We could examine whether data might reveal any differences between the three weeks' games played with replacement referees and the next three weeks' games that were played with regular referees. For example, did games generally take less or more time to play with replacement referees than with regular referees? The pair of dotplots below displays data about the duration of games (in minutes), separated by the type of referees officiating the game.

a) What proportion of the 48 games officiated by replacement referees lasted for at least 3.5 hours (210 minutes)? What proportion of the 43 games officiated by regular referees lasted for this long?

b) What proportion of the 48 games officiated by replacement referees lasted for less than 3 hours (180 minutes)? What proportion of the 43 games officiated by regular referees lasted for this long?

c) Would you say that either type of referee tended to have longer games than the other on average? If so, which type of referee tended to have longer games, and by about how much on average?

d) Would you say that either type of referee tended to have more variability in game durations? If so, which type of referee tended to have more variability?

e) Write a paragraph in which you compare the distributions of game durations between the two types of referees. Be sure to comment on shape, center, variability, and unusual observations.

Penalties with replacement refs

P.2.6 Refer to Exercise P.2.5. The following pair of dotplots displays data on the total number of penalties called in the game, again separated by the type of referee. Summarize what these reveal about whether the two types of referees differ with regard to the distributions of this variable.

Weather in San Luis Obispo*

P.2.7 The July 8, 2012 edition of the *San Luis Obispo Tribune* listed predicted high temperatures (in degrees Fahrenheit) for locations throughout San Luis Obispo (SLO; California) county on that date. The following dotplot displays the distribution of these temperatures:

a) Write a paragraph describing the distribution of these predicted high temperatures. Be sure to comment on shape, center, variability, and unusual observations.

b) Can you suggest an explanation for the bimodal (two clusters/peaks) nature of this distribution?

Weather in California

P.2.8 Refer to the previous exercise. The newspaper also listed predicted high temperatures for locations throughout the state of California. The following dotplots display the two distributions:

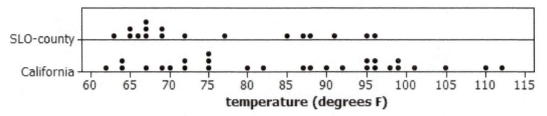

Compare and contrast the distributions of predicted high temperatures in the two locations.

Weather throughout the United States

P.2.9* Refer to the previous two exercises. The newspaper also listed predicted high temperatures for locations throughout the United States and for locations around the world. The following dotplots display the four distributions:

a) Without calculating, arrange the four groups from approximately the highest temperature (on average) to smallest.

b) Without calculating, arrange the four groups from approximately the most variability in temperatures to least.

c) Suggest an explanation for the low outlier in the distribution of the United States temperatures.

d) Suggest an explanation for the bi-modal shape in the distribution of the world temperatures.

Section P.3

P.3.1* It has been reported that the probability that a new business closes or changes owners within its first 3 years is about 0.6. Interpret what this probability means, beginning with: "About 60% of ...."

Ice cream shop discount

P.3.2 An ice cream shop offers a discount price on a small ice cream cone--the price is obtained by rolling two dice and then taking the larger number followed by the smaller number to be the price (in cents). So, rolling a 3 and a 2, or a 2 and a 3, would result in a price of 32 cents. If I visit the ice cream shop with only 50 cents in my pocket the probability that the ice cream cone costs no more than 50 cents is 4/9, which is about 0.444. How would you interpret this probability? (Circle all that apply.)

    A. If I go to this ice cream shop a very large number of times with only 50 cents each time, then I will be able to afford the ice cream cone in about 44.4% of all such visits.

    B. I will be able to afford the ice cream cone in 4 of my next 9 visits to this shop, if I always enter with two quarters.

    C. Among all customers who enter this ice cream shop tomorrow, about 4/9 will be able to afford the ice cream cone.

P.3.3* Which of the following numbers *cannot* represent a probability? Circle all that apply.

    A. 0.01

    B. -0.5

    C. 1.05

    D. 88%

    E. 107%

P.3.4 Explain in your own words what it means to say "the probability of..." for each statement below. Be clear about describing the random process that is repeated over and over again. Do ***not*** include the words "probability," "chance," "odds," or "likelihood" or any other synonyms of "probability."

a. The probability of getting a red M&M candy is 0.2.

b. The probability of winning at a "daily number" lottery game is 1/1000.

c. The probability of rain tomorrow is 0.3.

d. Pennies can be a nuisance. Suppose 30% of the population of adult Americans want to get rid of the penny. If I randomly select one person from this population, the probability this person wants to get rid of the penny is 0.30.

e. Suppose a polling organization takes a random sample of 100 people from the population of adults in a city, where 30% of this population wants to get rid of the penny). Then the probability is 0.015 that the sample proportion who want to get rid of the penny is less than 0.20.

P.3.5* Suppose that baseball team A is better than baseball team B. Team A is enough better that it has a 2/3 probability of beating team B in any one game, and this probability remains the same for each game, regardless of the outcomes of previous games.

a) Explain what it means to say that team A has a 2/3 probability of beating team B in any one game.

b) If team A plays team B for 3 games, is team A guaranteed to win exactly twice?

c) If team A plays team B for 30 games, is team A guaranteed to win exactly 20 times?

d) If team A plays team B for 30 games, do you think it's very likely that team A will win exactly 20 times? Explain.

P.3.6 Reconsider the previous exercise. Continue to assume that team A has a 2/3 probability of beating team B in any one game. Now suppose that team A and team B play a best-of-three series, meaning that the first team to win two games wins the series.

a) If you are a fan of team A, would you prefer to play a single game or a best-of-three series, or would you have no preference? Explain.

b) Describe how you could use a six-sided die to simulate one repetition of a best-of-three series between teams A and B.

c) Describe how you could use a six-sided die to approximate the probability that team A would win the best-of-three series against team B.

d) It turns out that the probability is 0.741 that team A would win this best-of-three series against team B. Interpret what this probability means.

P.3.7* Suppose that a birth is equally likely to be a boy or girl, and the outcome of one birth does not change this probability for future births.

    a) Describe how you could use a coin to approximate the probability that a couple with four children would have 2 boys and 2 girls.

    b) It turns out that the probability is 0.375 that a couple with four children would have 2 boys and 2 girls. Interpret what this probability means.

    c) Based on the probability given in part b), what is the probability that a couple with four children does not have 2 boys and 2 girls?

    d) Explain why it makes sense (at least in hindsight) that a couple with four children is more likely not to have 2 of each sex than to have 2 of each sex.

P.3.8 Suppose that there are 50 people in a room. It can be shown (under certain assumptions) that the probability is approximately 0.97 that at least two people in the room have the same birthday (month and date, not necessarily year).

    a) Explain what this value 0.97 means, as if to someone who has never studied probability or statistics.

    b) Now consider the question of how likely it is that at least one person in the room of 50 matches *your particular* birthday. (Do not attempt to calculate this probability.) Is this event more, less, or equally likely as the event that at least two people share *any* birthday? In other words, do you think this probability will be smaller than 0.97, larger than 0.97, or equal to 0.97? Explain your reasoning.

    c) Explain how you could use simulation (imagine you had a hat with 365 slips of paper in it, each with a number from 1 to 365 listed on it) to confirm that the probability is 0.97 in part a.

    d) Repeat part c), but now use simulation to confirm your answer to part b).

UNIT 1: FOUR PILLARS OF INFERENCE: STRENGTH, SIZE, BREADTH, AND CAUSE

The four chapters of this unit are devoted to four core concepts:

Significance – How *strong* is the evidence of an effect? (Chapter 1)

How *strong* is the evidence that the wording of default options affects recruitment of organ donors?

Estimation – How *large* is the effect? (Chapter 3)

How *much* will the neutral wording (forced choice) increase the donor pool?

Generalization – How *broadly* do the conclusions apply? (Chapter 2)

Do the conclusions apply only to that one set of volunteers? To other similar groups of volunteers? To all who apply for driver's licenses?

Causation – Can we say what *caused* the observed difference? (Chapter 4)

Can we conclude that it was the *default option wording* that was responsible for the increased rate of donor recruitment?

The four concepts have a natural pairing. The first two, significance and estimation, deal with the *logic* of inference. They correspond to Step 4 of our Six-Step Statistical Investigation Method. The last two concepts, generalization and causation, deal with the *scope* of inference. They correspond to Step 5.

In Unit 1 we present these four core concepts – *strength*, *size*, *breadth* and *cause* – one chapter at a time, in the simplest possible statistical setting: observed values for a single binary (Yes/No) variable. This setting is mathematically simple enough to study by flipping coins or dealing cards from a shuffled deck of red and black cards. However, even though the coins and cards are simple, they can be used for inference about a variety of interesting situations. Do dolphins communicate? Can dogs detect cancer? Do red uniformed Olympians win more often? Is the more competent looking politician more likely to win an election?

Chapter 1

Significance: How strong is the evidence?

Chapter Overview

This chapter is about statistical significance, the first of the four pillars of inference: strength, size, breadth, and cause. *Statistical significance* indicates the *strength* of the evidence. For example, how strong is the evidence that the form of the question about organ donation affects the chance that a person agrees to be a donor? Do we have only a suggestion of an impact or are we overwhelmingly convinced?

The goal of the chapter is to explain how statisticians measure strength of evidence. In the organ donation study, the researchers wanted to determine whether the wording (opt in, versus opt out, versus forced choice) really does make a difference. The data in our study pointed in that direction, but are we convinced? How strong is that evidence? Are there other explanations for the results found in the study? One way to approach our investigation is similar to a criminal trial where innocence is assumed and then evidence is presented to help convince a jury that the assumption of innocence and the actual evidence are at odds. Likewise, we will assume that the wording does *not* make a difference and then see whether the evidence (our data) is dramatic enough to convince us otherwise. But we know the outcomes for any one study are random in the sense that there is "chance variability." How, then, do we eliminate "they just got lucky" as an explanation? We have to compare our results to what we would expect to see, "by chance," if the wording did not have an effect. If the actual data and this "by chance" explanation are at odds, then which do you believe? Short of a fluke outcome, the actual data trumps the chance explanation and we then have strong evidence against the "wording does not have an effect" hypothesis and in favor of our research hypothesis that there is an effect. *Understanding this logic is the hardest challenge of the chapter.*

Section 1.1: Introduction to Chance Models

Introduction

A key step in the Statistical Investigation Method is drawing conclusions beyond the observed data. Statisticians often call this "statistical inference." There are four main types of conclusions (inferences) that statisticians can draw from data: significance, estimation, generalization, and causation. In the remainder of this chapter we will focus on *statistical significance*.

If you think back to the Organ Donor study from the Preliminaries, there were three groups: those in the neutral group were asked to make a yes/no choice about becoming a donor; those in the "opt in" group were told their default was not to be a donor, but they could choose to become a donor; and those in the "opt out" group were told their default was they were a donor, but they could choose not to be one if they wished. Let's further examine two of those groups. We saw that 23 of 55 or 41.8% in the "opt in" group elected to become donors, compared with 44/56 or 78.6% in the neutral group. A key question here is whether we believe that the 78.6% is far enough away from the 41.8% to be considered *statistically significant*, meaning unlikely to have occurred by random chance alone. True, 78.6% looks very different from 41.8%, but it is at least possible that the wording of the solicitation to donate actually makes no difference, and that the difference we observed happened by random chance.

To answer the question "Is our result unlikely to happen by random chance?," our general strategy will be to consider what we expect the results to look like if any differences we are seeing are solely due to random chance. Exploring the random chance results is critical to our ability to draw meaningful conclusions from data. In this section we will provide a framework for assessing random chance explanations.

Example 1.1: Can Dolphins Communicate?

A famous study from the 1960s explored whether two dolphins (Doris and Buzz) could communicate abstract ideas. Researchers believed dolphins could communicate simple feelings like "Watch out!" or "I'm happy," but Dr. Jarvis Bastian wanted to explore whether they could also communicate in a more abstract way, much like humans do. To investigate this, Dr. Bastian spent many years training Doris and Buzz and exploring the limits of their communicative ability.

During a training period lasting many months, Dr. Bastian placed buttons underwater on each end of a large pool—two buttons for Doris and two buttons for Buzz. He then used an old automobile headlight as his signal. When he turned on the headlight and let it shine steadily, he intended for this signal to mean "push the button on the right. "When he let the headlight blink on and off, this was meant as a signal to "push the button on the left." Every time the dolphins pushed the correct button, Dr. Bastian gave the dolphins a reward of some fish. Over time Doris and Buzz caught on and could earn their fish reward every time.

Then Dr. Bastian made things a bit harder. Now, Buzz had to push his button before Doris. If they didn't push the buttons in the correct order—no fish. After a bit more training, the dolphins caught on again and could earn their fish reward every time. The dolphins were now ready to participate in the real study to examine whether they could communicate *with each other*.

Dr. Bastian placed a large canvas curtain in the middle of the pool. (See Figure 1.1.)Doris was on one side of the curtain and could see the headlight, whereas Buzz was on the other side of the curtain and could not see the headlight. Dr. Bastian turned on the headlight and let it shine steadily. He then watched to see what Doris would do. After looking at the light, Doris swam near the curtain and began to whistle loudly. Shortly after that, Buzz whistled back and then pressed the button on the right—he got it correct and so both dolphins got a fish. But this single attempt was not enough to convince Dr. Bastian that Doris had communicated with Buzz through her whistling. Dr. Bastian repeated the process several times, sometimes having the light blink (so Doris needed to let Buzz know to push the left button) and other times having it glow steadily (so Doris needed to let Buzz know to push the right button).He kept track of how often Buzz pushed the correct button.

In this scenario, even if Buzz and Doris can communicate, we don't necessarily expect Buzz to push the correct button every time. We allow for some "randomness" in the process; maybe on one trial Doris was a bit more underwater when she whistled and the signal wasn't as clear for Buzz. Or maybe Buzz and Doris aren't communicating at all and Buzz guesses which button to push every time and just happens to guess correctly once in a while. Our goal is to get an idea of how likely Buzz is to push the correct button in the long run.

Figure 1.1: Depending whether or not the light was blinking or shown steadily, Doris had to communicate to Buzz as to which button to push.

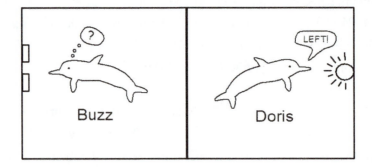

Let's see how Dr. Bastian was applying the Six-Step Statistical Investigation Method.

Step 1: Ask a research question. Can dolphins communicate in an abstract manner?

Step 2: Design a study and collect data. Notice Dr. Bastian took some time to train the dolphins in order to get them to a point where he could test a specific research conjecture. The research conjecture is that Buzz pushes the correct button more often than he would if he and Doris could not communicate. If Buzz and Doris could not communicate, Buzz would just be guessing which button to push. The *observational units* are Buzz's attempts and the *variable* for each attempt is whether or not on each attempt, Buzz pushes the correct button (a *categorical* variable).

Step 3: Explore the data. In one phase of the study, Dr. Bastian had Buzz attempt to push the correct button a total of 16 different times. In this *sample* of 16 attempts, Buzz pushed the correct button 15 out of 16 times. To summarize these results, we report the *statistic,* a numerical summary of the sample. For this example, we could report either 15, the number of correct pushes, or $15/16 = 0.9375$, as the statistic.

> **Definitions:** The set of observational units on which we collect data is called the *sample*. The number of observational units in the sample is the *sample size*. A *statistic* is a number summarizing the results in the sample.

The *sample size* in this example is 16. Note that the word "sample" is used as both a noun (the set of observational units being studied) and as an adjective, for example to mean "computed from the observed data," as, for example, "sample statistic."

Step 4: Draw inferences beyond the data. These 16 observations are a mere snapshot of Buzz's overall selection process. We will consider this a *random process*. We are interested in Buzz's actual long-run *probability* of pushing the correct button based on Doris' whistles. This unknown long-run probability is called a *parameter*.

> **Definition:** For a random process, a *parameter* is a long-run numerical property of the process.

Note that we are assuming this parameter is not changing over time, at least for the process used by Buzz in this phase of the study. Because we can't observe Buzz pushing the button forever, we need to draw conclusions (possibly incorrect, but hopefully not) about the value of the parameter based only on these 16 attempts. Buzz certainly pushed the correct button most of the time, so we might consider either:

- Buzz is doing something other than just guessing (his probability of a correct button push is larger than 0.50).
- Buzz is just guessing (his probability of a correct button push is 0.50) and he got lucky in these 16 attempts.

These are the two possible explanations to be evaluated. Because we can't collect more data, we have to base our conclusions only on the data we have. It's certainly *possible* that Buzz was just guessing and got lucky! But does this seem like a reasonable explanation to you? How would you argue against someone who thought this was the case?

> **Think about it:** Based on these data, do you think Buzz somehow knew which button to push? Is 15 out of 16 correct pushes convincing to you? Or do you think that Buzz could have just been guessing? How might you justify your answer?

So how are we going to decide between these two possible explanations? One approach is to choose a *model* for the random process (repeated attempts to push the correct button) and then see whether our model is consistent with the observed data. If it is, then we will conclude that we have a reasonable model and we will use that model to answer our questions.

The Chance Model

Scientists use models to help understand complicated real world phenomena. Statisticians often employ *chance models* to generate data from random processes to help them investigate such processes. You did this with the Monty Hall Exploration (P.3)to investigate properties of the two strategies, switching and staying with your original choice of door. In that exploration it was clear how the underlying chance process worked, even though the probabilities themselves were not obvious. But here we don't know for sure what the underlying real world process is. We are trying to decide whether the process could be Buzz simply guessing or whether the process is something else, such as Buzz and Doris being able to communicate.

Let us first investigate the "Buzz was simply guessing" process. Because Buzz is choosing between two options, the simplest chance model to consider is a coin flip. We can flip a coin to represent or *simulate* Buzz's choice *assuming he is just guessing* which button to push. To generate this artificial data, we can let "heads" represent the outcome that Buzz pushes the correct button and let "tails" be the outcome that Buzz pushes the incorrect button. This gives Buzz a 50% chance of pushing the correct button. This can be used to represent the "Buzz was just guessing" or the "random chance alone" explanation. The correspondence between the real study and the physical simulation is shown in Table 1.1.

Table 1.1: Parallels between real study and physical simulation

Coin flip	=	guess by Buzz
Heads	=	correct guess
Tails	=	wrong guess
Chance of Heads	= 1/2 =	probability of correct button when Buzz is just guessing
One repetition	=	one set of 16 simulated attempts by Buzz

Now that we see how flipping a coin can simulate Buzz guessing, let's flip some coins to simulate Buzz's performance. Imagine that we get heads on the first flip. What does this mean? This would correspond to Buzz pushing the correct button! But, why did he push the correct button? In this chance model, the only reason he pushed the correct button is because he happened to guess correctly—remember the coin is simulating what happens when Buzz is just guessing which button to push.

What if we keep flipping the coin? Each time we flip the coin we are simulating another attempt where Buzz guesses which button to push. Remember that heads represents Buzz guessing correctly and tails represents Buzz guessing incorrectly. How many times do we flip the coin? Sixteen, to match Buzz's 16 attempts in the actual study. After 16 tosses, we obtained the sequence of flips shown in Figure 1.2.

Figure 1.2: A sequence of 16 coin flips

Heads, Tails, Tails, Heads, Tails, Heads, Heads, Heads, Heads, Tails, Heads, Heads, Heads, Heads, Heads, Tails

Here we got 11 heads and 5 tails (11 out of 16 or 0.6875 is the simulated statistic). This gives us an idea of what could have happened in the study if Buzz had been randomly guessing which button to push each time.

Will we get this same result every time we flip a coin 16 times? Let's flip our coin another 16 times and see what happens. When we did this we got 7 heads and 9 tails as shown in the sequence of coin flips (7 out of 16 or 0.4375 is the simulated statistic) in Figure 1.3.

Figure 1.3: Another sequence of 16 coin flips

Heads, Tails, Heads, Heads, Tails, Tails, Tails, Heads, Tails, Tails, Heads, Tails, Heads, Heads, Tails, Tails

So can we learn anything from these coin tosses when the results vary between the sets of 16 tosses?

Using and evaluating the coin flip chance model

Because coin flipping is a random process, we know that we won't obtain the same number of heads with every set of 16 flips. But are some numbers of heads more likely than others? If we continue our repetitions of 16 tosses, we can start to see how the outcomes for the number of heads are distributed. Does the *distribution* of the number of heads that result in 16 flips have a predictable long-run pattern? In particular, how much *variability* is there in our simulated statistics between repetitions (sets of 16 flips) just by random chance?

In order to investigate these questions, we need to continue to flip our coin to get many, many sets of 16 flips (or many repetitions of the 16 choices where we are modeling Buzz simply guessing each time). We did this, and Figure 1.4 shows what we found when we graphed the number of heads from each set of 16 coin flips. Here, the process of flipping a coin 16 times was repeated 1,000 times—this number was chosen for convenience, but also appears to be large enough to give us a fairly accurate sense of the long-run behavior for the number of heads in 16 tosses.

Figure 1.4: A dotplot showing 1000 repetitions of flipping a coin 16 times and counting the number of heads

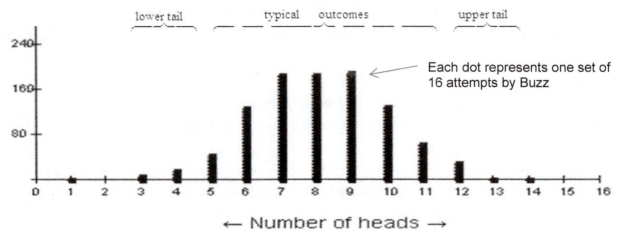

← Number of heads →

Let's think carefully about what the graph in Figure 1.4 shows. For this graph, each dot represents the number of heads in one set of 16 coin tosses. We see that the resulting number of heads follows a clear pattern: 7, 8, and 9 heads happened quite a lot, 10 was pretty common also (though less so than 8), 6 happened some of the time, 1 happened once. But we never got 15 heads in any set of 16 tosses! We might consider any outcome between about 5 and 11 heads to be typical, but getting fewer than 5 heads or more than 11 heads happened rarely enough we can consider it a bit unusual. We refer to these unusual results as being out in the 'tails' of the distribution.

Think about it: How does the analysis above help us address the strength of evidence for our research conjecture that Buzz was doing something other than just guessing?

What does this have to do with the dolphin communication study? We said that we would flip a coin to simulate what could happen if Buzz was really just guessing each time he pushed the button in 16 attempts. We saw that getting results like 15 heads out of 16 never happened in our 1000 repetitions. This shows us that 15 is a very unusual outcome— far out in the tail of the distribution of the simulated statistics—if Buzz is guessing. In short, even though we expect some variability in the results for different sets of 16 tosses, the pattern shown in this distribution indicates that an outcome of 15 heads is outside the typical chance variability we would expect to see when Buzz is simply guessing.

In the actual study, Buzz really did push the correct button 15 times out of 16, an outcome that we just determined would rarely occur if Buzz was just guessing. So, our coin flip chance model tells us that we have very strong evidence that Buzz was not just tossing a coin to make his choices. This means we have strong evidence that Buzz wasn't just guessing. Therefore, we don't believe the "by chance alone" explanation is a good one for Buzz. The results mean we have strong enough evidence to be *statistically significant*, not a result that happened by chance alone, and to convince us that something other than "random chance" was at play.

Definition: A result is *statistically significant* if it is unlikely to occur just by random chance.

If our observed result appears to be consistent with the chance model, we say that the chance model is *plausible* or believable.

What next? A glimpse into Steps 5 and 6

The steps we went through above have helped us evaluate how strong the evidence is that Buzz is not guessing (Step 4 of the Statistical Investigation Method). In this case, the evidence provided by this sample is fairly strong that Buzz isn't guessing. Still, there are some important questions you should be asking right now, such as: If Buzz isn't guessing what is he doing?

Step 5: Formulate conclusions. We should also ask ourselves: if Buzz wasn't guessing, does this prove that Buzz and Doris can communicate? And if so, what does this say about other dolphins? As we'll find in later chapters, the answers to these questions hinge mainly on how the study was designed and how we view the 16 attempts that we observed (e.g., we assume Buzz couldn't see the light himself, the light signal displayed each time was chosen randomly so Buzz couldn't figure out a pattern to help him decide which button to push; Buzz's 16 attempts are a good representation of what Buzz would do given many more attempts under identical conditions; but we might still wonder whether Buzz's behavior is representative of dolphin behavior in general or are there key differences among individual dolphins).

Step 6: Look back and ahead. After completing Steps 1-5 of the Statistical Investigation Method, we need to revisit the big picture of the initial research question. First, we reflect on the limitations of the analysis, and think about future studies. In short, we are now stepping back and thinking about the initial research question more than the specific research conjecture being tested in the study. In some ways, this is the most important step of the whole study because it is where we think about the true implications of the scientific study we've conducted. For this study, we would reflect on Dr. Bastian's methods, summarize the results for Buzz, and reflect on ways to improve the study to enhance the conclusions we can draw.

The 3S Strategy

Let us summarize the overall approach to assessing *statistical significance* that we have been taking in this section. We observed a sample statistic (e.g., the number of "successes" or the proportion of "successes" in the sample). Then we simulated 'could-have-been' outcomes for that statistic under a specific chance model (just guessing). Then we used the information we gained about the random variation in the 'by-chance' values of the statistics to help us judge whether the observed value of the statistic is an unusual or a typical outcome. If it is unusual–we say the observed statistic is *statistically significant*—it provides strong evidence that the chance-alone explanation is wrong. If it is typical, we consider the chance model plausible.

You may have noticed that we only simulated results for one specific model. When we saw that the sample statistic observed in the study was not consistent with these simulated results, we rejected the chance-alone explanation. Often, research analyses stop here. Instead of trying to simulate results from other models (in particular we may not really have an initial idea what a more appropriate model might be), we are content to say there is something other than random chance at play here. This might lead the researchers to reformulate their conjectures and collect more data in order to investigate different models.

We will call the process of simulating could-have-been statistics under a specific chance model the *3S Strategy*. After forming our research conjecture and collecting the sample data, we will use the 3S Strategy to weigh the evidence against the chance model. This 3S Strategy will serve as the foundation for addressing the question of statistical significance in Step 4 of the Statistical Investigation Method.

3S Strategy for Measuring Strength of Evidence

---

1. Statistic: Compute the statistic from the observed sample data.

2. Simulate: Identify a "by chance alone" explanation for the data. Repeatedly simulate values of the statistic that could have happened when the chance model is true.

3. Strength of evidence: Consider whether the value of the observed statistic from the research study is unlikely to occur if the chance model is true. If we decide the observed statistic is unlikely to occur by chance alone, then we can conclude that the observed data provide strong evidence against the plausibility of the chance model. If not, then we consider the chance model to be a plausible (believable) explanation for the observed data; in other words what we observed could plausibly have happened just by random chance.

Let's illustrate how we implemented the 3S Strategy for the Doris and Buzz example.

1. Statistic: Our observed statistic was 15, the number of times Buzz pushed the correct button in 16 attempts.

2. Simulate: If Buzz was actually guessing, the parameter (the probability he would push the correct button) would equal 0.5. In other words, he would push the correct button 50% of the time in the long run. We used a coin flip to model what could have happened in 16 attempts when Buzz is just guessing. We flip the coin 16 times and count how many of the 16 flips are heads, meaning how many times Buzz pressed the correct button ("success"). We then repeat this process many more times, each time keeping track of the number of the 16 attempts that Buzz pushed the correct button. We end up with a distribution of could-have-been statistics representing typical values for the number of correct pushes when Buzz is just guessing.

3. Strength of evidence: Because 15 successes of 16 rarely happens by chance alone, we conclude that we have strong evidence that, in the long-run, Buzz is not just guessing.

Notice that we have used the result of 15 out of 16 correct attempts to *infer* that Buzz's actual long-run probability of pushing the correct button was not simply 0.5.

Another Doris and Buzz study

One goal of statistical significance is to rule out random chance as a *plausible*(believable) explanation for what we have observed. We still need to worry about how well the study was conducted. For example, are we absolutely sure Buzz couldn't see the headlight around the curtain? Are we sure there was no pattern to which headlight setting was displayed that he might have detected? And of course we haven't completely ruled out random chance, he may have had an incredibly lucky day. But the chance of him being that lucky is so small, that we conclude that other explanations are more plausible or credible.

One option that Dr. Bastian pursued was to re-do the study except now he replaced the curtain with a wooden barrier between the two sides of the tank in order to ensure a more complete separation between the dolphins to see whether that would diminish the effectiveness of their communication.

Step1: Ask a research question. The research question remains the same: Can dolphins communicate in a deep abstract manner?

Step 2: Design a study and collect data. The study design is similar with some adjustments to the barrier between Doris and Buzz. The canvas curtain is replaced by a plywood board. The research conjecture, observational units, and variable remain the same.

In this case, Buzz pushed the correct button only 16 out of 28 times. The variable is the same (whether or not Buzz pushed the correct button), but the number of observational units (sample size) has changed to 28 (the number of attempts).

Think about it: Based on the results for this phase of the study, do you think that Doris could tell Buzz which button to push, even under these conditions? Or is it believable that Buzz could have just been guessing?

Step 3: Explore the data. So our observed statistic is 16 out of 28 correct attempts, which is $16/28 \times 100\% = 57.1\%$ of Buzz's attempts. This is again more than half the time, but not much larger than 50%. A simple *bar graph* of these results is shown in Figure 1.5.

Figure 1.5: Bar graph for Buzz's 28 attempts

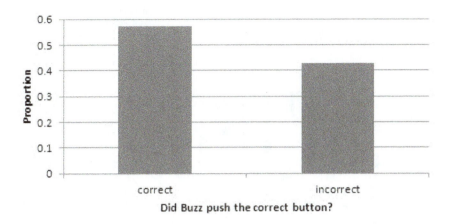

Step 4: Draw inferences. Is it plausible (believable) that Buzz was simply guessing in this set of attempts? How do we measure how much evidence these results provide against the chance model? Let's use the same chance model as we used earlier to see what could have happened if Buzz was just guessing. We will apply the 3S Strategy to this new study.

1. Statistic: The new observed sample statistic is 16 out of 28, or about 0.571.

> **Think about it:** Consider again our simulation of the chance model assuming Buzz is guessing. What do we need to change for this new phase of the study?

2. Simulation: This time we need to do repetitions of 28 coin flips, not just 16. A distribution of the number of heads in 1000 repetitions of 28 coin flips is shown in Figure 1.6. This models 1000 repetitions of 28 attempts with Buzz randomly pushing one of the buttons (guessing) each time.

Figure 1.6: A graph showing 1000 repetitions of flipping a coin 28 times and counting the number of heads. This models the number of correct pushes in 28 attempts when Buzz is guessing each time.

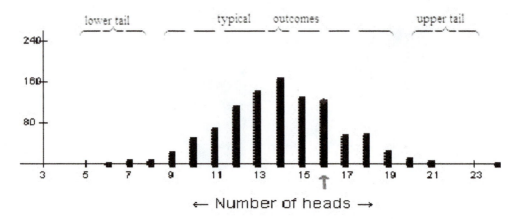

3. Strength of evidence: Now we need to consider the new observed statistic (16 out of 28 or 0.571). We see from the graph that 16 out of 28 is a fairly typical outcome if Buzz is just randomly guessing. What does this tell us? It tells us that the results of this study are something that could easily have happened if Buzz was just randomly guessing. So what can we conclude? We can say his 16 successes are not convincing evidence against the "by chance alone" model.

The graph in Figure 1.6 shows what happens for the hypothetical Buzz who just guesses. An actual outcome far out in the tail of that distribution would be strong evidence against the "just guessing" hypothesis. But be careful: The opposite result – an actual outcome near the center – is *not* strong evidence in support of the guessing hypothesis. Yes, the result is *consistent* with that hypothesis, but it is also consistent with many other hypotheses as well.

Bottom line: In this second study we conclude that there is not enough evidence that the "by chance alone" model is wrong. That model is still a *plausible* explanation for the statistic we observed in the study (16 out of 28). Based on this set of attempts, we do not have convincing evidence against the possibility that Buzz is just guessing, but other explanations also remain plausible. For example, the results are consistent with very weak communication between the dolphins. All we know from this analysis is that one plausible explanation for the observed data is that Buzz was guessing.

In fact, Dr. Bastian soon discovered that in this set of attempts, the equipment malfunctioned and the food dispenser for Doris did not operate and so Doris was not receiving her fish rewards during the study. Because of this malfunction, it's not so surprising that removing the incentive hindered the communication between the dolphins and we cannot refute that Buzz was just guessing for these attempts.

Dr. Bastian fixed the equipment and ran the study again. This time he found convincing evidence that Buzz was not guessing.

For a bit more discussion on processes and parameters, see FAQ 1.1.1.

FAQ 1.1.1: What is a random process?

Q: So a *parameter* is a numerical property of a process... but what is the process? All I have is Buzz's 16 attempts.

A: We can think of Buzz's 16 attempts as a sample from his long-run selection process. He has some underlying probability of pushing the correct button, but he's unlikely to be correct every time, even if the dolphins are communicating. There is *randomness* in his choices. His choices might be affected by Doris' squeaks, but also by how hungry he is, how tired he is, how warm the water is. We are trying to see how much Doris' communication is influencing him. Everything else gets lumped together into "random chance." Any one outcome from a random process is unknown to us in advance (like tossing a coin), but if we observe outcomes from the process for long enough, we should start to see some patterns amidst the randomness. So you can think of a random process as an unlimited source of potential observations for your sample.

Q: Can I use *any* old set of observations from a process as a sample?

A: Some samples are good and others are terrible.

Q: How do I tell the good apples from the rotten ones?

A: For a Yes/No process like Buzz's selections, we need to be willing to make certain assumptions along the way. For example, we assumed Buzz's probability of a correct guess was the same every time. In fact, under the null hypothesis, we assumed it was 50/50 every time. But even if the probability is not 0.5, we are assuming it is not changing. In other words, we assume that Buzz doesn't get better through learning and that he doesn't get tired or bored or quit trying. We are also assuming that his previous trials don't influence his future guesses – like a coin, he would have no memory of his past guesses.

Q: So how do I know if I have a good sample?

A: You need to believe that nothing about the process is changing and that each outcome does not depend on past outcomes. If you're a coin, your chance of heads doesn't change, and you have no memory of how you've landed in the past. If this is true about the process, then you will have a good sample! On the other hand, if Buzz's probability of success is different in the morning than in the afternoon, then only observing him in the morning will not give you a good representation of his overall probability of success.

Q: So then how does the parameter come into play?

A: Parameter is a hard concept, precisely because we never see it! For the dolphin example, we want to know whether Buzz is just guessing. We think of his guesses as a potentially never-ending process (like coin tossing), and the parameter is the *probability* that Buzz will be correct in his next attempt. As you saw in the Preliminaries, this means if we were to observe the random process forever, this probability is the long-run proportion of times that Buzz pushes the correct button.

Q: So I see why the parameter isn't just 15/16, that is his sample proportion and we may not have hit that long-run probability yet. So the parameter is just 0.5?

A: Not necessarily. The parameter equals 0.5 if he is just guessing. That's the chance model we simulated. But his probability may actually be different from 0.5.

Q: So the parameter is his actual probability of pushing the correct button. It's a number, but we don't know its value.

A: Correct! But keep in mind that we will define lots of different kinds of parameters in future chapters. The "probability of success" parameter only applies in this type of Yes/No random process.

Exploration 1.1: Can Dogs Understand Human Cues?

Dogs have been domesticated for about 14,000 years. In that time, have they been able to develop an understanding of human gestures such as pointing or glancing? How about similar non-human cues? Researchers Udell, Giglio, and Wynne tested a small number of dogs in order to answer these questions.

In this exploration, we will first see whether dogs can understand human gestures as well as non-human gestures. To test this, the researchers positioned the dogs about 2.5 meters from the experimenter. On each side of the experimenter were two cups. The experimenter would perform some sort of gesture (pointing, bowing, looking) toward one of the cups or there would be some other non-human gesture (a mechanical arm pointing, a doll pointing, or a stuffed animal looking) toward one of the cups. The researchers would then see whether the dog would go to the cup that was indicated. There were six dogs tested. We will look at one of the dogs in two of his sets of trials. This dog, a 4 year-old mixed breed, was named Harley. Each trial involved one gesture and one pair of cups, with a total of ten trials in a set.

We will start out by looking at one set of trials where the experimenter bowed toward one of the cups to see whether Harley would go to that cup.

Step 1. State the research question.

1.   Based on the description of the study, state the research question.

Step 2. Design a study and collect data. Harley was tested 10 times and 9 of those times he chose the correct cup.

2.   What are the observational units? *10*

3.   Identify the variable in the study. What are the possible outcomes of this variable? Is this variable quantitative or categorical?

Step 3. Explore the data.

> **Definitions:** The set of observational units on which we collect data is called the *sample*. The number of observational units in the sample is the *sample size*. A *statistic* is a number summarizing the results in the sample.

With categorical data, we typically report the number of "successes" or the proportion of successes as the statistic.

4.   What is the number of observational units (sample size) in this study?
     10

5.   Determine the observed statistic and produce a simple bar graph of the data (have one bar for the proportion of times Harley picked the correct cup, and another for the proportion of times he picked the wrong cup) *9/10*

6.   If the research conjecture is that Harley can understand what the experimenter means when they bow toward an object, is the statistic in the direction suggested by the research conjecture?

7.   Could Harley have gotten 9 out of 10 correct even if he really didn't understand the human gesture and so was randomly guessing between the two cups?
     Yes

8. Do you think it is likely Harley would have gotten 9 out of 10 correct if he was just guessing randomly each time? *No*

Step 4. Draw inferences beyond the data.

There are two possibilities for why Harley chose the correct cupnine out of ten times:

- He is merely picking a cup at random, and in these 10 trials happened to guess correctly in 9 of them. That is, he got more than half correct just by random chance alone.
- He is doing something other than merely guessing and perhaps understands what the experimenter means when they bow towards the cup.

The unknown long-run probability that Harley will choose the correct cup is called a ***parameter***.

Definition: For a random process, a ***parameter*** is a long-run numerical property of the process.

We don't know the value of the parameter, but the two possibilities listed above suggest two different possibilities.

9. What is the value of the parameter if Harley is picking a cup at random? Give a specific value.
    π
    50 %

10. What is the possible range of values (greater than or less than some value) for the parameter if Harley is not just guessing and perhaps understands the experimenter?
    $\pi > .50$

We will show you how statisticians use simulation to make a statement about the strength of evidence for these two possible statements about the parameter's value.

The Chance Model

Statisticians often use *chance models* to generate data from random processes to help them investigate the process. In particular, they can see whether the observed statistic is consistent with the values of the statistic simulated by the chance model. If we determine that Harley's results are not consistent with the results from the chance model, we will consider this to be evidence against the chance model and in favor of the research conjecture, that he understands the bowing gesture. In this case, we would say Harley's results are *statistically significant*, meaning unlikely to have occurred by chance alone.

We can't perform the actual study more times in order to assess the second possibility, but we can simulate the behavior of Harley's choices if we were to assume the first possibility (that he is simply guessing every time).

11. Explain how you could use a coin toss to represent Harley's choices if he is guessing between the two cups each time. How many times do you have to flip the coin to represent one set of Harley's attempts? What does heads represent?

12. If Harley was guessing randomly each time, on average, how many out of the 10 times would you expect him to choose the correct cup?

13. Simulate one repetition of Harley guessing randomly by flipping a coin 10 times (why 10?) and letting heads represent selecting the correct cup("success") and tails represent selecting the incorrect cup ("failure"). Count the number of heads in your 10 flips. Combine your results with the rest of the class to create a *dotplot* of the distribution for the number of heads out of 10 flips of a coin.

   a. Where does 9 heads fall in the distribution? Would you consider it an unusual outcome or a fairly typical outcome for the number of heads in 10 flips?

   b. Based on your answer to the previous question, do you think it is plausible (believable) that Harley was just guessing which cup to choose?

Using an applet to simulate flipping a coin many times

To really assess the typical values for the number of heads in 10 coin tosses (number of correct picks by Harley assuming he is guessing at random), we need to simulate many more outcomes of the chance model. Open the **One Proportion** applet from the textbook webpage.

Notice that the probability of heads has been set to be 0.50, representing the guessing between the two cups. Set the number of tosses to 10 and press the **Toss Coins** button. What was the resulting number of heads?

Notice that the number of heads in this set of 10 tosses is then displayed by a dot on the graph. Uncheck the **Animate** box and press the **Toss Coins** button 9 more times. This will demonstrate the how the number of heads varies randomly across each set of 10 tosses. Nine more dots have been added to your dotplot. Is a pattern starting to emerge?

Now change the **Number of repetitions** from 1 to 990 and press **Toss Coins**. The applet will now show the results for the number of heads in 1000 different sets of 10 coin tosses. So each dot represents the number of times Harley chooses the correct cup out of ten attempts assuming he is just guessing. Remember why we conducted this simulation: to assess whether Harley's result (9 correct in 10 attempts) would be unlikely to occur by chance alone if he were just guessing between the pair of cups for each attempt.

14. Locate the result of getting 9 heads in the dotplot created by the applet. Would you consider this an unlikely result, in the tail of the distribution of the number of heads?

15. Based on the results of 1000 simulated sets of 10 coin flips each, would you conclude that Harley would be very unlikely to have picked the correct cup in 9 times in 10 attempts if he was randomly guessing between the two cups each time? Explain how your answer relates to the applet's dotplot.   *Yes*

Definition: A result is ***statistically significant*** if it is unlikely to occur just by random chance. If our observed result appears to be consistent with the chance model, we say that the chance model is ***plausible*** or believable.

---

16. Do the results of this study appear to be statistically significant?

17. Do the results of this study suggest that Harley just guessing is a plausible explanation for Harley picking the correct cup 9 out of 10 times?

Summarizing your understanding

18. To make sure that you understand the coin flipping chance model fill in the following table
    indicating what parts of the real study correspond to the physical (coin-flipping) simulation.

Table 1.2: Parallels between real study and physical simulation

Coin flip	=	
Heads	=	
Tails	=	
Chance of Heads	=	
One repetition	=	one set of ___ simulated attempts by Harley

The 3S Strategy

We will call the process of simulating could-have-been statistics under a specific chance model the *3S Strategy*. After forming our research conjecture and collecting the sample data, we will use the 3S Strategy to weigh the evidence against the chance model. This 3S Strategy will serve as the foundation for addressing the question of statistical significance in Step 4 of the Statistical Investigation Method.

<u>**3S Strategy for Measuring Strength of Evidence**</u>

---

1. Statistic: Compute the statistic from the observed sample data.

2. Simulate: Identify a "by chance alone" explanation for the data. Repeatedly simulate values of the statistic that could have happened when the chance model is true.

3. Strength of evidence: Consider whether the value of the observed statistic from the research study is unlikely to occur if the chance model is true. If we decide the observed statistic is unlikely to occur by chance alone, then we can conclude that the observed data provide strong evidence against the plausibility of the chance model. If not, then we consider the chance model to be a plausible (believable) explanation for the observed data; in other words what we observed could plausibly have happened just by random chance.

Let's review how we have already applied the 3S Strategy to this study.

19. **Statistic.** What is the statistic in this study?

20. **Simulate.** Fill in the blanks to describe the simulation.

We flipped a coin _____ times and kept track of how many times it came up heads. We then repeated this process _____ more times, each time keeping track of how many heads were obtained in each of the _____ flips.

21. **Strength of evidence.** Fill in the blanks to summarize how we are assessing the strength of evidence for this study.

Because we rarely obtained a value of _____ heads when flipping the coin _____ times, this means that it is _____ (believable/unlikely) that Harley is just guessing, since if Harley was just guessing he _____ (rarely/often) would get a value like _____ correct out of _____ attempts.

Step 5: Formulate conclusions.

22. Based on this analysis, are you convinced that Harley can understand human cues? Why or why not?

Another Study

One important step in a statistical investigation is to consider other models and whether the results can be confirmed in other settings.

23. In a different study, the researchers used a mechanical arm (roughly the size of a human arm) to point at one of the two cups. The researchers tested this to see whether dogs understood non-human gestures. In 10 trials, Harley chose the correct cup 6 times.

a. Using the dotplot you obtained when you simulated 1000 sets of 10 coin flips assuming Harley was just guessing, locate the result of getting 6 heads. Would you consider this an unlikely result, in the tail of the distribution?

b. Based on the results of 1000 simulated sets of 10 coin flips each, would you conclude that Harley would be very unlikely to have picked the correct cup 6 times in 10 attempts, if he was randomly guessing between the two cups each time? Explain how your answer relates to the applet's dotplot.

c. Is this study's result statistically significant?

d. Do the results of this study suggest that Harley just guessing is a plausible explanation for Harley picking the correct cup 6 out of 10 times?

e. Does this study prove that Harley cannot understand the mechanical arm?

Step 6: Look back and ahead.

24. Compare the analyses between the two studies. How does the unusualness of the observed statistic compare between the two studies? Does this make sense based on the value of the observed statistic in the two studies? Does this make sense based on how the two studies were designed? Explain.(*Hint*: Why might the results differ for human and mechanical arms? Why would this matter?)

25. A single study will not provide all of the information needed to fully understand a broad, complex research question. Thinking back to the original research question, what additional studies would you suggest conducting next?

Section 1.1 Summary

The set of observational units on which we collect data is called a **sample**. The number of observational units is the **sample size**. A number computed to summarize the variable measured on a sample is called a **statistic**.

For a chance process, a **parameter** is a long-run numerical property of that process, such as a probability (long-run proportion).

A simulation analysis based on a chance model can assess the strength of evidence provided by sample data against a particular claim about the chance model. The logic of assessing statistical significance employs what we call the **3S Strategy**:

- **Statistic**: Compute an observed statistic from the data.
- **Simulate**: Identify a model for the "by chance alone" explanation. Repeatedly simulate values of the statistic that could have occurred from that chance model.
- **Strength of evidence**: Examine how unusual the observed value of the statistic would be under repeated application of the chance model.
  - If the observed value of the sample statistic is *unlikely* to have occurred from the chance model, then the data provide *strong* evidence against the chance model as the explanation.
  - If the observed value of the sample statistic is *not* unlikely to have occurred from the chance model, then the chance model is a *plausible* explanation for the observed data.

The **chance model** considered in this section involved tossing a fair coin. This chance model allowed for assessing whether an observed number of "successes" in a study provided strong evidence that the two outcomes of a categorical variable were not equally likely. In the next section you will consider other chance models, but the reasoning process will remain the same.

Section 1.2: Measuring the Strength of Evidence

Introduction

In the previous section, we discussed testing whether or not a chance model of *equally choosing between two options (e.g., flipping a coin)* was plausible (believable) based on some observed sample data. Not all situations call for a coin flipping model. For example, if you were guessing at the answers to a true-false test, you would expect to have a 50% chance of getting a question correct. However, if you were guessing at the answers to a multiple choice test where each question had four possible answers, you should expect to have only a 25% chance of guessing a particular question correctly. Other events, as you will soon see, can be a bit more complicated than this. In this section, we will apply the Six-Step Statistical Investigation Method in the more general setting of an arbitrary probability of "success."

We will also learn helpful terminology (null and alternative hypotheses, p-value) that are commonly used in statistical investigations to describe the process of drawing conclusions from data. These things will both help us formalize the procedure of a ***test of significance*** and give us some guidelines to help us determine when we have strong enough evidence that our chance model is not correct. We will also introduce some new symbols, for convenience, so try to keep the big picture in mind.

Example 1.2: Rock Paper Scissors

Have you ever played Rock-Paper-Scissors (or *Rochambeau*)?It's considered a "fair game" in that the two players are equally likely to win (like a coin toss). Both players simultaneously display one of three hand gestures (rock, paper or scissors), and the objective is to display a gesture that defeats that of your opponent. Official rules are available from the World RPS Society (www.worldrps.com), but the main gist is that rocks break scissors, scissors cut paper, and paper covers rock.

But is it really a fair game? Do some players exhibit patterns in their behavior that an opponent can exploit? An article published in *College Mathematics Journal* (Eyler, Shalla, Doumaux, and McDevitt, 2009) found that players, particularly novices, tend to not prefer scissors. Suppose you decide to investigate this tendency with a friend of yours who hasn't played the game before. You explain the rules of the game and play 12 rounds. Suppose your friend only shows scissors twice in those 12 plays.

Think about it:
- What are the observational units? What is the variable?
- What is the underlying process from which the researchers gathered their sample data?
- What is the parameter of interest?
- What is the observed statistic?

A similarity between this study and ones we've looked at previously (e.g., Buzz and Doris) is that we have repeated outcomes from the same random process. In this study, the individual plays of the game are the observational units, and the variable is whether or not the player chooses scissors. This categorical variable has just two outcomes (scissors or not scissors), and so is sometimes called a ***binary variable***.

Definition: A ***binary variable*** is a categorical variable with only two outcomes. Often we convert categorical variables with more than two outcomes (e.g., rock, paper, or scissors) into binary variables (e.g., scissors or not scissors).In this case we also define one outcome to be a "success" and one to be a "failure."

The underlying process here is choosing a hand gesture in each repeated play of the game. The parameter of interest is the long-run probability that any player picks scissors. The observed statistic is number of times scissors was chosen in the 12 plays (2) or the proportion of times that scissors was chosen (1/6).

Think about it: What is the research conjecture about the long-run probability?

We are told that players seem to pick scissors less than the other two gestures. This would imply that the long-run probability for picking scissors is less than one-third. And in this study, that's what you found. Your friend only chose scissors 1/6 of the time.

But perhaps that was just an unlucky occurrence in this study? Maybe your friend would play scissors 1/3 of the time if he played the game for a very long time, and, just by chance, you happened to observe less than one-third in the first 12 games? So, again, we have two possible explanations for why our statistic (1/6) is below what we think is the true parameter (1/3) if the random chance model is correct.

- Novice players, like your friend, pick equally among the 3 gestures and we just happened to observe fewer scissor choices in this study by chance alone.

- Novice players really do tend to pick scissors less than 1/3 of the time.

We can rewrite these two possible explanations as two competing hypotheses:

> Null hypothesis: Novice players pick equally between the 3 gestures in the long-run (picking scissors one-third of the time in the long run).

> Alternative hypothesis: Novice players pick scissors less than one-third of the time in the long run.

Definition:
- The *null hypothesis* typically represents the "by random chance alone" explanation. The chance model (or "null model") is chosen to reflect this hypothesis.
- The *alternative hypothesis* typically represents the "there is an effect" explanation that contradicts the null hypothesis. It represents what the researchers are hoping to gather evidence to support (the research conjecture).

Notice that the null and alternative hypotheses are statements about the parameter (probability of choosing scissors) and the underlying process (in the long-run), not just about what was observed in this study. In fact, we should state the hypotheses prior to conducting the study, before we ever gather any data! Our goal is to use the sample data to estimate the parameter and to infer whether this unknown value is less than one-third.

Use of Symbols

We can use mathematical symbols to represent quantities and simplify our writing. Throughout the book we will emphasize written explanations, but will also show you mathematical symbols which you are free to use as a short-hand once you are comfortable with the material. The distinction between parameter and statistic is so important that we always use different symbols to refer to them.

We will use the Greek letter for p which is π (pronounced "pie") to represent a parameter that is a probability. For example, the long-run probability that a noviceplayer picks scissors can be represented by π. We will then use the symbol $\hat{p}$ (pronounced "p-hat") to represent the proportion in the sample. In this example, $\hat{p} = 1/6 \approx 0.167$, the proportion of times that your friend chose scissors. In fact, one way to distinguish between the parameter and the statistic is verb tense! The statistic is the proportion of times that your friend did (past tense, observed) show scissors. The parameter is the long-run proportion he would throw scissors (future tense, unobserved) if he played the game forever.

We can also use symbols for the hypotheses. The null hypothesis is often written as H_0 and the alternative as H_a. Finally, the sample size, which is the same as the number of observational units, also has a symbol: the letter "n."

We can use these symbols to rewrite the null and alternative hypotheses.

$H_0: \pi = 1/3$

$H_a: \pi < 1/3$

Where π represents your friend's true probability of throwing scissors.

Notice that both of these hypothesis statements are claims about the parameter π. When testing a single probability as we are doing in this example, we compare the actual (but unknown) value of the parameter (π) to the same numerical value in both the null and alternative hypotheses. In this case that value is 1/3 (often called the "hypothesized" probability). What differs between the two statements is the inequality symbol. The null hypothesis will always contain an equals sign and the alternative hypothesis will contain either a (strictly) greater than sign (as it would have in the Doris and Buzz example from the previous question), a (strictly) less than sign (as it does in this example), or a not equal to sign like we will see in Section 1.4. Which in equality symbol to use in the alternative hypothesis is determined by the research conjecture.

Applying the 3S Strategy

Now, let's apply the 3S Strategy to evaluate the evidence against the null hypothesis in favor of the alternative hypothesis, that your friend is less likely to choose scissors than the other options.

1. Statistic: Your friend showed scissors 1/6 of the time in the first 12 plays.

2. Simulation: We will again focus on the chance alone explanation, to see whether we have strong evidence that your friend chooses scissors less than 1/3 of the time in the long run. So we will use a chance model that assumes the probability of your friend choosing scissors is one-third, and then examine the types of results we get for the sample proportion of times he chooses scissors to see whether our observed statistic is consistent with that chance variability.

> **Think about it:** Can we use a coin to represent the chance model specified by the null hypothesis like before? It not, can you suggest a different random device we could use? What needs to be different about our simulation this time?

We cannot directly use a coin to simulate the chance model for this scenario. The key difference is now we want to sample from a process where the probability of "success" is not 0.5 (as with coin tossing) but 1/3. This will model your friend choosing scissors one-third of the time. We could use dice (e.g., rolls of 1 or 2 represent scissors) or cards (like with Monty Hall), or the computer can do this for us with correspondences as outlined in Table1.2.

Table 1.2: Parallels between actual study and simulation.

Observational unit	=	One round of the game
Success	=	Plays scissors
Failure	=	Doesn't play scissors
Chance of success	= 1/3 =	Three gestures are equally likely; chance of scissors is one out of three
One repetition	=	A simulated set of 12 rounds of the game

For the first 12 plays of the game, your friend only showed scissors twice. If scissors were chosen as often as the other two options, we would expect four of the 12 plays to be scissors, so maybe this is not all that different, especially in the short run. Let's first evaluate how surprising it is for your friend to only play scissors twice, looking at results from a simulation that models your friend choosing scissors one-third of the time in the long-run.

In the One Proportion applet you can change the probability of success to any number between 0 and 1, and the applet will generate successful attempts at that rate. One way to think about it is as a spinner that will spin and land in one are a one-third of the time and the other area two-thirds of the time. So we now have a chance model that generates successes with a long-run probability equal to the null hypothesis probability of 1/3.

Figure 1.7 shows us the distribution of the simulated sample proportion of scissor choices (successes) that could have happened in 12rounds of the game assuming this chance model. We can use this distribution to assess the strength of evidence against the 1/3 chance model. The distributions of simulated statistics we have been looking at represent what could have happened in the study assuming the null hypothesis was true. For that reason, we will now refer to this distribution as the ***null distribution***.

Figure 1.7: The null distribution of simulated sample proportion of successes in 12 plays assuming scissors is chosen 1/3 (≈ 0.3333) of the time in the long run

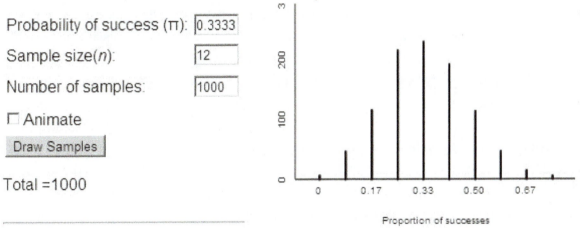

Notice that we are using the proportion of successes as the statistic instead of the number of successes, but these are equivalent. This distribution is centered around 1/3 and has a very slight skew to the right. But what does the distribution tell us about whether or not your friend tends to avoid scissors?

3. Strength of evidence

Think about it: How do we use this null distribution to evaluate whether the 1/3 model is appropriate for your friend's scissors selection process?

The key is to see whether the observed statistic (2 out of 12 ≈ 0.167) is consistent with the chance variability in the 1/3 process. So we need to decide whether or not 0.167 is a typical value or falls in the tail of the distribution. In Figure 1.7, we see that this is a bit of a difficult call. The outcomes at 0.17 are not that far in the tail of the distribution, but are not right in the middle either. What we need is a more systematic way of measuring how unusual an outcome of 0.167 is in this null distribution.

We could see how often our observed result occurs in the distribution, but that becomes difficult to compare across different distributions (see FAQ 1.2.1). A more common approach for measuring how unusual an observation is in a distribution is to determine how much of the distribution lies to the left of the value and how much to the right (think "percentiles" from when you took the SAT or ACT!). The probability of getting a result at least as extreme as the observed statistic, when the null hypothesis is true, is called the *p-value*. We can estimate the p-value from our simulated null distribution by counting how many (and what proportion) of the simulated sample proportions are as extreme, or more extreme (in the direction of the alternative hypothesis), than the observed sample proportion.

Definition: The *p-value* is the probability of obtaining a value of the statistic at least as extreme as the observed statistic when the null hypothesis is true. We can estimate the p-value by finding the proportion of the simulated statistics in the null distribution that are *at least as extreme* (in the direction of the alternative hypothesis) as the value of the statistic actually observed in the research study.

The p-value takes into account the could-have-been outcomes (assuming the null hypothesis is true) that are as extreme, or more extreme, than the one we observed. This provides a direct measure of our strength of evidence against the "by chance alone" or null model and allows for a standard, comparable value for all scientific research studies. Smaller p-values mean the value of the observed statistic, under the null model, is more unlikely by chance alone. Hence, smaller p-values indicate stronger evidence against the null model.

Let's reconsider the distribution from Figure 1.7. In Figure 1.8, we added a vertical line at the observed statistic, 0.167. The "proportion of repetitions" counts how many times 0.167 or smaller occurred under the specified chance model. In this case, we found 173 of the 1000 samples from a process with $\pi = 1/3$resulted in a simulated sample proportion $\hat{p}$ of 0.167 or smaller. So 0.173 is our estimate of the study's p-value.

Figure 1.8: The null distribution of simulated sample proportion of successes in 12 rounds of rock-paper-scissors for a novice that plays scissors one-third of the time in the long run with the observed proportion (0.167) or even smaller shown, yielding an approximate p-value of 0.1730.

Calculating a p-value from a simulation analysis is an approximation. Different simulations will have slightly different p-values based on 1000 repetitions. Performing more repetitions generally produces more accurate approximations. For our purposes, using 1000 repetitions is typically good enough to provide reasonable approximations for p-values.

Notice that the approximate p-value of 0.173 in Figure 1.8 is computed as the proportion of the 1000 simulated samples with a sample proportion of 0.167 *or smaller*. A p-value always computes "more extreme" in the direction of the alternative hypothesis. In this case the alternative hypothesis is "less than 1/3" and so we look at the lower tail to compute "more extreme than" for the p-value.

So how do we *interpret* this 0.173? As a probability; so we can say that in the long run, if we repeatedly generate random sets of 12 rounds of the game under identical conditions with the probability of scissors equal to 1/3, we expect to observe a sample proportion of 0.167 or smaller, by chance alone, in about 17.3% of those repetitions.

But what does 0.173 tell us about the strength of the evidence? As stated earlier, smaller p-values are stronger evidence against the null hypothesis in favor of the alternative hypothesis. Is 0.173 small enough? Although there is no hard and fast rule for determining how small is small enough to be convincing, we offer the following guidelines.

Guidelines for evaluating strength of evidence from p-values

0.10 < p-value	not much evidence against null hypothesis; null is plausible
0.05 < p-value ≤ 0.10	moderate evidence against the null hypothesis
0.01 < p-value ≤ 0.05	strong evidence against the null hypothesis
p-value ≤ 0.01	very strong evidence against the null hypothesis

> **The smaller the p-value, the stronger the evidence against the null hypothesis.**

Many researchers consider a p-value ≤ 0.05 to be sufficient to conclude there is convincing evidence against the null hypothesis (see FAQ 1.2.2 for some intuition on that number), but in some situations you may want stronger evidence. For now, just keep in mind that the smaller the p-value, the stronger the evidence against the null hypothesis (chance model is true) and in favor of the alternative hypothesis (typically the research conjecture).

So we would consider only 2 plays of scissors in the first 12 rounds of the game to not be much evidence against the null hypothesis your friend would play scissors 1/3 of the time in the long run. Why? Because a p-value of 0.173 indicates that getting 2 or fewer choices of scissors in 12 plays, if the probability of scissors was really 1/3, is *not surprising*. Hence, 1/3 is still a plausible value for your friend's long-run probability.

What if your friend had only played scissors once? Then our statistic becomes $\hat{p} \approx 0.083$. This is even further away from the expected one-third. So what will happen to the p-value? We can use the same null distribution but now we need to see how often we find a sample proportion as small as 0.083 or smaller. Such simulated statistics are at or below the tall vertical line shown in Figure 1.9.

> **Key idea:** Values of the statistic that are even further from the hypothesized parameter result in a smaller p-value and stronger evidence against the null hypothesis.

Conclusions

So what can we conclude from this study? We approximated a p-value (0.173) for the first 12 plays of the game that was not small and did not provide strong evidence against the null hypothesis. Have we *proven* the null hypothesis is true? No! In fact, we will never get to prove the null hypothesis true because we had to assume it was true to do the analysis. So the results of the simulation can never "support" the null hypothesis. What we can say is that these results are not inconsistent with the type of result we would expect to see when the null hypothesis is true. In other words, the null hypothesis is one plausible (or believable) explanation for the data. If we wanted to investigate the issue more closely, we could have our friend play more games (increase the sample size). A larger sample size would give us a better chance of detecting any tendency that might be there (we'll discuss the issue of sample size more fully in section 1.4).

Figure 1.9: The null distribution of simulated sample proportion of successes in 12 rounds of rock-paper-scissors for novices that play scissors one-third of the time in the long run.If the observed proportion would only have been 0.083, we can see the p-value would have been smaller and we would therefore have stronger evidence against the null hypothesis.

The approximate p-value has decreased to 0.077. This makes sense because if the null hypothesis is true ($\pi = 1/3$), it should be more surprising to get a sample proportion further from 1/3.

Key assumptions: Model vs. reality

Stepping back, we need to keep something in mind about the use of chance models in assessing strength of evidence. Chance models are models, and, thus, are not reality. They make key assumptions that we need to pay careful attention to, otherwise we may too quickly leap to an incorrect conclusion.

For example, with the Rock-Paper-Scissors game we simulated data from a chance model with a one-third success probability. This means each and every round has exactly the same probability of scissors. But, is that a reasonable assumption? What if your friend changes his strategy part way through? Our chance model ignores these aspects of "reality" and makes the situation much simpler than it really is.

Thus, if we could say "we have strong evidence against the chance model" are we saying that your friend subconsciously avoids scissors? Well, maybe. All we are saying is we think something else is going on other than your friend picking equally among the three gestures. When we think about Steps 4 and 5 of the Six-Step Statistical Investigation Method, we must be aware about these assumptions we're making---specifically about the ways in which our model does not match reality.

FAQ 1.2.1 Why do we need to include "or more extreme" when computing our p-value?

Q: So Buzz pushed the correct button 15 out of 16 times, why isn't the p-value the likelihood I would get exactly 15 heads out of 16 flips of a coin? Why do we have to also include the likelihood of getting more than 15 heads?

A: In flipping just 16 coins, the probability of getting 15 heads is very similar to the probability of getting 15 or more heads, so I can see how this would be confusing. I have an idea to help you understand this. Do you have any coins with you?

Q: I might, let me check. It looks like I have a few.

A: Good. Do you think you can toss the coin "fairly"?

Q: I don't see why not.

A: Start flipping it. As you do this, keep track of the number of heads you get as well as the total number of flips. Also think about whether or not your results indicate whether this is an unfair tossing process.

Q: Okay, I flipped the coin 10 times and got 6 heads. Because this is close to 50-50, I don't think I could conclude that my coin tossing is unfair.

A: Good. The probability of getting 6 heads out of 10 flips is about 0.20 whereas the probability of getting 6 or more heads (our actual p-value) is about 0.37. Either way you think of this, getting 6 heads out of 10 coin flips is not too unlikely. Let's keep flipping.

Q: Okay, I've flipped it 50 times and …
A Keep flipping.

Q: I've flipped it now 100 times and…

A: Keep flipping.
  (1 hour later)

Q: I think my arm is going to fall off!

A: I guess you can stop now. What are your results?

Q: I flipped the coin 1000 times and got 505 heads.

A: Do you think your results show that your coin tossing is not fair?

Q: Of course not. Getting 505 heads out of 1000 flips is close enough to 50% that the result is not unexpected.

A: Do you mean you expected to get 505 heads?

Q: Well no, not exactly 505, but I did expect to get something close to 500.

A: In fact, the probability of getting exactly 505 heads out of 1000 flips of a fair coin is only about 0.02. Because there are so many different outcomes possible, the probability of any one particular outcome is rather small. But we wouldn't want to look at the probability of 0.02 and consider this a surprising outcome - this outcome is definitely among the typical values in the distribution. This is better conveyed by noting that the probability of getting 505 heads or more in 1000 flips is about 0.39.

Q: And because this p-value is so high, I would not conclude my coin tossing is unfair!

A: Exactly!

FAQ 1.2.2: What p-value should make us suspicious?

Q: So, that table you gave us about strength of evidence and p-values seems pretty arbitrary. Where did that come from?

A: Did you ever google "Persi Diaconis"?

Q: No, why?

A: Try it sometime. He's unusual … even for a statistician he's unusual. He was one of the first people to win one of the MacArthur "genius" awards.

Q: Is that all? What else can he do?

A: Flip a coin and make it come up heads every single time.

Q: I don't believe anyone can do that. Isn't this whole dialog just a stat prof's geeky way to sneak in p-values?

A: Sorry, I can't help it. We stat profs are so transparent sometimes. That's just the data. But humor me: How many heads in a row would Persi have to get to make you begin to think maybe it's not just chance? Would heads on the first two tosses do it for you?

Q: Of course not. That happens 25% of the time.

A: You win a point for extra credit. What about heads on the first three tosses?

Q: You're making me get serious about this. Three heads in a row wouldn't happen very often, but it's still not unusual enough to be suspicious.

A: What about four in a row? Five in a row?

Q: Now it's really getting serious ... and suspicious. I find it hard to believe that you can get five heads in a row just by chance. Sure, it can happen, but five in a row is enough to make me think maybe there's something else going on.

A: As it turns out the chances of getting four heads in a row is about 6% and the chances of getting five heads in a row is about 3%. This means that most people start getting suspicious when something occurs less than ~5% of the time.

Q: That sounds familiar. Didn't you tell us that when the p-value is less than 0.05 that gives us strong evidence?

A: Exactly! So, just like getting four or five heads in a row makes us think something else is going on, getting a p-value less than 0.05 gives us strong evidence the null hypothesis may be incorrect.

Exploration 1.2: Tasting Water

People spend a lot of money on bottled water. But do they really prefer bottled water to ordinary tap water? Researchers at Longwood University (Lunsford and Dowling Fink, 2010) investigated this question by presenting people who came to a booth at a local festival with four cups of water. Three cups contained different brands of bottled water, and one cup was filled with tap water. Each *subject* (person) was asked which of the four cups of water they most preferred. Researchers kept track of how many people chose tap water in order to see whether tap water was chosen significantly less often than would be expected by random chance.

Step 1: Ask a research question.

1. What is the research question that the researchers hoped to answer?

Step 2: Design a study and collect data.

2. Identify the observational units in this study.

3. Identify the variable. Is the variable quantitative or categorical?

4. How many outcomes can the variable have?

> **Definition:** A *binary variable* is a categorical variable with only two outcomes. Often we convert categorical variables with more than two outcomes (e.g., four brands of water) into binary variables (e.g., tap water or not). In this case we also define one outcome to be a "success" and one to be a "failure."

5. Describe the parameter of interest (in words).(*Hint*: The parameter is the long-run probability of …?)

6. One possibility here is that subjects have an equal preference among all four waters and so are essentially selecting one of the four cups at random.In this case what is the long-run probability that a subject in this study would select tap water?

7. Another possibility is that the subjects are less likely to prefer tap water than the bottled water brands.In this case what can you say about the long-run probability that a subject in this study would select tap water?(*Hint*: You are not to specify a particular value this time; instead indicate a *direction* from a particular value.)

8.

> **Definition:**
> - The *null hypothesis* typically represents the "by chance alone" explanation. The chance model (or "null model") is chosen to reflect this hypothesis.
> - The *alternative hypothesis* typically represents the "there is an effect" explanation that contradicts the null hypothesis. It represents what the researchers are hoping to gather evidence to support (the research conjecture).

9. Your answers to #6 and #7 should be the null and alternative hypotheses for this study. Which is which?

The researchers found that 3 of 27 subjects selected tap water.

Step 3: Explore the data.

10. Calculate the value of the relevant statistic.

Use of symbols

We can use mathematical symbols to represent quantities and simplify our writing. Throughout the book we will emphasize written explanations, but will also show you mathematical symbols which you are free to use as a short-hand once you are comfortable with the material. The distinction between parameter and statistic is so important that we always use different symbols to refer to them.

When dealing with a parameter that is a long-run probability, such as the probability that a (future) subject in this study *would* choose tap water as most preferred, we use the Greek letter π (pronounced "pie").But when working with a statistic that is the proportion of "successes" in a sample, such as the proportion of subjects in this study who *did* choose tap water as most preferred, we use the symbol $\hat{p}$ (pronounced "p-hat").Finally, we use the symbol *n* to represent the sample size.

11. What is the value of $\hat{p}$ in this study?

$$3/27$$

12. What is the value of *n* in this study?

$$27$$

13. Hypotheses are always conjectures about the unknown parameter. π. You can also use H_0 and H_a as short-hand notation for the null and alternative hypotheses, respectively. A colon, ':', is used to represent the word 'is.' Restate the null and alternative hypotheses using π.

H_0: $1/4 = \pi$

H_a: $\pi < 1/4$

Step 4: Draw Inferences.

14. Is the sample proportion who selected tap water in this study less than the probability specified in the null hypothesis?

15. Is it *possible* that this proportion could turn out to be this small even if the null hypothesis was true (i.e., even if people did not really dislike the tap water and were essentially selecting at random from among the four cups)?

    Yes

As we did with Buzz and Doris in Section 1.1, we will use simulation to investigate how surprising the observed sample result (3 of 27 selecting tap water) would be, if in fact subjects did not dislike tap water and so each had a ¼ probability of selecting tap water. (Note also that our null model assumes the same probability for all subjects.)

Think about it: Can we use a coin to represent the chance model specified by the null hypothesis like before? It not, can you suggest a different random device we could use? What needs to be different about our simulation this time?

16. Explain why we cannot use a simple coin toss to simulate the subjects' choices, as we did with the Buzz/Doris study.

17. We could do the simulation using a set of four playing cards: one black and three red. Explain how the simulation would work in this case.

18. Another option would be to use a spinner like the one shown here , like you would use when playing a child's board game. Explain how the simulation would work if you were using a spinner. In particular:

    What does each region represent?

    How many spins of the spinner will you need to do in order simulate one repetition of the experiment when there is equal preference between the four waters (null hypothesis is true)?

19. We will now use the **One Proportion** applet to conduct this simulation analysis. Notice that the applet will show us what it would be like if we were simulating with spinners.

  a. First enter the **probability of success** value specified in the null hypothesis.

  b. Enter the appropriate **sample size** (number of subjects in this study).

  c. Enter 1 for the number of samples, and press **Draw Samples**. Report the number of "successes" in this simulated sample.

  d. Now, select the radio button for "Proportion of successes." Report the proportion of successes in this simulated sample. Use your answer to c. to verify how this value is calculated.

  e. Leaving the "Proportion of successes" radio button selected, click on **Draw Samples** four more times. Do you get the same results each time?

  f. Now enter 995 for the number of samples and click on **Draw Samples**, bringing the number of simulated samples to 1000.Comment on the center, variability, and shape of the resulting distribution of sample proportions.

This distribution of simulated sample proportions is called the ***null distribution***, because it is based on assuming the null hypothesis to be true.

20. Recall that the observed value of the sample proportion who selected tap water in this study was $\hat{p} = 3/27 \approx 0.1111$.Looking at the null distribution you have simulated, is this a very unlikely result, if the null hypothesis was true? In other words, is this value far in the tail of the null distribution?

You might very well find that #19 is a bit of a tough call. The value 0.1111 is not far in the tail of the distribution, but it's also not near the middle of the distribution. To help make a judgment about strength of evidence in this case, we can count how many (and what proportion) of the simulated sample proportions are as extreme, or more extreme, than the observed value.

21. Use the applet to count how many (and what proportion) of the simulated sample proportions are more extreme than the observed value.

To do this, first click on the $\geq$ inequality symbol to change it to $\leq$ (to match the alternative hypothesis).Then enter **0.1111** (the observed sample proportion who chose tap water) in the box to the left of the Count button.Then click on the **Count** button.Record the number and proportion of simulated sample proportions that are as extreme, or more extreme, than the observed value.

Definition: The *p-value* is estimated as the proportion of simulated statistics in the null distribution that are *at least as extreme* (in the direction of the alternative hypothesis) as the value of the statistic actually observed in the research study.

How do we *evaluate* this p-value as a judgment about strength of evidence provided by the sample data against the null hypothesis? One answer is: The smaller the p-value, the stronger the evidence against the null hypothesis and in favor of the alternative hypothesis. But how small is small enough to regard as convincing? There is no definitive answer, but here are some guidelines:

Guidelines for evaluating strength of evidence from p-values

$0.10 <$ p-value	not much evidence against null hypothesis; null is plausible
$0.05 <$ p-value ≤ 0.10	moderate evidence against the null hypothesis
$0.01 <$ p-value ≤ 0.05	strong evidence against the null hypothesis
p-value ≤ 0.01	very strong evidence against the null hypothesis

The smaller the p-value, the stronger the evidence against the null hypothesis.

22. Is the approximate p-value from your simulation analysis (your answer to #20) small enough to provide much evidence against the null hypothesis that subjects prefer tap water equally to the brands of bottled water? If so, how strong is this evidence? Explain.

23. When computing p-values, "more extreme" is always measured in the direction of the alternative hypothesis. Use this fact to explain why you clicked the $\leq$ earlier.

Step 5: Formulate conclusions.

24. Do you consider the observed sample result to be statistically significant? Recall that this means that the observed result is unlikely to have occurred by chance alone.

25. How broadly are you willing to generalize your conclusions? Would you be willing to generalize your conclusions to water drinkers beyond the subjects in this study? How broadly? Explain your reasoning.

Step 6: Look back and ahead.

26. Suggest a new research question that you might investigate next, building on what you learned in this study.

Alternate Analysis

Instead of focusing on the subjects who chose tap water, you could instead analyze the data based on the subjects who chose one of the three bottled waters. Because 3 of 27 subjects chose tap water, we know that 24 of the 27 subjects chose one of the brands of bottled water. Now let the parameter of interest (denoted by π) be the long-run probability that a subject selects one of the *bottled* water cups as most preferred.

27. Conduct a simulation analysis to assess the strength of evidence provided by the sample data.

    a. The research conjecture is that subjects tend to select bottled water (more or _less_) often than tap water. (Circle your answer.)

        More   Less

    b. State the null hypothesis in words and in terms of the (newly defined) parameter $\square$.

c. State the alternative hypothesis in words and in terms of the (new) parameter □.

d. Calculate the observed value of the relevant statistic.

e. Before you use the **One Proportion** applet to analyze these data, indicate what values you will input:

    Probability of success:

    Sample size:

    Number of samples:

f. Use the applet to produce the null distribution of simulated sample proportions. Comment on the center, variability, and shape of this distribution. Be sure to comment on how this null distribution differs from the null distribution in #16f.

g. In order to approximate the p-value, you will count how many of the simulated proportions are _____ or (larger or smaller) and then divide by _____.

h. Estimate the p-value from your simulation results.

i. *Interpret* this p-value.(*Hint*: This is the probability of what, assuming what?)

j. *Evaluate* this p-value: How much evidence do the sample data provide against the null hypothesis?

28. Does your analysis based on the number who chose bottled water produce similar conclusions to your previous analysis based on the number who chose tap water? Explain.

You should have found that it does not matter whether you focus on the number/proportion that chose tap water or the number/proportion that chose bottled water. In other words, it does not matter which category you define to be a "success" for the *preferred water* variable. Your findings should be very similar provided that you make the appropriate adjustments in your analysis:

- using 0.75 instead of 0.25 as the null value of the parameter,
- changing the alternative hypothesis to "$\pi > 0.75$" rather than "$\pi < 0.25$", and
- calculating the p-value as the proportion of samples with $\hat{p} \geq 0.8889$ rather than $\hat{p} \leq 0.1111$.

Section 1.2 Summary

The 3S Strategy for assessing statistical significance also applies to chance models other than a coin toss:

- Other probabilities of success, such as 1/3, can be analyzed.
- The strategy can assess whether a long-run probability is *less than* a conjectured value, as well as greater than a conjectured value.

Introducing some terminology can help to clarify the 3S Strategy when used to conduct a **test of significance**:

- The **null hypothesis** is the "by chance alone" explanation.
- The **alternative hypothesis** contradicts the null hypothesis.
- The **null distribution** refers to the simulated values of the statistic generated under the assumption that the null hypothesis ("by chance alone" explanation) is true.

Strength of evidence can be assessed numerically by determining how often a simulated statistic as or more extreme than the observed value of the statistic occurs in the null distribution of simulated statistics.

- The **p-value** is estimated by determiningthe proportion of simulated statistic values in the null distribution that are at least as extreme as the observed value of the statistic.
- The smaller the p-value, the stronger the evidence against the null hypothesis ("by chance alone" explanation).

Some guidelines for evaluating **strength of evidence** based on a p-value are:

$0.10 < \text{p-value}$	not much evidence against null hypothesis
$0.05 < \text{p-value} \leq 0.10$	moderate evidence against the null hypothesis
$0.01 < \text{p-value} \leq 0.05$	strong evidence against the null hypothesis
$\text{p-value} \leq 0.01$	very strong evidence against the null hypothesis

---

Notation check: Here is a quick summary of the symbols we've introduced in this section.

- π represents the parameter when it is a probability

- $\hat{p}$ represents the statistic when it is the observed proportion

- H_0 represents the null hypothesis

- H_a represents the alternative hypothesis

- n represents the sample size

Section 1.3: Alternative Measure of Strength of Evidence

Introduction

In the previous section, you learned the formal process of a *test of significance*: make a claim about the parameter of interest (through competing null and alternative hypothesis); gather and explore data; follow the 3S Strategy to calculate an observed statistic from the data, simulate a null distribution and measure the strength of evidence the observed statistic has against the null hypothesis; draw a conclusion about the null hypothesis. The p-value was introduced as a standard way to measure the strength of evidence against the null hypothesis. We found that the smaller the p-value, the stronger our evidence against the null hypothesis and in favor of the alternative hypothesis. However, if we find strong evidence against the null hypothesis, this does not mean we have proven the alternative hypothesis to be true. Similarly, if we don't have strong evidence against the null hypothesis, this does not mean we have proven the null hypothesis to be true. What we can conclude is whether or not the "by chance alone" explanation is reasonable. You also confirmed in the previous section that if an observed result is not as far away from the proportion under the null hypothesis, then it provides less evidence against the null hypothesis, meaning that the null hypothesis is one plausible explanation for the observed data. In this section, you will explore another method often used to measure how far away the observed statistic is from the parameter value conjectured by the null hypothesis.

The key to Step 4 of the Statistical Investigation Method, when we apply the 3S Strategy with one categorical variable, has been assuming some claim about the long-run probability of a particular outcome, π , and then seeing whether or not our observed sample proportion is consistent with that claim. One approach is to create the null distribution (the could-have-been simulated sample proportions assuming the null hypothesis to be true), and then use the p-value to measure how often we find a statistic at least as extreme as that of the actual research study. The p-value gives us a standard way of measuring whether or not our observed statistic fell in the tail of this distribution. In this section you will find that for many distributions there is another convenient way to measure how far the observed statistic is from the hypothesized value under the null.

Example 1.3: Heart Transplant Operations

In an article published in the *British Medical Journal* (2004), researchers Poloniecki, Sismanidis, Bland, and Jones reported that heart transplantations at St. George's Hospital in London had been suspended in September 2000, after a sudden spike in mortality rate. Of the last 10 heart transplants, 80% had resulted in deaths within 30 days of the transplant. Newspapers reported that this mortality rate was over five times the national average. Based on historical national data, the researchers used 15% as a reasonable value for comparison.

Think about it: What research question can we ask about these data? Identify the observational units, variable of interest, parameter, and statistic. What is the null hypothesis?

We would like to know whether the current underlying heart transplantation mortality rate at St. George's hospital exceeds the national rate. So the observational units are the individual heart transplantations (the sample size for the data above is 10) and the variable is whether or not the patient died.

When we conduct analyses with binary variables, we often call one of the outcomes a "success" and the other a "failure," and then focus the analysis on the 'success' outcome. It is arbitrary which outcome is defined to be a success, but you need to make sure you do so consistently throughout the analysis. In many epidemiological studies, death is the outcome of interest and so 'patient did not survive for 30 days post-operation' is called a 'success' in this case!

Knowing the 'success' outcome, allows us to find the observed statistic which is the number of successes divided by the sample size. In this case, the observed statistic is 8 out of 10 or $\hat{p} = 0.80$, and the parameter is the actual long-run, current probability of a death within 30 days of a heart transplant operation at St. George's. We don't know the actual value of this probability; we have just observed a small sample of data from the heart transplantation operation process; if a different set of 10 people had been operated on, the statistic would mostly likely differ. We will use the symbol π to refer to this unknown probability.

The null hypothesis is that the current death rate at St. George's hospital is no different from other hospitals, but the researchers want to know whether these data are convincing evidence that the death rate is actually higher at St. George's. So we will state the hypotheses as follows:

> Null hypothesis: Death rate at St. George's is the same as the national rate (0.15).
> Alternative hypothesis: Death rate at St. George's is higher than the national rate.

Symbols

You can also write your null and alternative hypotheses like this

$H_0: \pi = 0.15$

$H_a: \pi > 0.15$

where π is the actual long-run probability of death after a heart transplant at St. George's.

Applying the 3S Strategy

Using the 3S Strategy from the previous section, we observed 0.80 as our statistic and now we will simulate one thousand repetitions from a process where π = 0.15 (under the null hypothesis).We did this using the One Proportion applet as shown in Figure 1.10.

Figure 1.10: Null distribution (could-have-been simulated sample proportions) for 1000 repetitions of drawing samples of 10 "patients" from a process where the probability of death is equal to 0.15. "Success" has been defined to be patient death.

Even though the sample size for this study is quite small ($n = 10$), we can see that 0.80 is not even close to the could-have-been results from the simulation. In fact, the approximate p-value is less than 1/1000, as we never observed a value of 0.8 or larger by chance in these 1000 repetitions. This provides very strong evidence against the null hypothesis and in favor of the alternative hypothesis. Of course, it tells us nothing about *why* the death rate from heart transplantations is higher at St. George's, but it is pretty convincing that something other than random chance is at play.

Note that in Figure 1.10 the average of the 1000 proportion of successes values is 0.152, which is quite close to 0.15-- the probability of deaths if the chance model is correct. This makes sense. Also, notice that the proportions of success vary quite a lot from sample to sample, ranging from 0 to 0.6. In fact, the variability of the distribution, as measured by the standard deviation is 0.113. Remember from the Preliminaries we can think of standard deviation as the distance a typical value in the distribution is away from the mean of the distribution. In this case, 0.113 is the average distance that a simulated value of the statistic (proportion of patients who died) is from 0.152. We'll come back and use the value of the standard deviation in a moment.

Digging deeper into the St. George's mortality data

We might still wonder whether these most recent 10 operations, which caught people's attention, are truly representative of the process as a whole. One approach is to investigate whether there were any major changes at the hospital recently (e.g., new staff, new sterilization protocols).Another is to gather more data over a longer period of time, to better represent the underlying process. So the researchers decided to examine the previous 361 heart transplantations at St. George's hospital, dating back to 1986.They found that 71 of the patients died within 30 days of the transplant.

> **Think about it:** Now what is the value of the observed statistic? Predict how the simulated null distribution will change for this new situation. Do you think the observed statistic will still be highly statistically significant?

Figure 1.11 shows the results of 1000 repetitions of drawing samples of size $n = 361$ from a process with $\pi = 0.15$ (still assuming the null hypothesis to be true).

Figure 1.11: The null distribution of 1000 repetitions of drawing samples of 361 "patients" from a process where the probability of death is equal to 0.15

> **Think about it:** Where does $\hat{p} = 0.197$ fall in this distribution? In the tail or among the more typical values? How does the p-value tell us whether the observed statistic is in the tail or not?

First, we note that again the p-value is small (0.003), so we still have strong evidence against the null hypothesis and in favor of the conclusion that the mortality rate at St. George's is above 0.15, but not quite as strong as in the first case. In the first case, all we could say was that the estimated p-value was less than 1 in 1000, but visually appeared to be much smaller still. While one option to more precisely quantify the p-value would be to increase the number of repetitions (say 10,000 or 100,000) we'll now look at another commonly used option.

An alternative to the p-value: Standardized value of a statistic

Using p-values is the most common way of assessing the strength of evidence by indicating the probability, under the null hypothesis, of getting a statistic as or more extreme than the one observed. Another way to measure strength of evidence is to *standardize* the observed statistic by measuring how far it is from the mean of the distribution using standard deviation units.(Common notation: *z*.)

Definition: To *standardize* a statistic, compute the distance of the observed statistic from the (hypothesized) mean of the null distribution and divide by the standard deviation of the null distribution.

$$\text{standardized statistic} = z = \frac{\text{statistic} - \text{mean of null distribution}}{\text{standard deviation of null distribution}}$$

For the second study, a standardized value of the statistic would result in the following calculation since Figure 1.11 tells us that the standard deviation of the null distribution is 0.018 and we know that the mean of the null distribution will be 0.15:

$$\text{standardized statistic} = z = \frac{0.197 - 0.15}{0.018} = 2.61$$

So we would say that the observed statistic (0.197) falls 2.61 standard deviations above the mean of the distribution (0.15).

Key idea: Observations that fall more than 2 or 3 standard deviations from the mean can be considered in the tail of the distribution.

Because the observed statistic is more than 2 standard deviations above the mean, we know the statistic is in the tail of the distribution.

We can also apply this approach to the original sample of 10 patients. In that simulation, the standard deviation of the null distribution was 0.113 with a mean of 0.15. Thus, the standardized value of the statistic is computed as:

$$\text{standardized statistic} = z = \frac{0.800 - 0.15}{0.113} = 5.73$$

Thus, $\hat{p} = 0.800$ is 5.73 standard deviations above the mean of 0.15. This is another way of saying that 0.800 would be extremely unlikely to happen by chance alone if the true long-run, current mortality rate was 0.150.

There are guidelines for assessing the strength of the evidence against the null hypothesis based on thestandardized value, as given next.

Guidelines for evaluating strength of evidence from standardized values of statistics

Standardizing gives us a quick, informal way to evaluate the strength of evidence against the null hypothesis. For standardized statistics:

between -1.5 and 1.5:   **little or no** evidence against the null hypothesis;

below -1.5 or above 1.5:**moderate** evidence against the null hypothesis;

below -2 or above 2:   **strong** evidence against the null hypothesis;

below -3 or above 3:**very strong**evidence against the null hypothesis.

Figure 1.12 illustrates the basis for using a standardized statistic to assess strength of evidence.

Figure 1.12: Positions of standardized statistics for a bell-shaped distribution.

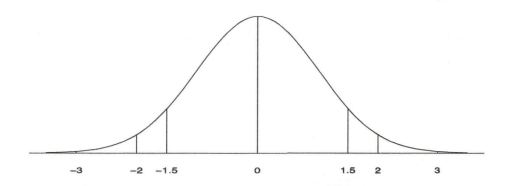

Notice that the farther from zero the standardized statistic is, the stronger the evidence against the null hypothesis.Like the p-value, we can directly compare the standardized values across datasets. We see the stronger evidence from the small dataset compared to the larger dataset (5.73 > 2.61).

Exploration 1.3: Do People Use Facial Prototyping?

A study in *Psychonomic Bulletin and Review* (Lea, Thomas, Lamkin, & Bell, 2007) presented evidence that "people use facial prototypes when they encounter different names." Participants were given two faces and asked to identify which one was Tim and which one was Bob. The researchers wrote that their participants "overwhelmingly agreed" on which face belonged to Tim and which face belonged to Bob, but did not provide the exact results of their study.

Step 1: Ask a research question. We will gather data from your class to investigate the research question of whether students have a tendency to associate certain facial features with a name.

Step 2: Design a study and collect data. Each student in your class will be shown the same two pictures of men's faces used in the research study. You will be asked to assign the name Bob to one photo and the name Tim to the other. Each student will then submit the name that he or she assigned to the picture on the left. Then the name that the researchers identify with the face on the left will be revealed.

1. Identify the observational units in this study.

2. Identify the variable. Is the variable categorical or quantitative?

The parameter of interest here is the long-run probability that a student in your class would assign the same name to the face on the left.

3. State the null and alternative hypotheses to be tested when the data are collected. Express these both in words and symbols.(*Hint*: Think about the parameter and the research question of interest here.)

Consider these two photos:

4. Do you think the face on the left is Bob or Tim? Collect the responses (data) for all the students in your class.

Step 3: Explore the data.

5. How many students put Tim as the name on the left? How many students participated in this study (sample size)? What proportion put Tim's name on the left?

    Number (Tim on left):   Sample size:

    Proportion (Tim on left):

When we conduct analyses with binary variables, we often call one of the outcomes a "success" and the other a "failure," and then focus the analysis on the 'success' outcome. It is arbitrary which outcome is defined to be a success, but you need to make sure you do so consistently throughout the analysis. In this case we'll call "Tim on left" a success because that's what previous studies have found to be a popular choice.

Step 4: Draw inferences. You will use the **One Proportion** applet to investigate how surprising the observed class statistic would be, if students were just randomly selecting which name to put with which face.

6. Before you use the applet, indicate what you will enter for the following values:

    a. Probability of success:

    b. Sample size:

    c. Number of repetitions:

7. Conduct this simulation analysis. Make sure the "Proportion of heads" button is selected in the applet and not the "Number of heads."

    a. Indicate how to calculate the approximate p-value (count the number of simulated statistics that equal _____ or _____).

    b. Report the approximate p-value.

    c. Use the p-value to evaluate the strength of evidence provided by the sample data against the null hypothesis, in favor of the alternative that students really do tend to assign the name Tim (as the researchers predicted) to the face on the left.

The p-value is the most common way to evaluate strength of evidence, but now we will explore a common alternative way to evaluate strength of evidence.The goal of any measure of strength of evidence is to use a number to assess whether the observed statistic falls in the tail of the null distribution (and is therefore surprising if the null hypothesis is true) or among the typical values we see when the null hypothesis is true.

8. Check the **Summary Stats** box in the applet.

    a. Report the mean (average) value of the simulated statistics.

    b. Explain why it makes sense that this mean is close to 0.5.

    c. Report the standard deviation (SD) of the simulated statistics.

    d. Report (again) the observed class value of the statistic.(What proportion of students in your class put Tim's name on the left?)

$$\hat{p} =$$

    e. Calculate how many standard deviations the observed class value of the statistic is from the hypothesized mean of the null distribution, 0.5.In other words, subtract the 0.5 from the observed value, and then divide by the standard deviation.In still other words, calculate:
      (observed statistic ($\hat{p}$)– 0.5) / SD of null distribution.

Your calculation in #8e is called standardizing the statistic. It is telling us how far above the mean the observed statistic is in terms of the 'how many standard deviations.'

Definition: To *standardize* a statistic, compute the distance of the statistic from the (hypothesized) mean of the null distribution and divide by the standard deviation of the null distribution.

$$\text{standardized statistic} = z = \frac{\text{statistic} - \text{mean of null distribution}}{\text{standard deviation of null distribution}}$$

Once you calculate this value, you interpret it as "how many standard deviations the observed statistic falls from the hypothesized parameter value."

The next question is how to evaluate strength of evidence against the null hypothesis based on a standardized value. Here are some guidelines:

Guidelines for evaluating strength of evidence from standardized values of statistics

Standardizing gives us a quick, informal way to evaluate the strength of evidence against the null hypothesis. For standardized statistics:

between -1.5 and 1.5:   **little or no** evidence against the null hypothesis;

below -1.5 or above 1.5: **moderate** evidence against the null hypothesis;

below -2 or above 2:   **strong** evidence against the null hypothesis;

below -3 or above 3:   **very strong** evidence against the null hypothesis.

The diagram in Figure 1.13 illustrates the basis for using a standardized statistic to assess strength of evidence.

Figure 1.13: Positions of standardized statistics for a bell-shaped distribution.

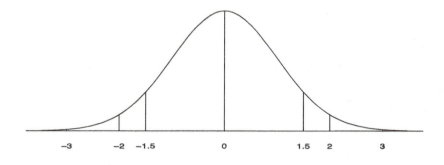

The figure can be summarized by the following key idea.

> **Key idea:** Observations that fall more than 2 or 3 standard deviations from the mean can be considered in the tail of the distribution.

Step 5: Formulate conclusions.

9. Let's examine the strength of evidence against the null.

    a. Based on the value of the standardized statistic, z, in #8e and the guidelines shown above, how much evidence do the class data provide against the null hypothesis?

    b. How closely does your evaluation of strength of evidence based on the standardized statistic compare to the strength of evidence based on the p-value in #7c?

Now, let's step back a bit further and think about the scope of inference. We have found that in most classes, the observed data provide strong evidence that students do better than random guessing which face is Tim's and which is Bob's. In that case, do you think that most students at your school would agree on which face is Tim's? Do you think this means that most people can agree on which face belongs to Tim? Further more, does this mean that all people do ascribe to the same facial prototyping?

Step 6: Look back and ahead.

10. Based on the limitations of this study, suggest a new research question that you would investigate next.

Extensions

11. In #5 you recorded the proportion of students in your class who put Tim's name with the photo on the left. Imagine that the proportion was actually larger than that (e.g., if your class was 60%, imagine it was 70%).

    a. How would this have affected the p-value:

    Larger   Same   Smaller

b.   How would this have affected the standardized statistic:

Larger   Same   Smaller

c.   How would this have affected the strength of evidence against the null hypothesis:

Stronger   Same   Weaker

12. Suppose that less than half of the students in your class had put Tim's name on the left, so your class result was in the opposite direction of the research conjecture and the alternative hypothesis.

a.   What can you say about the standardized value of the statistic in this case? Explain. (*Hint*: You cannot give a value for the standardized statistic, but you can say something specific about its value.)

b.   What can you say about the strength of evidence against the null hypothesis and in favor of the alternative hypothesis in this case?

Section 1.3 Summary

In addition to the p-value, a second way to evaluate strength of evidence numerically is to calculate a **standardized statistic**:

- Standardized statistic $= \dfrac{\text{statistic} - \text{mean of null distribution}}{\text{standard deviation of null distribution}}$

A standardized statistic provides an alternative to p-value for measuring how far an observed statistics falls in the tail of the null distribution. More specifically:

- A standardized statistic indicates how many standard deviations the observed value of the statistic is above or below the hypothesized process probability.
- Larger values of the standardized statistic (in absolute value) indicate stronger evidence against the null model.
- Values of a standardized statistic greater than 2 or less than -2 indicate strong evidence against the null model; values greater than 3 or less than -3 indicate very strong evidence against the null model.

Section 1.4: What Impacts Strength of Evidence?

Introduction

When we are conducting a test of significance for a single proportion we assume the null hypothesis is true (or that the long-run probability equals some number) and then determine how unlikely it would be to get a sample proportion that is as far away (or farther) from the probability assumed in the null hypothesis. The p-value and standardized scores are measures of how unlikely this is. Small p-values and large standardized scores (in absolute value) give us strong evidence against the null.

In this section we will explore some of the factors that affect the strength of evidence. You should have already seen that as the sample proportion moves farther away from the probability in the null hypothesis, we get more evidence against the null. We will review this factor and explore two more. We will see how sample size and what are called "two-sided tests" affect strength of evidence.

Example 1.4:Predicting Elections from Faces?

Do voters make judgments about a political candidate based on his/her facial appearance? Can you correctly predict the outcome of an election, more often than not, simply by choosing the candidate whose face is judged to be more competent-looking? Researchers investigated this question in a study published in *Science*(Todorov, Mandisodka, Goren, & Hall, 2005).Participants were shown pictures of two candidates and asked who has the more competent-looking face. Researchers then predicted the winner to be the candidate whose face was judged to look more competent by most of the participants. In particular the researchers predicted the outcomes of the 32 US Senate races in 2004.

> **Think about it:** What are the observational units? What is the variable measured? Is the variable categorical or quantitative? What is the null hypothesis?

The observational units are the 32 Senate races and the variable is whether or not this method correctly predicted the winner – a categorical variable. Because we are looking for evidence that this method works better than guessing we will state:

Null hypothesis: The probability this method predicts the winner between the two candidates equals 0.5 ($\pi = 0.5$)

Alternative hypothesis: The probability this method predicts the winner is greater than 0.5 ($\pi > 0.5$).

The researchers found the competent face method of predicting election outcomes to be successful in 23 of the 32 Senate races. Thus the observed statistic, or observed proportion of correct predictions is 23/32 ≈ 0.719, or 71.9%. We can use simulation to investigate whether this provides us with strong enough evidence against the null to conclude the competent face method is better than randomly choosing the winner. Figure 1.14 displays the results of 1000 simulated sets of 32 races, assuming that the "competent face" method is no better than flipping a coin, using the One Proportion applet.

Figure 1.14: The results of 1000 sets of 32 coin tosses from a process with probability 0.5 of success indicating which results are at least as extreme as 23 ($\hat{p} \geq 0.7188$)

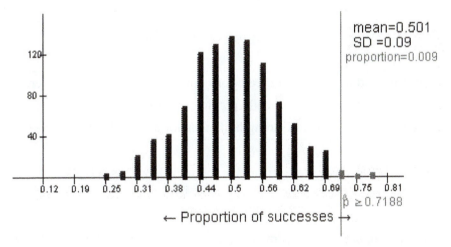

Only 9 of these 1000 simulated sets of 32 races show 23 or more correct predictions (or a proportion of 23/32=0.7188 or more), so the approximate p-value is 0.009. This p-value is small enough to provide strong evidence against the null hypothesis, in favor of concluding that the "competent face" method makes the correct prediction more than half the time in the long run. Alternatively, we can compute the standardized value of the statistic as

$$\text{standardized statistic} = \frac{0.7188 - 0.500}{0.09} = 2.43.$$

Thus, the observed statistic is 2.43 standard deviations away from hypothesized parameter value (the mean of the null distribution) specified by the chance model, and so this (being larger than 2), again, confirms the observation is in the tail of the null distribution, so there is strong evidence that the chance model is wrong.

What impacts strength of evidence?

In the previous two sections we've looked at two measures of the strength of evidence: p-value and standardized statistic, however we've not yet formally looked at what factors *impact* the strength of evidence. In other words, why is the strength of evidence (measured by p-value or standardized statistic) sometimes strong and sometimes weak or non-existent? We will look at three factors that impact the strength of evidence: the difference between the observed statistic ($\hat{p}$) and null hypothesis parameter value, the sample size, and whether we do a one or two-sided test.

1. Difference between statistic and null hypothesis parameter value

> **Think about it:** What if instead of 23 correct predictions out of 32, the researchers were able to correctly predict 26 elections? Or, what if they only correctly predicted 20 elections? How would the number of correct predictions in the sample impact our strength of evidence against the null hypothesis?

Intuitively, the more extreme the observed statistic, the more evidence there is against the null hypothesis. But, let's be a bit more precise.

If the researchers correctly predicted 26 elections, that is a success rate of $\hat{p} = 26/32 = 0.8125$, which is farther away from what would occur, in the long-run, if they were just guessing (0.50). Back in Figure 1.14 you can see that a value of 0.81 or larger never occurs just by chance, approximating a p-value <0.001. Similarly, the standardized statistic would be 3.47 (meaning 0.8125 is 3.47 standard deviations above the mean). In short, if researchers correctly predict 26 elections, this is extremely strong evidence against the null hypothesis because the observed statistic is farther out in the tail of the null distribution.

On the other hand, if the researchers correctly predicted only 20 elections, the success rate drops to $\hat{p} = 20/32 = 0.625$. In this case, the observed statistic (0.625) is something that is fairly likely to happen just by chance if the researchers were guessing who would win the election. The p-value increases to 0.115 (there were 115 out of 1000 times that 62.5% or more correct predictions occurred by chance), and the standardized statistic is closer to zero[(0.625 − 0.50)/0.09=1.39], suggesting that if the researchers correctly predicted 20 of the 32 elections, there is little evidence that the researchers' method performs better (in the long run) than guessing.

> **Key idea:** The farther away the observed statistic is from the average value of the null distribution, the more evidence there is against the null hypothesis.

2. Sample size

> **Think about it:** Do you think that increasing the sample size will increase the strength of evidence, decrease the strength of evidence, or have no impact on the strength of evidence against the null hypothesis (assuming that both the value of the observed statistic and the chance model do not change)?

Intuitively, it seems reasonable to think that as we increase the sample size, the strength of evidence will increase. Observing Doris and Buzz longer, playing rock paper scissors longer and looking at St. George's heart transplant patients outside of the initial set of 10 patients all intuitively suggest we'll have more knowledge about the truth. Let's dig a bit deeper into this intuition.

In Exploration P.3, we looked at how probability is the long-run proportion of times an outcome occurs. The key there is long-run. We realize that in the short-term, we expect more chance variability than in the long-term. For example, long-term investment strategies are typically good (variability is reduced). Whereas if you flip a coin just 3 times and get heads each time, you aren't convinced the coin isn't fair because with such a small sample size, almost "anything can happen."

In terms of statistical inference, our sample size (number of observational units) is what dictates how precise a measure of the parameter we have. In Figure 1.15, we can see how the null distribution changes as we increase the sample size from 32 senate races to 128 races (4 times as many) or 512 races (16 times as many).

Figure 1.15: As the sample size increases, the variability of the null distribution decreases

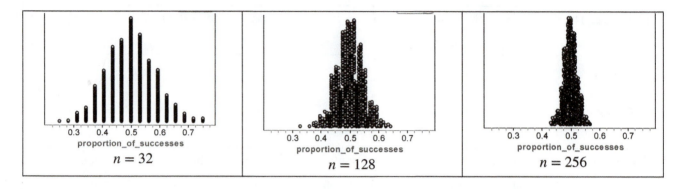

In each case the distribution is centered at 0.50, the null hypothesis value. What, then, is changing? What's changing is the variability of the distribution. With 32 races the variability of the distribution, as measured by the standard deviation is 0.088, with 128 races, it is 0.043, and with 256 races it is only 0.031. You can see this visually by noting that the distributions are getting squeezed closer and closer to the null hypothesis value. Stated yet another way, there is less sample to sample variability in the sample proportions as the sample size gets bigger.

What does this decrease in variability mean in terms of statistical significance? Remember from our observed data that in 71.9% of elections the competent face method predicted the winning candidate. This is strong evidence (p-value =0.009; standardized statistic=2.43) that the competent face method does better than guessing. But, what if the sample size was 128 elections and the competent face method correctly predicted 71.9% of the races? Now the p-value would be <0.001 and the standardized statistic would be $z=5.09$—the strength of evidence against the chance model of just guessing. The strength of evidence against the null hypothesis increases even more if the competent face method can get 71.9% correct in 256 elections (p<0.001; standardized statistic=7.06).

Key idea: As the sample size increases (and the value of the observed sample statistic stays the same) the strength of evidence against the null hypothesis increases.

Two more quick points: (1) If you are trying to pass a true/false test (let's say get a 60% or higher) but know NOTHING about what is going to be on the test, would you rather have more questions or fewer questions on the test? You, the student, would rather have fewer questions. If there was only one question on the test you'd have a 50% chance of passing! The teacher would rather have more questions on the test because the more questions they have on the test the more likely your outcome on the test will be close to 50% (just guessing!) and the less likely you would be to "get lucky" and pass the test by just guessing; (2) Importantly, we can't automatically assume that if we have Doris and Buzz do more trials or have your friend play more rounds of rock-paper-scissors or collect more data at St. George's Hospital that the strength of evidence will increase (smaller p-value, larger standardized statistic). Why not? Because when we collect more data our observed statistic will almost always change as well. If we have Doris and Buzz do more trials, they won't always get exactly 93.75% correct, your friend may not always throw scissors in exactly 1/6 of the rounds played, and the other heart transplant patients may not have 80% mortality after 30 days.

3. One sided versus two-sided tests

What if the researchers were wrong, and instead of the competent person being elected more frequently, it was actually the less competent person who was more likely to win the election?
Currently, as we've stated our null and alternative hypotheses, we haven't allowed for this possibility. The null hypothesis says that the competent face method predicts the winner 50% of the time, the alternative hypothesis says greater than 50%, but less than 50% doesn't appear.

> Null hypothesis: The probability this method predicts the winner equals 0.5.
> Alternative hypothesis: The probability this method predicts the winner is greater than 0.5.

These hypotheses can be written in symbols as:

> H_0: $\pi = 0.5$
> H_a: $\pi > 0.5$, where π = the probability this method predicts the correct winner

This type of alternative hypothesis is called "one-sided" because it only looks at one of the two possible ways that the null hypothesis could be wrong. In this case, it only considers that the null hypothesis could be wrong if the probability is more than 0.50.Many researchers consider this way of formulating the alternative hypothesis to be too narrow and too biased towards assuming the researchers are correct ahead of time. Instead, a more objective approach is to conduct a two-sided test, which can be formulated as follows.

> Null hypothesis: The probability this method predicts the winner equals 0.50.
> Alternative hypothesis: The probability this method predicts the winner is not 0.50.

These hypotheses can be written in symbols as:

> H_0: $\pi = 0.50$
> H_a: $\pi \neq 0.50$

In this case, the alternative hypothesis states that the probability the competent face method predicts the winner is not 0.50—might be more, might be less. This change to the alternative hypothesis, however, has ramifications on the rest of the analysis. Recall that p-values are computed as the probability under the null of obtaining a value that is equal to or more extreme than your observed statistic, where more extreme goes in the direction of the alternative hypothesis (greater than **or** less than). In the case of a ***two-sided test***, more extreme must go in both directions. The way this is operationalized is that the p-value is computed by finding out how frequently the observed statistic or more extreme occurred in one tail of the distribution, and adding that to the corresponding probability of being at least as extreme in the other direction - in the other tail of the null distribution.

For example, how often does 0.7188 or more occur by chance? Because 0.7188 is 0.2188 *above* 0.5, we need to look at what number is 0.2188 *below* 0.5. Calculating $0.5 - 0.2188 = 0.2812$, we need to also look at the proportion of outcomes at or below 0.2812. (See Figure 1.16.)

Figure 1.16: When determining a two-sided p-value we can look at outcomes in both tails for the null distribution that are the same distance above or below the mean

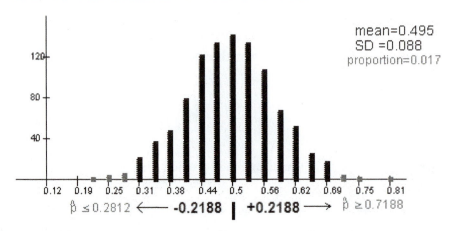

Figure 1.16 illustrates how a two-sided p-value is computed. In this case 0.7188 or greater was obtained 9 times by chance, and 0.2812 or less was obtained 8 times by chance. Thus, the two-sided p-value is approximately (8+9)/1000=0.017. Two-sided tests always increase the p-value (approximately doubling it from a one-sided test), and thus, using two-sided tests always decreases the strength of evidence. Two-sided tests are more conservative and are used most of time in scientific practice. However, because of their objectivity they require even stronger results before the observed statistic will be statistically significant.

> **Key idea:** Because the p-value for a two-sided test is about twice as large as that for a one-sided test, they provide less strength of evidence. However two-sided tests are used more often in scientific practice.

Technical note: When the hypothesized probability is something other than 0.5, there are actually a couple of different ways of defining an observation as being more extreme. The One Proportion applet determines an outcome is more extreme if it has a smaller individual probability than that of the observed outcome. In other words, if an observation is more unlikely, then it is considered more extreme. See FAQ 1.4.1 for more on one-sided vs. two-sided p-values.

Follow-up Study

As a way of applying what we just learned, consider the following. The researchers investigating the competent face method, also predicted the outcomes of 279 races for the U.S. House of Representatives in 2004, looking for whether the probability the "competent face" method predicts the correct winner (π) is different from 0.50. In these 279 races, the method correctly predicted the winner in 189 of the races, which is a proportion of $\hat{p} = 189/279 \approx 0.677$, or 67.7% of the 279 House races. Notice this sample percentage of 67.7% is bit smaller than the 71.9% correct predictions in 32 Senate races, however the sample size (279 instead of 32) is larger. We are also now considering a two-sided alternative hypothesis.

Think about it: Do you expect the strength of evidence for the "competent face" method to be stronger for the House results, weaker for the House results, or essentially the same for the House results as compared to the Senate results? Why?

Let's take the three factors separately:

1. **Distance of the observed statistic to the null hypothesis value.** In the new study the observed statistic is 0.677, whereas in the original study it was 0.719—this is a small change, closer to the hypothesized value of the parameter, which will slightly decrease the strength of evidence against the null hypothesis.

2. **Sample size.** The sample size is almost 10 times as large (279 vs. 32) in the new study which will have a large impact on the strength of evidence against the null hypothesis. This will increase the strength of evidence quite a bit, because the observed statistic didn't change much.

3. **One- or two-sided test.** When we looked at the Senate races, we obtained a p-value of 0.009 with a one-sided test and 0.017 with a two-sided test. Because we are asked to use a two-sided test here that means we will obtain a p-value about twice as large as what we would obtain with a one-sided test.

So, what's the p-value? Figure 1.17 shows the null distribution for a sample size of 279 with 0.5 as the probability under the null hypothesis.

Figure 1.17: A null distribution for the Senate sample of 279 races

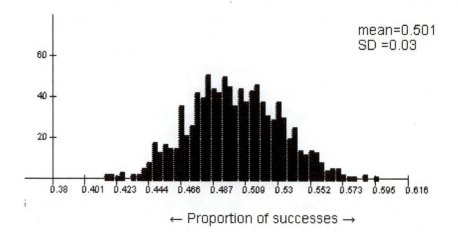

← Proportion of successes →

Notice that a proportion of 0.677 or larger does not occur just by chance in the null distribution. Neither do any values of 0.323 or smaller. (We need to look down there (0.677 − 0.500 ≈0.177, 0.500 − 0.177 ≈0.323) because this is a two-sided test.)Thus the p-value is approximately zero and the evidence against the null hypothesis in the House of Representatives sample is stronger than in the Senate sample. This is mainly due to the increased sample size and the fact that the observed sample statistic didn't change too much.

Looking at the standardized statistics, for the Senate races, we find $z = (0.719 − 0.500)/0.088 ≈ 2.49$and for the House of Representatives we find $z = (0.677 − 0.500)/0.03 ≈ 5.90$.We see that the numerator is a bit smaller in the second case, but the denominator (the standard deviation of the null distribution) is much smaller, providing much stronger evidence against the null hypothesis.

FAQ 1.4.1: When and why do we use two-sided alternative hypotheses?

Q: Why would we use a two-sided alternative?

A: Two-sided alternative hypotheses are common in scientific practice when researchers don't make a prior assumption about the direction of the outcome in the study. This can happen if the researchers don't have a prior suspicion about the direction, or if they are interested in finding out whether the parameter (in this case, the long-run probability) differs from the value hypothesized in the null hypothesis. In contrast, to use a one-sided test, one needs to have prior suspicion that the outcome will be in a certain direction.

Q: Why don't we look at the direction of the outcome in the data collected and state our alternative hypothesis to match that?

A: Because that would be considered cheating. Remember that hypotheses are always statements about the parameter(s) of interest, and are stated before we begin to explore the data. We use the data to test our hypotheses, not to form them.

Q: So, should we always do two-sided tests? Are there any drawbacks to using two-sided tests?

A: The reality is that most scientists do use two-sided tests the majority of the time. There are, however, downsides to using two-sided tests. The first is related to the size of the p-value. When we compute a p-value for a two-sided test, the p-value is approximately twice the size it would be had we computed it for the one-sided test. This means that two-sided tests always yield larger p-values than one-sided tests, making it harder to find strong evidence in support of your research conjecture. So, in some cases scientists will do one-sided tests if they are focused on only one type of result (and have prior evidence that that is the direction of the effect or that's the only direction of interest). For example, in some of the examples we've looked at in previous chapters (e.g., Doris and Buzz), we studied whether the dolphins could do better than just guessing, it makes less sense to study whether the dolphins did worse than just guessing (this would mean that they were purposely pushing the wrong button!). In these cases a one-sided test makes the most sense.

Q: And the second drawback?

A: The second drawback to two-sided tests is that the two-sided p-value is the strength of evidence for a difference, but does not tell us much about *how* (bigger or smaller?) the parameter (in this case, π) differs from the value hypothesized in the null hypothesis.

Exploration 1.4: Competitive Advantage to Uniform Colors?

In this exploration, we are going to explore three factors that influence the strength of evidence in a test of significance: (1) The difference between the observed sample statistic and the value of the parameter used in the null hypothesis;(2) The sample size; (3) One-sided tests versus two-sided tests. To do this we will look at a study conducted by Hill and Barton (*Nature*, 2005) to investigate whether Olympic athletes in certain uniform colors have an advantage over their competitors. They noticed that competitors in the combat sports of boxing, tae kwon do, Greco-Roman wrestling, and freestyle wrestling are randomly assigned red or blue uniforms. For each match in the 2004 Olympics, they recorded the uniform color of the winner.

> **Think about it:** What are the observational units? What is the variable measured? Is the variable categorical or quantitative?

The observational units in this study are the matches, and the variable is whether the match was won by someone wearing red or someone wearing blue – a categorical variable. Let's suppose that going into this study, the researchers wanted to see whether the color red had an advantage over blue. In other words, competitors that wear red uniforms will win a majority of the time.

1. State the null and the alternative hypotheses in words.

2. We will let π represent the probability that a competitor wearing a red uniform wins. Using this, restate the hypotheses using symbols.

3. Researchers Hill and Barton used data collected on the results of 457 matches and found that the competitor wearing red won 248 times, whereas the competitor wearing blue won 209 times. We will carry out a simulation to assess whether or not the observed data provide evidence in support of the research conjecture. This simulation will employ the **3S Strategy**: Determine the **s**tatistic, **s**imulate could-have-been outcomes of the statistic under the null model, and assess the **s**trength of evidence against the null model by estimating the p-value.

    a. What is the statistic we will use? Calculate the observed value of the statistic in this study.

    b. Describe how you could use a coin to develop a null distribution to test our hypothesis.

    c. Use the **One Proportion** applet to test our hypothesis. Based on your simulation, find the p-value and write a conclusion. Also write down the mean and standard deviation from your null distribution when the proportion of successes is used for the variable on the horizontal axis. You will need this later.

One sided vs. two-sided tests

One factor that influences strength of evidence is whether we conduct a two-sided or a one-sided test. Up until now we have only done one-sided tests. In a one-sided test the alternative hypothesis is either > or <. In a two-sided test, the alternative hypothesis is ≠. So, why would you do a two-sided test and what are the implications?

Suppose the researchers did not necessarily think that red would win more often, but they also didn't necessarily think that blue would win more often. They were just interested in whether one color would win more often than the other. A two sided alternative hypothesis (red wins at a rate other than 50% of the time (or at a rate not equal to 50%)) allows the researchers to be less sure of the anticipated location of the parameter than a one-sided test.

4. If we let π equal the probability that a competitor wearing a red uniform wins, state the hypotheses for this study in symbols using a two-sided alternative.

5. Return to the **One Proportion** applet to approximate the p-value for our original overall proportion of red winning 248 times out of 457 matches, but now select the "Two-sided" check box to find the "two-sided p-value."

    a. Describe how the portion of the null distribution that is shaded red is different than our first test done in #3c.

    b. Describe how the p-value is different from the p-value that was obtained in our original test done in #3c.

    c. To find the two-sided p-value, the applet is looking to see how often 0.543 or larger occurs and how often the comparable value on the other side of the null distribution (or smaller) occurs. To find this value, first compute how far 0.543 is from the center of the null distribution, and then go that same distance to the left (less than) the center of the null distribution. What is the comparable value?

d. Complete the following sentence: The two-sided p-value of _____ is the probability of obtaining _____ or larger plus the probability of obtaining _____ or smaller if the _____ is true.

e. You should have seen that when the alternative hypothesis is *two-sided*, the p-value is computed by looking at how extreme the observed data is in *both* tails on the null distribution. This makes the p-value about twice as large. Because of this, explain how switching from a one-sided test to a two-sided influences the strength of evidence against the null.

Key idea: Because the p-value for a two-sided test is about twice as large as that for a one-sided test, they provide less strength of evidence. However two-sided tests are used more often in scientific practice.

Difference between statistic and null hypothesis parameter value

6. A second factor that influences the strength of evidence against the null is how far apart the observed sample statistic and the value of the parameter specified under the null hypothesis are.For this study the null value was 0.5 and the observed sample statistic was about 0.543 (or 54.3% of the competitors wearing red won their matches).Suppose a larger proportion of competitors wearing red won their matches. If fact, suppose 57% of the 457 matches were won by a competitor wearing red.

a. Go back to the **One Proportion** applet and approximate the (one-sided) p-value for this situation where again we are testing to see whether the overall probability of winning is more than 0.5.

b. Is your p-value larger or smaller than your original one?Explain why this makes sense.

c.   Write a sentence explaining the relationship between the distance between the observed sample statistic and the value of the parameter specified under the null hypothesis to the strength of evidence against the null hypothesis.

Key idea: The farther away the observed statistic is from the average value of the null distribution, the more evidence there is against the null hypothesis.

Sample size

7.   The third factor we will look at that influences strength of evidence against the null hypothesis is the sample size. As we said earlier, the data for this study came from four combat sports in the 2004 Olympics. One of those sports was boxing. The researchers found that out of the 272 boxing matches, 150 of them were won by competitors wearing red. This proportion of $150/272 \approx 0.551$ is similar to the overall proportion of times the competitor wearing red won. Let's see what the smaller sample size does to the strength of evidence. Use the **One Proportion** applet to test the same hypotheses as we originally did, but just the boxing matches as our sample.

a.   Compare the null distribution you generate in this case to that generated in #3. In particular, how do the center and variability compare?

b.   What is your new p-value? Is it larger or smaller than your original p-value from #3c? Explain why this makes sense.

c.   Write a sentence explaining the relationship between sample size and the strength of evidence against the null hypothesis.

Key idea: As the sample size increases (and the value of the observed sample statistic stays the same) the strength of evidence against the null hypothesis increases.

Section 1.4 Summary

Three factors impact the strength of evidence provided by sample data against a null hypothesis:

- **Two-sided alternatives/tests** are used when the researcher does not have a prior suspicion about the direction that the parameter value is from the hypothesized value.

- A two-sided test produces a larger p-value than a one-sided test based on the same sample data.
  - Two-sided tests therefore require a higher "standard of proof" to provide convincing evidence against the null hypothesis.
  - The p-value for a two-sided test is generally twice as large as the p-value would have been from a one-sided test on the same sample data.
- The farther away the observed statistic is from the hypothesized value of the parameter, π, in the direction of the alternative hypothesis, the stronger the evidence against the null hypothesis.
- A larger sample size generally produces stronger evidence against the null hypothesis if the observed value of the statistic does not change (and if the observed result is in the direction of the alternative hypothesis).

Section 1.5: Inference on a single proportion: Theory-based approach

Introduction

The focus of this chapter has been on Step 4: Drawing Inferences. We have learned that we can draw inferences from data by comparing our observed statistic to a conjecture or claim about a long-run probability. In order to assess the strength of evidence for the claim we have always simulated the null distribution. However, simulation takes a lot of computer power, and historically that wasn't always possible. In this section we will see how, in many cases, we can predict what will happen when you simulate, thus avoiding the need to conduct a simulation, but still providing the ability to generate either a p-value or standardized statistic to assess strength of evidence. As we'll find out in this and later sections, in addition to being necessary historically, this method (we'll call it a theory-based approach) also gives insights into and advantages in standardization (Section 1.3) and confidence intervals (the theme of Chapter 3).

At the heart of this method is the fact that the null distributions of sample proportion sthat we have been simulating often exhibit a common and familiar shape. In particular:

(1) They are often, though not always, bell-shaped curves;
(2) They are centered at the null hypothesis value for π; and
(3) Their variability (spread, standard deviation) is determined by the sample size.

Figure 1.18 shows a few examples of null distributions from this chapter.

Figure 1.18: Null distributions for various studies we've explored so far. A bell-shaped distribution, called a normal distribution, can be used to nicely approximate some of them, but will not work well for others.

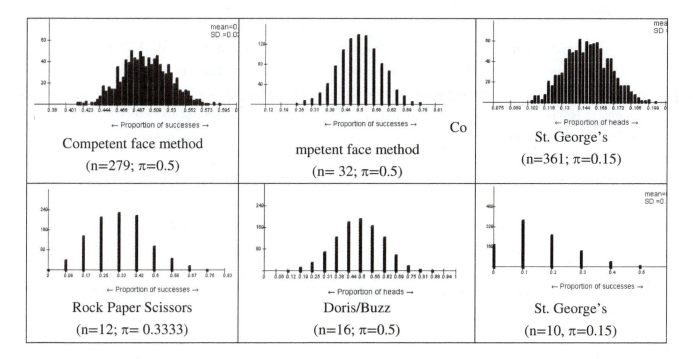

The competent face method and St. George's (n=279) simulations show bell-shaped curves. However, the Rock/Paper/Scissors study and the first sample of St. George's heart patients ($n = 10$) are not bell-shaped curves, in part because the distribution is not symmetric and in part because there are so few values that the sample proportion can be. For the Doris/Buzz dolphin study, even though the distribution of sample proportions is quite symmetric, an approximation based on a bell-shaped curve will not be very good because of the small number of possible values (i.e., because of the extreme discrete-ness, the gaps between the values) for the sample proportion in those simulations.

As we will learn in this section, in many, but not all, cases we can predict when the simulated distribution is bell-shaped (or **normally distributed**), where it will be centered, and how variable it will be. All of these predictions can be used to generate p-values and standardized statistics without simulating, in an approach called a one proportion z test, which is one example of Theory-based approach to statistical inference.

Example 1.5: Halloween Treats

Stemming from concern over the nation's obesity epidemic, researchers investigated whether children might be as tempted by toys as by candy for Halloween treats. Test households in five Connecticut neighborhoods offered children two plates: one with lollipops or fruit candy and one containing small, inexpensive Halloween toys, like plastic bugs that glow in the dark. The researchers observed the selections of 283 trick-or-treaters between the ages of 3 and 14 (Schwartz, Chen, and Brownell, 2003).

To investigate whether children show a preference for either the candy or the toys, we test the following hypotheses.

> Null hypothesis: The probability a trick-or-treater would choose candy is 0.5.
>
> Alternative hypothesis: The probability a trick-or-treater would choose candy is not 0.5.

Note that our null model assumes that the probability of choosing candy (π) is the same for all children. In symbols, these hypotheses translate to

H_0: $\pi = 0.5$

H_a: $\pi \neq 0.5$

The researchers collect the reactions of 283 children for the study. With a sample size of 283, under our null hypothesis, we simulated the null distribution (using 10,000 simulated samples) shown in Figure 1.19.

Figure 1.19: A null distribution representing 10,000 simulated samples of 283 children where the probability that an individual child would choose candy is 0.5

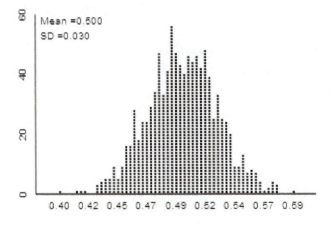

Proportion of heads

We notice that the distribution is quite bell-shaped, the average value is 0.5 (the null hypothesis value for π) and the standard deviation is 0.030. The theory-based approach could have predicted this would happen!

Theory-based approach (One proportion z test)

In the early 1900s, and even earlier, computers weren't available to do simulations, and as people didn't want to sit around and flip coins all day long, they focused their attention on mathematical and probabilistic rules and theories that could predict what would happen if someone did simulate.

They proved the following key result (often called the **Central Limit Theorem**):

> **Central Limit Theorem:** If the sample size (n) is large enough, the distribution of sample proportions will be bell-shaped (or normal), centered at the long-run probability (π), with a standard deviation of
>
> $$\sqrt{\frac{\pi(1 - \pi)}{n}} \ .$$

One bit of ambiguity in the statement is how large is large enough for the sample size? As it turns out, the larger the sample size is, the better the prediction of bell-shaped behavior in the null distribution is, but there is not a sample size where all of the sudden the prediction is good. However, some people have used the convention that you should have at least 10 successes and at least 10 failures in the sample.

> **Validity conditions:** The normal approximation can be thought of as a prediction of what would occur if simulation was done. Many times this prediction is valid, but not always, only when the validity condition (at least 10 successes and at least 10 failures) is met.

Let's see how this prediction compares to what actually happened in the simulation (back in Figure 1.19). Because the sample size is 283 (with 148 and 135 the two categories) the prediction should work. Looking at Figure 1.19 we see that:

1. The simulated distribution certainly looks bell-shaped.

2. The simulated distribution is centered at 0.5, the hypothesized value of π.

3. The simulated standard deviation of the sample proportions is 0.030, which is very close to the predicted standard deviation of $\sqrt{\dfrac{0.5(1 - 0.5)}{283}} = 0.0297.$

One advantage of using this result is we can determine the standard deviation of the sample proportion without having to conduct the simulation. Knowing the standard deviation allows us to calculate the standardized statistic, z:

$$z = \frac{0.523 - 0.5}{0.0297} = 0.77.$$

This tells us that our observed sample proportion is 0.77 standard deviations above the mean of 0.5. Because this is less than one standard deviation above the mean, it gives us little evidence against the null hypothesis.

To convert this standardized score to a two-sided p-value, we could look at the area under the (standard) normal curve to the right of 0.77 and to the left of -0.77. In the One Proportion applet, checking the box for **Normal Approximation** will overlay the theoretical normal curve and give you the p-value from the normal distribution. Figure 1.20 shows the results of this applet for the Halloween treat study where in the sample of 283 trick-or-treaters, 148 (52.3%) chose candy and 135 (47.7%) chose toys.

Figure 1.20: Simulated and Theory-Based p-values for Halloween study

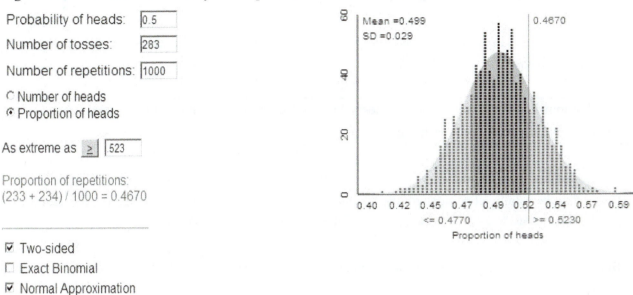

Probability of heads: 0.5

Number of tosses: 283

Number of repetitions: 1000

○ Number of heads
● Proportion of heads

As extreme as ≥ .523

Proportion of repetitions:
(233 + 234) / 1000 = 0.4670

☑ Two-sided
☐ Exact Binomial
☑ Normal Approximation

p-value = 0.4390 (Z = 0.77)

The normal distribution does a nice job of predicting the behavior of the null distribution of sample proportions in this case. The p-value from the theory-based test is 0.4390, compared to 0.4670 from the simulation. We interpret this p-value, and draw a conclusion from it, as always: If the probability each trick-or-treater prefers candy to toys is 0.5, then there's about a 44% chance that a random sample of 283 trick-or-treaters would have found 52.3% or more, or 47.7% or fewer, choosing the candy. Because this is not a small p-value, we do not have substantial evidence to suggest that trick-or-treaters prefer either type of treat. This was actually good news to the researchers, to learn that both types of treats are viable and we could probably distribute less candy at Halloween (at least in this Connecticut neighborhood).

The p-value of 0.4390 from the theory-based approach (One proportion z test) is very similar to that obtained from the simulation (see Figure 1.20) because the prediction of the shape, center, and variability of the null distribution is very good.[Note: We can improve the normal approximation by employing a "continuity correction" that would use a value just below 0.523 and just above 0.477 to include more area right at those cut-off values.]

A situation where a theory-based approach doesn't work

In Section 1.2, we looked at the Rock/Paper/Scissors game where novice players threw scissors two of the 12 rounds played.

> **Think about it:** Why do you think the Theory-based (One proportion z-test; Normal approximation) approach will not work well for this data?

In this case, the theory-based approach is not expected to work well because the sample size (12) is small. Recall that the validity conditions for the theory-based approach state the need for at least 10 successes and 10 failures, we have 2 scissors and 10 not scissors so this condition is not met.

But, let's see what happens when we use the theory-based approach anyway. The theory based approach predicts that:

1. The null distribution will be bell-shaped and approximately normal
2. The null distribution will be centered at 0.333
3. The standard deviation of the null distribution will be $\sqrt{\dfrac{0.333(1 - 0.333)}{12}} \approx 0.136.$

Figure 1.21 shows a picture of the simulated null distribution with the theory-based normal distribution overlaid.

Figure 1.21 shows that the approximate p-value from the simulation (0.1980) is not all that similar to the theory-based p-value (0.1108). Although both distributions (theoretical and simulation) are centered at nearly 0.3333, with similar standard deviations (0.134 vs. 0.136), the distribution is not "normal" in that it is not "filled in." In other words, the null distribution is too discrete (too much empty space between the possible observations) for it to be well-modeled by a normal distribution.

Figure 1.21: The null distribution for the Rock Paper Scissors example

Probability of success (π): 0.3333

Sample size(*n*): 12

Number of samples: 1000

○ Number of successes
◉ Proportion of successes

As extreme as ≤ 0.167 Count

Proportion of samples:
198 / 1000 = 0.1980

☐ Two-sided
☐ Exact Binomial
☑ Normal Approximation

p-value = 0.1108 (Z = 1.22)

With the slightly larger p-value, the simulation-based approach provides a bit less evidence against the null hypothesis. Although there is not a large difference between these two p-values, in this case we would have a bit more faith in the p-value obtained from the simulation. In summary, the theory-based approach should not have been used in this case because the sample size was too small, and, thus, the end result is that the theory suggested there was more evidence against the null than there was.

Exploration 1.5: Calling Heads or Tails

When asked to call the outcome of a coin toss, are people equally likely to choose heads or tails? Let's investigate this question by collecting some data from you and your classmates.

1. What would you call: heads or tails?

Before we even collect data from your classmates, let's think about what we want to test here. Conventional wisdom says that people tend to pick heads more often than tails, so that's the research hypothesis we'll investigate.

2. In the coin toss study with your class:

    a. What are the observational units in this study?

    b. What is the variable that is recorded?

    c. Describe the parameter of interest in words.(Use the symbol π to represent this parameter.)

    d. If people do *not* have a tendency to pick heads more often than tails (or tails more often than heads) what would you expect the numerical value of the parameter to be? Is this the null hypothesis or the alternative hypothesis?

    e. If people *do* have a tendency to pick heads more often than tails, what can you say about the numerical value of the parameter? Is this the null hypothesis or the alternative hypothesis?

3. Including yourself and your classmates, how many people participated in this study? How many picked heads? Calculate the sample proportion that picked heads.

4. To have a larger sample size to analyze, combine your class results with the results from one of the author's classes, in which 54 of 83 students picked heads. Now what are the sample size and the sample proportion that picked heads?

    Sample size: Sample proportion:

5. Use the **One Proportion** applet to test the hypotheses from #2d and #2e.

    a. Describe the shape of the null distribution of sample proportions. Does this shape look familiar? Where is the null distribution centered? Does this make sense? Check the **Summary Stats** box, and report the mean and standard deviation as reported by the applet.

Shape: Familiar?

Center?   Why does this make sense?

Mean: SD:

b. Approximate the p-value, and summarize the strength of evidence that the sample data
    provide regarding the research hypothesis.

c. Determine the standardized statistic, z, and summarize the strength of evidence. Confirm that
    the strength of evidence obtained using the standardized statistic is similar to that obtained
    using the p-value.

Theory-based approach (One proportion z test)

In question 5a, you probably described the shape of the null distribution using words such as bell-shaped,
symmetric, or maybe even normal. You have seen many null distributions in this chapter that have had
this same basic shape. You should have also noticed that the null distributions have all been centered at
the hypothesized value of the long-run probability used in the null hypotheses. You probably could have
predicted that your null distribution was going to be somewhat bell-shaped and centered at 0.5. You
probably would have a harder time predicting your null distribution's variability (standard deviation), but
this too can be predicted in advance, as we will see shortly.

We can use mathematical models known as normal distributions (bell-shaped curves) to approximate
many of the null distributions we have generated so far in this text. When rules and theories are used to
predict what the value of the standardized statistic and p-value would be if someone did simulate, we call
the approach a **theory-based approach**. The normal distribution provides a second way, in addition to
simulation, to approximate a p-value.

6. Check the box next to **Normal Approximation** in the applet. Does the shaded in blue region
    seem to be a good description (model) of what we actually got in the simulation?

Validity Conditions for Theory-Based Approach

The normal approximation to the null distribution is valid whenever the sample size is reasonably large. One convention is to consider the sample size to be large enough whenever there are at least 10 observations in each category.

7. According to this convention, is the sample size large enough in this study to use the normal approximation and theory-based inference? Justify your answer.

Validity conditions: The normal approximation can be thought of as a prediction of what would occur if simulation was done. Many times this prediction is valid, but not always, only when the validity condition (at least 10 successes and at least 10 failures) is met.

Formulas

The normal approximation will also give you values of the standardized statistic and p-value based on its mathematical predictions. As you learned in Section 1.3, the standardized score is calculated as:

$$z = \frac{\text{statistic} - \text{mean of null distribution}}{\text{standard deviation of null distribution}}$$

The mean of the null distribution is the hypothesized value of the long-run probability (π). The standard deviation can be obtained in two ways:

(1) Find the standard deviation of the null distribution by simulating.

(2) Predict the value of the standard deviation by plugging into this formula: $\sqrt{\dfrac{\pi(1-\pi)}{n}}$.

8. Use the formula to determine the (theoretical; predicted) standard deviation of the sample proportion. Then compare this to the SD from your simulated sample proportions, as recorded in #5a. Are they similar?

The predicted value of the standard deviation (using the formula) will be very close to the simulated standard deviation of the null distribution when the sample size is large enough. The Validity Condition mentioned earlier, says that a "large enough" sample size means at least 10 successes and at least 10 failures. This mathematical prediction is often called the "Central Limit Theorem."

Central Limit Theorem: If the sample size (n) is large enough, the distribution of sample proportions will be bell-shaped (or normal), centered at the long-run probability (π), with a standard deviation of

$$\sqrt{\frac{\pi(1-\pi)}{n}}.$$

9. Use the predicted value of the standard deviation from #8 to calculate the standardized statistic (z) by hand, and confirm that your answer is very close to what you found in #5c when using simulation.

In the applet, see that the predicted value of the standardized statistic, z, is given immediately below the button for "Normal approximation" in parentheses and should match your answer to #9.

10. The theory-based (Normal approximation) p-value is also now displayed. Compare this p-value to the one you got from simulation (#5b). Are they similar?

11. Why are the standard deviation (#8), standardized statistic (#9) and p-value (#10) similar when using the theory-based (one proportion z-test; normal approximation) to what you got in your simulation? When would they be different?

Follow-up Analysis #1

In his book *Statistics You Can't Trust*, Steve Campbell claims that people pick heads 70% of the time when they're asked to predict the outcome of a coin toss.

12. Use the theory-based approach to test Campbell's claim based on the sample data used above (your class combined with authors class) using a two-sided alternative. Report the null and alternative hypothesis, standardized statistic, and p-value. Summarize your conclusion, and explain the reasoning process by which it follows from your analysis.

Follow-up Analysis #2

In a small class of eight students, seven students picked heads when given the choice between heads and tails.

13. Use simulation to generate a two-sided p-value evaluating the strength of evidence that the long-run probability of students picking heads is different than 50% based on this small class's data alone.

14. Why can't you use the normal approximation in this case?

15. Use the normal approximation anyway. Compare and comment on the p-values obtained from the two methods.

Section 1.5 Summary

Most of the null distributions for a sample proportion that we have seen follow a common and familiar shape. This bell-shaped curve, known mathematically as a **normal distribution**, allows us to anticipate what a null distribution will look like, and to determine an approximate p-value, without bothering to conduct a simulation analysis.

All normal distributions have a bell-shaped curve, but they can differ with regard to center and variability. When working with the null distribution of a sample proportion:

- The center is described by the mean, which for a null distribution equals the hypothesized value of the long-run probability π.
- The variability is described by the standard deviation, which is determined primarily by the sample size.
    - The larger the sample size, the smaller the variability in sample proportions.
    - The standard deviation of the sample proportion equals $\sqrt{\dfrac{\pi(1-\pi)}{n}}$.

The **theory-based approach** (one-proportion z-test; normal approximation) standardizes the statistic based on the observed value of the statistic (sample proportion) and these theoretical mean and standard deviation values.

- The p-value is calculated by software as the area under the normal curve in the appropriate direction (as specified by the alternative hypothesis) from the standardized statistic.
- The p-value is interpreted and evaluated just as with the simulation-based method.

This theory-based method works well whenever the sample size is large enough for the normal curve to provide a good approximation to the null distribution.

- We consider the theory-based method to be valid when there are at least 10 observations of each category ("success" and "failure") in the sample.

- When this validity condition is not met, and even when it is met, an alternative is simply to use the simulation-based method.

The names, conditions and applets used for the simulation and theory-based tests are shown in the following table.

Figure 1.22 Summary of validity conditions and applets for one proportion tests

Type of data	Name of test	Applet	Validity conditions
Single binary variable	Simulation test for a single proportion	One proportion	---
	Theory-based test for a single proportion (1 Proportion z test)	One proportion	At least 10 successes and at least 10 failures

Chapter 1 Summary

This chapter has focused on Step 4 of the Statistical Investigation Method: Assessing the statistical significance of an observed result against some claim about the long-run probability of successes. This reasoning will stay the same as we consider other statistics (e.g., sample mean, difference in sample proportions) and other parameters (e.g., a difference in long-run probabilities) in other chapters. The basic approach is the 3S Strategy: Decide on an appropriate statistic, simulate values of that statistic under the assumption that the null hypothesis is true, and then assess the strength of evidence against the null hypothesis and in favor of the alternative hypothesis. Keep in mind that the null and alternative hypotheses are competing claims about the parameter value. We expect the observed sample statistic to differ from the parameter by chance. The goal is to see whether the difference observed in the study is larger than can be reasonably expected by chance alone. This unusualness (deciding whether the observed result in the tail of the null distribution) can be measured through the standardized statistic (e.g., z-statistic), which measures the distance between the observed statistic and the hypothesized parameter value in terms of number of standard deviations away, and/or the p-value, which measures how often a statistic at least as extreme would occur when the null hypothesis is true.

In the case of one categorical variable, when the sample size is large (e.g., at least 10 successes and at least 10 failures in the sample), then we can approximate the p-value using the normal distribution. This theory-based approach is referred to as a one-proportion z-test

You also considered factors that affect the strength of evidence against the null hypothesis:

- How far the observed sample statistic is from the hypothesized parameter value
- Sample size
- One-sided vs. two-sided alternatives.

The relationships you learned in this chapter will also apply in other scenarios as well.

Chapter 1 Glossary

p-hat ($\hat{p}$)

    The proportion or percentage of observational units that have a particular characteristic based on a measured variable. A statistic. 26

plausible

    A term used to indicate that the chance model is a reasonable/believable explanation for the data we observed 8, 19

p-value

    The probability of obtaining a value of the statistic at least as extreme as the observed statistic when the null hypothesis is true 29, 41

sample

    The set of observed values 4, 15

sample size

    The number of observational units in the sample 4, 15

standardize

    To standardize an observation, compute the distance of the observation from the mean and divide by the standard deviation of the distribution 50, 55

statistic

    A number computed from the sample 4, 15

statistical significance

    Results unlikely to have arisen by chance alone 1

statistically significant

    Unlikely to occur just by random chance 8, 19

strength

    How much evidence we have against the null hypothesis 1

subjects

    Study participants that are human 36

test of significance

    A procedure for measuring the strength of evidence against a null hypothesis about the parameter of interest 24

Theory-based approach

    mathematical approach which predicts the shape, center and variability of the null distribution instead of obtaining a null distribution by simulating 75

two-sided test

    Estimates the p-value by considering results that are at least as extreme as our observed result *in either direction* 64

Validity conditions

    Check to see that certain conditions are met that render the theory-based approach valid. Often these conditions deal with sample size and shape and variability of distributions 76

z-statistic

Chapter 1 Exercises

Section 1.1

Spinning tennis racquet*

Tennis players often spin a racquet to decide who serves first. The spun racquet can land with the manufacturer's label facing up or down. A reasonable question to investigate is whether a spun tennis racquet is equally likely to land with the label facing up or down. (If the spun racquet is equally likely to land with the label facing in either direction, we say that the spinning process is fair.) Suppose that you gather data by spinning your tennis racquet 100 times, each time recording whether it lands with the label facing up or down. Use this information to answer Exercises 1.1.1-4.

1.1.1
  a) Describe the relevant long-run probability of interest in words.
  b) What statistical term is given to the long-run probability you described in a?
  c) What value does the chance model assert for the long-run probability?
  d) Suppose that the spun racquet lands with the label facing up 48 times out of 100. Explain, as if to a friend who has not studied statistics, why this result does *not* constitute strong evidence against believing that the spinning process is fair.
  e) Is the result in d) statistically significant evidence that spinning is not fair, or is it plausible that the spinning process is fair?

1.1.2
  a) Suppose that the spun racquet lands with the label facing up 24 times out of 100. Explain, as if to a friend who has not studied statistics, why this result *does* constitute strong evidence against believing that the spinning process is fair.
  b) Is the result in a) statistically significant evidence that spinning is not fair, or is it plausible that the spinning process is fair

1.1.3
      To conduct a simulation analysis of this racquet spinning study for either **1.1.1** or **1.1.2**, you could flip a coin _____ times and repeat that process _____ times.
  A. 100, 1000
  B. 1000, 100

    C. 100, 1

    D. 1, 1000

1.1.4 Which of the following is the most important reason that a simulation analysis would repeat the coin-flipping process many times?

    A. To see whether the distribution of sample proportions follows a normal, bell-shaped curve

    B. To see whether the distribution of sample proportions is centered at 0.5

    C. To see how much variability results in the distribution of sample proportions

Shooting percentage

1.1.5 LeBron James of the Miami Heat hit 765 of his 1354 field goal attempts in the 2012/2013 season for a shooting percentage of 56.5%. Over the lifetime of LeBron's career, can we say he is more likely than not to make a field goal?

    a) Describe the parameter of interest.

    b) Is 56.5% a parameter or a statistic?

    c) What value would the chance model assign to the parameter from a)?

    d) Describe how you would use coin-flipping to simulate the 2012/1013 season under the assumption of the chance model

    e) What would be a typical value from a repetition of 1354 coin-flips? Justify your answer.

1.1.6 Dwayne Wade of the Miami Heat hit 569 of his 1093 field goal attempts in the 2012/2013 season for a shooting percentage of 52.1%. Over the lifetime of Dwayne's career, can we say that Dwayne is more likely than not to make a field goal?

    a) Is the long-run probability of Dwayne making a field goal a parameter or a statistic?

    b) Is 52.1% a parameter or a statistic?

    c). When simulating possible outcomes assuming the chance model, how many times would you flip a coin for one repetition of the 2012/2013 season?

    d) With each repetition, what would you keep track of?

    e) What would be a typical value from a repetition of 1093 coin-flips? Justify your answer.

1.1.7 If Dwayne Wade of the Miami Heat hits 52 out of his first 100 field goals in the 2013/2014 season, let's see how we might investigate if he is more likely than not to make a field goal?

    a) Based on these first 100 field goals, we want to find out what Dwayne's long-run probability of making a field goal is. Will this value be a statistic or a parameter?

    b) Is 52 out of 100 a statistic or a parameter?

    c) We can use the **One Proportion** applet to generate 1000 possible values of Dwayne's last 100 field goals under the chance model that he has a long-run probability of 50% of making a field goal. Match the aspects of the simulation in **Column A** to their equivalent aspects in the actual study listed in **Column B**.

Column A	Column B
Coin flip	Dwayne misses his field goal
Heads	Long-run probability that Dwayne makes a field goal
Tails	One set of 100 field goal shots by Dwayne
Chance of Heads	Dwayne shoots a field goal
One repetition	Dwayne makes his field goal

Minesweeper*

One of the authors sometimes likes to play *Minesweeper*, and of the last 20 times she played *Minesweeper*, she won 12 times. That is, she won 60% of the games. Use this information to answer Exercises 1.1.8-9.

1.1.8

a) Based on these 20 games, we would like to learn about her long-run probability of winning at Minesweeper. Is this value a parameter or a statistic?
b) Is 60% (12 wins in last 20 games) a parameter or a statistic?

To find out whether her performance on the last 20 games provides convincing evidence that her long-run probability of winning at *Minesweeper* is higher than 50%, she used the One Proportion applet to generate 100 possible values of what her number of wins could have been if her long-run probability of winning was *equal to 50%*.

c) In the table below, match the aspects of the simulation performed by the applet as listed in **Column A**, to their equivalent aspects in the actual study, as listed in **Column B**.

Column A	Column B
Coin flip	Author wins a game
Heads	Long-run probability that author wins
Tails	One set of 20 *Minesweeper* games played by author
Chance of Heads	Author loses a game
One repetition	Author plays a game of *Minesweeper*

1.1.9 Use the dotplot generated by the One Proportion applet, that assumed the long-run probability of winning was equal to 50%, to answer the following questions.

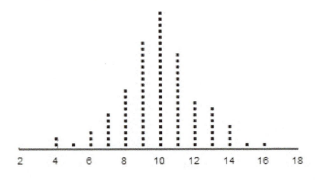

Number of heads

a) How many dots are in the dotplot? (*Hint:* you should reason it out, and *not* count the number of dots.)

b) What does each dot represent in terms of *Minesweeper* games and wins?

c) At what number is the dotplot centered? Could you have anticipated that? Why or why not?

d) Based on 12 wins on the last 20 games and the above dotplot, are you convinced that the author's long-run probability of winning at *Minesweeper* is higher than 50%? Explain how you are deciding.

e) Does this analysis prove that the author's long-run probability of winning at Minesweeper is 50%?

f) Suppose that it so happened that when the author played the last 20 games, she was also watching her favorite TV show. Does this information change your conclusion in part d)? If yes, how and why? If not, why not?

Buttered Toast

1.1.10 If you drop a piece of buttered toast on the floor, is it just as likely to land buttered side up as buttered side down? It sure seems like mine always lands buttered side down! Suppose that 7 of the last ten times I dropped toast it landed buttered side down. In order to carry out a statistical analysis, the One Proportion applet was used to produce 100 possible values for the number of times the toast could have landed butter side down if the long-run probability of landing butter side down is equal to 50%. Use the dotplot generated by the applet to answer the following questions.

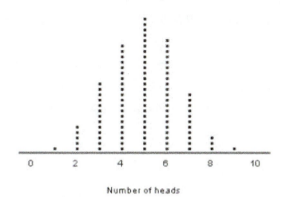

Number of heads

a) How many dots are in the dot plot?

b) What does each dot represent in terms of dropped toast and buttered side down?

c) At what number is the dotplot centered? Could you have determined that before running the applet simulation? Why or why not?

d) Based on 7 of the last 10 slices of toast landing buttered side down, and the above dotplot, are you convinced that the long 0 run probability of my toast landing buttered side down is greater than 50%? Explain how you are deciding.

e) Does this prove that my long-run probability of dropping toast buttered-side down is 50%?

Mark's Tennis Game*

1.1.11 Mark is practicing his tennis serves. He wants to be able to tell newspaper reporters the long-run probability of getting his first serve in. Mark gets 17 of his first 20 serves in.

a) Is 17 out of 20 (85%) a statistic or a parameter?

b) Is the long-run probability of Mark's first serve being in a statistic or a parameter?

c) Do you think that 17 out of 20 serves in is a *possible* outcome if Mark is just as likely to get his first serve in as to not get his first serve in?

d) Do you think that 17 out of 20 serves in is a *plausible* outcome if Mark is just as likely to get his first serve in as not to get his first serve in?

Lady tasting tea

A famous (in statistical circles) study involves a woman who claimed to be able to tell whether tea or milk was poured first into a cup. She was presented with 8 cups containing a mixture of tea and milk, and she correctly identified which had been poured first for all 8 cups. Use this information to answer Exercises 1.1.12-14.

1.1.12

a) Identify the observational units and variable in this study.

b) Identify the parameter for this study. (*Hint:* The long-run probability that…)

c) Identify the sample size. Also, identify the observed value of the statistic for this study.

d) Is it *possible* that the woman could get all 8 correct if she were randomly guessing with each cup?

e) Is it *unlikely* that the woman could get all 8 correct if she were randomly guessing with each cup? (Answer for now based on your intuition, without doing any analysis.)

1.1.13

a) Describe how you can use a coin to address the question "Is it *unlikely* that the woman could get all 8 correct if she were randomly guessing with each cup?" Be sure to include details such as, how many times you would toss the coin and why; what would "heads" and "tails" stand for; what you would record after each set of coin tosses; how many times you would repeat this process. Also, how is this repeated coin tossing going to help you address the question?

b) Now use an applet simulation to address the question "Is it *unlikely* that the woman could get all 8 correct if she were randomly guessing with each cup?" Then explain how your answer follows from your simulation results.

c) Based on your simulation analysis, would you conclude that the woman's result produces strong evidence that the she is not guessing as to whether milk or tea was poured first? Also explain the reasoning process behind your answer.

d) Is your result in c) statistically significant, or is it plausible she is just guessing?

1.1.14 Suppose that I try to discern whether tea or milk is poured first for 8 cups and make the correct identification 5 times.

a) I say: "5 out of 8 is more than half, so one must conclude that I'm doing better than random guessing." How would you respond? Use simulation results in your response.

b) With regards to my study on 8 cups of tea and milk, does this prove that I cannot tell whether milk or tea is placed in the cup first? Why or why not?

c) Now suppose that someone gets 14 correct out of 16 cups. Conduct a new simulation analysis to assess whether this result provides strong evidence that the person actually has ability better than random guessing to distinguish which was poured first. Be sure to describe which of the applet inputs needs to change and why. Also summarize your conclusion and explain your justification.

Zwerg*

An Austrian study was completed to determine if untrained sea lions and sea lionesses could follow various experimenter-given cues when given a choice of two objects. One experimenter-given cue was to point at one of the objects. One sea lioness, named Zwerg, successfully chose the pointed-at object 37 times out of 48 trials. Does this result show that Zwerg can correctly follow this type of direction by an experimenter more than 50% of the time? Use this information to answer Exercises 1.1.15-18.

1.1.15

a) What is the parameter of interest for this study? (*Hint:* The long-run probability that…)
b) What are two possible explanations for Zwerg's results of choosing the correct object 37 out of 48 times?
c) Which explanation of the two suggested in part b) do you think is the more plausible, based on Zwerg's results? Explain. Note: Provide an explanation based only on your intuition.
d) If Zwerg is just randomly choosing between the objects, what is the chance that she will choose the correct object?

1.1.16 In order to discern whether Zwerg is doing better than just guessing (50% correct) we will employ the 3S strategy:

a) *Statistic:* How many times did Zwerg pick the correct object? Out of how many attempts?
b) *Simulate:* Using an applet, simulate 1000 repetitions of having Zwerg choose between the two objects if she is doing so randomly.
c) Based on the value of the statistic from part a), do you think the chance model is wrong? Why or why not?
d) *Strength of evidence:* Based on your answers for (a)-(c), state your conclusions about the research question of whether the data provide convincing evidence that Zwerg can correctly follow this type of direction by an experimenter more than 50% of the time?
e) Are the results of this study statistically significant or is the chance model plausible? How are you deciding?

1.1.17 In a related study, another experimenter-given cue was to place a marker in front of the correct object. Zwerg successfully chose the object with the marker 26 out of 48 times. Does this result show that Zwerg can correctly follow this type of direction by an experimenter more than 50% of the time?

a) What are two possible explanations for Zwerg's results of choosing the correct object 26 out of 48 times?
b) Which explanation of the two suggested in part a) do you think is the more plausible, based on Zwerg's results? Explain based on your intuition.
c) If Zwerg is just randomly choosing between the objects, what is the chance that she will choose the correct object?

1.1.18. Refer to the previous questions data (26 out of 48) in order to discern whether Zwerg is doing better than just guessing (50% correct) we will employ the 3S strategy:

a) *Statistic:* How many times did Zwerg pick the correct object? Out of how many attempts?

b) *Simulate:* Using an applet, simulate 1000 repetitions of having Zwerg choose between the two objects if she is doing so randomly. Where is the distribution centered?

c) Based on the value of the statistic from part a), do you think the chance model is wrong? Why or why not?

d) *Strength of evidence:* Based on your answers for (a)-(c), state your conclusions about the research question of whether the data provide convincing evidence that Zwerg can correctly follow this type of direction by an experimenter more than 50% of the time?

e) Are the results of this study statistically significant or is the chance model plausible? How are you deciding?

f) Could you have anticipated this? Explain.

g) Does this prove that Zwerg is just guessing when the marker is used to indicate the correct object? Why or why not?

Janine's volleyball game

Janine is an ambidextrous volleyball player. She is practicing 'short serves' in volleyball because her team has an upcoming match against an opponent who does not pass short serves very well. Janine must land a majority of her 'short serves' in bounds or her coach will not let her use her short serve in the upcoming match. When serving left-handed Janine makes 23 out of 30 serves. Use this information to answer Exercises 1.1.19-22.

1.1.19

a) What is the parameter in this scenario? (*Hint:* the long-run probability that …)

b) What are two possible explanations for Janine getting 23 out of 30 left-handed short serves in bounds?

c) Which of the two explanations from part b) do you think is more plausible? No need to perform an analysis yet, explain your answer using only your intuition.

d) If Janine is just as likely to serve in as she is to not serve in, what is the chance she will serve in?

1.1.20 In order to discern whether or not Janine can land a majority of her short serves in bound, we will employ the 3S strategy.

a) *Statistic:* How many times did Janine land a short serve in bounds? Out of how many service attempts?

b) *Simulate:* Using an applet, simulate 1000 repetitions of having Janine serve 30 left-handed short serves if she is just as likely to serve in as she is to not serve in. Where is the distribution centered?

c) Based on the value of the statistic from part a), do you think the chance model is wrong? Why or why not?

d) *Strength of evidence:* Based on your answers for (a)-(c), state your conclusions about the research question of whether the data provide convincing evidence that Janine can get her short serve in with her left hand more than 50% of the time?

e) Are the results of this study statistically significant or is the chance model plausible? How are you deciding?

1.1.21 Janine serves 30 short serves with her right hand and gets 17 of the 30 in the court.

a) What are two possible explanations for Janine getting 17 out of 30 right-handed short serves in?

b) Which of the two explanations in part a) do you think is more plausible? No need to perform an analysis yet, explain your answer using only your intuition.

c) If Janine is just as likely to serve in as not, what is the chance she will serve in?

1.1.22 Using the data where Janine serves 17 out of 30 right-handed short serves into the court. In order to discern whether or not Janine can serve a majority of her short serves in, we will employ the 3S strategy.

a) *Statistic:* How many times did Janine get a short serve in? Out of how many service attempts?

b) *Simulate:* Using an applet, simulate 1000 repetitions of having Janine serve 30 right-handed short serves if she is just as likely to serve in as she is to not serve in.

c) Based on the value of the statistic from part a), do you think the chance model is wrong? Why or why not?

d) *Strength of evidence:* Based on your answers for (a)-(c), state your conclusions about the research question of whether the data provide convincing evidence that Janine can get her short serve in with her right hand more than 50% of the time?

e) Are the results of this study statistically significant or is the chance model plausible? How are you deciding?

f) Does this mean that Janine can serve a short serve in with her right hand 50% of the time? Explain.

Jerry's tennis game*

Jerry is a tennis player. He is working on a really tough first serve. While practicing his new tough serve, he gets 12 out of 20 or 60% of his serves in. Jerry wants to know if his long-run probability of getting his first serve in is greater than 50%. Use this information to answer Exercises 1.1.23-24.

1.1.23 What values would you enter into the **One Proportion** applet to run an analysis for Jerry?

a) Probability of heads _____

b) Number of tosses _____

c) Number of repetitions _____

d) Use the applet to conduct the simulation study. If Jerry's long-run probability of getting his new first serve in is truly 50%, is 12 out of 20 a likely value? Is it unlikely? How are you deciding?

1.1.24 Suppose Jerry continues to serve and gets 60 out of 100 serves in. Use the One Proportion applet again to test if Jerry's long-run probability of getting his first serve in is greater than 50%. State the values you would enter into the applet.

a) Probability of heads _____

b) Number of tosses _____

c) Number of repetitions _____

d) If Jerry's long-run probability of getting his new first serve in is truly 50%, is 60 out of 100 a likely value? Is it unlikely? How are you deciding?

e) Why do you suppose that you found different results in part (d) to this question as compared to part (d) to the previous question even though the sample statistic was 60% in both problems?

1.1.25 If an observed statistic from a study turns out to be a likely value under the chance model, then

a) we can say we have evidence against the chance model

b) we can say that the chance model is plausible

c) we can say that the chance model is true

d) we can't say anything about the chance model, because it isn't real data

Section 1.2

1.2.1* After you conduct a coin-flipping simulation, a graph of the _____ will be centered very close to 0.5. Choose from (A) – (D).

A. Process probability

B. Sample size

C. Number of heads

D. Proportion of heads

1.2.2 The graph of a null distribution will be centered approximately on:

A. The observed proportion

B. The observed count

C. The value of the probability in the null hypothesis

D. The number of repetitions performed

1.2.3* The p-value of a test of significance is:

A. The probability, assuming the null hypothesis is true, that we would get a result as extreme as the one that was actually observed.

B. The probability, assuming the alternative hypothesis is true, that we would get a result as extreme as the one that was actually observed.

C. The probability the null hypothesis is true.

D. The probability the alternative hypothesis is true.

1.2.4 Suppose a researcher is testing to see if a basketball player can make free throws at a rate higher than the NBA average of 75%. The player is tested by shooting 10 free throws and makes 8 of them. In conducting the related test of significance we have a computer applet do an appropriate simulation, with 1000 repetitions, and produce a null distribution. This distribution represents:

A. Repeated results if the player makes 80% of his shots in the long run.

B. Repeated results if the player makes 75% of his shots in the long run.

C. Repeated results if the player makes more than 75% of his shots in the long run.

D. Repeated results if the player makes more than 80% of his shots in the long run.

1.2.5* The simulation (flipping coins or using the applet) done to develop the distribution we used to find our p-values assumes which hypothesis is true?

A. Null hypothesis

B. Alternative hypothesis

C. Both hypotheses

D. Neither hypothesis

1.2.6 When using the coin-flipping chance model, the most important reason you repeat a simulation of the study many times is

A. To see whether the null distribution follows a symmetric, bell-shaped curve

B. To see whether the null distribution is centered at 0.5

C. To see whether your coin is really fair

D. To see how much variability there is in the null distribution

1.2.7* When we get a p-value that is very large, we may conclude that:

A. The null hypothesis has been proven to be true

B. There is strong evidence for the alternative hypothesis

   C. The null hypothesis is plausible

   D. The alternative hypothesis has been proven to be false

1.2.8 When we get a p-value that is very small, we may conclude that:

   A. The null hypothesis has been proven to be true

   B. There is strong evidence for the alternative hypothesis

   C. The null hypothesis is plausible

   D. The alternative hypothesis has been proven to be false

1.2.9* Suppose you are testing the hypotheses H_0: π 0.25 and H_a: $\pi < 0.25$ and the observed statistic, $\hat{p}$, is equal to 0.30 with a sample size of 100.

   a) If you are using a proportion as your statistic, where do you expect your null distribution to be centered?

   b) If you are using a count as your statistic, where do you expect your null distribution to be centered?

1.2.10 Explain the meaning of each of the following symbols.

   a) H_0

   b) H_a

   c) $\hat{p}$

   d) π

   e) n

1.2.11* What is the difference between $\hat{p}$ and p-value?

Sarah, the chimpanzee

1.2.12 A chimpanzee named Sarah was the subject in a study of whether chimpanzees can solve problems. Sarah was shown 30-second videos of a human actor struggling with one of several problems (for example, not able to reach bananas hanging from the ceiling). Then Sarah was shown two photographs, one that depicted a solution to the problem (like stepping onto a box) and one that did not match that scenario. Researchers watched Sarah select one of the photos, and they kept track of whether Sarah chose the correct photo depicting a solution to the problem. Sarah chose the correct photo in 7 of 8 scenarios that she was presented.

We want to run a test of significance to determine whether Sarah understands how to solve problems and will thus pick the photo of the correct solution more often that what would be done by random chance.

   a) Describe the parameter of interest in the context of this study, and assign a symbol to denote it.

b) What is the observed value of the statistic in this case?

c) Write out the null and alternative hypotheses for this first in words, and then in symbols.

d) We conducted a test of significance using the One Proportion applet, and got the following null distribution for the "number of heads." (Note this null distribution uses only 100 simulated samples and not the usual 1000 or more.) Based on the null distribution, what is the p-value for the test? How did you calculate it?

Probability of heads: 0.5

Number of tosses: 8

Number of repetitions: 100

☐ Animate

[Toss Coins]

Total = 100

Number of heads

e) Based on your p-value, do you have strong evidence that Sarah understands how to solve problems similar to those she was presented? How are you deciding? Justify.

f) Complete this sentence describing what the p-value means. If Sarah doesn't understand how to solve problems and is just guessing at which picture to select, the probability she would ...

g) What does a single dot in the null distribution represent in terms of Sarah and the photos?

Hope, the dog*

1.2.13 Suppose you are testing to see if your dog, Hope, understands pointing towards an object. You put Hope through 20 trials and 12 times (or 60%) she goes to the correct object when given a choice between two objects. You then conduct a test of significance and generate the following 100 simulations using an applet.

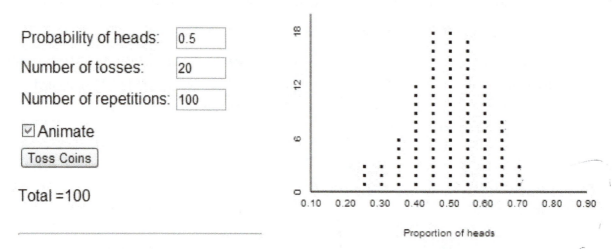

Probability of heads: 0.5

Number of tosses: 20

Number of repetitions: 100

☑ Animate

[Toss Coins]

Total = 100

a)   Write out the null hypothesis and alternative hypotheses for this study in words.

b)   If we let π be the long run probability Hope will go to the correct object, write out the null and alternative hypotheses for this study in symbols.

c)   Based on the null distribution shown, what is the p-value for your test?

d)   Based on your p-value, do you have strong evidence that Hope understands pointing?

e)   Using the null distribution shown, what is the smallest proportion of times Hope could go toward the correct object so that you would have strong evidence that she understands pointing toward an object?

f)   Which of the following best describes what a single dot represents in the null distribution shown?

   i)   One possible value of the proportion of times Hope goes to the correct object out of 20 if she goes to the correct object 50% of the time in the long run.

   ii)   One possible value of the proportion of times Hope goes to the correct object out of 20 if she goes to the correct object **more than** 50% of the time in the long run.

1.2.14 Suppose two researchers are conducting studies about animal behavior, and they both have the same null hypothesis and the same alternative hypothesis as each other. Also, suppose that researcher A's study results produce a p-value of 0.034, whereas researcher B's study results produce a p-value of 0.055. Which researcher has found stronger evidence against the null hypothesis and for the alternative hypothesis? Explain how you know.

1.2.15 Suppose you determine that in order to evaluate some data you need to conduct a simulation analysis.

a)   In your own words explain how to conduct a simulation using a six-sided die where the sample size is 20 and the null hypothesis probability is 1/6. Be specific.

b)   Referring to part a), explain how you would conduct the simulation using a regular deck of playing cards. Be specific.

c)   How would your answer to a) change if the sample size was 30?

d)   How would your answer to b) change if the sample size was 30?

e)   How would your answer to a) change if the null hypothesis probability is 2/3?

f)   How would your answer to b) change if the null hypothesis probability is 2/3?

Love, first

A recent study (Ackerman, Griskevicius, and Li, 2011) examined expressions of commitment between two partners in a committed romantic relationship. One aspect of the study involved 47 heterosexual couples who are part of an online pool of people willing to participate in surveys. These 47 couples were asked about which person was the first to say "I love you." For 7 of those couples, the two people disagreed about the answer to this question. But both people agreed for the other 40 couples, so those 40 responses were included in the analysis. Previous studies have suggested that males tend to say "I love you" first. Use this information to answer Exercises 1.2.16-17.

1.2.16

a)   Identify the observational units and variable in this study. Also classify the variable as categorical or quantitative.

b)   State the appropriate null and alternative hypotheses (in words) for testing whether males are more likely to say "I love you" first.

We can express these hypotheses with symbols as: Null: $\pi = 0.5$ Alternative: $\pi > 0.5$

c)   Describe (in words) what the symbol π stands for here.

It turned out that for 28 of the 40 couples in the sample (after the 7 couples who could not agree were excluded), the man said "I love you" before the woman did.

d)   Determine the sample proportion (written as a decimal) of couples for whom the man was the first to say "I love you." What symbol do we use to denote this proportion?

e)   Describe how you could use a coin-flipping model to find the p-value for these data.

f)   Use an applet to conduct a simulation analysis to assess the strength of evidence against the null hypothesis provided by the sample data. Report the values that you enter into the applet, and describe how you approximate the p-value. Also report the value of this (approximate) p-value.

g)   Write a sentence or two interpreting this p-value: the probability of ____, assuming _____.

h)   Summarize your conclusion from this p-value.

1.2.17 The researchers also interviewed 96 university students that had been or were currently involved in a romantic heterosexual relationship where at least one person said "I love you." The students were asked, "Think about your last or current romantic relationship in which someone confessed their love. In this relationship, who admitted love first?"

  a) Identify the observational units and variable in this study. Also classify the variable as categorical or quantitative.
  b) State the appropriate null and alternative hypotheses (in words) for testing whether males are more likely to say "I love you" first.

It turned out that for 59 of the 96 responses, the man said "I love you" before the woman did.

  c) Determine the sample proportion (written as a decimal) of couples for whom the man was the first to say "I love you." What symbol do we use to denote this proportion?
  d) Describe how you could use a coin-flipping model to find the p-value for these data.
  e) Use an applet to conduct a simulation analysis to assess the strength of evidence against the null hypothesis provided by the sample data. Report the values that you enter into the applet, and describe how you approximate the p-value. Also report the value of this (approximate) p-value.
  f) Write a sentence or two interpreting this p-value: the probability of _____, assuming _____.
  g) Summarize your conclusion from this p-value.

Rhesus Monkeys*

A recent article (Hauser, Glynn, and Wood, 2007) described a study that investigated whether rhesus monkeys have some ability to understand gestures made by humans. In one part of the study, the experimenter approached individual rhesus monkeys and placed 2 boxes an equal distance from the monkey. The experimenter then placed food in one of the boxes, making sure that the monkey could tell that one of the boxes received food without revealing which one. Finally, the researcher made eye contact with the monkey and then gestured toward the box with the food by jerking his head toward that box. This process was repeated for a total of 40 rhesus monkeys. It turned out that 30 of the monkeys approached the box that the human had gestured toward, and 10 approached the other box. The purpose is to investigate whether rhesus monkeys can interpret the head jerk better than random chance. Use this information to answer Exercises 1.2.18-19.

1.2.18

  a) Identify the observational units and variable in this study. Also classify the variable as categorical or quantitative.

b) Describe in words the parameter of interest for this study, and assign a symbol to it.

c) Determine the sample proportion of monkeys who picked the box towards which the human had gestured. Is this value a parameter or a statistic? What is the symbol you should use to denote this proportion?

d) State the appropriate null and alternative hypotheses in the context of this study, first in words, and then in symbols.

e) Describe how you could use a coin to conduct a simulation analysis of this study and its result. Give sufficient detail that someone else could implement this simulation analysis based on your description. Be sure to indicate how you would decide whether the observed data provide convincing evidence that rhesus monkeys can interpret human gestures better than random chance.

f) Use an applet to conduct a simulation analysis with at least 1000 repetitions. Based on null distribution produced by the applet, explain how you are deciding whether the observed data provide convincing evidence that rhesus monkeys can read human gestures better than random chance. Summarize the conclusion that you would draw about the research question of whether rhesus monkeys have some ability to understand gestures made by humans.

1.2.19. In one part of the study, the experimenter approached individual rhesus monkeys and placed 2 boxes an equal distance from the monkey. The experimenter then placed food in one of the boxes, making sure that the monkey could tell that one of the boxes received food without revealing which one. Finally, the researcher pointed and looked toward the box that contained the food. This process was repeated for a total of 40 rhesus monkeys. It turned out that 31 of the monkeys approached the box that the human had gestured toward, and 9 approached the other box. The purpose is to investigate whether rhesus monkeys can interpret human gestures better than random chance.

a) Identify the observational units and variable in this study. Also classify the variable as categorical or quantitative.

b) Describe in words the parameter of interest for this study, and assign a symbol to it.

c) Determine the sample proportion of monkeys who picked the box towards which the human had gestured. Is this value a parameter or a statistic? What is the symbol you should use to denote this proportion?

d) State the appropriate null and alternative hypotheses in the context of this study, first in words, and then in symbols.

e) Describe how you could use a coin to conduct a simulation analysis of this study and its result. Give sufficient detail that someone else could implement this simulation analysis based on your description. Be sure to indicate how you would decide whether the observed data provide convincing evidence that rhesus monkeys can interpret human gestures better than random chance.

f)   Use an applet to conduct a simulation analysis with at least 1000 repetitions. Based on null distribution produced by the applet, explain how you are deciding whether the observed data provide convincing evidence that rhesus monkeys can read human gestures better than random chance. Summarize the conclusion that you would draw about the research question of whether rhesus monkeys have some ability to understand gestures made by humans.

Minesweeper*

1.2.20 Recall that one of the authors liked to play the game *Minesweeper* (Exercise 1.1.8) and she won 12 of the last 20 games she played. Now that you know how to calculate p-values, use an applet to find the p-value to investigate whether her results provide convincing evidence that her long-run probability of winning at *Minesweeper* is higher than 50%? Report the p-value and use the p-value to state a conclusion about the strength of evidence.

Spider Solitaire

1.2.21 While the author in the previous question likes to play *Minesweeper*, another author likes to play *Spider Solitaire* on the computer. In his last 40 games, he won 24 of them. Use an applet to find the p-value to investigate whether his results provide convincing evidence that his long-run probability of winning at *Spider Solitaire* is higher than 50%? Report the p-value and use the p-value to state a conclusion about the strength of evidence.

Spinning a coin*

1.2.22 It has been stated that spinning a coin on a table will result in it landing heads side up fewer than 50% of the time in the long run. One of the authors tested this by spinning a penny 50 times on a table and it landed heads side up 21 times. A test of significance was then conducted with the following hypotheses.

$$\text{Null: } \pi = 0.5 \qquad \text{Alternative: } \pi < 0.5$$

a)   Describe what the symbol π stands for in this context.

b)   Use an applet to conduct a simulation with at least 1000 repetitions. What is your p-value? Based on your p-value is there strong evidence that the probability the spun coin will land heads up is less than 50%?

c)   Suppose you focused on the proportion of times the coin landed tails up instead of heads up. How would your hypotheses be different? What would you do differently to calculate your p-value?

Flipping a coin

1.2.23 According to Stanford mathematics and statistics professor Persi Diaconis, the probability a flipped coin that starts out heads up will also land heads up is 51%. Suppose you want to test this. More specifically, you want to test to determine if the probability that a coin that starts out heads up will also land heads up is more than 50%. Suppose you flip a coin (that starts out heads up) 100 times and find that it lands heads up 53 of those times.

a) If π stands for the probability a coin that starts heads up will also land heads up, write out the hypotheses for this study in symbols.

b) Use an applet to conduct a simulation with at least 1000 repetitions. What is your p-value? Based on your p-value is there strong evidence that the probability that a coin that starts heads up will also land heads up is more than 50%?

c) Even though the coin in the study landed heads up more than 50% of the time, you should not have found strong evidence that it will land heads up more than 50% of the time in the long run. Does this mean the Diaconis is wrong about his assertion that it will land heads up 51% of the time in the long run? Explain.

d) The newspaper article described the 51% probability by saying, "This means that if a coin is flipped with its heads side facing up, it will land the same way 51 out of 100 times." What is wrong with this statement?

Free throws*

1.2.24 Suppose your friend says he can shoot free throws as well as someone in the NBA and you don't think he is that good. You know that the NBA average for shooting free throws is 75% and decide to test your friend. You have him shoot 20 free throws and he makes 12 of them. Based on this, do you have strong evidence that he is worse than the average NBA player? Answer this by conducting a simulation and reporting your p-value.

1.2.25 Suppose you retest your friend from the previous question to see if he is a worse free throw shooter than the NBA average of 75%. He shoots 20 more free throws and again makes 12 of them. You combine the data from the two tests together so he made 24 of his 40 attempts. Based on this, do you have strong evidence that he is worse than the average NBA player? Answer this by conducting a simulation and reporting your p-value.

Section 1.3

1.3.1*Which standardized statistic (standardized sample proportion) gives you the strongest evidence against the null hypothesis?

    A. $z = 1$

    B. $z = 0$

    C. $z = -3$

    D. $z = 1.8$

1.3.2 Consider the output given below that was obtained using the One Proportion applet. Use information from the output to find the standardized statistic for a sample proportion value of 0.45.

1.3.3*Consider the two null distributions (A and B) given below, both for proportion of successes. For which null distribution will the standardized statistic for a sample proportion value of 0.60 be farther from 0? How are you deciding?

A.

B.

1.3.4 Suppose that your hypotheses are: $H_0: \pi = 0.25$, and $H_a: \pi < 0.25$. In the context of these hypotheses, which of the following standardized statistics would provide the strongest evidence against the null, and for the alternative hypothesis? Why?

    A. $z = -1$

    B. $z = 0$

    C. $z = 3$

    D. $z = -1.8$

1.3.5* Identify these statements as either true or false.

    a) A p-value can be negative.

    b) A standardized statistic can be negative.

    c) We run tests of significance to determine whether $\hat{p}$ is larger or smaller than some value.

    d) As a p-value gets smaller, its corresponding standardized statistic gets closer to zero.

1.3.6 Suppose that a standardized statistic (standardized sample proportion) for a study is calculated to be 2.45. Which of the following is the most appropriate interpretation of this standardized statistic?

    A. The observed value of the sample proportion is 2.45 SDs away from the hypothesized parameter value.

    B. The observed value of the sample proportion is 2.45 SDs above the hypothesized parameter value.

    C. The observed value of the sample proportion is 2.45 times the hypothesized parameter value.

    D. The study results are statistically significant.

Rock, Paper Scissors*

Refer to Example 1.2 which explored player's choices in the game Rock, Paper, Scissors when answering Exercises 1.3.7-9.

1.3.7 Suppose that you play the game with three different friends separately with the following results:

        Friend A chose scissors 100 times out of 400 games

        Friend B chose scissors 20 times out of 120 games

        Friend C chose scissors 65 times out of 300 games

Suppose that for each friend you want to test whether the long-run probability that the friend will pick scissors is less than one-third.

a) Determine the appropriate standardized statistic for each friend's results. (*Hint:* you will need to get the standard deviation of the simulated statistics from the null distribution produced by an applet.)

b) Based on the standardized statistics, which friend's data provides the strongest evidence that the long-run probability that the friend will choose scissors is less than one-third? Which friend's data provides the least strong evidence?

1.3.8 Suppose that you play the game, again, with two other friends separately with the following results:

Friend D chose rock 200 times out of 400 games

Friend E chose rock 20 times out of 40 games

Suppose that for each friend you want to test whether the long-run probability that the friend will pick rock is more than one-third.

a) Even though both of the friends played rock the same proportion of times, one of the two friends data provide more evidence the null hypothesis is wrong. Which friend's data do you think provides more evidence against the null hypothesis? Why?

b) Based on your answer to (a) which friend's data yields a smaller p-value? Why?

c) Based on your answer to (a) which friend's data yields a larger standardized statistic? Why?

d) Based on your answer to (c) which friend's null distribution has a smaller standard deviation? Why?

1.3.9 Suppose that you play the game, again, with two other friends separately with the following results:

Friend F chose rock 15 times out of 40 games

Friend G chose rock 30 times out of 40 games

Suppose that for each friend you want to test whether the long-run probability that the friend will pick rock is more than one-third. In which friend's case (F or G) will you find a larger (more extreme) standardized statistic? Explain your reasoning, without actually calculating the standardized statistics.

Minesweeper

1.3.10 Recall that one of the authors likes to play the game *Minesweeper* (Exercise 1.1.8) and in her last 20 games, she won 12. Use an applet to conduct appropriate simulations in order to calculate the standardized statistic for these data and investigate whether her results provide convincing evidence that her long-run probability of winning at *Minesweeper* is higher than 50%? Report the calculated standardized statistic, and use it to state a conclusion about the strength of evidence.

Doris and Buzz*

Refer to Example 1.1 involving two dolphins, Buzz and Doris, where the researcher Dr. Bastian was investigating whether dolphins can communicate to answer Exercises 1.3.11-12.

1.3.11 Recall that in the first study, out of 16 attempts, Buzz pushed the correct button 15 times. Calculate the standardized statistic for these data to investigate whether Buzz's results provide convincing evidence that his long-run probability of pushing the correct button is higher than 50%? Report the calculated standardized statistic, and use it to state a conclusion about the strength of evidence.

1.3.12 Recall that in the second study mentioned (where they later found out that the fish delivering equipment had been malfunctioning), out of 28 attempts, Buzz pushed the correct button 16 times. Calculate the standardized statistic for these data to investigate whether Buzz's results provide convincing evidence that his long-run probability of pushing the correct button is higher than 50%? Report the calculated standardized statistic, and use it to state a conclusion about the strength of evidence.

Right or left

1.3.13 Most people are right-handed, and even the right eye is dominant for most people. Molecular biologists have suggested that late-stage human embryos tend to turn their heads to the right. In a study reported in *Nature* (2003), German bio-psychologist Onur Güntürkün conjectured that this tendency to turn to the right manifests itself in other ways as well, so he studied kissing couples to see which side they tended to lean their heads while kissing. He and his researchers observed kissing couples in public places such as airports, train stations, beaches, and parks. They were careful not to include couples who were holding objects such as luggage that might have affected which direction they turned. For each kissing couple observed, the researchers noted whether the couple leaned their heads to the right or to the left. They observed 124 couples, age 13-70 years. Suppose that we want to use the data from this study to investigate whether kissing couples tend to lean their heads right more often than would happen by random chance.

a)   Define the parameter of interest in the context of the study, and assign a symbol to it.

b)   State the null hypothesis and the alternative hypothesis using the symbol defined in part a).

c)   Of the 124 kissing couples, 80 were observed to lean their heads right. What is the observed proportion of kissing couples who leaned their heads to the right? What symbol should you use to represent this value?

d)   Determine the standardized statistic from the data. (*Hint:* you will need to get the standard deviation of the simulated statistics from the null distribution.)

e) Interpret the standardized statistic in the context of the study. (*Hint:* You need to talk about the value of your observed statistic in terms of standard deviations assuming _____ is true.)

f) Based on the standardized statistic, state the conclusion that you would draw about the null and alternative hypotheses.

1.3.14 Refer to the previous exercise except suppose that instead of $H_0: \pi = 0.5$, our null hypothesis was $H_0: \pi = 0.60$.

a) In the context of this null hypothesis, determine the standardized statistic from the data where 80 of 124 kissing couples leaned their heads right. (*Hint:* you will need to get the standard deviation of the simulated statistics from the null distribution.)

b) How, if at all, does the standardized statistic calculated here differ from that when $H_0: \pi = 0.5$? Explain why this makes sense.

Love, first*

1.3.15 A previous exercise(1.2.16) introduced you to a study of 40 heterosexual couples. In 28 of the 40 couples the male said 'I love you' first. The researchers were interested in learning this data provided evidence that in significantly more than 50% of couples the male says I love you first.

a) State the null hypothesis and the alternative hypothesis in the context of the study.

b) Determine the standardized statistic from the data. (*Hint:* you will need to get the standard deviation of the simulated statistics from the null distribution.)

c) Interpret the standardized statistic in the context of the study. (*Hint:* You need to talk about the value of your observed statistic in terms of standard deviations assuming _____ is true.)

d) Based on the standardized statistic, state the conclusion that you would draw about the research question of whether males are more likely to say "I love you" first.

Rhesus Monkeys

Revisit exercise 1.2.18 about the study on Rhesus monkeys. When given a choice between two boxes, 30 out of 40 monkeys approached the box that the human had gestured toward, and 10 approached the other box. The purpose is to investigate whether rhesus monkeys can interpret human gestures better than random chance.

1.3.16 For this study…

    a) State the null hypothesis and the alternative hypothesis in the context of the study.

    b) Determine the standardized statistic from the data. (*Hint*: you will need to get the standard deviation of the simulated statistics from the null distribution in an applet.)

    c) Interpret the standardized statistic in the context of the study. (*Hint:* You need to talk about the value of your observed statistic in terms of standard deviations assuming _____ is true.)

    d) Based on the standardized statistic, state the conclusion that you would draw about the research question of whether rhesus monkeys have some ability to understand gestures made by humans.

Tasting tea*

Revisit exercise1.1.12 about the study on a lady tasting tea. When presented with 8 cups containing a mixture of milk and tea, she identified whether tea or milk was poured first, correctly for all 8 cups. Is she doing better than if she were just guessing?

1.3.17 For this study…

    a) Define the parameter of interest in the context of the study, and assign a symbol to it.

    b) State the null hypothesis and the alternative hypothesis using the symbol defined in part a).

    c) What is the observed proportion of times the lady correctly identified what was poured first into the cup? What symbol should you use to represent this value?

    d) Suppose that you were to generate the null distribution of the sample proportion of correct answers, that is, the distribution of possible values of sample proportion of correct identifications if the lady always guesses, where would you anticipate this distribution would center? Also, do you anticipate the SD of the null distribution to be negative, positive, or 0? Why?

    e) Use an applet to generate the null distribution of sample proportion of correct identifications, and use it to determine the standardized statistic.

    f) Interpret the standardized statistic in the context of the study. (*Hint:* You need to talk about the value of your observed statistic in terms of standard deviations assuming _____ is true.)

    g) Based on the standardized statistic, state the conclusion that you would draw about the research question of whether the lady does better than randomly guess.

Zwerg

Refer to a previous exercise, 1.1.15, involving Zwerg the sea lioness to answer Exercises 1.3.18-19.

1.3.18 In one study, Zwerg, when given two choices, successfully chose the object pointed at by the experimenter 37 times out of 48 trials. Does this result show that Zwerg can correctly follow this type of direction ("pointing at") by an experimenter more than 50% of the time?

a) Define the parameter of interest in the context of the study, and assign a symbol to it.

b) State the null hypothesis and the alternative hypothesis using the symbol defined in part a).

c) What is the observed proportion of times Zwerg picked the correct object? What symbol should you use to represent this value?

d) Suppose that you were to generate the null distribution of the sample proportion of correct answers, that is, the distribution of possible values of sample proportion of correct identifications if Zwerg always guesses, where would you anticipate this distribution would center? Also, do you anticipate the SD of the null distribution to be negative, positive, or 0? Why?

e) Use an applet to generate the null distribution of sample proportion of correct answers, and use it to determine the standardized statistic.

f) Interpret the standardized statistic in the context of the study. (*Hint:* You need to talk about the value of your observed statistic in terms of standard deviations assuming _____ is true.)

g) Based on the standardized statistic, state the conclusion that you would draw about the research question of whether Zwerg does better than randomly guess, when the cue is that the experimenter points at the correct object.

1.3.19 in a related study, Zwerg, when given two choices, successfully chose the object indicated by a 'marker' 26 out of 48 times. Does this result show that Zwerg can correctly follow this type of direction ("placing a marker") by an experimenter more than 50% of the time?

a) Define the parameter of interest in the context of the study, and assign a symbol to it.

b) State the null hypothesis and the alternative hypothesis using the symbol defined in part a).

c) What is the observed proportion of times Zwerg picked the correct object? What symbol should you use to represent this value?

d) Suppose that you were to generate the null distribution of the sample proportion of correct answers, that is, the distribution of possible values of sample proportion of correct identifications if Zwerg always guesses, where would you anticipate this distribution would center? Also, do you anticipate the SD of the null distribution to be negative, positive, or 0? Why?

e) Use an applet to generate the null distribution of sample proportion of correct answers, and use it to determine the standardized statistic.

f) Interpret the standardized statistic in the context of the study. (*Hint:* You need to talk about the value of your observed statistic in terms of standard deviations assuming _____ is true.)

g) Based on the standardized statistic, state the conclusion that you would draw about the research question of whether Zwerg does better than randomly guess, when the cue is that the experimenter places a marker in front of the correct object.

Helper, Hinderer*

Use the following information to answer Exercises 1.3.20-21. An investigation reported in the November 2007 issue of *Nature* (Hamlin, Wynn, and Bloom) aimed at assessing whether infants take into account an individual's actions towards others in evaluating that individual as appealing or aversive, perhaps laying the foundation for social interaction. In one component of the study, each of sixteen 10-month-old healthy infants (in New England) were shown a "climber" character (a piece of wood with "google" eyes glued onto it) that could not make it up a hill in two tries. Then they were shown two scenarios for the climber's next try, one where the climber was pushed to the top of the hill by another character ("helper") and one where the climber was pushed back down the hill by another character ("hinderer"). Each infant was alternately shown these two scenarios many times. Then the child was presented with both pieces of wood (the "helper" and the "hinderer") and asked to pick one to play with. In the study 14 out of the 16 babies picked the "helper" toy. Does this provide evidence that such 10-month old infants have a *genuine preference* for the helper toy?

1.3.20

a) Define the parameter of interest in the context of the study, and assign a symbol to it.
b) State the null hypothesis and the alternative hypothesis using the symbol defined in part a).
c) What is the observed proportion of times the infants picked the helper toy? What symbol should you use to represent this value?
d) Use an applet to generate the null distribution of the proportion of "successes. Determine the standardized statistic for the observed sample proportion of "successes." Show work.
e) Interpret the standardized statistic in the context of the study. (*Hint:* You need to talk about the value of your observed statistic in terms of standard deviations assuming _____ is true.)
f) Based on the standardized statistic, state the conclusion that you would draw about the research question of whether such 10-month old infants have a genuine preference for the helper toy.

1.3.21

a) Based on the value of the standardized statistic obtained in the previous question, do you anticipate the p-value to be small? Why or why not?
b) Use an appropriate applet to find and report the p-value. Also, interpret the p-value. (*Hint:* The p-value is the probability of ... assuming...)
c) Based on the p-value, state the conclusion that you would draw about the research question of whether such 10-month old infants have a genuine preference for the helper toy.
d) Do the p-value and standardized statistic provide similar strength of evidence? Do they both lead you to the same conclusion? Did you anticipate this? Why or why not?

Choosing numbers

Use the following information to answer Exercises 1.3.22-25. One of the authors read somewhere that it's been conjectured that when people are asked to choose a number from the choices 1, 2, 3, and 4, they tend to choose "3" more than would be expected by random chance.

To investigate this, she collected data in her class. Here is the table of responses from her students:

Chose 1	Chose 2	Chose 3	Chose 4
10	4	14	5

1.3.22

a) Define the parameter of interest in the context of the study, and assign a symbol to it.

b) State the null hypothesis and the alternative hypothesis using the symbol defined in part a).

c) What is the observed proportion of times students chose the number 3? What symbol should you use to represent this value?

d) Use an applet to generate the null distribution of the proportion of "successes." Report the mean and SD of this null distribution.

e) Determine the standardized statistic for the observed sample proportion of "successes."

f) Interpret the standardized statistic in the context of the study. (*Hint:* You need to talk about the value of your observed statistic in terms of standard deviations assuming _____ is true.)

g) Based on the standardized statistic, state the conclusion that you would draw about the research question of whether students tend to have a genuine preference for the number 3, when given the choices 1, 2, 3, and 4.

1.3.23

a) Based on the value of the standardized statistic obtained in the previous question, do you anticipate the p-value to be small? Why or why not?

b) Use an appropriate applet to find and report the p-value. Also, interpret the p-value. (*Hint:* The p-value is the probability of … assuming…)

c) Based on the p-value, state the conclusion that you would draw about the research question of whether when people are asked to choose from the numbers 1, 2, 3, and 4, they tend to choose "3" more than would be expected by random chance.

d) Do the p-value and standardized statistic provide similar strength of evidence? Do they both lead you to the same conclusion? Did you anticipate this? Why or why not?

1.3.24 Suppose that you wanted to investigate whether people tend to pick a "big" number (3 or 4) rather than a "small" number (1 or 2).

   a) In this context, define the parameter of interest, and assign a symbol to it.
   b) State the null hypothesis and the alternative hypothesis using the symbol defined in part a).
   c) Recall that of 33 students, 19 picked a "big" number (that is, 3 or 4). What is the observed proportion of times students picked a "big" number? What symbol should you use to represent this value?
   d) Use an applet to generate the null distribution of the proportion of "successes." Report the mean and SD of this null distribution.
   e) Determine the standardized statistic for the observed sample proportion of "successes." Show work.
   f) Based on the standardized statistic, state the conclusion that you would draw about the research question of whether people tend to pick a "big" number.

1.3.25 Suppose that you wanted to investigate whether people tend to pick a "big" number (3 or 4) rather than a "small" number (1 or 2).

   a) Based on the value of the standardized statistic obtained in the previous question, do you anticipate the p-value to be small? Why or why not?
   b) Use an appropriate applet to find and report the p-value. Also, interpret the p-value. (*Hint:* The p-value is the probability of … assuming…)
   c) Based on the p-value, state the conclusion that you would draw about the research question of whether when people are asked to pick a number from the choices 1, 2, 3, and 4, they tend to pick a "big" number.
   d) Do the p-value and standardized statistic provide similar strength of evidence? Do they both lead you to the same conclusion? Did you anticipate this? Why or why not?

Section 1.4

Racquet Spinning*

Use the following information to answer Exercises 1.4.1-3. Reconsider Exercise 1.1.1 where researchers investigated whether a spun tennis racquet is equally likely to land with the label facing up or down.

1.4.1   Does this racquet spinning study call for a one-sided or a two-sided alternative?

   A. One-sided, because there is only one variable: how the label lands
   B. Two-sided, because there are two possible outcomes: up or down

C. One-sided, because the researchers want to know whether the label is more likely to land face up

D. Two-sided, because the researchers want to know whether the spinning process is fair or biased in either direction

1.4.2 Which of the following will always be true about the standardized statistic for the racquet spinning study?

A. The standardized statistic increases as the sample proportion that land "up" increases.

B. The standardized statistic decreases as the sample proportion that land "up" increases.

C. The standardized statistic increases as the sample proportion that land "up" gets farther from 0.5.

D. The standardized statistic decreases as the sample proportion that land "up" gets farther from 0.5.

1.4.3 Which of the following will always be true about the p-value for the racquet spinning study?

A. The p-value increases as the sample proportion that land "up" increases.

B. The p-value decreases as the sample proportion that land "up" increases.

C. The p-value increases as the sample proportion that land "up" gets farther from 0.5.

D. The p-value decreases as the sample proportion that land "up" gets farther from 0.5.

Chess-Boxing

1.4.4 You have heard that in sports like boxing there might be some competitive advantage to those wearing red uniforms. You want to test this with your new favorite sport of chess-boxing. You randomly assign blue and red uniforms to contestants in 20 matches and find that those wearing red won 14 times (or 70%). You conduct a test of significance using simulation and get the following null distribution. (Note this null distribution uses only 100 simulated samples and not the usual 1000 or more.)

a) Suppose you want to see if competitors wearing red win more than 50% of the matches in the long run, so you test H_0: π = 0.5 versus H_a: π > 0.5. What is your p-value based on the above null distribution?

b) Suppose you now want to see if competitors wearing either red or blue have an advantage, so you test H_0: π = 0.5 versus H_a: π ≠ 0.5. What is your p-value now based on the above null distribution?

1.4.5 Suppose you are testing the hypotheses H_0: π = 0.5 versus H_a: π > 0.5. You get a sample proportion of 0.54 and find that your p-value is 0.08. Now suppose you redid your study with each of the following changes, will your new p-value be larger or smaller than the 0.08 your first obtained?

a) You increase the sample size and still find a sample proportion of 0.54.

b) Keeping the sample size the same, you take a new sample and find a sample proportion of 0.55.

c) With your original sample, you decided to test a two-sided alternative instead of H_a: π > 0.5.

1.4.6 Identify these statements as either true or false.

a) Using a simulation-based test, the p-value for a two-sided test will be about twice as large as the corresponding p-value for a one-sided test.

b) We run tests of significance to determine if $\hat{p}$ is larger, smaller, or different than some value.

Minesweeper*

1.4.7 Look back to question 1.1.3. Recall that in the last 20 games of *Minesweeper* she played, the author won 12.

a) What if she had played 20 games, and won 18. Would that provide stronger, weaker, or evidence of similar strength compared to 12 wins out of 20, to conclude that her long-run probability of winning at *Minesweeper* is higher than 50%? Explain how you are deciding.

b) What if she had played 100 games, and won 60. Would that provide stronger, weaker, or evidence of similar strength compared to 12 wins out of 20, to conclude that her long-run probability of winning at *Minesweeper* is higher than 50%? Explain how you are deciding.

c) What if she had played 30 games, and won 12. Would that provide evidence that her long-run probability of winning at *Minesweeper* is higher than 50%? Explain how you are deciding.

Divine Providence?

Use the following information to answer Exercises 1.4.8-12. Dr. John Arbuthnot (1667 – 1735) was physician to England's Queen Anne. He was also one of the first scientists to publish a use of p-values and to apply the logic of statistical inference.[1] His goal was to prove the existence of God, and he took as evidence the fact that for 82 years in a row, male christenings had outnumbered female christenings. His argument: The p-value is so extremely tiny that the extra male births cannot be due to chance. Therefore the excess males must be due to "Divine Providence." Here is a summary of his data:

Number of male christenings:	484,382
Number of female christenings:	453,841
Total number of christenings:	938,223
Proportion of males:	.516276

1.4.8 Hypotheses.

   a) What was Dr. Arbuthnot's research hypothesis?

   b) Let π be Arbuthnot's parameter. Tell in words what π refers to.

   c) State in words, and then in symbols, what Dr. Arbuthnot's null hypothesis was.

   d) Was Arbuthnot's alternative hypothesis one-sided or two-sided? Give your reasoning.

1.4.9 Based solely on the observed value of $\hat{p}$ and the null value for π would you expect the evidence against the null hypothesis to be strong or weak? Explain your reasoning.

1.4.10 Based solely on the sample size n (ignoring the value of distance $\hat{p} - \pi$) would you expect the evidence against the null hypothesis to be strong or weak? Explain your reasoning.

1.4.11 Suppose (against all historical evidence) Arbuthnot had been an early feminist, equally willing to accept an excess of female births as evidence of "Divine Providence." In that case, his p-value would have been _____ (half, twice) as large.

1.4.12 Summary. For the analysis based on the data above:

   a) The distance between $\hat{p}$ and π is _____ (tiny, small, large, huge).

[1] John Arbuthnot (1710). "An argument for Divine Providence, taken from the constant regularity observed in the births of both sexes". *Philosophical Transactions of the Royal Society of London* **27** (325–336): 186–190.

b) The sample size is _____ (tiny, small, large, huge).

c) The alternative hypothesis is _____-sided (one, two).

Buzz and Doris *

Use the following information to answer Exercises 1.4.13-15. Refer to Example 1.1. Recall that Buzz pushed the correct lever 15 out of 16 times when given the choice between two levers to push.

1.4.13 Buzz got 15 right out of 16. Suppose the numbers had been exactly double that: 30 right out of 32.

   a) Which of the three influences on strength of evidence (distance, sample size, one-sided or two) would change? Which would stay the same?

   b) Overall, the evidence against the null hypothesis would be _____ (stronger, weaker).

1.4.14 In the actual experiment, Buzz got 15 right out of 16. Suppose he had only guessed right 14 times out of 16.

   a) Which of the three influences on strength (distance, sample size, one-sided or two) would change? Which would stay the same?

   b) Overall, the evidence against the null hypothesis would be _____ (stronger, weaker).

1.4.15 Just for the sake of this exercise, imagine that the investigators wanted their statistical test to include the possibility that Buzz was a very sadistic dolphin, so malicious that he would willingly give up his own chance at a fish just to deprive Doris.

   a) How would this thinking affect the p-value based on the actual data (15 right out of 16): Double it or cut it in half? (Explain your reasoning.)

   b) Does this reframing of the alternative hypothesis makes the evidence against the null hypothesis stronger or weaker?

Harley

Use the following information to answer Exercises 1.4.16-17. Refer to Exploration 1.1.Researchers investigated whether Harley the dog could correctly choose between two objects when a researcher bowed towards one of the cups.

1.4.16 Harley got 9 right out of 10. Suppose instead the numbers had been 18 right out of 20.

a) Which of the three influences on strength of evidence (distance, sample size, one-sided or two) would change? Which would stay the same?

b) Overall, the evidence against the null hypothesis would be _____ (stronger, weaker).

1.4.17 Suppose that instead of the actual result (9 right out of 10) Harley had only been right 8 times out of 10.

a) Which of the three influences on strength (distance, sample size, one-sided or two) would change? Which would stay the same?

b) Overall, the evidence against the null hypothesis would be _____ (stronger, weaker).

Krieger*

Use the following information to answer Exercises 1.4.18-24. Can domestic dogs understand human body cues such as bowing, pointing or glancing? The experimenter presented a body cue toward one of two objects and recorded whether or not the dog being tested correctly chose the object indicated. A 4-year old male pit bull named Krieger participated in this study. He chose the correct object 6 out of 10 times when the experimenter turned and looked towards the correct object.

1.4.18

a) Identify the parameter of interest for this study, and assign a symbol to it.

b) If Krieger is just randomly choosing between the two objects, what is the chance that he will choose the correct object?

c) State your null hypothesis and an appropriate one-sided alternative hypothesis

1.4.19

a) *Statistic:* How many times did Krieger choose the correct object? Out of how many attempts? Thus, what proportion of the time did Krieger choose the correct object?

b) *Simulate:* Using an applet, simulate 1000 repetitions of having the dog choose between the two objects if he is doing so randomly. Report the null and standard deviation.

c) Based on the study's result, what is the p-value for this test?

d) *Strength of evidence:* What are your conclusions based on the p-value?

e) How would the evidence change if Krieger got 8 out of 10 correct?

1.4.20

a) Suppose that you decide to use a two-sided alternative hypothesis. What additional values for the parameter will now be part of the alternative?

b) Conjecture how, if at all, the p-value for the two-sided alternative will change compared to that reported in 1.4.19(c) for the one-sided alternative hypothesis. Explain your choice.

INCREASE   STAY THE SAME   DECREASE

c) Use an applet to find and report the p-value for the two-sided alternative. Did the p-value behave as you had conjectured in (b)?

d) Complete the following: A p-value corresponding to a one-sided alternative hypothesis provides

_____ (WEAKER/THE SAME/STRONGER) evidence against the null hypothesis, compared to a p-value calculated from the same data, but corresponding to a two-sided alternative hypothesis.

1.4.21 Suppose that we repeated the same study with Krieger, and this time he chose the correct object 12 out of 20 times.

a)   Conjecture how, if at all, the p-value would change from that reported in 1.4.19(c) (one-sided alternative)

INCREASE   STAY THE SAME   DECREASE

b)   Using an applet, find this p-value. Did it behave the way you conjectured in (a)?

c)   Complete the following: When the sample proportion of successes stays the same, as the sample size gets larger, then the strength of evidence against the null hypothesis gets _____ (WEAKER/THE SAME/STRONGER).

1.4.22 In another part of the study, instead of *looking* at the object, the experimenter kept her eyes on the dog and *leaned* toward the object. For this part of the study, Krieger got 9 right out of 10.

a)   Identify the parameter of interest for this part of the study, and assign a symbol to it.

b)   If Krieger is just randomly choosing between the two objects, what is the chance that he will choose the correct object?

c)   State your null hypothesis and an appropriate one-sided alternative hypothesis.

d)   Conjecture how, if at all, the p-value for these data will change compared to that from the data in exercise 1.4.18 – 1.4.19, (where Krieger got 6 out of 10 correct). Explain your choice.

INCREASE   STAY THE SAME   DECREASE

1.4.23 For the "leaning" version of the study,

a) *Statistic:* How many times did Krieger choose the correct object? Out of how many attempts? Thus, what proportion of the time did Krieger choose the correct object?

b) *Simulate:* Using an applet, simulate 1000 repetitions of having the dog choose between the two objects if he is doing so randomly. Report the null and standard deviation.

c) Based on the study's result, what is the p-value for this test?

d) Approximately what proportion of the 20 attempts would Krieger have needed to get correct in order to yield a p-value of approximately 0.05?

1.4.24

a) Based on the study's result (the 9 out of 10 from Exercise 1.4.22), what is the standardized statistic for the test?

b) *Strength of evidence:* What are your conclusions based on the p-value you found in Exercise 1.4.23c? Are the conclusions the same if you base them off the standardized-statistic you found in a)?

c) Revisit your conjecture in 1.4.22(d). Did the p-value behave the way you had conjectured?

The Sign Test

Exercises 1.4.25-29 use the ideas presented here. So far, the outcome has always been binary – Yes/No, Right/Wrong, Heads/Tails, etc. What if outcomes are quantitative, like heights or percentages? Although there are specialized methods for such data that you will learn in a later chapter, you can also use the methods and logic you have already learned for situations of a very different sort: (1) outcomes are quantitative, (2) you want to compare two conditions, A and B, and (3) your data come in pairs, one A and one B in each pair. To apply the coin toss model, you simply ask for each pair, "Is the A value bigger than the B value?" The resulting test is called the "sign test" because the difference (A-B) is either positive or negative. Here's a summary table where the notation P(A) means the probability of A.:

Coin toss	Heads	P(Heads)	Null hypothesis	Statistic
Buzz's guess	Right	= P(Right)	$\pi = 0.5$	$\hat{p}$
Each pair	A > B	= P(A > B)	$\pi = 0.5$	$\hat{p}$

1.4.25* Divine Providence Refer to Exercise 1.4.8-12. Dr. Arbuthnot's actual analysis was different from the analysis you saw earlier. Instead of using each individual birth as a coin toss, Arbuthnot used a sign test with each of the 82 years as a coin toss, and a year with more male births counted as a "success."

a) Complete the following table of comparisons:

Analysis method	Sample size n	Null value π_0	Value of $\hat{p}$
A: 1.4.8 – 1.4.12	_____	_____	_____
B: Sign Test	_____	_____	_____

b) For each method of anal...   ...gth of evidence against the null hypothesis, as one of:
...   ...ately strong, strong, or overwhelming.

1...   ...t the hypothesis that living in the country is better for
y...   ...ossible variation due to genetic differences, they
lo...   ...ber of each twin living in the country, the other in a
cit...   ...e of inhaled tracer particles remaining in the lungs
aft...   ...althy the lungs. They found that for six of the seven
twi...   ...r lungs.

...   ...sided?

b...   ...null value and observed proportion, estimate the
...of evidence: inconclusive, weak but suggestive, moderately strong, strong, or
overwhelming.

c) Use the probabilities for the number of heads in seven tosses of a fair coin given below to compute
the p-value and state a conclusion.

# Heads	0	1	2	3	4	5	6	7
Probability	0.0078	0.0547	0.1641	0.2734	0.2734	0.1641	0.0547	0.0078

1.4.27*Beestings Scientists gathered data to test the research hypothesis that bees are more likely to sting
a target that has already been stung by other bees. On eight separate occasions, they offered a pair of
targets to a hive of angry bees; one target in each pair had been previously stung, the other was pristine.
On six of the eight occasions, the target that had been previously stung accumulated more new stingers.

a) Is the alternative hypothesis one-sided or two-sided?

b) Based on the sample size and distance between null value and observed proportion, estimate the
strength of evidence: inconclusive, weak but suggestive, moderately strong, strong, or
overwhelming.

c) Use the probabilities for the number of heads in eight tosses of a fair coin given below to compute the
p-value and state a conclusion.

# Heads	0	1	2	3	4	5	6	7	8
Probability	0.0039	0.0313	0.1094	0.2188	0.2734	0.2188	0.1094	0.0313	0.0039

Presidential stature

In a race for US president, is the taller candidate more likely to win?

1.4.28 In the first election of the 20th century, Theodore Roosevelt (178 cm) defeated Alton B. Parker (175 cm). There have been 27 additional elections since then, for a total of 28. Of these, 25 elections had only two major party candidates with one taller than the other. In 19 of the 25 elections, the taller candidate won.

    a) Let $\pi = $ P(taller wins). State the research hypothesis in words and in symbols.

    b) State the null and alternative hypotheses in words and symbols.

    c) Compute the appropriate p-value using an applet.

    d) If you take the p-value at face value, what do you conclude?

    e) Are there reasons not to take the p-value at face value? Is yes, list them.

1.4.29 In this exercise, as in the one before, we eliminate elections with more than two major party candidates, and also eliminate elections with two candidates of the same height. In addition, we eliminate the two elections in which George Washington was unopposed and five elections with missing data. Consider four different data sets:

    A. Elections from 1960 (Kennedy) to the present: $n = 14$, $\hat{p} = 8/14 = 0.5714$.

    B. Elections from Theodore Roosevelt (1904) to the present: $n = 25$, $\hat{p} = 19/25 = 0.76$.

    C. Elections from John Adams (1796) through William McKinley (1900): $n = 16$, $\hat{p} = 5/16 = 0.3125$.

    D. Elections from John Adams (1796) to the present: $n = 41$, $\hat{p} = 24/41 = 0.5854$.

    a) The four p-values, from smallest to largest, are 0.007, 0.174, 0.395, 0.961. Match each data set (A – D) with its p-value.

    b) What do you conclude about the hypothesis that taller candidates are more likely to win?

Three features that affect p-values: An abstract summary*

1.4.30 The graph above shows six different curves labeled A – F. Each curve shows the relationship between p-value (y-axis) and the distance $\hat{p} - \pi$ (x-axis) for testing the null hypothesis $\pi = 0.5$.

Match each curve A - F with one of the descriptions:

    Sample size is $n = 25$, alternative hypothesis is right-sided: $\pi >$.

    Sample size is $n = 225$, alternative hypothesis is right-sided: $\pi >$.

    Sample size is $n = 25$, alternative hypothesis is left-sided: $\pi <$.

Sample size is $n = 225$, alternative hypothesis is left-sided: $\pi <$.

Sample size is $n = 25$, alternative hypothesis is two-sided: $\pi \neq$.

Sample size is $n = 225$, alternative hypothesis is two-sided: $\pi \neq$.

Exercises 1.4.30-35 use the graphs shown below. Exercise 1.4.30 is not an easy exercise, but don't let the surface mess of the picture put you off. The purpose of the exercise is to challenge you to think hard about very important issues in a new way, and at the same time to challenge you to become more skillful at connecting graphs with ideas. We hope you won't give up, and that you'll feel that what you learn from the time you spend is worth it. Give it a shot, see how far you can get, and if you get stuck, skip to Exercises 1.4.31-35 and do those first and then come back to Exercise 1.4.30.

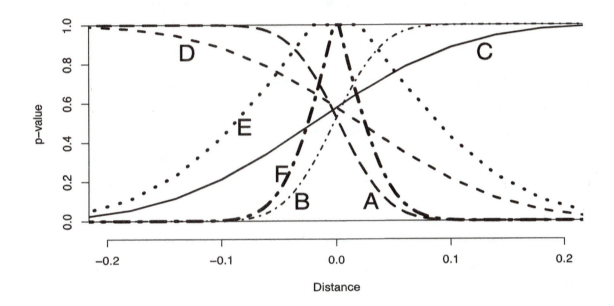

1.4.31 Notice that there are three basic shapes for the curves: increasing, decreasing, and "up-down".
Tell the letters for each shape:

    a) Increasing:

    b) Decreasing:

    c) Up-down:

1.4.32 Notice that the distance $\hat{p} - \pi$ (x-axis) can tell you percentage correct. Recall Example 1.1 For Buzz and Doris,

    a) a distance of 0.5 means Buzz is right _____% of the time.

    b) a distance of 0.0 means Buzz is right _____% of the time.

c) a distance of -0.5 means Buzz is right _____ %of the time.

1.4.33 Try to relate the shape of the curve to the alternate hypothesis.

a) Suppose Buzz really want to earn his fish and tries to guess right. Your alternative hypothesis is

that $\pi > 0.5$. The better Buzz does, the _____ (smaller, larger) the p-value. The graph should be

_____ (increasing, decreasing, up-down). The graphs that match this alternative are

_____.

b) Suppose you are studying sadism in dolphins and your research hypothesis is that Buzz is able to

guess right but he wants only to deprive Doris of her fish. Your alternative hypothesis is that

$\pi < 0.5$. The better Buzz does, the _____ (smaller, larger) the p-value. The graph should be

_____ (increasing, decreasing, up-down). The graphs that match this alternative are _____.

c) You are unsure about whether Buzz is fish-greedy or sadistic. Your more objective alternative

hypothesis is that $\pi \neq 0.5$. The graphs that match this alternative are _____ (increasing,

decreasing, up-down). Their letters are _____.

d) Complete the following summary table. For "Strongest evidence," choose from among $\hat{p} = 0$, $\hat{p} =$

1, and $\hat{p} = 0$ or 1. For "Shape of curve," choose from among increasing, decreasing, and up-

down.

Alternative hypothesis	Strongest evidence	Shape of curve
$\pi > 0.5$		
$\pi < 0.5$		
$\pi \neq 0.5$		

1.4.34 The six graphs are for two different sample sizes: one small (25) and one large (225). Your

challenge in this exercise is to figure out how to tell the difference between a small-sample curve and a

large-sample curve. Exercise 1.4.33 illustrates that the *form* of the alternative hypothesis determines the

shape of the curve.

a) Consider two sample sizes with the same distance: If Buzz is right 30 times out of 32, the p-value

will be _____ than if Buzz is right 15 times out of 16. (Choose from: higher, lower, can't tell

– it depends on the alternative)

b) More generally, for any given value of the distance $\hat{p} - \pi$ (x-axis), the p-value is smaller when the

sample size n is _____ (larger, smaller, can't tell – it depends on the alternative).

c) This means that compared to the graph for $n = 225$, the graph for $n = 25$ _____ (always

lies above, always lies below, neither: it depends on the alternative hypothesis.)

d) "Above/below" is not a useful comparison. Instead, use the steepness of the curve. Compared to

the graph for $n = 225$, the graph for $n = 25$ is _____ (steeper, less steep).

1.4.35 Look at the figure long enough to follow each curve with your eye, and recognize its shape and

steepness. Keep these two facts in mind – three shapes, two curves of each shape – and try to relate them

to the three main influences on p-values: distance, sample size, and one-versus-two sided alternatives.

Then fill in the following table indicating which letter (A-F) corresponds to combination of shape

descriptions.

	Increasing	Decreasing	Up-Down
Steeper			
Flatter			

Now go back to exercise 1.4.30 and use your table to do the matching.

Section 1.5

1.5.1*Which long-run probability of success, π, gives the largest standard deviation of the null

distribution when the sample size is 10?

   A. 0.05

   B. 0.25

   C. 0.50

   D. 0.90

1.5.2 Would your answer change in exercise **1.5.1** if the sample size were 100?

1.5.3*Which sample size, n, gives the smallest standard deviation of the null distribution where the long-

run probability, π, is 0.25?

   A. 30

   B. 40

   C. 50

   D. 60

1.5.4 What is calculated using the formula $\sqrt{\dfrac{\pi(1-\pi)}{n}}$?

1.5.5* What is calculated using the formula (statistic − mean of null distr.)/(SD of null distr.) ?

1.5.6 Suppose you are using theory-based techniques (e.g., a One proportion z test) to determine p-values. How will a two-sided p-value compare to a one-sided p-value (assuming the one-sided p-value is less than 0.5)?

    A. The two-sided p-value will be about the same as the one-sided.
    B. The two-sided p-value will be close to twice as large as the one-sided.
    C. The two-sided p-value will be exactly twice as large as the one-sided.
    D. The two-sided p-value will be half as much as the one-sided.

1.5.7*What is a z-statistic?

1.5.8 Suppose you ride to school with a friend and often arrive at a certain stop light when it is red. One day she states, "It seems like this light is green only 10% of the time when we get here." You think it is more often than 10% and want to test this. You keep track of the color (green/not green) the next 20 times you go to school and find that 4 times (4/20 = 20%) the light is green when you arrive. You wish to see if your sample provides strong evidence that the true proportion of times the light is green is greater than 10%. In other words, you are testing the hypotheses:

$$H_0: \pi = 0.1 \text{ versus } H_a: \pi > 0.1$$

where π = the long-run proportion of times the light is green.

Two different approaches were taken in order to yield a p-value and both are shown in the applet output.

    • **Option 1.** A simulation-based test was done and found a p-value of **0.148**, showing weak evidence against the null.
    • **Option 2.** A one proportion z test was conducted and found a p-value of **0.068**, yielding moderate evidence against the null.

a) Which test gives a more valid p-value?
b) Give one or two sentences why you picked the option you did.

1.5.9* According to statistician Persi Diaconis, the probability of a penny landing heads when it is spun on its edge is only about 0.2. Suppose you doubt this claim and think that it should be more than 0.2. To test this, you spin a penny 10 times and it lands heads side up 5 times. You put this information in the One proportion applet and determine a simulation-based p-value of 0.0770, but the One proportion z test p-value is 0.0303 as shown in the following applet output.

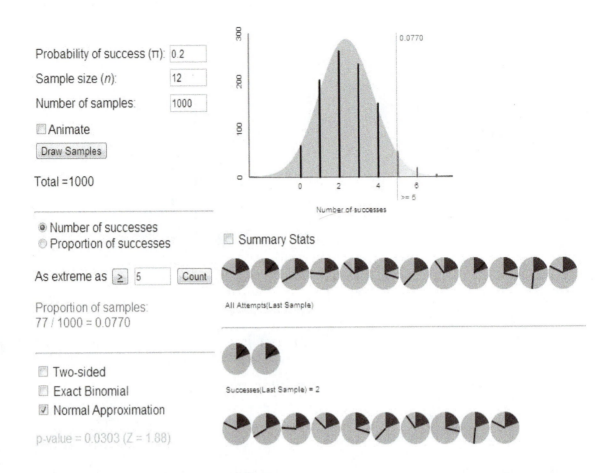

a) Write down the hypotheses for this study.

b) Which p-value would you say is the most valid? Explain.

c) Do you have strong evidence that a spun penny will land heads more that 20% of the time in the long run?

1.5.10 According to researchers, a coin flip may not have a 50% chance of landing heads and a 50% chance of landing tails. In fact, they believe that a coin is more likely to land the same way it started. So if it starts out heads up, it is more likely to land heads up. Suppose someone tests this hypothesis with 1000 flips of a coin where it starts out heads up each time.

a) Describe what the symbol π stands for in this context.

b) State your null and alternative hypotheses.

c) Suppose 52% of the sample of 1000 flips landed heads facing up. Verify the validity conditions that allow us to use a theory-based test.

d) A theory-based test reports a standardized statistic of 1.26. Interpret what this means.

e) A theory-based test reports a p-value of 0.1030. State your conclusion in terms of strength of evidence and what that means in the context of the study.

Heart Transplant Operations*

For Exercises 1.5.11-12 recall Example 1.3: **Heart Transplant Operations** which describes how in September 2000 based on the fact that of the last 10 heart transplant operations at St. George's hospital 8 had resulted in deaths, concerns arose and heart transplantations were suspended at this hospital. To investigate whether the data provide evidence that the long-run probability of dying within 30 days of a heart transplant at St. George's is higher than 15%, one can use the One Proportion applet to produce the null distribution of sample proportion of deaths as given below:

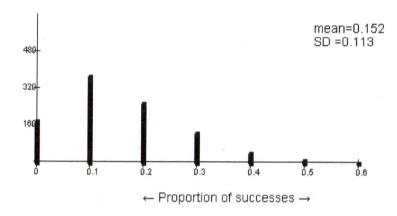

← Proportion of successes →

1.5.11

   a) According to the dotplot produced by the applet, what is the mean of the null distribution? Explain how you could have anticipated this.

   b) According to the dotplot produced by the applet, is the overall shape of the null distribution *bell-shaped*? That is, can the null distribution be described as being a normal distribution? Explain how you could have anticipated this.

1.5.12 Recall that to explore more deeply, the researchers went on to look at data from the most recent 361 heart transplantations at St. George's, and found that 71 had resulted in deaths within 30 days of the heart transplant procedure. Once again, one can use the One Proportion applet to produce the null distribution of sample proportion of deaths assuming the long-run probability of death is 0.15:

a) According to the dotplot produced by the applet, what is the mean of the null distribution? Explain how you could have anticipated this.

b) According to the dotplot produced by the applet, what is the SD of the null distribution? Explain how you could have anticipated this.

c) According to the dotplot produced by the applet, is the overall shape of the null distribution *bell-shaped*? That is, can the null distribution be described as being a normal distribution?

d) Explain why the normal approximation works better in the context of the null distribution in 1.5.12 compared to the null distribution in 1.5.11.

Psychic abilities

Use the following information to answer the next two questions. Statistician Jessica Utts has conducted extensive analysis of studies that have investigated psychic functioning. One type of study involves having one person (called the "sender") concentrate on an image while a person in another room (the "receiver") tries to determine which image is being "sent." The receiver is given four images to choose from, one of which is the actual image that the sender is concentrating on. This is a technique called Ganzfeld.

1.5.13

a) Describe what the symbol π stands for in this context.

b) State your null and alternative hypotheses.

c) If the subjects in these studies have no psychic ability, approximately what proportion will identify the correct image? Is this the null hypothesis or the alternative hypothesis?

Utts (1995) cites research from Bem and Honorton (1994) that analyzed 329 Ganzfeld sessions (http://www.ics.uci.edu/~jutts/air.pdf).

d) Use an applet to simulate 1000 repetitions of this study, assuming the null hypothesis to be true. Sketch a labeled graph of the results.

e) Utts reported that Bem and Honorton found a total of 106 "hits" in the 329 sessions. Does this result provide very strong evidence against the null hypothesis? Use the simulation that you carried out in part d) to find and report a p-value. Based on this p-value, summarize your conclusion in the context of this study, and explain your reasoning for having arrived at this conclusion. Be sure that you have used the 3S strategy for assessing the strength of evidence for the research conjecture.

f) Now carry out the same analysis, only instead of simulating the null, use the theory-based (normal approximation) approach for that null distribution. How does the p-value you found using the theory-based approach compare to the one you found using simulation? Does this surprise you? Why or why not?

1.5.14 In the same research article, Utts (1995) also cites research from Morris et al. (1995) where out of 97 Ganzfeld sessions, the receiver could correctly identify the image sent 32 times. You would like to know whether these data provide evidence that in Ganzfeld studies receivers will do better than just guess when picking which image is being sent.

a) Find the sample proportion of "hits" in Morris et al.'s study. How does this compare to the sample proportion of "hits" in Bem and Honorton's study: larger, smaller, or about the same?

b) Conjecture whether the p-value for Morris et al.'s study results will be larger, smaller, or about the same compared to that of Bem and Honorton's. Explain your reasoning.

c) Use simulation to find and report a p-value. Based on this p-value, summarize your conclusion in the context of this study, and explain your reasoning for having arrived at this conclusion.

d) Now carry out the same analysis, only instead of simulating the null, use the theory-based (normal approximation) approach for that null distribution. How does the p-value you found using the theory-based approach compare to the one you found using simulation? Does this surprise you? Why or why not?

e) Why do the p-values from the simulation versus theory-based approaches in the Bern and Honorton studynot differ as much from each other as do the p-values from the simulation versus theory-based approaches as here?

I love you, first*

1.5.15 Recall exercise 1.2.16 in which researchers asked 40 heterosexual couples which person said "I love you first." They found that the man said "I love you" in 28 of the 40 couples before the woman did. Since previous research indicated that the man tends to say "I love you" before the woman, we tested to see if the probability a man would say "I love you" before the woman is more than 50%.

a) State the appropriate null and alternative hypotheses (in symbols) for testing whether males are more likely to say "I love you" first.

b) Using an appropriate applet, find the p-value using a theory-based test (one-proportion z test; normal approximation).

c) Summarize your conclusion from your p-value.

d) What is the z-statistic for the test? What does this number represent in the context of this study?

e) Without using an applet, determine the value of the theory-based (normal approximation) p-value if we used a two-sided alternative hypothesis.

Rhesus monkeys

1.5.16 Recall exercise 1.2.19where researchers investigated whether rhesus monkeys have some ability to understand gestures made by humans. In one part of the study, they found that 31 of 40 monkeys correctly went to one of two boxes that researchers pointed towards.

a) State the appropriate null and alternative hypotheses in the context of this study, first in words, and then in symbols.

b) Using an appropriate applet, find the p-value using a theory-based test (one-proportion z test; normal approximation).

c) Summarize your conclusion from your p-value.

d) What is the z-statistic for the test? What does this number represent in the context of this study?

e) Without using an applet, determine the value of the theory-based (normal approximation) p-value if we used a two-sided alternative hypothesis.

Rock, paper, scissors*

Use the following information to answer the next two questions. In Example 1.2, we investigated some results of the game rock-paper-scissors. In the article referenced in the example, the researchers had 119 people play rock-paper-scissors against a computer. They found 66 players (55.5%) started with rock, 39 (32.8%) started with paper, and 14 (11.8%) started with scissors. We want to see if players start with scissors with a long-term probability that is different than 1/3.

1.5.17

a) State the appropriate null and alternative hypotheses in the context of this study, first in words, and then in symbols.

b) Using an appropriate applet, find the p-value using a theory-based test (one-proportion z test; normal approximation).

c) Summarize your conclusion from your p-value.

1.5.18 Based on the statistics from the previous question where people were playing rock-paper-scissors, we want to see if players start with *rock* with a long-term probability that is different than 1/3.

a) State the appropriate null and alternative hypotheses in the context of this study, first in words, and then in symbols.

b) Using an appropriate applet, find the p-value using a theory-based test (one-proportion z test; normal approximation).

c) Summarize your conclusion from your p-value.

Facial prototyping

1.5.19 Reconsider **Exploration 1.3: Do People Use Facial Prototyping?** where we explored whether the long-run probability that students in a class would correctly assign the names Bob and Tim to the pictures. We tested to see if this long-run probability is more than 0.5.

a) State the appropriate null and alternative hypotheses in the context of this study.

b) In a class of one of the authors 19 out of 30 students correctly identified that the face on the left belonged to Tim. Using an appropriate applet, find the p-value using a one-proportion z test (theory-based/normal approximation).

c) Summarize your conclusion from your p-value.

d) What is the z-statistic for the test? What does this number represent in the context of this study?

Predicting Elections from Faces*

1.5.20 In **Example 1.4: Predicting Elections from Faces?** we explored whether voters make decisions based on a candidate's facial appearance. The researchers described how they had people determine which of two candidates for the U.S. House of Representatives in 2004 had the more competent face. They then looked to see if the one with the more competent face won the election. We will let π = the probability this method predicts the winner.

a) State the appropriate null and alternative hypotheses in the context of this study.

b) This method predicted the winner in 189 of the 279 House races that year. Based on these results, what are the values of $\hat{p}$ and n?

c) Using an appropriate applet, find the p-value using a theory-based test (one-proportion z test; normal approximation).

d) Summarize your conclusion from your p-value.

Uniform Color

Use the following information to answer Exercises 1.5.21-22. Reconsider **Exploration 1.4: Competitive Advantage to Uniform Colors?** where we looked to see if an athlete's uniform color affects whether or not they win or lose. The athletes were randomly assigned red or blue uniforms in matches of four combat sports. We want to see if one color has an advantage over the other. If we focus on the red uniform, we can ask if the probability the competitor in the red uniform will win is different than 0.5.

1.5.21 The researchers found that of the 457 matches, the competitor in red won 248 times.

a) State the appropriate null and alternative hypotheses in the context of this study.

b) Using an appropriate applet, find the p-value using a theory-based test (one-proportion z test; normal approximation).

c) Summarize your conclusion from your p-value.

d) What is the z-statistic for the test? What does this number represent in the context of this study?

1.5.22 The researchers found that of the 272 boxing matches, the competitor in red won 150 times.

a) State the appropriate null and alternative hypotheses in the context of this study.

b) Using an appropriate applet, find the p-value using a theory-based test (one-proportion z test; normal approximation).

c) Summarize your conclusion from your p-value.

d) What is the z-statistic for the test? What does this number represent in the context of this study?

Yahtzee*

Use the following information to answer Exercises 1.5.23-24. One of the authors used to have an electronic Yahtzee game that he played frequently. The game would "roll" five virtual six-sided dice. But were the dice fair?

1.5.23 It seemed to the author that sixes showed up more than what they should if the dice were fair and he wanted to test this. He had the machine roll 500 dice and obtained 92 sixes.

a) Describe what the parameter is in the context of this problem.

b) State the appropriate null and alternative hypotheses in the context of this study.

c) Using an appropriate applet, find the p-value using a theory-based test (one-proportion z test; normal approximation).

d) Summarize your conclusion from your p-value.

1.5.24 It also seemed to the author that ones showed up less often than what they should if the dice were fair and he wanted to test this. In his 500 dice rolls, 72 resulted in ones.

a) Describe what the parameter is in the context of this problem.

b) State the appropriate null and alternative hypotheses in the context of this study.

c) Using an appropriate applet, find the p-value using a theory-based test (one-proportion z test; normal approximation).

d) Summarize your conclusion from your p-value.

Mario and Luigi

1.5.25 Suppose two brothers named Mario and Luigi like to compete by playing a certain video game. Mario thinks he is better at this game than Luigi and sets out to prove it by keeping track of who wins. After playing the game 30 times, Mario won 18 of them (or 60%). Mario then declares that this proves he is obviously the better player.Luigi, who just finished chapter 1 in his statistics class, realizes that perhaps he and his brother are evenly matched and just by chance Mario won 60% of the last 30 games. Luigi is going to test this by running a test of significance.

a) Describe what the parameter is that Luigi should be testing.

b) State the appropriate null and alternative hypotheses in the context of this study.
(*Hint*: Luigi likes two-sided tests because he knows that will make it harder to get strong evidence that his brother is better.)

c) Using an appropriate applet, find the p-value using a theory-based test (one-proportion z test; normal approximation).

d) Summarize your conclusion from your p-value.

e) The p-value you found in part c should have been greater than 0.05. Suppose the two brothers continue to compete and Mario continues to win 60% of the games. How many games will they have to play until he gets a p-value less than 0.05 after retesting?

End of Chapter

1.CE.1* A p-value is calculated assuming which hypothesis is true: null or alternative?

1.CE.2 The standard deviation of a null distribution is calculated using which value: the probability value in the null hypothesis, or the observed value of the sample proportion?

Fantasy Golf*

1.CE.3 A statistics professor is in a fantasy golf league with 4 friends. Each week one of the 5 people in the league is the winner of that week's tournament. During the 2010 season, this particular professor was the winner in 7 of the first 12 weeks of the season. Does this constitute strong evidence that his probability of winning in one week was larger than would be expected if the 5 competitors were equally likely to win?

 a) State the null hypothesis and the alternative hypothesis in the context of the study.

 b) Describe how you could use playing cards to conduct a tactile simulation analysis of this study and its result. Give sufficient detail that someone else could implement this simulation analysis based on your description. Be sure to indicate how you would decide whether the observed data provide convincing evidence that the statistics professor's probability of winning in one week was larger than would be expected if the 5 competitors were equally likely to win?

 c) Use an applet to conduct a simulation analysis with at least 1000 repetitions. Based on the null distribution produced by the applet, explain how you are deciding whether the observed data provide convincing evidence that the statistics professor's probability of winning in one week was larger than would be expected if the 5 competitors were equally likely to win. Summarize the conclusion that you would draw about the research question of whether the statistics professor's probability of winning in one week was larger than would be expected if the 5 competitors were equally likely to win.

 d) In answering (c), you should have reported a p-value. Interpret this p-value in the context of the study. (*Hint:* The p-value is the probability of what, assuming what?)

 e) Explain why it is not advisable to use a one proportion Z test (theory-based approach) to find a p-value for the data from this study.

Mythbusters

1.CE.4 On the television show *Mythbusters*, the hosts Jamie and Adam wanted to investigate which side buttered toast prefers to land on, when it falls through the air. To replicate a piece of toast falling through the air, they set up a specially designed rig on the roof of the *Mythbusters'* headquarters. In 48 attempts, 19 pieces of toast fell buttered side down. Do these data provide evidence that buttered toast falling through air, has a preference for any specific side (buttered or not buttered), when landing? Write a report describing all six steps of the Statistical Investigation process as they apply to this question and these data.

Cheating*

Use this information to answer Exercises 1.CE.5-6. The August 19, 2013 issue of Sports Illustrated included a brief article titled "Why Men Cheat," in which the author L. Jon Wertheim argued that economic facts led Major League Baseball (MLB) players from the United States are less likely than those from poorer countries to take performance-enhancing drugs (PEDs). The article mentioned that 57.3% of all professional baseball players between 2005–2013 were from the U.S., and 206 of the 595 (34.6%) players suspended for PED use in these years were from the U.S.

1.CE.5

a) One question is whether 0.346, the observed proportion of PED-suspended players from the U.S., is (statistically) significantly less than 0.573, the proportion of all players from the U.S. Before you attempt to answer this question, explain what the question means (i.e., what statistical significance means) in this context. Write as if to a friend who is a baseball fan but has not studied statistics.

b) Perform a simulation analysis to investigate the question in part a). Write a paragraph summarizing the results and conclusion that you draw from your simulation analysis, again as if to a friend who is a baseball fan but has not studied statistics.

c) Calculate the value of the standardized statistic (z-statistic) for comparing 34.6% to 57.3% in this context. Interpret what this value means, and summarize the conclusion that you can draw from it.

1.CE.6 Additional information supplied in the *Sports Illustrated* article were that 34.6% of all professional baseball players are from Latin-American countries, but 368 of the 595 (61.8%) players suspended for PED use between 2005–2013 are from Latin-American countries. Answer parts a)-c) of the previous exercise with respect to the question of whether 61.8% is (statistically) significantly greater than 34.6% in this study.

Bob or Tim?

Use this information to answer Exercises 1.CE.7-8. Gary Judkins is a teacher in New Zealand who asked his students the question about facial prototyping from Exploration 1.3. Mr. Judkins asked this question of 135 New Zealand students and found that 105 of the students associated the name Tim with the face on the left.

1.CE.7

a) Express the appropriate null and alternative hypotheses, in symbols and in words, for testing whether New Zealand students have a tendency to associate the name Tim with the face on the left.

b) Conduct a simulation analysis to test these hypotheses with data from Mr. Judkins' students. Summarize your conclusion.

c) Is it appropriate to use a theory-based approach with these data? Justify your answer.

d) Implement a theory-based approach to test the hypotheses in part a). Report the values of the test statistic and p-value. Summarize your conclusion.

1.CE.8 Reconsider the previous exercise. Suppose that you were to define the parameter π not as the probability that a New Zealand student would associate the name Tim with the face on the left but as the probability that a New Zealand student would associate the name *Bob* with the face on the left. Re-answer parts a)-d) of the previous exercise with this different set-up. Comment on how (if at all) your calculations and conclusions change.

Coin flipping*

1.CE.9 Suppose that Jose and Roberto both collect data to investigate whether people tend to call "heads" more often than "tails" when they are asked to call the result of a coin flip. If Jose uses a larger sample size than Roberto, is Jose guaranteed to have a smaller p-value than Roberto? Explain why or why not.

1.CE.10 Suppose that Sasha and Jayla both collect data to investigate whether people tend to call "heads" more often than "tails" when they are asked to call the result of a coin flip. If Sasha has a smaller p-value than Jayla, which of the following would you conclude? (There may be more than one correct conclusion.)

a) Sasha has stronger evidence than Jayla that people tend to call "heads" more often than "tails."
b) Jayla has stronger evidence than Sasha that people tend to call "heads" more often than "tails."
c) Both Sasha and Jayla have convincing evidence that people tend to call "heads" more often than "tails."
d) Sasha must have used a larger sample size than Jayla.
e) Sasha must have used a smaller sample size than Jayla.

Family structure

1.CE.11. Josephine wants to investigate whether more than one-third of families with two children have one child of each sex, so she collects data for 50 two-child families. Describe how she could use a die to perform a physical (by hand, without using a computer) simulation to approximate the null distribution.

1.CE.12 Alfonso wants to investigate whether fewer than one-fourth of families with two children have two girls, so he collects data for 25 two-child families. Describe how he could use coins to perform a physical (by hand, without using a computer) simulation to approximate the null distribution.

Athletic performance*

1.CE.13 Rick is a basketball player who wants to investigate whether he successfully makes more than 90% of his free throws by shooting with an underhand style.
a) Define (in words) the relevant parameter of interest for Rick.
b) State (in words and in symbols) the appropriate hypotheses to be tested by Rick.

1.CE.14 Lorena is a golfer who wants to investigate whether she successfully makes more than 60% of her 10-foot putts.
a) Define (in words) the relevant parameter of interest for Lorena.
b) State (in words and in symbols) the appropriate hypotheses to be tested by Lorena.

Odd numbers

Use this information to answer Exercises 1.CE.15-16. Many studies have investigated the question of whether people tend to think of an odd number when they are asked to think of a single-digit number (0 through 9). Combining results from several studies, Kubovy and Psotka (1976) used a sample size of 1770 people, of whom 741 thought of an even number and 1029 thought of an odd number. Consider investigating whether these data provide strong evidence that people have a tendency to think of an odd number rather than an even number in this situation.

1.CE.15
a) Describe the relevant parameter of interest.
b) State the appropriate hypotheses to be tested.
c) Calculate the observed value of the appropriate statistic.
d) Check whether a theory-based approach is appropriate for these data and hypotheses.
e) Calculate the standardized value of the statistic.
f) Determine an approximate p-value.
g) Summarize your conclusion.

1.CE.16 Refer to the previous exercise. The researchers also found that 503 of the 1770 people thought of the number 7. Do these data provide strong evidence that people have a tendency to select the number 7 disproportionately often? Investigate this question by answering parts a-g) of the previous exercise.

Racket spinning*

1.CE.17 One of the authors once spun his tennis racquet 100 times and found that it landed with the label up for 46 of those 100 spins (0.46). For testing whether the probability (that a spun tennis racquet lands with the label up) differs from 0.5, he calculated the p-value to be 0.484.

a) Summarize the conclusion that you would draw from this study.
b) Explain the reasoning process that leads to your conclusion, as if to someone with no knowledge of statistics.

1.CE.18 Reconsider the previous exercise. Notice that three different decimal numbers appear: 0.46, 0.484, and 0.5. Identify which is the null-hypothesize probability, which is the observed value of the statistic, and which is the p-value.

Your own random process

1.CE.19 Think of a random process of interest to you for which you would be interested in assessing evidence regarding a claim about the value of the long-run probability associated with the random process.
a) Describe the random process.
b) Describe the parameter of interest.
c) State the null and alternative hypotheses that you would be interested in investigating.
d) Describe how you could collect data to conduct this statistical investigation.

Investigation: Tire Story Falls Flat

PART I

A legendary story on college campuses concerns two students who miss a chemistry exam because of excessive partying but blame their absence on a flat tire. The professor allows them to take a make-up exam, and sends them to separate rooms to take it. The first question, worth five points, is quite easy. The second question, worth ninety-five points, asks: Which tire was it?

Step 1: Ask a research question

Do students pick which tire went flat in equal proportions? It has been conjectured that when students are asked this question and forced to give an answer (left front, left rear, right front, or right rear) off the top of their head, they tend to answer "right front" more than would be expected by random chance.

Step 2: Design a study and collect data

To test this conjecture about the right front tire, a recent class of 28 students was asked if they were in this situation, which tire would they say had gone flat. We obtained the following results:

Left front	Left rear	Right front	Right rear
6	4	14	4

1. What are the observational units?
2. What is the variable that is measured/recorded on each observational unit?
3. Describe the parameter of interest in words. (You can use the symbol π to represent this parameter.)
4. State the appropriate null and alternative hypotheses to be tested.

Step 3: Explore the data

5. What percentage of the students picked the right front tire? Is this more than you would expect if students randomly pick one of the four tires?
6. Is it *possible* that we could observe 14 students from this class of 28 students even if all of the students were just selecting randomly among the four tires?

Step 4: Draw inferences

Let's use our 3S strategy to help us investigate how much evidence the sample data provide to support our conjecture that there is something special about the right front tire.

Statistic

7. What is the statistic that you can use to summarize the data collected in the study and what symbol is associated with this statistic?

Simulate

8. Use the **One Proportion** applet to simulate 1000 repetitions of this study, assuming that every student in class selects randomly (equally) among the four tires. Report what values you input into the applet.

Probability of success (π)_____.25_____
Sample size (n) _____
Number of samples ____1000____

9. Using the proportion of successes for the values on the horizontal axis, what is the center of your null distribution? Does it make sense that this is the center? Explain.

Strength of evidence

10. Let's examine the strength of evidence with the three ways we used in Chapter 1.

   a) Determine the p-value from your simulation analysis. Also interpret what this p-value represents (i.e., the probability of what, assuming what?).

   b) Determine the standardized statistic from your simulation analysis. Also interpret what this standardized statistic represents.

   c) Use a theory-based test (or normal approximation) to determine the p-value for this test. Are the validity conditions met for this test? Why or why not?

11. Is there strong evidence against the null hypothesis for all three of the methods used in Question 10? Explain.

Step 5: Formulate conclusions

12. Summarize the conclusion that you draw from this study and your simulation analysis. Also explain the reasoning process behind your conclusion.

13. Now, let's step back a bit and think about the scope of our inference. What are the wider implications? Do you think that your conclusion holds true for people in general? (These are extremely important questions, that we'll discuss more when we talk about the scope of inference in Chapter 3.)

Step 5: Look back and ahead

14. If you were to repeat this study, what improvements might you make? What further research might you propose related to this topic in the future?

PART II

Now suppose another class conducts the same study with exactly half as many students, and suppose the proportional breakdown in the four categories is identical to the class of 28. In other words, 7 out of 14 students answered, "right front."

15. Before you analyze the data, would you expect to find stronger evidence for the research conjecture (that people pick the right front tire more than ¼ of the time), weaker evidence, or the same strength of evidence? Explain your thinking.

16. Conduct a simulation analysis to produce a simulated p-value. How does it compare to the p-value from the study of the class with 28 students? Is this what you expected? Explain.

PART III

Suppose we didn't have a preconceived notion that the right front tire would be chosen more often, but just wanted to find out if it was chosen at a rate that was different than one-fourth. Let's use our original data where 14 out of 28 chose the right front tire to test this.

17. Write the null and alternative hypotheses for this new question.

18. Use a theory-based test to find the p-value. How does this compare with the p-value you obtained back in question 10c.

19. Do you have strong evidence that the probability a student will choose the right front tire is different than one-fourth?

Research Article: Infants prefer to harm those who are different

Read "Not like me = Bad: Infants prefer those who harm dissimilar others" by Hamlin, Mahajan, Liberman and Wynn. Psychological Science, 2013. 24:589-594, and then answer the following questions.

Step 1. Ask a research question

1. In no more than 1-2 sentences state a broad research question (or related questions) that the researchers were trying to investigate in this study.

2. Identify at least two (cited) points of evidence the researchers cite as to why the research question(s) are important. Include the citation in your response.

The following questions (3- 13) refer to Experiment #1, which is described on page 590

Step 2. Design a study and collect data

3. What is the sample size of 14 month olds, not counting those eliminated from the final sample?

4. Why was Phase I of Experiment 1 necessary?

5. What variable is being measured in Phase 1 of Experiment 1? Is it categorical or quantitative? If categorical, what are the outcomes?

6. Explain the purpose of Phase 2 of Experiment 1.

7. Is there a variable being measured in Phase 2 of Experiment 1?

8. In Phase 3 of the experiment, of the babies saw the similar puppet (bunny who liked the same food they did), while the other of the babies saw the dissimilar puppet (bunny who liked different food than they did). Why was it necessary to report that in Phase 3 "Infants were permitted to observe the outcome of each event until they had looked away for 2 s or until 30 s elapsed?"

9. What variable is being measured in Phase 4 of Experiment 1? Is it categorical or quantitative? If categorical, what are the outcomes?

Step 3. Explore the data

10. In Figure 1a, explain how to generate the height of the white bar for 9-month olds? The black bar?[1]

11. What percent of 14-month olds who saw the similar target (bunny who like same food they did) chose the helper puppy? What percent who saw the dissimilar target chose the helper puppy?

Step 4. Draw inferences

12. In describing the results of Phase 1, the researchers state that "53% of 9-month olds chose graham crackers [when given a choice between graham crackers and green beans]." While no p-value is

---

[1]The sample size for 9-month olds is correctly stated in Figure 1a as 16 who saw the similar target, and 16 who saw the dissimilar target, for a total of 32 infants. The sample size is misstated in the first sentence of "Participants and procedure" for Experiment 1 on page 590.

provided in the article, name at least two reasons it's likely that there is little evidence (large p-value) against a null hypothesis that "50% of all 9-month olds will choose graham crackers"

13. Identify the null and alternative hypothesis for the p-value reported in the following sentence from the results: "…whereas infants who saw interactions involving the dissimilar rabbit puppet preferred the harmer dog puppet over the helper dog puppet (81% of 9 month olds, p=0.02)."

The following questions (14-16) refer to Experiment #2 which is described on page 593

Step 2. Design a study and collect data

14. In no more than 1-2 sentences summarize the main purposes of Experiment #2.

Step 3. Explore the data

15. Identify a statistic from Experiment #2 explain what it means.

Step 4. Draw inferences

16. State a null and alternative hypothesis from Experiment #2, and indicate the related p-value (any is fine). Indicate the strength of evidence provided by the p-value and provide an appropriate (in context) interpretation.

The remaining questions apply to the entire study

Step 5. Formulate conclusions

17. While the authors say little about how the infants were selected to participate in the study, why might this be important to the studies' conclusions?

Step 6. Look back and ahead

18. Identify at least two things that, if you were running the study and had a reasonable amount of financial resources, you would do differently. You must justify your answer.

19. On page 592, the researchers state that "Additionally, infants' evaluations did not differ according to which food the target liked…" Why is this statement important to the researchers' findings? Explain how it potentially undermines their conclusions if it is not true.

20. The final paragraph of the paper argues that they researchers "…did not demonstrate (a) that infants generated, either prior to or during the study, categories of "green-been lovers" and "graham cracker lovers" or even of "individuals like me" and "individuals not like me." How do the researchers argue that this is not important to their final conclusion?

Chapter 2

Generalization: How Broadly Do the Results Apply?

Chapter Overview

This chapter is about *generalization*, one of the four pillars of inference: strength, size, **breadth**, and cause. How **broadly** do the conclusions apply?

Polling organizations like Gallup and Harris take weekly polls on various issues. They claim to estimate how all adult Americans think. Have you ever been polled by Gallup? If not, how can Gallup think that your opinion has been represented? When pollsters ask only 1000 or 1500 people, can they get useful and reliable information about all Americans?

In Chapter 1, we looked at data from a potentially never-ending random process. For example, Buzz was making choices (e.g., which button to push) that were variable (he didn't always get it right). We set up a chance (null) model for his choices (making the correct choice had the same probability for each trial)---his probability of making a correct choice was 0.5 each time. We repeated the chance process to see how the statistic behaved over the long run, and used the could-have-been results to measure the strength of evidence against the null hypothesis that Buzz was just guessing.

So far, so good. But when we look at polling data, we begin to run into a different source of variability – the selection of the observational units. If Gallup chooses a different set of people to interview, each person's opinion won't change, but each different set of people will have its own value for the proportion voting Yes. Now the variability is in which group of people are selected, rather than in the response outcomes in repeated trials of a random process.

But our goal in such a study is not only to describe the *sample* that we see, but hopefully to *generalize* characteristics of the sample to a much larger *population* of individuals. When the population is too large for us to gather data on each and every individual, we rely on the sample to tell us about the population. But this only works when the sample is *representative* of the population.

So now we have two questions:

- How do we obtain a sample that is representative of the larger population?

- How do we model the randomness in the sampling process in order to make inferences about a larger population based only on the sample?

The main goal of this chapter is to explain when and why you can (sometimes) think of choosing your sample as a chance process. When you can, all of the techniques and theoretical results from Chapter 1 apply. When you can't, caution is the order of the day.

Chapter 2 has three sections. The first is about how random sampling works in practice and when you can draw inferences in the same way whether you are sampling from a process or from a population. The second section considers *quantitative data*, showing how random sampling suggests a way to make inferences about a population mean and that the reasoning process from Chapter 1 works for population means also. Finally, the third section reminds us that when we make inferences, there is always the chance that we will make the wrong decision, but we have some control over how often this happens.

Section 2.1: Sampling from a Finite Population

Introduction

Suppose you want to assess the opinion of students at your school on some issue (e.g., funding of sports teams). The observational units in such an investigation would be the individual students. The *population* of interest would be all students at your school. If you had the time and resources to interview all students at your school, you would be conducting a *census* and you would know exactly what proportion of students at your school agree on the issue (e.g., to increase funding). But what if you don't have such time and resources? If you only interview a subset of the students at your school, this subset is the *sample*, and the proportion or any number you calculate about this sample would again be considered a *statistic* as in Chapter 1. In this case, the corresponding *parameter* of interest can be defined as the proportion of *all* students at your school that agree on the issue. The key question is how can we obtain a statistic that we trust to be reasonably close to the actual (but unknown to us) value of the parameter? And just how close do we think we might be? It turns out that the key is how you select the sample from the population.

Example 2.1A: Sampling Students

Colleges and universities collect lots of data on their students. We have data from the registrar at a small midwestern college (hereafter known as "College of the Midwest") on all students enrolled at the college in Spring 2011. Two of the variables from the registrar are the student's cumulative GPA and whether or not the student is housed on campus. Table 2.1 shows a partial data table for these students.

Definitions: The *population* is the entire collection of observational units we are interested in, and the *sample* is a subset of the population on which we record data. Numerical summaries about a population are called *parameters*, and numerical summaries calculated from a sample are called *statistics*.

Table 2.1: Data table of College of the Midwest

Student ID	Cumulative GPA	On campus?
1	3.92	Yes
2	2.80	Yes
3	3.08	Yes
4	2.71	No
5	3.31	Yes
6	3.83	Yes
7	3.80	No
8	3.58	Yes
...	...	...

Notice that whether or not a student is housed on campus is the kind of variable that we focused on in Chapter 1 (specifically, a categorical variable with two categories), whereas cumulative GPA contains numeric data and is a *quantitative* variable, which we first discussed in the Preliminaries.

Definition: A *data table* (or statistical spreadsheet) is a convenient way to store and represent all data values. Typically in a data table, the rows correspond to observational units, columns represent variables, and the data are the table entries.

If we are only interested in students at this college, then we have data for the entire population of 2919 students. In this case, we can actually examine the distributions of these variables for the entire population (Figure 2.1).

Figure 2.1: Population distributions for cumulative GPA (quantitative) and whether or not student lives on campus (categorical)

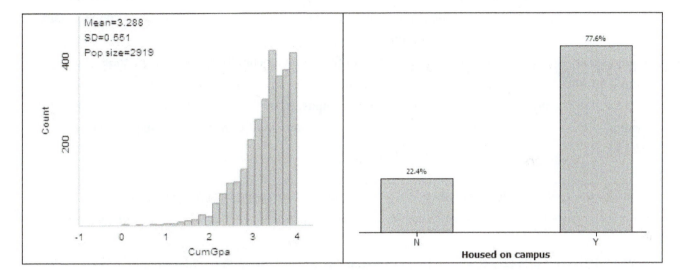

The figure on the left is called a ***histogram***. Like a *dotplot*, it displays the distribution of a quantitative variable. But when you have a large number of observations like this, it may be convenient to "bin" together similar observations into one bar. The height of each bar tells you how many of the values are in each bin (interval of values). The graph still helps us see features of the distribution, like that the GPA values are not symmetric but cluster around 3.30, with most GPAs falling between 2.00 and 4.00, and some falling below 2.00, but none above 4.00.

> **Definition:** A ***histogram*** is a graph that summarizes the distribution of a quantitative variable. Bar heights are used to represent the number of observational units in a particular interval of values.

However, we don't usually have the luxury of having access to the data on every member of the population. Instead, we would like to make inferences about the entire population based only on data from a sample of observational units. Now let's suppose that a researcher does not have access to our data for the population and, in fact, only has time to collect data on a sample of students. Suppose the researcher decides to ask the first 30 students he finds on campus one morning whether he or she lives on campus. This sampling plan is certainly convenient and allows him to select his sample quickly.

> **Think about it:**
> - For this context, describe the population, sample, parameter, and statistic.
> - Do you think this researcher's sampling plan is likely to obtain a sample proportion that live on campus that is close to the population proportion that live on campus?

A sample is only useful to us if the data we collect on our sample is similar to the results we would find in the entire population. In this sense, we say the sample is **representative** of the population. This researcher would like his sample of 30 students to have a similar proportion living on campus (the statistic) as the proportion of all students at this college that live on campus (the parameter). However, it's quite possible to think the first 30 students he sees on campus in the early morning are more likely to be on-campus students than those he might find later in the day. In fact, if he repeatedly applies this method every day, even if he finds different samples of students, we suspect that his samples are likely to consistently overestimate the proportion of students that live on campus. In this sense, we would say his sampling method is *biased* – likely to produce sample statistics that consistently overestimate the population parameter.

Definition: A sampling method is **biased** if, when using that sampling method, statistics from different samples *consistently* overestimate or consistently underestimate the population parameter of interest.

Note that bias is a property of a sampling *method*, not a property of an individual sample. Also note that the sampling method must *consistently* produce non-representative results in order to be considered biased. Sampling bias also depends on what is being measured. One could argue that this sampling method would also be biased in estimating the average (or mean) GPA of students at the college: students with lower GPAs may be more likely to be sleeping in and not present on campus first thing early in the morning. However, if we look at the proportion of students with black hair, it's hard to imagine why this early morning sampling method would consistently over or under-sample those students compared to the rest of the population. So, one could argue that the early morning sampling method might produce a *representative* sample with regard to hair color but not with respect to living on campus or GPA.

Think about it: So what can the researcher do instead? What would be a better way of selecting a sample of 30 students from this population to try to obtain a representative sample?

Simple Random Samples

The key to obtaining a representative sample is using some type of *random* mechanism to select the observational units from the population, rather than relying on **convenience samples** or any type of human judgment. The most common type of random sample is the **simple random sample**.

Key idea: A **simple random sample** ensures that every sample of size n is equally likely to be the sample selected from the population. In particular, each observational unit has the same chance of being selected as every other observational unit.

To take a simple random sample you need a list of all individuals (observational units) in the population, called the *sampling frame*. A computer is then used to randomly select some of the individuals in the sampling frame.

To understand the benefits of such an approach, we will see what happens when we take a simple random sample of students from the College of the Midwest. (This is of course a somewhat artificial example because we really do have access to the values for the entire population, so there's no need to examine only a sample. But this exercise of taking several simple random samples will allow us to explore properties of this approach.) Students are represented by ID numbers 1 to 2919. The computer then randomly selects four-digit numbers between 1 and 2919, the number of students in this population. Suppose that the computer chooses 827. So we find the student whose ID number is 827. He or she is selected for our sample. Then the computer chooses 1355. This student is also selected for our sample. Table 2.2 shows the IDs of the 30 people selected for our sample, along with their cumulative GPA and residential status.

Table 2.2: Results for a simple random sample of 30 students

ID	Cumulative GPA	On campus?	ID	Cumulative GPA	On campus?	ID	Cumulative GPA	On campus?
827	3.44	Y	844	3.59	N	825	3.94	Y
1355	2.15	Y	90	3.30	Y	2339	3.07	N
1455	3.08	Y	1611	3.08	Y	2064	3.48	Y
2391	2.91	Y	2550	3.41	Y	2604	3.10	Y
575	3.94	Y	2632	2.61	Y	2147	2.84	Y
2049	3.64	N	2325	3.36	Y	2590	3.39	Y
895	2.29	N	2563	3.02	Y	1718	3.01	Y
1732	3.17	Y	1819	3.55	N	168	3.04	Y
2790	2.88	Y	968	3.86	Y	1777	3.83	Y
2237	3.25	Y	566	3.60	N	2077	3.46	Y

Once we have our sample, we can calculate sample statistics. For example: What is the average cumulative GPA for these 30 students? What proportion of these students live on campus?

In Statistics, we use the symbol $\bar{x}$ ("xbar") to denote the sample average and the symbol $\hat{p}$ to denote the sample proportion. It is easy enough to calculate $\bar{x} = 3.25$ and $\hat{p} = 0.80$ from these sample data, but how do we know whether these values are at all close to the population values? If we had taken a different sample of 30 students we would probably have obtained completely different values for these statistics. So, how are these statistics of any use in estimating the population parameter values?

Note: Whereas $\overline{x}$ and $\hat{p}$ are common symbols used to denote the sample average and sample proportion, respectively, the corresponding population values are usually denoted by μ and π. Similarly, s is the commonly used symbol used to denote the sample standard deviation, with σ the typical symbol used to denote the population standard deviation. The use of Greek letters to denote population parameters helps remind us that the actual values of parameters are typically unknown to us.

Notation check: Here is a quick summary of the symbols we've introduced in this section.

Type of Variable	Quantitative	Categorical
Statistics	$\overline{x}$ = sample mean	$\hat{p}$ = sample proportion
	s = sample standard deviation	
Parameters	μ = population mean	π = population proportion
	σ = population standard deviation	

In order to understand the behavior of sample statistics calculated from simple random samples, let's take more simple random samples of thirty students to examine the distribution of the statistics across samples. Table 2.3 shows the values of the two statistics for five different simple random samples of 30 students.

Table 2.3: Observed statistics resulting from 5 different random samples of 30 students

Random sample	1	2	3	4	5
Sample average GPA ($\overline{x}$)	3.22	3.29	3.40	3.26	3.25
Sample proportion on campus ($\hat{p}$)	0.80	0.83	0.77	0.63	0.83

There are a few interesting things to note about these values. First, each simple random sample gave us different values for the statistics. That is, there is variability from sample to sample (*sampling variability*). This makes sense because the samples are chosen randomly and consist of different students each time, so we expect some differences from sample to sample. You also might notice that despite the variability, the values aren't changing that much: the average GPAs range from 3.22 to 3.40 and the sample proportions range from 0.63 to 0.83. Lastly, we should point out that the population parameters (that is, the true average cumulative GPA amongst all 2919 students and the true proportion of the 2919 students who live on campus) are $\mu = 3.288$ and $\pi \approx 0.776$ (2265/2919). (See Figure 2.1.) We can see that the statistics tend to be fairly close to the values of each of the parameters.

Let's see what happens if we take 1000 different simple random samples. The two histograms below (Figure 2.2) show the values of the sample average GPA and the sample proportion of students who live on campus, in 1000 different simple random samples of 30 students.

Figure 2.2: Results from 1000 simple random samples of 30 students from the College of the Midwest

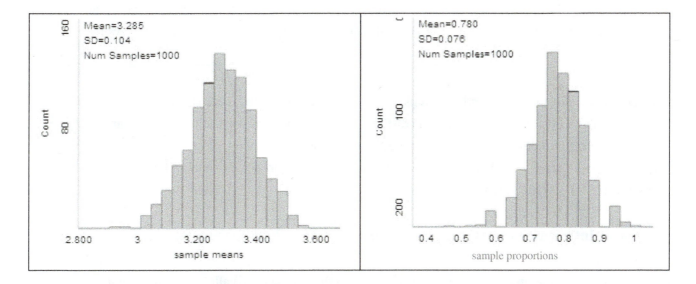

Taking 1000 random samples helps us see the long-run pattern to these results. In particular, with both statistics (sample mean and sample proportion), the middle of the distribution falls very near the value of the corresponding population parameter. In fact, the average of the 1000 sample averages on the left turns out to be 3.285---almost exactly the same as the population parameter (average GPA among all 2919 students is μ = 3.29). Similarly, the average of the 1000 sample proportions turns out to be 0.780, very close to the population proportion of 0.776 (proportion of all 2919 students who live on campus, π). In fact, if we took all possible random samples of 30 students from this population (rather than just 1000 samples as shown here), the averages of these statistics would match the parameters exactly. (Another interesting observation from Figure 2.2 is that both of these distributions have the familiar mound-shaped, symmetric pattern, even though the population distributions did not!)

This example illustrates that sample means and sample proportions computed from simple random samples cluster around the population parameter and the average of those could-have-been statistics equals the value of the population parameter. Thus, simple random sampling would be considered an *unbiased* sampling method because there is no tendency to over or underestimate the parameter. Whether a sampling method is biased or unbiased depends both on the sampling method and on the statistic you chose to calculate from the sample. Luckily, one can show that with simple random sampling, the sample average and the sample proportion both have this unbiased property. Although we don't know for sure whether the initially proposed convenience sample of selecting the first 30 people seen on campus provides a reasonable estimate of the parameter, this unbiased property gives us faith that a simple random sample is representative of the population.

Key idea: When we use *simple random sampling*, the average of the sample means from different samples equals the population mean and the average of the sample proportions from different samples equals the population proportion. (This means that these statistics are *unbiased* estimators for the corresponding parameter when we use simple random sampling.)

Although selecting a simple random sample is not always feasible (e.g., what if you don't have a complete sampling frame, what if it is too difficult to track down everyone selected?), it is an example of a broader class of probability sampling methods (e.g., stratified sampling, cluster sampling – see the Stratified and Cluster Sampling Appendix). The key is that probability, not human judgement, is responsible for selecting the sample.

Scope of Inference: Generalizing results from the sample to the population

Simple random sampling (and probability sampling methods in general) allow us to believe that our sample is representative of the larger population and that the observed sample statistic is "in the ball park" of the population parameter. For this reason, when we use simple random sampling we will be willing to *generalize* the characteristics of our sample to the entire population. But the next question is "how large is that ballpark?" That is, how far do we think any one observed statistic could be from the parameter? We will investigate that in the next example.

You may be asking: But, what if you don't have a simple random sample? Many people use convenience samples and other sampling designs where the observational units are not randomly selected from the population.

When we have convenience samples, we have no guarantee that the sample is representative of the population. For example, imagine we were trying to learn about how long baseball games take at the major league level. Suppose further that we think of this research question in April and decide to look at all games played within the previous week. Is this a bad idea? Well, this is not a random sample. A random sample would involve making a list of all of the major league baseball games over a certain time frame (e.g., an entire season or several seasons) and randomly choosing a subset of the games. Instead, we've just conveniently looked up the information for the last weeks' worth of games. But, do games in April really differ with regard to length as compared to games from other times of the year? If they don't, then you might argue that your convenience sample, although not random, is still representative of the population, at least with respect to game lengths. But…maybe the game lengths are different in April. Maybe there are more rain delays, or maybe managers use more relief pitchers. That's the problem with convenience samples. They are convenient, but we have no guarantees that they are representative—we're left to try to think about all the ways they might not be representative—a dangerous proposition.

Of course, if we were trying to use our sample of games to look at game time temperature, that would probably be a pretty bad idea and lead to some very misleading statistics! Instead, the effort to select a simple random sample may end up saving you time and effort later.

Impact of Increasing Sample Size

> **Think about it:** Suppose the researcher at the College of the Midwest decides 30 students may not be enough and decides to sample 75 students using his same early morning sampling method. Do you think this will make his sampling method any less biased?

If the researcher's sampling method tends to oversample students who live on campus, then selecting more students *in the same manner* will not fix this. A larger sample size does not imply our results are more representative or that the sampling method is any less biased. It is actually more effective to select a smaller but truly random sample than a large convenience sample!

Of course there are some benefits to using a larger sample size, as you saw in the previous chapter. With a larger sample size, there will be less sample to sample variability, and the statistics from different samples will cluster more closely around the center of the distribution (the statistical ball park will be smaller). But the random (unbiased) sampling method is what convinces us that the center of that distribution is in the right place.

Exploration 2.1A: Sampling Words

1.   Select a representative set of 10 words from the passage by circling them with your pen or pencil.

> Four score and seven years ago, our fathers brought forth upon this continent a new nation: conceived in liberty, and dedicated to the proposition that all men are created equal.
>
> Now we are engaged in a great civil war, testing whether that nation, or any nation so conceived and so dedicated, can long endure. We are met on a great battlefield of that war.
>
> We have come to dedicate a portion of that field as a final resting place for those who here gave their lives that that nation might live. It is altogether fitting and proper that we should do this.
>
> But, in a larger sense, we cannot dedicate, we cannot consecrate, we cannot hallow this ground. The brave men, living and dead, who struggled here have consecrated it, far above our poor power to add or detract. The world will little note, nor long remember, what we say here, but it can never forget what they did here.
>
> It is for us the living, rather, to be dedicated here to the unfinished work which they who fought here have thus far so nobly advanced. It is rather for us to be here dedicated to the great task remaining before us, that from these honored dead we take increased devotion to that cause for which they gave the last full measure of devotion, that we here highly resolve that these dead shall not have died in vain, that this nation, under God, shall have a new birth of freedom, and that government of the people, by the people, for the people, shall not perish from the earth.

The authorship of literary works is often a topic for debate. For example, researchers have tried to determine whether some of the works attributed to Shakespeare were actually written by Bacon or Marlow. The field of "literary computing" examines ways of numerically analyzing authors' works, looking at variables such as sentence length and rates of occurrence of specific words.

> **Definitions:** The *population* is the entire collection of observational units we are interested in, and the *sample* is a subset of the population on which we record data. Numerical summaries about a population are called *parameters*, and numerical summaries calculated from a sample are called *statistics*.

The passage is, of course, Lincoln's Gettysburg Address, given November 19, 1863 on the battlefield near

Gettysburg, PA. We are considering this passage a *population* of words, and the 10 words you selected are considered a *sample* from this population. In most studies, we do not have access to the entire population and can only consider results for a sample from that population, but to learn more about the process of sampling and its implications we will now deal with a somewhat artificial scenario where we sample from this known population.

2. Record each word from your sample, and then indicate the length of the word (number of letters) and whether or not the word contains at least one letter *e*.

Word	Length (no. of letters)	Contains *e*? (Y or N)
1.		
2.		
3.		
4.		
5.		
6.		
7.		
8.		
9.		
10.		

The table you filled in above is called a data table.

Definition: A *data table* (or statistical spreadsheet) is a convenient way to store and represent all data values. Typically in a data table, the rows correspond to observational units, columns represent variables, and the data are the table entries.

3. Identify the observational units and the variables you have recorded on those observational units. (Keep in mind that observational units do not have to be people!)

4. Is the variable "length of word" quantitative or categorical?

Note: When we are sampling from a finite population (e.g., all 268 words), we can consider the *parameter* to be a numerical summary of the variable in the population. With quantitative variables, we will often be interested in the population mean. With categorical variables, we will often be interested in the population proportion. To distinguish these values from the corresponding statistics, we will use

Greek letters, μ for a population mean, σ for the population standard deviation, and π for a population proportion. In contrast, common symbols for sample statistics are $\bar{x}$ for sample mean, s for the sample standard deviation, and $\hat{p}$ for sample proportion.

Notation check: Here is a quick summary of the symbols we've introduced in this section.

Type of Variable	Quantitative	Categorical
Statistics	$\bar{x}$ = sample mean	$\hat{p}$ = sample proportion
	s = sample standard deviation	
Parameters	μ = population mean	π = population proportion
	σ = population standard deviation	

5. Calculate the average length of the 10 words in your sample. Is this number a parameter or a statistic? Explain how you know. What symbol would you use to refer to this value?

6. The average length of the 268 words in entire speech equals 4.29 letters. Is this number a parameter or a statistic? Explain how you know. What symbol would you use to refer to this value?

7. Calculate the proportion of words in your sample that contain at least one *e*. Is this number a parameter or a statistic? Explain how you know. What symbol would you use to refer to this value?

8. The proportion of all words in the entire speech that contain at least one letter *e* is $125/268 \approx 0.47$. Is this number a parameter or a statistic? Explain how you know. What symbol would you use to refer to this value?

9. Do you think the words you selected are ***representative*** of the 268 words in this passage? Suggest a method for deciding whether you have a representative sample. (*Hint*: Whereas any one sample may not produce statistics that exactly equal the population parameters, what would we like to be true in general?)

10. Combine your results with your classmates' by producing a dotplot of the distribution of average word lengths in your samples. Be sure to label the axis of this dotplot appropriately.

    a. Explain what each dot on the dotplot represents by filling in the following sentence.

        Each dot on the dotplot is a single measurement of the value of

        _____ on a single

        _____.

    b. Describe the shape, center, and variability of the distribution of average word lengths as revealed in the dotplot.

11. Let's compare your sample statistics to the population parameters.

a. How many and what proportion of students in your class obtained a sample average word lengths larger than 4.29 letters, the average word length in the population?

b. How many and what proportion of students in your class obtained a sample proportion of *e*-words larger than 0.47, the proportion of *e*-words in the population? (You might use a show of hands in class to answer this question.)

Definition: A sampling method is ***biased*** if, when using that sampling method, statistics from different samples *consistently* overestimate or consistently underestimate the population parameter of interest.

Note that bias is a property of a sampling *method*, not a property of an individual sample. Also note that the sampling method must *consistently* produce non-representative results in order to be considered biased.

12. What do the answers to #11 tell you about whether the sampling method of asking students to quickly pick 10 representative words is biased or unbiased? If biased, what is the direction of the bias (tendency to overestimate or to underestimate)?

13. Explain why we might have expected that this sampling method (asking you to quickly pick 10 representative words) to be biased.

14. Do you think asking each of you to take 20 words instead of 10 words have helped with this issue? Explain.

Now consider a different sampling method: What if you were to close your eyes and point blindly at the page with a pencil 10 times, taking for your sample the 10 words that your pencil lands on.

15. Do you think this sampling method is likely to be biased? If so, in which direction? Explain.

A sample is only useful to us if the data we collect on our sample is similar to the results we would find in the entire population. In this sense, we say the sample is **_representative_** of the population.

16. Suggest another technique for selecting 10 words from this population in order for the sample to be representative of the population with regard to word length and *e*-words.

Taking a simple random sample

> **Key idea:** A **_simple random sample_** ensures that every sample of size *n* is equally likely to be the sample selected from the population. In particular, each observational unit has the same chance of being selected as every other observational unit.

The key to obtaining a representative sample is using some type of *random* mechanism to select the observational units from the population, rather than relying on **_convenience samples_** or any type of human judgment. Instead of having you choose "random" words using your own judgment, we will now ask you to take a simple random sample of words and evaluate your results. The first step is to obtain a **_sampling frame_** - a complete list of every member of the population where each member of the population can be assigned a number. Below is a copy of the Gettysburg address that includes numbers in front of every word. For example, the 43rd word is nation.

17. Go to the Random Numbers applet (or http://www.random.org).

Generate Random Number

Specify that you want 5 **Numbers per replication** in the range from 1 to 268.

Number of replications:	1
Numbers per replication:	5
Number range:	From: 1
	To: 268

Press **Generate** to view the 5 random numbers. Enter the random numbers in the table below:

Randomly generated five ID values from 1-268

Using your randomly generated values, look up the corresponding word from the sampling frame. Fill in the data table below.

	Word	Length (no. of letters)	*e*-word? (Y or N)
1.			
2.			
3.			
4.			
5.			

Notice that each simple random sample gave us different values for the statistics. That is, there is variability from sample to sample (***sampling variability***). This makes sense because the samples are chosen randomly and consist of different words each time, so we expect some differences from sample to sample.

Gettysburg Address Sampling Frame

1	Four	56	are	111	we	166	living	221	full
2	score	57	met	112	cannot	167	rather	222	measure
3	and	58	on	113	consecrate	168	to	223	of
4	seven	59	a	114	we	169	be	224	devotion
5	years	60	great	115	cannot	170	dedicated	225	that
6	ago	61	battlefield	116	hallow	171	here	226	we
7	our	62	of	117	this	172	to	227	here
8	fathers	63	that	118	ground	173	the	228	highly
9	brought	64	war	119	The	174	unfinished	229	resolve
10	forth	65	We	120	brave	175	work	230	that
11	upon	66	have	121	men	176	which	231	these
12	this	67	come	122	living	177	they	232	dead
13	continent	68	To	123	and	178	who	233	shall
14	a	69	dedicate	124	dead	179	fought	234	not
15	new	70	A	125	who	180	here	235	have
16	nation	71	portion	126	struggled	181	have	236	died

17	conceived	72	Of	127	here	182	thus	237	in
18	in	73	that	128	have	183	far	238	vain
19	liberty	74	field	129	consecrated	184	so	239	that
20	and	75	as	130	It	185	nobly	240	this
21	dedicated	76	a	131	far	186	advanced	241	nation
22	to	77	final	132	above	187	It	242	under
23	the	78	resting	133	our	188	is	243	God
24	proposition	79	place	134	poor	189	rather	244	shall
25	that	80	For	135	power	190	for	245	have
26	all	81	those	136	to	191	us	246	a
27	men	82	Who	137	add	192	to	247	new
28	are	83	here	138	or	193	be	248	birth
29	created	84	gave	139	detract	194	here	249	of
30	equal	85	their	140	The	195	dedicated	250	freedom
31	Now	86	lives	141	world	196	to	251	and
32	we	87	that	142	will	197	the	252	that
33	are	88	that	143	little	198	great	253	government
34	engaged	89	nation	144	note	199	task	254	of
35	in	90	might	145	nor	200	remaining	255	the
36	a	91	live	146	long	201	before	256	people
37	great	92	It	147	remember	202	us	257	by
38	civil	93	Is	148	what	203	that	258	the
39	war	94	altogether	149	we	204	from	259	people
40	testing	95	fitting	150	say	205	these	260	for
41	whether	96	and	151	here	206	honored	261	the
42	that	97	proper	152	but	207	dead	262	people
43	nation	98	that	153	it	208	we	263	shall
44	or	99	we	154	can	209	take	264	not
45	any	100	should	155	never	210	increased	265	perish
46	nation	101	do	156	forget	211	devotion	266	from
47	so	102	this	157	what	212	to	267	the
48	conceived	103	But	158	they	213	that	268	earth
49	and	104	in	159	did	214	cause		
50	so	105	a	160	here	215	for		
51	dedicated	106	larger	161	It	216	which		
52	can	107	sense	162	is	217	they		
53	long	108	we	163	for	218	gave		
54	endure	109	cannot	164	us	219	the		
55	We	110	dedicate	165	the	220	last		

18. Let's examine your sample and those of your classmates.

a. Calculate the average word length and proportion of *e*-words in your random sample.

     b.   Again produce a dotplot of the distribution of sample average word lengths for yourself and your classmates.

     c.   Comment on how this distribution compares to the one from #10 based on non-random sampling.

19. Let's compare your sample statistics to the population parameters.

     a.   How many and what proportion of students in your class obtained a sample average word length larger than the population average (4.29 letters)?

     b.   How many and what proportion obtained a sample proportion of *e*-words larger than the population proportion (0.47)? (You can use a show of hands.)

     c.   What do these answers reveal about whether simple random sampling is an unbiased sampling method? Explain.

You should see that the random samples do a better job than self-selected samples of centering around the population parameter value, *even though* we used a smaller sample size! Thus, simple random sampling would be considered an **unbiased** sampling method because there is no tendency to over or underestimate the parameter. Whether a sampling method is biased or unbiased depends both on the sampling method and on the statistic you chose to calculate from the sample. Luckily, one can show that with simple random sampling, the sample average and the sample proportion both have this unbiased property.

> **Key idea:** When we use **simple random sampling**, the average of the sample means from different samples equals the population mean and the average of the sample proportions from different samples equals the population proportion. (This means that these statistics are **unbiased** estimators for the corresponding parameter when we use simple random sampling.)

Although selecting a simple random sample is not always feasible (e.g., what if you don't have a complete sampling frame, what if it is too difficult to track down everyone selected?), it is an example of a broader class of probability sampling methods (e.g., stratified sampling, cluster sampling see the

Stratified and Cluster Sampling Appendix). The key is that probability, not human judgement, is responsible for selecting the sample

To really see the long-run pattern in these results, let's look at many more random samples from this population.

20. Open the **Sampling Words** applet. You will see three population distributions for three variables: the lengths of the words, whether or not a word contains at least one letter *e*, and whether or not the word is a noun. For now, use the pull-down menu to select the length variable. Before we draw random samples from this population, we want to point out that we often use dotplots to represent a set of quantitative values. Another option is a *histogram* like the one shown below.

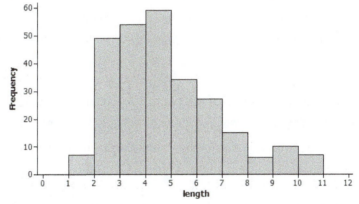

Like a dotplot, a histogram displays the distribution of a quantitative variable. But when you have a large number of observations like this, sometimes it may be convenient to "bin" together similar observations into one bar. The height of each bar tells you how many of the values are in each bin (interval of values). The graph still helps us see features of the distribution, like that the distribution of word lengths are not symmetric but cluster around 4 to 5 letters.

Definition: A *histogram* is a graph that summarizes the distribution of a quantitative variable. Bars are used to represent the number of observational units in a particular interval of values.

---

21. Check the **Show Sampling Options** box. Specify 5 in the **Sample Size** box and press the **Draw Samples** button. You will see the 5 selected words appear in the box below. You will also see 5 blue dots within the population distribution representing the 5 lengths that you have sampled. These 5 lengths are also displayed in a dotplot in the middle panel. The blue triangle indicates the value of this sample's mean, which is then added to the graph in the far right column. Press **Draw Samples** again. Did you get the same sample this time? Did you get the same sample mean?

You should now see two dots in the bottom graph, one for each sample mean (most recent is blue). The mean of these two sample means is displayed in the upper left corner of this panel. Confirm this calculation.

First sample mean: Second sample mean: Average:

22. Press the **Draw Samples** button 3 more times (for a total of 5 samples). Notice how the samples vary and how the sample mean varies from sample to sample. Now what is the average of your 5 sample means?

Change the **Number of Samples** to 995 (for a total of 1000 samples). Press the **Draw Samples** button. *(This should generate a dotplot of 1000 values of the sample mean for 1000 different simple random samples, each consisting of 5 words, selected randomly from the population of all 268 words in the Gettysburg Address.)*

23. Describe the resulting distribution of sample means: What are roughly the largest and smallest values? What is the average of the distribution? What is the standard deviation? How does the average of the 1000 sample means compare to the population mean μ = 4.29?

24. Now change the **Sample Size** from 5 to 20 and the **Number** of **Samples** to 1000. Press the **Draw Samples** button. The applet allows you to compare the two distributions ($n = 5$ to $n = 20$). How do these two distributions compare? Be sure to identify a feature that is similar and a feature that is different.

25. Did changing the sample size change the center of the distribution? If we used a biased sampling method, would increasing the sample size remove the bias? Explain.

You should see that whereas the convenience samples tended to consistently produce samples that overestimate the length of words and even the proportion of words that are *e*-words, the simple random sampling method does not have that tendency. In fact, the average of all the statistics from all possible simple random samples will exactly equal the population parameter value. For this reason, in the real world, where we don't have access to the entire population, when we select a simple random sample, we will be willing to believe that sample is representative of the population. In fact, we would prefer a small random sample over a large convenience sample. That doesn't mean our sample result will match the population result exactly, but we will be able to predict how far off it might be.

Simple random sampling (and probability sampling methods in general) allow us to believe that our sample is representative of the larger population and that the observed sample statistic is "in the ball park" of the population parameter. For this reason, when we use simple random sampling we will be willing to *generalize* the characteristics of our sample to the entire population. But the next question is "how large is that ballpark?" That is, how far do we think any one observed statistic could be from the parameter? We will investigate that in the next example and exploration.

Example 2.1B: Should Supersize Drinks be Banned?

In June of 2012, New York City Mayor Michael Bloomberg proposed banning the sale of any regular soda or other sugary drinks in containers larger than 16 oz. Diet drinks, those with more than 70% juice, or more than 50% milk would be exempt. Some people saw this as an effort to fight the obesity problem and others saw it as an example of government intrusion.

A poll of 1093 randomly selected New York City voters conducted by Quinnipiac University taken shortly after the mayor's announcement found that 46% of them supported the ban. Based on this sample proportion, we would like to decide whether less than half of all New York City voters support the ban.

So our parameter, represented by π, is the proportion of all New York City voters that favor the ban. But the poll didn't ask every New Yorker. Our statistic is the sample proportion $\hat{p} = 0.46$. We know that if we had taken a different sample we would likely get a different value for $\hat{p}$. So how can we use this observed $\hat{p}$ to make any conclusions about π ?

> **You might be thinking:** But didn't we do that in the last chapter?

In the last chapter you focused on the sample to sample variability in a sample proportion where the observations came from a never ending random *process* with a fixed probability of success. We stated a formula for the standard deviation of these sample proportions that helped us measure the amount of sampling variability:

$$\text{SD of } \hat{p} = \sqrt{\pi(1-\pi)/n}$$

Now, the subtle distinction is that we are looking at the variability in different samples from the same finite *population*.

But recall the distribution of sample proportions we found when we took samples of 30 students from the College of the Midwest (See Figure 2.2). The simulated standard deviation of those 1000 samples was 0.076, which, remembering that in that case $\pi = 0.776$, is also equal to $\sqrt{0.776(1-0.776)/30} \approx 0.076$! It can be shown that when the population is large relative to the size of the sample, the amount of sample to sample variability is not impacted by the population size.

> **Key idea:** When sampling from a large population, the standard deviation of sample proportions is estimated by the same formula $\sqrt{\pi(1-\pi)/n}$, where π represents the proportion of successes in the population and n represents the sample size. The population is considered large enough when it is more than 10 or 20 times that sample size.

As of 2009, there were more than 4 million registered voters in New York City. So this population size is definitely large enough for us to treat these as random samples from an infinite process. This means that we will consider a sample proportion to be far from a hypothesized population proportion if it is more than $2\sqrt{\pi(1-\pi)/n}$ away. So, if 50% of all New Yorker approve of the ban, a poll of 1093 New Yorkers should produce a sample proportion within $2\sqrt{0.5(0.5)/1093} \approx 2 \times .0151 = 0.0302$ of 0.5. In fact, we can use the One Proportion applet exactly as we did before. See Table 2.4 for the mapping of the simulation to the study.

Figure 2.3: Results from 1000 simple random samples of 1093 New York City voters, assuming 50% of all New York City voters favor a ban on super-sized drinks

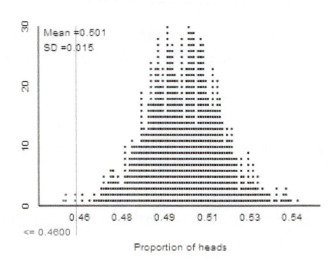

Table 2.4: Parallels between real study and simulation

One repetition	=	A random sample of 1093 NYC voters
Statistic	=	Sample proportion who favor the ban
Null model	=	Population proportion (π) = 0.5

Our result of 0.46 is more than 2 standard deviations away from 0.5 and a result at least this extreme would be expected to happen in less than 1% of random samples from the population. This gives us convincing evidence that random chance is not a plausible explanation for the lower sample proportion. We have strong evidence that less than half of all New York City voters favor the ban. We feel comfortable generalizing this sample result to the entire population of voters because the University selected a *random* sample from that population. We would not feel comfortable generalizing this result to all New Yorkers or to other states because the sampling frame only consisted of registered NY voters.

Furthermore, we can also use the Theory-Based approach to approximate the p-value under the same validity conditions as before, with the addition of the population size consideration.

> **Validity conditions:** The normal approximation can be used to model the null distribution of random samples from a finite population when the population size is more than 20 times the size of the sample, and if there are at least 10 successes and at least 10 failures in the sample.

Our sample size is quite large (503 successes and 590 failures) so we expect the normal approximation to be valid as shown in Figure 2.4. Read more about these validity conditions in FAQ 2.1.1.

Figure 2.4: The Theory-Based approach to the New York City ban on super-sized drinks

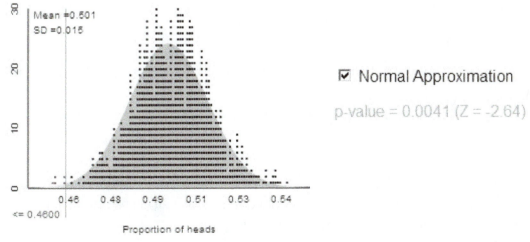

☑ Normal Approximation

p-value = 0.0041 (Z = -2.64)

Proportion of heads

Exploration 2.1B: Banning Smoking in Cars?

In Exploration 2.1A about the Gettysburg Address, you saw that values of statistics generated from simple random samples from a finite population follow a predictable pattern. In particular, the center of the distribution of sample statistics is equal to the population parameter we are trying to make inferences about. In this exploration, you will focus on the variability of sample statistics and see how we can use what we learned in Chapter 1 to predict the sample-to-sample variability for samples from finite populations.

In February of 2012, a bill was introduced into the Ohio State Senate that would outlaw smoking in vehicles if young children are present. That same month a poll was conducted by Quinnipiac University and asked randomly selected Ohio voters if they thought this bill was a good idea or a bad idea. Of the 1421 respondents, 55% said it was a good idea. We can use this sample proportion to investigate whether more than half of all Ohio voters in February of 2012 thought that this bill was a good idea.

1. Identify the population and sample in this survey.

   Population:

   Sample:

2. Is it reasonable to believe that the sample of 1421 Ohio voters is representative of the larger population? Explain why or why not.

3. Explain why 55% is a statistic and not a parameter. What symbol would you use to represent it?

4. Describe in words the corresponding parameter for this study. What symbol would you   use to represent this value?

5. Is it reasonable to conclude that exactly 55% of Ohio voters agree with the ban on smoking in cars when young children are present? Explain why or why not.

6. Describe how we could conduct a simulation to decide whether a simple random sample from this population could produce a sample proportion like 0.55 if, in fact, only 50% of the population supports ban on smoking in cars when young children are present.

    Each trial represents_____

    Number of trials =

    Probability of success =

7. Some estimates suggest that there are approximately 7.8 million registered voters in Ohio, suppose 50% of Ohio voters are in favor of this ban. If this is true, what is the probability that a randomly selected voter will be in favor of the ban?

8. If 50% of Ohio voters are in favor of the ban, how many people is this? Report your answer in millions.

9. Suppose I select a Ohio voter from the set of all 7.8 million voters, and that voter is in favor of the ban. What is the probability that the next randomly selected voter will also be in favor of the ban?

Your answer to the previous question suggests that when we are sampling from a finite population, without replacement, then we technically don't have a constant probability of success like we assumed in Chapter 1. However, if the population size is much, much larger than the sample size, then we can treat the probability of success as roughly constant. In other words, we will assume samples from a very large population behave just like samples from an infinite process.

> **Key idea:** When sampling from a large population, the standard deviation of sample proportions is estimated by the same formula $\sqrt{\pi(1-\pi)/n}$, where π represents the proportion of successes in the population and n represents the sample size. The population is considered large enough when it is more than 10 or 20 times that sample size.

This means that the inference methods that you learned in Chapter 1, both simulation-based and theory-based, can be applied to a random sample from a population. In such situations the parameter of interest is the population proportion with the characteristic, rather than the long-run process probability of success. (The population proportion and process probability can be considered to be equivalent, if you think of the process as randomly selecting an observational unit from the population.)

10. Use the **One Proportion** applet to estimate a p-value for testing whether more than 50% of Ohio voters are in favor of the ban for this survey. Begin by stating the null and alternative hypotheses.

    Be clear how many "coin tosses" you use and how many repetitions you use. Report the p-value you obtain.

11. Write a one-sentence interpretation of your p-value <u>and</u> summarize the strength of evidence against the null hypothesis provided by this p-value.

12. Report the standard deviation of the null distribution displayed by the applet. How does this compare to the formula for the standard deviation of a sample proportion discussed in Chapter 1($\sqrt{\pi(1-\pi)/n}$) when $\pi = 0.5$?

13. Are the validity conditions met for using a theory-based method to predict a p-value from this sample? Why or why not?

14. Check the box for **Normal Approximation** in the applet. Report and interpret the value of the standardized statistic.

15. Is the theory-based estimate of the p-value similar to the p-value from the simulation-based method?

Validity conditions: The normal approximation can be used to model the null distribution of the sample proportion for random samples from a finite population when the population size is more than 20 times the size of the sample, and if there are at least 10 successes and at least 10 failures in the sample.

Summary

An important question to ask about any statistical study is how broadly the results can be **generalized**. The key to conducting a study that allows us to generalize the results from the sample to a larger population is how the sample is selected from the population.

- Remember that a population is the entire group of interest; a sample is a part of the population on which data are gathered.
    - o   A parameter is a number that describes a population, typically unknown.
    - o   A statistic is a number calculated from a sample, often used to estimate the corresponding population parameter.

The primary goal of sampling is to select a sample that is **representative** of the population on all characteristics of interest.

- A sampling method based on **convenience** is often biased, and not likely to produce a representative sample.
    - o   A biased sampling method is one that systematically tends to over-represent some parts of the population and under-represent others.

- A biased sampling method leads to a biased statistic, meaning that the statistic systematically over-estimates or under-estimates the corresponding population parameter.

The best way to ensure that the sample will be representative is to take a **random sample**.

- A simple random sample gives every possible sample of the desired size the same chance of being the sample selected.
    - In fact, a simple random sample gives each observational unit in the population the same chance of being selected for the sample.
- With random sampling, the values of sample statistics (such as proportions and means) center around the corresponding population parameter.
- The sampling variability in sample statistics decreases as sample size increases.
    - But using a larger sample size is not helpful if the sampling method is biased in the first place.
    - The population size does not affect sampling variability, as long as the population is at least 10 to 20 times larger than the sample.

When a sample is not selected with random sampling, be cautious about generalizing results from the sample to a larger population.
- Depending on the characteristic being studied, generalizing may or may not be advisable.

When a random sample is selected from a population, the simulation- and theory-based approaches from Chapter 1 for determining p-values can be applied to make inferences about the population proportion based on the sample proportion. This is because sample-to-sample variability can be modeled the same way, whether we are taking random samples from populations or samples from random processes, provided that the population size is at least 10 to 20 times the sample size. More details about assumptions and contrasting a population and a process are provided in FAQ 2.1.2.

FAQ 2.1.1 Population size N and sample size n

Q: First you tell me n should be large, then you say N/n should be large. What gives? I can't change N: the population is what it is. So the only way to make N/n large is to make n small. Which is it, n large or n small?

A: Fair question. There are several issues, but I'll start with an answer to your question: larger n is better.

Q: So little n should be big. What about wanting N/n to be large?

A: Two points. First, the theory-based approximations were designed for really big populations, with N in the millions (registered voters in the US) or at least in the thousands (McDonald's restaurants). They don't work when your population is small, like your stat class. Mainly, they don't work because the SE for $\hat{p}$ is *smaller* than $\sqrt{\dfrac{\pi(1-\pi)}{n}}$ when N/n is smaller. Smaller SE is actually better. You just can't use the theory-based approximations. Second, if N is really big, N/n will be big, for a particular sample size n.

Q: But wait! If N is large, doesn't n also need to be large? I'm thinking that the sampling fraction n/N is what matters. A 5% sample from the US would be about a million and a half people.

A: A lot of stat students have that same intuition, but it's simply not right. Professional pollsters get by with samples a thousand times smaller, just 1,500 or so, not 1,500,000.

Q: How can that work? A sample as small as that almost guarantees that you won't get any people from small groups, like statistics professors.

A: True, but groups like statistics professors that are not in the sample tend to be groups that are also a very small part of the population.

Q: Say it another way.

A: If you want to know whether your soup is salty, a small taste is enough, *provided* the soup is thoroughly mixed. For sampling, randomizing is the way we mix the soup. With proper mixing, it's enough to taste a spoonful, whether your spoonful comes from a bowl or a barrel. But the mixing is critical.

Q: A spoonful of soup sounds like small n. Back at the beginning you said larger n is better. What does larger n get you?

A: Two things, variability and shape.

Q: Variability?

A: So long as N is really large, it's n that matters. With really large N, the formula for SE gives a good approximation to the true SE. Notice that in the formula, N isn't there at all. The sample size is there, as $\sqrt{n}$ in the denominator.

Q: That's why pollsters can get away with a sample size of just 1500?

A: Exactly. For Yes/No data, $n = 1500$ gets you an SE of about 0.013.

Q: OK. And shape?

A: "Shape" here refers to the shape of the distribution of the statistic from the random samples. When n is larger, that distribution is closer to normal.

Q: And?

A: The theory-based approximations *assume* the distribution is normal. So the larger your n, the better the approximations work.

Q: You said larger n gets you two things, variability and shape. You left out center.

A: I did. I meant to leave it out.

Q: Doesn't bigger n mean smaller bias in the sample?

A: No. If your sampling method is biased, doing more of it won't fix the problem. If you're in a hole, digging faster won't get you out.

Q: Can you trash-compact all this down?

A: Four points:

- Theory-based methods are designed for random samples from very large populations. The ratio N/n should be large because N is large, not because n is small.

- If N is very large, then what matters in predicting the behavior of your statistic is the sample size n. Little n should be big.

- If your sampling method is biased, larger n won't get rid of the bias …

- but larger n means the SE is smaller, and the statistic's distribution is closer to normal.

FAQ 2.1.2 The role of randomization

Q: Why should we let random chance dictate which units we choose for our sample?

A: Two main reasons, bias and inference. You know about bias already – just remember the Gettysburg Address. Didn't chance beat judgment?

Q: OK. But inference?

A: Just say after me, one, two, three:

    One: For inference you need a p-value.

    Two: For a p-value you need a chance model.

    Three: For a chance model you need random chance in your data.

Q: I can say it shorter if I go backwards:

    No chance, no model.

    No model, no p-value.

    No p-value, no inference.

A: Spoken like a statistician. But don't let mere slogans eat your brain. Real life, like your brain, is full of gray areas. Sometimes non-random samples behave "as if random." Evaluating "as if random" is where an educated brain can beat mere rules. To the degree that a chance-like model is reasonable, you can trust a p-value. To the extent not, then not.

Q: OK. So for choosing a sample from a population, I know the ideal way: make a complete list (the sampling frame), then use random numbers to choose the sample. What if I've got a process instead of a population?

A: Excellent question! Let's do a point-by-point comparison. Think coin toss. Inference depends on three key properties: chance, constant probability, and independence.

Q: Say it for a coin.

A:

    Chance: The outcome of each toss is chance-like.

    Constant probability: The chance of heads doesn't change.

    Independence: The coin has no memory. The outcome of the next toss doesn't depend on the outcome of the current toss.

Q: Okay. For a population,

- Chance: We get chance by using random numbers to choose the sample.
- Constant probability: We get constant probability because π is a property of the population, and the population doesn't change.
- Independence is approximate, and requires N/n large, but with really large populations that ratio will be large.

A: Excellent. So when we calculate a p-value, now we will be modelling the randomness in the sample selection process (that we introduced) rather than the randomness in individual outcomes of a process.

Q: So if we don't have a random sample, then it is difficult to interpret the p-value?

A: Correct. Now let's talk a bit more about the assumptions we actually had with a process but didn't dwell on in Ch. 1. With a process, you have less control over the sampling, so you have to rely more on assumptions.

Q: Assumptions? Which are?

A: Take our coin toss model for Buzz and Doris. The Yes/No model with fixed sample size n makes the same three assumptions as for a population:

    (1) Outcomes (his guesses) are chance-like.

    (2) The chance $\pi = P(\text{Success})$ is constant. (In fact, under the null model that Buzz is guessing, we assume his probability of picking the correct button is 0.5 every time.)

    (3) Outcomes are independent.

Q: OK. What's your point?

A: Our coin toss model depends on all three assumptions. If the p-value is tiny and we reject the model, it could be because any one of the three assumptions is wrong. It could also be that H_0 is false, that is, π

is not 0.5 for every trial. To reject H_0, we have to assume that all of (1) – (3) are true.

Q: Let's take them one at a time.

A: Our hypothesis is that Buzz is just guessing. That takes care of Assumption 1 that outcomes are chance-like.

Q: Agreed. And Assumption 2?

A: That's harder. If Buzz gets tired, his chance π of guessing right will go down over time. If Buzz learns from experience, his chance of guessing right goes up. Either way, π is not constant.

Q: That's true, but we can control for tiredness by giving Buzz and Doris a rest before we run the experiment, and we can control for learning by making sure Buzz and Doris have had lots of practice time before they take the test.

A: Good answer. But we still have to take it on faith that Buzz and Doris have had enough rest time and enough practice time. For this particular process, we can trust the experimenters, but maybe not so much for other data from a process.

Q: Last issue. Independence.

A: For a coin, that's easy. The coin has no memory and no volition. The coin doesn't say to itself, "I landed heads three times out of the last four; I'd better land tails this time to help even things out." Buzz and Doris are not coins. They want fish. They do have memories. They do have volition.

Q: You make it sound hopeless.

A: Not at all. Dr. Bastian can make the coin toss model apply to Buzz and Doris just by using a coin toss to choose whether the light will be on or flashing. For Marine, we can put the bags in a random order. So again, it is often helpful to our analysis to introduce randomness in the data collection process.

Q: That works when the experimenter controls the right answer. What if we're just observing a Yes/No process, like whether it rains day by day.

A: For that, because we have less control, we have to rely much more on assumptions. We have to assume (1) that outcomes are random. We know they're not, but for some purposes, the assumption might be OK. We have to assume also (2) that π is constant. We know it's not, but again, for some purposes, it might lead to a workable model. (3) For independence, we do have some control: If we think the current outcome might influence the next one, we can skip a few outcomes to let the influence die down. For example, "Did it rain today? (Yes/No)" depends on "Did it rain yesterday? (Yes/No)", but "Did it rain today? (Yes/No)" is largely independent of "Did it rain 10 days ago? (Yes/No)." If we record Yes/No for every tenth day, we can make the assumption of independence more reasonable.

Q: To sum up?

A: Randomization in sampling: 2 scenarios

- For sampling from a population:
  If you can, randomly select the sample to allow us to use a chance model. The theory assumes that N is very large. If so, then we consider the outcomes as independent because the conditional probability of success, based on the outcomes you have already selected, is essentially constant.

- For sampling from a Yes/No process:
  If you can, randomize the correct response. If you can't, your model depends on the assumptions that outcomes are random with constant probability. For independence, you can make the assumption more justifiable by taking every k^{th} value from the process.

Section 2.2: Inference for a Single Quantitative Variable

Introduction

In the previous section we saw that

- Selecting a random sample from a population allows us to *generalize* from our sample to the population we've sampled from, whether we have categorical or quantitative data. In particular, with categorical variables, we can make inferences about the population *proportion*; with quantitative variables, we can make inferences about the population *mean*.

- The simulation- and theory-based inference methods that we learned in Chapter 1 for analyzing sample proportions from random processes also work for random samples from large finite populations.

In this section we'll learn more about how we can draw inferences about populations when we have quantitative variables measured on random samples. We'll learn both simulation- and theory-based approaches for estimating p-values for a single population mean. We will also mention some of the differences between the mean and the median with regard to their ability to summarize the center of a distribution of a quantitative variable.

Example 2.2: Estimating Elapsed Time

Does it ever seem like time drags on (perhaps during one of your least favorite classes) or time flies by (like a summer vacation)? Perception, including that of time, is one of the things that psychologists study.

Students in a statistics class collected data on other students' perception of time. They told their subjects that they would be listening to some music and then after it was over they would be asked some questions. They played ten seconds of the Jackson 5's song "ABC." Afterwards, they simply asked the subjects how long they thought the song snippet lasted. They wanted to see whether students' could accurately estimate the length of this short song segment. Let's explore this study by working through our six step statistical investigation method.

Step 1. Ask a research question. Can people accurately estimate the length of a short song snippet?

Step 2. Design a study and collect data. The researchers asked 48 students on campus to be subjects in the experiment. The participants were asked to listen to a ten second snippet of a song, and after it was over, they were asked to estimate how long the song snippet lasted. (The subjects did not know in advance that they would be asked this question.) Their estimates of the song length are the data that will be analyzed. Because the true length of the song is 10 seconds, we are interested in learning whether people, on average, correctly estimate the song length as 10 seconds---or if they, on average, over- or under-estimate the song length. Thus, we have the following hypotheses.

> Null Hypothesis: People in the population accurately estimate the length of a 10 second-song snippet, on average.

> Alternative Hypothesis: People in the population do not accurately estimate the length of a 10-second song snippet, on average.

If we let μ represent the mean time estimate for the 10-second song snippet for everyone in the population, then we can write our hypotheses as follows.

> H_0: $\mu = 10$ seconds (on average, students estimate the song length correctly)
> H_a: $\mu \neq 10$ seconds (on average, students in the population over– or under–estimate the song length)

Step 3. Explore the data. Because the estimated length of time is a quantitative variable, we use different techniques to summarize the variable than in Chapter 1. Because quantitative variables can take many, many different values, there is typically more work to do to summarize the variable's distribution.

Recall from the Preliminaries that we can use a dotplot of the variable's distribution, and then describe the distribution's shape, center, variability, and identify unusual observations. The graph in Figure 2.5

displays the distribution of estimated times for the 48 subjects who participated in the study.

Figure 2.5: The results (in seconds) for 48 students trying to estimate the length of a 10 second song snippet

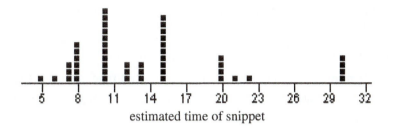

estimated time of snippet

Think about it: What does each dot on this dotplot represent?

In this graph the observational units are the 48 students and the variable is the estimated time of the snippet.

Shape: An important characteristic of the graph is that distribution of time estimates is not symmetric, but instead is *right skewed*.

Definition: A distribution of data is *skewed* if it is not symmetric, and, instead, the bulk of values tend to fall on one side of the distribution, with a "tail" on the other. Right skewed distributions have their tail on the right, and left skewed distributions have their tail on the left.

You can also see that responses in increments of 5 seconds were common (10, 15, 20, 30 sec).

Center: The average estimate of the ten second song snippet, $\bar{x}$, was 13.71 seconds. (We will refer to "seconds" here as the *measurement units* of the variable, an important description to include with numerical summaries of quantitative variables.)

Key Idea: When describing a quantitative variable, you should always be aware of the *measurement units* of the variable. Measurement units indicate information about the context and scaling of the variable (e.g., seconds vs. minutes). Always include measurement units when describing your variable.

However, when data are skewed, the mean may no longer be as appropriate a measure of center as with symmetric distributions. For example, more than half (28/48 = 58%) of the estimates in this sample are less than the mean estimate (13.7 seconds). With symmetric distributions approximately half of the data values will be less than the mean (and approximately half will be more).

Key Idea: The more *right skewed* the distribution is, the larger the percentage of data values that are below the mean. Similarly, the more *left skewed* the distribution is, the larger the percentage of data values that are above the mean. (The mean is pulled in the direction of the longer tail.)

For this reason, the *median* is sometimes a preferred way to measure the center of a skewed distribution. The median is the middle number when the data values are sorted from smallest to largest. This implies that the median is always the number which splits the data in half so that half the data values are larger and half are smaller.

Definition: The *median* is the middle data value when the data are sorted in order from smallest to largest. The location of the median can be found by determining $(n+1)/2$, where n represents the sample size. When there are an odd number of data values, the median is the $(n+1)/2^{th}$ observation. When there is an even number of data values, the median is the average of the middle two numbers.

To determine the median of the students' estimates of song length, first sort the 48 estimates in order from smallest to largest. We did this in Table 2.5 below.

Table 2.5. Student estimates of song length ordered from smallest to largest

Rank	1	2	3	4	5	6	7	8	9	10	11	12	13	14	15	16	17	18	19	20	21	22	23	24
Estimate	5	6	7	7	7	8	8	8	8	8	8	10	10	10	10	10	10	10	10	10	10	10	12	12

Rank	25	26	27	28	29	30	31	32	33	34	35	36	37	38	39	40	41	42	43	44	45	46	47	48
Estimate	12	13	13	13	15	15	15	15	15	15	15	15	15	15	20	20	20	20	21	22	30	30	30	30

In this case, because there are 48 data values (an even number), the median is the average of the 24th and 25th data values (12 and 12). So, the median is 12 seconds.

Think about it: Why is the median (12) lower than the mean (13.71) for this dataset?

Because the distribution is right skewed, the mean is "pulled" in the direction of the skewness; the median is not. It's common for the mean to be larger than the median with a right-skewed distribution and for the mean to be smaller than the median with a left-skewed distribution. With a stronger skew, the difference between the mean and median is typically larger than with a more modest skew.

Think about it: Which would change more – the mean or the median – if one of the 30 second values had instead been 120 seconds?

Because the median only considers the middle value, it is not affected by extreme values that do not fit the overall pattern of the distribution (sometimes called *outliers*). But because the mean considers every individual value, it can be greatly affected by outliers. In this case the mean would change considerably (from 13.7 to 15.6 seconds) if one of the 30 second values had instead been 120 seconds, but the median (12 seconds) would not change at all.

Definition: A statistic is *resistant* if its value does not change considerably when outliers are included in a dataset. The median is a resistant statistic, but the mean is not.

Variability: The third important way to describe a distribution is to discuss its variability. In this case, the standard deviation (s) of students' song length estimates was 6.50 seconds. This means roughly that, on average, students' song length estimates were about 6.50 seconds away from the average song length (13.71 seconds).

Think about it: Is standard deviation resistant to outliers?

No, the standard deviation is not resistant to outliers. Calculating a standard deviation involves comparing each data value to the mean (and then squaring that distance!), so extreme outlier values can have a very large impact on the standard deviation.

Variability can also be described by commenting on where the bulk of the data values are located. In this case we could say that all students in our sample estimated the song length to be between 5 and 30 seconds, with the majority of students estimating the song length to be between 5 and 20 seconds.

Unusual observations: Typically, the fourth part of summarizing a variable's distribution is to identify any unusual observations (outliers). Although there are some mathematical rules to define values that are outliers (e.g., a certain distance away from the bulk of the data), often a subjective description of unusual observations is sufficient. In this case, we can say that although the bulk of student estimates of song length are between 5 and 20 seconds, there were a few students who estimated the song length to be 30 seconds--substantially longer than most other students in the sample.

Step 4. Draw inferences beyond the data. Judging both from the graph and the mean and median, we can see that this group of subjects tended to overestimate the length of the snippet. Could this overestimate just come from random chance if, on average, the population is accurate when estimating the length of a short song snippet? Maybe the researchers were just "unlucky" in their sample, and the mean really would equal 10 seconds if the researchers had been able to ask this question of the entire population at their school.

We will now look at two approaches to evaluating hypotheses about a population mean: a simulation-based approach, and a theory-based approach.

Simulation-based approach: Applying the 3S strategy to hypotheses about a population mean

First, we note that the **statistic** is the sample mean of $\overline{x}_{observed} = 13.71$ seconds. But how will we simulate values of the statistic assuming the null hypothesis to be true? Clearly, we can no longer flip coins or spin spinners (Why not?). Instead, we need an alternative approach to simulate selecting a random sample of data assuming that the null hypothesis is true (namely, that the population mean, μ, is 10 seconds).

One option is to invent a "population" of values with a mean of 10 and then repeatedly take random samples (of this same size) from that population to simulate what samples could look like if the null hypothesis were true.

The histogram in Figure 2.6 shows a made-up population of 6215 estimates of the length of the song, where the mean estimated song length is 10 seconds, and the standard deviation is about 6.5 seconds.

Figure 2.6: Histogram of the distribution of song length estimates in a hypothetical population of 6215 individuals with mean 10 seconds

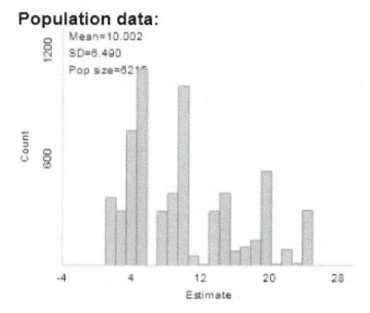

Think about it: What symbols would you use to represent the mean of 10 and standard deviation of 6.5 in the population? Why?

The mean of the population distribution is represented by the Greek letter, μ. Similarly, the standard deviation of 6.5 seconds is represented by the Greek letter σ as it is also is based on the population.

Notice that the distributions in Figures 2.5 (sample) and 2.6 (hypothetical population) are similar in many ways:

- Most individuals guessed whole numbers (1, 2, 3…) and not fractions for the length of the song.
- People were more likely to guess multiples of 5 seconds (like 5, 10, 15, 20, …) than other whole numbers.
- The sample standard deviation in Figure 2.5 is similar to the population standard deviation in Figure 2.6.
- Both distributions are right skewed, meaning that more people tended to guess shorter times, but a few people guessed very long times.

The main difference is that this population has a mean song length of 10 seconds, whereas our actual observed sample mean was larger (13.7 seconds). Thus, this distribution is a possible population if the null hypothesis is true. We can now apply the 3S strategy to evaluate the strength of evidence provided by the sample that the actual population did not have a mean of 10 seconds.

1. Statistic: The observed statistic is the average estimated song length among the 48 students in the sample, 13.7 seconds.

2. Simulate: We need to conduct a simulation which generates random samples of 48 students from a population where we know that the null hypothesis (the population mean, μ, is 10 seconds) is true.

One such approach is to take many random samples of size 48 from the population of 6215 estimates shown earlier in Figure 2.6. We use the One Mean applet to take a random sample of 48 students from this population. Figure 2.7 shows the results.

Figure 2.7: A random sample of 48 "students" from the hypothetical population of 6215 students. The dots represent the observed results (estimated song lengths) in the sample.

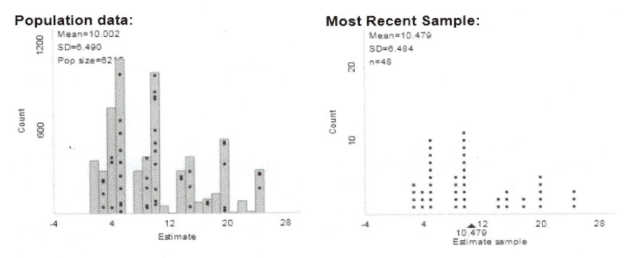

Table 2.6: Parallels between real study and simulation

One repetition	=	A random sample of 48 students
Null Model	=	Population mean (μ) = 10 seconds.
Statistic	=	Average time estimate of song by sample

Figure 2.7 shows that this simulated random sample of 48 students had an average estimated song length of 10.479 seconds, a "could-have-been" sample mean.

By now you know that when we simulate we need to repeat the simulation many, many times and keep track of the value of the statistic each time in order to build a null ("what if") distribution ("what if the population mean was 10 seconds") of the sample statistics. So next we simulated taking 1000 different random samples of 48 students from the same hypothetical population. Figure 2.8 illustrates the distribution of sample means in each of the 1000 samples.

Figure 2.8: Distribution of sample means across 1000 simulated random samples of 48 students from a population with μ = 10 seconds

Count Samples | Beyond | ▼ | 13.71

Count = 0/1000 (0.0000)

Notice that this null distribution of sample means is centered around 10 seconds (as was the population), has a symmetric shape (unlike the population), and has much less variability than the population (standard deviation of sample means = 0.926 seconds rather than the standard deviation of time estimates in the population of 6.490 seconds). Now we can use this null distribution to assess how surprising it would be to get a sample mean of 13.71 seconds from such a population by chance (random sampling) alone.

3. Strength of evidence: As shown in the histogram in Figure 2.8, sample means as large as 13.71 (the observed sample mean) rarely occur in either direction. In fact, out of 1000 random samples from this population, a sample mean as large as 13.71 or as small as 6.29 never occurred (see Figure 2.8). Thus, our estimated p-value from this simulation is 0, providing very strong evidence that our observed sample mean did not come from a population with a mean estimated song length of 10 seconds.

Alternatively, we could compute a standardized statistic to assess the strength of evidence by computing $(13.71-10)/0.926 = 4.01$. This reveals that the observed value of the sample mean is almost 4 standard deviations above the hypothesized value of 10 seconds. This large standardized statistic confirms our very strong evidence against the null hypothesis that the average estimated song length in the population is 10 seconds.

Thus, we see that the observed sample data provide very strong evidence that people in the population have a systematic tendency to, on average, over-estimate how much time elapses in a song snippet.

Theory-based approach: One-sample *t*-test

You may have noticed a familiar pattern in the null distribution simulated in Figure 2.8. As you saw in Chapter 1, when certain validity conditions are met, it is possible to predict the distribution of the statistic when the null hypothesis is true, instead of using simulation. This same option of using a theory-based test is available when testing hypotheses about a population mean, as we will now see.

In Chapter 1 we used a theoretical probability distribution called a normal distribution to predict the distribution of sample proportions. As it turns out people typically don't predict the distribution of sample means directly, instead they predict the distribution of the standardized statistic:

$$standardized\ statistic = \frac{sample\ mean - hypothesized\ population\ mean}{standard\ deviation\ of sample\ mean}$$

The question is what to use for the standard deviation of the sample mean. It can be shown mathematically that $SD(\bar{x}) = \sigma/\sqrt{n}$. However, we don't usually know the value of σ, the population standard deviation. So we substitute in s, the sample standard deviation, and compute the **standard error** of the sample mean: $s/\sqrt{n}$. For the above data set, $s = 6.5$, so the standard error is $6.5/\sqrt{48} \approx 0.938$ seconds. This is similar to what we saw in the simulation, which reported the standard deviation of the 1000 hypothetical sample means to be 0.926.

> **Key idea:** When sampling from a large population, the standard deviation of the sample means is estimated by $s/\sqrt{n}$. The population is considered large enough when it is more than 10 or 20 times that sample size. This is another version of the **Central Limit Theorem** we saw in Chapter 1 but for quantitative data.

Thus, for this data set we have: *standardized statistic* $= \frac{\bar{x} - \mu_0}{s/\sqrt{n}} = \frac{13.71 - 10}{6.5/\sqrt{48}} = \frac{3.71}{0.94} = 3.95$

(where μ_0 is a symbol used to represent the value of the population mean when the null hypothesis is true.) As above, this feels like a large value for a standardized statistic. However, our reference distribution for this standardized statistic is no longer the normal distribution, but rather the *t-distribution* (and the standardized statistic is often called a ***t-statistic*Error! Reference source not found.**). The shape of a *t*-distribution is still a mound-shape curve, and it is centered at zero. It looks a lot like a normal distribution but it is a bit more spread out than a normal distribution (more data in the 'tails,' less in the middle). Figure 2.9 shows the distribution of the standardized statistics for the 1000 random samples from Figure 2.8 with the *t*-distribution overlaid. You should notice that the *t*-distribution extends further in each direction than a normal distribution. The good news is that *t*-statistics can be interpreted very similarly to *z*-statistics. Namely, values roughly greater than 2 (or less than -2) can be considered strong evidence against the null hypothesis.

Figure 2.9: The 1000 simulated *t*-statistics with a *t*-distribution overlaid

Obtaining the p-value from the *t*-distribution (0.0003) gives us a similar result to the simulation. We find a *t*-statistic as large as 3.95 (in either direction) in only about 3 in 10,000 random samples from a population with μ = 10 seconds. Figure 2.10 shows the output from the Theory-based inference applet for this dataset.

Figure 2.10: Theory-Based Inference for time estimate study

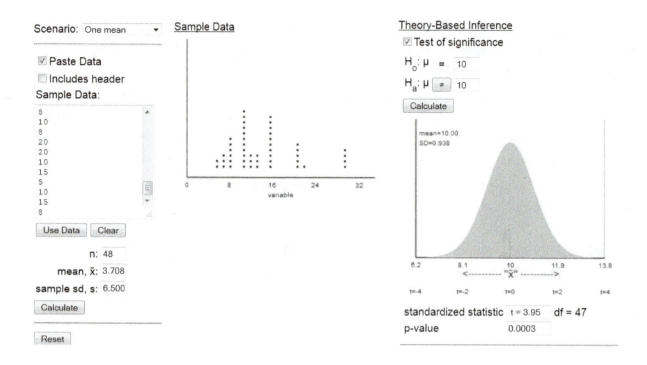

Validity Conditions

As we discussed in Chapter 1, theory-based tests are valid when certain conditions are met. In the case of a test on a single mean (also called a ***one-sample t-test***) the validity conditions are a little more complicated than they were for a single proportion.

> ***Validity Conditions***: The quantitative variable should have a symmetric distribution, or you should have at least 20 observations and the sample distribution should not be strongly skewed.

It's important to keep in mind that these conditions are rough guidelines and not a guarantee. All theory-based methods are approximations. They will work best when the distributions are symmetric, when the sample sizes are large, and there are no large outliers. When in doubt, use a simulation-based method as a cross-check. If the two methods give very different results you should probably consult a statistician. For more information see FAQ 2.2.1.

The sample size for the time estimates data is 48 and although the distribution is modestly skewed, it is not strongly skewed, so a theory-based approach should produce similar results to the simulation approach. As shown in Figure 2.10, the theory-based *t*-distribution matches the distribution of *t*-statistics from the simulation, and consequently the p-values are similar, and both are quite small.

So in general, the *t*-statistic for a test on a population mean can be found by plugging into this formula:

$$t = \frac{\bar{x} - \mu_0}{s/\sqrt{n}}$$

where μ_0 is a symbol used to represent the value of the population mean when the null hypothesis is true. Recall that *s* is the sample standard deviation and that *n* is the sample size. In particular, to find the *t*-statistic, subtract the hypothesized value (μ_0) from the sample statistic ($\bar{x}$) and then divide by the standard error of the statistic. When simulating we divide by the standard deviation of the simulated statistics, in the theory based approach shown here we divided by $s/\sqrt{n}$. (see FAQ 2.2.2 for more details).

FAQ 2.2.1 More about skewness and validity conditions

Q: You said that when my distribution is skewed, I need large *n* for the theory-based methods to work.

A: Right. The more skewed your distribution, the larger *n* needs to be.

Q: That's pretty vague. How skewed? How large?

A: You can summarize the skewness by how far apart the mean and median are: compute (mean – median)/SD.

Q: Why divide by the SD?

A: The SD adjusts distances according to the overall variability in the sample, similar to the idea of standardizing.

Q: OK. So what does (mean – median)/SD tell me?

A: Values near 0 come from symmetric distributions.

Q: And skewed distributions?

A: Values above 0.2 indicate extreme skewness.

Q: Let me guess: The 0.2 is arbitrary?

A: Yes, just a guideline.

Q: What do I do if my distribution is skewed?

A: You can often find a new scale that makes the data much more nearly symmetric. (For right-skewed data that can't be negative -- like incomes -- taking logarithms often works well.)

Q: Isn't it cheating to change the scale? I want a symmetric distribution but my actual data is skewed. So I change the scale. Lo and behold, I now have symmetric data.

A: Changing to a new scale doesn't change the facts, only the numbers that record those facts. Decibels, pH, and the Richter scale for earthquakes are examples of commonly used transformed scales.

Q: Will I get in trouble if I use the theory-based method anyway, without transforming?

A: There are other options. A statistical approach called bootstrapping is an option. But if you use the methods we show you here, then the answer depends on the accuracy you need. With extreme skewness and small samples, probabilities (p-values and coverage probabilities) may be off by as much as a factor of 2.

Q: Does this matter?

A: Sometimes not. Take p-values. If the theory-based method gives $p = 0.001$, you are safe to conclude that the evidence against H_0 very strong. Similarly, if $p = 0.20$, you are safe to conclude that there is no strong evidence against H_0. For p-values between 0.02 and 0.10, you should not trust the p-value with skewed data. Try a transformation, then use the theory-based method.

Q: And confidence intervals?

A: A confidence interval has two miss rates, one for the lower tail, one for the upper tail. For a 95% confidence intervals, both rates should be 2.5%. With highly skewed data, the rates could be off by a factor of 2 or more, with miss rates of 1% at one end and 5% at the other end. For some applications, this may not matter. If the theory-based interval claims to have a 95% coverage probability, you may be getting only 92%, but an approximate interval with high confidence may be good enough.

FAQ 2.2.2 More about z, t, and df

Q: Why do we use z for π but t for μ ?

A: The short answer is that we have two different kinds of data. For estimating π, we have binary data. For estimating μ, we have quantitative data.

Q: Is there a longer answer?

A: Here's part of it. For the coin toss model, there is only one unknown parameter, π. That single parameter determines both the center and the variability of the data distribution.

Q: (Oh, please, PLEASE, professor, call on me!) And for quantitative data, we have separate parameters, μ for the center, and σ for the variability.

A: Exactly. So we need separate estimates, $\bar{x}$ for μ, and s for σ.

Q: How is that related to z versus t?

A: If you don't know σ, you have to estimate it using s, and that means extra uncertainty about the variability of $\bar{x}$. That uncertainty makes your confidence intervals wider. The t-distributions have heavier tails than the normal.

Q: Heavier?

A: More probability away from the center.

Q: Is that why we need degrees of freedom?

A: Well done! The larger the df, the more the t-distribution looks like the normal. On the other hand, with only a few df, the tails are much heavier than for the normal, and the confidence intervals are much wider.

Q: Why do we subtract 1 to get the df? Why $n-1$?

A: Here's a short answer: It's just part of the definition of the t-distribution. The computer will take care of it. If you want, you can think of "df = (n-1)" as dorm food: Here it is; just eat it.

Q: And there's a longer answer, better than dorm food?

A: Glad you asked. There's always a longer answer: The df tells the number of "free" numbers.

Q: Glad I won't have to pay for them.

A: Not that kind of free. (And I don't blame you if the authors put corny words in your mouth. They do that to me, too.) To get serious, here's an example of free numbers. Take the numbers 1, 2 and 3 and compute the average.

Q: OK. The average is 2.

A: Now subtract the average from each of 1, 2, and 3.

Q: OK. I get -1, 0, and 1.

A: What do you notice about them?

Q: They add to zero.

A: Right. And that always happens: The deviations from an average always add to zero.

Q: What does this have to do with df?

A. We're almost there. Suppose I give you the first two deviations, the -1 and 0, and I tell you that all three deviations add to zero: $(-1) + (1) + (x) = 0$. There's only one possible value for x; it has to be 1.

Q: So once you know the first two deviations, the third one is fixed.

A: Exactly.

Q: And if I had n=5 deviations instead of n=3, there would be n-1 = 4 df?

A. Yes.

Q: And the bottom line?

A: It's the free numbers that tell you about σ. So it's the number of free numbers that tells you which t-distribution to use.

Q: Can you put that in a nutshell?

A: Here are the four key facts:
1. The df are df for using s to estimate σ.
2. We estimate σ using the set of n deviations from $\bar{x}$.
3. The deviations from $\bar{x}$ always add to zero
4. There are n deviations, but only $n-1$ df because the deviations add to zero. If you know all but one of the deviations, there's only one possible value for the last one.

Alternative Analysis: What about the median?

Think about it: Because we said that the mean may not be the best way to summarize the center of the distribution, explain how you could modify the 3S strategy to use the median.

If we wanted to assess whether the sample data provided convincing evidence that the population median was equal to 10 seconds, we could simply find the value of the sample median (in this case 12 seconds) and compare it to many, many simulated sample medians from a population where the median is actually 10 seconds and see how likely the sample median (12 seconds) or more extreme would be to occur by chance (random sampling). As this straightforward analysis suggests, our simulation is quite easy to modify by using a different statistic. Figure 2.11 shows a population distribution that we created with a population median of 10 seconds and the simulated null distribution of sample medians.

Figure 2.11: The distribution of song lengths in a hypothetical population of 6215 students alongside the simulated null distribution for the sample median

Table 2.7: Parallels between the real study and simulation

One repetition	=	A random sample of 48 students
Null Model	=	Population median = 10 seconds.
Statistic	=	Sample median estimate of song

We could use the simulated results shown above to count how many simulated sample medians are larger than 12 seconds or smaller than 8 seconds to estimate the two-sided p-value. As shown, our simulation indicates that we do not have very strong evidence against this null hypothesis (p-value = 0.1100); that is, we do not have strong evidence that the median is different than 10 in the population.

Another important point is that the simulated null distribution shown above has far from a 'nice' distribution, suggesting that theory-based approaches using mound-shaped, symmetric distributions may not be appropriate. This is, in fact, the case. Although there are some theory-based tests for a median, there is no question that the simulation approach is more straightforward and easy to implement!

Step 5. Formulate conclusions. The analysis of the sample mean gave strong evidence that the population mean differs from 10 seconds. But what is an appropriate population to generalize to from this study? This study was conducted on a convenience sample of 48 college students at a Midwestern college. First, we need to recognize that this was not a random sample, thus we cannot, with certainty, generalize to any particular population. However, even though this was not a random sample, it is doubtful that there was something systematically different about this group of 48 students in their time estimation ability than those of the entire student body at the college. In fact, with regards to estimating elapsed time, it might be reasonable to consider this sample as representative of all college students in the U.S., even though the students were selected from just one Midwestern college. Of course, we have no guarantee of this since we did not take a random sample.

As noted earlier, because the data are skewed, the mean may not be the best measure of center. In our simulation-based analysis, we did not find strong evidence that the population median differs from 10 seconds. Despite a different finding in terms of strength of evidence, the ability to generalize this conclusion to all U.S. college students is the same as for the mean, with the appropriate caveats described in the previous paragraph.

Step 6. Look back and ahead. Researchers are interested in learning more about why certain activities seem to pass faster than others. Interestingly, we showed that among this sample of 48 college students, the average perceived time of a song snippet was significantly longer than the actual song length (mean 13.71 seconds for a 10 second snippet). Although this finding is interesting, it suggests a number of additional questions and follow-ups. In particular,

    a.   The sample is a convenience sample, and so the conclusion we drew is about a population mean of college students well represented by our sample. Taking a random sample would be an improvement, potentially broadening the diversity in the population (beyond just college students).

    b.   It is interesting that the perception of elapsed time was longer than the actual time, but it would be interesting to compare the perception of time passed while listening to different songs, or completing different tasks to start suggesting why the perceived time was longer, on average, than reality. Relatedly, participants could be asked about how much they enjoyed the song/task if enjoyment is thought to be a factor impacting time perception.

    c.   Because the mean was statistically significant but not the median, we could further explore the skewness in the distribution. We also need to be cautious that we don't say "most students overestimate the length of the snippet." It's possible that a few long estimates are pulling up the value of the mean, whereas 10 seconds is still plausible for a "typical" student.

Exploration 2.2: Sleepless Nights?

Step 1. **Ask a research question.** How much do students at your school sleep on a typical night? Let's make the question more specific and ask about *last* night. Is the average less than 8 hours? How can we estimate this average?

1. Based on these questions, what is the population of interest? What is the parameter? What symbol do we use for this parameter?

Population:

Parameter:

Symbol:

Step 2. Design a study and collect data.

Let's investigate whether the mean amount of sleep last night for the population of all students at your school was less than 8 hours.

2. Express the null and alternative hypotheses for investigating this question.

Null hypothesis:

Alternative hypothesis:

To test these hypotheses we need to collect data

3. Ideally, how should you obtain a sample from the population of all students at your school?

4. What variable would you measure on each student?

Whereas obtaining a simple random sample of students from your school is an ideal way to attempt to get a representative sample, there is no question that it would take a fair bit of work to make that happen. So much so, that we might not want to do the study anymore! This happens frequently in statistics, and so researchers often opt to use a convenience sample instead. One choice of a convenience sample of students at your school is to use your class.

5. Identify at least one way that your class is likely a good representation of all students at your school in terms of hours slept last night

6. Identify at least one way that your class is likely not a good representation of all students at your school in terms of hours slept last night

7. Recall from the previous section that convenience sampling may be _Bias_ (biased/unbiased) whereas simple random sampling is _unBias_ (biased/unbiased).

We will revisit the implications of using your class as a sample of students from your college later when we evaluate the study. For now, let's gather the data.

8. How much sleep (to the nearest quarter hour) did you get last night? Combine your answer with those of your classmates.

9. Did you report a number of hours or a number of minutes slept?

Key Idea: When describing a quantitative variable, you should always be aware of the *measurement units* of the variable. Measurement units indicate information about the context and scaling of the variable (e.g., hours vs. days). Always include measurement units when describing your variable.

Step 3. Explore the data.

After you collect data, the next step is to explore the data. In the case of a quantitative variable (like we have here), exploring the data is a bit more involved than in Chapter 1. Recall from the Preliminaries that we can summarize a quantitative variable using a dotplot and then describe the shape, center, variability, and unusual observations.

Use the **Descriptive Statistics** applet to examine a dotplot of the sample data. To do this, press **Clear** to delete the existing data in the applet and then copy and paste your class data (with a one-word variable name) into the "Sample data" box and press **Use Data**.

Descriptive Statistics

☐ Stacked (Group Value)
☑ Includes header
Sample data:

```
SleepHrs
6
6.5
7
7.5
8
5
10
2.5
6
```

Use Data   Clear   Top/Bottom

Shape

Earlier in the book we described the shapes of distributions as either symmetric or not symmetric. When distributions are not symmetric, they can sometimes be described as skewed.

Definition: A distribution is called *skewed* if it is not symmetric, and, instead, the bulk of observations values tend to fall on one side of the distribution, with a "tail" on the other. Right skewed distributions have their tail on the right, and left skewed distributions have their tail on the left.

---

10. Describe the shape of the distribution of sleep times in the sample as symmetric, right-skewed, left-skewed, or something else.

Center

11. One way to summarize the center of a distribution is with the mean. Check the box next to "Actual" in the Mean row and record the value of the average sleep time for your class. What symbol can we use to represent this value? Also report the measurement units for the value.

When describing the center of a distribution, the mean is not always the best option. In particular, when the distribution is skewed, the tail of the distribution pulls the mean in that direction, meaning that, in some cases, a large proportion of the data values can be above or below the mean.

Key Idea: The more *right skewed* the distribution is, the larger the percentage of data values that are below the mean. Similarly, the more *left skewed* the distribution is, the larger the percentage of data values that are above the mean. (The mean is pulled toward the longer tail.)

---

12. For your class, how many students' sleep hours are above the mean? To find this count the dots on the dotplot above the red arrow, or simply look at the data you pasted into the applet.

13. What percentage of your class had sleep hours above the mean?

14. What does this suggest about the skewness of your class's sleep time distribution?

Because in skewed distributions a large percentage of data values may be above or below the mean, the *median* is sometimes a preferred way to measure the center of a distribution. The median is the middle number when the data values are sorted from smallest to largest. This implies that the median is always the number which splits the data in half so that half the data values are larger and half are smaller.

Definition: The *median* is the middle data value when the data are sorted in order from smallest to largest. The location of the median can be found by determining $(n+1)/2$, where n represents the sample size. When there are an odd number of data values, the median is the $(n+1)/2^{th}$ observation. When there is an even number of data values, the median is the average of the middle two numbers.

---

15. Before you use the applet to find the median sleep hours for your class, make sure you understand how to find the median by finding the median of another (small!) class of statistics students. Here are the sleep times of a small class of 6 statistics students: 6, 7.5, 5.5, 8, 6.5, 7.5

    What is the median for this small class? Find this by hand, don't use the applet!

16. Now, let's return to your class's data. Use the applet to find the median sleep time for your class by checking the "Actual" box in the Median row.

17. Do the mean and median for your class differ by much? Are the observed values of the sample mean and median preliminary evidence in favor of our alternative hypothesis?

Variability

The third way to describe a distribution is to discuss its variability. One way to summarize a distribution's variability is to report the standard deviation.

18. What is the standard deviation of sleep times for students in your class? Use the applet to find this by checking the "Actual" box next to Std dev.

19. Fill in the blanks to correctly interpret the standard deviation. "The standard deviation of _____ (sleep times/ average sleep times) for _____ (students in my class/all students at my school) is ___1,438___ (fill in the value) ___hous___ (give the measurement units), which means that, ___6.429___ (on average/typically/rarely) the sleep times of _____ (students in my class/all students at my school) were within _____ (fill in the value) ____ (give the measurement units) of the average (mean) sleep time of _____ (fill in the value)."

Unusual observations

The fourth typical part of summarizing a variable's distribution is to identify any unusual observations (outliers). Although there are some mathematical rules to define values that are outliers (e.g., a certain distance away from the bulk of the data), often times a subjective description of unusual observations is sufficient.

20. Are there any sleep hours in your class that you would characterize as unusual (outliers)? That is, sleep times that are far away from the bulk of the data?

2-hous

Resistance

Let's look at the mean vs. median choice in one more way.

21. In the "Sample data" box in the applet find the longest sleep hours value and change it to 20.

Find the new value of the mean: 6.857

How has this changed from what the mean was before? Why?

Find the new value of the median:_____

How has this changed from what the median was before? Why?

Find the new value of the standard deviation of sleep hours: 2.134

How has this changed from what the standard deviation was before? Why?

Definition: A summary statistic is *resistant* if its value does not change considerably when outliers are included in a dataset.

---

22. Circle which of the following three statistics are not resistant.

    Mean   Median   Standard Deviation

Step 4. Draw inferences beyond the data

Now that we have explored the data and better understand the distribution of last night's sleep hours in your class, we will learn how to use both simulation and theory-based approaches to evaluate the strength of the evidence against the null hypothesis.

Simulation-based approach: Applying the 3S strategy to hypotheses about a population mean

To evaluate the strength of evidence that your sample data provide against the null hypothesis we need to investigate how often a sample mean as small as the one observed in your sample would occur, if you were selecting a random sample from a population in which the mean sleep time really is 8 hours (null hypothesis). In essence, we will apply the 3S strategy as we have done before.

1. Statistic

> 23. Remind yourself of the average sleep hours last night in your class. $\bar{x}$ = _____

2. Simulate

First, let's *assume* that the **population** of sleep hours follows a normal distribution with mean $\mu = 8$ hours (as indicated under the null hypothesis) and standard deviation $\sigma = 1.5$ hours (probably in the ballpark to what you found in your sample). Note: we use the symbol σ to represent the population standard deviation.

We can then investigate whether the observed value of your sample mean is surprising to arise for a random sample from such a population.

To do this, we have manufactured a large, hypothetical population of many thousands of college students with these population parameter values (mean $\mu = 8$ hours and standard deviation $\sigma = 1.5$ hours). You will use the One Mean applet to take random samples of the same number of students from this population and evaluate how likely your class's mean would occur.

> 24. Open the **One Mean** applet. Notice that "Pop1" has been selected and a histogram displays its distribution. Verify that the population distribution looks to be normally distributed with mean very close to 8 hours and SD about 1.5 hours.
>
> Check the **Show Sampling Options** box. Keep **Number of Samples** set to one for now. Set the **Sample Size** to match your class survey. Press **Draw Samples**. Scroll down to see the ID numbers and sleep time values that were randomly selected from this hypothetical population.

Record the values of the sample mean for this sample from the 'Most Recent Sample' dotplot. Notice that this sample mean is added to the Statistic dotplot.

First simulated sample mean =

25. Press **Draw Samples** again. You will see the second sample mean added to the Statistic dotplot. The average of the two sample means is reported as the mean in that dotplot. Now change the **Number of Samples** to 998 (for 1000 samples total) and press **Draw Samples**.

    a. What is the shape of this null distribution?

    b. Where is it centered? Why does this make sense?

    c. What is the standard deviation (SD) of the null distribution?

3. Strength of evidence

26. Change the **Count Samples** pull-down menu to **Less Than** (to match the direction of the alternative hypothesis) and specify *the observed sample mean for your class* in the box. Press the **Count** button and report the approximate p-value.

    Based on this p-value, do your sample data provide strong evidence that the population mean sleep time at your school last night was less than 8 hours? Justify your answer, as if to a friend who has not studied statistics.

27. Obtain the value of a standardized statistic by using the same formula we used in Section 1.3:

$$\text{Standardized Statistic} = \frac{\text{Statistic} - \text{Mean of Null Distribution}}{\text{SD of Null Distribution}} =$$

28. Based on the value of the standardized statistic, summarize the strength of evidence provided by the sample data against the null hypothesis.

Theory-based Approach

In the previous two chapters you've seen that we can predict the shape of the null distribution using theory instead of using simulation, as long as certain validity conditions are met. We will now explore a theory-based approach for testing a single mean.

In Chapter 1 we used a theoretical probability distribution called a normal distribution to predict the distribution of sample proportions. As it turns out people typically don't predict the distribution of sample means directly, instead they predict the distribution of the standardized statistic. The mathematical distribution they found is called a *t-distribution*. The shape of a t-distribution is still a mound shape curve, but it is centered at zero. It looks a lot like a normal distribution but it is a bit more spread out than a normal distribution (more data in the 'tails' less in the middle). For this reason, the standardized statistic for testing hypotheses about a population mean is not a z-statitics (normal distribution), but is called a *t-statitics* (*t*-distribution).

Validity Conditions

As we discussed in Chapter 1, theory-based tests are valid when certain conditions are met. In the case of a test on a single mean (also called a *one-sample t-test*) the validity conditions are a little more complicated than they were for a single proportion.

> *Validity Conditions*: The quantitative variable should have a symmetric distribution or you should have at least 20 observations and the sample distribution should not be strongly skewed.

It's important to keep in mind that these conditions are rough guidelines and not a guarantee. All theory-based methods are approximations. They will work best when the distributions are symmetric, when the sample sizes are large and there are no large outliers. When in doubt, use a simulation-based method as a cross-check. If the two methods give very different results you should probably consult a statistician.

29. Would you consider the validity conditions met for population 1?

Formula for a *t*-statistic

30. The *t*-statistic for a test on a population mean can be found by plugging into this formula:

$$t = \frac{\bar{x} - \mu_0}{s/\sqrt{n}}$$

where μ_0 is a symbol used to represent the value of the population mean when the null hypothesis is true. Recall that s is the sample standard deviation and that n is the sample size.

Compute the value of the *t*-statistic for your class data.

31. The good news is that *t*-statistics can be interpreted very similarly to *z*-statistics. Namely, values greater than 2 (or less than -2) provide strong evidence against the null hypothesis. Interpret what the *t*-statistic is telling you about strength of evidence against the null hypothesis.

Notice that the formula for a *t*-statistic is very similar to how we computed the standardized statistic in the simulation. In particular, to find the *t*-statistic, subtract the hypothesized value (μ_0) from the sample statistic ($\bar{x}$) and then divide by something. When simulating we divide by the standard deviation of the simulated statistics, here we divide by $s/\sqrt{n}$. As it turns out $s/\sqrt{n}$ is a prediction of the standard deviation of the simulated statistics!

> **Key idea:** When sampling from a large population, the standard deviation of the sample means is estimated by $s/\sqrt{n}$. The population is considered large enough when it is more than 10 or 20 times that sample size and the validity conditions are met. This is another version of the ***Central Limit Theorem*** we saw in Chapter 1, but for quantitative data.

32. Compare the standard deviation of the null distribution you reported in #25 to the value of $s/\sqrt{n}$. Why should these values be similar?

33. In the **One Mean** applet, above the Statistic graph, change the choice of statistic from mean to *t*-statistic. Check the box to **Overlay *t* Distribution**. Does this theoretical model do a reasonable job of predicting the behavior of the simulated distribution of *t*-statistics? How does this relate to your answer to question #29?

34. Use the applet to count the number of simulated samples with a *t*-statistic less than (our alternative hypothesis) the observed value of your *t*-statistic to find the approximate p-value (based on *t*-statistics).

35. The theory-based p-value (one sample *t*-test) is also provided in the output. What is the theory-based p-value?

36. Why is the theory-based p-value similar to the p-value you got in #34 (using *t*-statistic) and in #26 (using the sample mean)?

Step 5. Formulate conclusions.

37. Think again about how the sample was selected from the population. Do you feel comfortable generalizing the results of your analysis to the population of all students at your school? Explain.

Step 6. Look back and ahead.

38. Looking back. Did anything about the design and conclusions of this study concern you?

Issues you may want to critique include:
- The match between the research question and the study design?
- How the observational units selected?
- How the measurements recorded?
- Whether what we observed is of practical value?

39. Looking ahead. What should the researchers' next steps be to fix the limitations or build on this knowledge?

<u>**Follow-up #1**</u>

To carry out the above analysis, we had to make up a hypothetical population to sample from. How sensitive is the analysis to that choice of population?

1. Simulate

40. Now let's assume a different shape for the population of sleep times. Use the radio button to select **Pop 2**.

    a. Describe how the shape of this population is different from the population in #24. Also report the population mean and standard deviation (SD).

b. Next determine an approximate p-value for your class results. Describe how you have done so and report this p-value. Based on this p-value, do your sample data provide strong evidence that the population mean sleep time at your school last night was less than 8 hours?

41. Does the theory-based p-value appear to be a valid approximation of the simulated p-value for population 2 and this sample size? Explain how you are deciding.

42. Did the assumed shape of the population (Pop 1 vs. Pop 2) make a substantial difference on the *distribution of sample means*? Justify your answer. (*Hint*: Think about the shape, center, and variability of the sample means.)

2. Strength of Evidence

43. Did the assumed shape of the population (Pop 1 vs. Pop 2) make a substantial difference on *the strength of evidence* against the null hypothesis? Explain.

So in fact, the simulation-based method may not be all that sensitive to the shape of the population that we use. One advantage of the *t*-procedure is that we don't have to generate a population to sample from at all, as long as the validity conditions are met.

Follow-up #2

Earlier in this section we talked about the potential advantages of the median compared to the mean in summarizing quantitative variables when the distribution of the variable is skewed. However, we analyzed the strength of evidence with regards to a population mean. What if we used the median instead?

44. Restate the null and alternative hypotheses in terms of a population median of 8 hours

45. Explain how you would modify the 3S strategy to test these hypotheses

    a) **Statistic.** What is the value of the new statistic?

    b) **Simulate.** How would you modify the simulation? (*Hint*: What information will you collect from each random sample?)

    c) **Strength of Evidence.** How would you modify your assessment of the strength of evidence?·

46. Use the **One Mean** applet to simulate 1000 random samples from population 1 using the **median** as the statistic. What do you conclude about the strength of evidence provided by your sample median against the null hypothesis?

47. Re-examine population 2. Could we use it as is to test whether there is evidence that our sample median came from such a population? Explain.

Whereas modifying the 3S strategy to test the median is not hard, a theory-based test for a median is more difficult and less common. Although there are some options to do a theory-based test for a median, there is no question that the simulation approach is more straightforward and easy to implement!

Section 2.2 Summary

In this section we explored simulation- and theory-based approaches for evaluating claims about a single population mean. These approaches are used to estimate the population mean or to assess whether the sample mean differs significantly from a particular hypothesized value for the population (or process) mean. We also explored the idea of skewness of a distribution, and the potential use of median as a measure of center which always divides the data in half (half the distribution larger/half the distribution smaller), unlike the mean which, for skewed distributions, gets pulled in the direction of the skew.

- The median is the middle value in an order set of data values; half of the values are larger and half are smaller.
- Skewed distributions (asymmetric distributions with a "tail") pull the mean away from the median in the direction of the skew.
- The median is resistant to outliers, but the mean is not resistant. In other words, the mean can be greatly affected by a single extreme value.
- Typically with quantitative data, the statistic is the sample mean, that is, $\bar{x}$, and the parameter of interest is the population mean, μ.
- The null hypothesis then asserts a particular conjectured value for the population mean.
- The theory-based approach for testing a single mean uses a t-statistic, which is a standardized statistic that follows a theoretical probability distribution known as a t-distribution; this theory-based approach is valid if the sample data are fairly symmetric OR if there are more than 20 observations and the sample data are not strongly skewed (see the table below).
- The formula for the standardized statistic is $t = \dfrac{\bar{x} - \mu_0}{s/\sqrt{n}}$.

Summary of validity conditions and applets for a test of a single mean

Type of data	Name of test	Applet	Validity conditions
Single quantitative variable	Theory-based test for a single mean (1 sample t-test)	One mean; Theory-based inference	The quantitative variable should have a symmetric distribution or you should have at least 20 observations and the sample distribution should not be strongly skewed.

Process vs. Population

In Section 2.1 we talked about the difference between an infinite process and a finite population. Here the sample data were selected from a finite population: students at a particular college. But, there are also times when quantitative variables can be measured on sample data from an infinite process. For example, the measurements on the eruption times of Old Faithful Geyser were from a process. Just like we saw in the previous section, although the parameter may be described slightly differently (the mean in the population vs. the long-run mean of the process), all the resulting inference tools (3S strategy; one-sample t-test, etc.) are the same!

Section 2.3: Errors and Significance

Introduction

We have seen that a test of significance assesses the *strength of evidence* that sample data provide against a null hypothesis, in favor of an alternative hypothesis. A slightly different approach is to use a test of significance to make a *decision* about whether the data provide sufficiently strong evidence to *reject* the null hypothesis. When we make such a decision, we naturally hope that we are making the correct one. But of course mistakes happen. There are two different types of errors that can be made in these situations. In this section, we will explore these two types of errors and see how the probability of making one of them can be determined.

Example 2.3: Heart Transplant Operations (continued)

In Example 1.3, we looked at a study published in the *British Medical Journal* (2004) in which it was reported that heart transplantations at St. George's Hospital in London had been suspended in September 2000, after a sudden spike in mortality rate. Of the last 10 heart transplants, 80% had resulted in deaths within 30 days of the transplant. Newspapers reported that this mortality rate was over five times the national average. Based on historical national data, the researchers used 15% as a reasonable value for comparison.

We tested whether the data observed at St. George's (8 deaths in 10 operations) provided strong evidence that the probability of death at St. George's exceeded the national value of 0.15. In other words, we tested the following hypotheses:

> Null hypothesis: Probability of dying within 30 days of a heart transplant at St. George's is the same as the national rate ($\pi = 0.15$).
> Alternative hypothesis: Probability of dying within 30 days of a heart transplant at St. George's is higher than the national rate($\pi > 0.15$).

Using the **One Proportion** applet, we simulated 1000 repetitions of 10 heart transplant cases, assuming 0.15 for the death probability. In none of those 1000 repetitions did we find 8 or more deaths. Our approximate p-value was therefore essentially zero, and we concluded that 8 deaths in 10 operations provided extremely strong evidence that the death rate at St. George's was higher than the national rate.

Notice that this conclusion is phrased in terms of strength of evidence. Another way to approach this situation is by considering whether the data provides sufficiently strong evidence to *reject* the null hypothesis (that the death rate at St. George's is 0.15) in favor of the alternative hypothesis (that the death rate at St. George's is greater than 0.15). In other words, this second approach is about making a *decision* rather than simply assessing strength of evidence. For example, suppose that some government agency is investigating another hospital and must decide whether or not to suspend the hospital from performing heart transplants or not. They need to make a yes or no decision, not simply characterize the results as moderately, strongly, or very strongly significant.

Table 2.8 gives the p-value for various possible outcomes for the observed numbers of deaths in 10 operations, investigating whether the probability of dying within 30 days of a heart transplant at St. George's is higher than the national rate of 15%.

Table 2.8 p-values for different numbers of deaths

Number of deaths	2	3	4	5	6	7	8
p-value	0.456	0.180	0.050	0.010	0.001	0.0001	0.00001

It seems clear that with only 2 or 3 deaths in 10 operations, there's not strong evidence to reject the hypothesis that the underlying death probability is 0.15. It also seems safe to conclude that the underlying death probability is indeed greater than 0.15 if the 10 operations were to result in 7 or 8 deaths. But for other values, the strength of evidence is less definitive. If we (or the government agency) need to make a yes or no decision, then we need to have a clear criterion in mind for deciding how small the p-value must be to provide convincing evidence against the null hypothesis, and we need to specify this criterion *before* collecting the data.

Think about it: Where would you draw the line? How small would the p-value have to be before you would be sufficiently convinced to reject the null hypothesis and conclude that the death rate is actually larger than 0.15?

Remember that the meaning of a p-value in this context is the probability of observing at least as many deaths in the sample, if in fact the probability of death at the hospital were 0.15. Let's say that you choose 0.05 as your cut-off value. Using 0.05 as the criterion for the p-value being small enough to reject the null hypothesis means that if the null hypothesis really were true, 5% of random samples would produce a sample proportion extreme enough to reject the null hypothesis (even though it actually is true). (From the above table, we see that getting 4 or more deaths in 10 operations happens about 5% of the time when π = 0.15.) So by specifying 0.05 as the cut-off, we are saying that we are willing to allow this type of error, where we reject the null hypothesis when it is true, to happen 5% of the time. This 0.05 or 5% criterion that we are using to determine whether or not to reject the null hypothesis is known as the **significance level**, often denoted by the Greek letter α (*"alpha"*).

Definition: The **significance level** is a value used as a criterion for deciding how small a p-value needs to be to provide convincing evidence to reject the null hypothesis.

Key idea: When the p-value is larger than the pre-specified significance level, we *do not have enough evidence to reject* the null hypothesis. When the p-value is less than or equal to the significance level, we *do have enough evidence to reject* the null hypothesis in favor of the alternative hypothesis.

Think about it: Suppose that you believe that a stricter standard is necessary before you would be convinced to reject the null hypothesis and conclude that St. George's death probability is actually larger than the national rate of 0.15. Would you choose a larger or smaller value for the significance level α? Why?

Achieving a stricter standard of evidence requires a *smaller* significance level. For example, if you want to insist on stronger evidence than at the 0.05 level before you'll be convinced to conclude that St. George's has a higher death rate than 0.15, then you could select 0.01 (or any other value less than 0.05) for the significance level.

Key idea: The significance level controls the probability that you mistakenly reject a null hypothesis that is true. To achieve a stricter standard for rejecting the null hypothesis, use a smaller significance level.

Type I and Type II errors

Two kinds of errors can occur with any test of significance. If the null hypothesis is actually *true* but the data lead a researcher to *reject* the null hypothesis, then the researcher has made an error. This error is sometimes called a *false alarm*, because the researcher believes that he/she has discovered an effect/difference when there is actually no effect/difference. On the other hand, if the null hypothesis is actually *false* but the data does *not* lead the researcher to reject the null hypothesis, then the researcher has made a different kind of error. This error is sometimes called a *missed opportunity*, because the researcher fails to detect an effect or difference that really is present. These errors are summarized in Table 2.9.

> **Definitions:** A *Type I error* (false alarm) occurs when a true null hypothesis is rejected. A *Type II error* (missed opportunity) occurs when a false null hypothesis is not rejected.

Table 2.9: A summary of Type I and Type II errors

		What is true (unbeknownst to us)	
		Null hypothesis is true	**Null hypothesis is false**
What we decide (based on data)	**Reject null hypothesis**	Type I error (false alarm)	Correct Decision
	Do not reject null hypothesis	Correct Decision	Type II error (missed opportunity)

To make a Type I error in the heart transplant example, we would conclude that the death rate within 30 days of a heart transplant at the hospital is higher than the national benchmark when in fact it is not. This is a false alarm, which would lead to an untrue allegation being propagated against the hospital. A consequence is that the hospital might not be allowed to continue performing heart transplant operations, even though its underlying death rate is no higher than the national rate. To make a Type II error we would not conclude the death rate at the hospital is higher than the national benchmark, when in reality this probability really is higher than the benchmark. This is a missed opportunity in the sense that researchers should have concluded that something unusual was happening at the hospital but they failed to do so. In real-world situations, we never know for sure whether we make an error or make a correct decision, because we never know for sure whether the null hypothesis is true or not. So it is important to consider the possible consequences of each type of error in advance.

Hopefully you can see that these are two very different errors. To get a better feeling for this, it may be helpful to relate them to a jury trial example. Perhaps you can even try to put yourself in the place of a juror or the defendant.

We mentioned briefly in Chapter 1 that significance testing can be thought of like a criminal trial. Suppose a person is charged with a crime and is given a trial. This person, the defendant, is either innocent or guilty. You can think of the null hypothesis being that the defendant is innocent, and the alternative being that the defendant is guilty. In the trial, we initially assume the defendant is innocent and the prosecution brings evidence to try to show that the defendant is guilty. In tests of significance, we initially assume the null hypothesis to be true and we try to bring evidence (in the form of data from a sample) against the null hypothesis and in favor of the alternative hypothesis. The jury then makes a conclusion. Hopefully the conclusion they make is correct. If it is not, then they have made one of two kinds of errors. Finding an innocent person guilty is equivalent to making a Type I error. Finding a guilty person not guilty is equivalent to making a Type II error. (See Table 2.10.)

Table 2.10: Type I and Type II errors summarized in context of jury trial

<table>
<tr>
<td rowspan="2" colspan="2"></td>
<td colspan="2">**What is true (unbeknownst to the jury)**</td>
</tr>
<tr>
<td>**Null hypothesis is true (defendant is innocent)**</td>
<td>**Null hypothesis is false (defendant is guilty)**</td>
</tr>
<tr>
<td rowspan="2">**What jury decides (based on evidence)**</td>
<td>**Reject null hypothesis (Jury finds defendant guilty)**</td>
<td>Type I error (false alarm)</td>
<td>Correct Decision</td>
</tr>
<tr>
<td>**Do not reject null hypothesis (Jury finds defendant not guilty)**</td>
<td>Correct Decision</td>
<td>Type II error (missed opportunity)</td>
</tr>
</table>

Our judicial system is set up to make the likelihood of making a Type I error (finding an innocent person guilty) as small as possible. We don't want to send innocent people to prison. However, by focusing on making the probability of a Type I error small, the likelihood of making a Type II error (finding a guilty person not guilty) can be large. The same holds true when doing a test of significance. When we make the probability of a Type I error smaller by selecting a smaller significance level, then we unfortunately make the probability of a Type II error larger.

Why does reducing the probability of a Type I error change the probability of a Type II error? Think of our criminal justice system, if we make it harder to convict someone, this will result in less innocent people in jail (lower significance level), however this will also result in more guilty people not being convicted because it is harder to be convicted. A guilty person going free is a Type II error. Hence, lowering the Type I error rate increases the Type II error rate.

Key idea: One tradeoff for selecting a lower significance level (probability of Type I error if null hypothesis is true), is that the probability of a Type II error will increase.

Power and the probability of a Type II error

Just like the significance level is the probability of making a Type I error, researchers can also find the probability of making a Type II error. However, typically, researchers don't talk about the probability of a Type II error directly, instead they choose to talk about the power of the test---which is the probability of rejecting the null hypothesis when the alternative hypothesis is true (a correct decision), which is simply one minus the probability of a Type II error.

Definition: The probability of rejecting a false null hypothesis is called the **power** of the test. Tests with higher values of power are preferred over tests with lower values of power. Power can also be thought of as one minus the probability of a Type II error.

For this reason, researchers would like their tests to be powerful, meaning the power of the test (probability of correctly rejecting the null hypothesis if the alternative hypothesis is true) is high. Later we'll talk more about how researchers can control the power of their test.

Exploration 2.3: Parapsychology Studies

Statistician Jessica Utts has conducted an extensive analysis of Ganzfeld studies that have investigated psychic functioning. Ganzfeld studies involve a "sender" and a "receiver." Two people are placed in separate acoustically-shielded rooms. The sender looks at a "target" image on a television screen (which may be a static photograph or a short movie segment playing repeatedly) and attempts to transmit information about the target to the receiver. The receiver is then shown four possible choices of targets, one of which is the correct target and the other three are "decoys." The receiver must choose the one s/he thinks best matches the description transmitted by the sender. If the correct target is chosen by the receiver, the session is a "hit." Otherwise, it is a miss. Utts reported that her analysis considered a total of 2124 sessions and found a total of 709 "hits" (Utts, 2010).

1. If the subjects in these studies have no psychic ability, what would be the long-run probability that the receiver would identify the correct target?

2. State the appropriate null and alternative hypotheses for testing whether the data provide strong evidence of psychic ability. Be sure to define the parameter clearly in your hypotheses.

3. Calculate the proportion of "hits" (successful transmissions of the target image) among the 2124 sessions that the researcher analyzed. Is this proportion larger than the null-hypothesized value for the probability of a successful transmission?

4. Use a simulation-based method to determine an approximate p-value for testing the hypotheses stated in #2).

5. Confirm that the validity conditions for a theory-based method are satisfied. Then use the theory-based method to calculate a standardized statistic and p-value for testing the hypotheses stated in #2).

6. Based on the p-values and standardized statistic, would you characterize the evidence against the null hypothesis as weak, moderate, strong, or very strong? Explain.

7. What if you had to make a yes or no decision about psychic ability based on these data? Would you decide to reject the null hypothesis of no psychic ability in favor of the alternative hypothesis that the subjects have some psychic ability? Why would you make this decision?

8. How small of a p-value would you need to see to convince you to reject the null hypothesis of no psychic ability in favor of the alternative hypothesis that the subjects have some psychic ability? In other words, what value would you use for a cut-off value for determining whether a p-value is small enough to provide convincing evidence to reject the null hypothesis?

Definition: The *significance level*, often denoted by the Greek letter α ("*alpha*"), is a value used as a criterion for deciding how small a p-value needs to be to provide convincing evidence to reject the null hypothesis.

> **Key idea:** When the p-value is larger than the pre-specified significance level, we *do not have enough evidence to reject* the null hypothesis. When the p-value is less than or equal to the significance level, we *do have enough evidence to reject* the null hypothesis in favor of the alternative hypothesis.

9. Using a 0.05 significance level, would you reject the null hypothesis of no psychic ability?

10. If your friend has a stricter criterion than you, requiring stronger evidence before deciding to reject the null hypothesis, would your friend use a smaller or a larger significance level than you? Explain.

> **Key idea**: The significance level controls the probability that you mistakenly reject a null hypothesis that is true. To achieve a stricter standard for rejecting the null hypothesis, use a smaller significance level.

Two kinds of errors can occur with any test of significance. If the null hypothesis is actually *true* but the data lead a researcher to *reject* the null hypothesis, then the researcher has made an error. This error is sometimes called a *false alarm*, because the researcher believes that he/she has discovered an effect/difference when there is actually no effect/difference. On the other hand, if the null hypothesis is actually *false* but the data does *not* lead the researcher to reject the null hypothesis, then the researcher has made a different kind of error. This error is sometimes called a *missed opportunity*, because the researcher fails to detect an effect/difference that really is present.

> **Definitions:** A *Type I error* (false alarm) occurs when a true null hypothesis is rejected. A *Type II error* (missed opportunity) occurs when a false null hypothesis is not rejected.

These types of errors can be represented in the Table 2.11:

Table 2.11: Type I and Type II errors summarized in context of jury trial

		What is true (unbeknownst to us)	
		Null hypothesis is true	**Null hypothesis is false**
What we decide	**Reject null hypothesis**	Type I error (false alarm)	Correct Decision

(based on data)	Do not reject null hypothesis	Correct Decision	Type II error (missed opportunity)

11. Using a 0.05 significance level, which type of error (false alarm or missed opportunity) could Utts and her colleagues conceivably have made in this ESP study? Explain your answer. (*Hint*: First ask whether the data led to rejecting the null hypothesis or not. Then ask what kind of error is possible with that decision.)

The article by Utts (2010) also refers to 56 individual studies that led to the combined total of 2124 sessions that you have analyzed here. Most of these studies involved a sample size of about 50 sessions. One particular study involved 50 sessions and resulted in 15 "hits."

12. Use a simulation or theory-based analysis to approximate the p-value for a study that produces 15 hits in 50 sessions. Is this p-value small enough to reject the null hypothesis of no psychic ability at the 0.05 significance level? What about the 0.10 level?

13. What type of error could you be making with your decision in #12? Explain what this error means in the context of this study.

14. What if the researchers used a very small significance level? Say, 0.0001. How does this decision affect the probability of making a Type I error? Explain this in context.

15. There is another consequence of the choice of a very small significance level as chosen in the previous question. Explain what you think will happen to the probability that the researchers find significant evidence of psychic ability if the true probability of a hit is greater than 25%? Why?

Key idea: One tradeoff for selecting a lower significance level (probability of Type I error if null hypothesis is true), is that the probability of a Type II error will increase.

Power and the probability of a Type II error

Just like the significance level is the probability of making a Type I error, researchers can also find the probability of making a Type II error. However, typically, researchers don't talk about the probability of a Type II error directly, instead they choose to talk about the power of the test---which is the probability of rejecting the null hypothesis when the alternative hypothesis is true (a correct decision), which is simply one minus the probability of a Type II error.

Definition: The probability of rejecting a false null hypothesis is called the **power** of the test. Tests with higher values of power are preferred over tests with lower values of power. Power can also be thought of as one minus the probability of a Type II error.

For this reason, researchers would like their tests to be powerful. In other words, researchers want the power of the test (probability of correctly rejecting the null hypothesis if the alternative hypothesis is true) to be high. Later we'll talk more about how researchers can control the power of their test.

16. If the probability of a hit is actually 33%, then which hypothesis is true? Which error is possible?

17. If the probability of a hit is actually 33%, the sample size is 50 and the researchers use a significance level of 5%, the Type II error rate is 77.5%. What is the power of this test? Name at least one thing the researchers could do to improve the power of this test.

Section 2.3 Summary

As opposed to assessing strength of evidence against a null hypothesis, another approach to tests of significance is to make a decision about whether or not the sample data provides sufficiently strong evidence to reject the null hypothesis. The key idea with this approach is to specify, before collecting the data, a **significance level** that will be regarded as a cut-off value for how small the p-value needs to be in order to provide convincing evidence to reject the null hypothesis. The smaller the significance level, the stronger the evidence that is required to reject the null hypothesis. See FAQ 2.3.1 for more details.

Two kinds of errors can occur with any test of significance. If the null hypothesis is actually *true* but the data lead a researcher to *reject* the null hypothesis, then the researcher has made a **Type I error**. We can call this a *false alarm*, because the researcher believes that he or she has discovered an effect or difference when there actually is none. On the other hand, if the null hypothesis is actually *false* but the data do *not* lead the researcher to reject the null hypothesis, then the researcher has made a **Type II error**.

This error is sometimes called a *missed opportunity*, because the researcher fails to detect an effect/difference that really is present.

- A **Type I error** (false alarm) occurs when a true null hypothesis is rejected.
- A **Type II error** (missed opportunity) occurs when a false null hypothesis is not rejected.

The significance level of the test determines the probability of committing a Type I Error. Selecting a smaller significance level makes it harder to reject the null hypothesis and produces a smaller probability of Type I error. Unfortunately, selecting a smaller significance level also increases the probability of making a Type II error.

FAQ 2.3.1   Hypothesis testing: p-values, α, and the two types of errors

Q: Didn't you say that "statistically significant" means "p-value ≤ 0.05?" That means from here on, I can just say "significant" if $p \leq 0.05$ and "not significant" if $p > 0.05$?

A: You're part right, but only part right. For many people, "statistically significant" does mean the same as $p \leq 0.05$, but **you should always report the actual p-value**.

Q: How come?

A: The p-value tells you more about the strength of evidence than just a simple Yes/No.

Q: Isn't the point just to decide whether the result is significant?

A: No. Scientific thinking is never as black-and-white as that. If *your* study gives $p = 0.05001$ and *my* study gives $p = 0.04999$ do you really think we should disagree about the strength of evidence?

Q: I get the point, but what's the right conclusion? Isn't there a rule?

A: Sorry, the only way I can give you a rule is by being arbitrary. You can decide to reject if $p \leq \alpha$, but your value of α is arbitrary. (Science is a way of thinking, not a machine.)

Q: First you tell me the rule says to reject H_0 if $p \leq 0.05$, then you tell me the rule is arbitrary.

A: Guilty as charged.

Q: How do I make sense of that?

A: The problem comes from trying to turn a quantitative variable (the *p*-value) into a binary variable (reject/don't reject). Would you call a 7 foot person tall?

Q: Of course.

A: What about 6'4"? or 5'11"? or 5'7"? You get the idea: "Tall" is a useful concept, but the actual height gives more information.

Q: OK. If I believe what you say, I have a new question: Why don't we just report the *p*-value and stop there? Why pick an arbitrary α?

A: Good question. Do you remember Type I and Type II errors?

Q: Sure. Type I is a false alarm, and Type II is a missed opportunity. What's that got to do with my question?

A: The Type I error rate is the *probability* that you reject H_0 when H_0 is true.

Q: I think I see. You can't talk about Type I error *rates* unless you have a rule that tells you when to reject.

A: Exactly. Those error rates, and power, are useful concepts, but they are based on having a rule, and any rule will be somewhat arbitrary.

Q: Can you wrap it up for me?

A: You can think in terms of two different goals.

Goals:

• If your goal is to study how a method works in the long run, you need a rule that tells when to reject.

• If your goal is to learn from a particular data set, the *p*-value tells the strength of evidence, but any rejection rule is arbitrary.

Q: While we're on the subject of errors, which is worse, Type I or Type II?

A: Remember the analogy: Significance testing is like our criminal justice system. It's worse to declare an innocent person guilty than let a guilty person go free. A false alarm is worse than a missed opportunity. You don't want to build a scientific theory on results that won't stand the test of time.

Q: Can you say more?

A: Here's another analogy. You can think of a significance test as a kind of spam filter for weeding out likely dead ends -- research hypotheses that are not supported by data. The logic is that you don't want to spend your time and resources investigating results that could be due purely to random chance.

Q: What about the missed opportunities? According to the court analogy, once a guilty person is found innocent, they're in the clear. Doesn't that mean lots of valid research hypotheses get overlooked?

A: That does happen. The court and spam analogies are useful, but they ignore an important difference between a decision and a scientific conclusion. When a court reaches its final decision, that's a decision; it's really final. In science, conclusions are always contingent. They get refined, modified, and woven into a larger context.

Q: That's the last step in the Six Step process, right? Look ahead to what's next.

A: Bingo. Go to the head of the class. Just don't let your head swell.

Chapter 2 Glossary

Exercises

Section 2.1

2.1.1.* The population will always be _____ the sample.

    A. At least as large as

    B. Smaller than

2.1.2. In most statistical studies the _____ is unknown and the _____ is known.

    A. Parameter/Statistic

    B. Statistic/Parameter

2.1.3.* The reason for taking a random sample instead of a convenience sample is that...

    A. ...random samples tend to be smaller and so take less time to collect

    B. ...random samples tend to represent the population of interest

    C. ...random samples always have 100% participation rates

    D. ...random samples tend to be easier to implement and be successful

2.1.4. True or false

    A. Larger samples are always better than smaller samples, regardless of how the sample was collected

    B. Larger random samples are always better than smaller random samples

    C. You shouldn't take a random sample of more than 10% of the population size

2.1.5.* True or false

    A. Random samples only generate unbiased estimates of categorical variables (not quantitative)

    B. Non-random samples are always biased

    C. There is no way that a sample of 100 people can be representative of all adults living in the United States

2.1.6. When stating null and alternative hypotheses, the hypotheses are

    A. Always about the parameter only

    B. Always about the statistic only

    C. Always about both the statistic and the parameter

    D. Sometimes about the statistic and sometimes about the parameter

2.1.7.* When using simulation or theory-based methods to test hypotheses about a proportion, the process of computing a p-value is

      A. different if the sample is from a process instead of from a finite population

      B. the same if the sample is from a process instead of from a finite population

      C. sometimes different and sometimes the same if the sample is from a process instead of a finite population

School Cafeteria

2.1.8. Argue whether or not you believe using a sample of students from your school's cafeteria (you recruit the next 100 people to visit the cafeteria to participate) may or may not yield biased estimates based on the variable being measured/research question being investigated in each of the following situations.

    a) Using the proportion of students with type O blood to learn about the proportion of U.S. adults with type O blood.

    b) Using the proportion of students who eat fast-food regularly to learn about the proportion of U.S. college students who eat fast-food regularly

    c) Using the proportion of students who have brown hair to learn about the proportion of all students at your school who have brown hair

    d) Using the proportion of students who have brown hair to learn about the proportion of all U.S. adults who have brown hair

Sample of birds

2.1.9.* Argue whether or not you believe using a sample of birds which visit a bird feeder in your yard (the next 50 birds that visit) may or may not yield biased estimates based on the variable being measured/research question being investigated in each of the following situations.

    a) Using the proportion of finches that visit the bird feeder to estimate the proportion of finches among all birds that live near your backyard

    b) Using the proportion of finches that are at the bird feeder when another finch is also at the bird feeder to estimate the proportion of the time that finches prefer to eat with other finches

    c) Using the proportion of male birds that visit the bird feeder to estimate the proportion of male birds among all birds that live near your backyard

    d) Using the proportion of male birds that visit the bird feeder to estimate the proportion of male birds that typically visit bird feeders

Class survey

In order to estimate the typical amount of TV watched per day by students at her school of 1000 students, a student has all of the students in her statistics class (30 students) take a short survey, finding that, on average, students in her statistics class report watching 1.2 hours of television per day.

2.1.10. Use the previous information to answer the following questions

    a) What is the population of interest?

    b) Do you believe that the average hours of television per day in the sample is likely less than, similar to, or greater than the average hours of television watched per day in the population? Why?

2.1.11. Use the previous information to answer the following questions about the class survey

    a) Identify the parameter

    b) Identify the statistic

2.1.12. Use the previous information to answer the following questions about the class survey

    a) Identify the variable measured on each student.

    b) Is the variable categorical or quantitative?

    c) Identify at least two statistics that the student could use to summarize the variable

    d) Identify at least one graph that the student could use to summarize the variable

2.1.13. In the survey the student also asked students whether or not they watched at least 10 minutes of TV yesterday. The student found that 21 of 30 students reported watching at least 10 minutes TV yesterday.

    a) Identify the variable measured on each student

    b) Is the variable categorical or quantitative?

    c) Identify at least one statistic that the student could use to summarize the variable

    d) Identify at least one graph that the student could use to summarize the variable

2.1.14. Using the information from the previous question, the student wishes to test whether there is evidence that more than 50% of student's at the school watched at least 10 minutes of TV yesterday.

    a) State the null and alternative hypotheses for this test

    b) Describe the parameter of interest for this research question

    c) What is the value of the statistic?

2.1.15. Evaluate the strength of evidence for the hypotheses in the previous question

    a) Find the p-value for the hypotheses in the previous question using a simulation-based approach.

    b) Based on the p-value evaluate the strength of evidence and state a conclusion about the TV watching among students yesterday.

    c) To which population, if any, are you comfortable drawing your conclusion. Why?

    d) If a theory-based approach would be reasonable for this data, find the p-value and comment on the similarity of the p-value from the theory-based approach to the p-value you found in A. If a theory-based based approach would not be reasonable for this data explain why not.

2.1.16. Among the students in her class, the information for three particular students is given here:

    (i) Alejandra reported watching 2 hours of TV a day, but not watching TV yesterday

    (ii) Ben reported watching 4 hours of TV a day and watching TV yesterday

    (iii) Cassie reported watching 30 minutes of TV a day and not watching TV yesterday

Use this information to fill in the blank data-table below, where the last row (…) indicates that additional information is present in the real data table since there were actually 30 students in the survey.

…	…	…

2.1.17. If the study was conducted by taking a random sample, explain (in enough detail that someone could follow your directions to obtain the sample) how it could have been done. If the study was not conducted by taking a random sample, explain how you could modify the study design to make it be a random sample, and explain (in enough detail that someone could follow your directions to obtain the sample) how it could have been done.

Political survey*

Problems 2.1.18 to 2.1.25 refer to the following study.

In order to estimate the proportion of all likely voters to vote for the incumbent in the upcoming city's mayoral race, a random sample of 267 likely voters is taken, finding that 65% state they will likely vote for the incumbent.

2.1.18. Use the previous information to answer the following questions

    a) What is the population of interest?

    b) Do you believe that the proportion of likely voters for the incumbent in the sample is likely less than, similar to, or greater than the proportion of likely voters who would say they will vote for the incumbent in the population? Why?

2.1.19. Use the previous information to answer the following questions about the political survey

    a) Identify the parameter

    b) Identify the statistic

2.1.20. Use the previous information to answer the following questions about the political survey

    a) Identify the variable measured on each likely voter.

    b) Is the variable categorical or quantitative?

    c) Identify at least one statistic that the polling agency could use to summarize the variable

    d) Identify at least one graph that the polling agency could use to summarize the variable

2.1.21. The polling agency wishes to test whether there is evidence that more than 50% of likely voters will likely vote for the incumbent.

    a) State the null and alternative hypotheses for this test

    b) Describe the parameter of interest for this research question?

    c) What is the value of the statistic?

2.1.22. Evaluate the strength of evidence for the hypotheses in the previous question

   a) Find the p-value for the hypotheses in the previous question using a simulation-based approach.
   b) Based on the p-value evaluate the strength of evidence and state a conclusion about the likely voting outcome in the mayoral race if the election were to take place today.
   c) To which population, if any, are you comfortable drawing your conclusion. Why?
   d) If a theory-based approach would be reasonable for this data, find the p-value and comment on the similarity of the p-value from the theory-based approach to the p-value you found in A. If a theory-based based approach would not be reasonable for this data explain why not.

2.1.23. The poll also asked voters to report the time the respondent has spent reading or learning about local politics over the last week. The poll finds that, on average, respondents have spent 42 minutes learning or reading about local politics over the last week.

   a) Identify the variable measured on each likely voter
   b) Is the variable categorical or quantitative?
   c) Identify at least two statistics that the polling agency could use to summarize the variable
   d) Identify at least one graph that the polling agency could use to summarize the variable

2.1.24. Among the sample, the information for three particular likely voters is given here:
   (i) Likely voter #1 reported spending 0 minutes in the last week learning/reading about local politics and likely voting for the incumbent
   (ii) Likely voter #2 reported spending 0 minutes in the last week learning/reading about local politics and likely voting for the incumbent
   (iii) Likely voter #3 reported spending 60 minutes in the last week learning/reading about local politics and likely voting for the challenger (not the incumbent)

Use this information to fill in the blank data-table below, where the last row (…) indicates that additional information is present in the real data table since there were actually 267 likely voters in the survey.

…	…	…

2.1.25. If the study was conducted by taking a random sample, explain (in enough detail that someone could follow your directions to obtain the sample) how it could have been done. If the study was not conducted by taking a random sample, explain how you could modify the study design to make it be a random sample, and explain (in enough detail that someone could follow your directions to obtain the sample) how it could have been done.

Television news survey

Problems 2.1.26 to 2.1.33 refer to the following study.

In order to understand more about how people in the United States feel about the outcome of a recent criminal trial in which the defendant was found not guilty, a television news program invites viewers to go the news programs website and indicate their opinion about the event. At the end of the show 82% of the people who voted in the poll indicated they were unhappy with the verdict.

2.1.26. Use the previous information to answer the following questions

   a) What is the population of interest?
   b) Do you believe that the proportion of people unhappy with the verdict in the sample is likely less than, similar to, or greater than the proportion of individuals unhappy with the verdict in the population? Why?

2.1.27. Use the previous information to answer the following questions about the news program survey

   a) Identify the parameter
   b) Identify the statistic

2.1.28. Use the previous information to answer the following questions about the news program

   a) Identify the variable measured on each participant in the survey.
   b) Is the variable categorical or quantitative?
   c) Identify at least one statistic that the news program could use to summarize the variable
   d) Identify at least one graph that the news program could use to summarize the variable

2.1.29. The news program wishes to test whether there is evidence that more than three-quarters of US adults are opposed to the verdict. Five hundred and sixty-two participants participated in the survey.

    a) State the null and alternative hypotheses for this test

    b) Describe the parameter of interest for this research question?

    c) What is the value of the statistic?

2.1.30. Evaluate the strength of evidence for the hypotheses in the previous question

    a) Find the p-value for the hypotheses in the previous question using a simulation-based approach.

    b) Based on the p-value evaluate the strength of evidence and state a conclusion about the opinions of US adults about the verdict.

    c) To which population, if any, are you comfortable drawing your conclusion. Why?

    d) If a theory-based approach would be reasonable for this data, find the p-value and comment on the similarity of the p-value from the theory-based approach to the p-value you found in A. If a theory-based based approach would not be reasonable for this data explain why not.

2.1.31. The survey also asked participants to report the time the respondent spent reading or watching news coverage about the trial during the last 3 days. The poll finds that, on average, respondents have spent 92 minutes reading or watching news coverage about the trial during the last 3 days.

    a) Identify the variable measured on each respondent

    b) Is the variable categorical or quantitative?

    c) Identify at least two statistics that the news program could use to summarize the variable

    d) Identify at least one graph that the news program could use to summarize the variable

2.1.32. Among the sample, the information for three particular respondents is given here:

    (i) Respondent #1 reported spending 240 minutes in the last three days reading/watching news coverage about the trial and not being happy with the verdict.

    (ii) Respondent #2 reported spending 90 minutes in the last three days reading/watching news coverage about the trial and not being happy with the verdict.

    (iii) Respondent #3 reported spending 30 minutes in the last three days reading/watching news coverage about the trial and being happy with the verdict

Use this information to fill in the blank data-table below, where the last row (…) indicates that additional information is present in the real data table since there were actually 562 likely voters in the survey.

…	…	…

2.1.33. If the study was conducted by taking a random sample, explain (in enough detail that someone could follow your directions to obtain the sample) how it could have been done. If the study was not conducted by taking a random sample, explain how you could modify the study design to make it be a random sample, and explain (in enough detail that someone could follow your directions to obtain the sample) how it could have been done.

Shark survey*

Problems 2.1.34 to 2.1.41 refer to the following study.

A zoologist at a large metropolitan zoo is concerned about a potential new disease present among the 243 sharks living in the large aquarium at the zoo. The zoologist takes a random sample of 15 sharks from the aquarium, temporarily removes the sharks from the tank and tests them for the disease. He finds that 3 of the sharks have the disease.

2.1.34. Use the previous information to answer the following questions

    a) What is the population of interest?

    b) Do you believe that the proportion of diseased sharks in the sample is likely less than, similar to, or greater than the proportion of diseased sharks in the population? Why?

2.1.35. Use the previous information to answer the following questions about the shark survey

    a) Identify the parameter

    b) Identify the statistic

2.1.36. Use the previous information to answer the following questions about the disease

    a) Identify the variable measured on each shark.

    b) Is the variable categorical or quantitative?

    c) Identify at least one statistic that the zoologist could use to summarize the variable

    d) Identify at least one graph that the zoologist could use to summarize the variable

2.1.37. The zoologist wishes to test whether there is evidence that less than ¼ of sharks in the aquarium are diseased.

    a) State the null and alternative hypotheses for this test

    b) Describe the parameter of interest for this research question?

    c) What is the value of the statistic?

2.1.38. Evaluate the strength of evidence for the hypotheses in the previous question

    a) Find the p-value for the hypotheses in the previous question using a simulation-based approach.

    b) Based on the p-value evaluate the strength of evidence and state a conclusion about diseased sharks.

    c) To which population, if any, are you comfortable drawing your conclusion. Why?

    d) If a theory-based approach would be reasonable for this data, find the p-value and comment on the similarity of the p-value from the theory-based approach to the p-value you found in A. If a theory-based based approach would not be reasonable for this data explain why not.

2.1.39. Because the disease affects a shark's ability to extract oxygen from the water, the zoologist also measures the shark's blood oxygen content. The zoologist finds that, on average, shark's blood oxygen content is 4.2 percent.

    a) Identify the variable measured on each shark

    b) Is the variable categorical or quantitative?

    c) Identify at least two statistics that the zoologist could use to summarize the variable

    d) Identify at least one graph that the zoologist could use to summarize the variable

2.1.40. Among the sample, the information for three particular sharks is given here:

    (i) Shark #1 had the disease and had a blood oxygen level of 1.2 percent

(ii) Shark #2 did not have the disease and had a blood oxygen level of 5.6 percent

(iii)   Shark #3 did not have the disease and had a blood oxygen level of 6.2 percent

Use this information to fill in the blank data-table below, where the last row (…) indicates that additional information is present in the real data table since there were actually 15 sharks in the sample

…	…	…

2.1.41. If the study was conducted by taking a random sample, explain (in enough detail that someone could follow your directions to obtain the sample) how it could have been done. If the study was not conducted by taking a random sample, explain how you could modify the study design to make it be a random sample, and explain (in enough detail that someone could follow your directions to obtain the sample) how it could have been done.

School survey

Problems 2.1.42 and 2.1.43 use the following information.

In order to investigate how many hours a day students at their school tend to spend on course work outside of regularly scheduled class time, a statistics student takes a random sample of 150 students from their school by randomly choosing names from a list of all full-time students at their school that semester. The student finds that the average reported daily study hours among the 150 students is 2.23 hours.

2.1.42. Identify each of the following by describing it or stating its value (where possible)

a) Population of interest

b) Sample

c) Parameter of interest

d) Statistic

2.1.43. In the previous question, explain whether or not you believe the sample is representative of the population of interest and why.

Business survey*

Problems 2.1.44 and 2.1.45 use the following information.

In order to understand reasons why consumers visit their store, a local business conducts a survey by asking the next 100 people who visit their store to fill out a short survey. The business finds that 40 of the 100 people state that the main reason they visited the store was because the store is running a sale on coats that week.

2.1.44. Identify each of the following by describing it or stating its value (where possible)

      a) Population of interest
      b) Sample
      c) Parameter of Interest
      d) Statistic

2.1.45. In the previous question, explain whether or not you believe the sample is representative of the population of interest and why.

Section 2.2

2.2.1* On January 28, 1986, the Space Shuttle Challenger broke apart 73 seconds into its flight, killing all seven astronauts on board. All investigations into reasons for the disaster pointed towards the failure of an O-ring in the rocket's engine. Given below is a dotplot and some descriptive statistics on O-ring temperature (°F) for each test firing or actual launch of the shuttle rocket engine.

a) The numeric values of two possible measures of center are calculated to be 65.86°F and 67.50°F. Which one of these is the mean, and which the median? How are you deciding?

b) If we removed the observation 31°F from the dataset, how would the following numerical statistics change, if at all?

Mean:	SMALLER	SAME	LARGER
Median:	SMALLER	SAME	LARGER
Standard deviation:	SMALLER	SAME	LARGER

Salaries

Problems 2.2.2 to 2.2.5 use the same data.

2.2.2. The monthly salaries of the 3 people working in a small firm are: $3500, $4000, and $4500. Suppose the firm makes a profit and everyone gets a $100 raise, how, if at all, would the **average** of the 3 salaries change?

   a) The average would stay the same.
   b) The average would increase.
   c) The average would decrease.
   d) Cannot be answered without doing calculations.

2.2.3 Reconsider the previous question. The monthly salaries of the 3 people working in a small firm are: $3500, $4000, and $4500. Suppose the firm makes a profit and everyone gets a $100 raise, how, if at all, would the **standard deviation** of the 3 salaries change?

   a) The standard deviation would stay the same.
   b) The standard deviation would increase.
   c) The standard deviation would decrease.
   d) Cannot be answered without doing calculations.

2.2.4 Reconsider the previous question where the monthly salaries of the 3 people working in a small firm are: $3500, $4000, and $4500. If instead of a $100 raise, everyone gets a 10% raise, how, if at all, would the **average** of the 3 salaries change?

   a) The average would stay the same.
   b) The average would increase.
   c) The average would decrease.

        d) Cannot be answered without doing calculations.

2.2.5   Reconsider the previous question where the monthly salaries of the 3 people working in a small firm are: $3500, $4000, and $4500. If instead of a $100 raise, everyone gets a 10% raise, how, if at all, would the **standard deviation** of the 3 salaries change?

        a) The standard deviation would stay the same.

        b) The standard deviation would increase.

        c) The standard deviation would decrease.

        d) Cannot be answered without doing calculations.

Exam scores*

2.2.6:

a) Suppose that an instructor decides to add 5 points to every student's exam score in a class. What effect would this have on the mean exam score? On the median exam score? What effect would this have on the standard deviation of exam scores? Explain.

b) Suppose that an instructor decides to add 5 points only to the exam score for the highest-scoring student in the class. What effect would this have on the mean exam score? On the median exam score? What effect would this have on the standard deviation of exam scores? Explain.

c) Suppose that an instructor decides to add 5 points only to the exam score for the lowest-scoring student in the class. What effect would this have on the mean exam score? On the median exam score? What effect would this have on the standard deviation of exam scores? Explain.

States visited

2.2.7: An instructor collected data on the number of states students in her class have visited, in the U.S. Here is the dotplot for the data collected.

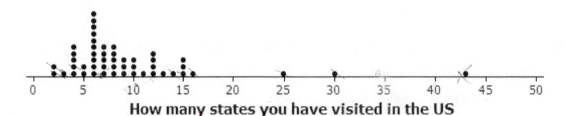

How many states you have visited in the US

a) Identify the observational units.

b) Identify the variable recorded, and whether it is categorical or quantitative.

c) Describe what the graph tells us about distribution of the variable recorded. Be sure to comment on shape, center and variability.

d) Use the dotplot to find and report the median value for the number of states visited by the students in this study.

e) Would the mean value for these data be smaller than, larger than, or the same as the median, as reported in (d)? Explain your reasoning

f) Suppose that the observation recorded as 43 states is a typo, and was meant to be 34. If we corrected this entry in the dataset, how would the following numerical statistics change, if at all?

Mean: ✗ SMALLER   SAME   LARGER

Median: SMALLER   ✗ SAME   LARGER

Standard deviation: ✗SMALLER   SAME   LARGER

Birth weights*

2.2.8 Birth weights of babies in the United States can be modeled by a normal distribution with mean 3250 grams and standard deviation 550 grams. Based on this information, which of following is *more unlikely*? Circle one.

A. That a randomly selected baby has a birth weight greater than 4000grams

B. That a random sample of 10 babies has an average birth weight greater than 4000grams

C. Both are equally likely

D. Cannot be answered without doing calculations

Heights

2.2.9 In which scenario would you expect to see more variability in the data: heights of a random sample of 100 college students, or heights of a random sample of 500 college students?

a) Random sample of 100 college students

b) Random sample of 500 college students

c) Both samples will have similar variability

d) Cannot be answered without doing calculations

Haircuts*

2.2.10 Students in a statistics class were asked the cost of their last haircut. The dataset can be found on the textbook website as **HaircutCosts**. Put this dataset into the **Descriptive Statistics** applet and answer the following questions.

    a) Is the distribution of haircut costs symmetric, skewed to the left, skewed to the right, or something else?

    b) Would you think the mean haircut cost is about the same as the median, higher than the median, or lower than the median? Why?

    c) Find and report both the mean and median haircut costs. Were you correct in your answer to the previous question?

    d) The highest haircut cost is $150 and there were three of these. Suppose one of them actually had a haircut cost of $300. If this value were changed, would it change the mean? Would it change the median? (Change one of the $150 to $300 and verify your answer.)

How much TV do you watch?

Problems 2.2.11 to 2.2.15 refer to the same study.

2.2.11 According to a 2011 report by the United States Department of Labor, civilian Americans spend 2.75 hours per day watching television. A faculty researcher, Dr. Sameer, at California Polytechnic State University (Cal Poly) conducts a study to see whether a different average applies to Cal Poly students.

    a)   Identify the variable of interest, and whether the variable is categorical or quantitative.

    b)   Describe Dr. Sameer's parameter of interest, and assign an appropriate symbol to denote it.

    c)   Write the appropriate hypotheses using symbols.

2.2.12 Reconsider Dr. Sameer's research question about the how much time Cal Poly students spend on watching television. Suppose that Dr. Sameer surveys a random sample of 100 Cal Poly students, and for this sample the mean number of hours per day spent watching TV turns out to be 3.01 hours.

    a) Is the number 3.01 a parameter or a statistic?

    b) Assign an appropriate symbol to the number 3.01.

    c) Describe how you can conduct a simulation-based test of significance to investigate whether the data provide evidence that the average number of hours per day Cal Poly students spend watching TV is different from 2.75 hours. Be sure to provide details on how one would find a p-value.

    d) For the following table, complete the column on the right to draw parallels between the simulation and the real study:

Simulation		Real study
One repetition	=	
Null model	=	
Statistic	=	

2.2.13 Reconsider Dr. Sameer's research question about the how much time Cal Poly students spend on watching television.

a) Suppose that on analyzing the data from the survey of a random sample of 100 Cal Poly students, the p-value for Dr. Sameer's study was computed to be 0.16. Interpret what this p-value means in the context of the study. (*Hint:* This question is *not* asking for a *conclusion* from the p-value.)

b) Another faculty researcher, Dr. Elliot, had hypothesized that Cal Poly students spend more than 2.75 hours/day watching TV, on average. How would this researcher's p-value compare to the p-value reported in (a)? Explain how you are deciding. (*Hint:* Bigger, smaller, or about the same? If bigger or smaller, by how much?)

2.2.14 Reconsider Dr. Sameer's research question about the how much time Cal Poly students spend on watching television. Suppose that for the random sample of 100 Cal Poly students the mean number of hours per day spent watching TV turns out to be 3.01 hours, and the standard deviation of the number of hours per day spent watching TV turns out to be 1.97 hours.

a) Is the number 1.97 a parameter or a statistic? Assign an appropriate symbol to this number.

b) Find the standardized statistic to investigate whether the data provide evidence that the average number of hours per day Cal Poly students spend watching TV is different from 2.75 hours.

c) What is your conclusion about the hypotheses, based on the calculated value of the standardized statistic? How are you deciding?

2.2.15 Reconsider Dr. Sameer's research question about the how much time Cal Poly students spend on watching television, and recall that for the random sample of 100 Cal Poly students' number of hours per day spent watching TV the mean and standard deviation turned out to be 3.01 hours and 1.97 hours, respectively.

a) In the context of this study, is the theory-based approach (one-sample t-test) valid to find a p-value to investigate whether the data provide evidence that the average number of hours per day Cal Poly students spend watching TV is different from 2.75 hours? Explain.

b) Regardless of your answer in (a), use a one-sample t-test to find and report a standardized statistic (*t*-statistic) and a p-value for Dr. Sameer's hypotheses.

    c)   State your conclusion from this study based on the p-value reported in (b), being sure to provide justification for your conclusion.

What's the SPF of your sunscreen?*

Problems 2.2.16 to 2.2.18 refer to the following sunscreen data.

2.2.16   Most dermatologists recommend using sunscreens that have a Sun Protection Factor (SPF) of at least 30. One of the authors wanted to find out whether the SPF of sunscreens used by students at her school (which is in a very sunny part of the United States), exceeds this value, on average?

    a)   Identify the variable of interest, and whether the variable is categorical or quantitative.
    b)   Describe the author's parameter of interest, and assign an appropriate symbol to denote it.
    c)   Write the appropriate hypotheses using symbols.

To collect data, the author surveyed students in her introductory Statistics course, and found that in a sample of 48 students, the average SPF was 35.29, and the standard deviation of SPF was 17.19.

    d)   Assign appropriate symbols to the numbers 48, 35.29, and 17.19.
    e)   Do the data come from a *random* sample? Explain.
    f)   Do you think the students in the introductory Statistics course are representative of all students at this school, with regard to SPF of their sunscreens? Why or why not?
    g)   Describe a population for which this sample is surely *not* representative, with regard to SPF of sunscreens used.

2.2.17   Reconsider the study about one of the authors wanting to find out whether the SPF of sunscreens used by students at her school (which is in a very sunny part of the United States), exceeds 30, on average?

    a)   Describe how you can conduct a simulation-based test of significance to investigate whether the data provide evidence that the average SPF of sunscreens used by students at this school exceeds 30. Be sure to provide details on how one would find a p-value.
    b)   For the following table, complete the column on the right to draw parallels between the simulation and the real study:

Simulation		Real study
One repetition	=	
Null model	=	
Statistic	=	

2.2.18 Reconsider the study about one of the authors wanting to find out whether the SPF of sunscreens used by students at her school (which is in a very sunny part of the United States), exceeds 30, on average?

a) In the context of this study, is the theory-based approach (one-sample t-test) valid to find a p-value to investigate whether the data provide evidence that the average SPF of sunscreens used by students at this school exceeds 30? Explain.

b) Regardless of your answer in (a), use a one-sample t-test to find and report a standardized statistic (*t*-statistic) and a p-value.

c) Interpret the p-value reported in (b).

d) State your conclusion from this study based on the p-value reported in (b), being sure to provide justification for your conclusion.

Needles!

Problems 2.2.19 to 2.2.21 refer to the following data on needles.

2.2.19 Consider a manufacturing process that is producing hypodermic needles that will be used for blood donations. These needles need to have a diameter of 1.65 mm – too big and they would hurt the donor (even more than usual), too small and they would rupture the red blood cells, rendering the donated blood useless. Thus, the manufacturing process would have to be closely monitored to detect any significant departures from the desired diameter. During every shift, quality control personnel take a random sample of several needles and measure their diameters. If they discover a problem, they will stop the manufacturing process until it is corrected.

a) Identify the variable of interest, and whether the variable is categorical or quantitative.

b) Write the appropriate hypotheses using appropriate symbols, to test whether the average diameter of needles from the manufacturing process is different from the desired value.

c) Suppose that the most recent random sample of 35 needles have an average diameter of 1.64 mm, and a standard deviation of 0.07mm. Assign appropriate symbols to these numbers.

d) Suppose that the diameters of needles produced by this manufacturing process have a bell-shaped distribution. Sketch the distribution of the average diameter of samples of 35 needles, assuming that the process is not malfunctioning. Be sure to clearly label the axis of the graph, and provide values for what you think the mean and standard deviation for this distribution should be.

2.2.20 Consider the investigation of the manufacturing process that is producing hypodermic needles.

a) Describe how you can conduct a simulation-based test of significance to investigate whether the data provide evidence that the average diameter of needles produced by this manufacturing process is different from 1.65mm. Be sure to provide details on how one would find a p-value.

b) For the following table, complete the column on the right to draw parallels between the simulation and the real study:

Simulation		Real study
One repetition	=	
Null model	=	
Statistic	=	

2.2.21 Consider the investigation of the manufacturing process that is producing hypodermic needles. Recall that the most recent random sample of 35 needles have an average diameter of 1.64 mm, and a standard deviation of 0.07mm.

a) Use the **Theory-Based Inference** applet to find and report a standardized statistic (t-statistic) and a p-value to investigate whether the average diameter of needles produced by this manufacturing process is different from 1.65mm.

b) Based on your p-value, write out a conclusion, being sure to include justification.

How cool are you?*

2.2.22 The dataset **MaleTemps** (found on the textbook website) consists of 65 body temperatures from healthy male volunteers aged 18 to 40 that were participating in vaccine trials. Put this dataset into the **Descriptive Statistics** applet and answer the following questions.

a) Is the distribution of temperatures symmetric, skewed to the left, skewed to the right, or something else?

b) Would you think the mean temperature is about the same as the median, higher than the median, or lower than the median? Why?

c) Find and report both the mean and median temperatures. Were you correct in your answer to the previous question?

d) Remember that standard deviation can be described as roughly the average distance between a data value in the distribution and the mean of the distribution. Check the **Guess** box next to the **Std dev** box in the applet. You should see a box centered on the mean. Adjust the "length" of this box with your mouse so it extends as wide as what you think is one standard deviation in each direction based on the description of standard deviation given above. Now check the **Show actual** box for the standard deviation. What is the actual standard deviation? Where you accurate in your guess?

2.2.23 Normal (or average) body temperature of humans is often thought to be 98.6° F. Is that number really the average? To test this, we will use the **MaleTemps** dataset (available from the textbook website) from the previous exercise.

a) Write out the null and alternative hypotheses for our test in words or in symbols.

b) You should have found (in the previous exercise) the sample mean body temperature for the 65 males in our sample to be 98.105°F and the standard deviation to be 0.699°F. Use the **Theory-Based Inference** applet to find and report a standardized statistic (t-statistic) and a p-value for the test.

c) Based on your p-value, write out a conclusion, being sure to include justification.

d) The data did not come from a random sample, but a convenience sample of healthy adults that were involved in a vaccine study. Given that information, to what population do you think we can generalize our results?

2.2.24 In the same study as was mentioned in the previous two exercises, data were also collected on 65 healthy female volunteers aged 18 to 40 that were participating in the vaccine trials. The dataset **FemaleTemps** consisting of body temperatures from the 65 females is available from the textbook website. You will use the data to investigate whether the average body temperature of healthy adult females is different from 98.6° F.

a) Write out the null and alternative hypotheses in words or in symbols.

b) A dotplot of the 65 body temperatures is shown below. Based on this dotplot, does it appear the average body temperature is different than 98.6° F? How are you deciding?

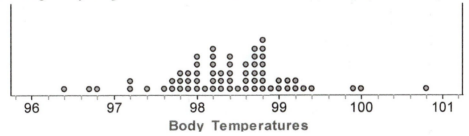

c) The sample mean body temperature for the 65 females in our sample is 98.39° F and the standard deviation is 0.743° F. Use the **Theory-Based Inference** applet to find and report a standardized statistic (*t*-statistic) and a p-value for the test.

d) Based on your p-value, write out a conclusion, being sure to include justification.

Commute time

2.2.25 According to a Pew Research Center report from 2012, the average commute time to work in California is 27.5 minutes. To investigate whether a different average is true for the small city that she lives in, a California high school student surveys 45 people she knows – her teachers, her parents and their friends and co-workers, and finds the average commute time for this sample to be 24.33 minutes, with a standard deviation of 9.53 minutes.

a) Write out the null and alternative hypotheses for our test in words or in symbols.

b) Use the **Theory-Based Inference** applet to find and report a standardized statistic (*t*-statistic) and a p-value for the test.

c) Based on your p-value, write out a conclusion, being sure to include justification.

d) The data did not come from a random sample, but a convenience sample of people to whom the student had access. Given that information, to what population do you think we can generalize our results?

Section 2.3

2.3.1* Suppose that we conduct a test of significance using 5% as our level of significance.

a) What is the probability of making a Type I error (if the null hypothesis is true)?

b) Suppose we conduct 10 different tests of significance each using a 5% significance level (and again where all the null hypotheses are true). How do you think the probability of making at least one Type I error in those ten tests will compare your answer to the previous question: will it be the same, be smaller, or be larger? Explain.

2.3.2 Indicate whether or not you would reject the null hypothesis, at the $\alpha = 0.05$ significance level, for the following p-values.

a) p-value = 0.078 – cannot reject

b) p-value = 0.045

c) p-value = 0.001

d) p-value = 0.051 reject

2.3.3* Indicate whether or not you would reject the null hypothesis, with a p-value of .064, for the following significance levels.

    a) $\alpha = 0.05$

    b) $\alpha = 0.10$

    c) $\alpha = 0.01$

    d) $\alpha = 0.065$

2.3.4 Suppose that you perform a significance test and, based on the p-value, decide to reject the null hypothesis at the $\alpha = 0.05$ significance level. Than suppose that your colleague decides to conduct the same test on the same data but using a different significance level. For each of the following significance levels, indicate whether you would reject the null hypothesis, fail to reject the null hypothesis, or do not have enough information to say. (Hint: First ask yourself what must be true about the *p*-value based on your decision to reject the null hypothesis at the $\alpha = 0.05$ significance level.)

    a) $\alpha = 0.10$ yes

    b) $\alpha = 0.01$

    c) $\alpha = 0.0001$

    d) $\alpha = 0.065$

2.3.5* Suppose that you perform a significance test and, based on the p-value, decide to reject the null hypothesis at the $\alpha = 0.01$ significance level. Than suppose that your colleague decides to conduct the same test on the same data but using a different significance level. For each of the following significance levels, indicate whether you would reject the null hypothesis, fail to reject the null hypothesis, or do not have enough information to say. (Hint: First ask yourself what must be true about the *p*-value.)

    a) $\alpha = 0.10$

    b) $\alpha = 0.05$

    c) $\alpha = 0.0001$

    d) $\alpha = 0.065$

2.3.6* Suppose that you perform a significance test using the $\alpha = 0.05$ significance level.

    a) For what *p*-values would you reject the null hypothesis?

    b) For what *p*-values would you fail to reject the null hypothesis?

2.3.7 Suppose that you perform a significance test and obtain a *p*-value of 0.036.

    a) For what significance levels α would you reject the null hypothesis?

    b) For what significance levels α would you fail to reject the null hypothesis?

2.3.8* A researcher decides to set the significance level to 0.001

      a. If the null hypothesis is true, what is the probability of a Type I error?

      b. If the researcher chooses a larger significance level, what will be the impact on the probability of a Type II error, assuming the alternative hypothesis is true?

2.3.9 A researcher decides to set the significance level to 0.10.

      a. If the null hypothesis is true, what is the probability of a Type I error?

      b. If the researcher chooses a smaller significance level, what will be the impact on the probability of a Type II error, assuming the alternative hypothesis is true?

Doris and Buzz*

Problems 2.3.10 and 2.3.11 refer to the Doris and Buzz study.

2.3.10 In Example 1.1, we looked at a study to investigate whether dolphins could communicate. In doing so, we tested whether Buzz, one of the dolphins, could push the correct button more than 50% of the time in the long run.

   a) Describe what a Type I error would be in this study.

   b) Describe what a Type II error would be in this study.

2.3.11 Reconsider dolphins Buzz and Doris from Example 1.1 and the previous exercise. We found that the observed data provided very strong evidence that Doris and Buzz were actually communicating better than random chance. Which type of error (I or II) could we possibly be making with this conclusion? What does this type of error mean in this context?

Harley the dog

Problems 2.3.15 and 2.3.16 refer to Harley the dog data.

2.3.12 In Exploration 1.1, we looked at a study to investigate whether Harley the dog could select the correct cup more than 50% of the time in the long run.

   a) Describe what a Type I error would be in this study.

   b) Describe what a Type II error would be in this study.

2.3.13 Reconsider Harley the dog from Exploration 1.1 and the previous exercise. We found that the observed data provided very strong evidence that Harley the dog was doing better than random chance. Which type of error (I or II) could we possibly be making with this conclusion? What does this type of error mean in this context?

Error tradeoff*

2.3.14 The significance level α determines the probability of making a Type I error. Errors are bad. So, why don't we always set alpha to be extremely small, such as .0001?

Needles

Problems 2.3.15 and 2.3.16 refer to needle data

2.3.15 Consider a manufacturing process that is producing hypodermic needles that will be used for blood donations. These needles need to have a diameter of 1.65 mm – too big and they would hurt the donor (even more than usual), too small and they would rupture the red blood cells, rendering the donated blood useless. Thus, the manufacturing process would have to be closely monitored to detect any significant departures from the desired diameter. During every shift, quality control personnel take a sample of several needles and measure their diameters. If they discover a problem, they will stop the manufacturing process until it is corrected.
   a)   Define the parameter of interest in the context of this study, and assign an appropriate symbol to it.
   b)   State the appropriate null and alternative hypotheses, using the symbol defined in a).
   c)   Describe what a Type I error would be in this study. Also, describe the consequence of such an error in the context of this study.
   d)   Describe what a Type II error would be in this study. Also, describe the consequence of such an error in the context of this study.

2.3.16 Recall that the two errors discussed in this chapter were regarded as "false alarm" and "missed opportunity." In some fields, these errors are also regarded as "consumer's risk" and "producer's risk." With regard to the previous exercise about the hypodermic needle manufacturing process, do you think the Type I error is the consumer's risk or the producer's risk? What about Type II error?

Campus newspaper*

2.3.17 Suppose that you are considering whether to publish a weekly alternative newspaper on campus. You decide to survey a random sample of students on your campus to ask if they would be likely to read such a newspaper. Your plan is to proceed with publication only if the sample data provide strong evidence that more than 10% of all students on your campus would be likely to read such a newspaper.
a) Identify the parameter of interest, in words.
b) Express the appropriate null and alternative hypotheses for conducting this test.
c) Specify what Type I error represents in this situation. Also indicate a potential consequence of Type I error.
d) Specify what Type II error represents in this situation. Also indicate a potential consequence of Type II error.

Diagnostic tests

2.3.18 As with a jury trial, another analogy to hypothesis testing involves medical diagnostic tests. These tests aim to indicate whether or not the patient has a particular disease. But the tests are not infallible, so errors can be made. The null hypothesis can be regarded as the patient being healthy. The alternative hypothesis can be regarded as the patient having the disease.
a) Describe what Type I error represents in this situation. Also indicate a potential consequence of Type I error.
b) Describe what Type II error represents in this situation. Also indicate a potential consequence of Type II error.
c) Which type of error would you consider to be more serious in this situation? Explain your thinking.

Lie detectors*

2.3.19 Lie detector tests are similar to hypothesis tests in that there are two possible decisions and two possible realities, and therefore two kinds of errors that can be made. The hypothesis can be considered as:

H_0: Subject is actually telling the truth.

H_a: Subject actually lying.
a) Describe what rejecting H_0 means in this context.
b) Describe what failing to reject H_0 means in this context.
c) Describe what Type I error means in this context.
d) Describe what Type II error means in this context.

Spam filters

2.3.20 Spam filters in an email program are similar to hypothesis tests in that there are two possible decisions and two possible realities, and therefore two kinds of errors that can be made. The hypothesis can be considered as:

H_0: Incoming email message is legitimate.

H_a: Incoming email message is spam.

a) Describe what rejecting H_0 means in this context.

b) Describe what failing to reject H_0 means in this context.

c) Describe what Type I error means in this context.

d) Describe what Type II error means in this context.

Treatments*

2.321 Later in the book you will encounter many hypotheses of the following type:

H_0: New treatment is *no* better than current treatment.

H_a: New treatment *is* better than current treatment.

a) Describe what Type I error means in this context.

b) Describe what Type II error means in this context.

Gender differences

2.3.22 Later in the book you will encounter many hypotheses of the following type:

H_0: Men and women do *not* differ on average with regard to the variable of interest.

H_a: Men and women *do* differ on average with regard to the variable of interest.

a) Describe what Type I error means in this context.

b) Describe what Type II error means in this context.

Organ donation*

2.3.23 Example P.1 describes a study that compared proportions of people who agreed to become organ donors, depending on the type of default option used. The null hypothesis is that all three default options have the same probability of a person agreeing to become an organ donor. The alternative hypothesis is that at least two of the default options have different probability of a person agreeing to become an organ donor.

a) Describe what Type I error means in this context.
b) Describe what Type II error means in this context.

Election winners

2.3.24 Reconsider Example 1.4 about whether winners of elections can be predicted better than random chance simply by considering the faces of the candidates.
   a) Describe what Type I error means in this context.
   b) Describe what Type II error means in this context.
   c) Based on the p-value in that study (which produced 23 successful predictions in 32 elections), would you reject the null hypothesis at the .01 significance level?
   d) Based on your answer to part c), which type of error could you possibly be making?

Trick or treat*

2.3.25 Reconsider Example 1.5 about whether trick-or-treaters have an overall preference between Halloween toys or candy.
   a) Describe what Type I error means in this context.
   b) Describe what Type II error means in this context.
   c) Based on the p-value in that study (in which 148 trick-or-treaters in the sample selected candy and 135 selected toys), would you reject the null hypothesis at the .10 significance level?
   d) Based on your answer to part c), which type of error could you possibly be making?

End of Chapter

Major league baseball games*
2.CE.1 Suppose that you select Major League Baseball (MLB) games played during the 2nd week in April as a sample of all MLB games played in the upcoming season.
   a) Would this be a simple random sample? Explain briefly.
   b) Following up on part a), consider two variables that you could record about each game: attendance and number of runs scored. With which of these two variables is the sample (of games played during the 2nd week in April) more likely to be representative of the population? Explain briefly.

Sampling students

2.CE.2 Suppose that you select a sample of students at your university by standing in front of the library and asking students who pass by to take a survey.

  a) Would this be a simple random sample from the population of all students at your university? Explain briefly.

  b) Name a variable for which you believe that this sample would be representative of the population of all students at your university. Explain briefly.

  c) Name a variable for which you believe that this sample would *not* be representative of the population of all students at your university. Explain briefly.

Big bang theory*

2.CE.3 A reader wrote in to the "Ask Marilyn" column in *Parade* magazine to say that his grandfather told him that in 3/4 of all baseball games, the winning team scores more runs in one inning than the losing team scores in the entire game. (This phenomenon is known as a "big bang.") Marilyn responded that this proportion seemed to be too high to be believable. To investigate this claim, an instructor examined the 45 Major League baseball games played on September 17–19, 2010.

  a) Was this a random sample of all Major League baseball games played in the 2010 season? Explain briefly.

  b) Restate the grandfather's assertion as a null hypothesis and Marilyn's response as an alternative hypothesis, in both symbols and in words.

The instructor found that 21 of the 45 games played on those dates contained a big bang.

  c) Calculate the sample proportion of games that had a big bang, and denote it with the appropriate symbol.

  d) Assume (for now) that the grandfather's claim is true, and use the One-Proportion applet to simulate 1000 samples of 45 games per sample. Describe how to calculate the approximate p-value from this simulation, and report its value.

  e) Would you conclude that the sample data provide strong evidence to support Marilyn's contention that the proportion cited by the grandfather is too high to be the actual value? Explain your reasoning, as if writing to the grandfather, who has never taken a statistics course.

Reading on the bowl

2.CE.4 A study published in 2009 (Goldstein et. al., 2009) involved a sample of 499 adults selected from various regions across the country of Israel. These adults were approached by interviewers in public gathering areas and asked to complete a survey about whether or not they read while using the toilet. The researchers found that 263 of the 499 adults responded that they do read while using the toilet.

    a) Calculate the sample proportion who responded that they read while using the toilet.

    b) Is this proportion a parameter or a statistic? Explain, and indicate the appropriate symbol to represent this proportion.

    c) Sketch a graph to display the data gathered from the study. What kind of graph did you sketch?

    d) Assuming half of all persons in Israel read while using the toilet, describe how to calculate a p-value from these data. Use the simulation-based One Proportion applet to calculate an approximate p-value. Based on this p-value, would you conclude that the population proportion who read while using the toilet differs from one-half? Explain.

    e) To what population do you feel comfortable generalizing the results of this study? For example, would you generalize the results to all adults in Israel? How about all adults in the world, or in another country such as the U.S.? Explain.

Keeping their names*

Problems 2.CE.5 and 2.CE.6 refer to the following data.

2.CE.5 What percentage of U.S. brides keep their own names after marriage, as opposed to taking their husband's name or using some modification (such as hyphenation) of her name and her husband's? Researchers investigated this question by selecting a sample of wedding announcements in the *New York Times* newspaper (Kopelman et. al., 2009). They found that 18% of a sample of 600 brides sampled between 2001 and 2005 kept their own names.

    a) Identify the observational units and variable in this study.

    b) Identify the population and sample in this study.

    c) Identify (in words) the parameter of interest.

    d) Based on how the sample was selected, do you have any concerns about generalizing the sample result to the population of all brides in the U.S. between 2001 and 2005? To what population would you feel comfortable generalizing the result? Explain.

2.CE.6 Refer to problem **2.CE.5**. Suppose you are testing to see if the population proportion of all brides who keep their own name is different from 15%.

a) If testing at a significance level of 5%, would you find evidence for the population proportion of brides between 2001 and 2005 who kept their own names to differ from 15%? Explain using the p-value you found.

b) If testing at a significance level of 10%, would you find evidence for the population proportion of brides between 2001 and 2005 who kept their own names to differ from 15%? Explain using the p-value you found.

c) If testing at a significance level of 1%, would you find evidence for the population proportion of brides between 2001 and 2005 who kept their own names to differ from 15%? Explain using the p-value you found.

Biased samples

Problems 2.CE.7 and 2.CE.8 refer to biased samples.

2.CE.7 Which samples do you think are non-biased samples of the population of all students at a college/four-year university? If it is a biased sample, is there a different population for which you believe the sample could be considered a non-biased sample?

a) A random sample of 100 students from the registrar's student list.
b) A random sample of 100 students in the library on a Thursday night.
c) A random sample of 100 students in the student center on a Friday night.
d) A random sample of 100 students at a basketball game on a Tuesday night.
e) A random sample of 10 students from 10 randomly chosen dorms on campus.
f) A random sample of 100 students from cars registered on campus.

2.CE.8 Biased samples either over-estimate or underestimate the parameter of interest. List the samples from exercise **2.CE.7**, you think might over-estimate the following parameters

a) Average student GPA
b) Proportion of male students
c) Proportion of female students
d) Proportion of out-of-state students

Body temperatures*

Problems 2.CE.9 and 2.CE.10 refer to body temperatures.

2.CE.9 Body temperatures were measured on a random sample of 20 females and 20 males applying for health insurance. Together, their average temperature was 98.8 with a standard deviation of 0.2.

    a) Are validity conditions met to carry out a theory-based test of significance to see if adult body temperature is different from 98.6? Explain why or why not. Is there any other information you would like to know?

    b) Carry out a theory-based test of a single mean.

    c) What is the value of the standardized test statistic & what does this tell you about your sample mean?

    d) Does your answer to past c coincide with your answer to part b? Explain.

2.CE.10 Refer to problem **2.CE.9**.

    a) Based on the test results found in **2.CE.9** part b), if you are testing at a significance level of 5% and you have made an error. What type of error have you made? (Type I or Type II)

    b) If you are testing at the 5% level of significance, and you have made a Type II error, what can you say about the p-value of the test?

    c) Name two ways in which you could decrease the probability of a Type I error.

Investigation: Faking Cell Phone Calls

Have you ever pretended to be talking on your cell phone in order to avoid interacting with people around you? Is faking cell phone calls a common practice among cell phone users? A recent survey conducted by the Pew Research Center during April 26 – May 22, 2011 asked cell phone users about this issue and many others regarding the respondents' cell phone usage in the past 30 days. The survey involved selecting a random sample of 1,858 American cell phone users.

    1. Was the sample random?

    2. Was the sampling method unbiased?

3. Do you feel comfortable generalizing your conclusions to all cell phone users? If not, is there a
    broader population you feel you can generalize your conclusions to?

Suppose instead your data came from a random sample of 1,858 college students.

4. Was the sample random?

5. Was the sampling method unbiased?

6. Do you feel comfortable generalizing your conclusions to all cell phone users? If not, is there a
    broader population you feel you can generalize your conclusions to?

Now suppose you asked 1,858 people walking on Wall Street throughout the day on a Monday.

7. Was the sample random?

8. Was the sampling method unbiased?

9. Do you feel comfortable generalizing your conclusions to all cell phone users? If not, is there a
    broader population you feel you can generalize your conclusions to?

A reporter for *International Business Times* took the Pew survey results and wrote that more than 1 in 10
cell phone users in the U.S. has engaged in such fake cell phone use in the past 30 days. Notice that the
reporter's claim is about the population of all cell phone users in the U.S. In the following questions, we
will investigate whether the survey results provide evidence that more than 1 in 10 cell phone users have
faked cell phone calls in the last 30 days.

Step 1: Ask a research question

   10. What is our research question?

Step 2: Design a study and collect data

   11. What are the observational units?

   12. What is the variable that is measured/recorded on each observational unit?

   13. Describe the parameter of interest in words. (You can use the symbol π to represent this
         parameter.)

   14. State the appropriate null and alternative hypotheses to be tested.

Of the 1,858 cell phone users, 13% admitted to faking cell phone call in the past 30 days.

   15. If another sample of 1,858 American cell phone users were surveyed, could the percent admitting
         to faking cell phone calls in the past 30 days, change? Explain your reasoning.

Step 3: Explore the data

   16. Is the 13% a statistic or a parameter? How are you deciding?

Step 4: Draw inferences

Let's use the 3S strategy to help us investigate how much evidence the sample data provide to support the conjecture more than 1 in 10 cell phone users fake cell phone calls.

Statistic

17.   What is the statistic that you can use to summarize the data collected in the study?

Simulate

18. If we assume that the population proportion of cell phone users who fake cell phone calls is actually 0.10. Is it possible that we could observe the statistic we did from this sample of 1858 cell phone users? Why?

19. Use the One Proportion applet to simulate 1000 repetitions of this study, assuming that the proportion of cell phone users who fake calls is 0.10. Report what values you input into the applet.

20. What is the center of your simulated null distribution? Does it make sense that this is the center? Explain.

21. Are there any values of *simulated* sample proportions that are less than 0.10? Are there any values of *simulated* sample proportions that are greater than 0.10? What does that tell you?

Strength of evidence

22. Based on the null distribution generated using the 1000 simulated values of the statistic, what values would you consider typical values and what would you consider atypical values of the statistic?

23. How do the actual study results compare to the null distribution obtained when simulating assuming 0.10 of the population faked cell phone calls? Do you believe the study results provide convincing evidence against the "1 in 10 cell phone users fake cell phone calls" null hypothesis and in favor of the "more than 1 in 10 cell phone users fake cell phone calls" alternative hypothesis? Why or why not?

24. Determine the approximate p-value from your simulation analysis. Also interpret what this p-value represents (i.e., the probability of what, assuming what?).

25. Use the Theory-Based Inference applet to find the 95% confidence interval for the true proportion of all American Cell phone users who faked a cell phone call within the last 30 days.

Step 5: Formulate conclusions

26. Now, let's step back a bit and think about the scope of our inference. What are the wider implications? Do you think that your conclusion holds true for people in general? What is the broader population we are able to generalize to? It is important to remember how we gathered our data to answer these questions.

Step 6: Look back and ahead

27. Summarize your findings. If you were to repeat this study, would you gather your sample in a different way? Are there things about the study you would change? What further research might you follow up with from what you have learned in this study?

Research Article: Social evaluation

Read the article "Social evaluation by preverbal infants," including the full (online) methods found in the journal *Nature* Vol. 450, p 557-559, November 22, 2007 by J.K. Hamlin, K. Wynn and P. Bloom and accessible here: http://www.nature.com/nature/journal/v450/n7169/full/nature06288.html, via your school's library or from your instructor. Embedded in the article are links to short video clips which may be useful in understanding the research design. After reading the article, answer the following questions.

Step 1: Ask a research question.

1. For each of the following experiments state the research question
   a. Experiment #1 with regards to choice between helper and hinderer
   b. Experiment #1 with regards to looking time between helper vs. hinderer
   c. Experiment #2 with regards to choice between pusher-up and pusher-down
   d. Experiment #3 with regards to choice between neutral and helper
   e. Experiment #3 with regards to choice between neutral and hinderer
   f. Experiment #3 with regards to looking time between neutral vs. helper
   g. Experiment #3 with regards to looking time between neutral vs. hinderer

2. In no more than one sentence, summarize all of the research questions being investigated in this study.

Step 2: Design a study and collect data.

3. Provide some demographic description of the infants in the sample (age, race, etc.) based on what you read in the article.

4. What other characteristics of the infants in the sample would you like to know that are not provided in the article?

5. How does the article indicate that the infants were selected? What do you think this means?

6. For each of the three experiments, what variable(s) were measured on the infants? Identify the variable(s) as categorical or quantitative. If categorical, indicate the outcomes.

Step 3: Explore the data,

7. For Experiment #1, when choosing between the helper and hinderer toy, what proportion of 6-month olds chose the helper toy? What proportion of 10-month olds chose the helper toy?

8. For Experiment #1, when choosing looking at the helper and hinderer toys, on average, how long did 10-month olds look at the helper toy interacting with the climbing toy? With the hindering toy? Answer the same questions for the 6-month olds.

9. Answer question #7 about Experiment #2

10. Answer question #7 about Experiment #3 comparing helper to neutral

11. Answer question #7 about Experiment #3 comparing hinderer to neutral

    Note: Looking time descriptive statistics were not provided for experiment #3

Step 4: Draw inferences beyond the data

12. In the paper, the authors use Binomial Probability tests to generate p-values. Binomial probability tests can be thought of as equivalent to what would result from millions of simulations. Report the p-value from the Binomial Probability test for the helper vs. hinderer comparison in 10-months olds. Write a conclusion about the strength of evidence based on this p-value.

13. Answer #12 for 6-month old infants.

14. Why would theory-based approaches not be an appropriate method to generate p-values for either of these two tests?

15. The researchers conducted paired t-tests to evaluate whether there was a significant difference in the average looking time of the infants at the climbing toy approaching the helper (no surprising) vs. the climbing toy approaching the hinderer (surprising). A paired t-test is equivalent to a one-sample t-test (Theory-based test of mean).

    The researchers measured the variable "difference in looking times" for each infant (hinderer minus helper). Is this variable quantitative or categorical?

16. The researchers tested the hypotheses that the average difference in looking times was equal to zero (null) or different than zero (alternative). The researchers report a p-value of 0.02 for the two-sided test on 10 month olds. Give a statement about the strength of evidence based on this p-value

17. Report and interpret the same p-value for 6-month olds.

18. Do you agree or disagree with the following statement made by the authors about the 6-month old p-value? "Six-month-olds, however, looked equally to both events."

Step 5: Formulate conclusions

19. To whom are you willing to generalize your conclusions about the 10-month old infants choices? 6-month infants choices? Why?

Step 6: *Look back* and ahead

20. The last paragraph of the Methods Summary says the following:

    "The following were counterbalanced across subjects in each experiment and age group: identities of helper/hinderer (experiment 1), pusher-up/pusher-down (experiment 2) and valenced/neutral characters (experiment 3); order of habituation events; order of choice and looking time measures (experiments 1 and 3); positions (left/right) of characters in choice and in

*looking time trials; order of climber's approach in looking time test trials to helper/hinderer
(experiment 1) and valenced/neutral character (experiment 3)."*

Why is this counter-balancing important? What if the helper was always a blue square and the
hinderer was always a yellow triangle?

21. Identify at least 2 conclusions the authors point out about their study that they consider very
important. Do you agree with the authors? Why or why not?

22. Identify at least 2 significant limitations of the authors' study.

23. Identify at least 2 "next steps" in the research that you would be interested to know the results of.

Chapter 3

Estimation: How Large is the Effect?

Chapter Overview

This chapter is about **estimation**, one of the four pillars of inference: strength, **size**, breadth, and cause. Often, estimation tells **how large** the effect is, not by giving us a single number, but by giving us an interval of values. For example, we might find in the Organ Donations study (Example P.1), that we are "95% confident that the 'true' effect of forcing people to make a choice between becoming an organ donor or not, rather than the opt-in setting where the default is that a person is not a donor, increases the long-run probability of people agreeing to be organ donors by somewhere between 20 percentage points and 55 percentage points."

The goal of this chapter is to help you understand the logic that statisticians use to compute an interval of plausible (i.e., believable) values for the size of the effect we want to know about. This set of values, called a **confidence interval**, comes with a measure of reliability, typically 95%, called the **confidence level**. For the Organ Donations study, the 95% confidence interval for the increase in long-run probability of organ donors when forcing people to make a choice between being a donor or not rather than giving them the choice to opt in, extends from 20 to 55 percentage points.

In this chapter, you will see three different ways to find confidence intervals, all based on methods from Chapters 1 and 2. The first method (Section 3.1) specifies an interval of **plausible** values to be those values of the parameter that are not rejected in a (two-sided) test of significance. The second method (Section 3.2) approximates an interval of plausible values for the parameter to be those values that fall within two standard deviations of the statistic. The third method, a theory-based approach (also in Section 3.2), uses similar logic but allows a more precise and flexible interval calculation. Although most of this chapter will focus on confidence intervals for a population proportion, we will take a look at confidence intervals for a population mean in Section 3.3. We show you all three methods to illustrate the logic of confidence intervals and how they are related to tests of significance, but the third method is the most commonly used in practice. This method also allows us to easily explore factors that impact the width of the confidence interval (e.g., our level of *precision*) in Section 3.4. The chapter ends with some final warnings and cautions about statistical inference in general in Section 3.5.

Section 3.1: Statistical Inference – Confidence Intervals

Introduction

Up to this point, you have been able to make conclusions such as "we have strong evidence that the long-run probability is larger than 0.5." A natural next question is "How much larger than 0.5?" Or our research question may have been initially phrased as "what is the probability that .." Our main goal may be to estimate the parameter value rather than having a single natural value to test. Up until now, we've not been able to answer these kinds of questions, which relate to the size of the parameter. Although the observed statistic gives us an estimate of the actual parameter value (e.g., 71.9% of elections are correctly predicted with the competent face method), this does not take into account the random variation associated with that statistic. What we would like is a method for saying something like "I believe 69% to 75% of all elections can be correctly predicted by the competent face method." This would give us very different information than if we were to say "I believe 55% to 89% of all elections can be correctly predicted by the competent face method." In both cases our estimate exceeds 0.5, but the interval is much wider in the second case and our estimate much less precise.

In this section, we will see how we can use a ***confidence interval*** to estimate a long-run probability or population proportion. In particular, we will see how one way to create a confidence interval is to consider many different significance tests about the parameter.

Example 3.1: Can Dogs Sniff Out Cancer?

Step 1. Ask a research question. We all know that dogs have an amazing sense of smell. They are used to help hunters find their prey, determine the location of missing people, and screen luggage for drugs. But can a dog's keen sense of smell also be used to determine whether a patient has cancer? This is a question that has been asked recently and studies have been conducted to try to answer it.

Step 2. Design a study and collect data. In 2011, an article published by the medical journal *Gut – An International Journal of Gastroenterology and Hepatology* (Sonoda, et al.) reported the results of a study conducted in Japan in which a dog was tested to see whether she could detect colorectal cancer. The dog used was an 8-year-old black Labrador named Marine. (As her name might suggest, she was originally trained for water rescues.)

The study was designed so that the dog first smelled a bag that had been breathed into by a patient with colorectal cancer. This was the standard that the dog would use to judge the other bags. Marine then smelled the breath in five different bags from five different patients, only one of which contained breath from a colorectal cancer patient (not the same as the original patient); the others contained breath from non-cancer patients. The dog was then trained to sit next to the bag which she thought contained breath from a cancer patient (i.e., had the cancer scent). If she sat down next to the correct bag, she was rewarded with a tennis ball.

Marine completed 33 attempts of this experimental procedure, with a different set of five patients each time: four non-cancer patients and one cancer patient. And, each time, whether or not she correctly identified the bag with the breath of the cancer patient was recorded.

> **Think about it:** What are the observational units in this scenario? What is the variable? Is the variable categorical or quantitative? If categorical, is it binary?

The observational units are Marine's 33 attempts. (Notice, $n = 33$, indicating that 33 measurements were taken. If you incorrectly say that "Marine" is the observational unit, this would imply $n = 1$.) The variable is whether or not she identifies the correct bag, and this is a categorical and binary variable. The order in which she sniffed the bags was randomly determined each time. We will use Marine's results to investigate whether dogs can successfully detect a patient with cancer by smelling their breath.

Let's first consider whether Marine performed significantly better than random chance if she was simply guessing among the five bags. We can state our hypotheses as follows.

Null hypothesis (H_0): Marine is randomly guessing among the five bags ($\pi = 0.20$)

Alternative hypothesis (H_a): Marine's ability to detect cancer is better than random guessing ($\pi > 0.20$)

where π represents her long-run probability of correctly identifying the cancer patient.

> **Think about it:** Why do we set π to be 0.20 under the null? Why don't we set the long run probability of correctly identifying the cancer patient to 0.50 to represent simply guessing?

Because Marine is choosing between 5 bags, she has a 1 in 5 chance or 0.20 probability of correctly identifying the cancer patient if she is simply guessing.

Step 3. Explore the data. In her 33 attempts, Marine correctly identified the bag containing the breath of the cancer patient 30 times ($\hat{p}$ = 30/33 ≈ 0.909). This result is definitely in the direction of the research conjecture that dogs can sniff out cancer!

Step 4. Draw inferences beyond the data. But what does this tell us about Marine's underlying probability of identifying the breath of the cancer patient? We could first consider that she was randomly guessing among the five bags. Figure 3.1 displays the results from the One Proportion applet.

Figure 3.1: Estimating a p-value from simulating samples of 33 attempts with π = 0.20 and counting the number of repetitions where the simulated sample proportion was 0.909 or higher

We see that 0.909 is not only in the tail of this null distribution, it is way beyond these 1000 simulated proportions! In fact, a sample proportion of 0.909 is more than 10 [z = (0.909-0.20)/0.07 ≈ 10.13] standard deviations away from the hypothesized probability of 0.20!

Therefore the sample provides extremely strong evidence that π , the probability that Marine correctly identifies the cancer patient, is greater than 0.20.

So what is her long-run probability? We suspect it's much closer to 0.90 than to 0.20, but what about other values? Is it plausible that Marine's probability is actually 0.70, but she had a lucky day? In other words, would it be surprising to find $\hat{p}$ ≈ 0.909 if π = 0.70? Well, let's test whether these sample data provide convincing evidence that π differs from 0.70. See Figure 3.2.

Think about it: Based on the two-sided p-value estimated in Figure 3.2, what would you conclude about whether or not Marine's long-run probability equals 0.70? How are you deciding whether the p-value is small enough?

Figure 3.2: Estimating a p-value from simulating samples of 33 attempts from a process with a long-run probability of success equal to 0.70

Most statisticians would consider a p-value of 0.009 to be convincing evidence against the null hypothesis. Therefore we have strong evidence that Marine's probability of correctly identifying the cancer patient is not equal to 0.70.

So what about 0.80? Do you think a sample proportion of 0.909 is unlikely to arise from a process with $\pi = 0.80$?

Think about it: What will be different about the null distribution now? Will we have stronger or weaker evidence against this null hypothesis ($\pi = 0.80$) than Figure 3.2?

Figure 3.3: Estimating a p-value from simulating samples of 33 attempts from a process with a long-run probability of success equal to 0.80

As you can see in Figure 3.3, when we increased the probability of success in the null hypothesis and moved it closer to our sample proportion, we got a larger p-value. (This is consistent with the effect of the difference between the observed proportion value and the conjectured probability value that we saw in Section 1.4.)

This p-value of 0.147 would not be considered strong evidence against the null hypothesis. From this we would conclude that 0.80 is a ***plausible value*** for Marine's long-run probability. We haven't *proven* that π equals 0.80, but we can say that we don't have convincing evidence against that value.

Think about it: Now that we have decided that 0.80 is a plausible value for Marine's probability of successfully detecting the cancer patient, are there other values that are plausible as well?

Note that 0.80 was smaller than our sample proportion of 0.909 and was found to be plausible. But 0.70 was not a plausible value for Marine's probability. Now we are interested in finding other plausible values for π. We would like some way of deciding which values are plausible and which are not. In this case, we are less interested in assessing the strength of evidence against one particular hypothesized probability, but we are more interested in making a decision as to which values we will consider plausible. To do this, we can specify a significance level. Remember that a significance level is the value used as a criterion for deciding how small a p-value needs to be to provide convincing evidence against the null hypothesis. The most common significance level used is 0.05. So any hypothesized probability that produces a two-sided p-value larger than 0.05 will be considered plausible, and any hypothesized probability with a p-value less than or equal to 0.05 will not be considered plausible.

Key Idea: We will consider a value of the parameter to be plausible if the two-sided p-value for testing that parameter value is larger than the level of significance.

So if we use 0.05 as our significance level, where is the boundary between plausible and not plausible for Marine's probability? In other words, for what hypothesized value of π does our two-sided p-value change from being larger than 0.05 to smaller than 0.05? From our work above, we know one of these boundaries is somewhere between 0.70 and 0.80.

Testing a null probability value of 0.79, we find a two-sided p-value of 0.131, so we include 0.79 in our list of plausible values. Testing the value 0.78, we find a two-sided p-value of 0.097 and include 0.78 in our list of plausible values. Testing the value of 0.77 we again get a p-value larger than 0.05 (0.066), and thus include 0.77 in our list of plausible values. Finally testing the value 0.76 we get a p-value of 0.044. Because this p-value is less than our significance value of 0.05 we conclude that the data provide convincing evidence that Marine's probability of success is *not* 0.76. Thus we have found 0.77 to be a lower bound for Marine's interval of plausible values.

What about values larger than 0.80? We know any probability close to 0.909 will also be plausible. What about values on the other side of 0.909? It's possible that Marine actually performed below her true long-run probability on this day. You can see the results for testing some probability values larger than 0.909 in Table 3.1.

Table 3.1: Results of testing different values of probabilities under the null hypothesis

Probability under the null	0.93	0.94	0.95	0.96	0.97	0.98	0.99
(Two-sided) p-value	0.501	0.447	0.227	0.144	0.076	0.028	0.004
Decision at a 0.05 significance level	Do not reject null	Do not reject null	Do not reject null	Do not reject null	Do not reject null	Reject null	Reject null
Plausible value of π ?	Yes	Yes	Yes	Yes	Yes	No	No

Putting these results together, we would say any value between 0.77 and 0.97 is a plausible value for Marine's long-run probability π . In other words, we think our sample proportion of 0.909 arose randomly from a process with a long-run probability between 0.77 and 0.97.

If we zoomed in even more closely on the two boundaries corresponding to our level of significance, we would find results as in Table 3.2.

Table 3.2: Results of testing different values of probabilities under the null

Probability under null	0.759	0.760	0.761	0.762	....	0.973	0.974	0.975	0.976
p-value	0.042	0.043	0.063	0.063		0.059	0.054	0.049	0.044
Plausible?	No	No	Yes	Yes	.... Yes	Yes	Yes	No	No

So, a more precise statement of the interval for estimating Marine's probability of success is 0.761 to 0.974, also written as (0.761, 0.974). This interval estimates the long-run probability that Marine correctly identifies the cancer patient. This interval is called a *confidence interval for* π▣ TA \l "*90% confidence interval for* π An interval of values of π that would not be rejected by a two-sided test of significance

with 0.10 level of significance." \s "90% confidence interval for π:" " \c 1 ▣ . Because we used 5% as our significance level to decide whether or not to consider the value under the null hypothesis to be plausible, our interval of 0.761 to 0.974 is called a 95% confidence interval, where the 95% comes from 100% minus the significance level of 5%. This 95% is called the **confidence level**.

Key idea: The interval of *plausible* values is also called a **95% confidence interval for π** TA \l "*90% confidence interval for π*

An interval of values of π that would not be rejected by a two-sided test of significance with 0.10 level

of significance." \s "90% confidence interval for π:" " \c 1 , the long-run probability Marine correctly

identifies the cancer patient. Why is this called *95% confidence*? This corresponds to the 5% (0.05) significance level that was used to decide whether there was enough evidence against a hypothesized value. Notice that 95% and 5% add up to give 100%. The 95% is called the **confidence level**, and is a measure of how confident we are about our interval estimate of the parameter, in this case, we are 95% confident that the long-run probability Marine correctly identifies the cancer patient is in the interval that we found.

Step 5. Formulate conclusions. To interpret these results, we state that we are 95% confident that Marine's probability of correctly picking the bag with breath from the cancer patient from among five bags is between 0.761 or 0.974. Notice that this is a much stronger statement than our initial significance test which concluded simply that Marine's probability was larger than 0.20.

Effect of Confidence Level

Think about it: If we increase the confidence level from 95% to 99%, what do you think would happen to the interval of plausible values for the parameter?

The confidence level of 95% gives some indication of how sure we are that we have captured the actual value of the parameter in our interval. If we want to be more sure (i.e., more confident) that our interval contains the parameter value, our interval of plausible values for the parameter should be wider. To obtain a wider interval of plausible values, we need to make it harder to reject the values of the parameter specified by the null hypothesis – so use a smaller significance level. More specifically, if we want the confidence level to be 99%, we should use a 1% significance level. Examining the results in Table 3.3 below for plausible values of Marine's long-run probability, we now look for values that correspond to two-sided p-values larger than 0.01.

Table 3.3: Results of testing different various values of probabilities under the null.

Probability under null	0.70	0.71	0.72	....	0.97	0.98	0.99
p-value	0.007	0.011	0.018	....	0.076	0.028	0.004

Thus a 99% confidence interval for Marine's long-run probability extends from 0.71 to 0.98. Notice that, as expected, when the confidence level is higher, the interval of plausible values for the parameter is wider.

> **Key idea:** As the confidence level increases, the *width* of the confidence interval also increases.

The correspondence between tests and intervals

Confidence intervals and tests of significance generally give complementary information. Values that are not considered plausible by the (two-sided) test of significance should not be contained in the corresponding confidence interval (confidence level = 100% − level of significance). Because 0.5 is not inside the confidence interval, this matches the test. It is not "further evidence against the null," but confirms our calculations agree. The confidence interval then gives us additional information in telling us how large the proportion is. (See FAQ 3.1.1).

Step 6. Look Back and Ahead. From our study we have strong evidence that Marine can correctly identify the cancer patient more than half the time, leading us to believe Marine can reliably detect cancer in a patient by smelling a bag of their breath. We are 95% confident the long-run probability that Marine can correctly identify the cancer patient by smelling their breath is between 0.761 to 0.974. It is important to keep in mind the limitations in this study. For one, the results really only apply to Marine. We should be cautious in generalizing this behavior to dogs in general, at least until we know more about Marine. Although if we did see similar success probabilities in other dogs, especially across breeds, this would be very compelling. Not only should further studies include different breeds of dogs, but also different kinds of cancer. It is also not clear how practical this method of smelling breaths for identifying cancer is in general and whether five bags is a reasonable comparison group.

FAQ 3.1.1 Interval or test or both?

Q: Did I get it right that confidence intervals are more informative than tests of significance?

A: Good question. The interval is more informative *provided* that when you say "testing" you mean answering Yes/No to the question "Is the p-value less than or equal to the significance level?" This "Reject/Don't reject" form of the test is just for the null value only. The interval tells you *all* the parameter values not rejected.

Q: So why bother with the test at all?

A: "Testing" can also mean computing the p-value. You should *always* report the p-value if you are testing a specific parameter value, because that number tells you the strength of the evidence against the null. The interval doesn't tell you that.

Q: So the p-value is additional evidence?

A: No, the p-value is another way of reporting the *same* evidence. The p-value and the interval are two different ways of summarizing the evidence from the same data.

Exploration 3.1: Kissing Right?

Most people are right-handed, and even the right eye is dominant for most people. Developmental biologists have suggested that late-stage human embryos tend to turn their heads to the right. In a study reported in *Nature* (2003), German bio-psychologist Onur Güntürkün conjectured that this tendency to turn to the right manifests itself in other ways as well, so he studied kissing couples to see which side they leaned their heads while kissing. He and his researchers observed kissing couples in public places such as airports, train stations, beaches, and parks. They were careful not to include couples who were holding objects such as luggage that might have affected which direction they turned. For each kissing couple observed, the researchers noted whether the couple leaned their heads to the right or to the left. They observed 124 couples, age 13-70 years.

1. Identify the observational units in this study.

    124

2. Identify the variable recorded in this study. Classify it as categorical or quantitative.

You will first use Güntürkün's data to test his conjecture that kissing couples tend to lean their heads to the right. Use the symbol π to denote the proportion of all kissing couples in these countries that lean their heads to the right.

3.   Is π a parameter or a statistic? Explain how you are deciding.

4.   Do we know the exact value of π based on the observed data? Explain.

no

5.   State the appropriate null and alternative hypotheses, both in words and in terms of the parameter π, for testing the conjecture that kissing couples tend to lean their heads to the right more often than by random chance.

Of the 124 couples observed, 80 leaned their heads to the right while kissing.

6.   Calculate the sample proportion of the observed couples who leaned their heads to the right while kissing. Also indicate the symbol used to denote this value.

$\hat{P} = 80/124$

7.   Conduct a simulation analysis (using the **One Proportion** applet) to assess the strength of evidence that the sample data provide for Güntürkün's conjecture that kissing couples tend to lean their heads to the right more often than they would by random chance. Report the approximate p-value and summarize your conclusion about this strength of evidence.

p-value = 0

Your simulation analysis should convince you that the sample data provide very strong evidence to believe that kissing couples lean their heads to the right more than half the time in the long run. That leads to a natural follow-up question: How much more than half the time? In other words, we have strong evidence that the long-run probability of leaning to the right is greater than one-half, but can we now estimate the value for that probability? We will do this by testing many different (null) values for the probability that a couple leans to the right when kissing.

8. Now test whether the data provide evidence that the long-run probability that a couple leans their heads to the right while kissing (π) is different from 0.60. Use the **One Proportion** applet to determine the p-value for testing the null value of 0.60. Report what you changed in the applet and report your p-value.

Think about it: What kind of test are we conducting if we want to determine whether the population parameter is different from a hypothesized value—a one-sided test or a two-sided test?

Remember from Section 1.4 that if we don't have a specific direction we are testing, greater than or less than, we need to use a two-sided test of significance. If you used a one-sided alternative in #8, determine the two-sided p-value now by checking the **Two-sided** box.

Recall from Section 2.3 that we can specify a *level of significance* in order to decide whether the p-value is small. For example, we can say a p-value of 0.05 or less as strong evidence against the null hypothesis and in favor of the alternative hypothesis. Thus, we can *reject* the null hypothesis when the p-value is less than or equal to 0.05. Otherwise, when the p-value is greater than 0.05, we do not have strong enough evidence against the null hypothesis and so we consider the null value to be *plausible* for the parameter.

Key Idea: We will consider a value of the parameter to be plausible if the two-sided p-value for testing that parameter value is larger than the level of significance.

---

9. Is the p-value for testing the null value of 0.60 less than 0.05? Can the value 0.60 be rejected, or is the value 0.60 plausible for the long-run probability that a couple leans their heads to the right while kissing?

The p-value you found in the previous question should not have been smaller than 0.05. Hence, you do not reject the null hypothesis at the 0.05 level of significance and therefore you do not reject 0.60 as a plausible value for π. Thus, it is plausible (i.e., believable) that the long-run probability that a kissing couple leans their heads to the right is 0.60.

10. Does this mean that you've *proven* that exactly 60% of kissing couples lean right? Why or why not?

Because there are still other plausible values, now we want to "zoom in" on which values for the long-run probability are plausible and which can be rejected at the 0.05 significance level.

11. Use the applet to test the probability values given in the following table.
    - Each time, change the **Probability of success** to match the value that you are testing (keeping the observed sample proportion that you count beyond the same).
    - Everything else should stay the same; press **Draw Samples** and then **Count** to see the new two-sided p-value (with the **Two-sided** box checked).

Probability under H_0	0.54	0.55	0.56	0.57	0.58	0.59	0.60
(Two-sided) p-value	.016	:03	.06	.114	.139	.22	.32
Reject or plausible?	no	no	Yes	Yes	Yes	Yes	Yes

Probability under H_0	0.70	0.71	0.72	0.73	0.74	0.75	0.76
(Two-sided) p-value	0.23	0.109	.069	.642			.003
Reject or plausible?	Yes	Yes	Yes	no	no	no	no

12. Using a 0.05 significance level and your results from #11, provide a list of plausible values for π, the long-run probability that a kissing couple leans their heads to the right.

This list of values represents an interval containing all values between two *endpoints*.

Key idea: The interval of *plausible* values that you wrote down in #12 is also called a **95% confidence interval for** π ◙ TA \l "*90% confidence interval for* π

An interval of values of π that would not be rejected by a two-sided test of significance with 0.10 level of significance." \s "90% confidence interval for π:" " \c 1 ◙ , the long-run probability that a kissing couple leans their heads to the right. Why is this called *95% confidence*? This corresponds to the 5% (0.05) significance level that was used to decide whether there was enough evidence against a hypothesized value. Notice that 95% and 5% add up to give 100%. The 95% is called the **confidence level**, and is a measure of how confident we are about our interval estimate of the parameter, in this case, we are 95% confident that the population proportion of all couples who lean their heads to the right while kissing is in the interval that you found.

---

13. Does your 95% confidence interval from #12 include the value of 0.5? Does it include the value 0.6? Explain how your answers relate to the significance test and p-value that you calculated in #7 and #8.

14. Now suppose we were to use a significance level of 0.01 instead of 0.05 to decide whether or not to reject the corresponding null hypothesis for each listed value of π. How would you expect the interval of plausible values to change: wider, narrower, or no change? Explain your reasoning.

15. Implement the 0.01 significance level to determine plausible values for the population proportion of all kissing couples that lean their heads to the right. (*Hint*: Start with the table in #11, although you might have to test more null values as well.) Report the interval of plausible values.

16. How confident are you about the interval of plausible values that you listed in #15?

17. What is the primary difference between the 95% confidence interval reported in #12 and the 99% confidence interval reported in #15? Which interval is wider? Is this consistent with what you predicted in #14?

18. How would you expect a 90% confidence interval to compare to the 95% and 99% confidence intervals? Explain.

Key idea: As the confidence level increases, the *width* of the confidence interval also increases.

You should have found very strong evidence, based on a small p-value, that a majority of kissing couples in the population lean their heads to the right. Moreover, your confidence intervals, which provide plausible values for of the population proportion that lean to the right, contain only values above 0.5, confirming the conclusion made from the p-value.

In fact confidence intervals and tests of significance generally give complementary information. Values that are not considered plausible by the (two-sided) test of significance should not be contained in the corresponding confidence interval (confidence level = 100% − level of significance). Because 0.5 is not inside the confidence interval, this matches the test. It is not "further evidence against the null" but confirms our calculations agree. The confidence interval then gives us additional information in how large this majority is.

Section 3.1 Summary

Once a sample proportion has been determined to differ significantly from a hypothesized value for the long-run process probability, a typical next step is to estimate the value of the long-run process probability with an interval of values, the **confidence interval** for the parameter. Or the research question may be phrased such that a confidence interval is the more appropriate tool (e.g., "what is the probability that…") in the first place.

- This interval can be thought of as containing plausible values of the parameter.
- A parameter value is considered to be plausible if it does not produce a small (two-sided) p-value when tested in a null hypothesis.
- With this approach, the **confidence level** of an interval is equal to one minus the significance level used for determining rejection/plausibility.
    o Using a larger confidence level produces a wider interval of plausible values.
- Confidence intervals and tests of significance generally give complementary information. That is, null hypothesis values that can be rejected will not be in the confidence interval and vice versa.

Section 3.2: 2SD and Theory-Based Confidence Intervals for a Single Proportion

Introduction

In the last section we explored a method for estimating a confidence interval for a long-run probability or population proportion. We found the confidence interval by conducting repeated tests of significance and changing the probability specified by the null hypothesis to see whether or not the hypothesized value was plausible. As suggested in the overview of this chapter, there are other ways to approximate a confidence interval. In this section, we will explore how we can use the idea of standardization to construct confidence intervals. In particular, we will see that a method we call the "Two Standard Deviation Method" (or 2 SD Method) gives us an approximate 95% confidence interval.

Whereas the 2SD Method is typically based on simulation, in this section we will also see a theory-based approach to generating confidence intervals, which avoids the needs to use simulation altogether. The theory-based approach also allows us to easily use any confidence level we would like, not just 95%.

Example 3.2: The Affordable Care Act

A November 23-24, 2013 Gallup poll based on a random sample of 1,034 adults asked whether the new healthcare law (the Affordable Care Act) had affected the respondents or their family. Sixty-nine percent responded that the law had no effect. Gallup (www.gallup.com) reports that interviews are "conducted with respondents on landline telephones and cellular phones, with interviews conducted in Spanish for respondents who are primarily Spanish-speaking … Landline and cell telephone numbers are selected using random-digit-dial methods. Landline respondents are chosen at random within each household on the basis of which member had the most recent birthday."

You will see numerous differences between this sampling method and that of the *Literary Digest* that will be discussed in Section 3.5. Organizations like Gallup are constantly revising methods to ensure representation of women, minority groups, and younger individuals. So we have some faith that this sample will indeed be representative of all American adults.

> **Think about it:** So is it sufficient to announce that 69% of adult Americans have not felt any impact from the Affordable Care Act?

Although we expect the population proportion (call this π) to be close to 0.69, we know that a different random sample would likely lead to a different sample proportion and that our sample proportion can't be expected to exactly match the population proportion, just from the random chance in the sampling process. Our goal with a confidence interval is to account for the likely size of this random sampling error. In this context, we aren't necessarily interested in deciding whether π is larger than say 0.5, but our goal is to give an interval of plausible values for π .

We could repeat the procedure we used in Section 3.1 to create an interval of plausible values for proportion of the population of adult Americans who would choose "no effect." Table 3.4 shows the results using a level of significance of 0.05.

Table 3.4: Finding an interval of plausible values using 0.05 as the significance level.

Probability under null	0.659	0.660	0.661	....	0.717	0.718	0.719
Two-sided p-value	0.0388	0.0453	0.0514	....	0.0517	0.0458	0.0365
Plausible value (0.05)?	No	No	Yes	....	Yes	No	No

This method gives us an interval of (0.661, 0.717), indicating that we are 95% confident that the proportion of all American adults who had not felt impact from the Affordable Care Act was between 0.661 and 0.717. In other words, we are about 95% confident that between 66.1% and 71.7% of all adult Americans had not felt any impact from the Affordable Care Act in November 2013. Notice that the *midpoint* of this interval, $(0.661 + 0.717)/2 = 0.689$, is essentially the observed value of the sample proportion who said no impact. The width of our interval is $0.717 - 0.661 = 0.056$. Half of the width of our confidence interval, $0.056/2 = 0.028$, is called the **margin-of-error**. We can use our sample proportion and the margin-of-error to write our confidence interval as 0.690 ± 0.028. Also notice that we are distinguishing between saying "95% confidence" and "95% chance" (See FAQ 3.2.1).

FAQ 3.2.1 Coverage probability

Q: You said there's a 95% chance that the actual value of the parameter is in my 95% confidence interval. Right?

A: Not quite. The issue is what's random and what's fixed. The way you put it makes it sound like the parameter is random and the interval is fixed. It's actually the other way around: The parameter value is fixed, and the interval is random. If you chose a different sample you'd get a different interval, but the parameter value would still be the same. (95% of all samples give an interval that covers the true value. So 95% is the "coverage probability" or "capture rate.")

Q: Say it a different way.

A: Think of the game of horseshoes. The parameter is the iron stake in the ground. It doesn't move. Each interval is a toss of a horseshoe. A really good player can get the horseshoe around the stake 95% of the time.

Q: That helps. It doesn't make sense for the horseshoe to stay fixed while the stake moves around. The confidence level tells the percentage of throws that are "ringers" and capture the true value.

A: Well said. See Exploration 3.4B on this issue. And remember the "plausible values" approach. The 95% interval is the set of values not rejected at the 5% level.

Q: Which is it: 95% coverage or not rejected at the 5% level?

A: Most of the time they're the same.

You will see this general form of writing a confidence interval as the observed statistic plus/ minus a measure of the accuracy of the statistic with different data types in future chapters.

The Plausible Values Method (Section 3.1) for determining a confidence interval requires conducting quite a few null distribution simulations to zoom in on the endpoints of the confidence interval. What we would like is a simpler method for estimating the margin-of-error.

A short-cut is provided by noticing that null distributions often have the familiar symmetric, bell-shaped distribution that we have seen before, regardless of the null probability used in the simulation. We saw in Chapter 1 that with such null distributions, most random samples (about 95%) generate a value of the statistic that falls within 2 standard deviations (2SDs) of the mean of the null distribution. Recall that we considered a statistic to be in the tail of the distribution if it was more than 2 standard deviations from the mean. (See Figure 3.4.)

Figure 3.4: When a null distribution is bell-shaped, about 95% of the statistics will fall within 2 standard deviations of the mean with the other 5% outside this region

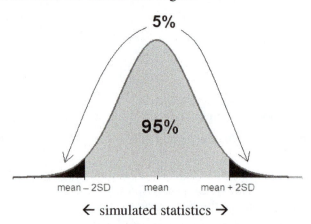

← simulated statistics →

Key Idea: When a distribution is bell-shaped, as your null distribution should be for this study, approximately 95% of the statistics in the null distribution will fall within two standard deviations of the mean. This implies that 95% of sample proportions will fall within two standard deviations of the long-run probability (π) , which means that π is within two standard deviations of the observed sample proportion for 95% of all samples.

We can extend this reasoning to generate a confidence interval for a long-run probability/population proportion. In particular, a parameter value is considered plausible if it is *not* more than two standard deviations from the observed value of the statistic. (This should roughly correspond to having a two-sided p-value less than 0.05.) Thus, another way to construct a 95% confidence interval is to include all values of π that are within two standard deviations of the observed sample proportion $\hat{p}$.

Key idea: We can construct a 95% confidence interval of plausible values for a parameter by including all values that fall within 2 standard deviations of the sample statistic. This method is only valid when the null distribution follows a bell-shaped, symmetric distribution. We call this the **2SD Method**. Thus we can present the 95% confidence interval for the long-run probability (or population proportion) 'π' in symbols as:

$$\hat{p} \pm 2 \times SD(\hat{p})$$

where $\hat{p}$ is the sample proportion and $SD(\hat{p})$ is the standard deviation of the null distribution of sample proportions. The value of 2×SD, which represents half the width of the confidence interval, is called the **margin-of-error** for 95% confidence.

But, how do we estimate the standard deviation of the distribution of sample proportions? There are actually several reasonable approaches. One is to simulate just one null distribution, read off its standard deviation, and use that value. But we need a value of π to conduct the simulation. In Chapter 1, you may have noticed that the standard deviation of the sample proportion is largest when $\pi = 0.5$. So using 0.5 in our simulation (Figure 3.5) should give us an upper bound on the standard deviation.

Figure 3.5: Null distribution of sample proportions for Affordable Care Act with $\pi = 0.5$

Proportion of heads

Using 0.016 as the standard deviation, we could construct our confidence interval to be

$$0.69 \pm 2(0.016) = 0.69 \pm 0.032$$

or approximately (0.658, 0.722). Notice this interval is similar (but wider) compared to what we found with the Plausible Values Method. The main distinction between the methods is the Plausible Values Method will allow the standard deviation to change when different values of π are utilized. When the sample size is large, you won't see much difference between these methods, with the 2SD Method being much more convenient.

Another method for estimating the standard deviation of sample proportions is to use the formula we saw in Chapter 1: $\text{SD}(\hat{p}) = \sqrt{\dfrac{\pi(1-\pi)}{n}}$. But again, what to use for the value of π ? We could substitute the conservative value of 0.5, giving us $0.5/\sqrt{n}$. In fact, when we apply the 2SD Method with this standard deviation, our margin-of-error simplifies to $1/\sqrt{n}$!

More generally, we can substitute the observed sample proportion in for π , giving us the "standard error of $\hat{p}$," $\text{SE}(\hat{p}) = \sqrt{\dfrac{\hat{p}(1-\hat{p})}{n}}$. This allows the standard deviation we use to slightly differ depending on the size of the sample proportion.

Definition: An estimate of the standard deviation of a statistic, based on sample data, is called the **standard error (SE)** of the statistic. In this case $\sqrt{\dfrac{\hat{p}(1-\hat{p})}{n}}$ is the standard error of a sample proportion $\hat{p}$.

Theory-Based Approach

Regardless of our choice for how to estimate the SD of the sample proportion, one limitation to the 2SD method it is only applies to 95% confidence. A more general method, valid for any confidence level, is to estimate the value of the parameter with

$$\text{statistic} \pm \text{multiplier x (SD of statistic)}$$

or, in this example of estimating a population proportion, with

$$\hat{p} \pm \text{multiplier} \times \sqrt{\dfrac{\hat{p}(1-\hat{p})}{n}} \; .$$

When we want a 95% confidence interval, the multiplier is approximately 2; hence the use of the name "2SD Method." If we want to change the confidence level, we change the multiplier. When the distribution of the sample proportion is approximately normal, this multiplier will come from the normal distribution. In Figure 3.6 we show corresponding multipliers for different confidence levels. For example, 90% of observations in a normal distribution should fall within 1.645 standard deviations of the mean of the distribution.

Figure 3.6: Change in confidence interval multiplier with change in confidence level

We learn several things from these graphs:

- You may notice that we can be a bit more precise than "2" with 95% confidence; that actually corresponds to a multiplier of 1.96.
- The multiplier increases as the confidence level increases. This agrees with what we learned in Section 3.1 about the effect of confidence level on the width of the interval.

So for 95% confidence, this formula gives us the interval: $0.69 \pm 1.96 \sqrt{\dfrac{.69(1 - .69)}{1054}}$ or $.69 \pm 0.029$ or $(0.662, 0.718)$. This is the formula used by the Theory-Based Inference applet by checking the Confidence interval box and pressing Calculate CI (Figure 3.7).

Figure 3.7: Using the Theory-Based Inference applet we can find a 95% confidence interval when the sample size is 1034 and the sample proportion is 0.690

As with all theory-based approaches, this method predicts the confidence interval you would have obtained from simulation. This approach uses the standard error formula to estimate the standard deviation of the sample proportions rather than carrying out a simulation to estimate that standard deviation. In fact, this approach makes no initial ballpark guesses for π. The advantage to using the Theory-Based approach over the 2SD Method is that we can change the confidence level to anything we would like, not just 95%. The output shown in Figure 3.8 shows the 90% and 99% confidence intervals.

Figure 3.8: Using a theory-based confidence interval allows us to easily change the confidence level

confidence level 90 % confidence level 99 %
(0.6663, 0.7137)   (0.6530, 0.7270)

Two things to observe about these intervals:
- The midpoint of each interval equals 0.690, the value of the observed sample proportion.
- The larger the confidence level, the wider the interval (smaller lower endpoint, larger upper endpoint).

In the next section, you will explore other factors that affect these interval calculations in general.

As we learned in Chapter 1, theory-based methods are only accurate when certain validity conditions are met.

> **Validity Condition:** The theory-based approach (called a one-sample z-interval) is considered valid if there are at least 10 observational units in each category of the categorical variable (i.e., at least 10 successes and at least 10 failures).

In this study, the validity condition is met because the sample included 713 who said "no effect" and 321 who cited an effect, which are both considerably greater than 10.

Exploration 3.2: American Exceptionalism

The Gallup organization conducted a survey with a random sample of 1019 adult Americans on December 10-12, 2010. They found that 80% of the respondents agreed with the statement that the United States has a unique character that makes it the greatest country in the world.

1. Identify the population and sample in this survey.

    Population:

    Sample:

2. Is it reasonable to believe that the sample of 1019 adult Americans is representative of the larger population? Explain why or why not.

3. Explain why 80% is a statistic and not a parameter. What symbol would you use to represent it?

4. Identify (in words) the parameter that the Gallup organization was interested in estimating.

5. Is it reasonable to conclude that exactly 80% of all adult Americans agree with the statement about American exceptionalism? Explain why or why not.

6. Although we expect π to be close to 0.80, we realize there may be other plausible values for the population proportion as well. First consider the value of 0.775. Is this a plausible value for π ? Use the **One Proportion** applet to simulate random samples of 1019 people from such a population. (*Hint*: Keep in mind that 0.775 is what we are assuming for the population proportion and 0.80 is the observed sample proportion.) What do you estimate for the two-sided p-value? Would you reject or fail to reject the null hypothesis at the 5% level of significance?

7. Also check the **Summary Stats** box and report the mean and standard deviation of this null distribution.

8. Now consider 0.5. Is this a plausible value for π ? Repeat #6 and record the mean and standard deviation for this null distribution as well.

Clearly 0.5 is going to be "too far" from $\hat{p} = 0.80$ to be plausible. But how far is too far? We could use the Plausible Values Method from Section 3.1 to produce a confidence interval for the proportion of all adult Americans who agree with the statement. But that approach is somewhat cumbersome and time-intensive, so we'll now learn some shortcut approaches.

9. Reconsider our first guess of $\pi = 0.775$. How many standard deviations is 0.80 from 0.775? (*Hint*: Standardize the value by looking at the difference between 0.775 and 0.80 and divide by the standard deviation you found in #7.)

You should notice that 0.775 and 0.80 are about 2 standard deviations apart AND that the two-sided p-value is around 0.05, so this value is close to the edge of values that can be considered plausible. Values between 0.80 and 0.775 are considered plausible and values smaller than 0.775, or more than two standard deviations below 0.80, will not be plausible values for the population proportion.

> **Key Idea:** When a distribution is bell-shaped, as your null distribution should be for this study, approximately 95% of the statistics in the null distribution will fall within two standard deviations of the mean. This implies that 95% of sample proportions will fall within two standard deviations of the long-run probability (π) , which means that π is within two standard deviations of the observed sample proportion for 95% of all samples.

We can then extend this idea to construct a 95% confidence interval.

> **Key idea:** We can construct a 95% confidence interval of plausible values for a parameter by including all values that fall within 2 standard deviations of the sample statistic. This method is only valid when the null distribution follows a bell-shaped, symmetric distribution. We call this the **2SD Method**. Thus we can present the 95% confidence interval for the long-run probability (or population proportion) π in symbols as:
>
> $$\hat{p} \pm 2\times\text{SD}(\hat{p})$$
>
> where $\hat{p}$ is the sample proportion and SD($\hat{p}$) is the standard deviation of the null distribution of sample proportions. The value of 2×SD, which represents half the width of the confidence interval, is called the **margin-of-error** for 95% confidence.

> **Think about it:** So how do we find the standard deviation to use for the 2SD Method?

10. How did the standard deviations you found in #7 (with $\pi = 0.775$) and in #8 (with $\pi = 0.5$) compare?

You should see that the standard deviation changes slightly when we change π , but not by much. We saw in Chapter 1, that the variability in the sample proportions is in fact largest when $\pi = 0.5$. So one approach would be to carry out one simulation (with lots of trials) using $\pi = 0.5$, and use that value of the standard deviation to calculate the margin-of-error.

11. Determine a 95% confidence interval using the 2SD Method:

    a. First calculate $2 \times$ (standard deviation for your null distribution of sample proportions) using 0.5 in the simulation to estimate the SD. (This is the margin-of-error.)

    b. Use this SD to produce a 95% confidence interval for π . (*Hint*: Subtract the margin-of-error from $\hat{p}$ to determine the lower endpoint of the interval and then add the margin-of-error to $\hat{p}$ to determine the upper endpoint of the interval.)

    c. Interpret the confidence interval: You are 95% confident that *what* is between what two values?

One limitation to this method is that it only applies to 95% confidence. What if we wanted to be 90% or 99% confident instead? We can extend this 2SD method to a more general theory-based approach. As we saw in Chapter 1, we don't always need to simulate a null distribution-not if we can accurately predict what would happen if we were to simulate. Instead, we can predict the standard deviation by the formula

$$\sqrt{\frac{\pi(1-\pi)}{n}}$$

and we compare the "number of standard deviations" to a normal distribution. But when constructing a confidence interval, we don't have a hypothesized value of π , so to estimate this standard deviation, we will substitute the observed sample proportion.

Definition: An estimate of the standard deviation of a statistic, based on sample data, is called the

standard error (SE) of the statistic. In this case $\sqrt{\frac{\hat{p}(1-\hat{p})}{n}}$ is the standard error of a sample proportion $\hat{p}$.

---

12. Calculate the standard error for this study. How does it compare to the standard deviations you found in #7 and #8?

So a more general formula for using the 2SD Method to estimate a population proportion would be:

$$\hat{p} \pm 2\sqrt{\frac{\hat{p}(1-\hat{p})}{n}}$$

But then how do we change the confidence level?

The 2SD Method was justified by saying 95% of samples yield a sample proportion within 2 standard deviations of the population proportion. If we want to be more confident that the parameter is within our margin-of-error, we can create a larger margin-of-error by increasing the multiplier. In fact a multiplier of 2.576 gives us a 99% confidence level, whereas a multiplier of 1.645 gives us only 90% confidence.

13. We will rely on technology to find the multiplier appropriate for our confidence level.

    a. In the **Theory-Based Inference** applet, specify the sample size (*n*) of 1019 and the sample proportion of 0.80 (or the sample count of 815) and press **Calculate**. (The applet will fill in the count.)

    b. Check the box for **Confidence Interval**, confirm the confidence level is 95% and press **Calculate CI** to generate a theory-based confidence interval. Report the 95% theory-based confidence interval.

14. Is this theory-based confidence interval similar to the one you obtained using the 2SD method?

Validity Condition: The theory-based approach (called a one-sample *z*-interval) is considered valid if there are at least 10 observational units in each category of the categorical variable (i.e., at least 10 successes and at least 10 failures).

Because we have a large sample size here, the theory-based approach should produce very similar results to the Plausible Values Method and the 2SD Method. In such a case, the theory-based approach is often the most convenient, especially if our confidence level is not 95%.

15. Change the confidence level in the applet from 95% to 99% and press the **Calculate CI** button again. Report the 99% confidence interval given by the applet. How does it compare to the 95% interval? (Compare both the midpoint of the interval = (lower endpoint + upper endpoint)/2 and the margin-of-error = (upper endpoint − lower endpoint)/2.)

Section 3.2 Summary

The Plausible Values Method to determining a confidence interval for a parameter reminds us that the confidence interval represents all possible values of the parameter that are considered "plausible" based on not putting the observed statistic too far in the tail of the distribution (two-sided p-value is not small, not below the significance level). However this method is fairly tedious, because it requires performing many significance tests to determine which parameter values are rejected and which are plausible.

An alternative is to produce a confidence interval using: **statistic ± margin-of-error**.
- This approach is valid when the null distribution is bell-shaped and symmetric.
- The **midpoint** of this confidence interval is the observed value of the sample statistic.
- The **margin-of-error** for 95% confidence can be approximated as 2 × (SD of the statistic)
  - o This corresponds to the prediction that 95% of samples produce a statistic that lies within two standard deviations of the underlying parameter.

In this section, our statistic was the sample proportion $\hat{p}$. One way to estimate the SD of the sample proportion is to simulate a null distribution (e.g., for $\pi = .5$).

A third approach for determining a confidence interval for a process probability or population proportion uses a **theory-based approach** based on the normal distribution.
- The theory-based approach also uses the form: **statistic ± margin-of-error**.
  - o The margin-of-error can be calculated as multiplier × standard error, which in the case of

    estimating a population proportion is *multiplier* × $\sqrt{\dfrac{\hat{p}(1-\hat{p})}{n}}$.
- This interval can be calculated with technology, without the need to perform a simulation analysis first.
- This method can be performed for any **confidence level**.

o   The most common confidence level is 95%, followed by 90% and 99%.

o   Increasing the confidence level increases the multiplier and widens the interval.

- The **validity condition** for applying the theory-based approach is that each category (successes, failures) includes at least 10 observations.

Validity conditions for a theory-based interval on a single proportion

Type of data	Name of test	Applet	Validity conditions
Single binary variable	Theory-based interval for a single proportion (1 Proportion z interval)	One proportion; theory-based inference	At least 10 successes and at least 10 in failures in sample

Section 3.3: 2SD and Theory-Based Confidence Intervals for a Single Mean

Introduction

In the previous sections we saw we can calculate confidence intervals for a population proportion by:

- Testing different values for the parameter to see whether they are plausible based on the observed proportion.

- Using a 2SD Method to extend two standard deviations (based on the null distribution of the statistic, the sample proportion in Section 3.2) from each side of the statistic

- Using a Theory-based Approach when certain validity conditions are met.

(You may want to review FAQ 3.3.1: Which interval do I use?)

FAQ 3.3.1 Which interval?

Q: I know you've said this already, but you've given us three different intervals! Which one should I use?

A. We used the "plausible values" approach to help you understand what the interval tells you—which

---

values are plausible (and not!) values of the unknown parameter. It's kind of tedious, though. Most people use the theory-based approximation if you have at least 10 responses of each kind (10 yes, 10 no).

Q. But what about the 2SD approach?

A: A short-cut to the one-sample z-interval is the 2SD Method, but it also needs a large sample size and it assumes 95% confidence.

Q: And if the validity conditions aren't met?

A: There are even more alternatives. Consult a statistician to talk more about these. Just keep in mind that no matter how you approximate the interval, you always interpret it the same way.

In Sections 3.1 and 3.2 we were focusing on inferences about a population *proportion* because our data consisted of a single categorical variable. In this section we will focus on data consisting of a single quantitative variable. Hence, we will make inferences about a population *mean* by creating confidence intervals. We will again consider a shortcut 2SD approach as well as a theory-based approach.

Example 3.3: Used Cars

Suppose you are in the market for a used car. You know you want a Honda Civic, but you have no idea how much you should expect to pay. The histogram in Figure 3.9 displays data collected on 102 used Honda Civics for sale on the internet in July, 2006.

Figure 3.9: Histogram of asking prices of 102 Honda Civics for sale on the internet

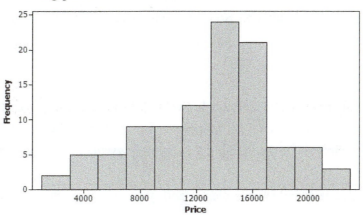

Think about it: The average price in this sample is $\bar{x}$ = \$13292 with standard deviation s = \$4535. Based on this information, what can you say about μ, the average price of all used Honda Civics?

First note that we should be cautious of whether the population of interest is well-represented by this sample. Was this a random sample of car on sale on the INTERNET? Do prices on the INTERNET differ from prices of cars sold, but not advertised on the INTERNET? Also keep in mind that this data was for the *asking price* and we do not have information on the actual selling prices.

Although we should be very cautious in doing so, let's treat this sample as representative of all such used Honda Civics for sale for this example. We don't presume that $\mu = 13{,}292$ exactly, yet it should be close. But how close? How much do we think the sample mean might wander away from the population mean? The answer to this question will help us find a confidence interval for μ.

To find a confidence interval for μ, we can use either a 2SD Method or we can use a theory-based method. Both methods we'll look at in this section will have similarities to the approaches we used in the previous section.

Let us revisit the idea behind how a confidence interval gets computed. The general form of a confidence interval for a parameter is

$$statistic \pm multiplier \times (SD\ of\ statistic)$$

where the choice of "*statistic*" should match the parameter being estimated. Meaning, if the parameter is the population proportion, the statistic should be a sample proportion. Similarly, if the parameter is a population average, the statistic should be a sample average, and so on. Recall that when we want a 95% confidence interval, the multiplier is approximately 2.

Hence the use of the name "2SD Method." So with quantitative data we could use

$$\bar{x} \pm 2 \times (SD \ of \ \bar{x})$$

It is critical to keep in mind the distinction between "SD of statistic" and "SD of sample." In this example, $s = \$4535$ tells us the amount of variability in asking price *from car to car*. What we need to know for our confidence interval is the variability in *average asking price* from *sample to sample*.

> **Key Idea:** There will be less sample to sample variability in the sample mean than car to car variability in prices.

In Section 2.2, we created a fake population and took repeated samples from that population to estimate this sample to sample variability. But that was more to convince us the variability in the sample means from different samples follows the formula $\sigma/\sqrt{n}$. And when we didn't know σ, (the population standard deviation) we could approximate the standard deviation σ with the standard error $s/\sqrt{n}$.

> **Key Idea:** An approximate 2SD Method for a 95% confidence interval for a population mean μ is $\bar{x} \pm 2$ $s/\sqrt{n}$.

For this example, with $\bar{x} = 13292$ and $s = 4535$, our 2SD interval would be:

$$13292 \pm 2 \times (4535/\sqrt{102}) \approx 13292 \pm 898.1$$

or approximately (12393.9, 14190.1). From this we would say we are about 95% confident that the average price of all Honda Civics for sale on the Internet around this time was between \$12,394 and \$14,190 (assuming the sample had been representative).

This approach should be reasonable as long as the null distribution of the sample means is reasonably symmetric. But what about a more general theory-based approach that allows us to change the multiplier and achieve different confidence levels?

Theory-based approach

In Section 2.2, we saw a theory-based method to find the p-value when testing hypotheses about a single population mean. Recall that the theory-based approach used a theoretical probability distribution called the *t* distribution to predict the behavior of the standardized statistic, and that this approach required certain validity conditions to be met. The implication here is that for a confidence interval for a population mean the multiplier will come from a *t* distribution rather than the normal distribution. And in fact this multiplier will change depending on the sample size (degrees of freedom) involved. We will let the technology figure that out, but do keep in mind that for 95% confidence, the multiplier is roughly 2, which is what we used as a shortcut earlier.

Using the Theory-Based Inference applet (see Figure 3.10), we can choose the One mean scenario and specify the summary statistics (or you can paste in the individual data values).

Figure 3.10: Theory-based inference confidence interval for mean used Honda asking price

This interval (12401, 14182) is similar to what we found with the 2SD Method, but is using a slightly different multiplier. The important thing to keep in mind is that this is an interval for the *population mean asking price*, not the price of one car. The rough sketch in Figure 3.10 shows that most cars are typically between $10,000 and $18,000, a much wider range than our confidence interval for the population mean asking price.

As before, the multiplier will increase as we increase our confidence level. (See Figure 3.11.)

Figure 3.11: 90% and 99% confidence intervals for mean used Honda asking price

confidence level 90 %	confidence level 99 %
(12546.5480, 14037.4520)	(12112.9546, 14471.0454)
df = 101	df = 101

Also keep in mind that this method should only be used when the validity conditions are met.

Validity Conditions: The theory-based interval for a population mean (called a *one-sample t-interval*) requires that the quantitative variable should have a symmetric distribution, or you should have at least 20 observations and the sample distribution should not be strongly skewed.

Keep in mind that the larger the sample size the more skewness in the sample you can tolerate.

In the context of the used Honda civic asking prices, the sample shows slight skewness to the left in Figure 3.9. Is this enough skewness to doubt the validity of our procedure? Perhaps, though a sample size of 102 should be enough to compensate for such slight skewness. If you are concerned, then you should use an alternative method like transforming the data first.

In summary, the formula for a theory-based confidence interval for a population mean (μ) is given by

$$\bar{x} \pm multiplier \times \frac{s}{\sqrt{n}},$$

where $\bar{x}$ denotes the sample mean, s denotes the sample standard deviation, and n is the sample size. This approach uses $s/\sqrt{n}$ to approximate the sample to sample variability in the sample means. When this is the case, we use a multiplier from the t distribution rather than the normal distribution as we did for proportions. Or, for 95% confidence we can continue to approximate this multiplier by 2, giving a quick approximation to that found by the Theory-Based Inference applet. The multiplier, and therefore the width of the confidence interval ,common will increase if the confidence level increases.

Exploration 3.3: Sleepless Nights? (continued)

Recall Exploration 2.2 where you investigated whether, on average, students at your school get less than 8 hours of sleep on a typical night. Even though 8 hours is often cited as the recommended amount, many people believe that college students will average less than 8 hours. But how much less? In this exploration, you will use the same data to estimate the average number of hours of sleep students at your school get.

1. Let μ represent the average number of hours of sleep all the students at your school got the previous night. The question we asked in Exploration 2.2 was, "How much sleep (to the nearest quarter hour) did you get last night?" If you still have the data from that exploration, give the summary statistics here. If not, collect a new set of data from your class and give those summary statistics here.

   Sample size, $n =$

   Sample mean, $\overline{x} =$

   Sample SD, $s =$

Similar to what you did in Exploration 3.1, we could conduct a simulation to generate a null distribution of the sample mean number of hours of sleep, assuming different values for μ to determine which are plausible. Recall that such an interval is called a confidence interval. But this would be rather tedious, so in this exploration we will consider some short-cuts to estimating a confidence interval for μ.

2. Based on your sample data, what is your best guess or estimate for the value of μ, that is, the average number of hours of sleep students at your school got the previous night?

3. Suppose you repeated the survey by asking a different group of students at your school. Will the average for this new sample be exactly the same as the original sample average reported in #2? Why or why not?

A natural guess or estimate for the population average number of hours of sleep students at your school got the previous night (μ) is the observed value of the sample average sleep time. But, we know that if we were to conduct this study again we would most likely observe a different value for the sample average due to *sampling variability* – we need to account for this in our estimate for μ. To do so, we create a confidence interval for μ, by using the formula

sample mean   multiplier x (SD of sample mean)

where the "*SD of sample mean*" accounts for the sample-to-sample variability that we expect in the sample average, the "*multiplier*" accounts for the confidence level, and the " " allows us to create an interval around the sample mean (these are the plausible parameter values that are not too far from the observed sample mean). Recall that when the confidence level is 95%, the multiplier is approximately 2.

To find a 95% confidence interval for μ we can use the 2SD Method that you saw in Section 3.2, but we need to know what to use for *SD of sample mean*.

4. In Question #25 of Exploration 2.2, when you simulated the distribution of sample means from the hypothetical population of sleep times (population 1), what did you find for the standard deviation of the null distribution? How did this value compare to the standard deviation of your sample?

Key Idea: There will be less sample to sample variability in the sample mean than person to person variability in sleep times.

---

5. In Exploration 2.2, we saw that $s/\sqrt{n}$ could be used to approximate the standard deviation of the sample means when the distribution of sleep times is reasonably symmetric or the sample size is reasonably large without extreme skewness. Report this value.

Key Idea: An approximate 2SD Method for a 95% confidence interval for a population mean is $\bar{x} \pm 2$ $s/\sqrt{n}$.

---

6. Use the value of $s/\sqrt{n}$ .and the 2SD Method to approximate a 95% confidence interval for μ in the context of the sleep study, and then interpret this interval. (*Hint*: When interpreting the interval, be clear about what the interval estimates. You are not estimating the individual sleep times of students, but how much sleep students at your school get on...)

7. Does the 95% confidence interval reported in #6 contain 8 (the population mean under the original null hypothesis)? Explain how you could have known this in advance based on your p-value from Exploration 2.2.

Theory-based approach

In Section 2.2, you saw a theory-based approach to find the p-value when testing hypotheses about a single population mean. This approach uses a more precise value of the multiplier in the confidence interval formula based on the *t*-distribution.

8. Go to the **Theory-Based Inference** applet, and select the **One Mean** option from the Scenario pull-down menu. Then, you can either copy and paste your class data, or just enter the sample statistics (as recorded in #2). Press **Calculate**, check the **Confidence Interval** box, and press **Calculate CI**. Report the 95% confidence interval and compare it to the interval you found in #6. Are they about the same? Same midpoint? Same width? Could you have anticipated this? Explain your reasoning.

Recall that the theory-based approach uses a theoretical probability distribution called the *t*-distribution to predict the behavior of the standardized statistic (denoted by *t*), and that this approach required certain validity conditions to be met.

> *Validity Conditions*: The theory-based interval for a population mean (called a *one-sample t-interval*) requires that the quantitative variable should have a symmetric distribution, or you should have at least 20 observations and the sample distribution should not be strongly skewed.

9. Are the validity conditions for using the 2SD or *t*-distribution based method to find a confidence interval met for your class data? Explain why or why not.

In summary, the formula for a theory-based confidence interval for a population mean (μ) is given by

$$\bar{x} \pm \text{multiplier} \times \frac{s}{\sqrt{n}},$$

where $\bar{x}$ denotes the sample mean, s denotes the sample standard deviation, and n is the sample size. We will rely on technology to find the appropriate multiplier for our sample size and our confidence level. Keep in mind that the multiplier is approximately 2 for 95% confidence and increases if we increase the confidence level.

Section 3.3 Summary

In this section we explored simulation- and theory-based approaches for finding confidence intervals for a single population mean.

- The general form of a confidence interval is statistic $\pm$ multiplier × (SD of statistic)
- When the parameter of interest is the population mean, μ, the statistic is the sample mean, that is $\bar{X}$.
- The general form of a confidence interval for a population mean (μ) is given by

  sample mean $\pm$ multiplier x (SD of sample mean)
    - o For 95% confidence, the multiplier will be roughly in the ballpark of 2.
    - o The standard deviation of the sample mean can be approximated by $s/\sqrt{n}$.
    - o The theory-based confidence interval is $\bar{x} + \text{multiplier} \times \frac{s}{\sqrt{n}}.$ where the multiplier comes from the *t* distribution with $n-1$ degrees of freedom.
    - o These methods are valid if the population distribution is normally distributed (make sure the sample is roughly symmetric) or the sample size is large (at least 20 or quite large if the sample is more heavily skewed).
- A confidence interval for a population mean estimates the value of the population mean, not the value of individual observations in the population.
- Notice we are not presenting a method for finding the confidence interval for a population median. Because the null distribution for medians did not follow the familiar symmetric, bell-shaped pattern, the approach outlined here would not be valid in most cases. If the parameter of interest is a median, you will need to consider more sophisticated approaches.

Summary of validity conditions and applets for a confidence interval of a single mean

Type of data	Name of test	Applet	Validity conditions
Single quantitative variable	Theory-based interval for a single mean (1 sample t-interval)	One mean; Theory-based inference	The quantitative variable should have a symmetric distribution or you should have at least 20 observations and the sample distribution should not be strongly skewed.

Section 3.4: Factors That Affect the Width of a Confidence Interval

Introduction

In the previous two sections we explored some different ways to construct confidence intervals for long-run probabilities or population proportions, and for population means. These different methods were more efficient than the Plausible Values Method seen in Section 3.1. In this section, we will explore how the sample size and the sample statistics can influence the width of confidence intervals. Finally, an interpretation of confidence *level* will be discussed in Exploration 3.4B.

Example 3.4: The Affordable Care Act (continued)

Recall the 2013 Gallup poll based on a random sample of 1,034 adults which asked whether the new healthcare law (the Affordable Care Act) had affected the respondents or their family (Example 3.2). Sixty-nine percent of the sample responded that the laws had had no effect.

We saw that changing the confidence level impacted the width of the interval. The theory-based confidence intervals for the three most common confidence levels are reproduced in Table 3.5.

Table 3.5: 90, 95, and 99% confidence intervals for the proportion of adult Americans not affected by the Affordable Care Act in November, 2013

90%	95%	99%
(0.6663, 0.7137)	(0.6618, 0.7182)	(0.6530, 0.7270)
width = 0.0474	width = 0.0564	width = 0.0740

Thus, as we saw in the previous sections, as the confidence level increases, the confidence interval gets wider. Many students find this, initially, counter-intuitive. Remember, this is an interval of plausible values for the parameter—we are hoping to "catch" the parameter in our interval. If you want to be more confident that you catch the parameter in your interval, then you need a wider interval. Read more on the idea of 'catching' the parameter in Exploration 3.4B.

Sample size

A key subgroup of interest in this survey is those currently without health insurance, one group the law is specifically designed to help. Interestingly, the poll found that 70% of people in this subgroup stated that the law had not affected them—very similar to the 69% found in the overall sample.

> **Think about it:** How will the confidence interval for the proportion of all uninsured Americans who feel the law has not impacted them differ from the original confidence interval?

To answer this question, we need to know the sample size – how many in the survey were not insured. Using the Theory-Based inference applet in Figure 3.12, we show 95% confidence intervals for three different sample sizes.

Figure 3.12: 95% confidence intervals for different subgroup sizes

n: 100	n: 200	n: 400
count: 70	count: 140	count: 280
sample p̂: 0.7000	sample p̂: 0.70	sample p̂: .70
confidence level 95 %	confidence level 95 %	confidence level 95 %
(0.6102, 0.7898)	(0.6365, 0.7635)	(0.6551, 0.7449)
0.70 ± 0.0898	0.70 ± 0.0635	0.70 ± 0.0449

Think about it: So what are the commonalities and differences among these intervals?

Because the sample proportion, $\hat{p} = 0.70$, is the same in all three cases, the midpoints of the intervals are all the same. However, the widths of the intervals change. The width for $n = 400$ is half the width with $n = 100$. (So we need to multiply the sample size by 4 to cut the margin-of-error in half.) See FAQ 3.4.1: The root n rule for more discussion of the role of sample size.

FAQ 3.4.1 The root n rule

Q: I was looking at the formulas for confidence intervals, and they have the form "estimate multiplier x SE."

A: Right.

Q: And the SE is either $\sqrt{\dfrac{\hat{p}(1-\hat{p})}{n}}$ for $\hat{p}$ or else $\dfrac{s}{\sqrt{n}}$ for $\bar{x}$.

A: Right again.

Q: Either way, there's a root n in the denominator.

A: You're three for three so far. You've got a question?

Q: Yes. That fact must tell us something about how the sample size and the width of a confidence interval are related. Can you say more about that?

A: We want narrow intervals, because they do a better job of pinning down the parameter value. Just to be concrete, suppose you have a quantitative variable with $s = 0.5$. You can check that the margin-of-error for a 95% interval for μ is roughly $\frac{1}{\sqrt{n}}$.

Q: OK.

A: So if $n = 1$, the margin-of-error is 1. Your interval is $\overline{x} \pm 1.0$. Suppose we double the sample size, to $n = 2$. Will that cut the margin-of-error in half?

Q: I thought I got to ask the questions. The margin-of-error goes from $\frac{1}{\sqrt{1}} = 1.0$ to $\frac{1}{\sqrt{2}} = 0.707$. Not in half.

A: Right. Because of the square root of n in the denominator, if you want to cut the margin-of-error by a factor of 2, you have to multiply the sample size by a factor of $2^2 = 4$.

Q: Ouch! That means that if $n = 1$ gives a margin-of-error of 1, I'll need 4 observations to cut that in half, down to 0.5.

A: And you'll need 16 observations to cut it by half again, down to 0.25. For a margin-of-error equal to 0.1, you'll need 100 observations.

Q: So how large should my sample size be?

A: Well, that depends on what you want your margin-of-error to be. Larger n is always better. But eventually you reach a point where it doesn't gain you much to increase the sample size.

Q: So a cheaper sample size may be "good enough"?

A: Yes, and what is good enough will depend on the context. You need to know about the variables involved to decide what is good enough.

Thus, we see that the width of confidence intervals is smaller when based on a larger sample size. This should make intuitive sense. When we have more information (based on a larger sample size), we can be more *precise* in our estimate of the long-run probability as evidenced by a narrower confidence interval.

> **Key idea:** The width of a confidence interval decreases as the confidence level decreases (for a fixed sample size) or as the sample size increases (for a fixed confidence level).

Do keep in mind that with the larger sample size we have made our estimate of the parameter more precise by specifying a smaller (narrower) interval of plausible values for the parameter, but we are still 95% confident in this interval.

With quantitative data, another factor that will impact the margin-of-error is the variability in the sample through the standard error $s/\sqrt{n}$. (See Exploration 3.4.) If our sample data are more precise (less variable), then our confidence interval will also be more precise (narrower). Though not as dramatic, the sample proportion is also related to the standard error used in the confidence interval as discussed next.

Optional: Effect of sample proportion

So far in this example we have focused on the proportion who said they had not been affected by the new health law. What if we focused on those who said they felt hurt by the new law? In the November 2013 study that sample proportion was 0.19.

> **Think about it:** Will the width of this confidence interval be larger or smaller than for $\hat{p} = 0.69$?

Returning to our sample size of $n = 1034$, we can still use the theory-based confidence interval because we will have 196 "successes" and 838 "failures" (Figure 3.13).

Figure 3.13: A 95% confidence interval for proportion of American adults who feel hurt by Affordable Care Act

The confidence interval of (0.1661, 0.2139) has a margin-of-error of $(.2139 - .1661)/2 = 0.0239$. This is actually a bit smaller than the margin-of-error we found with $\hat{p} = 0.69$ (0.0282). This is expected because the margin-of-error will be larger when $\hat{p}$ is closer to 0.50. So it is actually "easier" to (more precisely) estimate a population proportion that is not as close to 0.50.

Exploration 3.4: Holiday Spending Habits

A Nov. 7–10, 2013 Gallup poll asked 1,039 U.S. adults how much they planned to personally spend on Christmas gifts this year. The report cited an average of $704.

1. Identify the observational units and variable in this study. Classify the variable as quantitative or categorical.

2. Define the parameter of interest in this study. (*Hint*: Do you want to estimate a mean or a proportion? For what population?)

Because our variable of interest (dollar amount intending to spend) is quantitative, the parameter of interest here is the population mean for all adult Americans, call this μ . To construct a confidence interval for μ , we need to know the sample size, the sample mean, and the sample standard deviation.

3. Which two of these three quantities do you know? What symbols can be used to represent them?

What we are not told by Gallup is s, the sample standard deviation.

4. Explain in your own words what s represents in this context.

5. Suppose s was equal to $150. Use the 2SD Method to approximate and interpret a 95% confidence interval for μ . (Hint: You are 95% confident that *what* is between what two values?)

6. Suppose s was equal to $300. Use the 2SD Method to approximate the 95% confidence interval for μ . How does this interval compare to the interval from #5 (discuss both the midpoint and the width of the interval)? Explain why the changes you observe make intuitive sense.

The impact of sample size

7. The poll involved 562 men and 477 women. Suppose we looked at only one of these subgroups, how do you expect the width of the intervals to compare to the intervals based on the total sample (wider, narrower, or the same width as before)? Explain briefly.

8. Suppose the women reported planning to spend an average of $704 with standard deviation $150. Compare the margin-of-error of this interval (*Hint*: 2SD) to that in #5. Was your prediction correct?

9. In general, how does the width of a confidence interval change as the sample size increases? Why?

You should have seen that a smaller sample size produced a wider confidence interval. This makes sense because a smaller sample size conveys less information and therefore leaves more uncertainty about the actual value of the population mean. More directly, as we discussed in Chapters 1 and 2, increasing the sample size reduces the variability in the null distribution, thus the SD of the null distribution will be smaller, and so the confidence interval will also be narrower.

You also saw in Section 3.3 that the confidence level will impact the width of the interval as well. So with quantitative data, there are three main factors that impact the width of the confidence interval, the sample standard deviation (variability within the sample), the sample size (variability between samples), and the confidence level.

Key idea: The width of a confidence interval for a population mean, μ, decreases

a) as the confidence level decreases (for a fixed sample size and fixed sample standard deviation), or

b) as the sample size increases (for a fixed confidence level and fixed sample standard deviation), or

c) as the sample standard deviation decreases (for a fixed confidence level and fixed sample size).

The Gallup poll did give us a bit more information, see Table 3.6.

Table 3.6: Categorized responses to how much money respondents expected to spend on Christmas gifts 2013 (Gallup.com)

	$1,000 or more	$500-999	$250-499	$100-249	Under $100	No opinion	Median	Mean (w/ zero)	Mean (w/o zero)
2013 Nov 7-10	26	21	16	19	4	14	$500	$704	$773

10. Based on this information, which appears to be a more reasonable guess ($150 or $300) for the sample standard deviation? Explain your reasoning.

11. Based on this information, do you think the theory-based inference method is valid for these data? (*Hint*: Do you think the sample data are roughly symmetric?)

The information provided in Table 3.6 is enough to tell us that the sample distribution is skewed. We can consider the 2SD and Theory-Based approaches valid as long as the sample size is large. Our original sample size of 1,039 respondents should be sufficient to consider these methods valid. But with such a strong skewness in spending amounts, you may want to be cautious or consider first transforming your data.

Exploration 3.4B: Reese's Pieces

We have calculated 90%, 95%, and 99% confidence intervals, and we have seen that using a higher confidence level produces a larger margin-of-error and therefore a wider confidence interval. But what does it mean to say that we have "95% confidence" that an interval contains the parameter of interest? The following exploration leads you to investigate this question.

Reese's Pieces candies come in three colors: orange, yellow, and brown. Suppose that you take a random sample of candies and want to estimate the long-run probability that a candy is orange. Let's assume for now (although we would not know this when conducting the study) that this long-run probability, symbolized by π , is equal to 0.5.

1. Suppose that you take a random sample of 100 Reese's Pieces candies and find the sample proportion of orange. Is there any guarantee that the sample proportion will equal 0.5?

2. Suppose we calculate a confidence interval from this sample proportion. Is there any guarantee that the interval will contain the value 0.5?

3. Suppose that you select another random sample of 100 Reese's Pieces candies. Is there any guarantee that the sample proportion will be the same as for the first sample? Will the confidence interval based on the new sample necessarily be the same as the confidence interval based on the first sample?

4. To explore the behavior of confidence intervals arising from different random samples with a sample size of 100 from the same population with 50% orange, open the **Simulating Confidence Intervals** applet.

    - Make sure that **Proportions** and **Wald** are selected from the pull-down menus.
    - Set π to be 0.5, and *n* to be 100.
    - Let **Intervals** be set to 1.
    - Make sure that **Conf level** is set to 95%.
    - Press **Sample**.
    - Notice that this adds an interval (depicted by a horizontal line) to the graph in the middle with a dot corresponding to the observed sample proportion value to the graph labeled "*Sample Statistics*." Click on this dot. Doing this reveals the value of the sample proportion $\hat{p}$ and the endpoints of the 95% confidence interval based on the sample proportion (at the bottom).

    Record the following:

    a. Sample proportion:

    b. Endpoints of the 95% confidence interval

        Lower endpoint: Upper endpoint:

    c. Is the sample proportion within the confidence interval? Explain why this does not surprise you.

    d. Is the known long-run probability ($\pi = 0.5$) within the confidence interval?

Notice that 0.5 is either in the interval reported in #4b or not. Suppose the interval turned out to be (0.473, 0.537). It's not like 0.5 is sometimes between 0.473 and 0.537 and sometimes not between those two values. Similarly, even if we didn't know the value of π, the long-run probability is still some value, it doesn't change if you take a different sample. Therefore, it's technically incorrect to make a statement like "there is a 95% probability that π is between 0.473 and 0.537." This doesn't fit our "long-run proportion" interpretation of *probability*. So what is happening 95% of the time?

5.   Return to the **Simulating Confidence Intervals** applet.

   - Change **Intervals** from 1 to 99 for a total of 100 random samples from a process where $\pi = 0.5$.
   - Press *Sample*.
   - Notice that this produces a total of 100 possible values of a sample proportion which are graphed in the graph labeled "*Sample Statistics*" and the 100 confidence intervals (with 95% confidence) that are based on these sample proportions.

   a.   Study the graph labeled "*Sample Statistics*" closely. At about what number is this graph of sample proportions centered? Were you expecting this? Why or why not?

   b.   Notice that some of the dots on the "*Sample Statistics*" graph are colored red and some are colored green.

      - Click on any one of the *red* dots. Doing this will reveal the value of the sample proportion and the corresponding 95% confidence interval. Record the following:

      i   Sample proportion

      ii   Endpoints of the 95% confidence interval

      iii   Is the sample proportion within the confidence interval?

      iv   Is the probability you set for the overall process ($\pi = 0.5$) within the confidence interval?

      - Now click on any one of the *green* dots. Doing this will reveal the value of the sample proportion and the corresponding 95% confidence interval. Write down the following.

       i   Sample proportion

      ii   Endpoints of the 95% confidence interval

     iii Is the sample proportion within the confidence interval?

     iv   Is the long-run probability of orange ($\pi = 0.5$) within the confidence interval?

- The graph in the middle displays the 100 different 95% confidence intervals, depicting in green the ones that succeed in capturing the actual value of the long-run probability (which, you'll recall we assumed to be $\pi = 0.5$), and in red the intervals that fail to capture the actual value of π.

c. In the bottom left corner, the applet maintains a "*Running total*" for the percentage of the intervals produced so far that contain $\pi = 0.5$. Report this percentage here. That is, what percentage of the 100 random intervals you have generated succeeded in capturing the value of 0.5 inside the interval?

d. Now press the **Sample** button a few times, and watch how, if at all, the percentage reported under "*Running total*" changes. Now record what you think the percentage would be if you were to repeat this process forever.

e. Based on your observations from #5a – d, fill in the blanks:

Thus, 95% confidence means that if we repeatedly sampled from a process and used the sample statistic to construct a 95% confidence interval, in the long run, roughly _____% of all those intervals would manage to capture the actual value of the long-run probability, and the remaining _____% would not.

6. Now consider changing the confidence level to 90%.

a. Before you make this change, first predict what will happen to the resulting confidence intervals:

    i.   How will the widths of the confidence intervals change (if at all)?

ii.   How will the breakdown of successful/unsuccessful (green/red) intervals change? (What percentage of intervals do you suspect will be green in the long-run?)

b. Now change **Conf level** to 90% and (watch what happens as you) press **Sample**.

i.   How do the widths of the intervals change? Why does this make sense?

ii.   How does the running total change? Why does this make sense?

Key idea: The *confidence level* indicates the long-run percentage of confidence intervals that would succeed in capturing the (unknown) value of the parameter, if random samples were to be taken repeatedly from the population/process and a confidence interval produced from each sample. So, a 95% confidence level means that 95% of all samples would produce an interval that succeeded in capturing the unknown value of the parameter.

---

c. Now consider making the sample size four times larger: 400 candies, while keeping the confidence level at 90%.

d. Before you make this change, first predict what will happen to the resulting confidence intervals:

i.   How will the widths of the confidence intervals change (if at all)?

ii.   How will the percentage of successful (green) intervals change?

e. Now change *n* to 400 and (watch what happens as you) press **Sample**.

     i.   How do the widths of the intervals change? Why does this make sense?

     ii.   How does the running total change? Why does this make sense?

f.   To check your understanding about what 99% confidence means, suppose that you calculate a 99% confidence interval for the long-run probability of orange to be (.461, .589). For each of the following statements, indicate whether or not it is valid. If you think it is invalid, explain why.

     i.   There is a 99% chance that the long-run probability that a Reese's Pieces candy is orange is between 0.461 and 0.589.

     ii.   We are 99% confident that the long-run probability that a Reese's Pieces candy is orange is between 0.514 and 0.566.

     iii.   If we were to repeat the process of taking random samples of 100 Reese's Pieces, and then calculate a 99% confidence interval from the sample data for each of those samples, then in the long run, 99% of all those confidence intervals would contain the long-run probability that a Reese's Pieces candy is orange.

Section 3.4 Summary

Confidence intervals generally have the form: **statistic ± margin-of-error** where the margin-of-error is roughly approximated by two times the standard deviation of the statistic (for 95% confidence). For other confidence levels, we use a different multiplier. So the "accuracy" of our confidence interval depends on that multiplier and on the standard deviation of the statistic.

In estimating a population proportion (or process probability), the **midpoint** of the confidence interval is the observed value of the sample proportion, $\hat{p}$, and the **margin-of-error** depends on two main factors:

- Using a larger confidence level produces a larger margin-of-error and a wider confidence interval.
- A larger sample size produces a smaller margin-of-error and a narrower confidence interval.
- A factor we have less control over is the value of the sample proportion. Values closer to 0.5 generally correspond to wider intervals.

In estimating a population mean, the **midpoint** of this confidence interval is the observed value of the sample mean, $\bar{x}$, and the **margin-of-error** depends on three main factors:

- Using a larger confidence level produces a larger margin-of-error and a wider confidence interval.
- A larger sample size produces a smaller margin-of-error and a narrower confidence interval.
- A smaller sample standard deviation produces a smaller margin-of-error and a narrower confidence interval.

Notice that the confidence level and sample size are chosen separately; neither one determines the other. Exploration 3.4B also illustrated how the confidence level provides a measure of how reliable our confidence interval method is, how often, in the long run, we expect the confidence interval to successfully capture the actual parameter value of interest.

Section 3.5: Cautions When Conducting Inference

Introduction

You have now studied many of the concepts related to drawing inferences about populations and processes based on samples. In this section we will examine several cautions related to making such inferences. Some of these cautions are related to difficulties with selecting a random sample, but some difficulties arise even with a random sample (part A of this section). More issues are related to the difference between statistical significance and practical importance and to the kinds of errors that can occur with tests of significance (part B of this section).

Example 3.5A: The Controversial "Bradley Effect"

Tom Bradley, long-time mayor of Los Angeles, ran as the Democratic Party's candidate for Governor of California against Republican candidate George Deukmejian in 1982. Political polls of samples of likely voters showed Bradley with a substantial lead in the days before the election. Even exit polls favored Bradley significantly, with many media outlets projecting Bradley as the winner. However, Bradley ended up narrowly losing the overall race. After the election, research suggested that a smaller percentage of white voters had voted for Bradley than polls had predicted and a very large proportion of voters who, in the polls, claimed to be undecided, had voted for Deukmejian.

Think about it: What are some explanations for this discrepancy?

Some have argued that the reason for the discrepancies in the polling numbers and the outcome of the election was what has since come to be known as the "Bradley effect." Namely, that the way likely voters answered the questions was subject to "social desirability bias"—meaning that some people answered the polling questions in the way that they thought the interviewer wanted them to answer—the politically correct way. Tom Bradley is black and George Deukmejian is white, so many respondents may have indicated they would vote for the black candidate, but then, given the anonymity of the voting booth, they voted for the white candidate.

Since the 1982 election, the Bradley effect has been cited in numerous races and extended beyond race to include gender and other stances on political issues. More recently, in the 2008 presidential primaries for the Democrat candidate, the Bradley effect was cited as a possible explanation for unexpected polling versus voting results in the New Hampshire, and other primaries. In the 2008 New Hampshire democratic primary, Hillary Clinton received 112,404 votes compared to Barack Obama's 104,815 votes. As 287,527 voters participated in the primary, this meant that Obama received only 36.45% of the primary votes compared to 39.09% of the voters selecting Clinton. This result shocked many, because in the days leading up to the primary, Obama seemed to hold a commanding lead over Clinton. Specifically, in a *USA*

Today/Gallup poll of 778 likely voters, only days before the primary, 319/778 or 41% of likely voters said they would vote for Obama, compared to only 218/778 or 28% of likely voters saying they would vote for Clinton. How unlikely are the Clinton and Obama poll numbers given that 39.09% and 36.45% of actual primary voters voted for Clinton and Obama, respectively?

We will use a test of significance to evaluate this question, but let's talk about the assumptions our chance model will be making before we apply it.

We're assuming that the 778 people in the survey are a good representation of the people who will vote on primary day. But, there is no way that the 778 people in the survey are a simple random sample of primary voters. In order to take a simple random sample, pollsters would need to have a list of all voters in the election, and randomly choose some, but that list (sampling frame) is impossible to obtain.

Instead pollsters used **random digit dialing** and asked whether respondents were planning to vote in the upcoming Democratic primary. Only 9% of people (a total of 8644 people were contacted) agreed to participate in the survey. Of these, 319 and 218, respectively, said that they planned to vote for Obama and Clinton in the upcoming election.

So, our model makes the following assumptions:

1. Random digit dialing is a reasonable way to get a representative sample of likely voters
2. The 9% of individuals reached by phone who agree to participate are like the 91% who didn't
3. Voters who said they plan to vote in the upcoming Democratic primary will actually vote in the upcoming primary

4. Respondents' answers to who they say they will vote for matches who they actually vote for in the primary

The parameter of interest in this study, denoted by π, is the probability a poll respondent will claim he or she to vote for Obama. The question is whether this parameter value is the same as the proportion of all likely voters who actually vote for Obama (observed on primary day to be 0.3645). If the Bradley effect is at play, the value of our parameter will be larger than 0.3645, people are more likely to say they will vote for Obama than they actual are.

So the null and alternative hypotheses for our test of significance are:

H_0: The proportion of voters who will claim he or she plans to vote for Obama is 0.3645 ($\pi = .3645$)

H_a: The proportion of voters who will claim Obama is higher than 0.3645 ($\pi > .3645$)

Here are the results from a simulation where we simulated selecting 1000 samples of 778 individuals randomly chosen from a large population where 36.45% of the individuals in the population vote for Obama (with the population proportion equals 0.3645). We see in the output in Figure 3.14 that the probability of getting a sample proportion of at least $319/778 \approx 0.41$ successes is very small if the population proportion were .3645 (theory-based p-value is 0.0042 with $z = 2.64$). In other words, it's very surprising for 41% to say they plan to vote for Obama in our poll if in reality only 36.45% plan to vote for Obama.

Figure 3.14: Simulated-based p-value for testing whether significantly more respondents indicated they planned to vote for Obama vs. Clinton

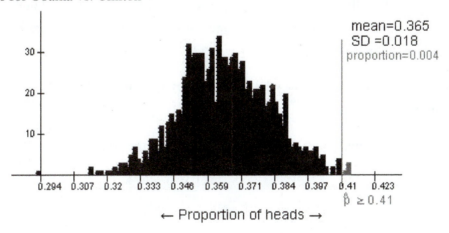

So, the Obama poll result of 41% is significantly different from what happened on primary day. We have convincing evidence that the discrepancy between what people said they would do and how they actually

voted is not explained by random chance alone (from the random sampling of the poll), suggesting that at least one of the four model assumptions is not true. Let's look at these explanations now.

Assumption #1. Random digit dialing is a reasonable way to get a sample of likely voters

Random digit dialing involves having a computer dial random phone numbers (in this case random digits after a New Hampshire area code) in order to select participants.

Random digit dialing is roughly equivalent to a simple random sample of all New Hampshire residents who have a landline or cell phone, except for slightly over-representing individuals who have more than one phone. Random digit dialing is a common survey technique in cases where a sampling frame (list of all members of the population) is unavailable.

Assumption #2. The 9% of individuals reached by phone who agree to participate are like the 91% who didn't

Only 9% of individuals reached by phone agreed to participate. The 91% of people who didn't participate includes people who didn't answer their phone and individuals who said they did not wish to participate. The assumption is that the respondents are like the non-respondents. Although the ***response rate*** was very low, it is in line with many polls and other surveys conducted by phone. So, though it is possible for non-respondents to be the cause of the bias observed, many other political surveys (of different candidates and other voting issues) conducted around the same time had similar response rates, but no bias. Of course, there is no guarantee that the 9% are representative.

Assumption #3. Voters who said they plan to vote in the upcoming Democratic primary will vote in the upcoming primary

It is typical to ask voters whether they plan to vote in the upcoming election/primary. But, there is no guarantee that they actually will. Sometimes the pollster will ask if the person voted in the previous election to gauge their commitment to voting in the upcoming election, but there is no guarantee they answer that question accurately either.

Assumption #4. Respondents answers to who they say they will vote for, matches who they actually vote for in the primary

There is no guarantee that people won't do something different in the voting booth than they say they will do when on the phone.

Explanation

Because of the wide disparity in the *USA Today/Gallup* poll, as well as nearly every other poll conducted prior to the New Hampshire primary, the AAPOR conducted an independent investigation and concluded the following were among the most likely explanations for the discrepancies:

1. People changed their opinion about who they were voting for at the last minute (Assumption #4)

2. People in favor of Hillary Clinton were more likely to be non-respondents. Non-respondents were more likely union households and those with less than a college education—groups that were more likely in favor of Clinton (Assumption #2)

3. Social desirability based on the race of the interviewer: black telephone interviewers were more likely to generate respondents who were in favor of Obama, than were white interviewers, even though the race of the interviewer was randomly assigned to survey participants (Assumption #4). This is an example of the Bradley effect.

4. Clinton was listed before Obama on every ballot (Assumption #4)

If Assumption 4 is not valid then even a pure random sample would have still exhibited this discrepancy in the results. Simple random samples should produce a representative sample, but they do nothing to control for the actions of the individuals in the sample. Having respondents change their minds or misrepresent their answers are examples of ***non-random errors***, reasons why the statistic may not be close to the parameter that are separate from random errors (sample to sample variability by chance alone).

Furthermore, in this case, with only 9% of those called choosing to participate, there is the potential for large differences between the sample and population on any number of characteristics. In particular, because the resulting sample willing to participate is not a random sample, there is no guarantee that it is representative of likely primary voters.

Key idea: Non-random (systematic) errors would still be a concern even if you could interview every single member of the population. In addition to the suspect memories and truthfulness of individuals when answering survey questions, other examples of non-random errors include the effects that the wording of a question can have on people's responses (e.g., how extreme the phrasing is), and the effects that the interviewer can have by virtue of his or her demeanor, gender, race, and other characteristics.

Exploration 3.5A: Voting for President

In the 1998 General Social Survey, a random sample of 2613 adult Americans revealed that 1783 claimed to have voted in the 1996 Presidential election.

1. Calculate the sample proportion who claimed to have voted. What symbol do we use for this value?

2. Describe (in words) the parameter that this sample proportion estimates.

3. Use the **Theory-Based Inference** applet to produce a 99% confidence interval for the population parameter.

4. Do you have 99% confidence that this interval captures the actual proportion of adult Americans who voted in the 1996 Presidential election? Explain. (*Hint*: Think about a subtle distinction between the parameter and what this question asks.)

5. The Federal Election Commission reported that 49.0% of those eligible to vote in the 1996 Presidential election actually voted. Is this value included within your 99% confidence interval?

6. Do not bother to conduct a significance test. But if you were to conduct a significance test of whether the sample data provide strong evidence to reject that the population proportion differs from 0.490, what can you say about how the p-value would turn out? Explain. (*Hint*: Base your answer on the 99% confidence interval.)

This example illustrates that a very difficult issue when conducting surveys is trying to elicit truthful responses from respondents. People may forget, or be unsure of how to answer an opinion question, or may outright lie in response to survey questions, especially if there is any "social expectation" to their answers. Statistical inference techniques such as confidence intervals and significance tests do not take these possible sources of ***non-random error*** into account.

> **Key idea**: Non-random (systematic) errors would still be a concern even if you could interview every single member of the population. In addition to the suspect memories and truthfulness of individuals when answering survey questions, other examples of non-random errors include the effects that the wording of a question can have on people's responses, and the effects that the interviewer can have by virtue of his or her demeanor, gender, race, and other characteristics.

Another famous case of problems in Presidential election polling

Founded in 1890, The *Literary Digest* was a popular magazine in the early 1900s. The magazine had correctly predicted the outcomes of the 1916, 1920, 1924, 1928, and 1932 presidential elections by conducting polls. With the Great Depression in full swing, the magazine ventured forth in 1936 to predict another presidential election outcome. Questionnaires were mailed out to more than 10 million adult Americans whose names and addresses were obtained from subscribers to the magazine and also from phone books and vehicle registration lists. More than 2.4 million responses were received, the largest survey that had ever been undertaken at that time, with 57% indicating that they planned to vote for Republican challenger Alf Landon over Democrat incumbent Franklin Roosevelt.

7. Identify the population and sample in this study. Also indicate (in words) what the relevant parameter and statistic are. If either of these values can be determined, perform the calculation(s).

> Population:
>
> Sample:
>
> Parameter: Known?
>
>
>
> Statistic: Known?

8. Do you expect a 99.9% confidence interval based on this sample result to be narrow or wide? Explain why.

With such a huge sample size, we should have no hesitation in using the theory-based inference procedures for one proportion.

9. Use the **Theory-Based Inference** applet to produce a 99.9% confidence interval for the proportion of all adult Americans who intended to vote for Alf Landon in 1936. Report the value of the statistic and margin-of-error, as well as the endpoints of the interval.

Estimate: Margin-of-error:

Endpoints:

10. Did your answer to #8 turn out to be correct?

11. Alf Landon actually received 36.5% of the votes cast in the election. Does your 99.9% confidence interval include this value? Does it come close?

12. Do you think the *Literary Digest*'s extremely misleading result can be blamed on selecting a non-representative sample or on non-sampling issues? Explain your answer.

The *Literary Digest* fiasco again illustrates the difficulty of selecting a representative sample from a population, even when a very large sample is selected. ***Voluntary response*** is an issue with many surveys, because those who choose to respond may be systematically different than those who choose not to respond.

13. An up-and-coming pollster named George Gallup made his first election prediction in 1936, the same year as the infamous *Literary Digest* poll. Gallup surveyed a sample of only 50,000 adult Americans and was able to predict the election result correctly. Do you think Gallup just got lucky, or how do you suspect that he did better than *Literary Digest*?

The secret to Gallup's success was ***random sampling***. His success revealed that a random sample, even of a smaller size, is much more informative and valuable than a large sample from a biased sampling method. For any study that fails to use random sampling methods to select the sample, you should be very cautious in generalizing the results to a larger population.

Also be aware that issues not associated with selecting a sample should be considered when drawing conclusions from survey results. People can often give inaccurate replies as a result of a poorly worded question, not being able to remember things correctly, or simply lying.

Key Idea: Consider two kinds of "errors":

(1) *Non-random errors are mistakes.* They come from a bad choice of sampling methods, such as voluntary response and convenience sampling. They can also occur from biased ways of asking a question, bias from non-response, etc. Theory can't help fix non-random errors after the data have been collected.

(2) *Random errors are not mistakes.* They are not errors in the everyday sense. These "errors" are planned; they come from the deliberate use of random chance to choose your sample. The margin-of-error tells the likely size of these errors. (Even though the term "error" is misleading. The word is still used because of its history in science.)

Summary

Non-random (systematic) errors describe reasons why sample results may not reflect the actual population for reasons other than poor sampling techniques. Examples of sources of non-random errors include:

- Wording of a question can make people respond differently.
- Order of questions can influence how a person responds.
- "Social desirability" can influence how a person responds to a question.
- The interviewer can affect how someone responds.
- People sometimes forget or misrepresent the truth.
- The measurement instrument used may not be accurate or valid or reliable.
- The sample is not taken randomly and may not be representative of the population
- The participants are selected randomly, but because people choose not to participate the sample is no longer representative.

Random errors are a way to describe sample-to-sample (chance) variability that is accounted for by the margin-of-error.

Example 3.5B: Parapsychology Studies (continued)

Recall the parapsychology study from Exploration 2.3. The receivers in the study tried to identify the target image that was being transmitted by the sender, from a group of four possible target images. Statistician Jessica Utts reported that her analysis considered a total of 2124 sessions and found a total of 709 "hits" (Utts, 2010).

For testing whether the sample data provide strong evidence that the long-run probability of a successful transmission exceeds 1/4, the *z*-statistic turns out to be 8.9. In other words, the observed sample proportion of 0.334 is 8.9 standard deviations above the null value of 0.25. The p-value is essentially zero, and the observed result is statistically significant at every reasonable significance level. The test leaves virtually no doubt that the long-run probability of a correct transmission is greater than 1/4.

> **Think about it:** But is the probability of a successful transmission much greater than 1/4? How is this question different from what the test and p-value address? How could you investigate that question?

The test result and p-value only reveal that the data provide extremely strong evidence that the long-run probability of a successful transmission is greater than 1/4. The test and p-value have nothing to say about *how much* greater than 1/4 this probability is. We can address this question (of how much greater) by producing a confidence interval to estimate the long-run probability of a successful transmission.

Using the theory-based method, which is justified by the large number of successes (709) an failures (1415) in the sample, a 95% confidence interval for this probability turns out to be (0.314, 0.354), and a 99% confidence interval is (0.307, 0.360).

> **Think about it**: Determine the margin-of-error for the 95% confidence interval. Why are these confidence intervals quite narrow? Which interval is wider?

The margin-of-error for the 95% confidence interval is ± 0.020, because the interval can be written as 0.338 ± 0.020. These confidence intervals are quite narrow because the sample size (2124 sessions) is fairly large. The 99% confidence interval is wider than the 95% confidence interval, because achieving more confidence requires a wider interval (more plausible values).

> **Think about it**: Would you say that these confidence intervals contain values that are *much* greater than 1/4?

Even with a confidence level of 99%, the interval extends only to the value 0.36. Is a 36% success percentage much greater than 25%?

On the one hand, these data suggest that there may be some validity to the claims of psychics; the significance test leaves little doubt that the long-run probability of a successful transmission in these studies is greater than 1/4. However, it is important to realize that although there is strong evidence that psychics *are* doing better than guessing, the confidence interval reveals that they are not doing *much* better than guessing.

How practically important is that fact? That's a judgment call. But, this brings up an important point: just because a result is statistically significant, does not mean that it is practically important. Of course, whether or not a result is of practical importance can be a subjective opinion left somewhat up to the person interpreting the research. The important thing to remember is that, especially when the sample size is large (so your confidence interval will be narrow), you may have a statistically significant result that doesn't necessarily provide a very practically important result.

> **Key idea:** *Significant is not the same as important.*
> (1) "Significant" means "statistically significant" and that means "too big to be likely due to chance."
> (2) "Important" means "practically important" and that means the difference is big enough to matter in the real world.
> To be clear, always use the adjective: *statistically* significant, *practically* important.

> **Key idea:** Especially with large sample sizes, even a fairly small difference between an observed and hypothesized value can be statistically significant. In such cases a confidence interval for the parameter can help to decide whether the difference is ***practically important***.

The article by Utts (2010) also refers to 56 individual studies that led to the combined total of 2124 sessions that you have analyzed here. Most of these studies involved a sample size of about 50 sessions.

> **Think about it**: Suppose that the actual value of the long-run probability of a successful transmission is 1/3 rather than 1/4. In this situation, do you think it's very likely that a sample of 50 sessions will lead to correctly rejecting the null hypothesis that the long-run probability is 1/4? Why or why not?

If π is equal to 1/3 and we fail to reject 1/4 as a plausible value, then we have committed a Type II Error (see Section 2.3). The question now is how often we can expect to make such an error. As we discussed in Section 2.3, researchers usually talk about one minus the probability of a type II error------the *power* of a study. Recall that power can be thought of as the probability of correctly rejecting a false null hypothesis.

Let's investigate this question with simulation.

The graph on the left in Figure 3.15 shows the could-have-been results (sample proportions) from simulating 10,000 samples of 50 sessions with a 0.25 probability of success per session. If we are using the $\alpha = 0.05$ significance level, then we will reject the null hypothesis that $\pi = 0.25$ whenever the sample proportion of successful transmissions is 0.38 or larger, because this is the smallest sample proportion that produces a p-value of 0.05 or less, as shown in the graph on the right.

Figure 3.15: Distributions of sample proportions when $\pi = 0.25$, with sample proportion corresponding to one-sided p-values below 0.05.

Now let's suppose that the actual long-run probability of success is 1/3 rather than 1/4. In this case the alternative hypothesis is true, so we want to reject the null hypothesis. But is it likely that a sample of 50 sessions would produce a sample proportion of 0.38 or greater for the proportion of successful transmissions?

Think about it: How often do you think we would obtain a sample proportion of 0.38 or more (with a sample size of 50), when the actual long-run probability of success is 1/3 rather than 1/4?

We can use another simulation to determine how likely it is to correctly reject the null hypothesis (of no psychic ability) when in fact the long-run probability of success is 1/3 (Figure 3.16). The graph on the left below shows the results of 10,000 repetitions. The corresponding graph on the right reveals that only about 30% of the 10,000 samples produced a sample proportion of 0.38 or greater. This means that there's only about a 30% chance of rejecting the null hypothesis ($\pi = 1/3$), even though the null hypothesis is false in this situation ($\pi = 1/4$). This means that the power of the test is only about 30%; there is approximately a 70% chance of making a type II error if the true success probability is 1/3.

Figure 3.16: Distributions of sample proportions when $\pi = 0.30$ and percentage that yield a sample proportion of at least 0.38

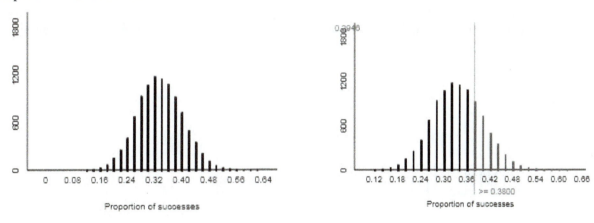

Think about it: Would increasing the *significance level* or decreasing the significance level result in greater power? Would increasing or decreasing the sample size result in greater power? Explain your answers.

The power of a test is influenced by the significance level used and by the sample size:

- Using a larger significance level results in greater power. One way to understand this is that using a larger significance level means that you don't require as strong of evidence to reject the null hypothesis. This means that it's easier to reject the null hypothesis, whether that null hypothesis is true or not. This means that you're more likely to reject the null hypothesis when it is false, which results in greater power.

      o   Greater power means a smaller probability of making a Type II error, so this is a good
           thing. The downside is that increasing the significance level means increasing the
           probability of making a Type I Error.

- Increasing the sample size results in greater power. One way to understand this is that with a larger sample size, sample proportions vary less about the true parameter value, so it's easier to obtain a sample proportion extreme enough to warrant rejecting the null hypothesis.
  - The downside is that using a larger sample size requires greater time, effort, and expense.

Power is also affected by a third factor:

- If the actual value of the parameter is farther away from the null value, then the test is more powerful.
  - So, for example, if the actual long-run probability of a successful transmission is 1/2, then the power of the test is greater than if the probability is actually 1/3. We will be much more likely to obtain a sample proportion that is far from 1/3 if the actual process probability is 1/2 rather than 1/4. (See Figure 3.17)

Figure 3.17: Distributions for sample proportions when $\pi = 0.50$ and proportion that yields a sample proportion of at least 0.38

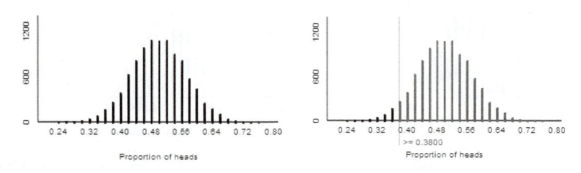

Exploration 3.5B: Cat Households

A survey of 47,000 American households in 2007 found that 32.4% own a pet cat.

1. Is the value 32.4% (or, as a decimal, 0.324) a parameter or a statistic? What symbol would you use to represent this value?

2. Describe (in words) the parameter of interest here. What symbol would you use to represent this parameter?

3. Conduct a theory-based significance test of whether the sample data provide strong evidence that less than one-third of all American households owns a pet cat. Specify the null and alternative hypotheses, and calculate the z-statistic and p-value. Also indicate whether the sample result is statistically significant at the $\alpha = 0.001$ level. Summarize your conclusion and explain how it follows from the test.

4. Determine a 99.9% confidence interval for the proportion of all American households that have a pet cat.

5. Explain why this confidence interval is so narrow.

Consider the following two possible conclusions:
- The sample data provide *very* strong evidence that less than 1/3 of all American households owns a cat.
- The sample data provide strong evidence that *much* less than 1/3 of all American households owns a cat.

6. Explain the difference between these two statements.

7. Which of these two statements can you use a p-value to assess? Which can you use a confidence interval to assess?

8. Which of the two conclusions would you draw from your analysis of these data? Explain your answer.

Key idea: Especially with large sample sizes, even a fairly small difference between an observed and hypothesized value can be statistically significant. In such cases a confidence interval for the parameter can help to decide whether the difference is *practically important*.

Now suppose that plan to take a random sample of only 100 American households and test whether the sample data provide strong evidence that less than one-third of all American households has a pet cat.

9. Use the **One Proportion** applet to simulate 1000 random samples of 100 households each, assuming that one-third of all households have a pet cat. Use trial-and-error to determine which values of the sample proportion would produce a p-value of less than 0.05. (Note that these values would lead to rejecting the null hypothesis at the 0.05 significance level.)

Now suppose that the 30% of all American households have a pet cat, so the actual value of the population parameter is 0.3.

10. Do you suspect that a test of significance, based on a random sample of 100 households, will be likely to reject the null hypothesis that one-third of all households have a pet cat, if in fact 30% of all households have a pet cat?

11. Use the **One Proportion** applet to simulate 1000 random samples of 100 households each, assuming that 30% of all households have a pet cat. Use the applet to count how many, and what proportion, of the 1000 samples produce a sample proportion of (your answer to #9) or smaller.

12. Re-answer #10 based on your answer to #11.

Recall that the probability of rejecting a false null hypothesis is called the **power** of the test. Tests with higher values of power are preferred over tests with lower values of power. Power can also be thought of as one minus the probability of a Type II Error.

13. Based on your simulation analysis in #9 and #11, what is the approximate power of this test with a sample size of 100, a significance level of 0.05, and an alternative value of 30%?

14. Do you suspect that increasing the *significance level* or decreasing the significance level result in greater power? Explain your thinking.

15. Use the **One Proportion** applet to investigate the issue in #14 by changing the significance level from 0.05 to 0.10. First re-answer #9 using 0.10 as the significance level. Then re-answer #11, again assuming that 30% of all households have a pet cat but using the sample proportions determined for the new significance level. Report the approximate power, and comment on whether power has increased or decreased from using a significance level of 0.05.

16. Do you suspect that increasing or decreasing the sample size result in greater power? Explain your thinking.

17. Use the **One Proportion** applet to investigate the issue in #16 by changing the sample size from 100 to 500. First re-answer #9 using 500 as the sample size. Then re-answer #11, again assuming that 30% of all households have a pet cat but for the sample proportions determined in #9 in a sample size of 100. Report the approximate power, and comment on whether power has increased or decreased from using a sample size of 100.

18. Do you suspect that increasing or decreasing the actual value of the population proportion with a pet cat would result in greater power? Explain your thinking.

19. Use the **One Proportion** applet to investigate the issue in #18 by changing the population proportion from 0.3 to 0.2. Report the approximate power, explaining how you found it, and comment on whether power has increased or decreased from using a sample size of 100.

Section 3.5 Summary

Keep in mind that a test of significance or a confidence interval is only as good as the data you give it. If you have sampling errors or non-sampling errors in your data collection, then the results from our simulation-based and theory-based procedures could be quite misleading.

Two other cautions to keep in mind with regard to drawing inferences concern statistical significance vs. practical importance and the power of a test.

A result can be statistically significant without being practically important. A statistically significant result is simply one that is unlikely to have occurred by chance alone (a small p-value). However, even a small p-value does not tell you how much the statistic varies from the parameter, and whether this is a meaningful difference. Practical importance can be assessed by producing a confidence interval.

Especially with large sample sizes, a sample result can be statistically significant without being practically important. Judging the practical importance of a result depends upon the context and is best judged by the researchers who know the context well.

The power of a test is the probability of rejecting the null hypothesis when the null hypothesis is false. Especially with a small sample size, a significance test may fail to detect an effect that may really be present because the test is not powerful enough. Factors that influence power are:

- Using a larger significance level results in greater power. One way to understand this is that using a larger significance level means that you don't require as strong of evidence to reject the null hypothesis. This means that it's easier to reject the null hypothesis, whether that null hypothesis is true or not. This means that you're more likely to reject the null hypothesis when it is false, which results in greater power.
    - o Greater power means a smaller probability of making a Type II error, so this is a good thing. The downside is that increasing the significance level means increasing the probability of making a Type I Error.
- Increasing the sample size results in greater power. One way to understand this is that with a larger sample size, sample proportions vary less about the null value, so it's easier to obtain a sample proportion extreme enough to warrant rejecting the null hypothesis.
    - o The downside is that using a larger sample size requires greater time, effort, and expense.
- If the actual value of the parameter is farther away from the null value, then the test is more powerful.

Chapter 3 Glossary

2SD Method
  Approximating a confidence interval by taking the statistic and the standard deviation of the statistic (from simulation or formula) and extending two standard deviations in each direction from the statistic.. 292, 298

confidence intervals
  An inference tool used to estimate the value of the parameter, with an associated measure of uncertainty due to the randomness in the sample data.. 274

confidence level
  A statement of reliability in the confidence interval method................................ 280, 286, 324,

margin-of-error
  The half-width of a confidence interval.. 292, 298

non-random errors
  Reasons why the statistic may not be close to the parameter that are separate from random errors which are accounted for by quantifying chance variability.. 334

one-sample z-interval
  Theory-based confidence interval for a population proportion 296

plausible value
  A parameter value tested under the null hypothesis where, based on the data gathered, we do not find strong evidence against the null 278

practically important
  A large enough discrepancy between the statistic and the hypothesized parameter to be of interest in context.. 336, 341

random digit dialing
  A common sampling technique when a sampling frame is unavailable. It involves a computer randomly dialing phone numbers within a certain area code by randomly selecting the digits to be dialed after the area code.. 327

response rate
  Of those selected to be in the sample, the percentage that respond. 329

validity condition for one-sample z-procedures

validity conditions for one-sample t-interval

A sample size of at least 20 or the distribution of the quantitative variable is not highly

Chapter 3 Exercises

Section 3.1

3.1.1* Let π denote some population proportion of interest and suppose a 95% confidence interval for π is calculated to be (0.6, 0.7). Also, suppose that we want to test

$$H_0: \pi = 0.63 \text{ vs. } H_a: \pi \neq 0.63.$$

What can you say about the corresponding p-value?
  A.   The corresponding p-value will be smaller than 0.05.
  B.   The corresponding p-value will be larger than 0.05
  C.   Need more information to answer this.

3.1.2 Let π denote some population proportion of interest and suppose a 99% confidence interval for π is calculated to be (0.6, 0.7). Also, suppose that we want to test

$$H_0: \pi = 0.74 \text{ vs. } H_a: \pi \neq 0.74$$

What can you say about the corresponding p-value?
  A.   The corresponding p-value will be smaller than 0.05, but larger than 0.01.
  B.   The corresponding p-value will be larger than 0.05.
  C.   The corresponding p-value will be smaller than 0.01.
  D.   I can't say anything about the corresponding p-value until I run the test.

3.1.3* Let π denote some population proportion of interest and suppose a 95% confidence interval for π is calculated to be (0.63, 0.73) and a 99% confidence interval for π is calculated to be (0.61, 0.75). Also, suppose that we want to test

$$H_0: \pi = 0.74 \text{ vs. } H_a: \pi \neq 0.74$$

What can you say about the corresponding p-value?
  A.   The corresponding p-value will be smaller than 0.05, but larger than 0.01.
  B.   The corresponding p-value will be larger than 0.05.
  C.   The corresponding p-value will be smaller than 0.01.
  D.   I can't say anything about the corresponding p-value until I run the test.

Cayle the cat

3.1.4 Suppose I am conducting a test of significance where the null hypothesis is my cat Cayle will pick the correct cancer specimen 25% of the time and the alternative hypothesis is that she will pick the cancer specimen at a rate different than 25%. I end up with a p-value of 0.002. I also construct 95% and 99% confidence intervals from my data. What will be true about my confidence intervals?

A. Both the 95% and the 99% intervals will contain 0.25.
B. Neither the 95% nor the 99% intervals will contain 0.25.
C. The 95% interval will contain 0.25, but the 99% interval will not contain 0.25.
D. The 95% interval will not contain 0.25, but the 99% interval will contain 0.25.

Grayce the cat*

3.1.5 Suppose I am conducting a test of significance where the null hypothesis is my cat Grayce will pick the correct cancer specimen 25% of the time and the alternative hypothesis is that she will pick the cancer specimen at a rate different than 25%. I end up with a p-value of 0.02. I also construct 95% and 99% confidence intervals from my data. What will be true about my confidence intervals?
A. Both the 95% and the 99% intervals will contain 0.25.
B. Neither the 95% nor the 99% intervals will contain 0.25.
C. The 95% interval will contain 0.25, but the 99% interval will not contain 0.25.
D. The 95% interval will not contain 0.25, but the 99% interval will contain 0.25.

Cayce the cat

3.1.6 Suppose I am conducting a test of significance where the null hypothesis is my cat Cayce will pick the correct cancer specimen 25% of the time and the alternative hypothesis is that he will pick the cancer specimen at a rate different than 25%. I end up with a p-value of 0.07. I also construct 95% and 99% confidence intervals from my data. What will be true about my confidence intervals?
A. Both the 95% and the 99% intervals will contain 0.25.
B. Neither the 95% nor the 99% intervals will contain 0.25.
C. The 95% interval will contain 0.25, but the 99% interval will not contain 0.25.
D. The 95% interval will not contain 0.25, but the 99% interval will contain 0.25.

Spinning rackets*

3.1.7 Tennis players often spin a racquet to decide who serves first. The spun racquet can land with the manufacturer's label facing up or down. A reasonable question to investigate is whether a spun tennis racquet is equally likely to land with the label facing up or down. (If the spun racquet is equally likely to land with the label facing in either direction, we say that the spinning process is fair.) Suppose that you gather data by spinning your tennis racquet 100 times, each time recording whether it lands with the label facing up or down. Further suppose that the 100 spins result in 46 with the label facing up and 54 with the label facing down. The resulting 95% confidence interval on the long-run probability of the racket landing face up is 0.36 to 0.56. This means that:

    A. There is strong evidence the racket lands face up less than 50% of the time

    B. There is not strong evidence that the racket lands face up less than 50% of the time

3.1.8 When conducting a two-sided test of significance, the value of the parameter under the null hypothesis is plausible and will be contained in a 95% confidence when the p-value is:

    A. Less than or equal to 0.05.

    B. Greater than 0.05

    C. There is no relationship between the p-value and the confidence interval.

3.1.9* When conducting a two-sided test of significance, the value of the parameter under the null hypothesis is not plausible and will not be contained in a 95% confidence when the p-value is:

    A. Less than or equal to 0.05.

    B. Greater than 0.05

    C. There is no relationship between the p-value and the confidence interval.

3.1.10 We use confidence intervals to estimate the value of:

    A. The population proportion.

    B. The sample proportion.

    C. The p-value.

    D. The standard deviation.

3.1.11* When we say that 0.5 is a plausible value for the population proportion, plausible means 0.5 will:

    A. Be in the 95% confidence interval

    B. Not be in the 95% confidence interval

3.1.12 Suppose a 95% confidence interval is constructed from a sample proportion and 0.5 is contained in the interval. Which of the following are true?

    A. A 90% confidence interval constructed from the same sample proportion will definitely contain 0.5. = no

    B. A 90% confidence interval constructed from the same sample proportion will definitely NOT contain 0.5.

    C. A 99% confidence interval constructed from the same sample proportion will definitely contain 0.5.

    D. A 99% confidence interval constructed from the same sample proportion will definitely NOT contain 0.5.

Dining out*

3.1.13 What proportion of San Luis Obispo (SLO) residents dine at restaurants at least once a week? To investigate, a local high school student, Deidre, decides to conduct a survey. She selects a random sample of adult residents of SLO, asks each participant whether he/she dines at restaurants at least once a week, and records their responses. Then, she uses her data to find a 95% confidence interval for the proportion of SLO adults who dine out at least once a week to be (0.38, 0.44).

    a) The 99% confidence interval based on the same data would be

        A. Narrower, because the more confident we are the narrower the interval

        B. Wider, because to be more confident we need to widen the interval

        C. Need more information to answer this

    b) According to the 95% confidence interval

        A. There is no evidence that less than half of SLO residents dine at restaurants at least once a week, because the interval does not contain 0.5.

        B. There is convincing evidence that less than half of SLO residents dine at restaurants at least once a week, because the interval does not contain 0.5.

        C. None of the above, because the confidence interval is unrelated to strength of evidence.

3.1.14 To determine a 95% confidence interval using the interval of plausible values method, what significance level should be used on the two-sided tests?

3.1.15* To determine a 99% confidence interval using the interval of plausible values method, what significance level should be used on the two-sided tests?

October 2013 Gallup poll

3.1.16 Based on an October 2013 Gallup poll, a 95% confidence interval for the proportion of American adults that think a degree from a well-respected university is more important for a young person to succeed than obtaining knowledge or skills to do a specific job is 0.43 to 0.51. Explain exactly what the confidence is estimating.

August 2013 Gallup poll*

3.1.17 Based on an August 2013 Gallup poll, a 95% confidence interval for the proportion of American adults that thought math was the most valuable subject they studied in school is 0.31 to 0.37. Explain exactly what the confidence is estimating.

3.1.18 Suppose we are constructing a confidence interval using repeated tests of significance to develop an interval of plausible values. Using two-sided tests each time with the following null hypotheses, we obtain the resulting p-values.

Null	p-value	Null	p-value
Proportion = 0.45	0.014	Proportion = 0.53	0.787
Proportion = 0.46	0.032	Proportion = 0.54	0.572
Proportion = 0.47	0.062	Proportion = 0.55	0.373
Proportion = 0.48	0.126	Proportion = 0.56	0.142
Proportion = 0.49	0.371	Proportion = 0.57	0.077
Proportion = 0.50	0.598	Proportion = 0.58	0.042
Proportion = 0.51	0.733	Proportion = 0.59	0.021
Proportion = 0.52	0.986	Proportion = 0.60	0.003

a) Using the results from the table, give a 90% confidence interval.

b) Using the results from the table, give a 95% confidence interval.

3.1.19* Suppose we are constructing a confidence interval using repeated tests of significance to develop an interval of plausible values. Using two-sided tests each time with the following null hypotheses, we obtain the resulting p-values.

Null	p-value	Null	p-value
Proportion = 0.45	0.007	Proportion = 0.53	0.602
Proportion = 0.46	0.012	Proportion = 0.54	0.124
Proportion = 0.47	0.045	Proportion = 0.55	0.084
Proportion = 0.48	0.079	Proportion = 0.56	0.052
Proportion = 0.49	0.121	Proportion = 0.57	0.034
Proportion = 0.50	0.254	Proportion = 0.58	0.019
Proportion = 0.51	0.643	Proportion = 0.59	0.012
Proportion = 0.52	0.986	Proportion = 0.60	0.004

a) Using the results from the table, give a 95% confidence interval.

b) Using the results from the table, give a 99% confidence interval.

3.1.20 Construct a 95% confidence interval for a population proportion using repeated tests of significance to develop an interval of plausible values based on a sample proportion of 0.52 from a sample of 300. Use two-sided tests with the following values under the null hypothesis to find the needed corresponding p-values to construct the interval.

Null	p-value	Null	p-value
Proportion = 0.45		Proportion = 0.53	
Proportion = 0.46		Proportion = 0.54	
Proportion = 0.47		Proportion = 0.55	
Proportion = 0.48		Proportion = 0.56	
Proportion = 0.49		Proportion = 0.57	
Proportion = 0.50		Proportion = 0.58	
Proportion = 0.51		Proportion = 0.59	
Proportion = 0.52		Proportion = 0.60	

3.1.21* Construct a 95% confidence interval for a population proportion using repeated tests of significance to develop an interval of plausible values based on a sample proportion of 0.52 from a sample of 600. Use two-sided tests with the following values under the null hypothesis to find the needed corresponding p-values to construct the interval.

Null	p-value	Null	p-value
Proportion = 0.45		Proportion = 0.53	
Proportion = 0.46		Proportion = 0.54	
Proportion = 0.47		Proportion = 0.55	
Proportion = 0.48		Proportion = 0.56	
Proportion = 0.49		Proportion = 0.57	
Proportion = 0.50		Proportion = 0.58	
Proportion = 0.51		Proportion = 0.59	
Proportion = 0.52		Proportion = 0.60	

July 2012 Gallup poll

3.1.22 In a July 2012 Gallup poll based on a representative sample of 1014 adult Americans, 48% reported drinking at least one glass of soda pop on a typical day. Now suppose that we test the null hypothesis $\pi = 0.5$ vs. the alternative hypothesis $\pi \neq 0.5$.

    a) Describe (in words) what the symbol π represents here.

    b) Use the **One Proportion** applet to determine the (approximate) two-sided p-value.

    c) Based on the two-sided p-value you reported in (b), does 0.5 appear to be a plausible value for π? Explain your reasoning.

    d) How would this two-sided p-value compare to a one-sided p-value: larger, smaller, or the same?

    e) Use the **One Proportion** applet for a single proportion, repeatedly testing other possible values for π, to determine plausible values and construct a 95% confidence interval for π.

    f) Interpret the 95% confidence interval you reported in (e).

3.1.23 The previous question gave the results of the Gallup poll where 48% of a sample of 1014 adult Americans reported drinking at least one glass of soda pop on a typical day. Use the **One Proportion** applet, repeatedly testing possible values for π, to determine plausible values and construct a 99% confidence interval for π.

May 2013 Gallup poll*

3.1.24 In a May 2013 Gallup poll based on a representative sample of 1535 adult Americans, 53% responded that they thought same-sex marriages should be legal and have the same rights as traditional marriages. Now suppose we test the null hypothesis $\pi = 0.5$ vs. the alternative hypothesis $\pi \neq 0.5$.

    a) Describe (in words) what the symbol π represents here.

    b) Use the **One Proportion** applet to determine the (approximate) two-sided p-value.

    c) Based on the two-sided p-value you reported in (b), does 0.5 appear to be a plausible value for π? Explain your reasoning.

    d) Use the **One Proportion** applet, repeatedly testing other possible values for π, to determine plausible values and construct a 95% confidence interval for π.

    e) Interpret the 95% confidence interval you reported in (d).

3.1.25 The previous question gave the results of the Gallup poll where 53% of a sample of 1535 adult Americans responded that they thought same-sex marriages should be legal and have the same rights as traditional marriages. Use the **One Proportion** applet, repeatedly testing possible values for π, to determine plausible values and construct a 99% confidence interval for π.

Section 3.2

May 2011 Gallup poll*

3.2.1 A Gallup survey of 1,001 randomly selected U.S. adults conducted May 2011, asked "*In your opinion, which one of the following is the main reason why students get education beyond high school?*"

Fifty three percent chose "to earn more money." Based on these data, I found the 95% confidence interval for the proportion of U.S. adults who choose "to earn more money" to be (0.499, 0.561). Which of the following is a valid statement?

A. A 99% confidence interval based on the same data would be narrower.

B. A 99% confidence interval based on the same data would be wider.

C. A 99% confidence interval based on the same data would have the same width as the 95% confidence interval.

Recent Gallup poll

3.2.2 A recent Gallup poll showed the president's approval rating at 60%. Some friends use this information (along with the sample size from the poll) and find theory-based confidence intervals for the proportion of all adult Americans that approve of the presidents performance. Of the following four confidence intervals, identify the ones that were definitely done **incorrectly**. (There may be more than one interval that is incorrect.)

    A. (0.57, 0.63)

    B. (0.60, 0.66)

    C. (0.58, 0.62)

    D. (0.47, 0.53)

3.2.3* Suppose a 95% confidence interval for a population proportion is found using the 2SD or theory-based method. Which of the following will *definitely* be contained in that interval?

    A. The population proportion

    B. The sample proportion

    C. The p-value

    D. All of the above

3.2.4 Which of the following is true of a 2SD confidence interval?

    A. They are approximate 95% confidence intervals.

    B. They can be used for any confidence level.

    C. They give exactly the same results as a theory-based interval.

    D. The SD represents the standard deviation of the sample data.

3.2.5* Which of the following is NOT true about theory-based confidence intervals for a population proportion?

    A. They should only be used when you have at least 10 success and 10 failures in your sample data.

    B. The process used to construct the interval relies on a normal distribution.

    C. They can be calculated using different confidence levels.

D. For a given sample proportion, sample size and confidence level, different intervals can be obtained because of their random nature.

3.2.6 Suppose a 95% confidence interval for a population proportion is (0.30, 0.60). Rewrite this interval in the form of $\hat{p} \pm$ margin-of-error .

3.2.7* Suppose a 95% confidence interval for a population proportion is (0.27, 0.49). Rewrite this interval in the form of $\hat{p} \pm$ margin-of-error .

Election poll

3.2.8 To estimate the proportion of city voters who will vote for the Republican candidate in the elections, two students Manny and Nina, each decide to conduct polls in the city. Manny selects a random sample of 50 voters, while Nina selects a random sample of 100 voters. Suppose both samples result in 48% of the voters saying they will vote for the Republican candidate. Whose 95% confidence interval will have the larger margin-of-error: Manny or Nina? How are you deciding?

Board games*

3.2.9 The *2011 Statistical Abstract of the United States* includes a table reporting that a national survey in fall of 2009 found that 18.2% of American adults played a board game in the past year. We can use this sample result to estimate the proportion of *all* adult Americans who had played a board game in the past year.

  a) What additional information do you need in order to determine a 95% confidence interval for the probability that a randomly selected American adult had played a board game in the past year?

  b) Consider three possible sample sizes for this study: 250, 1000, and 4000. Without performing any calculations, how do you expect three confidence intervals based on these sample sizes to compare to each other? Explain your reasoning.

  c) Carry out the simulation at each sample size to get the SD and use the 2SD rule to approximate the margin-of-error with each of these three sample sizes. Then produce the 95% confidence interval for each of these three sample sizes.

  d) How do the midpoints of these intervals compare?

  e) How do the widths of these intervals compare?

  f) By how many times must the sample size increase in order to cut the margin-of-error in half?

  g) Suppose that the sample size was really 1000. Based on the confidence interval, do the sample data suggest that fewer than 25% of all adult Americans had played a board game in the past year?

Hearing loss in teens

3.2.10 A recent study examined hearing loss data for 1771 U.S. teenagers. In this sample, 333 were found to have some level of hearing loss. News of this study spread quickly, with many news articles blaming the prevalence of hearing loss on the higher use of ear buds by teens. At MSNBC.com (8/17/2010), Carla Johnson summarized the study with the headline: "1 in 5 U.S. teens has hearing loss, study says." To investigate whether this is an appropriate or a misleading headline, you will conduct a test of significance with the following hypotheses.

$$\text{Null: } \pi = 0.2 \qquad \text{Alternative: } \pi \neq 0.2$$

a) Describe what the symbol π stands for in this context.
b) Use the **Theory-Based Inference applet** to determine a p-value.
c) Based on your p-value is there strong evidence that the proportion of all U.S. teens with some hearing loss is different than 1 in 5 (or 20%)?
d) Using the applet, find a 95% confidence interval for the proportion of U.S. teens that have some hearing loss.
e) What is the margin-of-error for you confidence interval from part (d) of this question?
f) Based on your confidence interval, is 0.2 a plausible value for the proportion of the population that has some hearing loss? Explain why or why not.
g) Based on your p-value, is 0.2 a plausible value for the proportion of the population that has some hearing loss? Explain why or why not.

Spinning dancer*

3.2.11 The spinning dancer (or silhouette illusion) is a moving image of a woman that appears to be spinning. Some people see her spinning clockwise and some see her spinning counterclockwise. A student showed other students this, and found that 30 out of 50 (or 60%) of them saw her spinning clockwise. The student researcher was interested in the proportion of people that would see the dancer spinning clockwise. He created the following null distribution for this.

a) Using the information provided above, determine a 2SD 95% confidence interval for the population proportion of people that would see the woman spinning clockwise.

b) Based on your confidence interval, can you conclude the population proportion of people that will see the woman spinning clockwise is greater than 50%? Explain.

3.2.12 Using the sample proportion of 30 out of 50 students that saw the spinning dancer spin clockwise (see Exercise 3.2.11), answer the following.

a) Using the **Theory-Based Inference** applet, determine a 95% confidence interval for the long-run probability a randomly chosen student would see the woman spinning clockwise.

b) Using the **Theory-Based Inference** applet, determine a 95% confidence interval for the long-run probability a randomly chosen student would see the woman spinning *counterclockwise*.

Competent faces

3.2.13 In Example 1.4 we looked to see if the competent face method could be used to predict the results of Senate races. We found this method worked in 23 out of 32 races. When testing to see if the probability of this method working is more than 0.5, we obtained a null distribution similar to that shown.

a) Using the information provided above, determine a 2SD 95% confidence interval for the long-run probability the competent face method will work.

b) Based on your confidence interval, can you conclude the long-run probability the competent face method will work is greater than 50%? Explain.

c) Why shouldn't a theory-based approach be used to compute the confidence interval for these data?

3.2.14 Recall that researchers also conducted a follow-up study predicting the outcomes of races in the House of Representatives with 189 correct predictions out of 279.

   a) Use a simulation and the 2SD method to compute a 95% confidence interval.

   b) Use a theory-based approach to generate the 95% confidence interval for this data.

   c) Compare the intervals you computed in a) and b), why are they similar?

Water taste test*

3.2.15 Students at Hope College were tested to see if they could determine the difference between tap water and bottled water. Of the 63 students tested, 42 correctly identified which was which. We will assume the students tested, are representative of all students at Hope College.

   a) Find a theory-based 95% confidence interval for the population proportion that can correctly distinguish tap water from bottled water.

   b) Find a theory-based 99% confidence interval for the population proportion that can correctly distinguish tap water from bottled water.

   c) Suppose π is the proportion of all Hope College students that can correctly distinguish tap water and bottled water and are testing the hypotheses Null: $\pi = 0.5$ and Alternative: $\pi \neq 0.5$. Based on just the confidence intervals you found in the previous questions, what can you say about the value of p-value for this test?

3.2.16 Referring to the data in the previous question. Use the 2SD rule to find the 95% confidence interval. Why is it similar to the one you obtained in part (a) of the previous question?

July 2012 Gallup poll

3.2.17 In July 2012 Gallup surveyed 1014 randomly selected U.S. adults, 5% said that they considered themselves vegetarians.

   a) Describe what the parameter is in this context.
   b) The standard deviation (SD) from the simulated null distribution is 0.007. Use the 2SD method to find a 95% confidence interval for the parameter.
   c) If we had 2020 in our sample, how would that change the width of our 95% confidence interval? Would it be wider, narrower, or have the same width?
   d) Using the **Theory-Based Inference** applet we find both 95% and 99% confidence intervals as shown below. Denote which is which.

$$(0.032, 0.068) \qquad\qquad (0.037, 0.063)$$

Coke or pepsi?*

3.2.18 Statistics students were conducting a test to see if people could taste the difference between Coke and Pepsi. They fill two cups with Coke and a third with Pepsi. They then asked their subjects which cola tasted different than the other two. Of the 64 people they tested, 22 were able to correctly identify which of the three cups of colas tasted different (i.e. which of the three cups was the Pepsi).

   a) Conduct a simulation-based test of significance to determine if the proportion of people that correctly identify which cola is different is greater than what would occur by just randomly guessing. Give the hypotheses, the p-value, and a conclusion.
   b) Using the standard deviation from your null distribution in part (a), determine a 2SD confidence interval for the population proportion that can correctly identify the cola that is different?
   c) Is the proportion under your null hypothesis from part (a) contained in your confidence interval from part (b)? Explain how this relates to the size of your p-value from part (a).

3.2.19 Statistics students conducted a test to see if people could taste the difference between Coke and Pepsi. They fill two cups with Coke and a third with Pepsi. They then asked their subjects which tasted different than the other two. Of the 64 people they tested, 22 were able to correctly identify which of the three cups of colas tasted different.
   a) Determine a 95% theory-based confidence interval for the population proportion that can correctly identify the cola that is different?
   b) What is the margin-of-error from your interval from part (a)?

Handedness

3.2.20 From a random sample of 97 male students at Hope College, 12 were left-handed. Determine a 99% confidence interval for the proportion of all male students at Hope College that are left-handed.

3.2.21 From a random sample of 92 female students at Hope College, 10 were left-handed. Determine a 99% confidence interval for the proportion of all female students at Hope College that are left-handed.

Facebook friend request*

3.2.22 In order to determine who would accept a Facebook friend request from someone they didn't know, a student researcher made up a phony Facebook profile that represented a female student at her college. She then sent out 118 friend requests and 61 of these accepted the request. Find a 95% confidence interval for the proportion of all students that would accept a Facebook request from a female at their college that they did not know.

3.2.23 In order to determine who would accept a Facebook friend request from someone they didn't know, a student researcher made up a phony Facebook profile that represented a male student at her college. She then sent out 101 friend requests and 18 of these accepted the request. Find a 95% confidence interval for the proportion of all students that would accept a Facebook request from a male at their college that they did not know.

Tasting Skittles

3.2.24 A sample of healthy students were blindly given a single Skittles candy to put in their mouth. They were told the five possible flavors and then were asked which flavor they had. Of the 154 students tested, 78 gave the correct answer. Find a 95% confidence interval for the proportion of all healthy students that could correctly identify a Skittles flavor blindly.

3.2.25 A sample of students that had stuffy or runny noses were blindly given a single Skittles candy to put in their mouth. They were told the five possible flavors and then were asked which flavor they had. Of the 118 students tested, 44 gave the correct answer. Find a 95% confidence interval for the proportion of all students with stuffy or runny noses that could correctly identify a Skittles flavor blindly.

3.2.26 Using the formula $\hat{p} \pm 1.96 \sqrt{\dfrac{\hat{p}(1-\hat{p})}{n}}$, compute a 95% confidence interval for a population proportion given the sample proportion is 0.35 and the sample size is 1000.

3.2.27 Using the formula $\hat{p} \pm 1.96 \sqrt{\dfrac{\hat{p}(1-\hat{p})}{n}}$, compute a 95% confidence interval for a population proportion given the sample proportion is 0.65 and the sample size is 500.

3.2.28 Show that the margin-of-error formula $1.96 \sqrt{\dfrac{\hat{p}(1-\hat{p})}{n}}$ simplifies to approximately $1/\sqrt{n}$ when $\hat{p} = 0.5$.

3.2.29 Using the simplified version for the margin-of-error from the previous exercise, $1/\sqrt{n}$, determine the smallest sample size needed so the margin-of-error is:
   a) 0.06
   b) 0.03
   c) 0.01

Section 3.3

3.3.1* True or False. When using the 2SD method to find a 95% confidence interval for the mean, multiply the standard deviation of the quantitative variable by 2 to obtain the margin-of-error.

3.3.2 True or False. In general, there will be more variability in the sample means across samples, than in the variability in the quantitative variable across observational units.

3.3.3* True or False. If you are concerned that the validity conditions aren't met, use a theory-based approach to computing a confidence interval for the mean.

How much TV do you watch?

3.3.4 According to a 2011 report by the United States Department of Labor, civilian Americans spend 2.75 hours per day watching television. A faculty researcher, Dr. Sameer, at California Polytechnic State University (Cal Poly) conducts a study to see whether a different average applies to Cal Poly students. Suppose that for a random sample of 100 Cal Poly students, the mean and standard deviation of hours per day spent watching TV turns out to be 3.01 hours and 1.97 hours. The data were used to find a 95% confidence interval: (2.619, 3.401) hours/day. Which of the following are valid interpretations of the 95% confidence interval? For each of the following, statements, say whether VALID or INVALID.

    a.   About 95% of all Cal Poly students spend between 2.619 and 3.401 hours/day watching TV.

    b.   There is a 95% chance that, on average, Cal Poly students spend between 2.619 and 3.401 hours/day watching TV.

    c.   We are 95% confident that, on average, these 100 Cal Poly students spend between 2.619 and 3.401 hours/day watching TV.

    d.   In the long run, 95% of the sample means will be between 2.619 and 3.401 hours.

    e.   None of the above.

3.3.5 Reconsider the previous question about hours spent watching TV every day. Suppose that the data had actually been collected from a sample of 150 students, and *not* 100, but everything else (mean and SD) was the same as reported earlier. How, if at all, would the *new* 95% confidence interval based on these data differ from the interval mentioned earlier (2.619, 3.401) hours?

    a)   The *new* interval would still be (2.619, 3.401) hours, because we are still 95% confident.

    b)   The *new* interval would be narrower than (2.619, 3.401) hours, because the sample size is bigger.

    c)   The *new* interval would be wider than (2.619, 3.401) hours, because the sample size is bigger.

    d)   More information is needed to answer this question.

Rattlesnakes*

3.3.6 Here is a dot plot for the ages of 21 male rattlesnakes captured at a single site. Assume that these 21 snakes can be regarded as a random sample of all male rattlesnakes at that site. The average age is 8.571, with a standard deviation of 2.942. An approximate 95% confidence interval is calculated to be: (7.287, 9.855).

    a) What is the observational unit for this data set? What is the population? What is the sample?

    b) Which, if any, of the validity conditions for using a theory-based interval are satisfied? Explain your answer.

    c) What is the parameter being estimated by the interval?

    d) True or false, and explain. The chance that the interval contains the parameter value is about 95%.

Sunscreen

3.3.7 Most dermatologists recommend using sunscreens that have a Sun Protection Factor (SPF) of at least 30. One of the authors wanted to find out whether the SPF of sunscreens used by students at her school, exceeds this value, on average? To collect data, the author surveyed students in her introductory Statistics course, and found that in a sample of 48 students, the mean SPF was 35.29, and the standard deviation of SPF was 17.19. The data was not strongly skewed.

   a) Identify the observational unit for this study.
   b) Identify the variable of interest, and whether it is categorical or quantitative.
   c) Use the **Theory-Based Inference** applet to find and report the 95% confidence interval for the parameter of interest, <u>and</u> provide an interpretation of the interval, in the context of the study.
   d) Assuming she has a representative sample, based on the confidence interval is the average SPF of sunscreens of the students at the author's school higher than 30? How are you deciding?
   e) Are the conditions required for the validity of the above results satisfied? State the condition(s) and explain your reasoning.

Studying statistics*

3.3.8 In a pre-course, anonymous survey of students in her introductory statistics course, one of the authors asked her students how many hours per week they expected to spend studying Statistics outside of class. Forty nine students responded to that question, with an average of 8.2 hours, and a SD of 3.79 hours. The data was not strongly skewed.

   a) Identify the observational unit for this study.
   b) Identify the variable of interest, and whether it is categorical or quantitative.
   c) Use the **Theory-Based Inference** applet to calculate and report the 95% confidence interval for the parameter of interest.
   d) In the context of this study, was it valid to use the theory-based (t-distribution) approach to find a confidence interval? Explain your reasoning.
   e) Interpret the 95% confidence interval reported in (c), in the context of the study.

How cool are you?

3.3.9 Normal (or average) body temperature of humans is often thought to be 98.6° F. Is that number really the average? To test this, we will use a dataset obtained from 65 healthy female volunteers aged 18 to 40 that were participating in vaccine trials. We will assume this sample is representative of a population of all healthy females.

a) The mean body temperature for the 65 females in our sample is 98.39° F and the standard deviation is 0.743° F. The data are not strongly skewed. Use the **Theory-Based Inference** applet to find a 95% confidence interval for the population mean body temperature for healthy females.

b) Based on your confidence interval, is 98.6° F a plausible value for the population average body temperature or is the average significantly more or less than 98.6° F? Explain how you are determining this.

c) In the context of this study, was it valid to use the theory-based (*t*-distribution) approach to find a confidence interval? Explain your reasoning.

3.3.10 As was explained in the previous question, normal (or average) body temperature of humans is often thought to be 98.6° F. Is that number really the average? To test this, we will use a dataset obtained from 65 healthy male volunteers aged 18 to 40 that were participating in vaccine trials. We will assume this sample is representative of a population of all healthy males.

a) The dataset can be found on the textbook website as **MaleTemps**. Use the **Theory-Based Inference** applet to find a 95% confidence interval for the population mean body temperature for healthy males.

b) Based on your confidence interval, is 98.6° F a plausible value for the population average body temperature or is the average significantly more or less than 98.6° F? Explain how you are determining this.

c) In the context of this study, was it valid to use the theory-based (*t*-distribution) approach to find a confidence interval? Explain your reasoning.

Traveling across the USA*

3.3.11 One of the authors came across an article (*USA Today*, 2008) that said that on average Americans have visited 16 states in the United States. In a survey of 50 students in her introductory statistics class, she found the average number of states the students had visited to be 9.48 and the standard deviation to be 7.13. The data was not strongly skewed.

a) Identify the observational unit for this study.

b) Identify the variable of interest, and whether it is categorical or quantitative.

c) Clearly the sample is not a randomly selected sample, but a convenience sample. With this in mind, would it still be okay to treat the sample as representative of the population of students at the author's school with regard to "number of states visited"? Explain.

d) Regardless of your answer to (c), state the null and the alternative hypotheses in symbols, to test whether the average number of states all students at the author's school have visited is different from 16.

e)   Use the 2SD approach to find a 95% confidence interval for the average number of states all students at the author's school have visited.

3.3.12 Reconsider the previous question about the average number of states all students at the author's school have visited. Recall that in the author's sample of 50 students the average number of states the students had visited was 9.48 and the standard deviation was 7.13.

a) Use the **Theory-Based Inference** applet to find a 95% confidence interval for the average number of states all students at the author's school have visited.

b) Does the 95% confidence interval that you have reported in (a) provide evidence that the average number of states all students at the author's school have visited is different from 16?

c) In the context of this study, was it valid to use the theory-based (*t*-distribution) approach to find a confidence interval? Explain your reasoning.

3.3.13 Reconsider the previous exercises about the average number of states all students at the author's school have visited. Recall that in the author's sample of 50 students the average number of states the students had visited was 9.48 and the standard deviation was 7.13.

a) Use the **Theory-Based Inference** applet to find and report a 88% confidence interval, and interpret this interval in the context of the study.

b) Explain why you cannot use a simulation-based 2SD method to find a 88% confidence interval for the average number of states all students at the author's school have visited.

3.3.14 Reconsider the previous exercises about the average number of states all students at the author's school have visited. Using the data from the author's sample, a 99% confidence interval for the average number of states all students at the author's school have visited is found to be (6.78, 12.18). For each of the following statements, say whether VALID or INVALID.

a) We are 99% confident that the average number of states visited by the 50 students the author surveyed is between 6.78 and 12.18.

b) Based on the 99% confidence interval, there is evidence that the number of states all students at the author's school have visited is different from 16.

c) Based on the 99% confidence interval, there is evidence that the number of states all students at the author's school have visited is 16.

d) We are 99% confident that the average number of states visited by all students at the author's school is between 6.78 and 12.18.

e) About 99% of the students at the author's school have visited between 6.78 and 12.18 states.

f) If we want to be less than 99% confident, we should take a smaller sample of students.

Needles!

3.3.15 Consider a manufacturing process that is producing hypodermic needles that will be used for blood donations. These needles need to have a diameter of 1.65 mm – too big and they would hurt the donor (even more than usual), too small and they would rupture the red blood cells, rendering the donated blood useless. Thus, the manufacturing process would have to be closely monitored to detect any significant departures from the desired diameter. During every shift, quality control personnel take a random sample of several needles and measure their diameters. If they discover a problem, they will stop the manufacturing process until it is corrected.

Suppose the most recent random sample of 35 needles have an average diameter of 1.64 mm, and a standard deviation of 0.07 mm. Also, suppose the diameters of needles produced by this manufacturing process have a bell-shaped distribution.

a) Identify the observational unit for this study.

b) Identify the variable of interest, and whether it is categorical or quantitative.

c) Use the 2SD approach to find a 95% confidence interval for the average diameter of needles produced by this manufacturing process.

3.3.16 Reconsider the investigation of the manufacturing process that is producing hypodermic needles. Recall that the most recent random sample of 35 needles have an average diameter of 1.64 mm, and a standard deviation of 0.07 mm.

a) In the context of this study, explain why it is valid to use the theory-based (t-distribution) approach to find a confidence interval?

b) Use the **Theory-Based Inference** applet to find and report a 95% confidence interval for the average diameter of needles produced by this manufacturing process.

c) Based on the above 95% confidence interval alone, is there evidence that the average diameter of needles produced by this manufacturing process is different from 1.65 mm? How are you deciding?

3.3.17 Reconsider the investigation of the manufacturing process that is producing hypodermic needles. Recall that the most recent random sample of 35 needles have an average diameter of 1.64

mm, and a standard deviation of 0.07 mm.

a) Use the **Theory-Based Inference** applet to find and report a 98% confidence interval, and interpret this interval in the context of the study.
b) Explain why you cannot use a simulation-based 2SD method to find a 98% confidence interval for the average diameter of needles produced by this manufacturing process.

3.3.18 Reconsider the investigation of the manufacturing process that is producing hypodermic needles. Using the data from the most recent sample of needles, a 90% confidence interval for the average diameter of needles is found to be (1.62 mm, 1.66 mm). For each of the following statements, say whether VALID or INVALID.

a) We are 90% confident that the average diameter of the sample of 35 needles is between 1.62 and 1.66 mm.

b) Based on the 90% confidence interval, there is evidence that the average diameter of needles produced by this manufacturing process is 1.65 mm.
c) Based on the 90% confidence interval, there is evidence that the average diameter of needles produced by this manufacturing process is different from 1.65 mm.
d) We are 90% confident that the average diameter of needles produced by this manufacturing process is between 1.62 and 1.66 mm.
e) About 90% of the needles produced by this manufacturing process have a diameter between 1.62 and 1.66 mm.
f) If we want to be more than 90% confident, we should take a larger sample of needles.

Facebook friends*

3.3.19 In a survey of her introductory statistics class, an instructor found the average number of Facebook friends the of 49 respondents had was 539.2, and the standard deviation was 298. The data on number of Facebook friends is not strongly skewed.

a) Identify the observational unit for this study.
b) Identify the variable of interest, and whether it is categorical or quantitative.
c) In the context of this study, explain why it is valid to use the theory-based (t-distribution) approach to find a confidence interval?
d) Use the **Theory-Based Inference** applet to find and report a 95% confidence interval for the parameter of interest. Also, interpret the interval in the context of the study.

3.3.20 Reconsider the previous question about number of Facebook friends. Using the data from the sample of 49 students, a 99% confidence interval for the average number of Facebook friends is found to be (425.0, 653.4). For each of the following statements, say whether VALID or INVALID.

    a) We are 99% confident that the average number of Facebook friends the 49 students who were surveyed have is between 425 and 653.4.

    b) We are 99% confident that the average number of Facebook friends all such people have is between 425 and 653.4.

    c) About 99% of all such people have between 425 and 653.4 number of friends, on average.

Textbook prices

3.3.21 Two Cal Poly freshmen gathered data on the prices for a random sample of 30 textbooks from the campus bookstore. They found the average price was $65.02, and the standard deviation of prices was $51.42. The data are not strongly skewed.

    a) Identify the observational unit for this study.

    b) Identify the variable of interest, and whether it is categorical or quantitative.

    c) Explain how you know from the sample average and sample standard deviation that the distribution of textbook prices is not bell-shaped.

    d) In the context of this study, explain why it is valid to use the theory-based (t-distribution) approach to find a confidence interval?

    e) Use the **Theory-Based Inference** applet to find and report a 95% confidence interval for the parameter of interest. Also, interpret the interval in the context of the study.

Mercury level in tuna*

3.3.22 Data were collected on random specimens of tuna from 1991 and 2010. The mercury level for each fish specimen was recorded. For the sample of 43 specimens of Albacore tuna, the average mercury level was found to be 0.358 parts per million (ppm), and the standard deviation of mercury level to be 0.138 ppm. The data are not strongly skewed.

    a) Identify the observational unit for this study.

    b) Identify the variable of interest, and whether it is categorical or quantitative.

    c) In the context of this study, explain why it is valid to use the theory-based (t-distribution) approach to find a confidence interval?

    d) Use the **Theory-Based Inference** applet to find and report a 95% confidence interval for the parameter of interest. Also, interpret the interval in the context of the study.

3.3.23 Data were collected on random specimens of Yellowfin tuna from 1991 and 2010. For the sample of 231 specimens of Yellowfin tuna, the average mercury level was found to be 0.354 ppm, and the standard deviation of mercury level to be 0.231 ppm. The data are not strongly skewed.

a) In the context of this study, explain why it is valid to use the theory-based (t-distribution) approach to find a confidence interval?

b) Use the **Theory-Based Inference** applet to find and report a 95% confidence interval for the parameter of interest. Also, interpret the interval in the context of the study.

Haircut prices

3.3.24 How much do you typically pay, including tips, for a haircut? In a survey of 50 students in her class, one of the authors found the average amount paid for haircuts to be $45.68 and the standard deviation to be $39.95. The data are not strongly skewed.

a) Explain how you know from the sample average and sample standard deviation that the distribution of price of haircuts is not bell-shaped.

b) In the context of this study, explain why it is valid to use the theory-based (t-distribution) approach to find a confidence interval?

c) Use the **Theory-Based Inference** applet to find and report a 95% confidence interval for the parameter of interest. Also, interpret the interval in the context of the study.

Close friends*

3.3.25 One of the questions asked of a random sample of adult Americans on the 2004 General Social Survey (GSS) was:

> *From time to time, most people discuss important matters with other people. Looking back over the last six months - who are the people with whom you discussed matters important to you? Just tell me their first names or initials.*

The interviewer then recorded how many names each person gave. The GSS is a survey of a representative sample of U.S. adults who are not institutionalized.

There were 1467 responses, and the average number of close friends reported was 1.987, with a standard deviation of 1.771. The data are not strongly skewed.

a) Explain how you know from the sample average and sample standard deviation that the distribution of number of close friends people tend to report is not bell-shaped.

b) In the context of this study, explain why it is valid to use the theory-based (t-distribution) approach to find a confidence interval?

c) Use the **Theory-Based Inference** applet to find and report a 99% confidence interval for the parameter of interest. Also, interpret the interval in the context of the study.

d) Based on the above 99% confidence interval alone, is there evidence that the average number of close friends people tend to report is different from 2? How are you deciding?

Section 3.4

3.4.1* Recall the form of a theory-based confidence interval for a population proportion π is:

$$\hat{p} \pm \text{multiplier} \times \sqrt{\frac{\hat{p}(1 - \hat{p})}{n}}$$

a) Remind me: What's the difference between $\hat{p}$ and π?

b) The midpoint of this interval is always equal to what?

c) What is $\sqrt{\dfrac{\hat{p}(1 - \hat{p})}{n}}$ called?

d) What is $\text{multiplier} \times \sqrt{\dfrac{\hat{p}(1 - \hat{p})}{n}}$ called?

3.4.2 Recall the form of a theory-based confidence interval for a population proportion π is:

$$\hat{p} \pm \text{multiplier} \times \sqrt{\frac{\hat{p}(1 - \hat{p})}{n}}$$

a) Increasing the confidence level has what effect, if any, on the midpoint of this interval?

b) Increasing the confidence level has what effect, if any, on the margin-of-error of this interval?

c) Increasing the sample size, if all else remains the same, has what effect, if any, on the margin-of-error of this interval?

3.4.3* Name three things that affect the margin-of-error in a theory-based confidence interval for a population proportion, π. Which of these three things is the hardest for an investigator to control?

3.4.4 Name three things that affect the margin-of-error in a theory-based confidence interval for a population mean, μ. Which of these three things is the hardest for an investigator to control?

Sudoku*

3.4.5 The *2011 Statistical Abstract of the United States* includes a table reporting that a national survey in fall of 2009 found that 11.6% of American adults had played a Sudoku puzzle in the past year. Suppose that these data are based on a sample of 1000 randomly selected American adults.

    a) Use this sample result and a theory-based approach to produce three confidence intervals (90%, 95%, 99%) for the probability that a randomly selected American adult had played a Sudoku puzzle in the past year.

    b) How do the midpoints of these three intervals compare?

    c) How do the widths of these three intervals compare? Is this what you expected? Explain.

    d) Based on the confidence intervals, does the sample result provide strong evidence that more than 10% of all American adults had played a Sudoku puzzle in the past year? Explain.

    e) Is it appropriate to use the theory-based approach to find the confidence intervals? Answer this by stating the validity conditions, and explain whether or not the condition is satisfied.

3.4.6 Refer to **3.4.5** to answer the following questions

    a) If the sample size had been 2000, how would the width of the 90%, 95%, and 99% confidence intervals be affected?

    b) If the sample size had been 100, how would the width of the 90%, 95%, and 99% confidence intervals be affected?

May 2011 Gallup poll

3.4.7 A Gallup survey of 1,001 randomly selected U.S. adults conducted May 2011, asked "In your opinion, which one of the following is the main reason why students get education beyond high school?" Fifty three percent chose "to earn more money." Based on these data, I found the 95% confidence interval for the proportion of U.S. adults who choose "to earn more money" to be (0.499, 0.561). Suppose that the sample size was 2,002 instead, while the proportion of respondents choosing "to earn more money" was still 53%.

    a) TRUE/FALSE: A 95% confidence interval based on these altered data would be narrower, because a larger sample size increases confidence which in turn makes the interval narrower.

    b) TRUE/FALSE: A 95% confidence interval based on these altered data would be narrower, because a larger sample size decreases the variability of the statistic.

    c) TRUE/FALSE: A 95% confidence interval based on these altered data would be wider, because a larger sample size increases confidence which in turn makes the interval wider.

    d) TRUE/FALSE: A 95% confidence interval based on these altered data would be wider, because a larger sample size increases the variability of the statistic.

3.4.8* Looking at the formula to calculate the SD of $\hat{p}$, the "root n relationship" means that to cut the SD of $\hat{p}$ in half, you need to

a)   multiply the sample size by $^1\!/_2$
b)   multiply the sample size by 2
c)   multiply the sample size by $\sqrt{2}$
d)   multiply the sample size by $^1\!/_4$
e)   multiply the sample size by 4

3.4.9 Gilbert uses a sample size of 25. Sullivan uses a sample size of 100. Gilbert's estimated SD for $\hat{p}$ will be _____ times as large as Sullivan's.

3.4.10* As n increases, the SD for $\hat{p}$ _____ (increases, decreases), more and more _____ (slowly, quickly).

Rolling dice

3.4.11 You have an ordinary six-sided die, but you suspect it may be loaded. If the die is fair, then the probability of rolling a 6 equals 1/6.

a)   Let π = the probability of rolling a 6. To estimate π to within 0.01 at the 95% confidence level, you need a sample size of _____.

b)   The SD for $\hat{p}$ is $\sqrt{\dfrac{\pi(1-\pi)}{n}}$. What values of π give a value of 0 for the SE? Explain why this result makes sense.

c)   Compute the value of $\pi(1-\pi)$ for values of $\square = 0, ^1\!/_4, ^1\!/_2, ^3\!/_4$. and 1. Plot the points with values of π on the x-axis and join the points with a smooth curve. What value of π gives the larges possible value for $SD(\hat{p})$?

3.4.12 Ordinary dice have six equally likely sides. You can buy dice online with (A) 4 sides, (B) 6 sides, (C) 8 sides, or (D) 10 sides. Suppose you want to estimate the probability of rolling a 1 for each of dice (A) – (D).

a)   Which die (A) – (D) will give an estimate for the probability of rolling a 1 with the largest SD?

b)   Explain your answer in a).

c)   Suppose the 10-sided die is in fact fair. If you roll 25 times and estimate the probability of rolling a 1, your SD for $\hat{p}$ will be _____.

d) Continue to suppose that the 10-sided die is fair. You want your estimate of π to have an SE of 0.01. How large should your sample size be?

Flipping coins*

3.4.13 You have a coin that you think is fair, with π = probability of heads = $\frac{1}{2}$. To check, you plan to flip the coin n times.

a) Compute the SD of $\hat{P}$ for each value of n in the table below, and fill in your SD values.

n	SD of $\hat{p}$
10	
20	
40	
100	
500	
1000	

b) Plot SD($\hat{P}$) on the y-axis versus n on the x-axis, and join your points with a smooth curve.

c) The SD _____ (increases/decreases) as n increases.

d) The curve becomes _____ (more, less) steep as n continues to increase.

3.4.14 You have a coin that you think is fair, with π = probability of heads = $\frac{1}{2}$.

a) Rogers has done 25 flips and plans to do another 25 flips. Hammerstein has done 100 flips and plans to do another 25 flips. Whose SD will go down more, Rogers or Hammerstein? Explain.

b) Rogers has done 25 flips and plans to do another 25. Hammerstein has done 100 and plans to do another 100. Whose SD for $\hat{P}$ will go down more?

School survey

In order to investigate how many hours a day students' at their school tend to spend on course work outside of regularly scheduled class time, a statistics student takes a random sample of 150 students from their school by randomly choosing names from a list of all full-time students at their school that semester. The student finds that the average reported daily study hours among the 150 students is 2.23 hours. The standard deviation of the hours studied is 1.05 hours.

3.4.15 A confidence interval is constructed for the population mean hours studied. Which confidence interval would be the widest?

   a) 99%

   b) 95%

   c) 90%

   d) 85%

3.4.16 Which confidence interval would be the narrowest?

   a) 99%

   b) 95%

   c) 90%

   d) 85%

3.4.17 If the standard deviation were 0.78 hours instead of 1.05, the width of a 95% confidence interval would (increase/decrease).

3.4.18 If the standard deviation were 1.25 hours instead of 1.05, the width of a 95% confidence interval would (increase/decrease).

3.4.19 If the sample size were 15 instead of 150, the width of a 95% confidence interval would (increase/decrease).

3.4.20 If the sample size were 1500 instead of 150, the width of a 95% confidence interval would (increase/decrease).

3.4.21 If you took repeated samples of size 150 and constructed a 95% confidence interval for the population mean hours slept from each sample, what percentage of these intervals would capture the population mean hours slept?

3.4.22 If you took repeated samples of size 150 and constructed a 99% confidence interval for the population mean hours slept from each sample, what percentage of these intervals would capture the population mean hours slept?

Business survey*

In order to understand how reasons why consumers visit their store, a local business conducts a survey by asking the next 100 people who visit their store to fill out a short survey. The business finds that 40 of the 100 people state that the main reason they visited the store was because of the store is running a sale on coats that week. Use this information to answer Exercises 3.4.23-31.

3.4.23 A confidence interval is constructed for the population proportion of consumers who would visit the store because of the coat sale. Which confidence interval would be the widest?

   a) 99%
   b) 95%
   c) 90%
   d) 85%

3.4.24 Which confidence interval would be the narrowest?

   a) 99%
   b) 95%
   c) 90%
   d) 85%

3.4.25 Instead of 40, if 30 of the 100 people had stated that the main reason they had visited the store was because of the store is running a sale on coats that week, the width of a 95% confidence interval would (increase/decrease).

3.4.26 Instead of 40, if 50 of the 100 people had stated that the main reason they had visited the store was because of the store is running a sale on coats that week, the width of a 95% confidence interval would (increase/decrease).

3.4.27 Instead of 40, if 60 of the 100 people had stated that the main reason they had visited the store was because of the store is running a sale on coats that week

   a) The width of the 95% confidence interval would (increase/decrease/remain same).
   b) The center of the 95% confidence interval would (increase/decrease/remain same).

3.4.28 If the sample size were 10 instead of 100, the width of a 95% confidence interval would (increase/decrease).

3.4.29 If the sample size were 1000 instead of 100, the width of a 95% confidence interval would (increase/decrease).

3.4.30 If you took repeated samples of size 100 and constructed a 95% confidence interval for the population proportion who would shop at the store because of the coat sale from each sample, what percentage of these intervals would capture the population proportion?

3.4.31 If you took repeated samples of size 100 and constructed a 99% confidence interval for the population proportion who would shop at the store because of the coat sale from each sample, what percentage of these intervals would capture the population proportion?

Heights of students

3.4.32 Suppose we have a collection of the heights of all students at your college. Each of the 250 people taking statistics randomly takes a sample of 40 of these heights and constructs a 95% confidence interval for mean height of all students at the college. Which of the following statements about the confidence intervals is most accurate?

    A. About 95% of the heights of all students at the college will be contained in these intervals.

    B. About 95% of the time, a student's sample mean height will be contained in his or her interval.

    C. About 95% of the intervals will contain the population mean height.

    D. About 95% of the intervals will be identical.

Mean household income*

3.4.33 A 95% confidence interval is computed to estimate the mean household income for a city. Which of the following values will definitely be within the limits of this confidence interval? Circle all that apply.

    A. The population mean.

    B. The sample mean.

    C. The sample standard deviation.

    D. The p-value.

    E. All the above.

Foot length

3.4.34 Suppose a random sample of 50 college students is asked to measure the length of their right foot in centimeters. A 95% confidence interval for the mean foot length for students at the college is found to be 21.709 cm to 25.091 cm. If a 99% confidence interval was calculated instead, how would it differ from the 95% confidence interval?

    A. The 99% confidence interval would be wider.

    B.   The 99% confidence interval would be narrower.

    C. The 99% confidence interval would have the same width as the 95% confidence interval.

    D. There is no way to tell if the 99% confidence interval would be wider or narrower than the 95% confidence interval.

Section 3.5 Exercises

Coffee Bars*

3.5.1 Suppose you want to compare waiting times, from getting in line to receiving the order, at two coffee bars on campus.

    a) Describe what might be wrong with looking at the first 40 people who come to each of the coffee bars on a Tuesday morning.

    b) Describe how you might find a more representative sample of customers at these coffee bars.

Literary Digest

3.5.2 Recall from *Exploration 3.5A* that the *Literary Digest* magazine conducted a very extensive public opinion poll in 1936 in order to predict the outcome of that year's presidential election. They mailed out over 10 million questionnaires to people whose names and addresses they obtained from telephone books and vehicle registration lists. They received more than 2.4 million responses, with 57% of the responses indicating an intention to vote for Republican challenger Alf Landon over the Democratic incumbent Franklin Roosevelt.

    a) Use the **Theory-Based Inference** applet to produce a 99.9% confidence interval. Also interpret this interval.

    b) Why is this interval so narrow, even with an extremely large confidence level of 99.9%?

    c) Would you really have a lot of confidence that this interval succeeds in capturing the population proportion of American voters who intended to vote for Landon in 1936? Why or why not?

Holocaust Denial*

3.5.3 In a November 1992 survey, the Roper Organization asked American adults: "Does it seem possible or does it seem impossible to you that the Nazi extermination of the Jews never happened?" The results were very surprising and widely discussed. Of the 992 adults surveyed, 22% responded that it's possible that this never happened, and 12% responded that they did not know. Therefore, only 66% of the respondents expressed certainty that the Holocaust *did* happen. What is wrong with the wording of this question that produced such surprising (and, it was later shown) erroneous responses?

3.5.4 Refer to the previous exercise. When the Roper Organization conducted another survey in March of 1994, they asked: "Does it seem possible to you that the Nazi extermination of the Jews never happened, or do you feel certain that it happened?" Do you think the wording of this question is more clear than the November 1992 version? Explain why or why not.

3.5.5 Refer to the previous question. The results from the 1994 sample of 991 adults were that 91% answered that they were certain the Holocaust happened, 8% did not know, and 1% thought it possible that it never happened.

   a) Produce an appropriate graph for comparing the results between the 1992 and 1994 surveys.

   b) Summarize how the results differ between the two surveys.

   c) For each survey, determine a 95% confidence interval for the population proportion of American adults who are certain that the Nazi extermination of the Jews happened.

   d) Are these confidence intervals similar? Do they overlap?

   e) Summarize the lesson to be learned from the two Roper polls.

Welfare

3.5.6 A survey question could ask for respondents' opinions about "welfare" or about "assistance to the poor"? Do you think that the wording used could influence people's responses? If so, which wording would you expect to produce a higher percentage of favorable responses towards these programs?

Military strength*

3.5.7 One version of a survey question asked whether subjects agree or disagree with the statement that "the best way to ensure peace is through military strength." Another version asked subjects to choose which of two choices they agree with: "the best way to ensure peace is through military strength" or "diplomacy is the best way to ensure peace." Which version would you expect to have a larger percentage favoring military strength? Explain.

Abortion

3.5.8 Answer the following about how questions can be worded in order to bias the results.

   a) How might you word a question about attitudes toward the government's relationship with abortion, if the goal were to bias the results in favor of making abortions more accessible?

   b) How might you word a question about attitudes toward the government's relationship with abortion, if the goal were to bias the results in favor of making abortions less accessible?

Healthcare*

3.5.9 If you support President Obama's health care policy, would you prefer to phrase a survey question in terms of "ObamaCare" or in terms of "Affordable Care Act"? Explain.

Biased responses

3.5.10 Think of a controversial issue about which you might be interested in asking people's opinions. Describe how you could ask a survey question about that issue in as clear and unbiased a manner as possible. Then describe how you could ask the question to bias responses in one direction, and then describe how you could ask the question to bias responses in the other direction.

Interviewer effects*

3.5.11 In an article titled "Interviewer Effects in Public Health Surveys," Davis et. al. (2010) argue that effects of interviewers on survey participant responses can be especially problematic when dealing with public health issues. For example, the article cited a study which found that black respondents report higher rates of alcohol and marijuana use to black interviewers than to white interviewers. Suggest another variable or issue for which you suspect that respondents might give different responses depending on the *race* of the interviewer. Justify your answer.

3.5.12 Refer to the previous exercise. The article also cited a study which found that female interviewers tended to obtain higher percentages of respondents reporting psychiatric symptoms. Suggest another variable or issue for which you suspect that respondents might give different responses depending on the *gender* of the interviewer. Justify your answer.

3.5.13 Suggest a variable or issue for which you suspect that respondents might give different responses depending on the *clothing worn* by the interviewer. Justify your answer.

3.5.14 If a survey is being conducted on the issue of whether to enact a smoking ban in a public place, how might responses be biased depending on whether or not the interviewer is smoking while asking the questions?

CNN.com polls

3.5.15

The news website CNN.com regularly posts a poll question that people who view the website can respond to. The following results were posted on January 10, 2012:

Do you surf the Web while on the job?

Often	62%	69,413
Seldom	25%	27,875
Never	13%	14652

Total votes: 111,938
This is not a scientific poll

a) Are the percentages reported in this graphic (62%, 25%, 13%) parameters or statistics? Justify your answer.

b) The margin-of-error for this survey is 0.003 (or 0.3%). Explain why the margin-of-error is so small. (*Hint*: Keep your response brief; you can use five words or less.)

c) Are you *really* confident that the proportion of all American workers who surf the Web while on the job is within 0.003 of 0.62? If not, do you suspect that the actual proportion is larger or smaller than 0.62? Explain.

3.5.16

On January 29, 2011, visitors to the CNN.com website were invited to answer a poll question. The results are shown below:

Are you exercising more in 2011?

No	51%	49,254
Yes	49%	47,232

Total votes: 96,486
This is not a scientific poll

A 99.9% confidence interval for the population proportion (of "yes" responses) turns out to be: (0.484, 0.495).

a) Explain (using no more than 10 words, preferably no more than 5 words) why this confidence interval is so narrow.

b) The standarized z-statistic for testing the null hypothesis that 50% of American adults claim to be exercising more in 2011 turns out to be -6.51, for a p-value of 2.5×10^{-10}. Explain what this indicates about the difference between statistical significance and practical importance.

c) Even if this question could have been asked of every American adult on January 29, 2011, the results would still be questionable. Describe one reason for this.

d) Explain why you think CNN.com adds the disclaimer that "this is not a scientific poll."

At your school*

3.5.17 For each of the following variables, indicate whether you think students at your school would generally be truthful or would tend to over- or under-state their honest responses. Briefly explain your answer for each.
   a) Amount of sleep per night
   b) Time spent on charity/volunteer work
   c) Church attendance
   d) Time spent studying
   e) How often a seat belt is worn

3.5.18 Answer the following questions about statistical significance and practical importance.
   a) Which is more appropriate to the issue of statistical significance: p-value or confidence interval?
   b) Which is more appropriate to the issue of practical importance: p-value or confidence interval?
   c) Explain the distinction between statistical significance and practical importance.

3.5.19 When is the distinction between statistical significance and practical importance more important to consider: with a large sample size, with a small sample size, or is sample size not relevant to this issue?

Penny flipping, spinning, tipping

3.5.20 Over several years, students of Professor Robin Lock have flipped a large number of coins and recorded whether the flip landed heads or tails. As reported in a 2002 issue of *Chance News*, these students had observed 14,709 heads in a total of 29,015 flips.
   a) What proportion of these flips resulted in heads?
   b) Calculate the standardized value of this statistic, for testing whether the long-run probability of heads differs from 0.5.
   c) Determine the theory-based p-value for testing whether the long-run probability of heads differs from one-half.
   d) What conclusion would you draw at the 0.05 significance level?
   e) Determine a 95% confidence interval for the long-run probability of heads.

f) Does the confidence interval include the value 0.5? Is this consistent with your test decision in part e)? Explain.

g) Based on the p-value and confidence interval, would you say that the difference between the sample proportion of heads and the value 0.5 is statistically significant? Would you say that the difference is practically important? Explain.

3.5.21 Refer to the previous exercise. Professor Lock's students also spun pennies on their side. The students found 9,197 heads in 20,422 spins. Answer parts a)-g) of the previous question with regard to penny *spinning* rather than flipping.

3.5.22 Refer to the previous two exercises. Professor Lock's students also "tipped" pennies by standing them on edge and then banging the table to see which way they would fall. The students found 10,087 heads in 14,611 tips. Answer parts a)-g) of the previous question with regard to penny *tipping* rather than flipping or spinning.

Pop, soda or coke?*

3.5.23 The website www.popvssoda.com invites people to indicate whether they refer to carbonated beverages as "pop" or 'soda" or "coke" or something else. As of December 30, 2013, the website had received 350,847 responses from the United States, broken down as follows:

Pop	Soda	Coke	Other	Total
131,403	149,377	54,072	15,995	350,847

a) Determine a 95% confidence interval for the population proportion who would answer "soda."

b) Based on this confidence interval, would you say that the sample proportion who answered "soda" is statistically significantly less than 0.5? Would you say that the difference is practically important? Explain your answers.

c) Based on how the sample was selected, do you have concerns about generalizing from this sample to any population? Explain.

Baseball batting power

3.5.24 Suppose that a baseball player has been a .250 hitter for his career, which means that his probability of a hit (success) has been 0.250. Then during one winter the player genuinely improves to the point that his probability of success improves to 0.333. You will investigate how likely the player is, in a sample of at-bats, to convince the manager that he has improved.

a) Open the **Power Simulation** applet. Enter 0.250 for the hypothesized value, 0.333 for the alternative value, and 20 for the sample size. Enter 200 for the number of samples, and click on

Draw Samples. Repeat four more times, to produce a total of 1000 simulated samples. Comment on how much overlap you see between the two distributions. Does this overlap suggest that it's easy or difficult to distinguish between a .250 hitter and a .333 hitter in a sample of 20 at-bats?

b) Choose the Level of Significance option, and enter 0.05 for the significance level. Click on Count. The top graph indicates how many hits the player needs in 20 at-bats in order to produce a p-value less than 0.05, to provide convincing evidence to reject the null hypothesis that his success probability is 0.250. How many hits are needed?

c) Based on the bottom graph, in what proportion of the 1000 samples did the player with a 0.333 probability of success will obtain enough hits to reject the hypothesis that he has a 0.250 probability of success? This is the approximate power of the test.

d) Would you say that the player has a good chance of convincing the manager that he has improved in a sample of 20 at-bats? Write a summary, as if to the player.

e) What factors could be changed in order to increase the power of the test?

3.5.25 Reconsider the previous exercise. Change the sample size to 100 at-bats, and answer questions a)-d). Then describe the effect of increasing the sample size on the power of the test.

3.5.26 Reconsider the previous two exercises. Change the significance level to 0.10 (with a sample size of 100 at-bats), and answer questions a)-d). Then describe the effect of decreasing the significance level on the power of the test.

3.5.27 Reconsider the previous three exercises. Now suppose that the player has actually become a .400 hitter. Make this change to the alternative value (with a significance level of 0.10 and a sample size of 100 at-bats), and answer questions a)-d). Then describe the effect of increasing the alternative value on the power of the test.

End of Chapter Exercises

Drinking water*

3.CE.1 An article "*Foul drinking water aboard airlines worsens*," in The *Seattle Times* (January 2005) reported that 29 out of 169 water specimens from randomly selected passenger aircraft carrying domestic and international passengers, tested positive for the presence of bacteria, thus failing to meet federal safety standards.

a) Use these data to estimate, with 95% confidence, the population proportion.

b) Interpret your answer from (a).

Talking while driving

3.CE.2 A Pew Research (2009) survey of nationally representative 242 cell phone users, ages 16 – 17 years, found that 52% had talked on the phone while driving.

   a) Use these data to estimate, with 95% confidence, the proportion of all 16 – 17 year-old cell phone users who talk on the phone while driving.

   b) Do these data provide evidence that a majority of 16 – 17 year-old cell phone users talk on the phone while driving? Explain your reasoning.

   c) Without performing any extensive calculations, find the 95% confidence interval for the proportion of all 16 – 17 year-old cell phone users who *do not* talk on the phone while driving.

   d) Suppose that you are planning to conduct a survey in your city to estimate, with 95% confidence, the proportion of 16 – 17 year-old cell phone users who talk on the phone while driving, and want the margin to be no bigger than 0.05. What should your sample size be? (*Hint*: Start by writing out the formula for the margin-of-error, and after plugging in appropriate values for the known quantities, solve for *n*.)

Elsa and Frank*

3.CE.3 Suppose that Elsa and Frank determine confidence intervals using the same confidence level, based on the same sample proportion. Elsa uses a larger sample size than Frank. How will midpoint and width of confidence intervals compare?

   A. Same midpoint and same width
   B. Same midpoint, Elsa has wider interval
   C. Same midpoint, Frank has wider interval
   D. Same width, Elsa has larger midpoint
   E. Same width, Frank has larger midpoint
   F. Elsa has larger midpoint, wider interval
   G. Frank has larger midpoint, wider interval

3.CE.4 Suppose that Elsa and Frank determine confidence intervals based on the same sample proportion and sample size. This time, Elsa uses a larger confidence level than Frank. How will midpoint and width of confidence intervals compare?

   A. Same midpoint and same width
   B. Same midpoint, Elsa has wider interval
   C. Same midpoint, Frank has wider interval

D. Same width, Elsa has larger midpoint
E. Same width, Frank has larger midpoint
F. Elsa has larger midpoint, wider interval
G. Frank has larger midpoint, wider interval

Margin-of-error

3.CE.5 Recall that a margin-of-error for a 95% confidence interval for process probability can be approximated by the expression $1/\sqrt{n}$.

a) Evaluate this expression (to three decimal places) for the following sample sizes: 100, 400, 500, 1000, 2000, 8000, 9000.
b) Based on your answers in (a), produce a graph using the following axes and scales. Describe how the margin-of-error changes with the sample size? Does it increase, decrease, or stay the same?

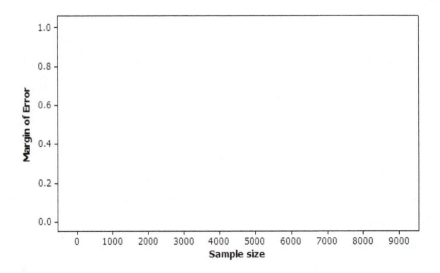

c) In order to cut the margin-of-error in half, by how many times must the sample size increase?
d) Which has the bigger impact on the margin-of-error: increasing the sample size from 100 to 500 (an increase of 400), or increasing the sample size from 8000 to 9000 (an increase of 1000)? Explain.

Blood donations*

3.CE.6 Suppose that researchers in three different countries collect sample data on whether or not a person has donated blood within the past year, with the following sample results:

A: 200 blood donors in a sample of 600 people
B: 50 blood donors in a sample of 300 people
C: 250 blood donors in a sample of 1000 people

Consider estimating the proportion of people in each country who donated blood in the previous year.

a) Explain why an interval of values is better than a single value for this estimate.

b) For which of the three countries would a 95% confidence interval be widest? For which would a 95% confidence interval be narrowest?

c) For which of the three countries would a 95% confidence interval include the value 0.2?

d) What does your answer to c) reveal about the p-values for testing whether proportion of people in the country who donated blood in the previous year differs from one-fifth? Explain.

3.CE.7. Suppose that researchers in three different countries collect sample data on whether or not a person has donated blood within the past year, with the following sample results:

A: 200 blood donors in a sample of 600 people
B: 50 blood donors in a sample of 300 people
C: 250 blood donors in a sample of 1000 people

Suppose that each researcher wants to test whether the proportion of people in the country who donated blood in the previous year differs from one-fifth.

a) Determine the appropriate standardized statistic for each researcher. (*Hint:* You will need to get the standard deviation of the simulated statistics from the null distribution produced by an applet.)

b) Based on the standardized statistics, which researcher has the strongest evidence that the population proportion differs from one-fifth? Which has the least strong evidence?

American exceptionalism

3.CE.8 Do Americans believe that the United States has a unique character that makes it the greatest country in the world? This belief has been called "American exceptionalism." A Gallup poll conducted on December 10-12, 2010 asked this question of a sample of 1019 adult Americans, with 80% responding that they do believe that the U.S. has a unique character that makes it the greatest country in the world.

a) Use the **Theory-Based Inference** applet to determine a 95% confidence interval based on this sample result. Also interpret what this interval says.

b) Would you expect values within this interval to be rejected by a two-sided significance test at the 0.05 significance level, or not to be rejected by such a test? Explain. (*Hint:* Think about whether the values within the confidence interval are *plausible* values of the parameter.)

3.CE.9 Use the data described in the previous question to answer the following questions.

a) Investigate your answer to 3.CE.8 (b) by using the **Theory-Based Inference** applet to test the following (null) values for the parameter: 0.76, 0.78, 0.80, 0.82, and 0.84. Report the z-statistic and (two-sided) p-value for each of these five tests.

b) Which of these five values in part a) are rejected at the 0.05 significance level?

c) Do the values that are not rejected correspond to values that are within the 95% confidence interval?

Follow-up on sub-groups*

3.CE.10 The Harris polling organization conducted a national survey in 2008, finding that 14% of the sample of 2302 American adults had a tattoo.

a) Use the **Theory-Based Inference** applet to determine a 99% confidence interval for the population proportion of adult Americans who had a tattoo in 2008.

b) Which possible values for the population proportion would not be rejected at the 0.01 significance level? Explain how you can tell based on the confidence interval, without performing any tests.

3.CE.11 A national survey of 47,000 American households in 2006 found that 32.4% of the households included a pet cat. This survey result was reported in the *2011 Statistical Abstract of the United States*, which listed the American Veterinary Medical Association as the source. No information was provided about how the sample households were selected.

a) Conduct a theory-based test of whether one-third of all American households include a pet cat. Report the null and alternative hypotheses and the values of the z-statistic and p-value. Also state your test decision using the 0.01 significance level.

b) Explain why the p-value turned out to be so small even though the observed sample proportion appears to be quite close to one-third.

c) Determine a 99% confidence interval for the population proportion of American households that include a pet cat.

d) Is the confidence interval *consistent* with the significance test result? How can you tell?

e) Does the confidence interval indicate that the population proportion is *much different* from one-third? Explain.

International students

3.CE.12 An instructor wants to find the proportion of international students at his school. He contacts the registrar's office, and obtains enrollment data for the year 2011. Using these data, he reports that "I am 95% confident that the proportion of international students at our school during 2011 is between 0.08 and 0.10." Is this an appropriate use of a confidence interval? Why or why not?

Confidence intervals*

3.CE.13 Higher confidence is better than lower confidence. So, why do statisticians not always use 99.99% confidence intervals?

3.CE.14 Narrower confidence intervals are better than wider confidence intervals. One way to achieve a narrower confidence interval is to use a larger sample size. So, why do statisticians interested in estimating a parameter value for the population of all adult Americans not always use a sample size of 1,000,000 people?

3.CE.15 Narrower confidence intervals are better than wider confidence intervals. One way to achieve a narrower confidence interval is to use a smaller confidence level. So, why do statisticians not always use a confidence level of 70%?

3.CE.16 Narrower confidence intervals are better than wider confidence intervals.
   a) What change to sample size produces a narrower confidence interval?
   b) What change to confidence level produces a narrower confidence interval?

Social networking

3.CE.17 In May of 2013, the Pew Research Center's Internet and American Life Project conducted a survey with a random sample of American adults. Of the 1,895 Internet users in the sample, 72% reported that they used social networking sites.
   a) Identify the population of interest for this study.
   b) Identify the parameter of interest.
   c) Determine a 95% confidence interval for the parameter.
   d) Interpret what this confidence interval says.
   e) Based on the confidence interval, does the sample result provide strong evidence against believing that 75% of all American adult Internet users in May of 2013 used social networking sites? Explain how your conclusion follows from the confidence interval.

3.CE.18 Refer to the previous exercise. The report from the Pew Research center also provided survey results for different age groups:
   - Among those aged 18-29, 89% of 395 used social networking sites.
   - Among those aged 30-49, 78% of 542 used social networking sites.

- Among those aged 50-64, 60% of 553 used social networking sites.
- Among those aged 65-older, 43% of 356 used social networking sites.

a) For each age group, determine and interpret a 95% confidence interval for the relevant parameter.

b) How do the margins-of-error for these age-specific confidence intervals compare to the margin-of-error based on the entire sample (larger, smaller, or the same)? Explain why this makes sense.

Halloween spending*

3.CE.19 The National Retail Federation conducted a national survey of 8,526 consumers on September 1-9, 2009. Among the findings reported were that:

- 29.6% of those surveyed said that the state of the U.S. economy would affect their Halloween spending plans;
- the average amount that the respondents said they expect to spend on Halloween is $56.31.

a) Are these numbers parameters or statistics? Explain.

b) Determine and interpret a 95% confidence interval based on the 29.6% value given in the article.

c) What additional information would you need to determine a 95% confidence interval based on the $56.31 value given in the article?

3.CE.20 Reconsider the previous exercise.

a) Suppose the sample standard deviation of the expected spending amounts was $30. Determine and interpret a 95% confidence interval for the population mean.

b) Now suppose the sample standard deviation was $50. Recalculate a 95% confidence interval for the population mean. Comment on what's different about the confidence interval based on the larger sample standard deviation.

c) What (if anything) must be true about the distribution of the expected spending amounts in order for these confidence intervals to be valid?

Hockey Goals

3.CE.21 Suppose you have very little knowledge about professional ice hockey, but you want to get a sense for how many goals are scored per game on average and for what a typical margin of victory is. You decide to select a sample of games from the current season and record the total number of goals and margin of victory in each game.

a) What are the observational units and variables in this study? Are the variables categorical or quantitative?

b) Describe how you could take a simple random sample of 44 National Hockey League (NHL) games from the current season.

Feeling too lazy to take a simple random sample of games, one of the authors recorded data on all 44 NHL games played on played on Thursday, December 19 – Monday, December 23, 2013. The data are in the file called **Hockey** on the textbook website.

c) Examine and describe a graph of the distribution of total number of goals in the sample.

d) Based on this graph, do you think it's valid to produce a theory-based 95% confidence interval based on these data? Explain.

e) Calculate the sample mean and sample standard deviation. Report these along with the appropriate symbols.

f) Determine a 95% confidence interval for the population mean number of goals scored in an NHL game.

g) Interpret what this confidence interval says.

3.CE.22 Reconsider the previous exercise. Answer questions c)-g) with regard to margin of victory rather than total number of goals.

Marriage Ages*

3.CE.23 A student went to the local county courthouse (in Cumberland County, Pennsylvania, in June and July of 1993) to gather data on ages of soon-to-be husbands and wives who had recently applied for marriage licenses. He gathered age data on a sample of 100 couples and calculated the difference in age (husband – wife) for each couple. The results, which can be found in the file **MarriageAges** on the textbook website, are displayed in the dotplot below:

a) Write a paragraph describing the distribution of differences.

b) Do you consider it valid to use a theory-based confidence interval for a population mean, based on these sample data? Explain.

c) Determine a 99% confidence interval for the population mean.

d) Interpret the interval that you calculated in part c).

e) Based on this confidence interval, do the sample data provide strong evidence for concluding that husbands tend to be older than their wives, on average? Explain how your answer follows from the confidence interval.

f) To what population of married couples do you feel comfortable generalizing the results of this study? Explain.

3.CE.24 Reconsider the previous exercise.

a) Based on the dotplot, determine the sample proportion of married couples for which the wife is older than the husband.

b) Is it valid to use a theory-based procedure to estimate the population proportion of married couples for which the wife is older than the husband? Justify your answer.

c) Determine and interpret a 90% confidence interval for the population proportion of married couples for which the wife is older than the husband.

d) Based on this confidence interval, do the sample data provide strong evidence that the wife is older than the husband in less than 50% of all marriages? Explain how your answer follows from the confidence interval.

3.CE.25 Reconsider the previous two exercises. A 95% confidence interval for the population mean difference in ages (husband – wife, in years) turns out to be (0.92, 2.92). For each of the following, indicate whether the conclusion is valid or invalid.

a) We are 95% confident that at least 92% of husbands are older than their wives.

b) 95% of all husbands are between 0.92 and 2.92 years older than their wives.

c) For about 95 the 100 married couples in this sample, the husband is between 0.92 and 2.92 years older than his wife.

d) We are 95% confident that the average difference in ages in the population is between 0.92 and 2.92 years.

Diameter of planets

3.CE.26 Suppose that you look up the diameters of the planets in our solar system and use the data to determine a 95% confidence interval. Would the resulting interval make any sense? Explain why or why not.

Emotional support*

3.CE.27 In the mid-1980s, sociologist Shere Hite undertook a study of American women's attitudes toward relationships, love, and sex by distributing 100,000 questionnaires in women's groups. One of the questions was: Do you give more emotional support to your husband or boyfriend than you receive from him? A total of 4500 women returned the questionnaire. An ABC News/*Washington Post* poll conducted at about the same time surveyed a random sample of 767 women, asking them the same question about emotional support.

   a)   Which survey would you expect to obtain a more representative sample of the population? Explain briefly.

Of the 4500 women who returned the Hite questionnaire, 96% said that they gave more emotional support than they received from their husbands or boyfriends. Of the 767 women interviewed in the ABC News/*Washington Post* poll, 44% claimed to give more emotional support than they receive.

   b)   Using *only* the poll corresponding to your answer to a), determine a 99% confidence interval for the relevant population parameter.
   c)   Write a sentence interpreting what your confidence interval reveals.
   d)   If you were to calculate the margin-of-error for both surveys (do not bother to actually do this calculation), which survey would have the smaller margin-of-error? Explain briefly.

Investigation: Cell phones while driving

Drivers today have many more distractions to deal with than drivers of a century ago. Let's focus on new drivers, aged 16 to 17. Do you think a majority of drivers aged 16 to 17 have ever talked on the cell phone while driving?

Step 1: Ask a research question

   1.   Identify the research question and state it as a conjecture.

Step 2: Design a study and collect data

To help answer this question a Pew Research (2009) conducted a survey of nationally representative 242 cell phone users, ages 16 to 17 years, and asked whether they had ever talked on the cell phone while driving.

2. What are the observational units?

3. What is the variable that is measured/recorded on each observational unit?

4. Describe the parameter of interest in words. (You can use the symbol □□to represent this parameter.)

5. State the appropriate null and alternative hypotheses to be tested.

The survey found that 52% of the 242 16 to 17 year-olds had talked on the phone while driving.

Step 3: Explore the data

6. How many of the 16 to 17 year-olds in the sample have ever talked on their cell phone while driving?

Step 4: Draw inferences

Let's use our 3S strategy to help us investigate how much evidence the sample data provide to support our conjecture that a majority of 16 to 17 year-olds have talked on their cell phone while driving.

Statistic

7. What is the statistic that you can use to summarize the data collected in the study?

8. Is the sample proportion who have talked on their cell phone while driving greater than 0.50? (If not, there's no need to conduct a simulation analysis. You will have no evidence that a majority of drivers aged 16 to 17 have ever talked on the cell phone while driving.)

Simulate

9. If we assume that each driver is equally as likely to have talked on the cell phone while diving as not, what is the chance that an individual will have talked on the cell phone while driving?

10. Is it possible that we could observe the statistic from this sample of 242 drivers even if all of the drivers were just as likely as not to have talked on the cell phone while driving? Why?

11. Use the **One Proportion** applet to simulate 1000 repetitions of this study, assuming that every driver is just as likely to talk on the cell phone as not while driving. Report what values you input into the applet.

12. What is the center of your simulated null distribution? Does it make sense that this is the center? Explain.

Strength of evidence

13. Based on the dotplot generated using the 1000 possible values of the statistic, what values would you consider typical values and what would you consider atypical values of the statistic when the chance model is true?

14. How do the actual study results compare to the null distribution obtained when simulating the chance model? Do you believe the study results provide convincing evidence against the "equally likely to talk on the cell phone as not" hypothesis? Why or why not?

15. Determine the approximate p-value from your simulation analysis. Also interpret what this p-value represents (i.e., the probability of what, assuming what?).

16. Summarize the conclusion that you draw from this study and your simulation analysis. Also explain the reasoning process behind your conclusion.

17. You should have seen that 0.50 is a plausible value for the true proportion of all 16 to 17 year-olds who've used their cell phones while driving. Does this prove that 0.50 is the true value?

18. Use the **One Proportion** applet to create an interval of plausible values with 95% confidence for the population proportion of drivers 16 to 17 who have ever talked on a cell phone while driving. Do this by putting different values in for π and seeing whether or not they are plausible.

19. Now use the **One Proportion** applet to create an interval of plausible values with 95% confidence for the population proportion of drivers 16 to 17 who have ever talked on a cell phone while driving, using the 2SD method.

    a) What is the interval?

    b) How does your confidence interval using the 2SD method and $\pi = 0.50$ compare to the confidence interval of plausible values you found in question 18?

c) Without doing extensive work, find the 95% confidence interval for the proportion of drivers aged 16 to 17 who have NOT talked on the cell phone while driving.

20. Finally, use the **Theory-Based Inference** applet to find a 95% confidence interval for the population proportion of proportion of drivers 16 to 17 who have ever talked on a cell phone while driving.

   a) What is the interval?

   b) How does this interval compare to the ones you found in 19a and 18? Why would you expect that?

21. What would happen to your 95% confidence intervals if the sample size had been 1000 instead of 242, but still 52% of your sample stated they had talked on a cell phone while driving? Use the applet to confirm your answer.

22. If you were to construct a 99% confidence interval like the 95% confidence interval you constructed in #18 by testing different values for π and seeing whether or not they are plausible, how would the width of the 99% confidence interval compare to that of the 95% confidence interval? Find this 99% confidence interval to back up your claim.

Step 5: Formulate conclusions

23. Now, let's step back a bit and think about the scope of our inference. What are the wider implications? Do you think that your conclusion holds true for people in general? (These are extremely important questions, that we'll discuss more when we talk about the scope of inference in Chapter 4.)

Step 6: Look back and ahead

    24. Summarize your findings. If you were to repeat this study, are there any changes or improvements you would make? What additional questions might you be interested in asking?

Research Article: TASERs

Read "Conducted Electrical Weapon (TASER) Use Against Minors: A Shocking Analysis" by Gardner, Hauda and Bozeman (Pediatric Emergency Care, 2012)

Step 1. Ask a research question

    1. In no more than 1-2 sentences state a broad research question (or related questions) that the researchers were trying to investigate in this study.

    2. Identify at least two (cited) points of evidence the researchers cite as to why the research question(s) are important. Include the citation in your response.

Step 2. Design a study and collect data

    3. What is the sample size of minor (under 18 years old) suspects who were in the study and received a CEW electrical discharge?

    4. From how many different law enforcement agencies were these minor suspects?

    5. How was medical information obtained on the minor suspects who received a CEW electrical discharge?

    6. Summarize the following characteristics of the sample of minors being studied by answering the following questions

        a. What was the average age of the minors in the sample? Median?

        b. What was the most common age?

        c. What was the youngest aged participant in the sample?

        d. What percentage of the minors were male?

        e. What percentage were suspected to be intoxicated via alcohol or drugs?

        f. What were the two leading causes of police involvement?

        g. What was the most common "mode" of use of the CEW? In what percentage of minors was that mode used exclusively?

7. When classifying injuries to minors resulting from CEW use, are injuries classified as a quantitative or categorical variable? If categorical, how many categories?

Step 3. Explore the data

8. What percentage of suspects in the sample received a moderate or severe injury?

9. What percentage of suspects in the sample received a mild injury?

Step 4. Draw inferences

Moderate or Severe injuries

10. Report the 97.5% confidence interval for the percentage of moderate or severe injuries reported.

11. What is the (presumed) population of interest for this confidence interval?

12. What is the parameter of interest (*Hint*: Do not give a specific value, just describe it.)

13. Provide a one-sentence interpretation of the confidence interval.

14. Assuming the sample is representative of the population of interest, is there significant evidence that the proportion of minors who will obtain moderate or severe injuries is different than 5%? Why?

15. The study states that the 97.5% confidence interval for the percent of minor suspects with no or minor injuries is 96.4% to 100%. Explain how this confidence interval relates to your answer to #10.

Mild injuries

16. Report the 97.5% confidence interval for the percentage of mild injuries.

17. What is the (presumed) population of interest for this confidence interval?

18. What is the parameter of interest (*Hint*: Do not give a specific value, just describe it.)

19. Provide a one-sentence interpretation of the confidence interval.

20. Assuming the sample is representative of the population of interest, is there significant evidence that the proportion of minors who will obtain moderate or severe injuries is different than 5%? Why?

Step 5. Formulate conclusions

21. Was random sampling used to obtain the sample used in this study? Explain the implications.

Step 6. Look back and ahead

22. Identify at least two things that, if you were running the study and had a reasonable amount of financial resources, you would do differently. You must justify your answer.

23. On page 876, the researchers state that "Lastly, an important limitation to this study, is the stature of our minor suspects…" Why is this statement important to the researchers' findings?

24. The final paragraph of the paper argues that the researchers "…support that CEW use during apprehension of adolescents does not seem to post unacceptable levels of risk." Why do you think the researchers used the term adolescents here despite using minors through much of the rest of the article?

Chapter 4

Causation: Can We Say What Caused the Effect?

Chapter Overview

This chapter is about **causation**, the last of the four pillars of inference: strength, size, breadth, and **cause**. What can we say about the cause of the pattern in the data? We saw in the study of organ donation in the Preliminaries that the proportion of "Yes" responses was larger when the question asked for a forced choice rather than just offering a chance to opt in. Can we conclude that it was the form of the question that *caused* the difference in proportion of "Yes" responses? The answer for that study is yes, but the conclusion is valid only because of careful planning by the investigators. (For contrast: Cities with more McDonald's restaurants record larger numbers of divorces, but it would be bad science to conclude that eating Big Macs increases the likelihood of divorce. Or: Year by year, the larger the number of licensed amateur radio operators, the larger the number of people who are mentally ill. But that doesn't mean … You get the idea.)

Whereas Chapters 1 and 3 were about the *logic* of statistic inference (e.g., how strong is the evidence that the difference in the proportion of "Yes" responses was not due to random chance alone and what do we estimate to be the size of the difference), cause, along with breadth (Chapter 2), are about the *scope* of inference. If we establish that the difference is "real," the breadth part of Step 5: Formulate conclusions tells us how broadly the conclusion applies. Finally, in this chapter, you will learn whether and when you can conclude that an observed difference is *caused* by a particular treatment.

Why do cities that sell more Big Macs have more divorces? A related variable is *city size*. The bigger the city, the more Big Macs are bought, and the more divorces there are. Section 4.1 explores such issues more formally. Section 4.2 tells how to choose a study design to avoid such issues. The key idea is to use a chance process to assign the conditions to be compared, as in the study of organ donation. This introduction of chance into the study design will play a different role than it has in previous chapters.

The main goal of this chapter is to explain when and why you can infer cause. When you can't, caution is the order of the day.

Section 4.1: Association and Confounding

Introduction

To oversimplify, but not by much, you can think of science as a search for cause-and-effect relationships and for theories that unite them.In this section we illustrate two things to look for as part of using data to find causal relationships.The first is **association** between two variables.Consider smoking and cancer. Initially, scientists found that smokers had higher rates of lung cancer than did non-smokers. Smoking and cancer were *associated*. Did that *prove* that smoking *caused* cancer? No. For example, some scientists thought there might be a gene that made people both likely to smoke and likely to get lung cancer. Presence of the gene (yes/no) could be a *confounding* variable. The confounder could explain the association. To conclude that smoking causes lung cancer, you must be able to rule out the effect of possible confounders.

Is association evidence of possible causation? Yes. If smoking *causes* cancer, there has to be an association. Cancer rates will be higher for smokers. If the two are associated, one *might* cause the other. Association is necessary, but association alone is not enough to prove cause-and-effect.

Example 4.1:Night Lights and Near-Sightedness

Myopia, or near-sightedness, typically develops during childhood years. Recent studies have explored whether there is an association between development of myopia and the use of nightlights with infants. One study (Quinn, et al., 1999) interviewed parents of 479 children who were seen as outpatients in a university pediatric ophthalmology clinic. One of the questions asked whether the child typically slept with the room light on, with a night light on, or in darkness before age 2. Based on the child's most recent eye examination, the children were also separated into two groups: near-sighted or not near-sighted. They found a higher percentage of near-sighted children among those using a room light (54.7%) or with a night light (33.6%) compared to children who slept in darkness (10.5%). Pediatricians and parents became concerned that the use of light in the infants' rooms was leading to near-sightedness.

For this study, the observational units are the 479 children. We have recorded two different variables about each child: *whether the child slept with the room light, a night light, or darkness* and *whether or not the child was near-sighted.* These are both categorical variables. The lighting variable has three categories and the eye condition variable has two. The percentages of children with near-sightedness presented earlier are calculated *conditional* on the amount of light present in the room at night. Table 4.1 displays the data used to calculate the conditional proportions which multiplied by 100 give to percentages.

Table 4.1: Data used to calculate conditional proportions of nearsightedness based on amount of light in bedroom

	Darkness	Night light	Room light	Total
Near-sighted	18	78	41	137
Not near-sighted	154	154	34	342
Total	172	232	75	479

Conditional proportions: $18/172 \approx 0.105$ (10.5%), $78/232 \approx 0.336$ (33.6%), and $41/75 \approx 0.547$ (54.7%). (These calculations will be discussed in more detail in Chapter 5.)

Notice that as the amount of light present in the room at night increases, so does the percentage of children that are near-sighted. Thus we can say that there exists an association between whether or not a child is nearsighted and amount of light used in the child's room before the age of 2.

Definition: Two variables are *associated* or related if the value of one variable gives you information about the value of the other variable. When comparing groups this means that the proportions or means take different values in the groups.

Here the two variables(amount of light and eye condition) are associated:Knowing the amount of light gives you information about (helps you predict) a child's later eye condition. For example, children who slept with a night light are much more likely to be nearsighted than are children who slept in the dark.

Explanatory and Response Variables

Association often raises questions about possible causal relationships. After seeing the data above we might wonder, "Does sleeping with a night light *cause* a child's chance of being near sighted to increase?"As you'll soon see, this study does *not* allow us to answer this question. As the saying goes, "Association is not causation. "All the same, it is useful to have labels for the two different roles of the variables:The light condition is called the*explanatory*variable, because in this context we think of it as "explaining" the values of the other, the*response* variable, in this case, eye condition.

Definitions: The *explanatory variable* is the variable we think is "explaining" the change in the response variable and the *response variable* is the variable we think is being impacted or changed by the explanatory variable. The explanatory variable is sometimes called the independent variable and the response variable is sometimes called the dependent variable.

Think about it:Imagine that you are planning a study to look for association between overall health (good, fair, not good) and having a pet. Which variable would be explanatory, and which would be the response?

If you think having a pet increases people's happiness and likelihood to exercise, then you are considering pet ownership to be the explanatory variable and health to be the response variable. On the other hand, you may want to consider the health of the individual as an explanation for why some people are more likely to have pets than others. This would reverse the role of the explanatory and the response variable.

In another study, diet could be the explanatory variable and whether or not the person has heart disease could be the response. However, in some cases, there is not a clear distinction between the roles of the variables, such as looking at hair color and eye color – it's not as though one of these variables precedes and perhaps influences the other.

Cause-and-effect

In this study, there is an association because the percentage of near-sighted children increases as the amount of light used increases. But does this convince us that the room light and night light use is *causing* the children to develop near-sightedness? Drawing such a "***cause-and-effect***" conclusion like this is much different from establishing an *association* between the variables.

Think about it: Can you suggest another explanation for the higher percentage of near-sightedness in children using the room light and night light compared to those with no light?

Keep in mind that the association between amount of light in the room and eyesight condition appears to be real. If we wanted to predict whether or not a child was near-sighted, it would be worthwhile to know what type of lighting they used before the age of 2. But that is still a very different statement from saying that it was in fact the lighting condition itself that *directly impacted* the children's eye conditions.

One alternative explanation is that children who are already near-sighted are more likely to prefer to sleep with light in their rooms. This reverses the roles of the explanatory and response variables, though keep in mind that the researchers asked about light conditions before the age of two (when children don't always have as much say about their environment!) and later eye condition.

Another alternative explanation is that near-sightedness is often inherited from parents (genetics). But does this explain the observed association?It does if parents who are near-sighted are more likely to use a room light or a night light with their infants, perhaps needing the extra light to more easily navigate the room when they enter at night to check on their child. This appears to be a plausible explanation. The problem is we have no way of deciding whether this explanation is better or worse than the cause-and-effect explanation. Both explanations are consistent with the data we have and the association we observed. If the parents' eye conditions do tend to differ among the children with well-lit rooms, night lights, and darkness, then we have no way of separating out the "effects" of this variable from those of the lighting condition. In such a case, *parents' eyesight* is considered a ***confounding variable***. The diagram in Figure 4. 1 illustrates the confounding. The top panel shows the design of the study:Children (units) are sorted into groups according to the explanatory variable (light condition), and the response (nearsightedness) is measured. The bottom panel shows the potential confounding effect of the parents' own vision.

Figure 4.1: Potential confounding in the study of light and nearsightedness

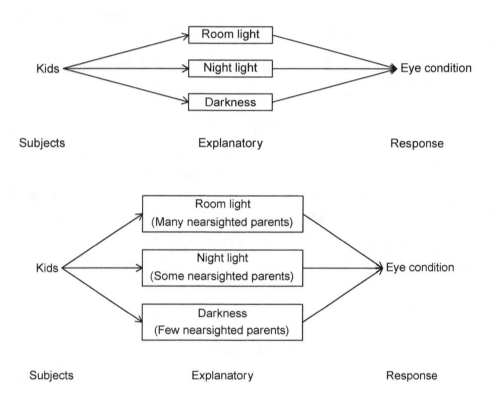

Definition: A ***confounding variable*** is a variable that is related both to the explanatory and to the response variable in such a way that its effects on the response variable cannot be separated from the effects of the explanatory variable.

Keep in mind that there will always be "other variables" floating around in any study. What we are most concerned about is these potential confounding variables that prevent us from isolating the explanatory variable as the only influence on the response variable. For example,*eye color* is another variable, but it is not likely related to eye condition or lighting use. Or maybe parents use a room light or nightlight more with girls than with boys. But since near-sightedness affects males and females equally, this wouldn't explain the higher percentages of near-sightedness for the children using light.

The key to discounting a cause-and-effect explanation is to identify a potential confounding variable and to explain how it is linked to <u>both</u> the explanatory variable and the response variable in a way that also explains the observed association.

Of course, this doesn't mean our explanatory variable is *not* influencing our response variable. Sometimes the explanation really is causation, as with the association between smoking and lung cancer. But we just can't feel comfortable jumping to a cause-and-effect conclusion based solely on these kinds of studies (we'll discuss this more in the next section!).In short, association does not imply causation, but sometimes it can be a pretty big hint.

Exploration 4.1: Home Court Disadvantage?

Sports teams prefer to play in front of their own fans rather than at the opposing team's site. Having a sell-out crowd should provide even more excitement and lead to an even better performance, right? Well, consider the Oklahoma City Thunder, a National Basketball Association team, in its second season (2008-09) after moving from Seattle. This team had a win-loss record that was actually worse for home games with a sell-out crowd (3 wins and 15 losses) than for home games without a sell-out crowd (12 wins and 11 losses). (These data were noted in the April 20, 2009 issue of *Sports Illustrated* in the Go Figure column.)

1. Identify the observational units and variables in this study. Also classify each variable as categorical (also binary?) or quantitative.

2. When did the Thunder have a higher winning percentage: in front of a sell-out crowd or a smaller crowd? Support your answer by calculating the proportion of sell-out games that they won and also the proportion of non-sell-out games that they won. (Write both these proportions as decimals.)

   Sell-out crowd:

Smaller crowd:

Definition: Two variables are *associated* or related if the value of one variable gives you information about the value of the other variable. When comparing two groups this means that the proportions or means take different values in the two groups.

---

3. Do the two variables appear to be associated?

Often when a study involves two associated variables, it is natural to consider one the *explanatory variable* and the other the *response variable*.

Definitions: The *explanatory variable* is the variable we think is "explaining" the change in the response variable and the *response variable* is the variable we think is being impacted or changed by the explanatory variable. The explanatory variable is sometimes called the independent variable and the response variable is sometimes called the dependent variable.

---

4. Which would you consider the explanatory variable in this study? Which is the response? (That is, what are the *roles* of these variables in this study?)

There are two possible explanations for this odd finding that the team had a better winning percentage with a smaller crowds:

- The sell-out crowd *caused* the Thunder to play worse, perhaps because of pressure or nervousness.
- The sell-out crowd did *not* cause a worse performance, and some other issue (variable) explains why they had a worse winning percentage with a sell-out crowds. In other words, a third variable is at play, which is related to both the crowd size and the game outcome.

(Of course, another explanation is random chance. Using methods you will learn later, we've determined that you can essentially rule out random chance in this case.)

5. Consider the second explanation. Suggest a plausible alternative variable that would explain why the team would be less likely to win in front of a sell-out crowd than in front of a smaller crowd. (Make sure it's clear not just that your explanation would affect the team's likelihood of winning, but that the team would be less likely to win in front of a sell-out crowd compared to a smaller crowd.)

> **Definition:** A *confounding variable* is a variable that is related both to the explanatory and to the response variable in such a way that its effects on the response variable cannot be separated from the effects of the explanatory variable.

6. Identify the confounding variable based on your suggested explanation in #5. Explain how it is confounding– what is the link between this third variable and the response variable, and what is the link between this third variable and the explanatory variable?(*Hint*: Remember that this variable has to be recorded on the observational units: home games for the Thunder.)

Another variable recorded for these data was whether or not the opponent had a winning record the previous season. Of the Thunder's 41 home games, 22 were against teams that won more than half of their games. Let's refer to those 22 teams as strong opponents. Of these 22 games, 13 were sell-outs. Of the 19 games against opponents that won less than half of their games that season (weak opponents), only 5 of those games were sell-outs.

7. Was the Thunder more likely to have a sell-out crowd against a strong opponent or a weak opponent? Calculate the relevant (conditional) proportions to support your answer.

When the Thunder played a strong opponent, they won only 4 of 22 games. When they played a weak opponent, the Thunder won 11 of 19 games.

8. Was the Thunder less likely to win against a strong opponent than a weak one? Again calculate the relevant (conditional) proportions to support your answer.

9. Explain how your answers to #7 and #8 establish that strength of opponent is a confounding variable that prevents drawing a cause-and-effect conclusion between crowd size and game outcome.

10. Summarize your conclusion about whether these data provide evidence that a sell-out crowd *caused* the Thunder to play worse. Write as if to a friend who has never studied statistics. Be sure to address the fact that the Thunder had a much smaller winning percentage in front of a sell-out crowd.

Confounding explains why you cannot draw a cause-and-effect conclusion from association alone: The groups defined by the explanatory variable could differ in more ways than just the explanatory variable when confounding is present. The diagram in Figure 4.2 illustrates the confounding in the Thunder study. The top panel shows the study design: Observational units (home games) are sorted into groups according to the explanatory variable (whether or not the arena was sold out). Then the response (win/lose) was observed. The bottom panel shows the confounding:sell out crowds tended to be against stronger opponents; weaker opponents tended not to sell out the arena.

Figure 4.2: Confounding in the study of the home court disadvantage

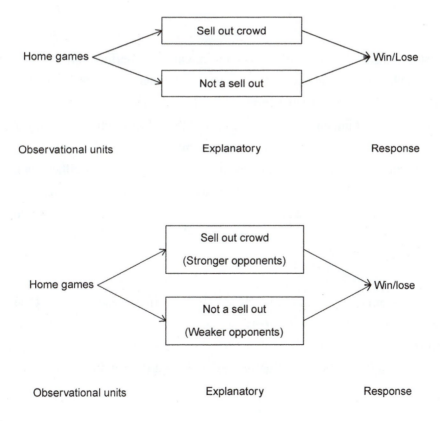

Section 4.1 Summary

Two variables are **associated** (related) if the values of one variable provide information about (help you predict) the values of the other variable.

Studies that involve two variables often distinguish between the roles played by the variables.
- The **explanatory variable** is the one that is suspected of possibly affecting the other.
- The **response variable is** the one that may be affected by the other.

A **confounding variable** is one whose effects on the response cannot be separated from those of the explanatory variable.

The possible presence of confounding variables is the reason that association alone does not justify a conclusion that differences in the explanatory variable *cause* differences in the response variable.

Section 4.2: Observational Studies Versus Experiments

Introduction

Many, if not most, scientific studies involve a search for cause-and-effect relationships between variables. Studies often start by looking for an association between explanatory and response variables, but the possible presence of confounding variables prevents us from being able to draw conclusions about cause. The main challenge in designing a good study is to avoid the effect of confounding variables. If you know in advance about a particular confounding variable, you may be able to choose a design that balances the effects of that confounder. But many confounders are "lurking" variables – either not known or known but not measured. To deal with these unseen confounders, we rely on a strategy called randomization or random assignment. We use a chance device like coin tosses or drawing names from a hat to decide which subjects go into which groups. Random assignment in this way gives approximate balance for all possible confounding variables.

The importance of random assignment to avoid confounding effects leads to one of the most important distinctions in all of science: the difference between an observational study and an experimental study. In an observational study, the values of the explanatory variable are simply observed – you don't get to choose. In an experiment, you – the investigator – get to assign the conditions to the different subjects.

Example4.2: Lying on the Internet

Are people more likely to tell lies with email than with pencil and paper? A study reported at the August 2008 meeting of the Academy of Management involved 48 graduate students in business who participated in a "bargaining" game. Researchers kept track of whether or not students misrepresented (lied about) the size of the pot when they were negotiating with other players.

Step 1: Ask a research question.

Does the Internet encourage lying? Investigators thought that the Internet might encourage users to feel anonymous, making them more likely to lie. To narrow their hypothesis, they decided to compare two groups of students playing a competitive game. Students in one group would play on the Internet by using e-mail as their method of communication. Students in the other group would use pencil and paper as their method of communication. The researchers hypothesized that those who played on the Internet would lie more often.

Step 2: Design a study and collect data.

> **Definition:** In an *observational study*, the groups you compare are "just there," that is, they are defined by what you *see* rather than by what you *do*. In an *experiment*, you actively create the groups by what you choose to do to the people or experimental units. More formally, you assign the conditions to be compared. These conditions may be one or more "treatments" or a "control" (a group you do nothing to).

> **Think about it:** Identify the explanatory and response variables in this study. Is the study observational or experimental? How can you tell?

The explanatory variable is the communication method. The response variable is whether or not the person lied. Both variables are categorical. This was an experiment rather than an observational study because the researchers actively intervened to determine the communication method used by the students.

> **Think about it.** Suppose there were 30 male and 18 female subjects. Imagine if you had all of the 30 men in the sample play on the Internet, and the 18 women used pencil and paper. If you were concerned that males might be more competitive and therefore more likely to lie in order to win, how would this limit your ability to draw conclusions about the initial research question from this study?

If all of the men played on the Internet and all of the women used pencil and paper, then if we saw differences in the two groups in terms of the proportion who lied, you would not be able to say whether it was gender or the format of the game (Internet vs. paper and pencil) that was causing the difference.

Think about it. How could you address this concern?

One strategy would be to assign half the men and half the women to each group. That way, both groups will have 15 male and 9 female subjects. The groups would be balanced with respect to gender.

Key idea: A comparative study is *balanced with respect to a possible confounding variable* if the distribution of the variable is the same for each group in the study.

Note that if you have many confounders, it will be hard to balance them all. Worse yet, if you have "lurking" variables – unmeasured or unrecognized potential confounders, the challenge might seem impossible. How can you balance what you can't even see?

Fortunately, random assignment offers a solution. If you know about and measure a potential confounder in advance, you may be able to design a study that gives exact balance for that variable. If you use a chance device to assign units to groups, you may not get exact balance. In fact, most of the time you won't. But – this is the key – you will get *approximate* balance for *all* potential confounders, including the ones you don't know about. If you toss a fair coin, you tend to get half heads, half tails, or at least an outcome pretty close to that. If you use a coin toss to assign subjects to two groups, you tend to get half the men in each group, and half the women in each group. You also, at the same time, tend to get half the young people in each group, half the vegetarians in each group, half the pet owners in each group, and so on.

Key idea: Randomly assigning experimental units to groups tends to balance out all other variables between the groups. Any variables that could have an effect on the response should be equalized between the two groups and therefore should not be confounding.

With a randomized experiment, if you observe a difference in lying between the Internet players and the pencil and paper players, there are only two explanations: Either there is a cause-and-effect relationship, or else the difference was just due to random chance, stemming from the random assignment process. We will be willing to assume there are no potential confounding variables to explain the difference.

The diagram in Figure 4.3 illustrates this. The top panel shows the possible confounding effect if the study design had not been randomized. We want to know whether differences in the explanatory variable cause differences in the response (dotted arrow). However, a confounding variable would affect the response variable and be associated with the explanatory variable, making it impossible to separate the effects of explanatory and confounding variable on the response. The bottom panel shows how randomization removes the effect of the potential confounders.

Figure 4.3: Randomization creates approximate balance and tends to eliminate confounding effects.

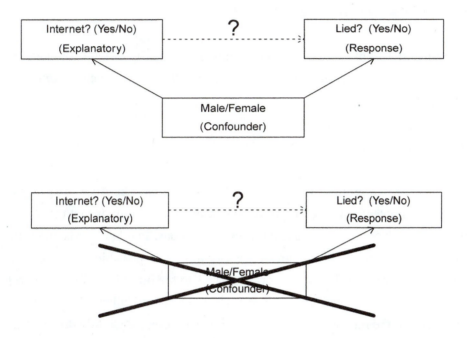

Definitions: In a *randomized experiment*, you use a chance device to make the assignments. The units in an experiment are often called *experimental units*. The role of the random assignment is to balance out potentially confounding variables among the explanatory variable groups, giving us the potential to draw cause-and-effect conclusions.

Note our change in terminology: we now will sometimes call the individuals in the study *experimental units* instead of *observational units*, because they are in an experiment instead of an observational study.

For this study each student was randomly assigned to one of two groups. Students in Group 1 used the Internet. Those in Group 2 used pencil and paper. You can think of the random assignment like drawing names out of a hat: Put all the names on slips of paper, put them in a hat, mix thoroughly, take them out one at a time, and put each of them into one of the two groups. (Internet, paper, Internet, paper, etc.)

Compare this study with the two studies in the last section, Night Lights and Home Court Disadvantage. Notice that in both of those studies, it was not possible for researchers to *assign* the conditions to be compared. By the time the children were seen at the eye clinic, it was too late to tell their parents what kind of light to use. For the study of sleep condition and vision, each child was an observational unit.

Values of the explanatory variable (level of light) were observed. They could not be chosen by the investigators. For the Thunder study, each home game was an observational unit. The strength of the opponent was observed, not assigned. If the home game is against the LA Lakers, there's no way you can make them be a weak team just for the sake of your research.

Step 3: Explore the data.

The researchers found that 24 of 26 (92%) who used email were guilty of lying about the potsize, compared to 14 of 22 (64%) who used paper and pencil. (We will discuss such *conditional* percentages in more detail in Chapter 5.)

Step 4: Draw inferences beyond the data.

When you are comparing two groups like this, if you find a difference between the groups, you again need to consider several potential explanations. One possible explanation is that the people feel more comfortable lying by email, perhaps because it seems anonymous. In other words, the format of the game *influenced* how people responded. But of course another important possible explanation to consider is random chance: Maybe it was just luck of the draw that led to the big difference in the proportions of liars between the two groups. We claimed that random assignment should balance out the two groups, but there is still a chance that the random assignment process itself created slightly different groups. Notice the randomness we are considering here stems from the random assignment, not random choices or random sampling. As it turns out, for this study we can rule out this explanation. Using methods you will learn in the next chapter, it's highly unlikely that the emailing players would lie this much more frequently, relative to the pencil and paper players, just by chance alone (i.e., the p-value is small, 0.015).

Step 5: Formulate conclusions.

Now it is time to formulate our final conclusions. We found a statistically significant difference between the two groups. Does this allow us to conclude that the format of the game *influenced* how people responded? Maybe those who use email tend to be younger and younger individuals are more likely to lie. Or there could be other personality traits that differ between the individuals in the email group and the paper and pencil group. Are these potential confounding variables?

But if you believe what we told you in the previous section, because the study used random assignment to form the email group and the paper and pencil group, we don't expect any systematic differences between the Internet group and the pencil-and-paper group. This is because the random assignment process should create groups that are similar to each other on all respects (e.g., gender, age, propensity for lying, familiarity with technology) apart from the explanatory variable of interest.

This means that when you observed a statistically significant difference in how the groups respond or behave (too large to be expected to happen by the random chance inherent in the random assignment process), then a plausible explanation is *because of* the group the experimental units were assigned to.

In formulating our conclusions, we still want to consider the *breadth* of our study. Because the subjects in this study business school graduate students, and volunteers, we should be very cautious in generalizing these results to a larger population. See FAQ 4.2.1 and Figure 4.5 at the end of this section for more details and a helpful table.

FAQ 4.2.1- Randomness and its implications

Q: You make it nice and clear. If a research study doesn't randomize the assignment, I can't conclude anything about cause. If it doesn't use random sampling, I can't generalize. No chance, no inference. Right?

A: Sorry, that's much too black and white.

Q: But wait. The theory is simple. List your units, then use chance to assign treatments. Isn't that what you said?

A: Yes, that's the ideal way to learn about cause-and-effect. But you can't always do that in practice. Think about smoking and lung cancer.

Q: Those were observational studies, not randomized experiments. How did scientists conclude that smoking causes cancer?

A: Remember that a main purpose of randomization is to protect against confounding. It's possible to design observational studies that also protect against confounding. It's a lot harder, but it's possible. It was over many, many of these more complicated observational studies, with more advanced statistical methods, and with support from other types of studies linking carcinogens to cigarettes, that it became an accepted scientific truth that smoking causes lung cancer.

Q: And random sampling?

A: Here, too, it's not black and white. Randomizing the selection protects against bias in your sampling method, so if you don't randomize, you have to be very careful about what larger group your sample represents, but that's where judgment comes in. As in all of statistics, the mathematical theory is an ideal, but reality is almost always more complicated. You don't want to throw out the baby (the data) just because the bath water (design) is not ideal.

Types of Experiments

We just saw that randomized experiments (experiments where the explanatory variable is manipulated by the researchers and assignment of observational units to the explanatory variable's groups is determined randomly) can potentially lead to cause-and-effect conclusions between the explanatory and response variables. What about experiments which manipulate the explanatory variable, but not randomly? These are often called *quasi-experiments*. For example, suppose we wanted to compare student learning gains when using a new curriculum to that of an old curriculum. It would be difficult to assign students *randomly* to which class they take. So the instructors might take the results from pre- and post-tests from students that used the old curriculum one year and compare these to the results from pre- and post-tests from students that used the new curriculum another year. Although this type of experiment may be the best option when random assignment is not feasible, there may still be reasons other than the explanatory variable (confounding variables) for the difference in outcomes between the two groups (e.g., time of day class is offered, percentage of upper-division students in the course) Therefore, even though this is considered an experiment, it is not a randomized experiment, and we should be cautious before drawing any cause-and-effect conclusions.

Also keep in mind that some explanatory variables of interest don't lend themselves to randomized experiments. For example, gender can't be randomly imposed on individuals, and other variables, such as smoking behavior, would not be ethical to manipulate!

Other Considerations

Think about it: In this study, the subjects were randomly assigned to the Internet group and the paper and pencil group. Do you think the researchers told the subjects in advance that they would be assigned to one of these two groups?

In most research studies, subjects are "blind" to the other treatment conditions. In fact, they often don't know that there are multiple conditions or exactly which condition they are in. For example, suppose I want to test a new diet pill. If I give you a pill and tell you it's going to give you more energy, you may feel a positive effect even if there is no active ingredient in the pill. If one group had this psychological suggestion and not the other, then that would be a confounding variable. So the second group, the "control group" would often be given a "placebo pill" (an empty treatment) so none of the subjects knew which group they were in. In fact, for something like measuring energy which may require subjective judgment (as opposed to heart rate), often the person making the measurements wouldn't know which group the subjects were in, to prevent any bias creeping into their judgments as well, making the study 'double-blind.' A "randomized, double-blind, placebo controlled experiment" have been considered by some to be the gold standard in scientific studies.

Definition: In a *double-blind* study, neither the subjects nor those evaluating the response variable know which treatment group the subject is in.

Exploration 4.2: Have a Nice Trip

An area of research in biomechanics and gerontology concerns falls and fall-related injuries, especially for elderly people. Recent studies have focused on how individuals respond to large postural disturbances (e.g., tripping, induced slips). One question is whether subjects can be instructed to improve their recovery from such disturbances.

Suppose researchers want to compare two such recovery strategies, lowering (making the next step shorter, but in normal step time) and elevating (using a longer or normal step length with normal step time). Subjects will have first been trained on one of these two recovery strategies, and they will be asked to apply it after they feel themselves tripping. The researchers will then induce the subject to trip while walking (but harnessed for safety), using a concealed mechanical obstacle.

Suppose the following 24 subjects have agreed to participate in such a study:

> Females: Alisha, Alice, Betty, Martha, Audrey, Mary, Barbie, Anna

> Males: Matt, Peter, Shawn, Brad, Michael, Kyle, Russ, Patrick, Bob, Kevin, Mitch, Marvin, Paul, Pedro, Roger, Sam

1. One way to design this study would be to assign the eight females to use the elevating strategy and the sixteen males to use the lowering strategy. Would this be a reasonable strategy? Why not?

2. One way to deal with this issue is to assign 4 females and 8 males to each group. Show how the proportion of males in each group is the same.

3. Now, if you saw a difference in the proportion of trips in the two groups, could it be because of gender? Why or why not? Could it be due to other variables, distinct from the recovery strategy? Why or why not?

4. Because there will always be more potential confounding variables which are distributed unevenly between the groups being compared, identify a better method for deciding who uses which strategy.

Definition: In an *observational study*, the groups you compare are "just there," that is, they are defined by what you *see* rather than by what you *do*. In an *experiment*, you actively create the groups by what you choose to do to the people or experimental units. More formally, you assign the conditions to be compared. These conditions may be one or more "treatments" or a "control" (a group you do nothing to).

Definitions: In a *randomized experiment*, you use a chance device to make the assignments. The units in an experiment are often called *experimental units*. The role of the random assignment is to balance out potentially confounding variables among the explanatory variable groups, giving us the potential to draw cause-and-effect conclusions.

Note our change in terminology: we now will sometimes call the individuals in the study *experimental units* instead of *observational units*, because they are in an experiment instead of an observational study.

5. Let's explore the process of random assignment to determine whether it does "work. "First, let's focus on the gender variable. Suppose we put each person's name on a slip, put those slips in a hat and mix them up thoroughly, and then randomly draw out 12 slips for names of people to assign to the elevating strategy. What proportion of this group do you expect will be male? What proportion of the lowering strategy do you expect will be male? Do you think we will always get an 8/8 split (8 males in each treatment group)?

6. To repeat this random assignment a large number of times to observe the long-run behavior, we will use the **Randomizing Subjects** applet. Open the applet, and press the **Randomize** button. What proportion of subjects assigned to Group 1 are men?Of Group 2?What is the difference in these two proportions?

You will notice that the difference in proportions of males is shown in the dotplot in the bottom graph. In this graph, each dot represents one repetition of the random assignment process where we are recording the difference in proportions of men between the two groups.

7. Press the **Randomize** button again. Was the difference in proportions of men the same this time?

8. Change the number of replications from 1 to 198 (for 200 total), uncheck the **Animate** option, and press the **Randomize** button. The dotplot will display the difference between the two proportions of men for each of the 200 repetitions of the random assignment process. Where are these values centered?

9. Does random assignment *always* equally distribute/balance the men and women between the two groups?Is there a tendency for there to be a similar proportion of men in the two groups? Explain.

Definition: A comparative study is *balanced with respect to a possible confounding variable* if the distribution of the variable is the same for each group in the study.

---

10. Prior research has also shown that the likelihood of falling is related to variables such as walking speed, stride rate, and height, so we would like the random assignment to distribute these variables equally between the groups as well. In the applet, use the pull-down menu to switch from the gender variable to the height variable. The dotplot now displays the differences in *average* height between Group 1 and Group 2 for these 200 repetitions. In the long-run, does random assignment tend to equally distribute the height variable between the two groups? Explain.

11. Suppose there is a "balance gene" that is related to people's ability to recover from a trip. We didn't know about this gene ahead of time, but if you select the "Reveal gene?" button and then select "gene" from the pull-down menu, the applet shows you this gene information for each subject and also how the proportions with the gene differ in the two groups. Does this variable tend to equalize between the two groups in the long run? Explain.

12. Suppose there were other "x-variables" that we could not measure such as stride rate or walking speed. Select the "Reveal both?" button and use pull-down menu to display the results for the x-variable (x-var). Does random assignment generally succeed in equalizing this variable between the two groups or is there a tendency for one group to always have higher results for the x-variable? Explain.

> **Key idea:** Randomly assigning experimental units to groups tends to balance out all other variables between the groups. Any variables that could have an effect on the response should be equalized between the two groups and therefore should not be confounding.

The diagram in Figure 4.4 illustrates this. The top panel shows the possible confounding effect if the study design had not been randomized. We want to know whether differences in the explanatory variable cause differences in the response (dotted arrow). However, a confounding variable would affect both the explanatory and response variables, making it impossible to separate the effects of explanatory and confounding variables on the response. The bottom panel shows how randomization removes the effect of the potential confounders.

Figure 4.4: Randomization creates approximate balance and tends to eliminate confounding effects

13. Suppose this study finds a statistically significant difference between the two groups. What conclusion would you draw? For what population? What additional information would you need to know?

14. As in #13, if you obtain a statistically significant result, what does that suggest about the potential for a cause-and-effect relationship? Why?

Another study: Cursive Handwriting

An article about handwriting appeared in the October 11, 2006 issue of the *Washington Post*. The article mentioned that among students who took the essay portion of the SAT exam in 2005-6, those who wrote in cursive style scored significantly higher on the essay, on average, than students who used printed block letters.

15. Identify the observational units in this study, as well as the explanatory and response variables. Also classify each variable as categorical or quantitative.

   Observational units:

   Explanatory variable: Type:

   Response variable: Type:

16. Explain how you know that this was an observational study.

17. Is it reasonable to conclude that using a cursive writing style *caused* higher scores on the essay, or can you think of an alternative explanation for why students who wrote in cursive style scored higher on average than students who write with block letters? In other words, can you think of other ways in which the cursive and block letter groups might have systematically differed and identify a potential confounding variable?

The same *Washington Post* article also mentioned a different study in which the identical essay was shown to many graders, but some graders were randomly chosen to see a cursive version of the essay and the other graders were shown a version with printed block letters. The average score assigned to the essay with the cursive style was significantly higher than the average score assigned to the essay with the printed block letters.

18. How does this study differ from the original one? Explain.

19. Would you be willing to draw a cause-and-effect conclusion from this study, as opposed to the original study? Explain why.

See the Section summary for a helpful table (Figure 4.5) which summarizes the types of conclusions you can draw with and without random sampling and with and without random assignment.

The "placebo effect" has been found in numerous studies: when subjects are told something good is going to happen, they often have a positive response even if nothing is actually done to them. For this reason, subjects are often kept "blind" as to which treatment group they are placed in, for example by giving one group the actual treatment and the other group a "fake" treatment, like a placebo (an empty pill). Placebo treatments have even been used in studies about knee surgery!

20. Discuss why it was important in this second study for the scorers of the essays to not know that there was no difference in the essays given to different individuals other than the writing style.

21. Discuss how this study also reveals that those evaluating the response variable in a study, if it requires any subjective judgment, should be "blind" to which treatment group individuals are in.

Definition: In **double-blind** studies, neither the subjects nor those evaluating the response variable know which treatment group the subject is in.

Section 4.2 Summary

A well-designed **experiment** uses random assignment to determine which observational (experimental) units go into which explanatory variable groups.

- The goal of **random assignment** is to produce groups that are as similar as possible in all respects except for the explanatory variable.
- Then if the groups can be shown to differ significantly on the response variable, the difference can be attributed to a **cause-and-effect relationship** between the explanatory and response variables.

Random assignment is a very different use of randomness from random sampling, with different implications for scope of conclusions.

- Random sampling aims to select a representative sample from a population, so that results about the sample can be generalized to the larger population.
- Random assignment aims to produce similar treatment groups, so that a significant difference in the response variable between groups can be attributed (causally) to the explanatory variable.

Figure 4.5 summarizes how these two types of randomness lead to different impact on the scope of conclusions.

In a **double-blind** experiment, neither the subjects nor those evaluating the response variable know which group the subject is in. This is to guard against biases such as the placebo effect, where the mere power of suggestion can influence subjects' responses.

Figure 4.5: Study design factors and their impact on scope of conclusions (adapted from Ramsey and Schafer, *The Statistical Sleuth*)

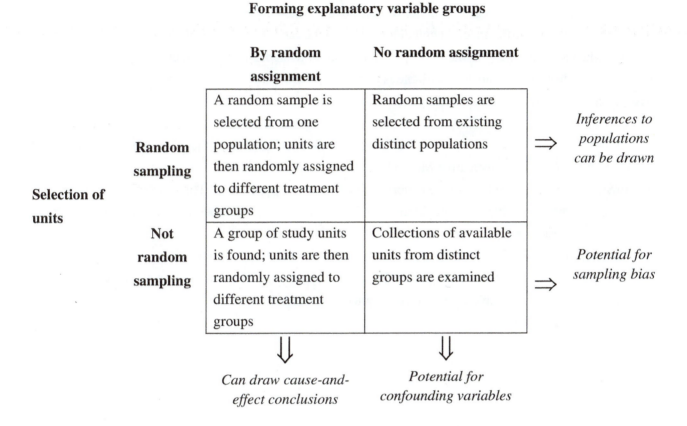

Chapter 4 Summary

In the first section, we saw that two variables are associated when the value of one variable provides information about (helps us predict) the value of the other variable. In a search for causal relationships, we call the possible cause variable *explanatory* and the outcome variable the *response*. However, association alone does not prove causation because of the potential for confounding variables(variables related to both the explanatory and response variables in such a way that their effects on the response variable are intertwined with those of the explanatory variable).

In Section 2we distinguished between observational and experimental studies. In an observational study, you simply record what's there. You don't get to choose which group each observational unit belongs to. In an experiment you choose and assign the conditions that define the groups to subjects or other units. Ideally, this is done using a chance device like coin tosses or drawing names out of a hat. Random assignment has the advantage that it tends to ensure balance for all possible confounding variables, both known and unknown. For a properly randomized experiment, there are only two explanations for any observed difference between groups:either the assigned explanatory variable conditions caused the difference, or else the difference is a coincidence due to the random assignment. (In the next chapter you will see when and how we can find evidence against random chance as an explanation.)Not all experiments where the explanatory variable is manipulated are randomized. Without randomized assignment, conclusions about cause are not justified.

You have now seen that there are two main "Scope of inference" questions that should be asked as part of Step 5 of the Statistical Investigation Method.

- First, is the result from the sample generalizable to a larger population? The answer is yes, if the sample is representative of the population, and we feel more comfortable believing such if the sample is a random sample.

- Second, can we conclude that there is a cause-and-effect relationship between the explanatory and response variable? The answer is yes, if the study is a randomized experiment, meaning that the value of the explanatory variable is randomly assigned, guarding against potential confounding variables

Chapter 4 Glossary

association

Two variables are associated or related if the distribution of the response variable differs across the

values of the explanatory variable 4, 7

cause-and-effect

In well-designed studies (randomized experiments), can conclude the explanatory variable is causing

the effect seen in the response variable... 4

confounding variable

A confounding variableis a variable that is related to both the explanatory and response variable in such

a way that its effects on the response variable cannot be separated from the explanatory variable. 5, 8

controltreatment

A *do nothing* treatment in an experiment.. 12, 19

double-blind

A study design where neither the subjects nor those evaluating the response know which treatment

group the subject is in 18, 25

experiment

 A study in which researcher actively assign subjects to treatment groups 12, 19

experimental units

 What observational units are called in an experiment study.. 14, 19

explanatory variable

The variable that, if the alternative hypothesis is true, is explaining changes in the response variable;

sometimes known as the independent or predictor variable 3, 7

observational study

Studies in which researchers observe individuals and measure variables of interest, but do not intervene

in order to attempt to influence responses. 12, 19

quasi-experiments

Experiments that manipulate the explanatory variable, but not randomly... 17

randomizedexperiment

An experiment where experimental units are randomly assigned to two or more *treatment* conditions

and the explanatory variable is actively imposed of the subjects. 14, 19

Section 4.1

GPAs and pulling all-nighters*

4.1.1 Exercises 4.1.1-3 refer to the following study. A study published in the January 2008 issue of the journal *Behavioral Sleep Medicine* involved a survey of 120 students at St. Lawrence University, a small liberal arts college in upstate New York. Researchers found that students who claimed to have never pulled an all-nighter had average GPAs of 3.1, compared to 2.9 for those students who do claim to have pulled all-nighters.

  a) In this context, what is the observational unit?

      A. Researcher

      B. Student

      C. GPA

      D. Whether or not have pulled an all-nighter

  b) Which of the following is the explanatory variable?

      A. Whether or not have pulled an all-nighter

      B. Student

      C. GPA

      D. Those students who have pulled all-nighters

  c) Which of the following is the response variable?

      A. GPA

      B. Average GPA

      C. Whether or not have pulled an all-nighter

      D. Students who have pulled all-nighters tend to have lower GPAs, on average

  d) The explanatory variable is _____ (pick one: CATEGORICAL   QUANTITATIVE)
  e) The response variable is _____ (pick one: CATEGORICAL   QUANTITATIVE)

4.1.2 Reconsider the previous study about GPAs and pulling all-nighters. Suppose that the difference between these two averages is shown to be statistically significant. Which of the following is a potential confounding variable, and provides an alternative explanation for why the all-nighter group would have a significantly lower average GPA?

  A. Students

  B. Type of lifestyle choices made by student

C. Difficulty level of classes offered at St. Lawrence

D. Class sizes

4.1.3 Reconsider the previous exercise about the potential confounding variable that provides an alternative explanation for why the all-nighter group would have a significantly lower average GPA. Sketch a well-labeled diagram showing how this explanation works.

Using candy cigarettes

4.1.4 In a study published in a 2007 issue of the journal *Preventive Medicine*, researchers found that smokers were more likely to have used candy cigarettes as children than non-smokers were. When hearing about this study, John responded: "But isn't the smoking status of the person's parents a confounding variable here?" When Karen asked what he meant, John said: "Children whose parents smoke are more likely to become smokers themselves when they become adults." What else does John need to say in order to explain how the parents' smoking status could be a confounding variable in this study?

Marriage views*

4.1.5 Do different generations view marriage differently? A 2010 survey conducted by the Pew Research Center asked the following question of each participant, "*Is marriage becoming obsolete?*" Here is how the data from Pew Research Center's study worked out:

		Generation				Total
		Millennial (ages 18-29)	**Gen X (ages 30-45)**	**Boomers (ages 46-65)**	**Age 65+**	
Marriage obsolete?	**Yes**	236	313	401	68	1018
	No	300	416	745	143	1604
	Total	536	729	1146	211	2622

a) Identify the observational units.

b) Identify the explanatory and response variables. Also identify the type (categorical or quantitative) of each.

c) The conditional proportions of people answer Yes, marriage is become obsolete in each generation of the sample are: 0.44 (Millenial), 0.429 (Gen X), 0.350 (Boomers) and 0.322 (Age 65+). Based on these conditional proportions calculated in does it appear that different generations tend to view marriage differently? Explain.

Heart attacks

4.1.6 Studies conducted in New York City and Boston have noticed that more heart attacks occur in December and January than in all other months. Some people have tried to conclude that holiday stress and overindulgence causes the increased risk of heart attack.

Identify a confounding variable whose effect on heart attack rate might be confounded with that of the month, providing an alternative explanation for the increased risk of heart attack in December and January. Also, include a diagram showing how this explanation works.

Number of TVs and life expectancy*

4.1.7 Exercises 4.1.7-8 contain questions about the following study. Analysis of data on a sample of 22 countries shows a strong positive association between the average life expectancy in the country and the number of TVs per 1000 people in the country. Meaning, countries that have more TVs per 1000 people also have higher average lifetimes. Based on this, is it reasonable to conclude that by sending TVs to countries with lower life expectancies, we can increase their inhabitants' lifetimes? If yes, explain how. If no, give an alternative explanation for the association between number of TVs and life expectancy.

4.1.8 Regarding the previous question about number of TVs and life expectancy.

a) Identify the observational units.

      A. People

      B. TVs

      C. Countries

      D. Life expectancy

b) Identify the explanatory variable.

      A. People

      B. TVs

      C. Countries

      D. Life expectancy

c) The explanatory variable is _____ (pick one: CATEGORICAL   QUANTITATIVE)

d) Identify the response variable.

      A. People

      B. TVs

      C. Countries

      D. Life expectancy

e) The response variable is _____ (pick one: CATEGORICAL   QUANTITATIVE)

Mediterranean diet

4.1.9 Based on a four-year (2003 – 2007) study of over 30,000 people who were 45+ years, where all individuals were followed up regularly for diet and health changes, researchers Tsivgoulis et al. (*Neurology,* 2013) reported that the Mediterranean diet was linked to better memory and cognitive skills.

a) Identify the explanatory and response variables.

b) Identify a confounding variable whose effect on memory and cognitive skills might be confounded with that of the explanatory variable identified in (a), providing an alternative explanation for the improved memory and cognitive skills among those on the Mediterranean diet. Also, include a diagram showing how this explanation works.

VCR Ownership*

4.1.10 A Gallup poll conducted in December 2013 found that 74% of respondents who were 65 years or older owned a VCR compared to 41% of 18 to 29 years old.

a) Identify the explanatory variable.

       A. Respondents

       B. Age group

       C. Whether own a VCR

       D. Are older folks more likely to own a VCR?

b) Identify the response variable.

       A. Respondents

       B. Age group

       C. Whether own a VCR

       D. Are older folks more likely to own a VCR?

c) Do the results from Gallup's poll indicate that there is an association between owning a VCR and age? Explain.

Hormone Therapy

4.1.11 In a study published in *Preventive Medicine* (1991), researchers Stampfer and Colditz observed that women who underwent hormone replacement therapy (HRT) showed a lower risk of coronary heart diseases.

a) Identify the explanatory and response variables.

b) Identify a confounding variable that provides an alternative explanation for the reduction in CHD risk among women who underwent HRT, compared to those who didn't undergo HRT. Also, include a diagram showing how this explanation works.

Eating Breakfast*

4.1.12 Based on a survey of almost 3000 adults, researchers Wyatt et al. (*Obesity Research*, 2002) reported that those who ate breakfast regularly tended to be more successful at maintaining their weight loss.

a) Identify the explanatory and response variables.

b) Identify a confounding variable that provides an alternative explanation for the improved success rate of weight loss among regular breakfast eaters, compared to those who don't eat breakfast regularly. Also, include a diagram showing how this explanation works.

Dinner and drugs

4.1.13 In June 2012, *Time* magazine ran an article titled *"Do Family Dinners Really Reduce Teen Drug Use?"* Answer the following questions in this context.

a) Identify the observational units.

b) Identify the explanatory and response variables.

c) Identify a confounding variable that provides an alternative explanation for the lower drug use among teens whose families ate dinner together, compared to those whose didn't. Also, include a diagram showing how this explanation works.

Smoking while pregnant*

4.1.14 Many studies have shown that women who smoke while pregnant tend to have babies that weigh significantly less, on average, than women who do not smoke while pregnant.

- a) Identify the population(s) of interest in these studies.
- b) Identify the explanatory variable in these studies. Also classify this variable as categorical or quantitative.
- c) Identify the response variable in these studies. Also classify this variable as categorical or quantitative.

d) Socio-economic status is potentially a confounding variable in these studies. Explain what it means for this to be a confounding variable in this context, and describe how this could provide an alternative explanation to concluding that smoking while pregnant causes lower birth weight in babies.

Spanking and IQ

4.1.15 Studies have shown that children in the U.S. who have been spanked (yes/no) have a significantly lower IQ score on average than children who have not been spanked.

a) Identify the explanatory variable in these studies. Also classify this variable as categorical or quantitative.

b) Identify the response variable in these studies. Also classify this variable as categorical or quantitative.

c) Identify a confounding variable that provides an alternative explanation for the lower average IQ score for children who have been spanked, compared to those who haven't.

Overweight friends*

4.1.16 In July 2013, Gallup surveyed 2027 randomly selected U.S. adults. They found that of the 921 people who described themselves as overweight, 424 reported having some or many friends/family member who were overweight (424/921 = 0.46). Also, of the 1106 people who described their weight as being "about right," 332 reported having some or many friends/family member who were overweight (0.30). Use the conditional proportions to explain whether the data indicate that there is an association between whether one perceives oneself as being overweight and how many friends/family members one perceives as being overweight.

Children and lifespan

4.1.17 Do men with children tend to live longer than men without children? To investigate, a group of students at Cal Poly randomly sampled men from the obituaries page on the *San Luis Obispo Tribune*'s webpage between June – November 2012. For each man selected, they noted the age at which the person died, and number of children under "survived by."

a) Identify the explanatory variable in this study. Also classify this variable as categorical or quantitative.

b) Identify the response variable in this study. Also classify this variable as categorical or quantitative.

Given below is a set of dotplots displaying the data:

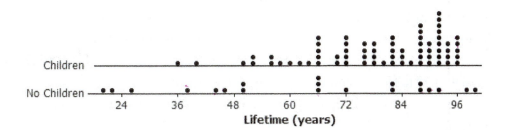

c)   Based on the above dotplots, is there any indication of an association between the explanatory variable and the response variable? Explain how you are deciding.

d)   Identify a confounding variable that provides an alternative explanation for the higher lifetime values among men who have children, compared to those who don't.

Happiness and income*

4.1.18 To investigate whether there is an association between happiness and income level, we will use data from the 2002 General Social Survey (GSS), cross-classifying a person's perceived happiness with their family income level. The GSS is a survey of randomly selected U.S. adults who are not institutionalized. Here are the data:

		Income			Total
		Above average	**Average**	**Below average**	
Happy?	**Very happy**	110	221	83	414
	Pretty happy	159	372	249	780
	Not too happy	21	53	94	168
Total		290	646	426	1362

a)   Identify the explanatory variable. Is it a categorical variable or a quantitative variable?

b)   Identify the response variable. Is it a categorical variable or a quantitative variable?

c)   Among above average income individuals 0.379 are Very happy (110/290), among average income individuals 0.342 are Very happy (221/646) and among below average individuals 0.195 are very happy (83/426). Do the data provide any indication of an association between happiness and income level? Why or why not?

d)   Is it okay to conclude that income affects happiness? If yes, explain why. If no, then identify a confounding variable that provides an alternative explanation for the association between happiness and income level.

Gender and body image

4.1.19 A 2013 Gallup poll asked randomly selected U.S. adults whether they wanted to stay at their current body weight or change. One purpose was to investigate whether there was any difference between men and women with regard to this aspect.

a) Identify the explanatory variable. Is it a categorical variable or a quantitative variable?

b) Identify the response variable. Is it a categorical variable or a quantitative variable?

c) Of the 562 men surveyed, 0.43 \wanted to stay at their current weight, whereas of the 477 women surveyed,0.36 wanted to stay at their current weight. Do the data provide any indication of a difference between how men and women perceive their current body weight? Explain using appropriate numbers.

Political party and evolution*

4.1.20 Is there an association between political party affiliation and beliefs about human evolution? A survey of a random sample of U.S. adults by *The Pew Research Center for the People and the Press*, conducted in March – April 2013 recorded each participant's political party affiliation and belief about human evolution. Here is how the data from Pew Research Center's study worked out:

		Political party affiliation		
		Republican	**Democrat**	**Independent**
Belief about human evolution	**Humans have evolved over time**	196	472	438
	Humans have existed in present form	218	190	189
	Don't know	41	43	47

a) Identify the explanatory and response variables. Also identify the type (categorical or quantitative) of each.

b) Based on the table, 0.43 Republicans believe Humans have evolved over time, 0.67 Democrats and 0.65 Independents. Do the data provide any indication of an association between political party affiliation and beliefs about human evolution?

Education and the census

4.1.21 A survey of a random sample of U.S. adults by *The Pew Research Center for the People and the Press*, conducted in early January 2010 recorded each participant's highest level of education completed, and whether they knew that responding to the Census was required by law. Of the 973 participants who had some college or less, 27.9% (271/973) knew that responding to the Census was required by law. Of the 526 participants who had a college degree or more, 37.1% (195/526) knew that responding to the Census was required by law.

a) Identify the explanatory variable. Is it a categorical variable or a quantitative variable?

b) Identify the response variable. Is it a categorical variable or a quantitative variable?

c) Do the data provide any indication of an association between education level and the awareness that responding to the Census is required by law? Explain.

Prayer and blood pressure*

4.1.22 In August 1998, an article titled "Prayer can lower blood pressure" appeared in the *USA Today*. The article was based on the findings of a study by the National Institutes of Health Initiatives that followed 2391 people aged 65 years or more. The article said,

"People who attended a religious service once a week and prayed or studied the Bible once a day were 40% less likely to have high blood pressure than those who don't go to church every week and prayed and studied the Bible less."

a) Identify the explanatory variable in this study. Also classify this variable as categorical or quantitative.

b) Identify the response variable in this study. Also classify this variable as categorical or quantitative.

c) Identify a confounding variable that provides an alternative explanation for the lower blood pressure among those who attended religious services regularly and studied the Bible regularly, compared to those who didn't.

Colds and exercise

4.1.23 In November 2010, an article titled "Frequency of Colds Dramatically Cut with Regular Exercise" appeared in the *Medical News Today*. The article was based on the findings of a study by researchers Nieman et al. (*British Journal of Sports Medicine,* 2010) that followed 1002 people aged 18 – 85 years for 12 weeks, asking them to record their frequency of exercise (5 or more days a week? Yes or No), as well as incidences of upper respiratory tract infections (Cold during last week? Yes or No).

a) Identify the explanatory variable in this study. Also classify this variable as categorical or quantitative.

b) Identify the response variable in this study. Also classify this variable as categorical or quantitative.

c) Identify a confounding variable that provides an alternative explanation for the lower frequency of colds among those who exercised 5 or more days per week, compared to those who were largely sedentary.

Politics and gun ownership*

4.1.24 In February 2013, the Pew Research Center surveyed randomly selected U.S. adults about their opinions on gun ownership and related issues. One of the questions they asked was, "Should states be allowed to ignore federal gun laws?" Of the 366 Republicans, 57.9% (212/366) responded "yes," compared to 18.1% (85/470) Democrats, and 38.1% (230/604) Independents.

a) Identify the explanatory variable in this study. Also classify this variable as categorical or quantitative.

b) Identify the response variable in this study. Also classify this variable as categorical or quantitative.

c) Do the data provide any indication that there is an association between political party affiliation and opinion of whether states should be allowed to ignore federal gun laws? Explain.

Red meat and heart disease

4.1.25 In March 2012, an article titled "Eating red meat regularly 'dramatically increases the risk of death from heart disease,'" appeared in the *Daily Mail* (www.dailymail.co.uk). The article was based on the findings of a study that followed over 120,000 men and women for almost 30 years, the data for which were analyzed by the Harvard School of Public Health in Boston.

a) Identify the explanatory variable in this study. Also classify this variable as categorical or quantitative.

b) Identify the response variable in this study. Also classify this variable as categorical or quantitative.

c) Identify a confounding variable that provides an alternative explanation for the higher risk of death from heart disease among those who ate red meat regularly, compared to those who didn't.

Chocolate and Nobel prizes*

4.1.26 Researcher Messerli published an article titled "Chocolate Consumption, Cognitive Function, and Nobel Laureates," in the October 2012 issue of the *New England Journal of Medicine*. The article shows a positive association between a country's chocolate consumption and the number of Nobel prizes won by the country (adjusted for population size).

a) Identify the observational units.

b) Identify the explanatory variable and the response variable in this study.

c) Identify a confounding variable that provides an alternative explanation for the higher number of Nobel prizes won by countries with higher chocolate consumption, compared to those who with lower chocolate consumption.

Million Women study

4.1.27 The Million Women study in England followed more than 1,000,000 women aged 50 to 64 years from 1996 to 2001, tracking their living habits, and maintaining records on their medical and social factors as well as cancer data. The researchers found that after around 12 years, 3.7% (217/5,877) of the South Asian women developed breast cancer, compared to 3.6% (180/4,919) of the black women, and 4.4% (45,191/1,038,144) of the white women.

a) Identify the explanatory and response variables in this context. Also, identify whether each variable is categorical or quantitative.

b) Use the conditional proportions to explain whether the data indicate that there is an association between one's race and whether or not one has breast cancer?

c) Identify a possible confounding variable that provides an alternative explanation for the observed differences in breast cancer rates among the different races.

Section 4.2

4.2.1* Which of the following is the primary purpose of randomly assigning subjects to treatments in an experiment?

A. To produce similar (experimental) groups so any differences in the response variable can be attributed to the explanatory variable
B. To give each subject a 50-50 chance of obtaining a successful outcome
C. To produce a representative sample so results can be generalized to a larger population
D. To simulate what would happen in the long run
E. Both A and C

4.2.2 A randomized experiment allows for the possibility of drawing a cause-and-effect conclusion between _____ and _____.

A. The subjects and the treatments
B. The observational units and the variables
C. The explanatory variable and the response variable
D. Statistical significance and statistical confidence

4.2.3* Which of the following is true of experiments?

A. The researchers assign the *explanatory* variable to subjects.
B. The researchers assign the *response* variable to subjects.
C. The researchers assign *both* the explanatory and response variables to subjects.
D. The researchers assign *neither* the explanatory nor response variable to subjects.

4.2.4 Which of the following must happen in a study to allow us to determine cause and effect?

A. Taking a random sample from a population.
B. Randomly assign the observational units to different treatment groups.
C. Simulation-based inference techniques.
D. Theory-based inference techniques.

4.2.5* Is random sampling or random assignment the more important consideration if the research question is whether faculty tend to drive older cars than students drive on your campus?

4.2.6 Is random sampling or random assignment the more important consideration if the research question is whether Facebook users tend to have lower grade point averages than students who do not use Facebook?

4.2.7* Is random sampling or random assignment the more important consideration if the research question is whether students tend to receive higher scores on essays if they are encouraged to submit a draft than if they are not so encouraged?

4.2.8 Is random sampling or random assignment the more important consideration if the research question is whether members of one political party tend to donate more to charities than members of another political party?

4.2.9* Is random sampling or random assignment the more important consideration if the research question is whether a waitress generates higher tips by giving her name when she first greets customers?

4.2.10 Reconsider previous exercises (**4.2.5** through **4.2.9**). Notice that the phrase "tend to" appears in many of the research questions. Explain what this phrase means and why it is important in these questions.

4.2.11* Can a study have both random sampling and random assignment? If so, explain what can be determined from such a study if significance is found.

4.2.12 Researchers could design an experiment where there is a gender balance between two experimental groups by putting half the females in one group and half in the other and do the same for the males. Why don't researchers always just force variables to be balanced out between groups, but often use random assignment?

4.2.13* Does random assignment *always* equally balance the variables between two groups? Is there a tendency for there to be a balance? Explain.

4.2.14 From which of the following studies can cause and effect conclusions potentially be drawn? If cause and effect can be determined, explain what may cause what.

a) From a random sample of city residents, it was found that those with higher incomes utilize the recycling services significantly more than those with lower incomes.

b) Students were randomly assigned to two groups. One group listened to music while taking a math test and one group did not. The one that did not listen to music scored significantly higher on the math test than the other.

c) A teacher gave one of his classes a math test while music was playing and another without music playing. The class that did not listen to music scored significantly higher on the math test than the other.

4.2.15* From which of the following studies can cause and effect conclusions potentially be drawn? If cause and effect can be determined, explain what may cause what.

a) Subjects were randomly assigned to watch one of two videos, one that was about a sad situation and one that was about a happy situation. The group that watched the sad video scored significantly lower on a quiz about their mood.

b) Students collected data on city residents shopping in their city's downtown stores and another sample from city residents that were shopping at the mall on the outskirts of the city. They found that those shopping downtown were significantly more likely to vote for the school millage than those shopping at the mall.

c) From a random sample of students, it was found that those who are members of Greek organizations have significantly lower grade point averages.

4.2.16 What is the difference between random sampling and random assignment and what types of conclusions can be drawn from each?

Smoking while pregnant*

4.2.17 Many studies have shown that babies born to women who smoked while pregnant tend to weigh less at birth than babies born to mothers who did not smoke while pregnant.
  a) Are these studies observational studies or experiments?
  b) Will a cause-and-effect conclusion be possible?
  c) Identify the observational units in these studies.
  d) Identify the explanatory variable in these studies.
  e) Identify the response variable in these studies.

4.2.18 Refer to the information provided in the previous question.
  a) While it's possible in principle to conduct a randomized experiment to investigate this issue, it would be unethical to do so. Explain why, as if to someone who has never studied statistics.
  b) Identify some possible confounding variables in these studies.

Why is your baby a boy?

4.2.19 Many studies have looked at the sex of babies, and then surveyed the mothers and fathers about their behaviors (diet, health, etc.) at or immediately before the time of conception attempting to find connections between these behaviors and the sex of the baby.

    a)   Are these studies observational studies or experiments?

    b)   Will a cause-and-effect conclusion be possible?

    c) Identify the observational units in these studies.

    d) Identify the explanatory variables in these studies.

    e) Identify the response variable in these studies.

Classical music and intelligence*

4.2.20 Many studies have found that children who listen to classical music while in the womb or when younger tend to have higher IQs later in life. These studies are typically done by surveying parents of older children about whether or not they played classical music when the child was younger, and conducting an IQ test of the child.

    a) Are these studies observational studies or experiments?

    b) Will a cause-and-effect conclusion be possible?

    c) Identify the observational units in these studies.

    d) Identify the explanatory variable in these studies.

    e) Identify the response variable in these studies.

Consumer attitudes

4.2.21 A team of researchers (Singer, et al., 2000) used the Survey of Consumer Attitudes to investigate whether incentives would improve the response rates on telephone surveys. A national sample of 735 households was randomly selected, and all 735 of the households were sent an "advance letter" explaining that the household would be contacted shortly for a telephone survey. However, 368 households were randomly assigned to receive a monetary incentive along with the advance letter, and the other 367 households were assigned to receive only the advance letter. Here are the data on how many households responded to the telephone survey.

		Received an incentive?		
		Yes	**No**	**Total**
Responded to the	**Yes**	286	245	531
telephone survey?	**No**	82	122	204

Total		368	367	735

a) Was this an observational study or an experiment? How are you deciding?

b) What are the observational units?

c) What are the variables recorded? For each variable, identify the type of the variable (categorical or quantitative), and the role of the variable (explanatory or response).

d) Did the study involve random *sampling*? If yes, what is the advantage of a randomly selected sample? If no, what is the disadvantage?

e) Did the study involve random *assignment* to either receive an incentive or not? If yes, what is the advantage? If no, what is the disadvantage?

f) An appropriate analysis of the data shows there is evidence that people receiving an incentive are more likely to respond to the telephone survey compared to those receiving no incentive. Is it appropriate to conclude that incentives *improve* response rates? Why or why not? Explain your reasoning.

Facebook friends*

4.2.22 Is the sex of a person making a friend request on Facebook to someone they do not know associated with whether or not the request is accepted? To answer this question, student researchers at Hope College made up phony Facebook profiles, one representing a Hope College female student and one representing a Hope College male student. The profiles were made to look as similar as possible except for a couple of pictures of the phony students, their names, and of course their sex. From a group of 219 students from one class at the college, 118 were randomly assigned to receive the friend request from the female "student" and 101 were randomly assigned to receive the friend request from the male "student." The results of the acceptance of these requests are shown in the following table.

		Facebook Profile		
		Female	**Male**	**Total**
Accepted friend request?	**Yes**	61	18	79
	No	57	83	140
Total		118	101	219

a) Was this an observational study or an experiment? How are you deciding?

b) What are the observational units?

c) What are the variables recorded? For each variable, identify the type of the variable (categorical or quantitative), and the role of the variable (explanatory or response).

d) Did the study involve random *sampling*? If yes, what is the advantage of a randomly selected sample? If no, what is the disadvantage?

e) Did the study involve random *assignment* to either get the friend request from either the female or male? If yes, what is the advantage? If no, what is the disadvantage?

f) An appropriate analysis of the data shows there is strong evidence that people receiving the request from the female are more likely to accept. Is it appropriate to conclude that the sex of the requestor is affecting the acceptance rates? Why or why not? Explain your reasoning.

Cuteness and aggression

4.2.23 Are cuteness and aggression related? A study done at Yale University tested this by showing people pictures of cute animals (like kittens and puppies) or pictures of older more serious looking animals. They tested the aggression of the subjects by giving them bubble wrap and let them pop the bubbles. The subjects in the group shown the cute animals popped an average of 120 bubbles compared 100 for the group seeing the pictures of the older animals.

a) Identify the explanatory variable in this study and tell whether it is categorical or quantitative.

b) Identify the response variable in this study and tell whether it is categorical or quantitative.

c) What should the researches do so they have the possibility of determining that the type of picture affects how many bubbles are popped?

Power poses*

4.2.24 A research article "Power Posing: Brief Nonverbal Displays Affect Neuroendocrine Levels and Risk Tolerance," published in *Psychological Science,* September 2010, describes a study involving 42 volunteer participants (male and female), where participants were randomly assigned to hold either low-power poses (contractive positions, closed limbs), or high-power poses (expansive positions, open limbs) for 2 minutes. All participants were told that the aim of this exercise was to see if their heart rate changed. After the exercise, each participant was given $2, and told that they could keep the money, or roll a die for a double or nothing. Of the 20 participants who held low-power poses, 3 took the "double or nothing" bet, whereas of the 22 participants who held high-power poses, 18 took the bet. Suppose we want to know, *Are people who hold "high-power poses" more likely to take risks (such as, the double or nothing bet) compared to those hold "low-power poses"?*

a) Was this an observational study or an experiment? How are you deciding?

b) What are the observational units?

c) What are the variables recorded? For each variable, identify the type of the variable (categorical or quantitative), and the role of the variable (explanatory or response).

d) Did the study involve random *sampling*, random *assignment,* or both? How are you deciding?

e) An appropriate analysis of the data shows that there is strong evidence that people in the high-power are more likely to take the double or nothing bet. Is it appropriate to conclude that the pose affects likelihood of taking the bet? Why or why not? Explain your reasoning.

Skittles taste test

4.2.25 Students recruited at the cafeteria at Hope College were blindly given a single Skittles candy to put in their mouth. They were told the five possible flavors and then were asked which flavor they had. Of the 154 healthy students tested, 78 gave the correct answer. Of the 118 students that had stuffy or runny noses tested, 44 gave the correct answer.

a) Was this an observational study or an experiment? How are you deciding?
b) What are the observational units?
c) What are the variables recorded? For each variable, identify the type of the variable (categorical or quantitative), and the role of the variable (explanatory or response).
d) Did the study involve random *sampling*, random *assignment,* both or neither? How are you deciding?
e) An appropriate analysis of the data shows that there is strong evidence that healthy students were more likely to give the correct answer. Is it appropriate to conclude that the health of the student affects ability to taste? Why or why not? Explain your reasoning.

End of Chapter Exercises

4.CE.1* When can you legitimately draw a cause-and-effect conclusion from a randomized experiment?
   A. When the p-value is small
   B. When the p-value is large
   C. Always, regardless of the p-value
   D. Never, regardless of the p-value

4.CE.2 Answer the following questions about explanatory and response variables.

a) Must an explanatory variable always be categorical?
b) Must a response variable always be quantitative?

Animal therapy*

4.CE.3 In a recent study of animal-assisted therapy (Cole, et. al., 2007), researchers investigated whether patients hospitalized with heart failure could be helped by a visit from a dog. The 76 patients in the study were randomly assigned to one of three groups: one group received a 12-minute visit from a volunteer and a dog, another group received a 12-minute visit from a volunteer only, and the third group received no visit. Many variables, including heart rate, blood pressure, and anxiety levels, were measured on the patients prior to and after the visit.

a) Identify the explanatory variable(s) in this study.

b) Identify the response variable(s) in this study.

c) Is this an experiment or an observational study? Explain.

d) Describe the purpose of randomly assigning patients to treatment groups.

e) Researchers also reported summary information on a variety of background variables such as age, gender, and smoking status. Do you think the researchers were hoping to find significant differences among the three groups on these variables? Explain why or why not.

Walking in socks

4.CE.4 In a study conducted in New Zealand, researchers Parkin et al. randomly assigned volunteers to either wear socks over their shoes (intervention), or wear usual footwear (control), as they walked downhill on an inclined icy path. Researchers standing at the bottom of the inclined path would then rate how confident the participant appeared ("confident," "cautious but did not hold onto supports," or "held onto supports"), and how much time (seconds) the participant took to walk down the path.

a) Was this an experiment or an observational study?

b) Identify the variables recorded. For each variable, identify the type (categorical or quantitative) and role (explanatory or response).

c) Did the study involve randomness? If so, how: random sampling, or random assignment, or both?

Cost of drugs*

4.CE.5 In a randomized, double-blind study reported in the *Journal of American Medical Association*, researchers Waber et al. (2008) administered a pill to each of the 82 healthy paid volunteers from Boston, Massachusetts, but told half of them that the drug had a regular price of $2.50 per pill, whereas the remaining participants were told that the drug had a discounted price of $0.10 per pill, mentioning no reason for the discount. In the recruitment advertisement, the pill was described as an FDA approved opioid analgesic, but in reality both groups were administered placebo pills. To simulate pain, the participants were administered electric shocks to the wrist, and the researchers recorded the proportion of subjects who reported a reduction in pain after taking the pill.

a) Identify the explanatory variable and response variable. For each variable, identify the type (categorical or quantitative).

b) Explain what "randomized" means in this study.

c) Explain what "double-blind" means in this study.

d) The researchers report "pain reduction was greater for the regular-price pill (p-value < 0.001)." Explain what the researchers mean by this. Be sure to comment, with correct justification, on

- Whether the results are statistically significant.
- Whether a cause-and-effect conclusion can be drawn.
- To whom the results of the study can be generalized.

Rebates and bonuses

4.CE.6 An article in a 2006 issue of *Journal of Behavioral Decision Making* reports on a study involving 47 undergraduate students in a class at Harvard. All of the participants were given $50, but some (chosen at random) were told that this was a "tuition rebate," while the others were told that this was "bonus income." After one week, the students were contacted again and asked how much of the $50 they had spent and how much they had saved. Those in the "rebate" group had spent an average of $22.04, while those in the "bonus" group had spent an average of $9.55.

   a) Identify the explanatory and response variables in this study. Also classify each variable as categorical or quantitative.
   b) Is this an observational study or an experiment? Explain.
   c) Did this study make use of random *sampling*, random *assignment*, both, or neither?
   d) If the difference in average spending amounts between the two groups is determined to be statistically significant, would it be legitimate to draw a cause-and-effect conclusion between what the money was called and how much was spent? Justify your answer.
   e) If the difference in average spending amounts between the two groups is determined to be statistically significant, would it be legitimate to generalize the result of this study to all adult Americans? Justify your answer.

Satisfaction with attractiveness*

4.CE.7 A poll conducted by the Gallup organization asked American adults whether or not they are generally satisfied with their physical attractiveness. One goal of the study was to investigate whether men and women differ with regard to responses on this issue.

   a) Identify the explanatory and response variables in this study.
   b) Is this an observational study or an experiment? Explain briefly.
   c) If this study were conducted well, would it have involved random *sampling*, random *assignment*, or both? Explain.

Drinking at college

4.CE.8 A study found that college students who live off-campus are significantly more likely to drink alcohol than those who live on-campus.

   a) Do you suspect that this was an observational study or an experiment? Explain.
   b) Is it appropriate to conclude that living off-campus causes a student to be more likely to drink alcohol? Explain why or why not.
   c) Is it appropriate to conclude that being a drinker causes a student to be more likely to live off-campus? Explain why or why not.

How to name your book*

4.CE.9 Ian Ayers, the author of a popular book titled *Super Crunchers*, conducted a study to help him decide what to name his book. He placed an ad on google.com, with the ad sometimes giving the title as *Super Crunchers* and other times giving the title as *The End of Intuition*. Google provided Ayres with data on how often a person clicked on the ad to obtain more information, depending on the title given in the ad. It turned out that viewers were significantly more likely to click through on the *Super Crunchers* ad than on *The End of Intuition* ad, so that's the title that Ayres chose.

a) Was this an observational study or an experiment? Explain how you know.
b) Identify the explanatory and response variables.
c) Was Ayres justified in concluding that the name *Super Crunchers* caused a higher click-through rate than the other name? Explain why or why not.

Spanking and IQ

4.CE.10 Studies have shown that children in the U.S. who have been spanked have a significantly lower IQ score on average than children who have not been spanked.

a) Is it legitimate to conclude from this study that spanking a child causes a lower IQ score? Explain why or why not.
b) Explain why conducting a randomized experiment to investigate this issue (of whether spanking causes lower IQs) would be possible in principle but ethically objectionable.

Reading Harry Potter*

4.CE.11 You want to investigate whether teenagers in the United Kingdom (UK) tend to have read more *Harry Potter* books, on average, than teenagers in the United States (US).

a) Identify and classify (as categorical or quantitative) the explanatory and response variable.
b) Would you ideally use random sampling for this study, or random assignment, or both? Explain.

Restaurant customer behavior

4.CE.12 Do restaurant customers tend to order more expensive meals when classical music is playing in the background than when other kinds of music are playing in the background? Describe how you could design a randomized experiment to investigate this question.

Pen color and grades*

4.CE.13 Do college professors who use a red pen to grade student essays tend to assign lower scores, on average, than professors who use a blue pen to grade student essays? Describe how you could design a randomized experiment to investigate this question.

Video games and kids

4.CE.14 Are American adults who have children significantly more likely to play video games than American adults who do not have children? Is it possible to design a randomized experiment to investigate this question? If so, describe how you would design such an experiment. If not, explain why not.

Smoking and church attendance*

4.CE.15 Are people who attend church regularly less likely to smoke than people who do not attend church regularly? Is it possible to design a randomized experiment to investigate this question? If so, describe how you would design such an experiment. If not, explain why not.

Investigation: High Anxiety and Sexual Attraction

Social Psychologists throughout the years have shown that an aggression-sexuality link exists in not only various animal species, but also in humans. Dutton and Aron (*Journal of Personality and Social Psychology*, 1974) set out to show that a more general link exists in humans, namely emotional arousal of all kinds and sexual attraction. They set up their study to compare men in a high emotional arousal situation to men in a low emotional arousal situation.

Researchers wanted to test the notion that an attractive female is seen as more attractive by males who encounter her in a fear-arousing situation than by males who encounter her in a non fear-arousing situation. In the high emotional arousal group, men deemed to be between the ages of 18 and 35 who crossed a suspension bridge 230 feet above rocks and shallow rapids in Capilano Canyon, North Vancouver, British Columbia, Canada and were not accompanied by a female were approached on the bridge by an attractive female interviewer. The same interviewer also approached men who fit the same criteria but crossed a solid wood bridge 10 feet above a rivulet that ran into the main river. Both groups of men were interviewed on the bridge and were told that the interview was for a Psychology class project on the effects of exposure to scenic attractions on creative expression. The men filled out a short questionnaire after which the interviewer wrote her phone number down on a slip of paper and said that if they were interested in the results of the experiment they could call her. The researcher talked to 18 men on the suspension bridge and 16 men on the wooden bridge.

Step 1: Ask a research question

The researcher is now interested in seeing whether a higher proportion of men in the group that crossed the Capilano Canyon bridge will call her versus the proportion of calls she will receive from the group that crossed the solid wooden bridge.

Step 2: Design a study and collect data

1.   Is this an observational study or an experiment? How are you deciding?

2.   What are the observational units?

3.   We have two variables: bridge type and whether the subject called or not. Which of these variables is explanatory and which is response?

You will learn in Chapter 5 how to state null and alternative hypotheses for studies like this. Here they are for this study:

Null: The long-run probability of calling the researcher from men in the study is the same in both groups

Alt: The long-run probability calling the researcher from men in the study is larger in the group of men crossing the Capilano suspension bridge

Of the 18 men on the suspension bridge that accepted her phone number, 9 called her. Of the 16 men on the solid wooden bridge that accepted her phone number, 2 called her.

Step 3: Explore the data

4.   Fill in the two-way table with the data from the study.

	Suspension bridge	Wooden bridge	Totals
Subject called interviewer			11
Subject did not call interviewer			23
Totals	18	16	34

5. What proportion of men on the suspension bridge called the interviewer?

6. What proportion of men on the wooden bridge called the interviewer?

Step 4: Draw inferences

In Chapter 5, we will learn how to use the 3S strategy to help us investigate how much evidence the sample data provide to support our conjecture that the long-run probability of men who call is greater for those on the suspension bridge than for those on the wooden bridge. As a result of the simulation analysis method used in Chapter 5 we received a p-value of 0.023.

7. Use the null and alternative hypotheses stated on the previous page and the p-value of 0.023 to write a conclusion about strength of evidence in this study. *Hint: Even though the study design is different, the way you interpret p-values in terms of the null and alternative hypothesis is still the same.*

8. How would the p-value have changed if the alternative hypothesis was stated as: *Alt: The long-run probability of calling the researcher from men in the study is different between the two groups*

9. Write a conclusion about the strength of evidence in the study based on the p-value from #8.

10. In studies comparing two groups, the parameter of interest is often stated as the true difference in the long run probabilities in the two groups. In this study, this is the long-run difference in the probabilities of men calling the researcher comparing the men crossing the Capilano suspension bridge to the men crossing the wooden bridge. A 95% CI for this parameter is 0.375 ± 0.35. Explain how the fact that the confidence interval does not include 0 corresponds to the size of the p-value obtained in #9.

Step 5: Formulate conclusions

11. *Generalization.* Now, let's step back a bit and think about the scope of our inference. Was this a random sample? Does this study represent all people? How about all men? Is there any population to which we can generalize our conclusion?

12. *Causation.* What about cause and effect conclusions? Did the researchers use random assignment to determine which of the 34 men crossed the suspension bridge and which crossed the wooden bridge? Are we able to conclude that the cause of the difference between the two groups of men was the type of bridge they crossed?

Step 6: Look back and ahead

The researchers acknowledged the fact that there were confounding variables involved in this study. They tried to minimize these by doing other studies to try to answer the same research question. In another study, they again interviewed subjects on the suspension bridge just as before. As a control, they interviewed subjects that had already crossed the suspension bridge at least 10 minutes earlier and were sitting or walking in the nearby park.

13. Describe the confounding variable they are trying to eliminate by completing this new study.

Let's see how this second study turned out. Of those on the suspension bridge, 13 of the 20 men called the female interviewer. Of those that had already crossed the bridge, 7 out of 23 men called the female interviewer.

14. In this study, what proportion of those on the suspension bridge called the interviewer? Of those who had already crossed the bridge?

The hypotheses are modified slightly for this new study.

Null: The long-run probability of calling the researcher from men in the study is the same in both groups

Alt: The long-run probability of calling the researcher from men in the study is larger in the group of men who are on the Capilano suspension bridge than the in the group of men who recently crossed.

15. Write a conclusion about the strength of evidence in this new study based on obtaining a p-value of 0.024.

16. *Generalization.* Have the researchers addressed the concerns about *generalization* with this new study?

17. *Causation.* Is a cause-effect conclusion possible in this new study? Why or why not?

Research Article: Impacting People's Willingness to Pay

Read "The Effect of Red Background Color on Willingness-to-pay: The Moderating Role of Selling Mechanism" and then answer the following questions.

Step 1. Ask a research question

1. In no more than 1-2 sentences state a broad research question (or related questions) that the researchers were trying to investigate in this study.
2. Identify at least two (cited) points of evidence the researchers cite as to why the research question(s) are important. Include the citation in your response.

The following questions (3- 12) refer to Study #1.

Step 2. Design a study and collect data

3. In the first part of study 1, the researchers sold 28, identical, Nintendo Wii bundles on eBay. There were over 900 bid jumps measured across the 28 auctions, in your own words explain what a bid jump is.

Step 3. Explore the data

4. Explain how the mean (20.10) and median (10.00) of the bid jumps data across the 28 auctions
    confirm the researchers conclusion that the data was right-skewed.

Step 4. Draw inferences

5. The average bid jump for Wii's sold on red backgrounds was \$20.82 and was \$19.22 on blue
    backgrounds, for a two-sided p-value of < 0.02. Complete the following sentence "We have
    _____ that the average bid jump is different (higher) for Wii's sold on a red
    background compared to blue.

Step 5. Formulate conclusions

6. In order to be a randomized experiment observational units (bid jumps) must be randomly
    assigned to the treatment groups (red or blue background). Is this study a randomized
    experiment? What are the implications on the conclusions from this study?
7. Why was it important to keep seller and product budget identical across auctions?
8. The researchers state that "Although the results support our expectations, we acknowledge that
    the red and the blue auctions may have differed on other factors (for instance, the number of
    individuals who viewed the listing or self-selection)." What is the term for these factors?

Step 2. Design a study and collect data. Step 3. Explore data. Step 4. Draw inferences

9. The researchers followed up by conducting a randomized experiment on 78 undergraduate
    students. Provide a one sentence summary of the strength of evidence in this study based on the
    p-value ($p < 0.01$), comparing the average bid jump for red (\$63.17) to the average bid jump for
    blue (\$35.13).

Step 5. Formulate conclusions

10. Because this study involved random assignment what should be true about how similar the 39
    students are in each of treatment groups (red and blue)?
11. Because of the random assignment, what type of conclusion is possible?
12. How were the participants for this study selected? What are the implications on the conclusions
    that can be drawn from this study?

Step 2. *Questions 13-16 refer to Study #2.* **Design a study and collect data. Step 3. Explore data. Step 4. Draw inferences**

13. In study #2, the researchers randomly assigned 89 individuals to make offers on a hotel where the pictures are placed on a blue or red background. Make a statement about the strength of evidence for a difference in the average bid price between the red and blue groups.

Step 5. Formulate conclusions

14. The researchers state that they measured how much participants liked the vacation (1 to 7 scale). How should the number of people who like the vacation a lot (6 or 7) compare between the two treatment groups (red and blue backgrounds)? Why?

15. Explain the impact of random assignment on the conclusions that can be drawn from study #2.

16. How were the participants for this experiment selected? What are the implications on conclusions you can draw from this study?

Step 6. Look back and ahead

17. Identify at least two things that, if you were running the study and had a reasonable amount of financial resources, you would do differently. You must justify your answers.

UNIT 2: COMPARING TWO GROUPS

In Unit 1 you saw the four pillars of statistical inference developed in the simplest possible situation. In almost every example, the data took the form of observed values on a single binary variable. It was possible to summarize the data using a single proportion, and to base inference on that one number. We also extended this case to another simple case: a single quantitative variable. But again, we were focused on just one outcome variable.

Now that you have experience with the basics, you are ready to apply the same ideas to more complicated situations. The chapters of Unit 2 will show you how to compare two groups using a single response variable. In Chapter 5, the response is binary, and we end up comparing two proportions, one for each group. In Chapter 6 the response is quantitative, and we end up comparing two means, one for each group. As you might guess, in many ways these two chapters are quite similar.

Chapter 7 is different. In both Chapters 5 and 6, the observational units in the two groups have nothing to do with each other, in other words they are *independent* of each other, and we get one outcome value for each unit. In Chapter 7, we get a *pair* of measurements on each observational unit. So there is a dependency built into the design of the study. The resulting paired structure has important advantages that you will learn about in Chapter 7.

Although the data structures of this unit will be somewhat more complicated, the basic ideas and methods remain the same as in Unit 1. As before, inference takes place in the framework of the Six-Step Statistical Investigation Method. The 3S Strategy also applies. In particular, the basic idea of the simulation step remains the same as before. You ask, "What if the null hypothesis was true. How could my statistic behave by chance alone?" Just as before, the answer requires investigating the null ("what if") distribution, and you use a p-value and/or standardized statistic to assess the strength of evidence against the null hypothesis. The main difference will be in the null hypothesis itself, which will typically take the form "No difference between parameters." You will also see how to construct a confidence interval to estimate the value of the population parameter, a difference in proportions or a difference in means.

Units 1 and 2 Data Structures

Unit 1: One binary or one quantitative variable
Unit 2: Comparing two groups

	Response	Explanatory
Chapter 5	Binary	Binary
Chapter 6	Quantitative	Binary
Chapter 7	Quantitative (paired)	Binary

Chapter 5

Comparing Two Proportions

Chapter Overview

In the studies of Unit 1 there was only one group and only one parameter. In this chapter there will be two groups. In Chapter 1, the only parameter for Doris and Buzz was π , the chance that Buzz would be right. Our null hypothesis was $\pi =$ .The value $\pi =$ came from the null hypothesis that Buzz was just guessing equally between two choices.

In this chapter, one of the studies is also about dolphins: Can swimming with dolphins help patients who have been diagnosed with depression? This study uses two groups: a treatment group and a control group. For each treatment, π is the long-run probability of substantial improvement, π_1 for the dolphin (treatment) group, π_2 for the control group. The null hypothesis is that $\pi_1 - \pi_2 = 0$. Or in other words, $\pi_1 = \pi_2$:swimming with dolphins makes no difference in the probability a subject improves. If we obtain a small p-value, and so have strong evidence against this null hypothesis, then we conclude that the difference we observed between the groups in the study did not arise from the random assignment process alone. As we saw in Chapter 4, in situations where we randomize observational units to groups we are comparing, cause-and-effect conclusions may be possible.

Even though having two groups is new, a lot of what we do will be familiar. The statistical investigation process follows the usual six steps. We start with a research question, design a study, and then explore the data we get. We will use the 3S Strategy to measure the strength of evidence against the null hypothesis, although the statistic will be new. We will get an interval estimate for the parameter, now a difference ($\pi_1 - \pi_2$), using the same methods as in Unit 1.First, we simulate and use the 2SD shortcut. Then we use a theory-based method.

Section 5.1 is about exploration and description of the sample data. The main new idea: *conditional* proportions and something called the relative risk. Section 5.2 explains the simulation-based approach to comparing proportions. The chapter ends with a third section on the theory-based approach.

Section 5.1 Comparing Two Groups: Categorical Response

Introduction

As we are now going to focus on drawing conclusions from studies involving two variables, we will start by focusing on how to summarize two categorical variables in order to make group comparisons.

Example 5.1: Good and Bad Perceptions

Student researchers were interested in how the phrasing of a question affects people's responses. In particular, they asked some students whether they were "having a good year" and other students whether they were "having a bad year." Then they recorded whether each of the 30 subjects indicated a "good perception of their year" (meaning they said either "Yes, I'm having a good year" or "No, I'm not having a bad year") or a "bad perception" (meaning they said either "No, I'm not having a good year" or "Yes, I'm having a bad year").Which question was posed to each subject was determined in advance by randomly assigning each of the 30 subjects to get one of the two questions.

> **Think about it:** What are the observational units in this study? How many variables were recorded on each observational unit? Identify and classify the roles of the variables in this study. Is this study an observational study or a randomized experiment?

The observational units in this study are the 30 students, but now we have two variables: *whether the question wording was "having a good year" or "having a bad year"* and *whether the respondent's perception of their year was positive or negative.* In this study, the student researchers thought the wording of the question would help explain why some people indicated a positive perception of their year and others indicated a negative perception of their year. Thus, the roles of these variables are wording of question as the explanatory variable and perception of year as the response variable. Because the explanatory variable was randomly assigned to each subject, this is a randomized experiment.

The student researchers recorded the data by noting which version of the question was asked and how the student responded. Table 5.1 shows the *data table* that they obtained.

Random experiment

Table 5.1: Data obtained by the student researchers

Individual	Question wording	Perception of Their Year	Individual	Question wording	Perception of Their Year
1	"Good year"	Positive	16	"Good year"	Positive
2	"Good year"	Negative	17	"Bad year"	Positive
3	"Bad year"	Positive	18	"Good year"	Positive
4	"Good year"	Positive	19	"Good year"	Positive
5	"Good year"	Negative	20	"Good year"	Positive
6	"Bad year"	Positive	21	"Bad year"	Negative
7	"Good year"	Positive	22	"Good year"	Positive
8	"Good year"	Positive	23	"Bad year"	Negative
9	"Good year"	Positive	24	"Good year"	Positive
10	"Bad year"	Negative	25	"Bad year"	Negative
11	"Good year"	Negative	26	"Good year"	Positive
12	"Bad year"	Negative	27	"Bad year"	Negative
13	"Good year"	Positive	28	"Good year"	Positive
14	"Bad year"	Negative	29	"Bad year"	Positive
15	"Good year"	Positive	30	"Bad year"	Negative

Notice that in the data table, each observational unit is one row, and the different columns (apart from the ID number) represent the two variables.

However, this table does not make it simple to notice any trends or patterns in the responses. One way to organize the data is to produce a *two-way table* of counts. As the name suggests, two-way tables summarize *two* categorical variables. You may also see these tables called "contingency tables." In such a table, the entries in the table (sometimes called *cells*) indicate the number of observational units that have the specified combination of categories for the two variables. For example, Table 5.2 is a two-way table of the survey data. It shows how many students were in each of the four possible combinations (e.g., 15 students gave a "good" response to the "Are you having a good year" question), and also shows the total for each category of the individual variables (e.g., 18 students were given the "Are you having a good year" version of the question).

Think about it:

- Verify that there were four individuals in the data table (Table 5.1) who answered with a positive perception to the "Are you having a bad year" question.

- Is it reasonable to conclude from this table that there is a tendency for more positive perceptions to the "Are you having a good year" question than to the "Are you having a bag year" question?

Table 5.2: A two-way table summarizing the number of good and bad responses to the two different phrasings of the question. This is a *"two by two (2 ×2)"* table because each variable has two outcomes.

<table>
<tr><td></td><td></td><td colspan="3" align="center">Question wording</td></tr>
<tr><td></td><td></td><td align="center">**"Good year"**</td><td align="center">**"Bad year"**</td><td align="center">**Total**</td></tr>
<tr><td>**Perception of**</td><td>**Positive**</td><td align="center">15</td><td align="center">4</td><td align="center">19</td></tr>
<tr><td>**their year**</td><td>**Negative**</td><td align="center">3</td><td align="center">8</td><td align="center">11</td></tr>
<tr><td></td><td>**Total**</td><td align="center">18</td><td align="center">12</td><td align="center">30</td></tr>
</table>

An important reminder is that comparing the *counts* in individual categories may not be useful if the group sizes differ. In this case, a few more people were asked the good year version of the question. So we might expect a slightly higher number of positive responses to the "good year" question than to the less often asked "bad year" question for that reason alone. To take into account the sample sizes, it is much more useful to look at the ***conditional proportions*** of positive responses, where we compute the proportion of "successes" separately within each explanatory variable group. For these data, we see that $15/18 \approx 0.833$ or about 83% of the individuals asked the "Are you having a good year" version of the question indicated a positive perception. We want to compare this to the proportion of individuals who indicated a positive perception to the "Are you having a bad year" version of the question: $4/12 \approx 0.333$.(Try to also be careful to say "higher *proportion* of positive responses" instead of the more ambiguous "more positive responses.") Keep in mind that these conditional proportions (0.833 and 0.333) are sample proportions because they have been calculated from sample data.

We can produce a picture of these data so that comparison becomes even more readily apparent. Previously, with just one categorical variable, we looked at a bar graph of the categorical variable outcomes. When evaluating two categorical variables, ***segmented bar graphs*** quickly illustrate the differences among the groups. In a segmented bar graph, the total bar heights are always 100%, but each bar is broken into categories and shows the breakdown of the proportions of observational units in each category. For example, in the segmented bar graph shown in Figure 5.1, the positive response segments go to 83% and 33%.

Figure 5.1: A segmented bar graph showing the percentage of people with good perceptions to each question

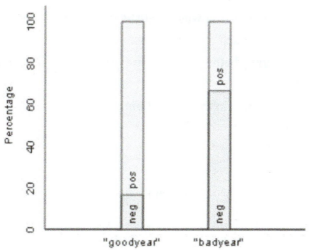

Summarizing the data

These data definitely support the claim that people will have a positive perception more often to the "Are you having a good year" question than to the "Are you having a bad year" question. In other words, there is an *association* between the phrasing of the question and how people respond. We can say the proportion of positive responses was $0.833 - 0.333 = 0.500$ higher for the "good year" phrasing than the "bad year" phrasing (Note: We can say there was a difference of 50 percentage points, but we would not say a difference of 50% because saying one number is 50% higher than another implies multiplying by 1.50.)

Another way we can compare the two conditional proportions is to take the ratio: $0.833/0.333 \approx 2.49$, so the proportion with a positive perception is 2.49 times higher (or 149% higher) in the "good year" group than the "bad year" group. This ratio is called the ***relative risk***. This ratio is typically used in medical fields to look at the relative risk of disease or death from one treatment compared to another. Convention puts the larger of the two values in the numerator and the smaller of the two values in the denominator to yield values larger than one.

Definition: *Relative risk* is the ratio of two conditional proportions. It indicates how many times greater the risk of an outcome is for one group compared to another.

Exploration 5.1: Murderous Nurse?

For several years in the 1990s, Kristen Gilbert worked as a nurse in the intensive care unit (ICU) of the Veteran's Administration hospital in Northampton, Massachusetts. Over the course of her time there, other nurses came to suspect that she was killing patients by injecting them with the heart stimulant epinephrine. Gilbert was eventually arrested and charged with these murders. Part of the evidence presented against Gilbert at her murder trial was a statistical analysis of 1641, eight-hour shifts during the time Gilbert worked in the ICU. For each of these shifts, researchers recorded two variables: whether or not Gilbert worked on the shift, and whether or not at least one patient died during the shift.

1. Identify the observational units in this study.(*Hint*: The observational units are not people. Think about what the sample size is.)

2. Classify each variable as categorical or quantitative.

   Whether or not Gilbert worked on the shift:

   Whether or not a patient died during the shift:

3. Which variable would you regard as explanatory and which as response?

   Explanatory: Worked

   Response: Deaths or No Deaths

4. Is this an observational study or an experiment? Justify your answer.

So what did the researchers find when they analyzed the 1,641 eight-hour shifts? They found that Gilbert worked on 257 of the shifts, and she did not work on 1384 of them. They also found that a patient died during 74 of the shifts, and no death occurred on the other 1,567 shifts.

5. Explain why the information in the previous paragraph does not provide any clues about whether there is an association between Gilbert working on a shift and a patient dying during the shift. What additional information do you need to know?

Here is some additional information: Among the 74 shifts during which a patient died, 40 were shifts on which Gilbert worked, and 34 were shifts on which Gilbert did not work.

6. Explain why simply comparing only the numbers 40 and 34 is not very informative.

When analyzing data on two categorical variables, a useful first step is to organize the data into a *two-way table* of counts. In this case both variables have only 2 categories, so we produce a 2×2 table.

7. Fill in the missing counts in the following ***two-way table***.(*Hint*: Start by entering 40 and 34 into the appropriate cells. Then make sure that both rows and both columns add up to the indicated totals.)

	Gilbert working on shift	Gilbert not working on shift	Total
Patient died during shift			74
No patient died during shift			1567
Total	257	1384	1641

8. Suggest a better way to compare the outcomes of the shifts on which Gilbert worked and the shifts on which she did not work than simply comparing 40 to 34. Explain why your answer is more informative.

A next step in analyzing categorical data in a two-way table is to calculate **conditional proportions**. A conditional proportion simply means that you consider the counts in only one category of the explanatory variable at a time.

9. Now let's calculate the conditional proportions.
    a. Calculate the conditional proportion of Gilbert shifts on which a patient died. (*Hint*: Put the total number of Gilbert shifts in the denominator of your fraction.)

    b. Calculate the conditional proportion of non-Gilbert shifts on which a patient died.

A ***segmented bar graph*** is an appropriate display for graphing data in a two-way table. Such a graph contains rectangles of total height 100% for each category of the explanatory variable. Then segments divide up each rectangle according to the conditional proportions of the response variable categories.

10. Fill in the following segmented bar graph to display the data from this study. First draw two rectangles, one for Gilbert shifts and one for non-Gilbert shifts, each with total height 100%. Then draw a line in each rectangle corresponding to the conditional proportion of shifts in which a patient died, as you calculated in #9.

11. Do the conditional proportions and segmented bar graph appear to provide evidence that a patient was more likely to die on a Gilbert shift than on a non-Gilbert shift? In other words, does there appear to be a tendency for Gilbert shifts to have a death more often than non-Gilbert shifts? Explain.

12. Without doing any further analysis, do you consider the difference between the conditional proportions to be striking, worth reporting to a jury?

A next step is to produce a single number (statistic) to summarize the data.

13. What arithmetic operation might you perform on the two conditional proportions to obtain a single statistic?

14. Calculate the *difference* in conditional proportions of death between Gilbert and non-Gilbert shifts. Does the value of this difference strike you as noteworthy?

15. Now calculate a different statistic: the *ratio* of conditional proportions of death between Gilbert and non-Gilbert shifts. This ratio is called a **relative risk**. Write a sentence interpreting this ratio value, as the prosecution might present to a jury. Does the value of this ratio strike you as noteworthy?

> **Definition:** *Relative risk* is the ratio of two conditional proportions. It indicates how many times greater the risk of an outcome is for one group compared to another.

16. Even if the difference between these conditional proportions of death is found to be highly statistically significant, is it reasonable to conclude that Kristen Gilbert was responsible for (i.e., *caused*) the increased chance of death on her shifts? Explain your answer.(*Hint*: Remember your answer to #4 about what kind of study this is.)

17. Suggest an alternative explanation that the defense attorneys might use to explain the large difference in death proportions between the two groups.

Further Analysis

18. Suppose that the data had turned out as follows:

	Gilbert working on shift	Gilbert not working on shift	Total
Patient died during shift	100	357	457
No patient died during shift	157	1027	1184
Total	257	1384	1641

    a.   Calculate the conditional proportions of death for Gilbert shifts and for non-Gilbert shifts.

    b.   Calculate the *difference* in these conditional proportions. How does this difference compare to the one in #14? Are they very similar?

    c.   Calculate the *relative risk* of these conditional proportions. How does this ratio compare to the one in #15? Are they very similar?

    d.   For which table of data, this new one or the original (real) data, would you consider the evidence against Gilbert to be more persuasive/incriminating? Explain.

19. Produce a hypothetical 2×2 table with the same marginal totals, but with the property that the data reveal virtually *no association* between the variables. In other words, produce a table so that the conditional proportions are very similar between the two groups.

	Gilbert working on shift	Gilbert not working on shift	Total
Patient died during shift			457
No patient died during shift			1184
Total	257	1384	1641

20. What would the segmented bar graph look like for a 2×2 table with virtually no association between the variables? Explain your answer.

Section 5.1 Summary

When a study involves two categorical variables, the data can be organized into a **two-way table** of counts.

A **segmented bar graph** displays the association between two categorical variables by showing the **conditional proportions** of success and failure across the explanatory variable groups.
- If the variables have absolutely no association, then the breakdown of segments is identical for across all of the bars.
- If the variables have absolutely no association, the conditional proportions are identical for all explanatory variable groups.

Comparing counts of successes between two groups is not a valid comparison, because the sample sizes in the two groups could differ substantially.
- Comparing conditional proportions is more appropriate.
- The difference between conditional proportions in the two groups is one statistic for measuring how different the groups' responses are.
- The ratio of the conditional proportions, called the **relative risk**, can also be used as a statistic to compare the two groups. The relative risk indicates how many times more likely an outcome is in one group compared to the other.

Section 5.2: Comparing Two Proportions: Simulation-Based Approach

Introduction

In the previous section we saw that when we have two categorical variables, our interest is less on the overall "success rate" and more on comparing the success rates (conditional proportions) between groups. The previous section focused on descriptively comparing conditional proportions between two groups. We will now assess whether the difference between two sample proportions is statistically significant. We will use the same reasoning process (3SStrategy) that we introduced in Chapter 1 to assess whether the two sample proportions differ enough to conclude that something other than random chance is responsible for the observed difference in groups. Once again we will evaluate the strength of evidence provided by the data based on how unlikely the observed data are to have occurred by chance alone. In this section, we will develop a new simulation approach to approximate the p-value for the group comparison. The next section, Section 5.3, will focus on a theory-based approach.

Example 5.2: Swimming with Dolphins

Step 1: Ask a research question. Swimming with dolphins can certainly be fun, but is it also therapeutic for patients suffering from clinical depression? To answer this question, researchers wanted to focus on whether or not the presence of dolphins helped some depression patients improve at a higher rate than other individuals in similar circumstances but without dolphins.

Step 2: Design a study and collect data. Researchers recruited 30 subjects aged 18-65 with a clinical diagnosis of mild to moderate depression. Subjects were required to discontinue use of any antidepressant drugs or psychotherapy four weeks prior to the experiment, and throughout the experiment. These 30 subjects went to an island off the coast of Honduras, where they were randomly assigned to one of two treatment groups. Both groups engaged in one hour of swimming and snorkeling each day, but one group did so in the presence of bottlenose dolphins and the other group did not. (It's important to note participants in the dolphin group and the control group experienced identical conditions except for the presence of dolphins.) At the end of two weeks, each subject's level of depression was evaluated, as it had been at the beginning of the study (Antonioli and Reveley, 2005). The response variable is defined as whether or not the subject achieved substantial reduction (improvement) in their depression.

> **Think about it:** Identify the observational units and the explanatory variable in this study. Classify the explanatory and response variables as quantitative or categorical. If categorical, how many categories does each variable have? Is this study an observational study or an experiment? Are the subjects in this study a random sample from a larger population?

The 30 subjects are the observational units, all with mild to moderate depression. It is difficult to imagine that they were randomly selected from some larger population and all happened to be available to go to Honduras. However, the subjects were randomly assigned to either the dolphin therapy or the control group (the explanatory variable). We also suspect that the subjects did not know there were two different groups, though at the end of the study everyone in both groups then had the opportunity to swim with the dolphins. So, this is a randomized, comparative experiment. The response variable has been coded into a binary categorical variable: whether or not there was substantial improvement in depression symptoms.

The researchers hoped to show the study results provide convincing evidence that swimming with dolphins is more beneficial for mild to moderately depressed subjects than simply going to Honduras. We will apply the "assume innocence" logic again from Chapter 1 and determine whether the results these researchers found are typical or not typical of what we would find if the dolphin therapy was *not* more beneficial. Let's write down the null and alternative hypotheses remembering the concept of association from Chapter 4.

> Null hypothesis: Presence of dolphins does not help; that is, whether or not someone swims with dolphins has ***no association*** with whether or not someone shows substantial improvement.

> Alternative hypothesis: The presence of dolphins is helpful; that is, swimming with dolphins increases the probability of substantial improvement in depression symptoms (there is an association).

If the variables are associated, this means that knowing which group someone is in (swim with dolphins yes or no) helps us predict whether or not they will reduce their depression symptoms. This suggests another way to state the hypotheses, which focuses on the parameters in the study (long-run probability of improvement given the dolphin therapy treatment and the probability of improving given the control treatment):

> Null: The probability of exhibiting substantial improvement after swimming with dolphins (denoted by π_{dolphins}) is the same as the probability of exhibiting substantial improvement after swimming without dolphins (denoted by π_{control}).

Alternative: The probability of exhibiting substantial improvement after swimming with dolphins is higher than the probability of exhibiting substantial improvement after swimming without dolphins.

We can, thus, define our parameter of interest to be the difference in probabilities of substantial improvement between the two groups, which we can symbolize with $\pi_{dolphins} - \pi_{control}$. Meaning that we can restate the null and alternative hypotheses again as:

$$H_0: \pi_{dolphins} - \pi_{control} = 0$$

$$H_a: \pi_{dolphins} - \pi_{control} > 0$$

Note that the parameter of interest is the difference in the two probabilities, not the individual probabilities themselves. Stated another way, we will be analyzing a potential difference between the two treatments, not the overall probability of depression patients improving their symptoms.

The results that the researchers found are shown in Table 5.3 (a 2×2 table), with the explanatory variable as the columns of the table.

Table 5.3: A summary of the results of the experiment

	Dolphin therapy	Control group	Total
Showed substantial improvement	10	3	13
Did not show substantial improvement	5	12	17
Total	15	15	30

Step 3: Explore the data. A segmented bar graph comparing the response variable outcomes between the two explanatory variable groups is shown in Figure 5.2.

Figure 5.2: A segmented bar graph comparing the improvement in depression symptoms between the dolphin therapy group and the control group

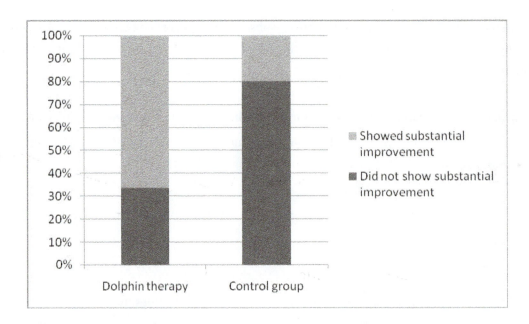

We see that subjects in the dolphin therapy group were more likely to show substantial improvement. In the dolphin therapy group, 10 of 15 or 66.67% showed substantial improvement compared to 3 of 15 or 20% in the control group.

Think about it: What are two possible explanations for the observed difference in these conditional proportions?

This difference of 46.67 percentage points (66.67% - 20% = 46.67%) appears quite substantial, and may reflect a genuine tendency for depressed individuals to be more likely to improve if their activities include dolphins. Still, the overall sample size (30 subjects) is rather small. Perhaps 13 subjects were going to show substantial improvement (and 17 were not) just by being participants in this experiment, regardless of which treatment group they were assigned to and we just happened, by random chance alone, to assign more of the improvers to the dolphin therapy group. We are not considering any other explanations or confounding variables because of the randomized blind nature of the study design. So, there are two possible explanations for the observed difference in these conditional proportions. Either the dolphins really helped (Alternative hypothesis)or that random chance alone was responsible for the observed difference (Null hypothesis). We need to decide which of these explanations is more plausible based on the data from the study.

Step 4: Draw inferences beyond the data.

As in Chapter 1, our strategy will be to determine how surprising the observed result would be if the null hypothesis were true. We can determine this by using simulation to generate a large number of "could-have-been" results under that null hypothesis and examine the null distribution of our statistic. If the observed result from our study falls in the tail of that distribution, then we have strong evidence against the null hypothesis.

Applying the 3S Strategy

1. Statistic: One possible choice of statistic for comparing these two groups is the difference in the observed sample proportions who improved. In this study, $\hat{p}_{\text{dolphins}} - \hat{p}_{\text{control}} = 0.4667$.

FAQ 5.2.2 Which Statistic?

Q: Which summary should I use, the difference in conditional proportions, or the ratio (relative risk)?

A: For describing the data, you can use whichever you like, or better yet, use both.

Q: Why both?

A: Suppose a new vaccine cuts the chance of disease by 0.01. If that difference is from 0.51 to 0.50, it's no big deal. But if the difference is from 0.02 to 0.01, you've cut the chance in half. Using both statistics gives a more complete description.

Q: Which one should I use to compute a p-value?

A: It doesn't matter. Use either one. You'll get the same p-value either way.

Q: Either way? No way. You'll have to convince me.

A: Fair enough. But I get to ask you two questions first.

Q: OK.

A: Think dolphin therapy. First question. Suppose I shuffle 13 black ("improver") cards and 17 red ("non-improver") cards, then deal out 15 for the dolphin pile. How many does that leave for the other pile?

Q. Duh. It has to be 15. You started with 30 cards and dealt out 15, so you've got 15 cards left

---

for the no-dolphin pile.

A: Will that always happen?

Q: Sure. The total number of cards always equals number of black plus number of red. Number of cards left always equals total number (black plus red) minus number in the dolphin pile. Next question?

A. It's a little harder. Suppose my dolphin pile has 10 black cards. How many red in the pile?

Q: Five: 15 cards in the pile, 10 of them black, so $15 - 10 = 5$ red. That was supposed to be hard? Who am I, Homer?

A: We're not done yet, Bart. What's in the other pile?

Q: Well, there were 13 black in all. Of those 13, 10 ended up in the dolphin pile. That leaves $13 - 10 = 3$ for the other pile.

A: Right. And how many red in that other pile?

Q: It's gotta be 12. The other pile has 15 cards, and 3 are black. Even Homer can do $15 - 3 = 12$.

A: Can Homer also do $17 - 5 = 12$?

Q: (Huh?) Oh I get it. We can get the answer two different ways [Think about it.]

A: And the overall lesson is?

Q: If I know the row totals and I know the column totals, I only need to know *one* more number to fill in the whole table.

A: Excellent! That's a really deep insight. (Hang onto it: It's the key to understanding degrees of freedom.)

Q: Payback time. I've answered your questions, but you haven't answered mine. How come I get the same p-value whether I choose the difference or the relative risk?

A: We're almost there. Back to the dolphin therapy. What's the difference in conditional proportions of improving?

Q: 10/15 – 3/15. That works out to 0.467.

A: Remember the margins are fixed. Can any other table give you that difference? Check it out: Try 0/15 – 15/15, 1/15 – 14/15, etc.

Q. I get the idea. There's one difference for each table, and one table for each difference.

A: And what's the relative risk ratio for the data?

Q: Just divide instead of subtract (10/15)/(3/15) = 10/3. That works out to 3.33.

A: Can any other table give you that ratio?

Q: No. Same idea as before.

A: There you have it. For each table there's exactly one difference and one ratio. I can tell you the table or the difference or the relative risk. If you know one, you can figure out the rest.

Q: So if the margins are fixed, the set of extreme tables is the same, regardless of whether you use the difference or the relative risk.

A: Congratulations. One last thing: I've been using "difference" and "ratio" as a kind of shorthand. Remember that in the context of 2 × 2 tables "difference" means "difference in conditional proportions" and "ratio" means "relative risk."

2. Simulation: If the null hypothesis is true, then the dolphin therapy is no more effective than the control therapy and we were going to have 13 improvers and 17 non-improvers regardless of which group they were assigned to. If this is true, then any differences we do see between the two groups arise solely from the randomness in the assignment to the explanatory variable groups.

Key idea: To evaluate the significance of the observed difference in our groups, we will investigate how large the difference in conditional proportions tends to be just from the random assignment of outcomes to the explanatory variable groups.

Think about it: What difference in conditional proportions would we expect to see if the null hypothesis were true?

We can perform this simulation analysis with index cards. We will let 13 blue cards represent the improvers and 17green cards represent the non-improvers. We are assuming these outcomes were going to happen no matter which treatment group subjects ended up in. We will simulate the random assignment process of the subjects to the two groups by shuffling the cards and putting 15 in one pile (for the dolphin therapy) and 15 in another pile (for the control group). This way an improver is equally likely to be assigned to the dolphin therapy as the control group. Note: We aren't saying subjects are equally likely to improve or not as we often did in Chapter 1, only that the long-run probability of being an improver is the same for both treatments. Figure 5.3 shows the results of one such reshuffling, with Group A arbitrarily representing the dolphin therapy group and Group B the control group.

Figure 5.3: The results of shuffling 30 cards with 13 improvers (blue cards) and 17 non-improvers (green cards) into 2 groups of 15

	Group A	Group B	Total
Blue cards (success)	6	7	13
Green cards (failure)	9	8	17
Total	15	15	30

Think about it: For the simulation, why do the row totals of 13 improvers (successes) and 17 non-improvers (failures) stay the same as the original study? Why do the column totals of 15 in the dolphin therapy group (Group A) and 15 in the control group (Group B) stay the same?

Notice the row totals are the same as in the actual study because we haven't changed who will improve or not. Similarly, the column totals have not changed because we are still assigning 15 subjects to each group, matching the study. But we have changed how those improvers and non-improvers are distributed between the two groups. As expected with the random shuffling we see a more equal distribution between the two groups. This is reflected in the segmented bar group for this "could-have-been" result (Figure 5.4).

Figure 5.4: Segmented bar graph for shuffled results

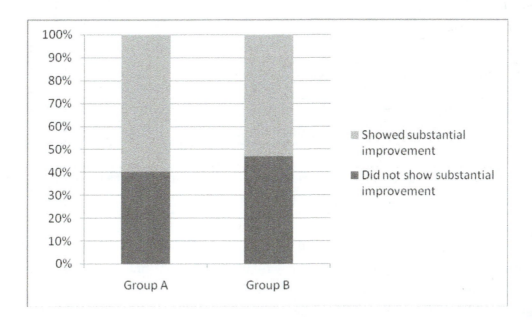

The conditional proportion of improvers in the dolphin therapy group is $6/15 = 0.40$ and the conditional proportion of improvers in the control group is $7/15 \approx 0.467$, for a difference of $0.40 - 0.467 = -0.067$. In fact, in this simulated trial, Group B (the control group) had a higher rate of improvers. But this can happen by random chance when there is no underlying impact from the treatment.

Think about it: When the null hypothesis is true, how often do you expect to see Group B with a larger conditional proportion of improvers than Group A?

If there is really no effect of dolphin therapy, then we expect Group B to have a larger proportion of improvers just as often as Group A (50% of the time). We repeated this process three more times and found 8, 5, and then 6 of the improvers in the dolphin therapy group.These results are shown in Figure 5.5.

Figure 5.5: The results of three more repetitions. This time we found 8, 5, and 6improvers were randomly placed in the dolphin therapy group. These yielded differences in proportions of improvers of 0.20, -0.20, and -0.067 respectively.

Simulated Results				Simulated Results				Simulated Results		
	Group A	Group B			Group A	Group B			Group A	Group B	
Success	8	5	13	Success	5	8	13	Success	6	7	13
Failure	7	10	17	Failure	10	7	17	Failure	9	8	17
	15	15	30		15	15	30		15	15	30

$8/15 - 5/15 = 0.20$ $5/15 - 8/15 = -0.20$ $6/15 - 7/15 = -.067$

Mapping this simulation to the study we find:

Null hypothesis = no difference in probabilities of success
One repetition = random assignment of response outcomes to two groups
Statistic = difference in conditional proportions

Think about it: Do any of these repetitions result in a difference in proportions of improvers that are at least as extreme as our observed study result of $0.667 - 0.20 \approx 0.467$?

None of the differences in proportions obtained in the four reshuffles are as extreme as or more extreme than the observed difference of 0.467, because they (-0.067, 0.20, -0.20) are all closer to 0, the difference expected under the null hypothesis, compared to 0.467. From just these four repetitions, it appears that the actual result of a difference in proportion of improvers in the dolphin group minus the control group of 0.467 is unlikely to happen by random chance alone when the null hypothesis is true. But we can't see much of a long-run pattern with only four repetitions so we repeated this process, using the Two Proportions applet, until we had 1000 total repetitions, each time computing the difference in proportion of improvers in the dolphin therapy group minus the proportion of improvers in the control group. The results are shown in Figure 5.6.

Think about it: What does the distribution in Figure 5.6 represent?

Figure 5.6: The results of 1000 repetitions of simulating the dolphin therapy experiment under the null hypothesis that dolphin therapy is not more effective than the control. Each dot in this dotplot represents a difference in conditional proportions.

Think about it: Does this distribution behave as you expected? In particular, where is the center and what is the shape of the distribution?

The mean of this distribution is approximately zero which we would expect because it represents the could-have-been outcomes when there is no underlying difference in the probability of improving between the two treatments. In other words, $\pi_{dolphin} - \pi_{control} = 0$, so we expect our difference in sample proportions, generated assuming the null hypothesis is true, to reflect that and cluster around zero as well. When the null hypothesis is true, we should see that Group A has the higher proportion of improvers about half the time, and Group B has the higher proportion of improvers about half the time. We do realize there will be some random variation but because the labeling of Group A and Group B was arbitrary, we expect that pattern of that variation to be symmetric around the mean of zero.

Think about it: How can we use the distribution shown in Figure 5.6 to evaluate whether dolphin therapy is effective?

3. Strength of evidence: However, the distribution of the difference in sample proportions, generated assuming the null hypothesis is true, does not tell us much until we see where our observed statistic falls in the distribution. We see that the statistic of 0.467 is in the tail of this null distribution. This shows that even though it's possible that the random assignment process alone could lead to such a large difference between the two groups, it's quite unlikely. In the 1000 trials in Figure 5.6, there were 13 simulated results with a difference of 0.47 or higher, so our estimate of the p-value is 13/1000 = 0.013. This small p-value provides strong evidence against the null hypothesis. We can also use the simulation results to standardize the statistic, as in Section 1.3. This indicates that 0.47 lies (0.47–0)/0.178 ≈ 2.654 standard deviations above zero. Because this standardized statistic is greater than 2, this again demonstrates that the researcher's result is an unusual outcome when the null hypothesis is true and provides strong evidence that the data in this study did not arise from a process where there was no difference in the probability of improving between the two treatments. We have strong evidence that dolphin therapy is more effective than the control for relieving depression symptoms.

Estimation

In fact, if we used a 2SD method to estimate a 95% confidence interval for the difference in the probability of substantial improvement between these two conditions using the standard deviation from the null distribution, we would find:

$$0.467 \pm 2(0.178) = 0.467 \pm 0.356 = (0.111, 0.823)$$

In other words, we are 95% confident that when allowed to swim with dolphins, the probability of improving is between 0.111 and 0.823 higher than when no dolphins are present. Notice that this interval is entirely positive, complementing our earlier analysis suggesting that the probability of improving is higher when swimming with dolphins than when not.

Step 5: Formulate conclusions. When we first looked at these data, we found a substantial difference in the proportions of improvers between the two groups. After considering how much variation we might see in the difference in conditional proportions from the random assignment process we confirmed that a difference this large was unlikely to occur by chance alone (p-value ≈ 0.013).Thus, the difference in the improvement rates between the two groups in this study is statistically significant and appears to be practically important as well[95% CI: (0.111, 0.823)].

Can we say that the presence of the dolphins *caused* this improvement? Because this was a randomized experiment, and assuming everything else was identical between the two groups other than the presence of dolphins (e.g., weather, amount of time in water, diet), then we do have compelling evidence that the imposition of the explanatory variable was the cause.

However, we have some cautions in generalizing this to a larger population. We must at least restrict ourselves to mild to moderately depressed patients in this age range willing to volunteer for such a study, and even that would be risky because we don't have any evidence that random selection was used to find the original 30 subjects. See FAQ 5.2.1 for more discussion on interpreting p-values and scope of conclusions when comparing two proportions.

FAQ 5.2.1 Where's the randomness?

No randomness

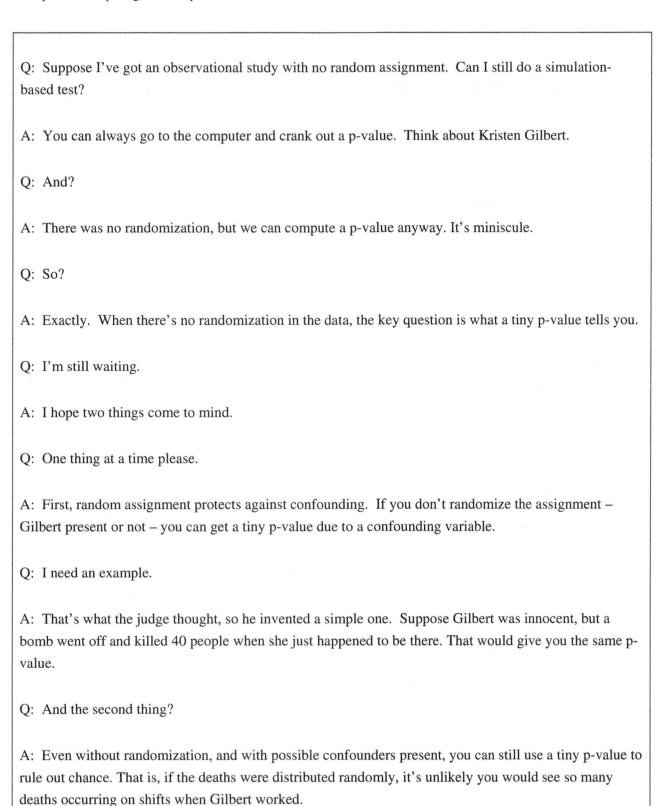

Q: Suppose I've got an observational study with no random assignment. Can I still do a simulation-based test?

A: You can always go to the computer and crank out a p-value. Think about Kristen Gilbert.

Q: And?

A: There was no randomization, but we can compute a p-value anyway. It's miniscule.

Q: So?

A: Exactly. When there's no randomization in the data, the key question is what a tiny p-value tells you.

Q: I'm still waiting.

A: I hope two things come to mind.

Q: One thing at a time please.

A: First, random assignment protects against confounding. If you don't randomize the assignment – Gilbert present or not – you can get a tiny p-value due to a confounding variable.

Q: I need an example.

A: That's what the judge thought, so he invented a simple one. Suppose Gilbert was innocent, but a bomb went off and killed 40 people when she just happened to be there. That would give you the same p-value.

Q: And the second thing?

A: Even without randomization, and with possible confounders present, you can still use a tiny p-value to rule out chance. That is, if the deaths were distributed randomly, it's unlikely you would see so many deaths occurring on shifts when Gilbert worked.

Q: Make it concrete, please.

A: Think Gilbert again. We don't know what *caused* the extreme result, but the tiny p-value tells us *we can rule out coincidence.* We're justified in *looking* for a cause.

Random sampling versus random assignment

Q: What if the chance comes from sampling instead of assigning?

A: Good question. Remember the shorthand: random sampling lets you generalize to the population; random assignment let's you detect a causal relationship.

Q: Can you make it concrete?

A: Think dolphin therapy again. What was randomized, the sampling or the assignment?

Q: Assignment.

A: And we concluded that swimming with dolphins *caused* the improvement, but only for the subjects in the experiment.

Q: Because there was no random sampling, the conclusion applies only to the subjects in the experiment. Right?

A: Strictly speaking, yes, but it doesn't have to be yes/no. To decide how representative those subjects might be is a matter of judgment.

Q: And if we don't randomize the assignment?

A: Suppose we let the subjects choose whether they wanted to swim with dolphins.

Q: If the dolphin group got better, we wouldn't know how much of the improvement was due to the dolphins themselves, and how much might be due to the kind of person who chose the dolphins.

A: Exactly. And with many observational studies you can't randomize. People make their own choices. You can ensure unbiased selection by using random sampling, but that kind of randomizing does not protect against confounding in the way you need for inference about cause. You have to use your knowledge of the context to make judgments about possible confounding.

Chance from a random process

Q: Suppose I take a simple random sample of cable subscribers and find an association between minority status (Yes/No) and news preference (FOX/MSNBC). My p-value is way below sea level. What can I conclude?

A: You can safely *generalize* from your sample to all cable subscribers in your sampling frame. But you have to be careful about *cause*. Of course you don't need statistics to rule out the hypothesis that watching MSNBC *causes* you to be a person of color, or that watching FOX causes you to be white, but you do need to be careful about the meaning of the association between ethnicity and cable preference. Just for example, maybe the MSNBC ads feature more people of color, and the political leanings are irrelevant.

Q: And if chance comes from a process instead of from sampling?

A: OK. Assume the process is chance-like. Then a lot comes down to whether you assign first and then observe (active plan), or whether you just observe (passive plan). The active plan is like random assignment; the passive plan is like random sampling.

Q: I've been waiting, and now it's my turn. Back at you: make it concrete.

A: Fair enough. For an example of an active plan, go back to Buzz and Doris. Each try was the result of a random process, but on top of that Dr. Bastian also randomized the assignment of conditions – whether to make the light blink or not. His randomizing the assignment of conditions justified inference about cause.

Q: And the passive plan, just observing, without assigning?

A: That's your cable TV example. You can be confident that the association applies to the whole population, but cause and effect remains tricky and a matter for judgment.

Step 6: Look back and ahead. There is strong evidence that the presence of the dolphins increases depression patients' probability of demonstrating substantial improvement in their depression symptoms over this time period (11 to 83 percentage points).Broad generalizations were difficult to make due to the convenience sample of participants. Researchers should look into a better design for selecting participants. Because this may not be a feasible long-term strategy to reach a large number of depression sufferers, the researchers may now want to investigate other animal care programs to see whether the effects are similar.

Follow-up Analysis

Think about it: Suppose the actual results had found 8 successes in the dolphin therapy group and 5 successes in the control group for these 30 subjects, what would change, if anything, about your analysis and conclusion?

Testing the same null hypothesis, we would use the same null distribution as in Figure 5.6 but this time we would see where a difference of 8/15 − 5/15 =0.20.Such an outcome does not fall in the tail of the distribution. In fact we can use the Two Proportions applet to approximate the p-value from 1000 repetitions of the shuffling process. This applet (with the 2 × 2 box checked) allows you to specify the values of your own two-way table by clicking in each cell, deleting the old value and typing in the new value, and then pressing Use Table. The applet (see Figure 5.7) reports an approximate p-value of 0.2320, giving us little to no evidence against the null hypothesis. This large p-value, suggesting that it would not be surprising to see a difference in sample proportions of 0.20, if the null hypothesis of no difference in the underlying probabilities was true, should make sense because the observed difference in proportions is smaller, so there is less evidence against the null hypothesis. As we saw in Chapter 1, the closer the observed value of the statistic is to the hypothesized value, the less evidence against the null hypothesis. Here the observed difference in conditional proportions is closer to zero.

Figure 5.7: Results for new data table shading the one-sided p-value

Relative Risk

Recall from Section 5.1, that the relative risk is another way to summarize the association between two binary variables. For this study, the observed relative risk is 0.667/0.15 = 3.33, meaning that individuals who swam with dolphins were 3.33 times more likely than those who did not swim with dolphins to show substantial improvement on depression symptoms. We can modify the 3S strategy in a minor way to conduct a test of significance for the relative risk.

Think about it: What would the value of the relative risk be if there was no association between swimming with dolphins and showing improvement on depression symptoms?

To state the null and alternative hypotheses, think about what value of the relative risk would mean there was no association. A relative risk of one means that the proportions are equal, and there is no association. Thus, we can write the null and alternative hypotheses as:

Null: The relative risk of improvement on depression symptoms (computed as the ratio of the probability of exhibiting substantial improvement after swimming with dolphins to as the probability of exhibiting substantial improvement after swimming without dolphins) is one

Alternative: The relative risk of improvement on depression symptoms (computed as the ratio of the probability of exhibiting substantial improvement after swimming with dolphins to as the probability of exhibiting substantial improvement after swimming without dolphins) larger than one

To apply the 3S strategy, we use the observed relative risk as the statistic, and use the same method of simulation.

Statistic: The observed relative risk is $0.667/0.15 = 3.33$

Simulate: We use the same simulation strategy described earlier, except that now we compute the relative risk with every shuffle. Figure 5.8 shows the results of 1000 shuffles using the relative risk as the statistic. Notice that the null distribution is no longer symmetric, but is now right skewed. However, the median of this distribution will be approximately equal to one – half the time group 1 has a larger proportion and half the time group 2 has a larger proportion (with a mean larger than one due to the skewness.

Strength of evidence: To evaluate the strength of evidence determine how many times the observed statistic (3.33) or larger (in the direction of our alternative hypothesis) occurred in the simulations. In this, it happened 13 out of 1000 times, yielding a p-value of 0.013. Notice that the p-value for this test using the relative risk is the same as the one we got when using the difference in proportions. That is no accident! In fact, it will always happen. See FAQ 5.2.2 for more details.

We can also calculate a confidence interval for the population relative risk, but this is a bit more complicated.

Figure 5.8 Distribution of simulated relative risks

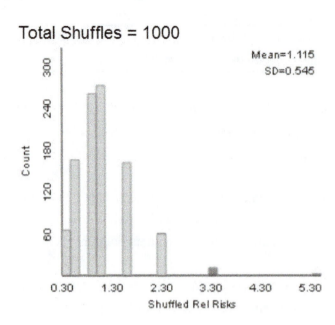

Count Samples Greater Than ▾ 3.33   [Count]
Count = 13/1000 (0.0130)

Exploration 5.2: Is Yawning Contagious?

Step 1: Ask a research question. Is yawning contagious? Conventional wisdom says yes: When we see someone else yawn, we're prone to feel sleepy and let out a yawn ourselves. Has this happened to you, or have you noticed it in others? Will data support this claim if we put it to a scientific test?

Step 2: Design a study and collect data. The folks at *Myth Busters*, a popular television program on the Discovery Channel, investigated this issue by using a hidden camera. Fifty people attending a local flea market were recruited to participate. Subjects were ushered, one at a time, into one of three rooms by co-host Kari. She yawned (planting a yawn "seed"), as she ushered subjects into two of the rooms, and for the other room she did not yawn. The researchers decided in advance, with a random mechanism, which subjects went to which room. As time passed, the researchers watched to see which subjects yawned.

1.   Think about why the researchers made the decisions they did.

a. Why did the researchers include a group that didn't see the yawn seed in this study? In other words, why didn't they just see how many yawned when presented with a yawn seed?

b. Why did the researchers use random assignment to determine which subjects went to the yawn seed group and which to the control group?

c. Is this an observational study or a randomized experiment? Explain how you are deciding.

d. The researchers clearly used random assignment to put subjects into groups. Do you suspect that they also use random sampling to select subjects in the first place? What would random sampling entail if the population was all flea market patrons?

2. Identify the explanatory and response variables in this study. Also classify them as categorical or quantitative.

Explanatory: Type:

Response: Type:

3. The two competing hypotheses to be tested are stated below. Identify which is the null hypothesis and which is the alternative hypothesis remembering that the null hypothesis is typically a statement of no effect.

Yawning is not contagious.   Yawning is contagious.

4. If yawning is not contagious what does this say about whether or not there is an association between the explanatory and response variables?

5. If yawning is contagious, what does this say about whether or not there is an association between the explanatory and response variables?

6. State the null and alternative hypothesis in terms of association between the explanatory and response variables in this study.

An equivalent option to stating the null and alternative hypotheses for this study in terms of association is to talk about long-run probabilities.

7. There are two long-run probabilities (parameters) in this study. What are they?

8. Write the null and alternative hypothesis in terms of the two long-run probabilities in words and using appropriate symbols.

The researchers found that 11 of 34 subjects who had been given a yawn seed actually yawned themselves, compared with 3 of 16 subjects who had not been given a yawn seed.

9. Organize this information into the following 2×2 table:

	Yawn seed planted	Yawn seed not planted	Total
Subject yawned			
Subject did not yawn			
Total			

Step 3: Explore the data.

10. Calculate the conditional proportion of subjects who yawned in the "yawn seed" group, and then do the same for the control group. Then calculate the *difference* between these proportions, subtracting in the direction indicated below.

    Proportion who yawned in "yawn seed" group:

    Proportion who yawned in control group:

    Difference in conditional proportions ("yawn seed" – control):

Notice that the difference in conditional proportions is the statistic of interest now that we are comparing two groups.

11. Produce a segmented bar graph to compare the distributions of whether or not the subject yawned between the treatment (yawn seed) and control group.

12. Comment on what the graph and calculations reveal about the research question.

    a. Did a larger proportion of subjects yawn who were exposed to the yawn seed, as compared to the control group?

    b. Based on your analysis thus far, do you think these data provide much evidence that yawning is contagious? Explain.

Step 4: Draw inferences beyond the data. We see that the proportion of subjects in the "yawn seed" group who yawned was greater than the proportion who yawned in the control group. But does this provide convincing evidence of a genuine difference in the long-run probabilities? To determine whether the *MythBusters'* results provide convincing evidence that yawning is contagious, we will apply the same logic that we have used previously: We will use a simulation analysis to determine whether their results are typical or surprising for what we would find if yawning is not contagious.

13. As with earlier studies, there are two possible explanations for the *MythBusters'* results. What are they? (*Hints*: These correspond to the null and alternative hypotheses stated earlier.)

The key to our simulation analysis is to assume that if yawning is not contagious (null hypothesis), then **the 14 yawners would have yawned regardless of whether or not they had seen the yawn seed**. Similarly, we'll assume that the 36 non-yawners would not have yawned, no matter which group they had been assigned. In other words, our simulation assumes the null hypothesis is true—that there is no association, no connection, between the yawn seed and actual yawning.

> **Key idea:** To evaluate the significance of the observed difference in our groups, we will investigate how large the difference in conditional proportions tends to be just from the random assignment of response outcomes to the explanatory variable groups.

We cannot use coins to conduct this simulation analysis, because we have two variables to consider: whether or not the yawn seed was planted, and whether or not the subject yawned. Instead of coins, we will use index cards. Here's our strategy:

- Take a set of 50 cards, with 14 blue cards (to represent the yawners) and 36 green cards (to represent those who did not yawn).
- Shuffle the cards well, and randomly deal out 34 to be the yawn seed group (the rest go to the control group)
- Count how many yawners (blue cards) you have in each group and how many non-yawners (green cards) you have in each group.
- Construct the two-way table to show the number of yawners and non-yawners in each group. Clearly nothing different happened to those in "group A" and those in "group B" and any differences between the two groups that arise are due purely to the random assignment process.

14. Do this shuffling and dealing once.
   a. Report the (simulated) 2×2 table that your shuffling and dealing produces:

	Yawn seed planted	Yawn seed not planted	Total
Subject yawned			14

Subject did not yawn			36
Total	34	16	50

b. Calculate the conditional proportions who yawned for your simulated data, and also the difference in those proportions:

(Simulated) Proportion who yawned in treatment group:

(Simulated) Proportion who yawned in control group:

(Simulated) Difference in conditional proportions (treatment – control):

c. Is your simulated statistic (difference in conditional proportions) at least as large as the observed value of the statistic from the *MythBusters'* study?

We need to perform a large number of repetitions (say, 1000 or more) in order to assess whether the *MythBusters'* result is typical or surprising when yawning is not contagious. To do this we will use an applet specifically designed for this purpose: the **Two Proportions** applet.

15. Open this applet and verify the two-way table, segmented bar graph, and observed difference in simulated proportions you found above.
   a. Check the **Show Shuffle Options** box. Notice how the cards have been set up: 14 blue cards, with 10 in the yawners group. Press **Shuffle** to shuffle the cards and redistribute them to the two groups. How many yawners ended up in the seeded group? What is the corresponding difference in the conditional proportions for the shuffled data?

b. Press the **Shuffle** button four more times. Record the difference in proportions each time. Was it always the same number?

Notice that each of these values has been added to the dotplot on the right.

(*Technical detail*: In this case, it's equivalent to look at the *number of blue cards in the yawn group* or the *difference in conditional proportions* as the statistic of interest.)

16. Use this applet to conduct 1000 repetitions of this simulation. Change the **Number of Shuffles** from 1 to 995 (for a total of 1000) and press **Shuffle**. The applet produces a dotplot showing the null distribution for the difference in proportions of yawners between the two groups.

    Map this simulation to the research study:

Null hypothesis =	
One repetition =	
Statistic =	

17. Look more closely at the simulated null distribution for the difference in proportions of yawners between the two groups.

    a. Is this null distribution centered around 0? Explain why this makes sense. (*Hint*: Think about the choice of statistic and about the null hypothesis, which was the basis of the simulation analysis.)

    b. Is the observed value of the statistic from the *MythBusters'* study (0.136) out in the tail of this null distribution, or not so much? In other words, does the observed result appear to be typical or surprising when the null hypothesis (that yawning is not contagious) is true?

    c. As you know, you can assess the strength of evidence against the null hypothesis by calculating a p-value. To calculate a p-value from this null distribution, you will count the number of repetitions that produced a simulated difference in proportions equal to _____ or _____ <choose more or less>.

18. Enter the observed difference in the sample proportions in the **Count Samples** box and choose the direction from the pull-down menu and press **Count**. Check whether the shaded region of the dotplot corresponds with your answer to part #17c. Then report this approximate p-value.

19. *Interpret* this p-value by filling in the following blanks:

Under the assumption that _____, if we repeatedly _____ many, many times, the probability we would obtain results as or more extreme than _____ is about _____.

20. *Evaluate* this p-value: Using guidelines for assessing the strength of evidence from p-values, would you conclude that the *MythBusters'* result provides much evidence that yawning is contagious?

We have a small confession to make. We fudged the data from this study a bit. The actual data showed that 10 of the 34 subjects in the yawn seed group yawned themselves, and 4 of the 16 in the control group yawned.

21. Recalculate the conditional proportions who yawned for these data:

Proportion who yawned in the "yawn seed" group =

Proportion who yawned in the control group =

Difference in conditional proportions ("yawn seed" – control) =

22. How has your observed statistic changed from the originally presented data?

smaller   same   larger

Before you carry out inference using this post-confession observed statistic, let's make some predictions: How do you expect the p-value and strength of evidence to change from your earlier analysis? Circle your choices below and then explain why you circled what you did.

p-value:

      smaller   same   larger

strength of evidence:

      weaker   same   stronger

Explanation:

23. Conduct a simulation analysis to check your answer predictions from the previous question. To do this, and enter the new values into the 4 boxes of the table and press **Use Table**. Then ask for 1000 repetitions in your simulation analysis. Enter the new observed value of the statistic and then press the **Count** button. Report the approximate p-value and summarize your conclusion in terms of strength of evidence that yawing is contagious.

Estimation

Because the *MythBusters'* data did not reveal a significant difference between the groups, it's less important than usual to estimate the parameter value. But we can still use the simulation results to estimate a 95% confidence interval for the relevant parameter. Keep in mind that the relevant parameter is now the difference in the long-run probabilities between the yawn seed treatment and the no-yawn seed treatment.

24. The parameter of interest in this yawning study can be expressed in symbols as: $\pi_{\text{yawn seed}} - \pi_{\text{control}}$.

    a.   Describe what this parameter means in words.

b. Report (again) the observed value of the statistic from the *MythBusters* actual data. (*Hint*: Recall that the statistic is the difference in conditional proportions between the two groups.)

The observed value of this statistic is your best estimate of the unknown parameter value. But we should produce an interval estimate centered on this observed value. To do that we again need to consider the *chance variability* in the statistic.

c. From the simulation analysis that you just conducted on the post-confession *Mythbusters'* data, what was the standard deviation of the null distribution?

d. Determine endpoints of a 95% 2SD confidence interval for the difference by taking the observed value of the statistic $\pm$ 2 standard deviations.

e. Does this confidence interval include only positive values, only negative values, or does the interval include zero as well as positive and negative values?

f. Explain the importance of your answer to part e) in terms of whether the data provide evidence that subjects who experience a yawn seed are more likely to yawn than those who do not experience a yawn seed?

Step 5: Formulate conclusions.

25. The *MythBusters* hosts concluded from their study that there is "little doubt, yawning seems to be contagious." Based on your simulation analysis of their data, considering the issue of statistical significance, do you agree with this conclusion? Explain your answer, as if to the hosts, without using statistical jargon. Be sure to include in your answer an explanation for

why you conducted the simulation analysis and what the analysis revealed about the research question.

26. If you had decided that the two groups differed significantly, would you have been justified in drawing a cause-and-effect conclusion between the yawn seed and an increased probability of yawing? Explain, based on how the study was conducted.

27. Based on how the sample was selected, to what larger population would you feel comfortable generalizing the results of this study? Justify your answer.

28. Think about the difference between these two possible conclusions:

      i.   The data strongly suggest that yawning is not contagious.

      ii.   The data provide no evidence to suggest that yawing is contagious.

Explain why these conclusions are not the same. Also indicate which one is the more appropriate conclusion from your analysis, and explain why.

29. As we discussed in Chapter 2, in every statistical analysis it is possible to make a Type I or a Type II error. Identify which error is possible here, describe it in context and briefly identify the consequences.

Step 6: Look back and ahead.

30. Critique the design and conclusion of this study. Were there any limitations, such as how the subjects were selected, or how the measurements were recorded? Was the sample size large enough? Did you observe a result of practical value? Discuss how you might address a few of these limitations. Suggest how you might design a follow-up study to investigate whether yawning is contagious. Address details such as how to ensure that the yawn seed is noticed, how long to wait for the subject's yawn to be connected to the seed, and how to select/recruit subjects.

Effect of Sample Size

Now let's suppose that this yawning study had involved 500 people, 10 times as many as the actual study. Let's also suppose that the conditional proportions who yawned in each group were identical to the actual study. The 2×2 table would therefore be:

	Yawn seed planted	Yawn seed not planted	Total
Subject yawned	100	40	140
Subject did not yawn	240	120	360
Total	340	160	500

31. Before you conduct a simulation analysis, indicate how you expect the following to change:

a. Difference in conditional proportions

     Smaller   Same   Larger

b. p-value

     Smaller   Same   Larger

c. Strength of evidence that yawning is contagious

     Weaker   Same   Stronger

d. Test decision at the 0.05 significance level

     Reject null (unlike before)   Fail to reject null (as before)

32. Use the **Two Proportions** applet, again entering the appropriate table of data first, to conduct a simulation analysis. Report the approximate p-value and summarize your conclusion in

terms of strength of evidence that yawing is contagious. Also indicate whether you would find strong evidence against the null hypothesis at the 0.05 significance level.

33. Summarize how the ten-fold increase in the sample size has changed your conclusions. Do you have stronger evidence that yawning is contagious? Is the evidence now very strong that yawning is contagious?

Relative risk

In Section 5.1 we introduced you to the relative risk as an alternative statistic to summarize the relationship between two binary variables.

34. What value of the relative risk indicates no association between the two variables in this study?

35. State the null and alternative hypothesis for the yawning study in terms of the relative risk.

We will now apply the 3S strategy using the relative risk. Use the sample size of 500 people you used in #31 through #33 for the following questions.

36. **Statistic.** Find the observed value of the relative risk by dividing the conditional proportion of yawners in the yawn seed group compared to the conditional proportion of yawners in the no yawn seed group.

In the **Two proportions** applet, change the statistic to "Relative risk" in the pull down menu. Confirm that the statistic matches your answer to #36.

37. **Simulate.** If you still have your shuffles from #32 use them, otherwise generate 1000 new shuffles. Find the p-value using the relative risk. Compare this p-value to the p-value you obtained in #32.

If you used the same simulated tables, you should have found that the p-value for the relative risk is exactly the same as for the difference in proportions! This is because there is a "one-to-one" correspondence between these statistics. See FAQ 5.2.2 for more details.

Section 5.2 Summary

The 3S Strategy can be used to assess whether two sample proportions differ enough to conclude that there is a genuine difference in the population proportions/process probabilities. The reasoning process of **statistical significance** is the same as with inference for a single proportion.

- The null hypothesis can be expressed as no association between the two variables.
    - o Equivalently, the null hypothesis asserts that the population proportions (or process probabilities) are the same in the two groups.
- The alternative hypothesis says that there is an association between the two variables.
    - o The alternative hypothesis can be one-sided or two-sided, depending on whether the researchers have a particular direction in mind prior to collecting data.
- The difference in conditional proportions between the two groups is the statistic of interest.
- The key idea in simulating the null distribution of the difference in sample proportions is to shuffle the values of the response variable.
    - o This shuffling simulates values of the statistic under the assumption of no association.
    - o With a randomized experiment, this shuffling is equivalent to re-randomizing the subjects into the two groups, assuming that the subjects' responses would have been the same regardless of which group they were assigned to.
- As always, the p-value is calculated as the proportion of repetitions in which the simulated value of the statistic is at least as extreme as the observed value of the statistic.
    - o Also as always, a small p-value provides evidence against the null hypothesis.

The **2SD Method** can be used to produce a confidence interval for estimating the size of the difference between the two success probabilities (or population proportions).

- The most important question is whether the interval is entirely negative, entirely positive, or includes zero.
  - o If the interval does not include zero, that provides evidence that the two population proportions/process probabilities are not the same.
  - o Whether the interval contains positive or negative values indicates which group has the larger population proportion, and by how much.
  - o If the interval contains zero, there is no evidence against the null hypothesis that the two population proportions/process probabilities are equal.

Some **factors that affect p-valuesand confidence intervals** are similar to before:

- The larger the difference between the sample proportions, the stronger the evidence that the population proportions differ.
- Larger sample sizes produce stronger evidence that the population proportions differ.
- The difference in sample proportions is the midpoint of the confidence interval.
- Larger sample sizes produce narrower confidence intervals.

Section 5.3: Comparing Two Proportions: Theory-Based Approach

Introduction

In Chapter 1 we saw that we could often predict the results obtained via simulation using a theory-based approach that uses normal distributions. We could do this because, in many situations, the null distribution followed a very predictable pattern: an approximately normal distribution with a (mathematically) known average and standard deviation. Furthermore, we saw how theory-based approaches also gave us much simpler ways to generate a confidence interval for the parameter of interest. In this chapter, you may have noticed that your simulated distributions were often symmetric, bell-shaped distributions again it's no coincidence. In this section we will, once again, see how we can use theory-based approaches instead of simulation to predict the outcome of tests comparing two group proportions and to generate related confidence intervals.

Example 5.3: Smoking and Birth Gender

An area of research that generates a lot of media coverage is examining how parents' behavior may be associated with the sex of their children. One such recent study (Fukuda et al., 2002) found the following:
- Of 565 births where both parents smoked more than a pack of cigarettes a day, 255 (45.1%) were boys.

- Of 3602 births where both parents did not smoke, 1975 (54.8%) were boys.

Figure 5.9: Asegmented bar graph comparing the proportions of boys and girls born to parents that smoked with parents that didn't smoke

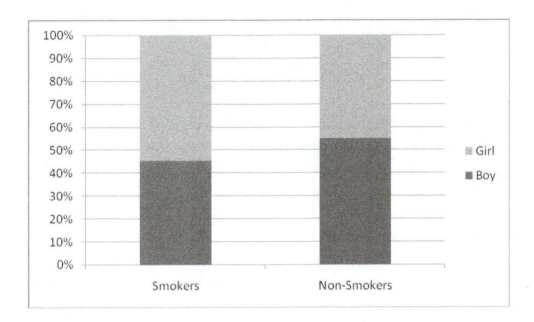

The proportion of boys born to parents who smoke (0.451) and to parents whodon't smoke (0.548) seem like they are considerably different.Let's comparethese two proportions to see whether the difference isstatistically significant.We start by stating the hypotheses.

> Null hypothesis: The probability of having a boy is the same for parents who smoke at least one pack per day as it is for parents who do not smoke. There is no association between smoking status of parents and sex of child.

> Alternative hypothesis: The probability of having a boy is not the same for parents who smoke at least one pack per day as it is for parents who do not smoke. There is an association between smoking status of parents and sex of child.

Think about it: Identify the two variables in this study. Would you consider one an explanatory variable and the other the response variable? What is different about how this study was conducted compared to those in Section 5.2?

The researchers are interested in whether the smoking habits of the parents might help explain the sex of the newborn. It is natural to think of smoking habit as the explanatory variable and sex as the response variable. This study was an observational study and not a randomized experiment, because the researchers did not randomly determine which parents would smoke and which would not. Therefore, even if we find the difference in these conditional proportions to be statistically significant, we will not draw any cause-and-effect conclusions from the association. Instead, we are more interested in what these sample data can tell us about the populations of smoking and non-smoking parents. So we can state these hypotheses in terms of population parameters. Let's define $\pi_{smoking}$ as the proportion of boys born into the population of all double smoking parents and $\pi_{nonsmoking}$ as the proportion of boys born into the population of all non-smoking parents. The null states $\pi_{smoking} = \pi_{nonsmoking}$. That is

$$H_0: \pi_{smoking} - \pi_{nonsmoking} = 0, \text{ versus } H_a: \pi_{smoking} - \pi_{nonsmoking} \neq 0.$$

Because these populations are quite large, applying what you learned in Chapter 2, we will consider these population proportions to be equal to the probability that a randomly selected set of parents has a boy from each population. But does this change how we assess the statistical significance of the difference in sample proportions?

Shuffling without random assignment

In Section 5.2 we "re-randomized" the subjects to the treatment groups, shuffling the response variable outcomes to the two groups mimicking what was done in the research study (assuming those outcomes wouldn't change based on the group assignment). But here the subjects weren't randomized to their explanatory variable groups, because it is an observational study. So, can we use the procedure as in Section 5.2? As it turns out, we can. To see this, note that we stated our null hypothesis as there was "no association" between parents' smoking status and the gender of the child. If we shuffle and deal the cards into two stacks (smoking and non-smoking) then the color of the card (blue or green=girl or boy) will have no association (relationship) with which stack the card is in. In other words we will have simulated data assuming the null hypothesis is true. Thus, regardless of whether the study design involves random assignment, random sampling, both, or neither, you can simulate data assuming the null hypothesis of no association between explanatory variable and response variable using shuffling (see FAQ 5.3.1 for more details).

Key idea: Shuffling is an appropriate way to estimate a p-value for comparing groups regardless of the study design (observational study or experiment, random assignment or random sampling or neither). Of course, the study design is very important in determining the appropriate scope of conclusions. (Can you generalize to a larger population? Can you infer a cause-and-effect conclusion?)

FAQ 5.3.1. Which Way To Simulate

Setting up the simulation

Q: If we do a simulation test for the study of dolphin therapy, how do we set it up?

A: The null hypothesis tells the colors. The experimental plan tells the piles. In short, null tells colors, plan tells piles.

Q: One at a time, please.

A: What's the null hypothesis for dolphin therapy?

Q: The treatment makes no difference. The outcome will be the same regardless of which group you get assigned to.

A. Exactly. So what's fixed?

Q: Each person's response, improve or not.

A: Right. So the null says the outcomes are fixed. They should be the colors. Seventeen black for the improvers, thirteen red for the non-improvers. These stay the same regardless of which group they go to, just as the null hypothesis says. What's random?

Q: The group you get assigned to, dolphins or no dolphins.

A: So that assignment – the group – is what we randomize. Shuffle and deal 15 for the dolphin pile, with the leftovers for the no-dolphin pile.

Q: That's clear enough. Just as you said, *the null tells what to fix; the plan tells what to randomize.*

A: Excellent. But prepare for a surprise. Suppose you get it backwards, and fix the groups and shuffle the outcomes.

Q: Then I'm toast?

A: Wrong. Believe it or not, you get exactly the same p-value either way.

Q: How can that be?

A: You can prove it with mathematical logic (google "Wiki Fisher's exact test") but there is also an intuitive way to think about it.

Q: Yes?

A: Either way, you've got 30 cards. Think of them set out in a row. You've got two sets of labels, one for colors (17 "black" and 13 "red") and one for piles (15 "one" and 15 "two"). Instead of shuffling the cards, you shuffle the labels.

Q: OK.

A: You can shuffle and deal the color labels first, then shuffle and deal the pile labels after. Call this Method 1.

Q: Or the other way: Shuffle and deal the pile labels first, then shuffle and deal the color labels after. That's Method 2.

A: Once you've done the first shuffle and deal, that's fixed. Method 1 is like fixing the colors and shuffling the piles. Method 2 is like fixing the piles and shuffling the colors.

Q: Both ways, the cards don't change, and the chances don't change. Order doesn't matter.

A: It's tricky, but I think you've got it.

Simulating for observational studies

Q: So far I'm OK. But let's go back to smoking and birth gender.

A: I'm with you: Example 5.3.

Q: There was no random assignment.

A: Right.

Q: But for the study of dolphin therapy, they used chance to assign people to groups.

A: Right again.

Q: Wasn't that randomness the reason we shuffled the cards? Shuffling was a way to mimic the random assignment.

A: Keep going. You're on a winning streak.

Q: (Pouncing) So if there's no random assignment, the shuffling makes no sense.

A: I like the way you think. The connections that link your data to your simulation to your inference are critical. No connections, no inference.

Q: Thanks for the complement, but you ducked my question.

A: Guilty as charged. Ask me again.

Q: If I don't have random assignment, what's the justification for shuffling?

A: The families in the sample could be regarded as chosen randomly. The chance is in the choice, not the assignment.

Q: So?

A: What's the null hypothesis?

Q: No relationship between smoking and boy/girl.

A: Right. Suppose we choose at random and get 565 smokers and 3602 non-smokers. What does the null tell us to expect about boys and girls?

Q: They're random. Boys and girls are always random.

A: How many of each?

Q: What we got: 2230 boys and 1937 girls.

A: Yes. So we have the same set-up as we have for random assignment. Under the null hypothesis it doesn't matter which label each person gets, the response is going to stay the same regardless of which group we assign people to. This time the randomness comes from sampling, not assignment, but the logic is the same.

Q: Last question. Because we randomized, we can conclude that smoking causes girls to be more likely.

If you want a girl, smoke two packs a day.

A: Nice try, but I know you know better. You need random assignment for inference about cause. Random selection lets you generalize to the population. Simulating the random assignment of group labels to get a p-value does not create randomness in the actual study design.

Q: Yeah. I knew that. I just wanted to get your goat.

Now let's apply the 3S Strategy to these data.

1. Statistic: The observed difference in conditional proportions of boys born to nonsmokers and boys born to smokers is $0.548 - 0.451 = 0.097$.

2. Simulate: We used the **Two Proportions** applet to do our simulation. First we can type in the two-way table and press **Use Table**. Then we select the **Show Shuffle Options** box. We conducted 5000 repetitions of randomly shuffling the 2230 boys and 1937 girls and "dealing them" randomly to the 565 smoking parents and 3602 nonsmoking parents, modeling the null hypothesis to be true, and calculating the difference in proportions of boys between the two groups for all 5000 repetitions. This random shuffling simulates the null hypothesis of no association, because shuffling and dealing ensures there is no relationship between smoking status (the pile the card is dealt to) and gender of child (the color of the card). The results are shown in Figure 5.10.

Figure 5.10: Simulated results of the difference in proportions of

Total Shuffles = 5000

Mean=0.000
SD=0.022

Count Samples | Beyond ▼ | .097

Count

boys   Count = 0/5000 (0.0000)

3. Strength of evidence: We see from the graph in Figure 5.9 that nothing as extreme as our observed statistic (larger than 0.097 or smaller than -0.097) ever occurred in these 5000 repetitions, thus the estimated two-sided p-value is 0. The sample data provide extremely strong evidence that the probability of boys is different between smoking parents and non-smoking parents.

But notice how the null distribution is centered at zero, the value of the population parameter specified by the null hypothesis, and is bell-shaped. The applet shows a checkbox to **Overlay Normal distribution** leading to the picture in Figure 5.11. The close agreement between these two distributions in Figure 5.10 suggests that we could have calculated a p-value using the normal distribution. As we saw in Chapter 1, there are situations where the null distribution can be predicted using theoretical models and we don't have to go to the trouble of shuffling cards (by hand or using technology) to develop our null distribution. The normal distribution also reports the theory-based p-value to be approximately zero.

Figure 5.11. A normal distribution overlaid on a simulated null distribution of differences in sample proportions.

You can also use the **Theory-Based Inference** applet to determine the theory-based p-value as shown in Figure 5.12. We first chose "Two proportions" from the top pull-down menu and then used a two-sided alternative ($\neq$).Notice that the hypotheses use the notation $\pi_1 - \pi_2$ for the difference in the population proportions. Because we input the non-smoker data for group 1 and the smoker data for group 2, we are letting π_1 represent the proportion of babies that are boys in our population of nonsmokers and π_2 represent proportion of babies that are boys in our population of smokers. We entered the two sample sizes (n) and the number of boys in each sample (count).The applet calculated the conditional proportions, the standardized statistic (z-statistic), and the corresponding (two-sided) p-value. From the standardized statistic of $z = 4.30$, we can see that our observed difference in sample proportions is more than four standard deviations above the hypothesized mean of zero. This is so far out in the tail that we don't see it showing up on the graph and we get a p-value of 0.0000 when it is rounded to four decimal places. So, just like the simulation-based result, we have very strong evidence that the population proportion of boys born to smokers is different than the population proportion of boys born to nonsmokers. Note: Formulas for computing z are provided at the end of this section. From our data we can also see that the sample proportion of boys is smaller for smokers.

Figure 5.12: The results of comparing the sample proportion of boys born to smokers to the sample proportion of boys born to nonsmokers using the Theory-Based Inference applet

Estimation

As we discussed in Section 5.2, a natural follow-up question is "how much smaller is the population proportion of boys born to parents who smoke, compared to that born to parents who do not smoke?" In other words, if the difference in the proportions is not 0, what is it? A confidence interval will help answer this question.

Remember that when we are comparing two proportions, the population parameter we are trying to estimate is the difference in population proportions or process probabilities ($\pi_1 - \pi_2$). Because our sample difference in proportions of boys (subtracting the smoking parents' proportion from the non-smoking parents' proportion) is 0.097, it makes sense for the interval to be centered on that number. The 2SD Method (Section 5.2) tell us that a 95% confidence interval is 0.097 ± 0.044, as the simulated SD for $\hat{p}_1 - \hat{p}_2$ is 0.022 (Figure 5.10).

Of course, a theory-based approach is also available. Figure 5.13 shows the result of using the theory-based inference applet to find a 95% confidence interval on the difference in population proportions. Note: Formulas for how this theory-based interval is computed are provided at the end of this section

Figure 5.13: Results from the Theory-Based Inference applet showing a confidence interval for the difference in population proportions

☑ Confidence interval

confidence level 95 % Calculate CI

(0.0529, 0.1411)

We used the same inputs (group 1= nonsmokers) for computing the confidence interval as we did for the test of significance. The only thing we did differently was to check the box to display the Confidence Interval and press Calculate CI. This gives us our interval of (0.053, 0.141), which is essentially the same as we obtained using the 2SD Method.

Thus, we are 95% confident that the difference in population proportions ($\pi_{nonmoker} - \pi_{smoker}$) is between 0.053 and 0.141. Stated more clearly, we are 95% confident that the population proportion of males born is 0.053 to 0.141 higher for families where neither parent smokes as compared to families where both parents smoke. Because 0 is not in this interval, 0 is not a plausible value for the difference in the two population proportions—a conclusion that it is consistent with the result of our significance test above (p-value < 0.001).

Think about it: How would our interval change if we changed the confidence level to 99%?

One advantage of the theory-based approach to finding confidence intervals is that it allows us to easily change the confidence level. We saw in earlier chapters how the level of confidence affects the width of confidence intervals for a single proportion. The same holds true for the difference in two proportions. To be more confident that we have captured the population parameter value (in this case, the difference in population proportions), we would need to make our confidence interval wider, thus including more values that are plausible for the parameter of interest. This would make our confidence level higher. So if we increase the confidence from 95% to 99%, we should see a wider interval. We have done that and the results are shown in Figure 5.14.

Figure 5.14: When we increase the confidence level to 99% we get a wider confidence interval

confidence level [99]   %

(0.0390, 0.1550)

Our 99% confidence interval for the difference in the population proportion of boys born to smokers and boys born to nonsmokers is 0.039 to 0.155. We can rewrite this as 0.097 ± 0.058. This margin of error of 0.058 is larger than the margin of error of 0.044 we had for our 95% confidence interval, and the interval is wider.

Think about it: How would our 95% confidence interval change if we were estimating $\pi_{smoker} - \pi_{nonsmoker}$, instead of $\pi_{nonsmoker} - \pi_{smoker}$?

The 95% confidence interval for $\pi_{smoker} - \pi_{nonmoker}$, based on these data, would be (-0.141, -0.053). Even though the endpoints of the confidence interval are now flipped around and are both negative, they are giving the exact same information as they did in Figure 5.12. We are still 95% confident that the population proportion of a boys born to parents who both smoke is *lower* than that to parents who do not smoke by somewhere between 0.053 and 0.141.

Validity Conditions for Theory-Based Approach (Two Proportions)

Just as we saw in earlier chapters, theory-based approaches that predict the characteristics of the null distribution work well most, but not all, of the time. In Chapter 1 we said that you needed a fairly large sample size in order for the theory-based approaches to work well—a similar fact is true here as well. Conservatively, the theory-based significance test and confidence interval work well if there are at least 10 observations in each of the four cells of the two-way table.

Validity Conditions: The theory-based test and interval for the difference in two proportions (called a *two-sample z-test or interval*) work well when there are at least 10 observations in each of the four cells of the 2×2 table.

For example, we have the smoking data summarized in Table 5.4. Notice that all four values in the cells (255, 310, 1975 and 1627, not any of the totals) are quite a bit larger than 10 (smallest is 255), indicating that the theory-based approaches will generate results very similar to those obtained using simulation and the reported confidence interval should be accurate.

Table 5.4: A two-way table of the smoking data

	Parents smoked at least a pack a day	**Parents both non-smokers**	**Total**
Male	255	1975	2230
Female	310	1627	1937
Total	565	3602	4167

Think about it: What should you do if you are concerned about the validity of the theory-based approach?

Because the theory-based approach is an approximation (prediction) of what would happen if you simulated, if the validity conditions are not met you should simulate to generate a p-value.

Follow-up

As mentioned earlier, the strongly significant result in this study yielded quite a bit of press when it came out (as have other studies about variables related to parental behaviors and children's gender, like one on breakfast cereal choice). Shortly thereafter, however, other studies came out (Parazinni et al., 2004, Obel et al., 2003), which found no such relationship between smoking and gender. Subsequently James (2004) wrote an article arguing that confounding variables such as social factors, diet, environmental exposure or stress were the reason why some were seeing an effect and others weren't. This, of course, is possible, because this wasn't a randomized experiment—just an observational study!

Formulas

Earlier, we saw that we obtained a value of the standardized statistic of 4.30. But how is this value obtained? As we've seen before the value of the standardized statistic, z, is equal to the number of standard deviations the observed statistic is away from the mean (in this case zero). However, in this case, unlike Chapters 1 and 2, the observed statistic is a difference in two proportions, instead of a single proportion, and so the resulting standardized statistic has the form:

$$z = \frac{\text{observed statistic-hypothesized value}}{\text{standard error of statistic}} = \frac{\hat{p}_1 - \hat{p}_2 - 0}{\sqrt{\hat{p}(1 - \hat{p})\left(\frac{1}{n_1} + \frac{1}{n_2}\right)}}$$

where $\hat{p}$ is the overall proportion of successes for the two groups combined (total number of successes divided by the total sample size). We use this overall proportion of success because tests of significance are carried out under the assumption that the null hypothesis (that the two population proportions are the same) is true, so we combine the two groups in order to estimate the(common) population proportion of success. Sometimes you will see this overall proportion of success called a *pooled* proportion of success.

In the example shown earlier, the combined proportion of success is

$\hat{p} = (255+1975)/(3602+565)=2230/4167=0.535$

and so the standardized statistic is:

$$z = \frac{\hat{p}_1 - \hat{p}_2}{\sqrt{\hat{p}(1 - \hat{p})\left(\frac{1}{n_1} + \frac{1}{n_2}\right)}} = \frac{0.548 - 0.451}{\sqrt{0.535(1 - 0.535)\left(\frac{1}{565} + \frac{1}{3602}\right)}} = 4.30$$

This says that the two sample proportions are more than 4 standard errors apart, which is a very large difference, producing a p-value of approximately zero.

Notice that the value for z changes as we would it expect it to. Namely, as the value of the difference between $\hat{p}_1$ and $\hat{p}_2$ increases, z gets farther from zero. Similarly, as the sample size in either (or both) groups increases, the denominator gets smaller and so the value of z gets farther from zero. This means that the strength of evidence for the alternative hypothesis increases as the difference in in the conditional proportions increases and as the sample size increases.

Also, we saw that a 95% confidence interval for the difference in proportions is 0.097 ± 0.044, where 0.097 is the difference in the proportion of successes between the two groups and 0.044 is the margin-of-error. The formula for the margin-of-error is shown below:

$$\text{margin} - of - error = multiplier \times \sqrt{\left(\frac{\hat{p}_1(1 - \hat{p}_1)}{n_1} + \frac{\hat{p}_2(1 - \hat{p}_2)}{n_2}\right)}$$

where the value of the multiplier is dependent upon the confidence level. For 95% confidence intervals we typically use 2 (though 1.96 would be a bit more precise), whereas for 99% confidence intervals the multiplier is 2.576, and for 90% confidence intervals is 1.645. Notice that as the confidence level increases the value of the multiplier also increases, making the width of the confidence interval increase.

Note that, in general, the equation for margin-of-error for most statistical situations is defined as the multiplier $\times$ SE, where SE denotes the standard error of the statistic (the estimated standard deviation of the statistic). This is slightly different than the denominator of the standardized z statistic because we are estimating the size of the difference between the population proportions under the assumption that there is a difference (rather than pooling them together as we did when we assumed the null hypothesis to be true). Thus, when finding a confidence interval for a difference in population proportions,

$$SE = \sqrt{\left(\frac{\hat{p}_1(1-\hat{p}_1)}{n_1} + \frac{\hat{p}_2(1-\hat{p}_2)}{n_2} \right)}.$$

Exploration 5.3: Donating Blood

Blood donations from volunteers serve a crucial role in the nation's health care system. Are Americans any more or less generous about donating blood in some years than others? Are women any more or less willing to give blood than men? To investigate these questions we can analyze data from the General Social Survey, which is a national survey conducted every two years on a nationwide random sample of adult Americans.

1. Did this study make use of random sampling, random assignment, both, or neither? Also explain what your answer means in terms of the scope of conclusions that can be drawn from these data.

Let's start by comparing the years 2002 and 2004.

2. State the appropriate null and alternative hypotheses, in words and in symbols, to address the research question of whether Americans were more generous with blood donations in one of these years than the other.(*Hints*: Remember that hypotheses are always about population parameters. And think about whether the alternative should be one- or two-sided before you see the data.)

Data from the 2002 and 2004 surveys are summarized in the following table:

	2002	2004
Donated blood in previous 12 months	210	230
Did not donate blood in previous 12 months	1152	1106
Total	1362	1336

3. For each year, calculate the (conditional) proportion of adult Americans sampled who donated blood in the previous 12 months. Use appropriate symbols to represent them. Also calculate the difference between these sample proportions (taking the 2002 proportion minus the 2004 proportion).

   2002 proportion:

   2004 proportion:

   Difference (2002 – 2004):

4. Produce a segmented bar graph to display the conditional proportions who gave blood in these two years.Comment on what the graph and your calculations from #3 reveal about these two samples.

5. Use the **Two Proportions** applet to conduct a simulation analysis to approximate a p-value for testing the hypotheses that you stated in #2. You can enter the table of counts in the 2×2 table or type in the table in the Sample data box and press **Use Table**. Or you can copy the data from the text webpage and paste into the Sample data box and press **Use Data**. Verify the segmented bar graph that you created.

Sample data: (2x2: ☑)

	GroupA	GroupB	Totals
Success	210	230	440
Failure	1152	1106	2258

Sample data: (2x2: ☐)

(explanatory,response)

```
          2002 2004
donated   210  230
didn't    1152 1106
```

6. Check the **Show Shuffle Options** box, ask for 1000 repetitions and press **Shuffle**.

    a. Indicate how to find the p-value from the null distribution.(*Hint*: Remember whether the alternative hypothesis is one-sided or two-sided.)

    b. Report the approximate p-value.

    c. Based on this p-value, do the sample data provide much evidence that the population proportion who gave blood in 2002 differs from the population proportion who gave blood in 2004?

7. Does the null distribution of the simulated $\hat{p}_{2002} - \hat{p}_{2004}$ values appear to follow an approximately normal distribution? Centered around what value? Also report the standard deviation of these values.

    Approximately normal?   Mean: SD:

Key idea: Shuffling is an appropriate way to estimate a p-value for comparing groups regardless of the study design (observational study or experiment, random assignment or random sampling or neither). Of course, the study design is very important in determining the appropriate scope of conclusions. (Can you generalize to a larger population? Can you infer a cause-and-effect conclusion?)

---

8. Let's see what the standardized statistic tells us.

    a. Report the observed value in the sample for the statistic $\hat{p}_{2002} - \hat{p}_{2004}$.

    b. How many standard deviations (based on your answer to #7) above or below 0 (the hypothesized value) is the observed value of the statistic?

    c. Is the strength of evidence against the null hypothesis based on this standardized statisitic consistent with what the p-value revealed in #6? Explain.

Just as you learned in Chapters 1 and 2 a theory-based approach for making inferences about a single proportion, so too will you now learn a theory-based approach for making inferences about comparing proportions between two groups.This theory-based approach is again based on an approximation from a normal distribution. Note: Formulas for these methods are given at the end of Example 5.3.

9.   Check the **Overlay Normal distribution** box. Is the behavior of the two-distributions similar?How does the p-values compare?

10. Check the box to Compute 95% CI(s) for proportions. [This box appears in the left hand panel when the difference is selected as the statistic.]
     a.   Report the endpoints of this interval.

     b.   Interpret this interval. Be sure to mention clearly what parameter is being estimated by this interval. (*Hint*: In other words, you are 95% confident that this interval contains what?)

     c.   Does this interval contain only negative values, only positive values, or does the interval include the value zero as well as both positive and negative values? Comment on the importance of what this means.

11. Another applet that can compute p-values and confidence intervals for tests of two proportions is the **Theory-Based Inference** applet. Use the pull-down menu to select **Two proportions** and enter the relevant summary information, using the 2002 sample as group 1. Check the **Test of Significance** box and press on the **inequality** sign button until the alternative hypothesis becomes two-sided. Press **Calculate**. Verify the p-value you obtained earlier (#9), as well as the confidence interval (#10).

Validity Conditions for Theory-Based Approach (Two Proportions)

As with a single proportion, the theory-based inference approach for comparing two proportions is only valid when a normal distribution provides a good approximation to the distribution of the statistic $\hat{p}_1 -$ $\hat{p}_2$ .Conservatively, the theory-based significance test and confidence interval work well when there are at least 10 observations in each of the four cells of the two-way table. In other words, we require at least 10 "successes" and at least 10 "failures" in each sample in order to use this theory-based approach.

Validity Conditions: The theory-based test and interval for the difference in two proportions (called a *two-sample z-test or interval*) work well when there are at least 10 observations in each of the four cells of the 2×2 table.

---

12. Is it valid to use the theory-based approach for two proportions in this study?Justify your answer.

13. Is it reasonable to generalize your conclusions from this analysis to the population of all adult Americans? Explain why or why not.

14. Let's see what will happen if we switched groups 1 and 2 around.
    a. Before doing any calculations, what do you think would happen to the standardized statistic, p-value, and confidence interval if you entered the 2004 sample as group 1 and the 2002 sample as group 2?

        Standardized statistic: p-value:

        Confidence interval:

    b. Make this change in the **Theory-Based Inference** applet (i.e., switch the roles of groups 1 and 2), and then press **Calculate**. Report the new values of the standardized statistic, p-value, and confidence interval.

        Standardized statistic: p-value:

        Confidence interval:

  c. Did these change as you predicted in part a?

Using the General Social Survey data to classify respondents by gender, and combining the years 2002 and 2004, produces the following table:

	Male	Female
Donated blood in previous 12 months	239	201
Did not donate blood in previous 12 months	1032	1226
Total	1271	1427

15. Analyze these data to address the question of whether American men and women differ with regard to donating blood. Include a descriptive analysis (graphs and numbers) as well as an inferential analysis (p-value and confidence interval). Use theory-based approaches, if you decide that the validity conditions are satisfied, for the inferential analysis. If you decide that men and women differ significantly, be sure to estimate by how much they differ. Write a paragraph or two summarizing your analysis and conclusions.

Section 5.3 Summary

Under certain circumstances, essentially when sample sizes are large enough, a theory-based approach can be used to approximate p-values and confidence intervals for the difference between two proportions.

- The standardized statistic again subtracts the hypothesized value from the observed statistic and then divides by the standard deviation (or standard error) of the statistic.

o   The statistic in this case is the difference in sample proportions, that is, $\hat{p}_1 - \hat{p}_2$.

o $$z = \frac{\hat{p}_1 - \hat{p}_2}{\sqrt{\hat{p}(1-\hat{p})\left(\frac{1}{n_1} + \frac{1}{n_2}\right)}}$$ where $\hat{p}$ is the overall observed proportion of successes, both samples combined.

- The confidence interval again takes the form statistic ± margin-of-error
  - o   Statistic $= \hat{p}_1 - \hat{p}_2$
  - o   Margin-of-error = multiplier × standard error of statistic
  - o   Standard error $$SE = \sqrt{\left(\frac{\hat{p}_1(1-\hat{p}_1)}{n_1} + \frac{\hat{p}_2(1-\hat{p}_2)}{n_2}\right)}$$

- Validity conditions: for the theory-based approach to be valid each explanatory variable group should have at least 10 observations in each category of the response variable.

- It does not matter which group is labeled as group 1 and which as group 2, provided that the labeling is consistent throughout the analysis, including interpretations and conclusions.

Type of data	Name of test	Applet	Validity conditions
Two binary variables	Simulation test or interval for two proportions	Two proportions	---
	Theory-based test or interval for two proportions (two sample z test; interval)	Two proportions; Theory-based inference	At least 10 observations in each of the four cells of the two-way table

Chapter 5 Summary

After this chapter, you can now apply the six steps of the Statistical Investigation Method to comparing two groups on a binary response. In analyzing the sample data, examine and describe conditional proportions and segmented bar graphs. If you go on to an inferential analysis, state hypotheses in terms of the difference in the underlying process probabilities or population proportions. To estimate a p-value, use either the Two Proportions applet or, if the sample sizes are large (e.g., at least 10 successes and at least 10 failures in each of the groups), the Theory-Based Inference applet using the Two Proportions pull-down menu. The latter can also be used to find confidence intervals for any confidence level. Remember to interpret this parameter, the difference in population proportions, carefully. When using either applet, remember which category of the explanatory variable is called group 1 and which group 2 is arbitrary; just remember to be consistent and interpret the results accordingly. Finally, consider how the data were collected to decide whether you can draw a cause-and-effect conclusion when the difference is statistically significant, and to what populations you are willing to generalize the results.

In the next chapter, you will repeat this process to compare two groups on a quantitative response variable rather than a categorical response variable. In other words, the next chapter will involve comparing two means rather than two proportions.

Chapter 5 Glossary

Exercises

Section 5.1

5.1.1* Four 2x2 tables, numbered (1) – (4), are shown below. For each one the response is Yes/No and the explanatory variable is A/B.

1	A	B
Yes	10	20
No	10	20

2	A	B
Yes	10	10
No	20	20

3	A	B
Yes	20	10
No	10	20

4	A	B
Yes	20	10
No	20	10

a) Which two tables have the same pair of conditional proportions

b) For which table(s) is the difference (A versus B) in conditional proportions the largest?

c) ... the smallest?

5.1.2 Consider the following two 2×2 tables:

	Treatment A	Treatment B
Success	90	98
Failure	10	2
Total	100	100

	Treatment C	Treatment D
Success	40	48
Failure	60	52
Total	100	100

a) Calculate the proportion of *failures* for all four treatments.

b) Calculate the *difference* in proportions of failure between Treatments A and B. Then do the same calculation to compare Treatments C and D.

c) Calculate the *relative risk* of failure between Treatments A and B. Then do the same calculation to compare Treatments C and D.

d) Which pair of treatments do you think produce responses that are more different: A and B, or C and D?

e) Which statistic reflects your answer to part d): difference in proportions, or relative risk? Explain.

Gender and playing sports*

5.1.3 Fill in the cells in the following table so there is no association between the explanatory and response variables.

		Male	Female	Total
Played sports in high school	Yes			200
	No			160
Total		150	210	360

Drinking and Greek Life

5.1.4 Fill in the cells in the following table so there is no association between the explanatory and response variables.

		Member of Fraternity or Sorority		
		Yes	No	Total
Has one or more alcoholic drinks per week	Yes			152
	No			230
Total		42	340	382

School spirit*

5.1.5 Think about the proportion of students at your college who are wearing clothing that displays the college name or logo today. Also suppose that a friend of yours attends a different college and the two of you have a recurring discussion about which college displays more school pride. You decide to measure school pride by the proportion of students at the college who wear clothing that displays the college name or logo on a particular day. You want to investigate whether this proportion differs between your college (call it Exemplary Students University, ESU) and your friend's college (call it Mediocre Students University, MSU).

a) What are the observational units in this study?

b) What is the response variable in this study? Is it categorical or quantitative?

c) Would you use random sampling to select the observational units? If so, explain why and how.

d) Would you use random assignment to create the two groups to compare? If so, explain why and how.

e) Draw and label a (blank) 2×2 table suitable for summarizing the data.

Medicare audits

5.1.6 The U.S. government authorizes private contractors to audit bills paid by Medicare and Medicaid. The contractor audits a random sample of paid claims and judges each claim to be either fully justified or an overpayment. Here is a 2×2 table that summarizes data from one such audit. (One of the authors served as a statistical consultant in connection with this audit. For reasons of confidentiality we cannot identify the health care provider.) For this audit, all claims were divided into two sub-populations according to amount of the claim, small or medium. Two simple random samples were chosen, 30 small claims and 30 medium claims. We want to answer the question, "Does the chance that a claim is judged to be an overpayment depend on the size of the claim?"

	Small	Medium	Total
Fully justified	16	22	38
Overpayment	14	8	22
Total	30	30	60

a) What is the response variable? What is the explanatory variable?

b) What are the observational units for this study?

c) As for any 2×2 table, there are two pairs of conditional proportions. For this table, the two pairs are 14/30 versus 8/30 and 16/38 versus 14/22. Which pair corresponds to the question of interest?

Rattlesnakes*

5.1.7 The table below shows the numbers of male and female rattlesnakes caught at two different sites, B and G. Assume that the snakes caught at a site can be regarded as a random sample from the population of all snakes at the site.

	Site B	Site G	Total
Female	11	12	23
Male	10	21	31
Total	21	33	54

We want to know whether there is an association between site and sex. Site is the explanatory variable, and sex is the response variable.

    a) Explain why we use conditional proportions to evaluate the association.

    b) Explain why we look at the proportion of females and proportion of males, conditional on the site, rather than the proportion from each site conditional on the sex.

    c) Reasoning informally, do you think the data show evidence of an association between Site and sex? Explain.

U.S. Senators

5.1.8 The 2014 U.S. Senate consists of 80 men and 20 women. The 2014 U.S. Senate consists of 45 Republicans and 55 Democrats (counting 2 Independents as Democrats because they vote with Democrats more than with Republicans). What additional information do you need in order to produce a 2×2 table and investigate whether there is an association between sex and party?

5.1.9 Reconsider the previous exercise, along with this additional information: 16 Senators are women Democrats.

    a) Produce a 2×2 table that classifies Senators in the 2014 U.S. Senate according to party and gender.

Provide the relevant proportion(s) to answer the following questions:

    b) Are most Senators Republicans?

    c) Are most Democrat Senators women?

    d) Are most women Senators Democrats?

5.1.10 Reconsider the previous two exercises.

    a) Produce a well-labeled segmented bar graph to display the association between party and gender.

    b) Does the graph reveal that there is an association between party and gender in the 2014 U.S. Senate? If so, describe the association.

5.1.11 Reconsider the previous three exercises, but now ignore the real data. Invent your own fictional data that shows *no association* between party and gender. Provide the 2×2 table, calculate conditional proportions, and produce a segmented bar graph. Also explain how you invented the data to make sure that there would be no association between the variables.

Church and Smoking*

5.1.12 A Gallup poll conducted in 2013 examined whether there is an association between church attendance and smoking status among American adults. The following table reports the percentage of people who classify themselves as smokers, conditional on self-described level of church attendance.

At least once per week	Almost every week	About once a month	Seldom	Never
12%	14%	22%	25%	30%

  a) Is this an observational study or an experiment? Explain how you can tell.
  b) Identify the explanatory variable and the response variable.
  c) Produce an appropriate graph to display these percentages.
  d) Summarize what these percentages reveal with regard to the question of whether there is an association between church attendance and smoking status among American adults.
  e) Would you conclude from these data that attending church more often *causes* a decrease in smoking? Explain.

5.1.13 Reconsider the previous exercise.
  a) Calculate the relative risk of being a smoker, comparing those never attend church to those who attend church at least once per week. Also write a sentence interpreting what this calculation reveals.
  b) Repeat a) for the other three categories of church attendance, comparing each to the "at least once a week" category.

Breaking up digitally

5.1.14 A Pew Research study in April and May of 2013 asked single American adults whether they have ever broken up with someone by email, text, or online message. Consider the following 2×2 table of counts:

	Female	Male	Total
Has broken up with someone by digital means	52	55	107
Has NOT broken up with someone by digital means	237	309	546
Total	289	364	653

a) Notice that there are more males (55) than females (52) who have broken up with someone by digital means. Explain why this comparison is not very useful.

b) Suggest a better comparison for investigating whether men or women are more likely to break up with someone by digital means.

c) Perform the calculation that you suggest in b).

d) Comment on what your calculation in c) reveals.

5.1.15 Reconsider the previous exercise. The Pew study also considered age as an explanatory variable, producing the following 2×2 table of counts:

	Age 18-29	Age 30-64	Total
Has broken up with someone by digital means	54	47	101
Has NOT broken up with someone by digital means	191	316	507
Total	245	363	608

Analyze these data with appropriate numerical summaries and graphical displays. Write a paragraph summarizing what you learn about whether there is an association between age and whether or not a person has broken up with someone by digital means.

HIV and AZT*

5.1.16 In a 1994 study, 164 pregnant, HIV-positive women were randomly assigned to receive the drug AZT during pregnancy and 160 such women were randomly assigned to a control group that received a placebo. It turned out that 40 of the mothers in the control group gave birth to babies who were HIV-positive, compared to only 13 in the AZT group.

a) Is this an observational study or an experiment?

b) Identify the explanatory and response variables.

c) Produce a 2×2 table of counts, with the explanatory variable in columns.

d) Calculate the proportion of babies who were born HIV-positive in each group.

e) Calculate the relative risk of a baby being born HIV-positive, comparing those in the placebo group to those in the AZT group.

f) Produce a segmented bar graph to display these data.

g) Summarize what these data reveal about the question of whether AZT is effective for reducing the HIV-positive proportion of new-born babies?

Pet CPR

5.1.17 According to a survey conducted by the Associated Press and petside.com in 2009, 63% of dog owners and 53% of cat owners would be at least somewhat likely to give CPR to their pet in the event of a medical emergency. The survey involved a nationwide sample of 1166 pet owners.

a) Produce a well-labeled segmented bar graph to display these percentages.
b) Identify the two categorical variables displayed in the graph.
c) What additional information would you need in order to construct a 2×2 table?

5.1.18 Reconsider the previous exercise. The same survey found that 65% of women and 50% of men would give CPR to their pets. Answer questions a)-c) from the previous exercise for these data.

Student Data*

5.1.19 College students were asked to indicate their gender, whether or not they drink coffee, and what letter they would use to complete the four-letter word beginning with F A I __ . Results are given in the following data table (* indicates a missing value):

ID	sex	Coffee?	Letter?		ID	sex	Coffee?	Letter?
1	Female	Yes	R		14	Male	Yes	R
2	Female	Yes	R		15	Female	No	R
3	Male	No	L		16	Male	No	R
4	Female	No	*		17	Male	Yes	L
5	Female	Yes	R		18	Female	No	L
6	Male	Yes	R		19	Female	No	L
7	Male	No	L		20	Male	Yes	R
8	Male	Yes	R		21	Male	No	R
9	Female	No	*		22	Female	No	L
10	Female	No	*		23	Female	Yes	R
11	Female	No	R		24	Female	No	R
12	Female	Yes	R		25	Female	No	L
13	Male	No	L		26	Female	Yes	L

a) Produce a 2×2 table of counts for investigating whether the data suggest any association between gender and coffee drinking.

b) Calculate and compare the proportions of coffee drinkers between male and female students.

c) Produce a segmented bar graph to display the data from part a).

d) Summarize what you conclude about whether the data suggest any association between gender and coffee drinking.

5.1.20 Reconsider the data from the previous exercise. Now investigate whether there is any association between gender and the letter chosen to complete the four-letter word.

5.1.21 Reconsider the data from the previous two exercises. Now investigate whether there is any association between coffee drinking and the letter chosen to complete the four-letter word.

Trusting People

5.1.22 The following segmented bar graph was produced from data collected in the 2010 General Social Survey, with regard to the question "Can people be trusted?"

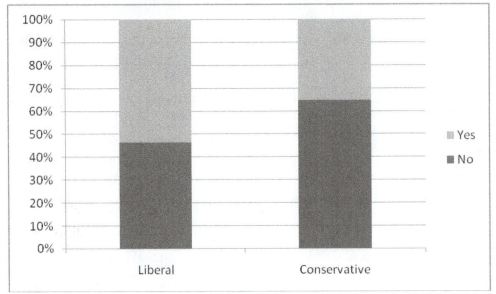

a) About what proportion of the liberals responded that people can be trusted?

b) About what proportion of the conservatives responded that people can be trusted?

c) Does the graph provide any information about what proportion of respondents were liberals?

d) Does the graph suggest that there is a relationship between political leanings and trust? Explain.

Simpson's Paradox at the hospital

5.1.23 The following two-way table classifies hypothetical hospital patients with a certain disease, according to the hospital that treated them and whether they survived or died:

	Hospital A	Hospital B
Survived	800	900
Died	200	100
Total	1000	1000

a) Calculate the proportion of hospital A's patients who survived and the proportion of hospital B's patients who survived. Which hospital saved the higher percentage of its patients?

Suppose that we further classify each patient according to a third variable: whether they were in fair condition or poor condition prior to treatment. We obtain the following two-way tables:

fair condition:

	Hospital A	Hospital B
Survived	590	870
Died	10	30

poor condition:

	Hospital A	Hospital B
Survived	210	30
Died	190	70

b) Show that when the fair and poor condition patients are combined, the totals are indeed those given in the table above.

c) Among those in *fair* condition, compare the recovery rates for the two hospitals. Which hospital saved the greater percentage of its patients who had been in fair condition?

d) Among those in *poor* condition, compare the recovery rates for the two hospitals. Which hospital saved the greater percentage of its patients who had been in poor condition?

This phenomenon is called **Simpson's paradox**: an association or comparison that holds for all of several groups can reverse direction when the data are merged to form a single group.

e) Explain (arguing from the data given) the apparent paradox here. [*Hints*: Do fair or poor patients tend to survive more often? Does one hospital tend to treat one type of patient?]

f) Which hospital would you rather go to if you were sick with this disease? Explain.

Simpson's paradox at Berkeley*

5.1.24 In 1973, a lawsuit was filed against the University of California at Berkeley alleging sex discrimination in its graduate admissions policies. The following table pertains to two of the graduate programs at the university. For each program, it lists the number of men accepted, the number of men denied, the number of women accepted, and the number of women denied:

	Men accepted	Men denied	Women accepted	Women denied
Program A	511	314	89	19
Program F	22	351	24	317
Combined	533	665	113	336

a) Start with the combined data. Determine the proportion of men applicants who were accepted, and compare this to the proportion of women applicants who were accepted. Which proportion is larger?

b) Now consider only Program A. Did men or women have a higher proportion of applicants who were accepted? Provide calculations to justify your answer.

c) Repeat b) for Program F.

d) Is this an example of Simpson's paradox (see the previous exercise)? Explain.

e) Explain how this paradox happens, based on the data, as if to someone who doesn't know much about Statistics.

Simpson's paradox in Softball

5.1.25 Show that it is possible for Simpson's paradox (see Exercise 5.1.23) to occur with two softball players. Produce a hypothetical example where one player (Amy) has a higher percentage of hits than another (Barb) in June, and Amy also has a higher percentage of hits than Barb in July. Yet when June and July are combined, Barb has a higher percentage of hits than Amy. (Hints: Be sure to give the two players different numbers of at-bats in the two months, and make one month easier to get a hit than the other.)

Section 5.2 Exercises

Medicare Audits*

5.2.1: Reconsider the exercise in section 5.1 about Medicare audits (5.1.6). Of the 30 small claims, 14 were judged to be overpayments; of the 30 medium claims, 8 were judged to be overpayments. Our research question is this "Does the chance that the claim is judged to be an overpayment tend to differ by the size of the claim?"

a) Which of the following is the appropriate null hypothesis? **Circle one.**

    → A. Small claims are more likely to be judged overpayments compared to medium claims

      B. Small claims are less likely to be judged overpayments compared to medium claims

      (C.) Small and medium claims are equally likely to be judged overpayments

    ×D. Small and medium claims are not equally likely to be judged overpayments

b) Which of the following is the appropriate alternative hypothesis? **Circle one.**

      A. Small claims are more likely to be judged overpayments compared to medium claims

      B. Small claims are less likely to be judged overpayments compared to medium claims

    ⁺ C. Small and medium claims are equally likely to be judged overpayments

    (D.) Small and medium claims are not equally likely to be judged overpayments

c) In this context, which of the following is an (are) appropriate statistic value(s) to compare small to medium claims? **Circle all that apply.**

    A. $14/60 - 8/60 = 0.10$

    (B.) $14/30 - 8/30 = 0.20$

    C. $14/22 - 16/38 = 0.22$

    D. $(14/30)/(8/30) = 1.75$

Suppose that we want to use cards to carry out a randomization-test of the appropriate hypotheses.

d) How many cards will we need? **Circle one.**

    A. 30

    B. 52

    (C.) 60

    D. Doesn't matter

e) We need cards of two different colors. Let's say we decide to use red and black cards. Which of the following is an appropriate combination of red and black cards? **Circle all that apply.**

    A. 14 red and 8 black

    B. 30 red and 30 black

    C. 22 red and 38 black

    D. 38 red and 22 black

f) You shuffle the stack of red and black cards, and deal them into 2 piles. How many cards should you place in each pile? Why?

g) What statistic should you record after you have shuffled and dealt the cards into two piles? (*Hint:* There is more than one correct answer here.)

h) Suppose that you have repeated the shuffle-and-deal many times, and recorded the appropriate statistic every single time. What should you do next to find the p-value?

5.2.2 Reconsider the previous question about Medicare audits. A statistics student decides to use the Two Proportions applet to find a p-value to investigate whether there is an association between the size of a claim and whether it is judged to be an overpayment. Here is the output:

a) The above graph is a dotplot of 1000 possible values of difference in proportion of claims judged to be overpayments between small and medium sized claims. At what numeric value does this graph center? Explain why that makes sense.

b) The p-value is computed to be 0.184. Shade the region on the above graph that represents the p-value.

c) Interpret the p-value reported in part b) in the context of the study.

d) Based on the p-value what would you conclude about strength of evidence?

e) Calculate an appropriate standardized statistic, and interpret the standardized statistic in the context of the study.

f) Do the p-value and standardized statistic both lead you to the same conclusion?

g) Use the 2SD Method to find a 95% confidence interval for the parameter of interest. Report this interval.

h) Interpret the interval reported in part g) in the context of the study.

i) Is your conclusion from using either the p-value or the standardized statistic consistent with your finding from the 95% confidence interval? How are you deciding?

j) State a complete conclusion about this study including Significance, Estimation (confidence interval), Causation and Generalization. Be sure to explain how you are arriving at your conclusion.

5.2.3: Reconsider the exercise in section 5.1 about school spirit (exercise 5.1.5). Suppose that you find that 24% of a random sample of students at ESU wear clothing with the school name or logo, compared to 16% of a random sample of students at MSU. Our research question: *Is the observed difference in percentage of students wearing school clothing at ESU versus MSU statistically significant?*

a) Which of the following is the appropriate null hypothesis? **Circle one.**

    A. In general, a greater percentage of ESU students wear school clothing compared to MSU students.

    B. In general, a smaller percentage of ESU students wear school clothing compared to MSU students.

    C. In general, the percentage of ESU students who wear school clothing is the same as that of MSU students.

    D. In general, the percentage of ESU students who wear school clothing is the different from that of MSU students.

b) Which of the following is the appropriate alternative hypothesis? **Circle one.**

    A. In general, a greater percentage of ESU students wear school clothing compared to MSU students.

    B. In general, a smaller percentage of ESU students wear school clothing compared to MSU students.

    C. In general, the percentage of ESU students who wear school clothing is the same as that of MSU students.

    D. In general, the percentage of ESU students who wear school clothing is the different from that of MSU students.

c) In this context, calculate an appropriate statistic value to compare ESU students to MSU students, with regard to wearing school clothing.

d) What additional information do you need in order to assess whether this difference calculated in part c) is statistically significant?

e) Describe a scenario, involving the additional information you cited in part d), in which you believe that this difference would be statistically significant.

f) Describe a scenario, involving the additional information you cited in part d), in which you believe that this difference would *not* be statistically significant.

Surgery or not

5.2.4 ResearchersWilt et al. (*New England Journal of Medicine* (2012)) investigated whether surgery, compared to just observation, was (more) effective in improving men's survival chances after being diagnosed with prostate cancer. The researchers identified 731 men with localized prostate cancer who volunteered to participate. They randomly assigned 364 men to surgery, and the remaining 367 to observation. All participants were followed for about 10 years. In those 10 years, 21 surgery recipients died of prostate cancer related reasons compared to 31 observation recipients.

a) Identify the observational units. **Circle one.**

   A. Men

   B. Men with prostate cancer

   C. Men with prostate cancer who underwent surgery

   D. Men with prostate cancer who received observation

b) What type of study is this? **Circle one.**

   A. Experiment

   B. Observational study

c) What is the <u>primary purpose</u> of random assignment in this type of study? **Circle one.**

   A. To ensure that subjects are representative of the population of interest

   B. To ensure that the groups are of equal sizes

   C. To create treatment groups are alike in all aspects except for the treatment administered

   D. To improve accuracy of results

d) Identify the explanatory variable. **Circle one.**

   A. Whether or not man dies of prostate cancer related reasons

   B. Whether or not man undergoes surgery

   C. Percentage of men who die of prostate cancer related reasons

   D. The number of men who undergo surgery, and the number of men who are just observed

e) Identify the response variable. **Circle one.**

   A. Whether or not man dies of prostate cancer related reasons

   B. Whether or not man undergoes surgery

   C. Percentage of men who die of prostate cancer related reasons

   D. The number of men who undergo surgery, and the number of men who are just observed

f) The explanatory variable is _____ (pick one: CATEGORICAL   QUANTITATIVE)

g) The response variable is _____ (pick one: CATEGORICAL   QUANTITATIVE)

h) In this context, which of the following is an (are) appropriate statistic value(s)? **Circle all that apply.**

   A. $21/731 - 31/731 = -0.014$

   B. $21 - 31 = -10$

   C. $21/364 - 31/367 = -0.027$

   D. $(21/364)/(31/367) = 0.68$

i) Give two possible explanations for the observed statistic value(s) as selected in (h).

5.2.5 Reconsider the data from the previous exercise about investigating whether there is a relationship between undergoing surgery and whether a man dies due to prostate cancer related reasons.

a) Which of the following is the appropriate null hypothesis? **Circle one.**

   A. Surgery recipients are equally likely as those just under observation to die of prostate cancer related reasons

   B. Surgery recipients are less likely than those just under observation to die of prostate cancer related reasons

   C. Surgery recipients are more likely than those just under observation to die of prostate cancer related reasons

   D. Surgery recipients are *not* equally likely as those just under observation to die of prostate cancer related reasons

b) Which of the following is the appropriate alternative hypothesis? **Circle one.**

   A. Surgery recipients are equally likely as those just under observation to die of prostate cancer related reasons

   B. Surgery recipients are less likely than those just under observation to die of prostate cancer related reasons

   C. Surgery recipients are more likely than those just under observation to die of prostate cancer related reasons

   D. Surgery recipients are *not* equally likely as those just under observation to die of prostate cancer related reasons

Suppose that we want to use a tactile simulation to carry out a randomization-test of the appropriate hypotheses.

c) In total, how many cards will we need?

d) We need cards of different colors. How many colors do we need? How many cards of each color do we need? Explain.

e) We will shuffle the stack of cards, and deal them into multiple piles. How many piles should we make, and how many cards should we place in each pile? Why?

f) What statistic should we record after we have shuffled and dealt the cards into two piles? (*Hint:* There is more than one correct answer here.)

g) Suppose that we have repeated the shuffle-and-deal many times, and recorded the appropriate statistic every single time. What should we do next to find the p-value?

h) Which of the following is the purpose of shuffling and dealing cards into two piles repeatedly, and recorded a chosen statistic value for each repetition?

   A. To generate more samples

   B. To improve accuracy of the study

   C. To generate values of the statistic that could have happened by chance alone

   D. All of the above

5.2.6 Reconsider the data from the previous exercise about investigating whether there is a relationship between undergoing surgery and whether a man dies due to prostate cancer related reasons.

a) Organize the counts in a well-labeled 2×2 table.

b) Use an appropriate randomization-based applet to find a p-value. Report <u>and</u> interpret the p-value, in the context of the study.

c) Based on this p-value, how much evidence do you have against the null hypothesis?

d) Calculate an appropriate standardized statistic, and interpret the standardized statistic in the context of the study.

e) Do the p-value and standardized statistic both lead you to the same conclusion?

f) Use the 2SD Method to find a 95% confidence interval for the parameter of interest. Report this interval, and interpret the interval in the context of the study.

g) Is your conclusion from using either the p-value or the standardized statistic consistent with your finding from the 95% confidence interval? How are you deciding?

h) State a complete conclusion about this study including Significance, Estimation (confidence interval), Causation and Generalization. Be sure to explain how you are arriving at your conclusion.

Chest-Compression-only or Standard Cardiopulmonary Resuscitation*

5.2.7 To investigate whether giving chest-compression-only (CC) instructions rather than standard cardiopulmonary resuscitation (CPR) instructions, to the witness of a heart attack will improve the victim's chance of surviving, researchers Hupfl et al. (*The Lancet,* 2010) combined the results from three randomized experiments. In each experiment, the emergency services dispatcher randomly assigned either CC or CPR instructions to the bystander who was at the site where a person had just experienced a heart attack. They found that of the 1500 cases where CC instructions had been given, 211 people had survived, whereas out of the 1531 cases where standard CPR instructions had been given, the number was 178.

Here are the data organized in a 2x2 table:

		CC	CPR	Total
Survived?	Yes	211	178	389
	No	1289	1353	2642
Total		1500	1531	3031

a) Identify the observational units. **Circle one.**

    A. Whether given CC or CPR instructions

    – B. Whether heart attack victim survived or not

    (C.) Cases of heart attacks where emergency services were contacted

    D. Those people who survived

b) What is the <u>primary purpose</u> of random assignment in this type of study? **Circle one.**

    A. To ensure that subjects are representative of the population of interest

    B. To ensure that the groups are of equal sizes

    (C.) To create treatment groups are alike in all aspects except for the treatment administered

    D. To improve accuracy of results

c) Identify the explanatory variable. **Circle one.**

    A. Whether given CC or CPR instructions

    – B. Whether heart attack victim survived or not

    (C.) Cases of heart attacks where emergency services were contacted

    D. Those people who survived

d) Identify the response variable. **Circle one.**

    –A. Whether given CC or CPR instructions

B. Whether heart attack victim survived or not

C. Cases of heart attacks where emergency services were contacted

D. Those people who survived

e) The explanatory variable is _____ (pick one: CATEGORICAL   QUANTITATIVE)

f) The response variable is _____ (pick one: CATEGORICAL   QUANTITATIVE)

g) In this context, which of the following is an (are) appropriate statistic value(s)? **Circle all that apply.**

A. $211/1500 - 178/1531 = 0.024$

B. $211/3031 - 178/3031 = 0.011$

C. $211/389 - 178/389 = 0.085$

D. $(211/1500)/(178/1531) = 1.21$

h) Give two possible explanations for the observed statistic value(s) as selected in (g).

5.2.8 Reconsider the data from the previous exercise about investigating whether giving chest-compression-only (CC) instructions rather than standard cardiopulmonary resuscitation (CPR) instructions, to the witness of a heart attack will improve the victim's chance of surviving.

a) Which of the following is the appropriate null hypothesis? **Circle one.**

A. Giving CC instructions is just as effective as giving CPR instructions at saving lives of heart attack victims.

B. Giving CC instructions is more effective than giving CPR instructions at saving lives of heart attack victims.

C. Giving CC instructions is less effective than giving CPR instructions at saving lives of heart attack victims.

b) Which of the following is the appropriate alternative hypothesis? **Circle one.**

A. Giving CC instructions is just as effective as giving CPR instructions at saving lives of heart attack victims.

B. Giving CC instructions is more effective than giving CPR instructions at saving lives of heart attack victims.

C. Giving CC instructions is less effective than giving CPR instructions at saving lives of heart attack victims.

Suppose that we want to use cards to carry out a tactile simulation to investigate whether giving chest-compression-only (CC) instructions rather than standard cardiopulmonary resuscitation (CPR) instructions, to the witness of a heart attack will improve the victim's chance of surviving.

c) In total, how many cards will we need?

d) We need cards of different colors. How many colors do we need? How many cards of each color do we need? Explain.

e) We shuffle the stack of cards, and deal them into multiple piles. How many piles should we make, and how many cards should we place in each pile? Why?

f) What statistic should you record after we have shuffled and dealt the cards into two piles? (*Hint:* There is more than one correct answer here.)

g) Suppose that we have repeated the shuffle-and-deal many times, and recorded the appropriate statistic every single time. What should we do next to find the p-value?

h) Which of the following is the purpose of shuffling and dealing cards into two piles repeatedly, and recorded a chosen statistic value for each repetition?

    A. To generate more samples

    B. To improve accuracy of the study

    C. To generate values of the statistic that could have happened by chance alone

    D. All of the above

5.2.9 Reconsider the data from the previous exercise about investigating whether giving chest-compression-only (CC) instructions rather than standard cardiopulmonary resuscitation (CPR) instructions, to the witness of a heart attack will improve the victim's chance of surviving.

a) Use an appropriate randomization-based applet to find a p-value. Report <u>and</u> interpret the p-value, in the context of the study.

b) Based on this p-value, how much evidence do you have against the null hypothesis?

c) Calculate an appropriate standardized statistic, and interpret the standardized statistic in the context of the study.

d) Do the p-value and standardized statistic both lead you to the same conclusion?

e) Use the 2SD Method to find a 95% confidence interval for the parameter of interest. Report this interval, and interpret the interval in the context of the study.

f) Is your conclusion from using either the p-value or the standardized statistic consistent with your finding from the 95% confidence interval? How are you deciding?

g) State a complete conclusion about this study including Significance, Estimation (confidence interval), Causation, and Generalization. Be sure to explain how you are arriving at your conclusion.

5.2.10 Reconsider the context of investigating whether giving chest-compression-only (CC) instructions rather than standard cardiopulmonary resuscitation (CPR) instructions, to the witness of a heart attack will improve the victim's chance of surviving. One of the randomized experiments that the researchers Hupfl et al. (*The Lancet,* 2010) looked at was by Halls trom et al. (*New England Journal of Medicine,* 2000) where there were 35 survivors in the CC group of 240, and 29 survivors in the CPR group of 278.

a) Find the observed difference in proportion of survivors comparing those received CC to those who received CPR.

b) Recall that in Hupfl et al. study where data from three studies were combined, there were 211 survivors in the CC group of 1500, and 178 survivors in the CPR group of 1531. Note that such a study is called a meta-analysis. Find the observed difference in proportion of survivors comparing those received CC to those who received CPR in Hupfl et al.'s meta-analysis.

c) Compare the answers from part a) and part b). Which is bigger?

d) Use an appropriate randomization-based applet to find a p-value for Hallstrom et al.'s data. Report and interpret the p-value, in the context of the study.

e) Based on this p-value, how much evidence do you have against the null hypothesis?

f) Compare the p-value from part d) to that from Hupfl et al.'s study. Which provides stronger evidence against the null hypothesis? How are you deciding?

g) You should have found that even though the observed difference in proportion of survivors is greater for Hallstrom et al.'s study compared to Hupfl et al.'s, the p-value is smaller for Hupfl et al. Explain why that makes sense.

h) Use the 2SD Method to find a 95% confidence interval for the parameter of interest based on Hallstrom et al.'s data. Report this interval, and compare the midpoint and width of this interval to that based on Hupfl et al.'s data. Do the differences between the midpoints and widths of the two intervals make sense? Why?

Prenatal care for mothers with gestational diabetes

5.2.11 A baby weighing more than 4000g at birth is considered to be Large for Gestational Age (LGA). Gestational diabetes in the mother is believed to be a common risk factor for LGA. In an article published in the *New England Journal of Medicine* (October 2009), researchers Landon et al. reported a study of 958 women diagnosed with mild gestational diabetes between 24 and 31 weeks of pregnancy who volunteered to participate in the study, and then were randomly assigned to one of two groups: 473 to usual prenatal care (control group), and 485 to dietary intervention, self-monitoring of blood glucose, and insulin therapy, if necessary (treatment group). Of the 473 women in the control group, 68 had babies who were LGA, and of the 485 women in the treatment group, 29 had babies who were LGA.

a) Was this an experiment or an observational study? Explain how you are deciding.

b) Identify the observational units.

c) Identify the explanatory and the response variables. Also, for each variable, be sure to identify the kind of variable (categorical or quantitative).

d) Organize the data in a well-labeled 2×2 table. Also create a segmented bar chart with well-labeled axis to display the data. Discuss what the segmented bar chart shows with regard to a relationship between the type of prenatal care a woman with gestational diabetes receives, and whether or not she has an LGA baby.

e) Find the observed difference in the conditional proportion of mothers who had LGA babies between those who received the usual prenatal care and those received specialized care.

f) Give two possible explanations for the observed difference in conditional proportions as reported in (e).

5.2.12 Reconsider the data from the previous study about the effect of specialized prenatal care for women with gestational diabetes. Do the data provide evidence of a relationship between the type of prenatal care a woman with gestational diabetes receives, and whether or not she has an LGA baby?

a) Define (in words) the parameters of interest of this study. Also, assign symbols to the parameters.

b) State the appropriate null and the alternative hypotheses in words.

c) State the null and the alternative hypotheses in symbols.

d) Give detailed, step-by-step instructions on how one could conduct a tactile simulation to generate a p-value to test whether the observed difference in the conditional proportion of mothers who had LGA babies between those who received the usual prenatal care and those received specialized care is statistically significant. Be sure to include details on the following:

- Would the simulation involve coins, or dice, or index cards?
- How many tosses, or rolls, or cards would be used?
- How many sets of tosses, or rolls, or shuffles would you observe?
- What would you record after every repetition?
- How would you compute the p-value?

5.2.13 Reconsider the data from the previous study about the effect of specialized prenatal care for women with gestational diabetes. Do the data provide evidence of a relationship between the type of prenatal care a woman with gestational diabetes receives, and whether or not she has an LGA baby?

   a) Use an appropriate randomization-based applet to find a p-value to test whether the observed difference in the conditional proportion of mothers who had LGA babies between those who received the usual prenatal care and those received specialized care is statistically significant. Report <u>and</u> interpret the p-value, in the context of the study.

   b) Based on this p-value, how much evidence do you have against the null hypothesis?

   c) Calculate an appropriate standardized statistic, and interpret the standardized statistic in the context of the study.

   d) Do the p-value and standardized statistic both lead you to the same conclusion?

   e) Use the 2SD Method to find a 95% confidence interval for the parameter of interest. Report this interval, and interpret the interval in the context of the study.

   f) Is your conclusion from using either the p-value or the standardized statistic consistent with your finding from the 95% confidence interval? How are you deciding?

   g) State a complete conclusion about this study including Significance, Estimation (confidence interval), Causation, and Generalization. Be sure to explain how you are arriving at your conclusion.

5.2.14 Reconsider the data from the previous study about the effect of specialized prenatal care for women with gestational diabetes. Do the data provide evidence of a relationship between the type of prenatal care a woman with gestational diabetes receives, and whether or not she has an LGA baby?

   a) Calculate the observed value of the relative risk of having an LGA baby for the pregnant women who received usual prenatal care compared to those who received specialized prenatal care.

   b) State, the null and alternative hypotheses in the context of relative risk of having an LGA baby.

   c) Use an appropriate randomization-based applet to find a p-value to test whether the observed relative risk (as calculated in part a)) is statistically significant. Report <u>and</u> interpret the p-value, in the context of the study. (*Note*: Make sure to change the statistic to "Relative risk" when using the applet.)

d) If you have a p-value from using difference in conditional proportions as your statistic (from the previous exercise), compare that p-value to the one reported in part c). Are they close? Explain why that makes sense.

Risk taking*

5.2.15: In the Chapter 4 exercises you read about a research article "Power Posing: Brief Nonverbal Displays Affect Neuroendocrine Levels and Risk Tolerance," that was published in *Psychological Science,* September 2010, and describes an experiment involving 42 participants (male and female), where 20 participants were randomly assigned to hold low-power poses (contractive positions, closed limbs), and 22 to hold high-power poses (expansive positions, open limbs) for 2 minutes. All participants were told that the aim of this exercise was to see whether their heart rate changed. After the exercise, each participant was given $2, and told that they could keep the money, or roll a die for a double or nothing. Here are the data:

		Low power pose	High-power pose	Total
Took the "double or nothing" bet?	Yes	3	18	21
	No	17	4	21
Total		20	22	42

Research question: Are people who hold "high-power poses" more likely to take risks (such as, the double or nothing bet) compared to those hold "low-power poses"?

a) Find the observed difference in the conditional proportion of bet takers between those who held low power poses and those who held high power poses.

b) Give two possible explanations for the observed difference in conditional proportions as reported in (a).

5.2.16 Reconsider the data from the study investigating whether people who hold "high-power poses" more likely to take risks (such as, the double or nothing bet) compared to those hold "low-power poses." For a statistic use the difference in the conditional proportion of bet takers between those who held low power poses and those who held high power poses.

a) Define (in words) the parameters of interest of this study. <u>Also, assign symbols</u> to the parameters.

b) State the null and the alternative hypotheses in words.

c) State the null and the alternative hypotheses in symbols.

d) Give detailed, step-by-step instructions on how one could conduct a tactile simulation to generate a p-value to test the hypotheses stated in (a) and (b). Be sure to include details on the following:

- Would the simulation involve coins, or dice, or index cards?
- How many tosses, or rolls, or cards would be used?
- How many sets of tosses, or rolls, or shuffles would you observe?
- What would you record after every repetition?
- How would you compute the p-value?

5.2.17 Reconsider the data from the study investigating whether people who hold "high-power poses" more likely to take risks (such as, the double or nothing bet) compared to those hold "low-power poses." For this exercise, for a statistic use the difference in the conditional proportion of bet takers between those who held low power poses and those who held high power poses.

a) Use an appropriate randomization-based applet to find a p-value. Report <u>and</u> interpret the p-value, in the context of the study.

b) Based on this p-value, how much evidence do you have against the null hypothesis?

c) Calculate an appropriate standardized statistic, and interpret the standardized statistic in the context of the study.

d) Do the p-value and standardized statistic both lead you to the same conclusion?

e) Use the 2SD Method to find a 95% confidence interval for the parameter of interest. Report this interval, and interpret the interval in the context of the study.

f) Is your conclusion from using either the p-value or the standardized statistic consistent with your finding from the 95% confidence interval? How are you deciding?

g) State a complete conclusion about this study including Significance, Estimation (confidence interval), Causation and Generalization. Be sure to explain how you are arriving at your conclusion.

5.2.18 Reconsider the data from the study investigating whether people who hold "high-power poses" more likely to take risks (such as, the double or nothing bet) compared to those hold "low-power poses."

a) Calculate the observed value of the relative risk of taking the bet for those who held high-power poses compared to those who held low-power poses.

b) State your null and alternative hypotheses in the context of relative risk of taking a bet.

c) Use an appropriate randomization-based applet to find a p-value to test whether the observed relative risk (as calculated in part a)) is statistically significant. Report <u>and</u> interpret the p-value, in the context of the study. Note: Make sure to change the statistic to "Relative risk" when using the applet.

d) If you have a p-value from using difference in conditional proportions as your statistic (from the previous exercise), compare that p-value to the one reported in part c). Are they close? Explain why that makes sense.

Want change?

5.2.19 A 2013 Gallup poll asked randomly selected U.S. adults whether they wanted to stay at their current body weight or change. One purpose was to investigate whether there was any difference between men and women with regard to this aspect. Of the 562 men surveyed, 242 wanted to stay at their current weight, whereas of the 477 women surveyed, 172 wanted to stay at their current weight.

a) Is this an experiment or an observational study? Explain how you are deciding.

b) Organize the data in an appropriate well-labeled 2×2 table.

c) Find the observed difference in the conditional proportion of those wanted to stay at their current weight between men and women.

d) Give two possible explanations for the observed difference in conditional proportions as reported in (c).

5.2.20 Reconsider the data from the 2013 Gallup poll that asked randomly selected U.S. adults whether they wanted to stay at their current body weight or change.

a) State the appropriate null and alternative hypotheses *in words*.

b) State the appropriate null and alternative hypotheses *in symbols*. Be sure to specify what the symbols denote.

c) Give detailed, step-by-step instructions on how one could conduct a tactile simulation to generate a p-value to test the hypotheses stated in (b) and (c). Be sure to include details on the following:

- Would the simulation involve coins, or dice, or index cards?
- How many tosses, or rolls, or cards would be used?
- How many sets of tosses, or rolls, or shuffles would you observe?
- What would you record after every repetition?
- How would you compute the p-value?

5.2.21 Reconsider the data from the 2013 Gallup poll that asked randomly selected U.S. adults whether they wanted to stay at their current body weight or change.

  a) Use an appropriate randomization-based applet to find a p-value. Report <u>and</u> interpret the p-value, in the context of the study.

  b) Based on this p-value, how much evidence do you have against the null hypothesis?

  c) Calculate an appropriate standardized statistic, and interpret the standardized statistic in the context of the study.

  d) Do the p-value and standardized statistic both lead you to the same conclusion?

  e) Use the 2SD Method to find a 95% confidence interval for the parameter of interest. Report this interval, and interpret the interval in the context of the study.

  f) Is your conclusion from using either the p-value or the standardized statistic consistent with your finding from the 95% confidence interval? How are you deciding?

  g) State a complete conclusion about this study including Significance, Estimation (confidence interval), Causation, and Generalization. Be sure to explain how you are arriving at your conclusion.

Census awareness*

5.2.22 A survey of a random sample of U.S. adults by *The Pew Research Center for the People and the Press*, conducted in early January 2010 recorded each participant's highest level of education completed, and whether they knew that responding to the Census was required by law. Of the 973 participants who had some college or less, 271 knew that responding to the Census was required by law. Of the 526 participants who had a college degree or more, 195 knew that responding to the Census was required by law. Is there an association between a person's level of education and her/his Census awareness?

a) Organize the data in an appropriate well-labeled 2×2 table.

b) Find the observed difference in the conditional proportion of those knew that responding to the Census was required by law between those who had a college degree or more and those with some college or less.

c) Give two possible explanations for the observed difference in conditional proportions as reported in (b).

d) Is this an experiment or an observational study? Explain how you are deciding.

5.2.23 Reconsider the data from *The Pew Research Center for the People* survey of U.S. adults to investigate whether there is an association between a person's level of education and her/his awareness of her/his response to the Census being required by law.

a) State the appropriate null and alternative hypotheses in words.

b) State the appropriate null and alternative hypotheses in symbols. Be sure to specify what the symbols denote.

c) Give detailed, step-by-step instructions on how one could conduct a tactile simulation to generate a p-value to test the hypotheses stated in (a) and (b). Be sure to include details on the following:

- Would the simulation involve coins, or dice, or index cards?
- How many tosses, or rolls, or cards would be used?
- How many sets of tosses, or rolls, or shuffles would you observe?
- What would you record after every repetition?
- How would you compute the p-value?

5.2.24 Reconsider the data from *The Pew Research Center for the People* survey of U.S. adults to investigate whether there is an association between a person's level of education and her/his awareness of her/his response to the Census being required by law.

a) Use an appropriate randomization-based applet to find a p-value. Report <u>and</u> interpret the p-value, in the context of the study.

b) Based on this p-value, how much evidence do you have against the null hypothesis?

c) Calculate an appropriate standardized statistic, and interpret the standardized statistic in the context of the study.

d) Do the p-value and standardized statistic both lead you to the same conclusion?

e) Use the 2SD Method to find a 95% confidence interval for the parameter of interest. Report this interval, and interpret the interval in the context of the study.

f) Is your conclusion from using either the p-value or the standardized statistic consistent with your finding from the 95% confidence interval? How are you deciding?

g) State a complete conclusion about this study including Significance, Estimation (confidence interval), Causation, and Generalization. Be sure to explain how you are arriving at your conclusion.

Eating habits

5.2.25 Suppose three high school students separately conduct polls in their city to investigate there is an association between being a vegetarian and whether people like to eat at home or eat at restaurants.

- Sally finds that 35 out of 45 vegetarians preferred to eat at home whereas 20 out of 105 non-vegetarians preferred to eat at home.

- Tara finds that 70 out of 90 vegetarians preferred to eat at home whereas 40 out of 210 non-vegetarians preferred to eat at home.

- Uma finds that 30 out of 45 vegetarians preferred to eat at home whereas 25 out of 105 non-vegetarians preferred to eat at home.

a) Comparing Sally's study to Tara's study: Who will find stronger evidence of a difference between vegetarians and non-vegetarians with regard to preference to eat at home? Answer without doing any extensive calculations. **Circle one.**

   A. Sally

   B. Tara

   C. The strength of evidence will be similar.

   D. Cannot be answered without finding a p-value

b) Comparing Sally's study to Uma's study: Who will find stronger evidence of a difference between vegetarians and non-vegetarians with regard to preference to eat at home? Answer without doing any extensive calculations. **Circle one.**

   A. Sally

   B. Uma

   C. The strength of evidence will be similar.

   D. Cannot be answered without finding a p-value

Blood donations*

5.2.26 Have you ever donated blood? If so, how often do you donate blood? Studies have looked at who is more likely to donate blood: men or women add. Suppose three students Eddie, Francisco, and Gina each decide to take random samples of men and women, and ask each of the participants whether they have donated blood in the past 12 months.

- Eddie finds that 20 of 100 men and 15 of 100 women in his study have donated blood in the past 12 months.

- Francisco finds that 25 of 100 men and 23 of 100 women in his study have donated blood in the past 12 months.

- Gina finds that 40 of 200 men and 30 of 200 women in her study have donated blood in the past 12 months.

   a) Comparing Eddie's study to Francisco's study: who will find stronger evidence of a difference between men and women with regard to probability of donating blood? Answer without doing any extensive calculations. **Circle one.**

Eddie

Francisco

C. The strength of evidence will be similar.

D. Cannot be answered without finding a p-value

b) Comparing Eddie's study to Gina's study: who will find stronger evidence of a difference between men and women with regard to probability of donating blood? Answer without doing any extensive calculations. **Circle one.**

A. Eddie

B. Gina

C. The strength of evidence will be similar.

D. Cannot be answered without finding a p-value

Sexual Discrimination

5.2.27 In July 2011, an Italian firm Ma-Vib was in the news for sexual discrimination for having chosen only its female employees to be dismissed. Before the layoffs, Ma-Vib employed 18 women and 12 men; it then fired 15 of the women and none of the men.

a) What proportion of the men were fired?

b) What proportion of the women were fired?

c) Find the difference in proportion of men and women fired.

Suppose that we want to investigate whether the proportion of females fired was significantly higher than the proportion of males fired by Ma-Vib. That is, was the difference in proportions as reported in (c) unlikely to have happened by chance alone, if nothing suspicious was going on.

d) Give detailed, step-by-step instructions on how one could conduct a tactile simulation to generate a p-value to investigate whether the proportion of females fired was significantly higher than the proportion of males fired by Ma-Vib.. Be sure to include details on the following:

   • Would the simulation involve coins, or dice, or index cards?

   • How many tosses, or rolls, or cards would be used?

   • How many sets of tosses, or rolls, or shuffles would you observe?

   • What would you record after every repetition?

   • How would you compute the p-value?

5.2.28: Recall previous exercise which described the Italian firm Ma-Vib that was in the news for sexual discrimination for having chosen only its female employees to be dismissed. Before the layoffs, Ma-Vib employed 18 women and 12 men; it then fired 15 of the women and none of the men. Using the **Two Proportions** applet for 1000 shuffle and deals, the following output was obtained.

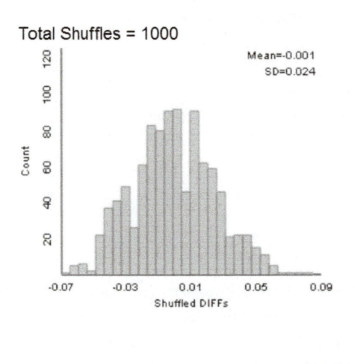

Count Samples Beyond   ▼   [   ]   [Count]

a) The above graph is a dotplot of 1000 possible values of difference in dismissals between men and women. At what numeric value does this graph center? Explain why that makes sense.

b) What value should be entered in the box next to "Beyond"?

c) Estimate <u>and</u> report the p-value from the graph. <u>Also</u>, shade the region of the graph that represents the p-value.

d) Based on the p-value reported in (e), would it be okay to say that the proportion of women fired by Ma-Vib was significantly higher than that of men? How are you deciding?

e) Based on your findings above, would it be okay to say that Ma-Vib fired the 15 women *because* they were women? How are you deciding?

f) Explain why it does not make sense to calculate a confidence interval is this context. (*Hint:* What would the confidence interval be estimating?)

Do incentives work?*

5.2.29 A team of researchers (Singer et al., 2000) used the Survey of Consumer Attitudes to investigate whether incentives would improve the response rates on telephone surveys. A national sample of 735 households was randomly selected, and all 735 of the households were sent an advance letter explaining that the household would be contacted shortly for a telephone survey. However, 368 households were randomly assigned to receive a monetary incentive along with the advance letter, and of these 286 responded to the telephone survey. The other 367 households were assigned to receive only the advance letter, and of these 245 responded to the telephone survey.

a) Is this an experiment or an observational study? Explain how you are deciding.

b) Organize the counts in a well-labeled 2×2 table.

c) Find the observed difference in the conditional proportion of households that responded to the telephone survey between those who received an incentive and those with didn't receive an incentive.

d) Give two possible explanations for the observed difference in conditional proportions as reported in (c).

5.2.30 Reconsider the data from Singer et al.'s study about whether incentives improve response rates on telephone surveys.

a) State the null and the alternative hypotheses in words.

b) State the null and the alternative hypotheses in symbols.

c) Give detailed, step-by-step instructions on how one could conduct a tactile simulation to generate a p-value to test the hypotheses stated in (a) and (b). Be sure to include details on the following:

- Would the simulation involve coins, or dice, or index cards?
- How many tosses, or rolls, or cards would be used?
- How many sets of tosses, or rolls, or shuffles would you observe?
- What would you record after every repetition?
- How would you compute the p-value?

5.2.31 Reconsider the data from Singer et al.'s study about whether incentives improve response rates on telephone surveys.

  a) Use an appropriate randomization-based applet to find a p-value. Report and interpret the p-value, in the context of the study.

  b) Based on this p-value, how much evidence do you have against the null hypothesis?

  c) Calculate an appropriate standardized statistic, and interpret the standardized statistic in the context of the study.

  d) Do the p-value and standardized statistic both lead you to the same conclusion?

  e) Use the 2SD Method to find a 95% confidence interval for the parameter of interest. Report this interval, and interpret the interval in the context of the study.

  f) Is your conclusion from using either the p-value or the standardized statistic consistent with your finding from the 95% confidence interval? How are you deciding?

  g) State a complete conclusion about this study including Significance, Estimation (confidence interval), Causation, and Generalization. Be sure to explain how you are arriving at your conclusion.

For a song

5.2.32: In an article published in *Psychology of Music* (2010), researchers reported the results of a study conducted to investigate the effects of "romantic lyrics on compliance with a courtship request." The researchers recruited undergraduate female students who were studying social and managerial science, and told them that "the purpose of the study was to discuss organic products with another participant." When a participant came in for the study, she was randomly assigned to listen to either a romantic song or a neutral song. After three minutes, she was greeted by a male "confederate" (chosen and trained by the researchers to pretend to be another participant), who while discussing the organic products, also asked for her phone number so that he could call her up to ask her out.

The male confederate was kept unaware of the aim of the study, and was trained to use the same script when talking to all the females.

Of the 44 women who listened to the romantic song, 23 gave their phone numbers, whereas of the 43 who listened to the neutral song, only 12 did.

  a) Was this an experiment or an observational study? Explain how you are deciding.

b) Notice that neither the participants nor the male confederate were aware of the aim of the study, and thus did not know who was getting what treatment. This is called double-blinding. Explain why the researchers used double-blinding.

c) Identify the observational units.

d) Identify the two variables recorded. Also, for each variable, be sure to identify the kind of variable (categorical or quantitative), and the role of the variable (explanatory or response).

e) Organize the counts in a well-labeled 2×2 table.

f) Find the observed difference in the conditional proportion of females who gave their phone number to the male confederate between those who listened to the romantic song and those who listened to the neutral song.

g) Give two possible explanations for the observed difference in conditional proportions as reported in (f).

5.2.33 Reconsider the data from the previous study about the effects of "romantic lyrics on compliance with a courtship request."

a) Define (in words) the parameters of interest of this study. Also, assign symbols to the parameters.

b) State the null and the alternative hypotheses in words.

c) State the null and the alternative hypotheses in symbols.

d) Give detailed, step-by-step instructions on how one could conduct a tactile simulation to generate a p-value to test the hypotheses stated in (a) and (b). Be sure to include details on the following:

- Would the simulation involve coins, or dice, or index cards?
- How many tosses, or rolls, or cards would be used?
- How many sets of tosses, or rolls, or shuffles would you observe?
- What would you record after every repetition?
- How would you compute the p-value?

5.2.34 Reconsider the study about the effects of "romantic lyrics on compliance with a courtship request." Recall that of the 44 women who listened to the romantic song, 23 gave their phone numbers, whereas of the 43 who listened to the neutral song, only 12 did.

a) Use an appropriate randomization-based applet to find a p-value. Report and interpret the p-value, in the context of the study.

b) Based on this p-value, how much evidence do you have against the null hypothesis?

c) Calculate an appropriate standardized statistic, and interpret the standardized statistic in the context of the study.

d) Do the p-value and standardized statistic both lead you to the same conclusion?

e) Use the 2SD Method to find a 95% confidence interval for the parameter of interest. Report this interval, and interpret the interval in the context of the study.

f) Is your conclusion from using either the p-value or the standardized statistic consistent with your finding from the 95% confidence interval? How are you deciding?

g) State a complete conclusion about this study including Significance, Estimation (confidence interval), Causation, and Generalization. Be sure to explain how you are arriving at your conclusion.

Section 5.3 Exercises

Hearing loss*

5.3.1 Researchers investigated whether the proportion of American teenagers with some level of hearing loss was different in 2005-06 than in 1988-94. They collected data on random samples of American teenagers in those two time periods. Let the symbol $\pi_{05\text{-}06}$ denote the population proportion of American teenagers with some level of hearing loss in 2005-06 and similarly for $\pi_{88\text{-}94}$. A 95% confidence interval for the parameter $\pi_{05\text{-}06} - \pi_{88\text{-}94}$ turns out to be (0.0015, 0.0467). Which of the following is an appropriate conclusion to draw?

A. The sample data provide strong evidence that a higher proportion of American teenagers had hearing loss in 2005-06 than in 1988-94.

B. The sample data provide strong evidence that a higher proportion of American teenagers had hearing loss in 1988-94 than in 2005-06.

C. The sample data provide strong evidence that the proportion of American teenagers with hearing loss was the same in 1988-94 as in 2005-06.

D. The sample data provide little evidence to doubt that the proportion of American teenagers with hearing loss was the same in 1988-94 as in 2005-06.

5.3.2 Reconsider the previous exercise. If a 95% confidence interval had instead been produced from the same sample data for the parameter $\pi_{88\text{-}94} - \pi_{05\text{-}06}$, how would this confidence interval have turned out?

A. Same as above: (0.0015, 0.0467)

B. Negative of above: (-0.0467, -0.0015)

C. Impossible to say

5.3.3 Reconsider the previous exercise. If a 99% confidence interval had instead been produced from the same sample data for the parameter $\pi_{05\text{-}06} - \pi_{88\text{-}94}$, how would this confidence interval have turned out?

    A. Wider than the 95% confidence interval: (0.0015, 0.0467)
    B. Narrower than the 95% confidence interval: (0.0015, 0.0467)
    C. Impossible to say

5.3.4 Reconsider the previous exercise. If a 99% confidence interval had instead been produced from the same sample data for the parameter $\pi_{05\text{-}06} - \pi_{88\text{-}94}$, what would the midpoint of this confidence interval be? If the midpoint can be calculated, find it. If the midpoint cannot be calculated using only the information given, explain why not.

5.3.5 Reconsider the previous exercise. If a 99% confidence interval had instead been produced from the same sample data for the parameter $\pi_{05\text{-}06} - \pi_{88\text{-}94}$, what would the width of this confidence interval be? If the width can be calculated, find it. If the width cannot be calculated using only the information given, explain why not.

Medicare audits

5.3.6 Reconsider a previous exercise from Section 5.1 about Medicare audits. Of the 30 small claims, 14 were judged to be overpayments; of the 30 medium claims, 8 were judged to be overpayments. Our research question is this "Does the chance that the claim is judged to be an overpayment tend to differ by the size of the claim?"

Explain why it is not advisable to use a theory-based approach to answer this research question.

Do incentives work?*

5.3.7 Recall a previous exercise from Section 5.2 about the use of incentives to encourage people to participate in a telephone survey. A team of researchers (Singer et al., 2000) used the Survey of Consumer Attitudes to investigate whether incentives would improve the response rates on telephone surveys. A national sample of 735 households was randomly selected, and all 735 of the households were sent an "advance letter" explaining that the household would be contacted shortly for a telephone survey. However, 368 households were randomly assigned to receive a monetary incentive along with the advance letter, and the other 367 households were assigned to receive only the advance letter. Here are the data on how many households responded to the telephone survey.

		Received an incentive?		
		Yes	No	Total
Responded to the telephone survey?	Yes	286	245	531
	No	82	122	204
Total		368	367	735

a) Define the parameters of interest. Assign symbols to these parameters.

b) State the appropriate null and the alternative hypotheses in words.

c) State the appropriate null and the alternative hypotheses in symbols.

d) Explain why it would be okay to use the theory-based method (that is, normal distribution based method) to find a p-value and confidence interval, for this study.

e) Use an appropriate applet to find and report the following from the data.
   - The standardized statistic.
   - The theory-based p-value.
   - The theory-based 95% confidence interval.

f) Include a screenshot of the applet output in your report.

g) Interpret the p-value in the context of the study.

h) Interpret the 95% confidence interval, in the context of the study.

i) State a complete conclusion about this study including Significance, Estimation (confidence interval), Causation, and Generalization. Be sure to explain how you are arriving at your conclusion.

Study of Tamoxifen and Raloxifene (STAR)

5.3.8 Researchers Vogel et al. (*JAMA*, 2006) reported the following findings about the Study of Tamoxifen and Raloxifene (STAR), a study involving postmenopausal women who were at an increased risk for invasive breast cancer. Of the 9726 women randomly assigned to use tamoxifen daily, 163 developed invasive breast cancer sometime during the next 5 years, compared to 168 in the group of 9745 who were randomly assigned to use raloxifene daily.

a) Is this an experiment or an observational study? Explain how you are deciding.

b) Identify the explanatory and response variables. Also, identify whether each is categorical or quantitative.

c) Organize the counts in a well-labeled 2×2 table.

d) Find the observed difference in proportion of breast cancer cases between the tamoxifen and raloxifene users.

e) State the appropriate null and the alternative hypotheses *in words*.

f) Explain why it would be okay to use the theory-based method (that is, normal distribution based method) to find a p-value for this study.

g) Use an appropriate applet to find and report the following from the data.
- The standardized statistic.
- The theory-based p-value.

h) State a complete conclusion about this study including Significance, Estimation (confidence interval), Causation, and Generalization. Be sure to explain how you are arriving at your conclusion.

5.3.9 Recall the data from the STAR study: Of the 9726 women randomly assigned to use tamoxifen daily, 163 developed invasive breast cancer sometime during the next 5 years, compared to 168 in the group of 9745 who were randomly assigned to use raloxifene daily.

a) Define the parameters of interest. Assign symbols to these parameters.

b) Explain why it would be okay to use the theory-based method (that is, normal distribution based method) to find a confidence interval, for this study.

c) Use an appropriate applet to find and report the theory-based 99% confidence interval.

d) Interpret the 99% confidence interval, in the context of the study.

e) Relatively speaking, is the 99% confidence interval narrow or wide? Explain why that makes sense.

5.3.10 Recall the data from the STAR study: Of the 9726 women randomly assigned to use tamoxifen daily, 163 developed invasive breast cancer sometime during the next 5 years, compared to 168 in the group of 9745 who were randomly assigned to use raloxifene daily.

a) Calculate the observed value of the relative risk of developing invasive breast cancer comparing women who took tamoxifen daily to those who took raloxifene daily.

b) Suppose that you were to calculate a 95% confidence interval for the long-run relative risk of developing invasive breast cancer comparing women who will take tamoxifen daily to those who will take raloxifene daily. Will the number 1 lie within this interval? Explain your reasoning.

Solitaire*

5.3.11 Enamored with the solitaire game on his new computer, author A sets out to estimate his probability of winning the game and wins 25 games while losing 192 games. Anxious to outperform author A, author B plays 444 games of solitaire and wins 74. Author B wants to know if she is performing significantly different than Author A.

a) Do these data arise from sampling from two processes or sampling from two populations?
b) Staying consistent with your answer in part a) define the parameters of interest. Also, assign symbols to these parameters.
c) State the appropriate null and the alternative hypotheses in words.
d) State the appropriate null and the alternative hypotheses in symbols.
e) Use an appropriate applet to use a simulation-based approach to find a p-value to test the hypotheses stated.
f) Explain why it would be okay to use the theory-based method (that is, normal distribution based method) to find a p-value and confidence interval, for this study.
g) Use an appropriate applet to find and report the following from the data.
    - The standardized statistic.
    - The theory-based p-value.
    - The theory-based 95% confidence interval.
h) Interpret the p-value in the context of the study.
i) How much evidence is there based on this p-value?
j) Interpret the 95% confidence interval, in the context of the study.
k) Does the 95% confidence interval contain 0? How does this correspond to the p-value you obtained earlier?
l) State a complete conclusion about this study including Significance, Estimation (confidence interval), Causation, and Generalization. Be sure to explain how you are arriving at your conclusion.

Romantic lyrics

5.3.12: Recall the exercise from Section 5.2 about the study conducted to investigate the effects of "romantic lyrics on compliance with a courtship request."

Of the 44 women who listened to the romantic song, 23 gave their phone numbers, whereas of the 43 who listened to the neutral song, only 12 did.

a) Define (in words) the parameters of interest of this study. <u>Also, assign symbols</u> to the parameters.

b) State, in symbols, the null and alternative hypotheses, to test whether women are <u>more</u> likely to comply with a "courtship request" when they have been listening to a romantic song rather than a neutral song.

c) Explain why it is okay to use theory-based methods to test the hypotheses stated above.

To analyze the data, the **Theory-based Inference applet** was used. Here is the output.

d) From the applet output, identify the following:

    i. The standardized statistic.

    ii. The p-value.

    iii. The 95% confidence interval.

e) Interpret the p-value in the context of the study. (*Hint:* Not the conclusion, but what the p-value is a probability of…)

f) State your conclusion, in the context of the study. (*Remember to address Significance, Estimation, Causation, and Generalizability!*)

g) Interpret the 95% confidence interval in the context of the study.

h) Recall that the formula for the standardized statistic in this scenario is

$$z = \frac{\hat{p}_1 - \hat{p}_2}{\sqrt{\hat{p}(1-\hat{p})\left(\frac{1}{n_1} + \frac{1}{n_2}\right)}}$$

. Plug in relevant numbers into this formula, and verify the value of the standardized statistic obtained using the applet.

i) Recall that the formula for the confidence interval in this scenario is

$$\hat{p}_1 - \hat{p}_2 \pm \text{multiplier} \times \sqrt{\left(\frac{\hat{p}_1(1 - \hat{p}_1)}{n_1} + \frac{\hat{p}_2(1 - \hat{p}_2)}{n_2} \right)}$$. Plug in relevant numbers into this

formula, and verify the value of the 95% confidence interval obtained using the applet.

j) Suppose a different research team wants to investigate whether type of music affects the likelihood of compliance with a courtship request. If this research team uses the same dataset to explore this, what would their p-value be? Answer without carrying out the test of significance. Explain your reasoning.

Surgery and prostate cancer*

5.3.13 Recall the exercise in section 5.2 where researchers Wilt et al. (*New England Journal of Medicine,* 2012) investigated whether surgery, compared to just observation, was (more) effective in improving men's survival chances after being diagnosed with prostate cancer. The researchers identified 731 men with localized prostate cancer and randomly assigned 364 to surgery, and the remaining 367 to observation. All participants were followed for about 10 years. In those 10 years, 21 surgery recipients died of prostate cancer related reasons compared to 31 observation recipients.

   a) Define (in words) the parameters of interest of this study. <u>Also, assign symbols</u> to the parameters.
   b) State the appropriate null and the alternative hypotheses in words.
   c) State the appropriate null and the alternative hypotheses in symbols.
   d) Explain why it is valid to use the theory-based approach to test the hypotheses stated above.
   e) Use an appropriate applet to find a theory-based p-value to test the above hypotheses. Report the standardized statistic as well as the p-value.
   f) Use an appropriate applet to find a theory-based 99% confidence interval, and interpret the resulting interval in the context of the study.
   g) Based on your findings, state a complete conclusion about the study. Be sure to address: significance, estimation (confidence interval), causation, and generalizability.

CPR

5.3.14 Recall the exercise in Section 5.2 about investigating whether giving chest-compression-only (CC) instructions rather than standard cardiopulmonary resuscitation (CPR) instructions, to the witness of a heart attack will improve the victim's chance of surviving, researchers Hupfl et al. (*The Lancet*, 2010) combined the results from three randomized experiments. In each experiment, the emergency services dispatcher randomly assigned either CC or CPR instructions to the bystander who was at the site where a person had just experienced a heart attack. They found that of the 1500 cases where CC instructions had been given, 211 people had survived, whereas out of the 1531 cases where standard CPR instructions had been given, the number was 178.

a) Define (in words) the parameters of interest of this study. <u>Also, assign symbols</u> to the parameters.

b) State the appropriate null and the alternative hypotheses in words.

c) State the appropriate null and the alternative hypotheses in symbols.

d) Explain why it is valid to use the theory-based approach to test the hypotheses stated above.

e) Use an appropriate applet to find a theory-based p-value to test the above hypotheses. Report the standardized statistic as well as the p-value.

f) Use an appropriate applet to find a theory-based 95% confidence interval, and interpret the resulting interval in the context of the study.

g) Based on your findings, state a complete conclusion about the study. Be sure to address: significance, estimation (confidence interval), causation, and generalization.

Prenatal care for mothers with gestational diabetes

5.3.15 Recall the exercise in section 5.2 about the link between babies being Large for Gestational Age (LGA), and gestational diabetes in the mother. In an article published in the *New England Journal of Medicine* (October 2009), researchers Landon et al. reported a study of 958 women diagnosed with mild gestational diabetes between 24 and 31 weeks of pregnancy who were randomly assigned to one of two groups: 473 to usual prenatal care (control group), and 485 to dietary intervention, self-monitoring of blood glucose, and insulin therapy, if necessary (treatment group). Of the 473 women in the control group, 68 had babies who were LGA, and of the 485 women in the treatment group, 29 had babies who were LGA.

a) Define (in words) the parameters of interest of this study. <u>Also, assign symbols</u> to the parameters.

b) State the appropriate null and the alternative hypotheses in words.

c) State the appropriate null and the alternative hypotheses in symbols.

   d) Explain why it is valid to use the theory-based approach to test the hypotheses stated above.

   e) Use an appropriate applet to find a theory-based p-value to test the above hypotheses. Report the standardized statistic as well as the p-value.

   f) Use an appropriate applet to find a theory-based 95% confidence interval, and interpret the resulting interval in the context of the study.

   g) Based on your findings, state a complete conclusion about the study. Be sure to address: significance, estimation, causation, and generalizability.

   h) Recall that the formula for the standardized statistic in this scenario is

$$z = \frac{\hat{p}_1 - \hat{p}_2}{\sqrt{\hat{p}(1-\hat{p})\left(\frac{1}{n_1}+\frac{1}{n_2}\right)}}$$

. Plug in relevant numbers into this formula, and verify the value of the standardized statistic obtained using the applet.

   i) Recall that the formula for the confidence interval in this scenario is

$$\hat{p}_1 - \hat{p}_2 \pm \text{multiplier} \times \sqrt{\left(\frac{\hat{p}_1(1-\hat{p}_1)}{n_1}+\frac{\hat{p}_2(1-\hat{p}_2)}{n_2}\right)}$$

. Plug in relevant numbers into this formula, and verify the value of the 95% confidence interval obtained using the applet.

5.3.16 Reconsider the data from the previous study about the effect of specialized prenatal care for women with gestational diabetes. Do the data provide evidence of a relationship between the type of prenatal care a woman with gestational diabetes receives, and whether or not she has an LGA baby?

   a) Calculate the observed value of the relative risk of having an LGA baby for the pregnant women who received usual prenatal care compared to those who received specialized prenatal care.

   b) Suppose that you were to calculate a 95% confidence interval for the long-run relative risk of having an LGA baby for pregnant women who will receive usual prenatal care compared to those who will receive specialized prenatal care. Will the 1 lie within this interval? Explain your reasoning.

Risk taking*

5.3.17: Recall a study from the exercises in Section 5.2, called "Power Posing: Brief Nonverbal Displays Affect Neuroendocrine Levels and Risk Tolerance," that was published in *Psychological Science,* September 2010, and describes an experiment involving 42 participants (male and female), where 20 participants were randomly assigned to hold low-power poses (contractive positions, closed limbs), and 22 to hold high-power poses (expansive positions, open limbs) for 2 minutes. All participants were told that the aim of this exercise was to see whether their heart rate changed. After the exercise, each participant was given $2, and told that they could keep the money, or roll a die for a double or nothing. Here are the data:

		Low power pose	High-power pose	Total
Took the "double or nothing" bet?	Yes	3	18	21
	No	17	4	21
Total		20	22	42

Is it valid to use the theory-based approach to examine whether the proportion of risk takers was significantly higher in the high-power pose compared to the low-power pose? Explain your reasoning.

Change your weight

5.3.18 Recall from section 5.2 exercises, the results from a 2013 Gallup poll that asked randomly selected U.S. adults whether they wanted to stay at their current body weight or change. Of the 562 men surveyed, 242 wanted to stay at their current weight, whereas of the 477 women surveyed, 172 wanted to stay at their current weight.

   a) Define (in words) the parameters of interest of this study. Also, assign symbols to the parameters.
   b) State the appropriate null and the alternative hypotheses in words.
   c) State the appropriate null and the alternative hypotheses in symbols.
   d) Explain why it is valid to use the theory-based approach to test the hypotheses stated above.
   e) Use an appropriate applet to find a theory-based p-value to test the above hypotheses. Report the standardized statistic as well as the p-value.
   f) Use an appropriate applet to find a theory-based 95% confidence interval, and interpret the resulting interval in the context of the study.
   g) Based on your findings, state a complete conclusion about the study. Be sure to address: significance, estimation (confidence interval), causation, and generalization.

Census awareness*

5.3.19 Recall from section 5.2 exercises the survey of a random sample of U.S. adults by *The Pew Research Center for the People and the Press*, conducted in early January 2010 recorded each participant's highest level of education completed, and whether they knew that responding to the Census was required by law. Of the 973 participants who had some college or less, 271 knew that responding to the Census was required by law. Of the 526 participants who had a college degree or more, 195 knew that responding to the Census was required by law. Is there an association between a person's level of education and her/his Census awareness?

  a) Define (in words) the parameters of interest of this study. Also, assign symbols to the parameters.

  b) State the appropriate null and the alternative hypotheses in words.

  c) State the appropriate null and the alternative hypotheses in symbols.

  d) Explain why it is valid to use the theory-based approach to test the hypotheses stated above.

  e) Use an appropriate applet to find a theory-based p-value to test the above hypotheses. Report the standardized statistic as well as the p-value.

  f) Use an appropriate applet to find a theory-based 95% confidence interval, and interpret the resulting interval in the context of the study.

  g) Based on your findings, state a complete conclusion about the study. Be sure to address: significance, estimation (confidence interval), causation, and generalization.

Sexual discrimination

5.3.20: Recall from section 5.2 exercises that in July 2011, an Italian firm Ma-Vib was in the news for "sexual discrimination" for having chosen only its female employees to be dismissed. Before the layoffs, Ma-Vib employed 18 women and 12 men; it then fired 15 of the women and none of the men.

Is it valid to use the theory-based approach to examine whether the proportion of women fired was significantly higher than the proportion of men fired? Explain your reasoning.

Cardiac disease and depression*

5.3.21 Researchers Penninx et al. (*Archives of General Psychiatry,* 2001) looked at the relationship between depression and a person's ability to survive cardiac disease. The researchers identified 450 men and women with cardiac disease, evaluated them for depression, and followed them for 4 years. In those 4 years, of the 89 patients diagnosed with minor or major depression, 26 died of cardiac disease. Of the 361 patients with no diagnosis of depression, 67 died of cardiac disease.

a) Define (in words) the parameters of interest of this study. Also, assign symbols to the parameters.

b) State the appropriate null and the alternative hypotheses in words.

c) State the appropriate null and the alternative hypotheses in symbols.

d) Explain why it is valid to use the theory-based approach to test the hypotheses stated above.

e) Use an appropriate applet to find a theory-based p-value to test the above hypotheses. Report the standardized statistic as well as the p-value.

f) Use an appropriate applet to find a theory-based 95% confidence interval, and interpret the resulting interval in the context of the study.

g) Based on your findings, state a complete conclusion about the study. Be sure to address: significance, estimation (confidence interval), causation, and generalizability.

Cheap pill to swallow?

5.3.22 In a randomized, double-blind study reported in the *Journal of American Medical Association*, researchers Waber et al. (2008) administered a pill to each of the 82 healthy paid volunteers from Boston, Massachusetts, but told half of them that the drug had a regular price of $2.50 per pill, whereas the remaining participants were told that the drug has a discounted price of $0.10 per pill, mentioning no reason for the discount. In the recruitment advertisement the pill was described as an FDA approved "opioid analgesic," but in reality both groups were administered placebo pills.

To simulate pain, the participants were administered electric shocks to the wrist, and the researchers recorded whether subjects reported a reduction in pain after taking the pill. Of the 41 people in the regular-price pill, 35 reported a pain reduction, whereas of the 41 participants in the discount-price pill, 25 reported a pain reduction.

Is it valid to use the theory-based approach to examine whether the proportion of subjects reporting reduction in pain among those who received the regular-price pill is significantly different from the proportion of subjects reporting reduction in pain among those who received the discount-price pill? Explain your reasoning.

The Physicians' Health Study (heart attacks)*

5.3.23 The Physicians' Health Study is a very large, randomized study designed to "test the effects of low-dose aspirin… in the prevention of cardiovascular disease (CVD)." The subjects were 22,071 U.S. male physicians (aged 40 to 84 years, in the year 1982), who were randomly assigned to be either in the low-dose aspirin group or the placebo group. Each participant was required to take the assigned pill every other day, for 5 years. The study was double-blind. Of the 11,034 physicians who took the placebo, 189 suffered heart attacks during the study. Of the 11,037 physicians who took aspirin, 104 had heart attacks.

a) Is this an experiment or an observational study? Explain how you are deciding.

b) Identify the explanatory and response variables. Also, identify whether each is categorical or quantitative.

c) Explain what it means for a study to be "double-blind." Also, explain the purpose of using "double-blinding" in this study.

d) Organize the counts in a well-labeled 2×2 table.

e) Find the observed difference in proportion of heart attack cases between the physicians who took aspirin and those who took the placebo.

f) State the appropriate null and the alternative hypotheses *in words*.

g) Use an appropriate applet to use a simulation-based approach to find a p-value to test the hypotheses stated.

h) Evaluate the strength of evidence based on the p-value.

i) Explain why it would be okay to use the theory-based method (that is, normal distribution based method) to find a p-value, for this study.

j) Use an appropriate applet to find and report the following from the data.
   - The standardized statistic
   - The theory-based p-value

k) How do the simulation-based and theory-based p-values compare?

5.3.24 Recall the data from the Physicians' Health Study: Of the 11,034 physicians who took the placebo, 189 suffered heart attacks during the study. Of the 11,037 physicians who took aspirin, 104 had heart attacks.

a) Define the parameters of interest. Assign symbols to these parameters.

b) State the appropriate null and the alternative hypotheses in symbols.

c) Explain why it would be okay to use the theory-based method (that is, normal distribution based method) to find a confidence interval, for this study.

d) Use an appropriate applet to find and report the theory-based 99% confidence interval. Interpret the 99% confidence interval, in the context of the study.

e) Also, use the appropriate applet to find and report the theory-based p-value.

f) Is there *very strong evidence* of a difference in the probability of heart attacks between aspirin takers and non-aspirin-takers? How are you deciding?

g) Based on your findings, state a complete conclusion about the study. Be sure to address: significance, estimation (confidence interval), causation, and generalizability.

h) Relatively speaking, is the 99% confidence interval narrow or wide? Explain why that makes sense.

i) Is there evidence of a *very large difference* in the probability of heart attacks between aspirin takers and non-aspirin-takers? How are you deciding?

5.3.25 Recall the data from the Physicians' Health Study: Of the 11,034 physicians who took the placebo, 189 suffered heart attacks during the study. Of the 11,037 physicians who took aspirin, 104 had heart attacks.

a) Recall that the formula for the standardized statistic in this scenario is

$$z = \frac{\hat{p}_1 - \hat{p}_2}{\sqrt{\hat{p}(1-\hat{p})\left(\frac{1}{n_1} + \frac{1}{n_2}\right)}}$$. Plug in relevant numbers into this formula, and verify the value of the standardized statistic obtained using the applet.

b) Recall that the formula for the confidence interval in this scenario is

$$\hat{p}_1 - \hat{p}_2 \pm multiplier \times \sqrt{\left(\frac{\hat{p}_1(1-\hat{p}_1)}{n_1} + \frac{\hat{p}_2(1-\hat{p}_2)}{n_2}\right)}$$. Plug in relevant numbers into this formula, and verify the value of the 95% confidence interval obtained using the applet.

5.3.26 Recall the data from the Physicians' Health Study: Of the 11,034 physicians who took the placebo, 189 suffered heart attacks during the study. Of the 11,037 physicians who took aspirin, 104 had heart attacks.

a) Calculate the observed value of the relative risk of heart attacks comparing physicians who took aspirin daily to those who took the placebo daily.

b) Suppose that you were to calculate a 95% confidence interval for the long-run relative risk of a heart attack comparing aspirin-taking men (like those in the study) to non-aspirin-taking men. Will the number 1 lie within this interval? Explain your reasoning.

The Physicians' Health Study (ulcers)

5.3.27 Another outcome of interest in the Physicians' Health Study was whether the subjects developed ulcers or not. Of the 11,034 physicians who took the placebo, 138 developing ulcers during the study. Of the 11,037 physicians who took aspirin, 169 developed ulcers.

a) Identify the explanatory and response variables. Also, identify whether each is categorical or quantitative.

b) Organize the counts in a well-labeled 2×2 table.

c) Find the observed difference in proportion of ulcer cases between the physicians who took aspirin and those who took the placebo.

d) State the appropriate null and the alternative hypotheses in words.

e) Use an appropriate applet to use a simulation-based approach to find a p-value to test the hypotheses stated.

f) Based on this p-value, how strong is the evidence against the null hypothesis.

g) Explain why it would be okay to use the theory-based method (that is, normal distribution based method) to find a p-value, for this study.

h) Use an appropriate applet to find and report the following from the data.
   - The standardized statistic
   - The theory-based p-value

i) How do the simulation-based and theory-based p-values compare?

5.3.28 Recall the data from the Physicians' Health Study: Of the 11,034 physicians who took the placebo, 138 developing ulcers during the study. Of the 11,037 physicians who took aspirin, 169 developed ulcers.

a) Define the parameters of interest. Assign symbols to these parameters.

b) State the appropriate null and the alternative hypotheses in symbols.

c) Explain why it would be okay to use the theory-based method (that is, normal distribution based method) to find a confidence interval, for this study.

d) Use an appropriate applet to find and report the theory-based 95% confidence interval.

e) Does the 95% confidence interval contain 0? Were you expecting this? Explain your reasoning.

f) Interpret the 95% confidence interval, in the context of the study.

g) Use the 95% confidence interval to state a conclusion about the strength of evidence in the context of the study.

h) Relatively speaking, is the 95% confidence interval narrow or wide? Explain why that makes sense.

5.3.29 Recall the data from the Physicians' Health Study: Of the 11,034 physicians who took the placebo, 138 developing ulcers during the study. Of the 11,037 physicians who took aspirin, 169 developed ulcers.

a) Recall that the formula for the standardized statistic in this scenario is

$$z = \frac{\hat{p}_1 - \hat{p}_2}{\sqrt{\hat{p}(1 - \hat{p})\left(\frac{1}{n_1} + \frac{1}{n_2}\right)}}$$

. Plug in relevant numbers into this formula, and verify the value of the standardized statistic obtained using the applet.

b) Recall that the formula for the confidence interval in this scenario is

$$\hat{p}_1 - \hat{p}_2 \pm multiplier \times \sqrt{\left(\frac{\hat{p}_1(1 - \hat{p}_1)}{n_1} + \frac{\hat{p}_2(1 - \hat{p}_2)}{n_2}\right)}$$

. Plug in relevant numbers into this formula, and verify the value of the 95% confidence interval obtained using the applet.

5.3.30 Recall the data from the Physicians' Health Study: Of the 11,034 physicians who took the placebo, 138 developing ulcers during the study. Of the 11,037 physicians who took aspirin, 169 developed ulcers.

a) Calculate the observed value of the relative risk of ulcers comparing physicians who took aspirin daily to those who took the placebo daily.

b) Suppose that you were to calculate a 95% confidence interval for the long-run relative risk of developing ulcers comparing aspirin-taking men (like those in the study) to non-aspirin-taking men. Will the number 1 lie within this interval? Explain your reasoning.

End of Chapter Exercises

5.CE.1* Suppose you collect data on two categorical variables and summarize the results in a 2×2 table of counts such as the following:

	Group A	Group B
Success		
Failure		

Your goal is to decide if the data provide strong evidence that Group A and Group B differ with regard to the long-run probability of success.

a) Is it possible to investigate this question even when the sample sizes of the two groups are different? Explain your answer, as if to a friend who is skeptical.

b) Is it possible to investigate this question even when the sample sizes of the two groups are both less than 10? Explain your answer, as if to a friend who is skeptical.

5.CE.2 Reconsider the previous exercise.

a) If the observed proportion of successes turns out to be the same for both groups, would you expect the p-value to be very small, somewhat small, or not small at all? Explain.

b) If Group A results in only successes, and Group B results in only failures, would you expect the p-value to be very small, somewhat small, or not small at all? Explain.

5.CE.3* Explain why the methods of this chapter are not appropriate for investigating the following research questions:

- Have teenagers in the United Kingdom read more Harry Potter books, on average, than teenagers in the United States?
- Do cows tend to produce more milk, on average, when they are spoken to by name than when they are not spoken to by name?

Praising children

5.CE.4 Psychologists investigated whether praising a child's intelligence, rather than praising his/her effort, tends to negative consequences such as undermining their motivation (Mueller and Dweck, 1998). Children participating in the study were given a set of problems to solve. After the first set of problems, half of the children were randomly assigned to be praised for their intelligence, while the other half was praised for their effort. The children were then given another set of problems to solve and later told how many they got right. They were then asked to write a report about the problems for other children to read, including information about how many they got right. Some of the children misrepresented (i.e., lied about) how many they got right, as shown in the table below. Researchers were interested in learning whether there was a difference in the proportion of children who lied depending on how they were praised.

	Praised for intelligence	Praised for effort	Total
Misrepresented their score (lied)	11	4	15
Did not misrepresent (did not lie)	18	26	44
Total	29	30	59

a) Identify the explanatory and response variables in this study.

b) For each group, determine the proportion who lied, and identify them with appropriate symbols.

c) Describe how you could use index cards to conduct a simulation analysis for determining whether the difference between these proportions is statistically significant. Include the following information in your description:

  i) how many cards you would use

        ii) how the cards should be marked/colored and how many of each type

        iii) how many you would deal out

        iv) which kinds of cards you would count

        v) what you would compare the results to, after you conducted a large number of repetitions

d) Use the appropriate applet to conduct a simulation with 1000 repetitions. Sketch the resulting histogram, labelling the axes appropriately and report the p-value from the applet.

e) Provide a complete, detailed interpretation (in one or two sentences) of what this p-value means in this context (i.e., probability of what, assuming what?)

f) Summarize your conclusion about whether the data provide evidence that praising a child's intelligence leads to more negative consequences than praising his/her effort. Be sure to address the issue of causation as well as the issue of significance.

Minority coaches*

5.CE.5 Professional baseball teams have one coach at first base and one coach at third base, with third base regarded as the more important and prestigious position. An article in the August 11, 2010 *New York Times* raised a concern that minority coaches are under-represented at third base compared to first base. The article cited the following data:

- 27 of 60 base coaches in Major League Baseball (MLB) are members of minority groups
- 20 of the 27 minority coaches are first base coaches
- 10 of the 33 non-minority coaches are first base coaches

a) Is this an observational study or an experiment?

b) Identify the explanatory and response variables.

c) Organize these counts in a 2×2 table, with the explanatory variable in columns.

d) Conduct a simulation analysis, and report the p-value.

e) Summarize your conclusion about whether the data provide string evidence that minority coaches are more likely to be first base coaches than non-minorities.

f) Is drawing a cause-and-effect conclusion warranted in this study? Explain.

Nicotine lozenge

5.CE.6 A study conducted in 2002 by Shiffman, Dressler, Hajeh, and Gilburt investigated whether a nicotine lozenge is helpful for smokers trying to quit smoking. The researchers recruited smokers who were interested in quitting through advertisements near four sites in the United Kingdom and 11 sites in the United States. Those smokers who met the screening qualifications were randomly assigned to one of two groups: one group received nicotine lozenges and the other group received placebo lozenges. The subjects were compared on various background variables at the beginning of the study, and at the end of the study they were compared on whether or not they successfully abstained from smoking. Of the 459 subjects in the nicotine group, 42.9% were male. Of the 458 subjects in the placebo group, 40.2% were male.

a) Is this an observational study or an experiment?

b) Produce a 2×2 table of counts to represent these data.

c) Create a relevant graph to display the data.

d) Comment on whether the percentages seem to suggest a strong association between gender and treatment.

e) Conduct an appropriate theory-based test to investigate whether the percentages of males differ significantly between the two groups. Report the test statistic and p-value, and summarize your conclusion.

f) Do you think the researchers would be pleased with the result of this significance test? Explain why or why not.

5.CE.7 Reconsider the previous exercise. At the end of the 52-week study, 17.9% of those in the nicotine group had successfully abstained from smoking, compared to 9.6% of those in the placebo group.

a) Produce a 2×2 table of counts to represent these data.

b) Create a relevant graph to display the data.

c) Calculate and interpret the *difference* in the conditional proportions who had successfully abstained from smoking.

d) Calculate and interpret the *ratio* of the conditional proportions who had successfully abstained from smoking.

e) Conduct an appropriate theory-based test to investigate whether the percentages who successfully abstained from smoking differ significantly between the two groups. Report the test statistic and p-value, and summarize your conclusion.

f) Do you think the researchers would be pleased with the result of this significance test? Explain why or why not.

g) Do you think a cause-and-effect conclusion (between using the nicotine lozenge and being more likely to abstain from smoking) is justified based on this study? Explain.

5.CE.8 Reconsider the previous two exercises.

a) Determine a 95% confidence interval for the difference in long-run probabilities of successfully abstaining from smoking between the two groups.

b) Interpret what this interval reveals. Be sure to address the question of whether the data provide strong evidence that the nicotine lozenge is more effective than the placebo.

5.CE.9 Reconsider the previous three exercises. Now consider only the results for the nicotine lozenge, not for the placebo.

a) Determine a 95% confidence interval for the long-run probability that a smoker who uses a nicotine lozenge will successfully abstain from smoking for 52 weeks.

b) Based on this confidence interval, would you conclude that a smoker has a very good chance of successfully quitting if he/she uses a nicotine lozenge? Explain your answer.

Attractiveness*

5.CE.10 A survey conducted by the Gallup organization in 1999 asked American adults whether they are satisfied with their physical attractiveness or wish they could be more attractive. The survey found that 71% of women and 81% of men responded that they were satisfied with their attractiveness.

a) Could this study have involved random sampling, random assignment, or both? Explain your answer.

b) Produce a (well-labeled) segmented bar graph to display these results.

c) State the relevant null and alternative hypotheses for testing whether these proportions differ significantly.

d) What additional information do you need in order to conduct a test of these hypotheses?

5.CE.11 Reconsider the previous exercise.

a) Suppose the study included 100 people of each sex. Determine the test statistic and p-value for testing whether the proportions who are satisfied with their attractiveness differ significantly between the two sexes using a theory based test

b) Now suppose that the study had included 500 people of each sex. Re-answer a).

c) Summarize what this reveals about the role of sample sizes in a two-proportion z-test.

Credit Cards

5.CE.12 The Nellie Mae organization conducts an extensive annual study of credit card usage by college students. For their 2004 study, they analyzed credit bureau data for a random sample of 1413 undergraduate students between the ages of 18 and 24. They found that 76% of the students sampled held a credit card. Three years earlier they had found that 83% of undergraduate sampled held a credit card. (Assume that the sample size was similar in 2001.)

    a) State the appropriate null and alternative hypotheses for testing whether the proportion of undergraduate students who held a credit card differed between these two years.

    b) Conduct an appropriate test of these hypotheses. Report the test statistic and p-value, and summarize your conclusion.

5.CE.13 Reconsider the previous exercise.

    a) Determine a 90% confidence interval for the difference in proportions of undergraduates who held a credit card between the years 2001 and 2004.

    b) Repeat a), with a 95% confidence interval.

    c) Repeat a), with a 99% confidence interval.

    d) Which (if any) of these intervals include the value 0? Summarize what you conclude from this.

5.CE.14 Reconsider the previous two exercises.

    a) Describe what Type I error would mean in this context.

    b) Describe what Type II error would mean in this context.

5.CE.15 Reconsider the previous three exercises.

    a) Describe what power means in this study.

    b) Would you expect that you could increase the power of a test by using a larger sample size or a smaller sample size? Explain.

    c) Would you expect that using a smaller (more strict) significance level would increase or decrease the power of a test? Explain.

    d) Would you expect that a larger difference between the groups would increase or decrease the power of a test? Explain.

Exciting Life*

5.CE.16 The General Social Survey (GSS) is a large-scale national survey conducted every two years with a representative group of American adults. The 2×2 table below shows results from the 2010 survey regarding sex and whether the respondent considers life to be exciting:

	Male	Female
Exciting	303	336
Routine or dull	271	356

Analyze these data to investigate the question of whether men and women differ significantly with regard to whether or not they consider life to be exciting. Write a report of your findings that includes statistics, graphs, a test of significance, and a confidence interval.

5.CE.17 Reconsider the previous exercise. Using the same data but grouping "routine" with "exciting" rather than with "dull" produces the following 2×2 table:

	Male	Female
Exciting or routine	540	660
Dull	34	32

Analyze these data to investigate the question of whether men and women differ significantly with regard to whether or not they consider life to be *dull*. Write a report of your findings that includes statistics, graphs, a test of significance, and a confidence interval.

Investigation: **Does Vitamin C improve your health?**

In 1970 Linus Pauling, a well-known chemist and Nobel Prize winning scientist, published *Vitamin C and the Common Cold* (1970) creating a great deal of public and scientific interest. In short, Pauling argued that taking Vitamin C would reduce one's risk of the common cold. This book almost singlehandedly made Vitamin C one of the most widely used dietary supplements, a status it retains to this day (Nutritional Supplement Review, 2009). Subsequent to the publishing of his book, Pauling wrote a paper that appeared in the Proceedings of the National Academy of Sciences in 1971. In this paper he describes a study conducted by a physician in Basel, Switzerland in the early 1960s.

Here is an excerpt from the paper explaining the study design:

"The study was carried out in a ski resort with 279 skiers during two periods of 5-7 days. The conditions were such that the incidence of colds during these short periods was large enough (about 20%) to permit results with statistical significance to be obtained. The subjects were roughly of the same age and had similar nutrition during the period of study. The investigation was double-blind, with neither the participants nor the physicians having any knowledge about the distribution of the ascorbic-acid tablets (1000 mg) and the placebo tablets. The tablets were distributed every morning and taken by the subjects under observation, so that the possibility of interchange of tablets was eliminated. The subjects were examined daily for symptoms of colds and other infections. The records were largely on the basis of subjective symptoms, partially supported by objective observations (measurement of body temperature, inspection of the respiratory organs, auscultation of the lungs, and so on). Persons who showed cold symptoms on the first day were excluded from the investigation.

After the completion of the investigation, a completely independent group of professional people was provided with the identification numbers for the ascorbic-acid tablets and placebo tablets, and this group performed the statistical evaluation of the observations." **Although not stated explicitly in the paragraph above, the participants were randomly assigned to take ascorbic-acid (vitamin C) tablets or the placebo.**

Step 1: Ask a research question

1. State the research question.

Step 2: Design a study and collect data

2. Is this study an experiment or an observational study? How are you deciding?

3. What are the observational/experimental units?

4. What are the variables that are measured/recorded on each unit?

5. Describe the parameters of interest in words. (You can use the symbol π to represent these parameters.)

6. State the null and alternative hypotheses for a test of significance to see whether this experiment provides statistically significant evidence that Vitamin C prevented colds in the skiers. Use a one-sided test.

7. Of the 139 skiers assigned to take a vitamin C, 17 developed a cold. Of the 140 skiers assigned to take a placebo, 31 developed a cold. Fill in the 2×2 table provided.

	Vitamin C	Placebo	Total
No cold			
Cold			
Total			

8. *Incidence* is the term used to describe the percent of the sample who onset with an illness over a certain time period. In this case, the incidence rates are the same as the conditional proportions. Find the incidence rates for individuals who received Vitamin C and those who received the placebo.

Step 3: Explore the data

9. What is the difference in incidence rates in the two groups? Before doing a test of significance, comment on whether you think this is evidence that Vitamin C prevents colds.

Step 4: Draw inferences

Let's use our 3S strategy to help us investigate how much evidence the sample data provide to support our conjecture that Vitamin C prevents colds.

10. What is the statistic that you can use to summarize the data collected in the study?

11. Use the **Two Proportions** applet to carry out a test of significance.

12. When you shuffle, you are simulating one of the two hypotheses (null or alternative) to be true, which one are you simulating?

13. What is the shape of your null distribution and where is its center located? Why does it make sense for the center to be located where it is?

14. Find the p-value of your test. Is this strong enough evidence for you to conclude that Vitamin C prevented colds in the skiers? Write a sentence stating your conclusion as it applies to the question. Remember to use words pertaining to the research question.

Step 5: Formulate conclusions

15. Can we generalize our conclusion to the entire human population? What should be true of the 279 skiers in order for your conclusion to be drawn to the entire human population? Discuss how the sample was obtained. Are there characteristics about the sample subjects that make this an unreasonable generalization?

16. Can we make a cause and effect conclusion? Why or why not?

Step 6: Look back and ahead

17. Summarize your findings from the study. What did you like about the study design? What would you change to improve the study? What further research might you want to do to follow up or expand upon the results in this study?

There continues to be a lot of controversy regarding the use of vitamin C both in the prevention of colds and other negative health outcomes. Recently, the Physicians Health Study II was completed. In November 2008, a paper (Sesso, et al. 2008) was published in the Journal of the American Medical Association which reported on some of the results of the Physicians Health Study II. Consider the following excerpt from the abstract of the paper:

The Physicians' Health Study II was a randomized, double-blind, placebo-controlled trial of vitamin C that began in 1997 and continued until its scheduled completion on August 31, 2007. There were 14,641 US male physicians enrolled, who were initially aged 50 years or older.

18. Explain what it means for the Physicians' Health Study II to be….

a) …randomized.

b) …double-blind.

c) …placebo-controlled.

19. Later in the paper we find out that 619 out of 7329 men who got vitamin C had a "Major cardiovascular event" compared to 626 out of 7312 men who got the placebo. We also read that the p-value for a *difference* in incidence rates is 0.91. Find the incidence rates, state the null and alternative hypotheses for a test of the effectiveness of vitamin C draw inferences about the effectiveness of vitamin C in preventing major cardiovascular events in this sample.

20. In the paper it states that "In this large, long-term trial of male physicians…vitamin C supplementation [did not] reduce the risk of major cardiovascular events. These data provide no support for the use of [vitamin C] for the prevention of cardiovascular disease in middle-aged and older men." Why do you think the authors specifically qualified their conclusion to say "middle-aged and older men?"

Final Comment: There are actually well over 1000 different experiments that have been done on vitamin C for cold prevention, heart attack prevention and a multitude of other health benefits. The studies reported here are merely two of more well-known studies. Neither study is conclusive, nor is there general consensus in the scientific community as to the health benefits of supplementing your diet with vitamin C to attain levels beyond the minimal levels necessary to prevent scurvy

Research Article: The price of fame

Read "Death in The New York Times: the price of fame is a faster flame" by Epstein and Epstein QJ Med, 2013, 106:517-521 and then answer the following questions.

Step 1. Ask a research question

1. In no more than 1-2 sentences state a broad research question (or related questions) that the researchers were trying to investigate in this study.

2. Identify at least two (cited) points of evidence the researchers cite as to why the research question(s) are important. Include the citation in your response.

Step 2. Design a study and collect data

3. What is the sample size of obituaries?

4. Aside from gender and age, identify the other two main variables of interest and indicate which variable is explanatory and which one is response.

Step 3. Explore the data

5. What percent of the sample of obituaries is male?

6. In the sample, what is the average age at death for males? Females?

Step 4. Draw inferences

The researchers conducted a series of tests in the paper as described in the following three statements.

> **Statement #1.** P. 518. "More women than men died before age 70 (24 vs. 17%; p<0.02)"
>
> **Statement #2.** P. 518. "Significant distributional differences were apparent between males and females in terms of...performance/sports (18.4% of males vs. 38.0% of females; $p<10^{-5}$)"
>
> **Statement #3.** P. 519. "Lung cancer deaths were commonest in performers and significantly less common in professionals/academics (7.4 vs. 1.4%, p=0.005).

The reported tests are all two-sided tests. As you will learn in a later chapter (Chapter 8), the chi-squared (χ^2) tests they used are equivalent (yield the same p-value) to the theory-based approach for comparing two proportions that you learned in section 5.3.

7. For Statement #1, state a null and alternative hypothesis, indicate the strength of evidence provided by the p-value and provide an appropriate (in context) interpretation.

8. Repeat question 7 for Statement #2.

9. Repeat question 7 for Statement #3.

10. The article also states that "The pattern of deaths attributable to...cancer (25 vs. 28%;) was comparable" but no p-value is given. Based on this statement what do you think is true about the size of the p-value?

Step 5. Formulate conclusions

11. Is the sample of obituaries a random sample? If yes, what is the population of interest. If no, explain how the sample was taken clarifying how it is not a random sample.

Step 6. Look back and ahead

12. Identify at least two things that, if you were running the study and had a reasonable amount of financial resources, you would do differently. You must justify your answer.

13. In your own words, explain why the researchers think they found a higher average age of death for males in their study than females, even though in population based samples the reverse is true

14. Explain the researchers' statement that "our study sheds no light on whether other complex variables predispose to risk taking…" in light of your understanding of observational studies and confounding variables.

Chapter 6

Comparing Two Means

Chapter Overview

In the previous chapter, we switched our focus to comparisons of two groups. The studies in Chapter 5 involved a binary response variable and the parameters were either long-run probabilities or population proportions. To summarize the data numerically, we computed conditional proportions, and for graphical displays we used segmented bar charts. In this chapter, we will continue to compare groups, but for a quantitative response variable. We will expand the study of quantitative variables that we began in Chapters 2 and 3, using dotplots, histograms or boxplots (something new!) to explore the data, and means and medians to summarize the data numerically.

Despite all that is new, the big picture remains the same as in Chapter 5. Following the usual Six-Step Method, we start with the research hypothesis and study design, then display, explore, and summarize the data. To measure strength of evidence, we simulate the null distribution of the statistic and compute a p-value. To obtain an interval estimate for a parameter, we rely on the usual approaches (e.g., statistic $\pm$ 2SD). Finally, we assess the scope of inference: Can we generalize to some larger group? Is a conclusion about cause-and-effect justified?

The three sections of Chapter 6 echo the three sections of Chapter 5. The first section is about displays and summaries of our sample data. The second section sets out a simulation-based approach to inference about the larger populations or processes (p-values and confidence intervals). The final section presents theory-based shortcuts.

Section 6.1: Comparing Two Groups: Quantitative Response

Introduction

Recall some of the research questions you have seen so far. *Do people perceive their year differently depending on how you ask the question? Is dolphin therapy beneficial to depression patients? Is yawning contagious?* Notice that all of these scenarios involved *categorical* variables. But, as we saw in the

Preliminaries and in Chapters 2 and 3, when we have a quantitative variable, we summarize our data differently.

In this section, you will learn about comparing groups of data on a *quantitative* response variable. We still want to reveal patterns exhibited by the sample data and will do so by focusing on the shape, center, and variability of the distribution(s) of quantitative data. We will also investigate any observations that do not appear to follow the general patterns in the data (e.g., outliers).

Example 6.1: Geyser Eruptions

Recall the Old Faithful data we explored in Example P.2. Back in the Preliminaries, we investigated times between eruptions. We even compared the times between eruptions depending on whether the previous eruption had been long or short. Now we will examine data on the times between eruptions from two different years to see whether the distributions of these times appears to have changed between the years. In particular, you will examine data on 107 eruptions occurring between 6am and midnight between August 1 and August 8, 1978 (from Weisberg, 1985) and data on 95 eruptions from the same week in 2003 (gathered from http://www.geyserstudy.org/geyser.aspx?pGeyserNo=OLDFAITHFUL). Thus, variables (and potential sources of variability) such as season and weather should be consistent for the two data sets. Scientists have speculated that some geological event, such as an earthquake, might have changed Old Faithful's eruption patterns between these two years.

Think about it: What are the observational units in this study? What are the two variables? What types of variables are these?

The observational units in this study are the geyser eruptions. The variables are the year (a categorical variable) and the time until the next eruption (a quantitative variable). The dotplots in Figure 6.1 display the distributions of inter-eruption times for the two years.

Figure 6.1: "Parallel" dotplots (using the same scale along the horizontal axis) of the distributions of inter-eruption times at Old Faithful in 1978 and 2003

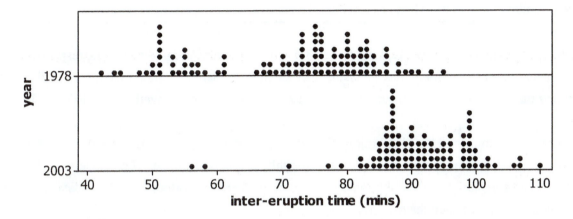

What do we see in these graphs? One difference in the distributions of inter-eruption times between the two years concerns shape: The distribution in 2003 has a few low outliers. Apart from the outliers, we might say the distribution is slightly skewed to the right. In 1978 however, the distribution is clearly bi-modal. Something was going on with this process in 1978 that led to inter-eruption times lasting between roughly 50 to 60 minutes or roughly 70 to 90 minutes, with very few times between 60 and 70 minutes. Also note that the outliers of less than 60 minutes we noted in the 2003 distribution would not have been unusual in 1978.

Perhaps the most obvious difference between these two years is that the inter-eruption times were generally longer in 2003 than in 1978. A typical inter-eruption time in 2003 was about 90 to 100 minutes, whereas in 1978 most inter-eruption times were between 50 and 60 minutes or between 75 and 85 minutes. The variability in the times also differs noticeably between these two years: The times in 2003 had much less variability than the times in 1978.

> **Think about it:** How do you expect the means of the times between eruptions to differ between the two years? Based on the dotplots, estimate the values of the means and medians.

The means of the times between eruptions are 71.0 minutes in 1978 and 91.2 minutes in 2003. So, the average wait time until the next eruption was slightly more than 20 minutes longer in 2003 than in 1978. The medians are 75 and 91 minutes in 1978 and 2003, respectively. Notice that the mean time in 1978 is 4 minutes smaller than the median, because the mean is pulled lower by the lower (but smaller) clump of observations.

Think about it: How do you expect the standard deviations of times between eruptions to compare between the two years? Which year do you expect to have the larger standard deviation, or do you expect the two standard deviations to be quite similar?

The standard deviations of times turn out to be 13.0 and 8.5 minutes in 1978 and 2003, respectively. Notice that the 2003 distribution has a considerably smaller standard deviation than the 1978 distribution, indicating that the times between eruptions were more consistent (closer together) in 2003 than in 1978.

Another way to summarize the distribution of a quantitative variable such as time until the next eruption is by dividing the distribution into four pieces of roughly equal size (number of observations). In other words, we can summarize the distribution by determining where the bottom 25% of the data are, the next 25%, the next 25%, and then the top 25%.

Definitions: The value for which 25% of the data lie below that value is called the **lower quartile** (or 25th percentile). Similarly, the value for which 25% of the data lie above that value is called the **upper quartile** (or 75th percentile). Quartiles can be calculated by determining the median of the values above/below the location of the overall median. The difference between the quartiles is called the **inter-quartile range** (IQR), another measure of variability along with standard deviation. The **five-number summary** for the distribution of a quantitative variable consists of the minimum, lower quartile, median, upper quartile, and maximum.

The five-number summaries for the times between eruptions turn out to be:

	Minimum	Lower quartile	Median	Upper quartile	Maximum
1978 times	42	58	75	81	95
2003 times	56	87	91	98	110

An important characteristic of the IQR is that, like the median, it is not sensitive to extreme values—unlike the SD. For example, if the two lowest wait times (56 and 58 minutes) in 2003 are removed—they are a lot different than the rest of the wait times—the IQR (11) and the median (91) don't change.

However, both the mean (91.19 before, 91.9 after) and standard deviation (8.46 before, 6.87 after) do change.

> **Key Idea:** IQR is a resistant measure of variability, whereas the standard deviation is sensitive to extreme values and skewness.

Notice that all of the values in the five-number summary are smaller in the 1978 distribution than in the 2003 distribution, reflecting that 1978 generally had smaller wait times than 2003. A visual display of the five-number summary called is a *boxplot*.

> **Definition:** A *boxplot* is a visual display of the five-number summary. The box displays the middle 50% of the distribution and its width (the IQR) helps us see the spread of the bulk of the distribution; the whiskers extend to the smallest and largest values in the data set.

Figure 6.2 shows the parallel boxplots for these two distributions.

Figure 6.2: Parallel boxplots (using the same scale along the horizontal axis) of the distributions of inter-eruption times at Old Faithful in 1978 and 2003

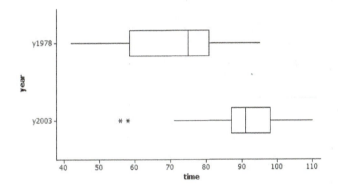

Notice in Figure 6.2 that the outliers we visually identified in 2003 are plotted separately and the whisker extends to the next observation which is not considered an outlier. (The most common criteria to identify an observation as an outlier is if it falls more than $1.5 \times$ IQR from the edge of the box. See FAQ 6.1.1. for details).

These boxplots give us a very quick comparison of the two distributions. We immediately see that the quickest 75% of times in 1978 were below the longest 75% of times in 2003. We also see that the length of the box (IQR) is much smaller in 2003 reflecting the smaller variability in those times. But you may also notice one disadvantage of boxplots: you can lose some important details about the shape of your distribution. For example, we wouldn't notice the two clusters of observations in 1978 that we saw in the dotplots. For this reason, we recommend examining boxplots in combination with dotplots to histograms.

> **Think about it:** Based on the distributions of times between eruptions, in which year would you have preferred to be a tourist waiting for the next eruption of Old Faithful? Can you think of one way in which 1978 would have been preferable, and a different way in which 2003 would have been preferable?

If you are a tourist who cares most about having a shorter wait for the next eruption, you would have preferred 1978 to 2003 because the average wait time in 1978 was about 20 minutes less than in 2003. But if your primary concern is being able to predict accurately when the next eruption would occur, you would have preferred 2003 to 1978 because the variability in wait times was much smaller in 2003.

FAQ 6.1.1 Mean, median, IQR and Outliers

Mean versus median

Q: When should you use the median rather than the mean?

A: It depends.

Q: Is that supposed to be helpful?

A: It depends in part on your data. If your distribution is perfectly symmetric, then the mean and median will be exactly equal. Of course that never happens in practice.

Q: But for distributions that are roughly symmetric, it doesn't matter whether you use the mean or the median?

A: That's basically right. I should state for a theory-based shortcut, statisticians prefer the mean, in part because the theory is much simpler. But both mean and median can be used in simulation-based analyses.

Q: Now what if your distributions are skewed, or have outliers?

A: Then the right summary to choose depends on what you want to find out.

Q: There you go again with "It depends."

A: Suppose you care mainly about the centers of your distributions, and you don't want your summary numbers to be influenced by outliers or values out in the tail of a skewed distribution. Then the median is better than the mean. Remember, the median is resistant to outliers.

Q: Why not just use the median every time you have outliers or skewness?

A: Most of the time that's a good way to go, but there are exceptions. If you want to estimate the sum of all values for a population, you can compute the sum from the mean, but not from the median. The sum of all values is simply the mean times the sample size.

Q: So the bottom line would be to use means for symmetric distributions and use medians if you have long tails or outliers and care mainly about the centers of the distributions.

A: Right.

IQR and outliers

Q: OK, follow-up question. What about outliers? What exactly is an outlier?

A: A value that doesn't fit the overall pattern of the data.

Q: That sounds like a textbook. How can I tell? I mean, what if I think it fits the overall pattern of the data, but you don't?

A: This is another one of those things about Statistics that makes it more art than science sometimes. I'd worry less about a black-and-white rule about outliers, and more about thinking of the implications of potential extreme values—sometimes they matter and sometimes they don't.

Q: And if I still wanted a black and white rule, you could make one up for me?

A: I don't have to make one up. There's already a rule. It's worth knowing, because a lot of people use it. But remember that in statistics, any rule is at best a *guide* to thinking, never a *substitute* for thinking.

OK, OK. What's the rule?

Q: An outlier is any value that is more than1.5 IQRs below the first quartile or more than 1.5 IQRs above the third quartile.

A: Q: So I find the first quartile Q1 and third quartile Q3, then subtract to get the inter-quartile range Q3-Q1, then multiply by 1.5.

A: Right. Any value less than Q1-1.5*IQR or larger than Q3 + 1.5*IQR is an outlier, according to the rule. A picture may help:

Q: Thanks. I promise to remember that the rule is just a guide.

Is the SD a resistant measure of spread?

Q: You said the median is a resistant measure of center (location). Right?

A: Yes. The mean is sensitive to extreme values; the median is not.

Q: Is the standard deviation resistant?

A: Not at all. First of all, the SD looks at deviations from the mean, so SD is based on a non-resistant statistic. Worse, because of the squaring, the SD is even more influenced by outliers than is the mean.

Q: How should you measure the size of variability when you have outliers or skewed distributions with long tails?

A: It depends.

Q: Not again.

A: If your goal is to describe the variability in your data, one standard choice is the inter-quartile range, or IQR, which you've just seen. There are other choices, but the IQR is the most common.

Q: OK, I'll use IQR to describe the variability in the data. What about for test statistics like the
t- statistics? It has the SD in its denominator.

A: If your sample size is large enough, you can still use the *t*-statistic even though it relies on the SD.
The reason is that the *t*-distribution is based on the distribution of the sample mean, and that
distribution tends to be a lot more symmetric than the distribution of the data values.

Exploration 6.1A: Haircut Prices

Do women pay more than men for haircuts? Is this a statistical tendency, or always true? By how much
do women spend more than men, on average? How much do haircut prices vary within a gender as well as
between genders?

To investigate these questions a professor asked students in her class to report the cost of their most
recent haircut, along with their gender.

1. Which would you consider to be the explanatory variable, and which the response? Also classify
   the type (categorical or ~~quantitative~~) for each variable.

   Explanatory: Type:
   Gender   *Categorical Binary*
   Response: Type:
   Price   *quantitative*
                   quant.

2. Is this an experiment or an observational study? Explain briefly.
   Observational

3. Did the professor who collected the data make use of random sampling, random assignment,
   both, or neither?   *Neither*

The following "parallel" dotplots (using the same scale along the horizontal axis) reveal the sample
distributions of haircut prices, by gender:

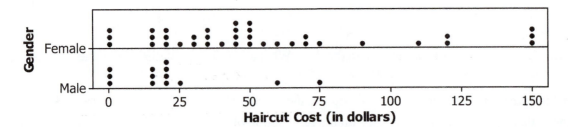

4. Compare and contrast the distributions of haircut prices between men and women in this class. (*Hint*: As you learned in the Preliminaries, comment on center, variability, shape, and unusual observations. You should give enough detail that someone reading your comments could recreate the overall pattern of the graphs from your description. Also be sure to relate your comments to the context.) • more women getting hair cuts. women: 150
   • women have a bigger range.   men : 60
   • women average (median) is larger
                      women: 50   variable: womens
5. Further explore the data. men : 20ish   more var.
   a. Explain why the right-skewness of these distributions makes sense in this context.
      No one wants to spend a lot of money

   b. Explain why the unusual behavior at the lower end of the dotplots – several values at $0 and then a gap to the next smallest prices – makes sense in this context.
      Friends or family ~~mother~~ did it.
             Beauty schools.

6. Based on the dotplots (without performing any calculations), make a guess for each gender's mean haircut price.

   Men: ~~XXX~~   Women: 50/55
        20/25

7. Based on the dotplots (without performing any calculations), which gender do you think has the larger standard deviation of haircut prices? Explain your answer.
      women. much more spread out.

8. Copy the haircut/gender data, which you can access from the book's website, to the clipboard. Open the **Descriptive Statistics** applet. Check the **Stacked** check box (notice the applet assumes the explanatory variable is the first column and the response variable is the second column.) Also keep the **Includes header** box checked and press **Clear**. Paste the data into the Sample data box and press **Use Data**. You should see that the dotplots are similar to the ones previously shown. Check the **Actual** boxes to show the means and standard deviations (Std dev).

a. Report the sample size, mean haircut price, and standard deviation (SD) of haircut prices for each gender. (Include appropriate symbols and measurement units.)

	Sample size	Sample mean	Sample SD
Men	13	21.85	22.14
Women	37	54.10	41.51

b. Which gender has the larger mean haircut price? Is this what you predicted in #6?

women; we were close (yes)

c. Which gender has the larger SD of haircut prices? Is this what you predicted in #7?

women; yes

9. Would you conclude that these data show an *association* between haircut price and gender? If so, describe the nature of this association.   *yes; women care more & more complicated.*

10. Let's return to the research questions that we started with. Address these questions based on the previous graphs and statistics.

a. Do women pay more than men for haircuts? If so, is this a statistical tendency (i.e., true on average), or always true?

yes; not always true so statistical tendency

b. By how much do women spend more than men, on average?

54 - 21 = 32 dollars more

11. Do the unequal sample sizes between the two genders lead you to doubt whether any conclusions can be drawn from these data? Explain.

Probably not; there are things that could change the study.

12. Based on how these data were collected, would you feel comfortable with generalizing your results to the population of all college students in the U.S.? How about the population of all college students at the professor's university? Explain your answers.

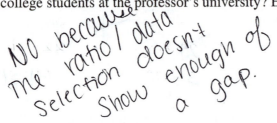

NO because the ratio/data selection doesn't show enough of a gap.

13. Based on both the differences in centers and the amount of overlap in the distributions of haircut prices between men and women, do you predict that the difference will turn out to be statistically significant? (You will learn how to assess statistical significance in the next section.)

yes

Further Analyses

The individual values of the haircut prices that you have been analyzing are:

Women ($n = 37$):   0, 0, 0, 15, 15, 15, 20, 20, 20, 25, 30, 30, 35, 35, 35, 40, 45, 45, 45, 45
50, 50, 50, 50, 55, 60, 65, 70, 70, 75, 90, 110, 120, 120, 150, 150, 150

Men ($n = 13$):   0, 0, 0, 14, 15, 15, 20, 20, 20, 22, 23, 60, 75

As you learned in Section 3.2, the median is the middle value in a dataset, once they are arranged in order. The location of the median can be found by calculating $(n+1)/2$.

$$\left(\frac{13+1}{2} \right)$$

7th position

14. Now let's look at the median haircut prices and compare them to the mean haircut prices
   a. Determine (by hand) the median haircut price for each gender. (You can verify your calculation by checking the **Actual** box for the **Median** in the applet.)

   Median haircut price for women: 45

   Median haircut price for men: 20

   b. Which gender has the larger median haircut price? Is this what you expected?

   women   yes

c.   Are the medians less than or greater than the means? Is this consistent with the right-skewed distributions of haircut prices? Explain.   *Less*   ~~Mean~~

yes.

One way to summarize the distribution of a quantitative variable such as haircut price is by dividing the distribution into four pieces of roughly equal size (number of observations). In other words, summarize the distribution by determining where the bottom 25% of the data are, the next 25%, the next 25%, and then the top 25%.

> **Definitions:** The value for which 25% of the data lie below that value is called the *lower quartile* (or 25th percentile). Similarly, the value for which 25% of the data lie above that value is called the *upper quartile* (or 75th percentile). Quartiles can be calculated by determining the median of the values above/below the location of the overall median. The difference between the quartiles is called the *inter-quartile range* (IQR), another measure of variability along with standard deviation. The *five-number summary* for the distribution of a quantitative variable consists of the minimum, lower quartile, median, upper quartile, and maximum.

15. Explore the data using quartiles and the 5-number summary.

a.   Calculate the lower quartile for the women's haircut prices. First note that the median is the $(37+1)/2 = 19^{th}$ ordered value. So, the lower quartile is the median of the bottom 18 values, which is found in position $(18+1)/2 = 9.5$. So, the lower quartile is the average of the 9th and 10th ordered values from the bottom.

b.   Similarly, calculate the upper quartile for the women's haircut prices.

lower **22.5**   **70**

c.   Calculate the lower and upper quartiles for the men's haircut prices.

7   **22.5**

16. Report the five-number summary for the women's haircut prices and for the men's haircut prices:

	Minimum	Lower quartile	Median	Upper quartile	Maximum
Women's haircut prices	0	7	20	~~22.25~~ 22.50	~~145~~ 75
Men's haircut prices	0	22-5	45	70	155

Definition: A *boxplot* is a visual display of the five-number summary. The box displays the middle 50% of the distribution and its width (the IQR) helps us compare the spread of the distribution; the whiskers extend to the smallest and largest values in the data set.

---

girls are willing to pay more. & its almost double.

17. In the applet, check the box for **Boxplot** to overlay the two boxplots as well. Describe what the boxplots reveal about how the distributions of haircut prices compare between male and female students in the professor's class.

MEN   0   7   20   22 -50   75

WOMEN   0   22-50   45   70   150

18. Find the IQR for both the men's and women's haircut prices.

20 & 45   2.

One advantage to the IQR is that it is not sensitive to extreme values/outliers like the standard deviation is. Just like the median is not sensitive to extreme values/outliers, but the mean is.

Key Idea: IQR is a resistant measure of variability, whereas the standard deviation is sensitive to extreme values and skewness.

---

19. Remove the male haircut price of $75 from the data and (in the applet) find the new SD and IQR. Compare them to the values you had earlier and confirm that this illustrates that the IQR is more resistant to extreme values than is the SD

its alot lower than before.

More Data

20. Collect data from yourself and your classmates on most recent haircut price and gender. Enter the data into the applet, and examine dotplots, boxplots, and summary statistics. Write a paragraph or two summarizing what the data reveal about whether haircut price is *associated* with gender among your classmates. Also address the research questions with which this exploration began. Finally, comment on how broadly you would be willing to *generalize* your findings.

Exploration 6.1B: Cancer Pamphlets

Researchers in Philadelphia investigated whether pamphlets containing information for cancer patients are written at a level that the cancer patients can comprehend (Short, Moriarty, and Cooley, 1995). They applied tests to measure the reading levels of 63 cancer patients and also the readability levels of 30 cancer pamphlets (based on such factors as the lengths of sentences and number of polysyllabic words). These numbers correspond to grade levels, but patient reading levels of under grade 3 and above grade 12 are not determined exactly.

The following tables indicate the number of patients at each reading level and the number of pamphlets at each readability level:

Table 6.1: Comparing of pamphlet readability and patient reading levels

Patients' reading levels	< 3	3	4	5	6	7	8	9	10	11	12	> 12	Total
Count (number of patients)	6	4	4	3	3	2	6	5	4	7	2	17	63

Pamphlets' readability levels	6	7	8	9	10	11	12	13	14	15	16	Total
Count (number of pamphlets)	3	3	8	4	1	1	4	2	1	2	1	30

1. Explain why the way these data are presented do not allow for you to calculate the *mean* reading level of a patient.

2. Determine the *median* reading level of a patient and the median readability level of a pamphlet. (*Hint*: First note that there are 63 patients and 30 pamphlets.)

   Patient: Pamphlet:

3. How do these medians compare? Are they fairly close?

4. Does the closeness of these medians indicate that the pamphlets are well matched to the patients' reading levels? Explain.

5. What proportion of the patients in the study do not have the reading skill level necessary to read even the simplest pamphlet in the study?

6. Do you want to re-think your answer to #4 in light of your answer to #5?

This exploration reveals that measures of center (or even variability) do not always tell the whole story when you are analyzing data to address a particular research question. In this case the research question of whether pamphlets' readability levels are well-aligned with patients' reading levels requires looking at the entire distributions, not simply at measures of center.

Section 6.1 Summary

With **quantitative** as well as categorical variables, the first steps in exploring data are to produce graphs and calculate summary statistics.

- When comparing two groups with a quantitative response variable, *parallel dotplots* can display the distributions.
    - Examining the centers of the dotplots and calculating means (averages) can reveal whether one group has a tendency toward larger or smaller values than the other.
    - Examining the variability of the distributions and calculating standard deviations (SDs) can reveal whether one group has more variability (less consistency) than the other.
    - Comparing medians between the groups can be more informative than comparing means when the distributions are skewed or when outliers are present.
    - The **inter-quartile range** (IQR) provides a resistant alternative to standard deviation as a measure of variability
- The **five-number summary** can be used to summarize the distribution of a quantitative variable. This summary consists of the minimum, lower quartile, median, upper quartile, and maximum.

o **Boxplots** provide a visual display of the five-number summary. Comparing the overlap and width of the boxplots helps illuminate differences in center and spread of two or more distributions.

Section 6.2: Comparing Two Means: Simulation-Based Approach

Introduction

In Section 6.1 we saw how to describe and explore data when comparing two groups on a quantitative variable. In this section, we will dig into the 3S Strategy to compare two groups with a quantitative response variable and address questions about how we can assess the statistical significance of the observed difference between two groups. Instead of using proportions (as was done with categorical variables) we will be using means (averages) for our comparisons.

Example 6.2: Bicycling to Work

Step 1. Ask a research question. Does type of bicycle frame affect commuting time?

Step 2. Design a study and collect data. An article that appeared in the *British Medical Journal* (2010) presented the results of a study conducted by researcher Jeremy Groves, whose objective was to determine whether the type of frame of his bicycle could affect his travel time to work. On each of 56 days (from mid-January to mid-July 2010), Dr. Groves tossed a coin to decide whether he would ride the 27 miles to work on his carbon frame bicycle that was light in weight, weighing 20.9 lbs. or on his steel frame bicycle that was heavier, weighing 29.75 lbs. He then recorded the commute time for each trip.

Think about it: What are the experimental units in this study? Identify and classify the roles of the variables in this study.

The experimental units in this study are the trips made to work on 56 different days. The explanatory variable is which bike Dr. Groves rode, and the response variable is his commute time (in minutes). Notice that the explanatory variable is categorical (binary) with two outcomes: carbon frame bicycle or steel frame bicycle, whereas the response variable is quantitative.

The data table that Dr. Groves used to collect his data started like that in Table 6.2.

Table 6.2: Data collected by Jeremy Groves on commuting times for his two different bikes

Date	Bike Frame	Time (min)
Jan. 20	Steel	115.50
Jan. 21	Carbon	115.25
Jan. 22	Steel	116.50
Jan. 23	Carbon	114.00
Jan. 24	Carbon	119.00
...	...	...

Remembering the research question for this study, we can state our hypotheses as follows:

Null hypothesis: There is no association between which bike is used and commute time. That is, commute time is not affected by which bike is used.

Alternative hypothesis: There is an association between which bike is used and commute time. That is, commute time is affected by which bike is used.

Recall that in Chapter 5, when the response variable was categorical, we compared the *proportions* of successes between the two samples. Now that the response variable is quantitative, we can compare *means* between the two samples. Accordingly, we can define our parameter of interest to be the underlying difference in the long-run mean commute time for the two types of bikes. This can be represented in symbols as $\mu_{carbon} - \mu_{steel}$ where

μ_{carbon} represents the long-run mean commute time using carbon frame (lighter) bike,
μ_{steel} represents the long-run mean commute time using steel frame (heavier) bike.

Note that the Greek letter "mu" (μ) is used to denote the parameter "long-run mean." So we are distinguishing the *long-run means* that would result from Dr. Groves riding to work forever from the *sample means* that he determined in his study (see FAQ 6.2.1 for details). Using the symbols μ_{carbon} and μ_{steel} we can restate our hypotheses to be,

$H_0: \mu_{carbon} = \mu_{steel}$ OR $\mu_{carbon} - \mu_{steel} = 0$
$H_a: \mu_{carbon} \neq \mu_{steel}$ OR $\mu_{carbon} - \mu_{steel} \neq 0$

Keep in mind that hypotheses are always about populations or processes, not about the sample data. We don't need to hypothesize or make inferences about the samples, because we have the data from the sample and we can see that the sample means are not identical. But we want to use these sample data to draw inferences about the long-run process or population the data came from. That is why, as you may have already noticed, the hypotheses stated for this bike study are statements made about the general association between commute time and which bike is used, not about the sample of 56 trips.

Think about it: Summarize the study design used here: Is this an observational study or an experiment?

This was a randomized experiment because the researcher used random assignment (coin flip) to decide which bike he would ride on a particular day.

Step 3. Explore the data. To graphically display the distributions of sample commute times, separated by which bike was used, we can create parallel dotplots, as shown in Figure 6.3.

Figure 6.3: Graphical display of the distribution of commute times for the two bikes

To numerically summarize the commute times, for each bike type, we can report the mean commute time, the standard deviation (SD), and the five number summary. Recall that larger values of the standard

deviation indicate more variability about the mean in the sample data. Table 6.3 reports these summary statistics for Dr. Groves' data.

Table 6.3: Summary statistics for commute time (in minutes), separated by bike type

Bike type	Sample size	Mean	SD	Min.	First Quart.	Med.	Third Quart.	Max.
Carbon frame (lighter)	26	108.34	6.25	100.1	102.75	107.7	112.5	123.3
Steel frame (heavier)	30	107.81	4.89	97.7	104.65	107.5	111.3	117.7

Step 4. Draw inferences beyond the data. As can be seen from the dotplots and the numerical summaries, both the sample mean commute time and the sample variability were higher for the carbon frame (lighter) bike. This indicates that there was a tendency for longer commute times and less consistency in commute times with the carbon frame. But this difference doesn't appear to be large. Maybe he was unlucky in his random assignment and just happened to have heavier traffic on the route on the days he used the carbon frame bike? Or does this reflect a difference in the overall mean commute times between the two routes that is not typical of random chance? If so, can we estimate how much longer we think the average commute would be with the carbon frame? Is it enough longer to be willing to switch to the heavier bike permanently?

Think about it: Is it *possible* to have obtained a difference in sample mean commute times of 0.53 minutes (108.34 – 107.81) if, in general, commute time is not affected by type of bike used? Do you think it's *very unlikely* for such a large difference in sample means to have occurred by chance alone, due to random assignment, if commute time is not affected by type of bike used? How might you decide?

Notice that these are the same questions that we asked repeatedly in Chapter 5 when we analyzed studies with a categorical response variable. The only difference now is that the response variable is quantitative. Again, the answer is yes. Of course it's *possible* to obtain such an extreme difference in group means by chance alone. But is it *plausible*? Again, as we have done since Chapter 1, we will use simulation to investigate whether such a difference is very unlikely to have occurred by chance alone. In this case, we can consider the "chance" as arising from the random assignment process – each trip would have had the same commute time and it was simply random chance that labeled some trips as carbon frame and others as steel frame.

3S Strategy: We will now apply the 3S Strategy to the sample data to evaluate how likely such an extreme difference in group means is to occur by chance alone. As we've seen before, the first step in the 3S process is to choose a statistic. In this case, because the distributions of commute times are reasonably symmetric in both groups, a reasonable choice of statistic is the difference in sample means.

1. Statistic: We calculate that the observed difference in sample mean commute times ($\overline{x}_{carbon} - \overline{x}_{steel}$) is 0.53 minutes.

2. Simulation: To investigate the key question of whether the observed difference in sample mean commute times is unlikely to have occurred by chance alone, we will conduct a simulation analysis. We can use $26 + 30 = 56$ index cards representing his 26 trips on the lighter carbon frame bike and 30 on the heavier steel frame bike. But now instead of color coding the cards to represent having a success/failure response as in Chapter 5, we need to incorporate the numerical values of the commute times into our

analysis. We can do this by writing the 26 commute times for the lighter bike on 26 of the cards, and do the same for the 30 commute times with the heavier bike on the other 30 cards.

Now suppose the time of the commute is not associated with which bike is used. (This is what the null hypothesis says.) We can simulate this null hypothesis by shuffling all 56 cards and randomly redistributing them into two stacks: one with 26 cards (to represent the carbon frame bike) and another with 30 cards (to represent the steel frame bike), regardless of the response values (or group membership) written on the cards. We can then calculate the difference in the mean times between our two stacks of cards. By repeating this over and over again, we can develop a null distribution for the difference in the mean commute times. The table below summarizes the key aspects of the simulation:

Null hypothesis = Average times on each of the two bikes are the same (bike type is not associated with commute time)
One repetition = Re-randomizing commute times to bike type
Statistic = Difference in two group means

Figure 6.4(a) shows the dotplots of the data observed in the original study, with the top dotplot showing the 26 observations of carbon (lighter) bike commute times as circles, and the dotplot below it showing the 30 observations of steel (heavier) bike commute times as solid squares. These are equivalent to the dotplots from Figure 6.3, just with overlapping dots. Figure 6.4(b) shows one possible shuffle (or re-randomization) of the commute times into the two groups. Notice that the shuffling and redistribution has resulted in some of the solid square (heavier bike) values ending up in the re-randomized lighter bike

group, and vice versa. This shuffle produced a difference of 2.42 minutes (109.35 – 106.93), which is more extreme than the observed difference of 0.53 minutes, because it is farther away from 0.

Figure 6.4: (a) Data distribution in *Bicycling to Work,* (b) After shuffling (re-randomizing) the commute times.

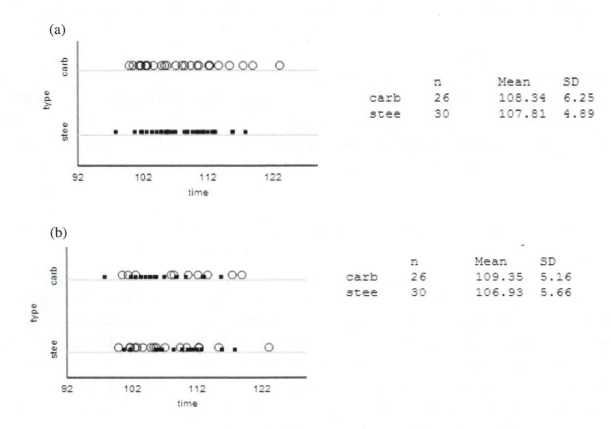

Figure 6.5 shows the results of a second shuffle. Notice that the difference in group means here was -0.70 minutes (= 107.68 – 108.38), a value that (like the result of the previous shuffle) is more extreme than the observed difference of 0.53 minutes, because -0.7 is farther away from the null hypothesized difference of 0, compared to 0.53.

Figure 6.5: After shuffling the commute times in *Bicycling to Work* a second time.

	n	Mean	SD
carb	26	107.68	6.43
stee	30	108.38	4.67

Think about it: What does the value -0.7 minutes imply? How can we have a negative number when we are talking about time?

Keep in mind that our statistic here is the *difference* in the two sample means. Therefore, a difference of −0.70 tells us that the mean commute time for the carbon (lighter) bike was smaller than that for the steel (heavier) bike, after the shuffling. Under the null hypothesis, we expect both positive and negative values because the longer commute times are just as likely to end up in either group. In fact, we would expect a positive difference roughly half the time and a negative difference roughly half the time.

Figure 6.6 shows the results from a third shuffle. This shuffled difference in means turned out to be 1.05 minutes (108.62 – 107.57). Is this difference more or less extreme than the observed difference of 0.53 min? It is more extreme because 1.05 is farther from the null hypothesized difference of 0 than 0.53 is.

Figure 6.6: After shuffling and re-randomizing the commute times in *Bicycling to Work* a third time

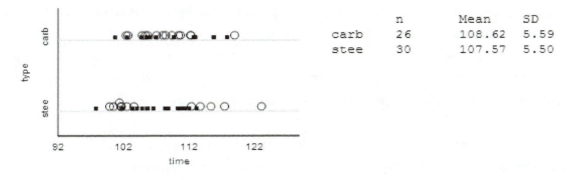

	n	Mean	SD
carb	26	108.62	5.59
stee	30	107.57	5.50

To this point we've seen three separate shuffled differences in means between the two groups: 2.42, -0.70, and 1.05. In all three cases we've obtained a difference in means larger (in absolute value) than the observed value of the statistic (0.53). Of course, we really need more shuffles (or re-randomizations) to get a better sense of the long-run pattern of the difference in means by chance (random assignment) alone. To see the long-run pattern of the could-have-been differences in sample means, if the null hypothesis of no association is true, we want to repeat the shuffling many more times. Figure 6.7 shows a histogram of 1000 possible values of the difference in sample means that could have happened if commute time was not affected by which bike was used.

Figure 6.7: Distribution of 1000 values of difference of group means that could have been observed if long-run mean commute time was the same for both bikes

> **Think about it:** At what numeric value is this histogram in Figure 6.7 centered? Why does that make sense?

Notice that the histogram in Figure 6.7 is centered at approximately 0. Remember that these values of the difference in sample means were generated assuming the null hypothesis is true, that is, the long-run means of the commute times are the same for the two bikes. If the null hypothesis is true, the difference in sample means should be around 0 (sometimes more than 0, and sometimes less than 0). The key question is how far the differences would get away from 0, which will depend on how large the sample sizes were and how variable the sample data were (the commute times for each bicycle). Notice in Figure 6.7 that all of the values of the difference in sample means fall between roughly -5 and 5 minutes.

In Dr. Groves' experiment, the observed difference in mean commute times was 0.53 min, and we want to find out whether this value is far enough away from 0 to provide convincing evidence that this result did not happen by chance alone and that there really is a difference between the underlying long-run mean commute times for the two bikes.

Think about it: Looking at the histogram in Figure 6.7, where is the observed difference of 0.53 located? If the null hypothesis were true, would a difference of 0.53 min be surprising? How are you deciding?

If the null hypothesis were true, and commute time is not associated with the bike chosen, it would not be at all surprising to see a difference of 0.53 min in the sample mean commute times just by chance. Notice that in Figure 6.7, 0.53 is not out in the tail, but is quite close to the null hypothesized difference of 0. As we have done before, we can count how often the simulation produced a statistic at least as extreme as the observed value (0.53 minutes). In other words, we can use the simulation results to approximate the p-value for this study.

Think about it: Which values will we count in order to approximate the p-value for this study?

Because our alternative hypothesis is two-sided, we approximate the p-value by looking in *both* tails of the distribution of the differences in sample means. Figure 6.8 shows that out of the 1000 simulated random shuffles, 732 resulted in a difference in sample means that was less than or equal to -0.53 or that was greater than or equal to 0.53, making our approximate two-sided p-value 0.732.

Figure 6.8: Histogram showing the 732 repetitions in the two tails where the re-randomization resulted in a difference in sample means that was at least as extreme as the observed difference of 0.53

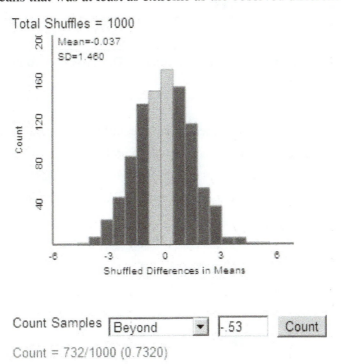

> **Think about it:** How would you interpret this p-value: the probability of what? Assuming what?

We interpret this p-value just as we have interpreted all p-values: the probability of obtaining such an extreme (or more) value for the observed statistic if the null hypothesis were true. Applied to this study: Assuming that the underlying long-run (process) mean commute times for the two bikes are the same, if we were to repeat the random assignment of the carbon frame (lighter) bike to 26 days and the steel frame (heavier) bike to 30 days, we would find the difference in sample mean commute times to differ by 0.53 minutes or even more (in either direction) in about 73.2% of the random assignments.

> **Think about it:** Based on the p-value, what is your conclusion about Dr. Groves' research question of whether the type of bike affects his commute time? Explain the reasoning process behind this conclusion.

3. Strength of evidence. A p-value of 0.732 is quite large, indicating that the observed result would not be surprising at all if the null hypothesis were true. Thus, we do not have evidence to say that type of bike affects commute time. In other words, we do not have evidence that the underlying long-run mean commute times are different for the two bikes. Even though the study was a randomized experiment, the large p-value does not let us conclude that commute time is affected by which bike is used. Of course we are assuming that these 56 trips are representative of the underlying commuting process (vs. all occurring in winter).

> **Think about it:** Based on the p-value, have we proven that Dr. Groves' two bicycles have no effect on commute time?

Remember, a large p-value is not "strong evidence that the null hypothesis is true" (See FAQ 6.2.1 for details). This large p-value merely suggests that the null hypothesis is plausible—we shouldn't rule out random chance, because it is a reasonable explanation for the observed difference in sample mean commuting times. But it's certainly possible there really is an effect of the bicycle on commute time that was not detected in this study.

FAQ 6.2.1 Remind me: What does "long run" mean?

Q: I've got a few questions about this study with Dr. Groves. To start, when you told us about Dr. Groves and his two bicycles, you said that μ_{carbon} represented the mean of an infinite random process. Right?

A: Right.

Q: But Groves can't ride forever.

A: Right. The phrase "infinite random process" is a metaphor. It's the same idea as flipping a coin. You can't flip the coin forever either. Eventually, it would get worn down to nothing.

Q: So can you be clearer about what you really mean?

A: I'll try. (You may want to refer back to Section 3 of the Preliminaries.) In order to talk about a long-run probability or long-run average, our simulation-based inferences require three things: (1) Dr. Groves' times using his carbon bike are *random*. (2) The times are *independent* of each other. His time today does not depend on any of his other times. (3) All the times with the carbon bike come from the *same distribution*.

Q: But none of these three things are true.

A: Not exactly true, but perhaps approximately true. How true will depend on the situation, and that is often a matter of judgment.

A:

        Q: I'm still struggling.

        The key point is to understand that there are two different means. The easy one is the mean that we actually see. The hard one is the parameter, the one we can't see. We use the metaphor as a way to describe it. But we can also think of the observed mean as the true but unseen mean plus a random error.
        Q: Can you say it in coin-speak?
        A: If I toss a coin 100 times and get 53 heads, my observed proportion of heads is 0.53. If the coin is fair, then the true probability of heads is 0.5. The 0.53 is the observed value, the 0.5 is the parameter value, which we describe as the long-run probability.

Why should I care whether studies are blind?

Q: The bicycle study by Dr. Groves was a randomized experiment, right?

A: Yes.

Q: And you need a randomized experiment for conclusions about cause, right?

A:	Yes.
Q:	So if you do a randomized experiment and reject the null hypothesis of no difference, you can conclude that the treatment you assigned was the cause of the difference you observed. Right?
A:	Not so fast. You have to be very careful about the meaning of two phrases, the "treatment" you assigned, and the difference you "observed."
Q:	I need an example.
A:	Think about Dr. Groves. It may seem that the "treatment" assigned was the bicycle, but that's not the whole story. Groves *knew* which bike he was riding, so part of the assigned treatment was that knowledge. The same issue is critically important in medical studies. Patients are not allowed to know which treatment they are getting. Otherwise, it could be their belief in the medicine, and not the medicine itself, that is responsible if they get better.
Q:	And the "observed" part?
A:	In medical studies, the person who measures or records or diagnoses the outcome is not allowed to know which treatment a patient got. Otherwise that knowledge might influence the outcomes and so be responsible for any observed differences. Studies that keep both the patients and the evaluators in the dark are called "double blind." We first talked about this idea back in Chapter 4.

What does it tell you if you don't have enough evidence to reject H_0?

Q:	Because we didn't reject the null hypothesis in Dr. Groves' case, we accept it, right? The bikes both give him the same commute time?
A:	Sorry, that's wrong. The data do not *prove* that the two means are equal. But in stat-speak, if you "accept" the null you are concluding that the means are equal. In statistics, there's a big difference between "do not reject" and "accept."
Q:	How can that be? Isn't "accept" the opposite of "reject?"
A:	No, at least not in statistics. Is "red" the opposite of "blue?"
Q:	Of course not.
A:	"Blue" is not "red," but "not red" won't always be "blue." It might be "green" or "yellow."
Q:	Can we please get back to statistics?
A:	I thought you'd never ask.
Q:	Suppose I test the null hypothesis that $\mu_1 = \mu_2$, and I get a large p-value. What can I conclude?
A:	It's certainly possible that $\mu_1 = \mu_2$, but there are other possibilities as well.
Q:	Like what?
A:	Perhaps the two means are really not equal, but your samples are too small to detect the

	difference.
Q:	Anything else?
A:	Perhaps your study was badly planned, and/or your data values show a lot of variability?
Q:	Anything else?
A:	Perhaps the two means are not exactly equal, but are so close to equal that the true difference has no practical importance.
Q:	I'm sure you could go on, but please don't.
A:	OK, I'll stop. But promise me you'll never use the phrase "accept the null hypothesis."

Estimating a confidence interval

As we've seen in previous chapters, a natural follow-up question to a test of significance is to estimate the size of the underlying, long-run difference in group means. We use a confidence interval to generate a range of plausible values for this unknown value.

> **Think about it:** How would you apply the 2SD Method to generate a confidence interval for the difference in long-run mean commute times between the two bikes?

The 2SD Method says that we can generate an approximate 95% confidence interval by taking two times the standard deviation of the simulated null distribution (as shown in Figure 6.8) and adding and subtracting the resulting value to the observed statistic. In this case, the standard deviation of the null distribution is 1.46 minutes (Figure 6.8). Thus, the endpoints of the 95% confidence interval for the difference in long-run mean commute times are roughly $0.53 - 2 \times 1.46 = -3.45$ to $0.53 + 2 \times 1.46 = 2.59$. In short, we are 95% confident that the difference in long-run mean commute times between the two bikes ($\mu_{carbon} - \mu_{steel}$) is between -3.45 minutes and 2.59 minutes. This means that the mean commute time with the carbon bike could be as much as 3.45 minutes longer than with the steel bike, or as much as 2.59 minutes shorter.

> **Think about it:** What are the implications of having a confidence interval that contains 0?

The fact that the confidence interval contains zero means that it is plausible that the difference in long-run means between the two groups is zero—in other words that the null hypothesis is plausible. Thus, the confidence interval yields a result that is consistent with the result of the test of significance.

Step 5. Formulate conclusions.

> **Think about it:** Can we generalize our conclusion to a larger population? Can we draw a cause-effect conclusion?

As we've seen before, there are two key questions to ask with regards to the scope of conclusions possible in this study: Was the sample randomly selected from a larger population? Were the observational units randomly assigned to treatments?

The answer to the first question is no, the sample was not randomly selected from a population. In this case, Jeremy Groves commuted on consecutive days and so if he, say, did the experiment in winter as opposed to summer, he might obtain different results. And, of course, because this is only one person, and two different bikes, we certainly shouldn't generalize any further than this particular individual and his two bikes.

The answer to the second question is yes, the observational units were randomly assigned to the treatments. So, had the p-value been small, we might be inclined to say a cause-effect conclusion is possible here (namely, that bike choice is causing a difference in his commute times). However, a serious limitation of this study is that it was not *double-blind*.

In this case, the researcher knew which bike he was on for each of the commutes. Could this have influenced his performance? It could have. He stated in the article that his steel frame bike was more comfortable. Perhaps this comfort helped him pedal faster. Perhaps he likes the steel frame bike better and this prompted him to pedal faster. This implies there could have been factors apart from the type of frame that were affecting his commute time differently for the two bikes.

Step 6. Look back and ahead. Bicyclists are willing to pay a lot of money to reduce the weight of their bikes. Presumably they do this to increase their speed or reduce their effort in riding. Does this study show that this added expense is worth it?

No, for a number of reasons. In addition to only being a single individual on two particular bikes, the lack of blinding in the study is a serious flaw in the design which could impact results. And, of course, our large p-value doesn't provide convincing evidence against the null hypothesis (that choice of bike does not affect commute time), only that it's plausible that there is no difference on average. Further studies looking at a variety of individuals, terrains, weather conditions, and bikes would be needed to make a stronger conclusion. Though difficult, a randomized experiment with blinding (e.g., make identical looking bikes except for the material of the frames; see FAQ 6.2.1 for details) would allow for stronger, more definitive, conclusions. Finally, even if there is a difference in the commute times on the two different bicycles, the 95% confidence interval suggests that the difference is likely only to be 2-3 minutes, on average, suggesting that the difference may not be worth the extra expense.

Exploration 6.2: Lingering Effects of Sleep Deprivation

Step 1: Ask a research question. Many students pull "all-nighters" when they have an important exam or a pressing assignment. Concerns that may arise include: *Can you really function well the next day after a sleepless night? What about several days later: Can you recover from a sleepless night by getting a full night's sleep on the following nights?*

1. What is your research conjecture about whether or not one can recover from a sleepless night by getting a full night's sleep on the following nights? What are some other related questions that you would be interested in investigating related to this issue?

Step 2: Design a study and collect data. Researchers Stickgold, James, and Hobson investigated delayed effects of sleep deprivation on learning in a study published in *Nature Neuroscience* (2000). Twenty-one volunteers, aged 18 to 25 years, were first trained on a visual discrimination task that involved watching stimuli appear on a computer screen and reporting what was seen.

Figure 6.9. Subjects were flashed the screen on the left and then it was masked by the screen on the right. Then subjects were asked whether they had seen an L or a V and whether the slanted lines were placed vertically or horizontally.

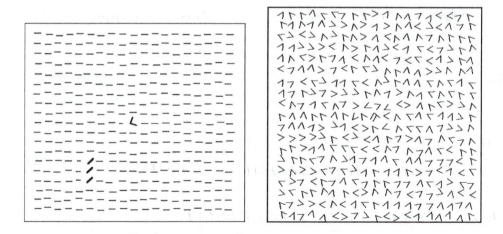

After the training period, subjects were tested. Performance was recorded as the minimum time (in milliseconds) between the appearance of stimuli and an accurate response. Following these baseline measurements, one group was randomly assigned to be deprived of sleep for 30 hours, followed by two full nights of unrestricted sleep, whereas the other group was allowed to get unrestricted sleep on all three nights. Following this, both groups were retested on the task to see how well they remembered the training from the first day. Researchers recorded the *improvement* in performance as the decrease in time required at retest compared to training. (Note: For example, if someone took 5 milliseconds (ms) to respond at the beginning of the study and then 2 ms, to respond at the end, the improvement score is 3 ms. But if someone took 2 ms at the beginning and then 5 ms at the end, the improvement score is -3 ms.)

The goal of the study was to see whether the *improvement scores* tend to be higher for the unrestricted sleep treatment than for the sleep deprivation treatment.

2. Identify the explanatory and response variables in this study. Also classify them as either categorical or quantitative.

    Explanatory: Type:
       Sleep   *Cat. Binary*

    Response: Type:
       difference/
       Performance time in
       milliseconds

3. Was this an experiment or an observational study? Explain how you are deciding.

4. Let $\mu_{unrestricted}$ be the long-run mean improvement on this task three days later when someone has had unrestricted sleep and let $\mu_{deprived}$ denote the long-run mean improvement when someone is sleep deprived on the first night.

    $\mu_{restricted} \neq \mu_{sleep\ deprivation}$

    In words and symbols, state the null and the alternative hypotheses to investigate whether sleep deprivation has a negative effect on improvement in performance on visual discrimination tasks. (Hint: For the alternative hypothesis: Do you expect the people to do better or worse when sleep deprived? Based on your answer, what sign/direction should you choose for the alternative hypothesis?)

    $\bar{X}$ unrestricted
    $- \bar{X}$ sleep depr.
    _____
    16.92

    H0: No association Between Sleep Deprivation & Improvement scores

    Ha: There [is] Absolution Between sleep deprivation & improvement scores.

Here are the data, with positive values indicating better performance at retest than at training, and negative values indicating worse performance at retest than at training:

Unrestricted-sleep group's improvement scores (milliseconds):
25.2, 14.5, -7.0, 12.6, 34.5, 45.6, 11.6, 18.6, 12.1, 30.5

Sleep-deprived group's improvement scores (milliseconds):
-10.7, 4.5, 2.2, 21.3, -14.7, -10.7, 9.6, 2.4, 21.8, 7.2, 10.0

Step 3: Explore the Data.

5. To look at graphical and numerical summaries of the data from the study, go to the **Multiple Means** applet. The sleep deprivation data have already been entered into the applet.

    a. Notice that the applet creates parallel dotplots, one for each study group. Based on these dotplots alone, which group (unrestricted or deprived) appears to have had the higher mean improvement? How are you deciding?

    b. Based on the dotplots alone, which group (unrestricted or deprived) appears to have had more variability in improvement? How are you deciding?

    c. Notice also that the applet also computes numerical summaries of the data, such as, the mean and standard deviation (SD) for the improvements in each group.

        i.   For the *unrestricted* group, record the sample size (n), mean, and SD.

             $n_{unrestricted} =$ $\bar{x}_{unrestricted} =$ $SD_{unrestricted} =$

        ii. For the *deprived* group, record the sample size (n), mean, and SD.

             $n_{deprived} =$ $\bar{x}_{deprived} =$ $SD_{deprived} =$

Recall from earlier in the course, that standard deviation is a measure of variability. Relatively speaking, smaller standard deviation values indicate less variability and a distribution whose data values tend to cluster more closely together, compared to a distribution with a larger standard deviation.

    d. Based on the numerical summaries reported in #5c, which group (unrestricted or deprived) had the higher mean improvement?

e.   Based on the numerical summaries reported in #5c, which group (unrestricted or deprived) had the higher variability in improvement?

f.   Notice that the applet also reports the observed difference in means for the improvements of the two groups. Record this value (and its measurement units).

$$\overline{x}_{unrestricted} - \overline{x}_{deprived} = 15.92$$

g.   Before you conduct an inferential analysis, does this difference in sample means (as reported in #5f) strike you as a meaningful difference? Explain your answer.

yes

Step 4: Draw Inferences.

6.   What are two possible explanations for why we observed the two groups to have different sample means for improvement in performance?

There is really no difference.

7.   Describe how you might go about deciding whether the observed difference between the two sample means is statistically significant. (*Hint*: Think about how you assessed whether an observed difference between two sample proportions was statistically significant in Chapter 5. Use the same strategy, with an appropriate modification for working with means instead of proportions.)

Once again the key question is how often random assignment alone would produce a difference in the groups at least as extreme as the difference observed in this study, if there really were no effect of sleep condition on improvement score. You addressed similar questions in Chapter 5 when you analyzed the dolphin therapy and yawing studies. The only change is that now the response variable is *quantitative* rather than *categorical*, so the relevant statistic is the difference in group *means* rather than the difference in group *proportions*. Also once again, we use *simulation* to investigate how often such an extreme difference would occur by chance (random assignment) alone (if the null hypothesis of no difference / no effect / no association were true).

In other words, we will again employ the 3S Strategy.

1. Statistic:

8. A natural statistic for measuring how different the observed group means are from each other is the difference in the mean improvement scores between the two groups. Report the value of this statistic, as you did in #5f. *15.92 - Sample.*

2. Simulate: You will start by using index cards to perform a tactile simulation of randomly assigning the 21 subjects between the two groups, *assuming* that sleep condition has no impact on improvement. Because the null hypothesis asserts that improvement score is not associated with sleep condition, we will assume that the 21 subjects would have had exactly the same improvement scores as they did, *regardless* of which sleep condition group (unrestricted or deprived) the subject had been assigned.

9.

    a.   How many index cards do you need to conduct this simulation?

    b.   What will you write on each index card?

To conduct *one repetition* of this simulation:

- Shuffle the stack of 21 cards well, and then randomly distribute cards into two stacks: one stack with 10 cards (the unrestricted group) and one with 11 (the sleep deprived group).

- Calculate and report the sample means for each re-randomized group:

    Re-randomized unrestricted group's mean:

    Re-randomized deprived group's mean:

- Calculate the difference in group means, subtracting unrestricted mean minus sleep deprived mean. Report this value.

- Combine this result with your classmates' to create a dotplot that shows the distribution of several possible values of difference in sample means that could have happened due to pure chance, if sleep condition has no impact on improvement. Sketch the dotplot below, being sure to label the horizontal axis.

Label:

c. At about what value is the dotplot centered? Explain why this makes sense. (*Hint*: What are we assuming to be true when we conduct the simulation?)

d. Where is the observed difference in means from the original study (as reported in #8) on the dotplot? Did this value happen often, somewhat rarely, or very rarely? How are you deciding?

10. As before with simulation-based analyses, you would now like to conduct many, many more repetitions to determine what is typical and what is not for the difference in group means, assuming that sleep condition has no impact on improvement score. We think you would prefer to use a computer applet to do this rather than continue to shuffle cards for a very long time, calculating difference of group means by hand. Go back to the **Multiple Means** applet, check the **Show Shuffle Options** box, select the **Plot** display, and press **Shuffle Responses**.

a. Describe what the applet is doing and how this relates to your null hypothesis from #4. (Note: You can check the Show Original Colors box to see which groups the observations originally belonged to.)

b. Record the simulated difference in sample means for the re-randomized groups, as given in the applet output. Is this difference more extreme than the observed difference from the study (as reported in 9 to 8)? How are you deciding?

c. Click on **Shuffle Responses** again, and record the simulated difference in sample means for the re-randomized groups. Did it change from #10b?

d. Click on **Re-Randomize** again, and record the simulated difference in sample means for the re-randomized groups. Did it change from #10b and #10c?

e. Now to see many more possible values of the difference in sample means, assuming sleep condition has no impact on improvement, do the following in the **Multiple Means** applet:

    • Change **Number of Shuffles** from 1 to 997.
    • Press **Shuffle Responses**, to produce a total of 1000 shuffles and re-randomized statistics.

f. Consider the histogram of the 1000 could-have-been values of difference in sample means, assuming that sleep condition has no effect on improvement.

    i.   What does one dot on the dotplot represent? (*Hint*: Think about what you would have to do to put another dot on the graph.)

    ii.   Describe the overall shape of the null distribution displayed in this dotplot.

    iii. Where does the observed difference in sample means (as reported in 9 to 8) fall in this dotplot: near the middle or out in a tail? Are there a lot of dots that are even more extreme than the observed difference, assuming sleep condition has no impact on improvement? How are you deciding?

g. To estimate a p-value, continue with the **Multiple Means** applet.

- Type in the observed difference in group means (as reported in #8) in the Count Samples box (for the one-sided alternative hypothesis), and press **Count**.
- Record the approximate p-value.

h. Fill in the blanks of the following sentence to complete the interpretation of the p-value.

The p-value of _____ is the probability of observing _____

_____ assuming _____.

3. Strength of Evidence:

11. Based on the p-value, evaluate the strength of evidence provided by the experimental data against the null hypothesis that sleep condition has no effect on improvement score: not much evidence, moderate evidence, strong evidence, or very strong evidence?

12. *Significance*: Summarize your conclusion with regard to strength of evidence in the context of this study.

13. *Estimation*:
    a. Use the 2SD Method to approximate a 95% confidence interval for the difference in long-run mean improvement score for subjects who get unrestricted sleep minus the long-run mean improvement score for subjects who are sleep deprived. (*Hints*: Remember the observed value of the difference in group means, and obtain the SD of the difference in group means from the applet's simulation results. The interval should be observed difference in means ± 2 SD, where SD represents the standard deviation of the null distribution of the difference in group means.)

    b. Interpret what this confidence interval reveals, paying particular attention to whether the interval is entirely positive, entirely negative, or contains zero. (*Hint*: Be sure to mention "direction" in your interpretation by saying how much larger improvement scores are on average for the treatment you find to have the larger long-run mean: I'm 95% confident that

the long-run mean improvement score is _____ to _____ higher with the
_____ treatment as opposed to the _____ treatment.)

Step 5: Formulate Conclusions.

14. *Generalization*: Were the participants in this study randomly selected from a larger population?
Describe the population to which you would feel comfortable generalizing the results of this
study.

15. *Causation*: Were the participants in the study randomly assigned to a sleep condition? How does
this affect the scope of conclusion that you can draw?

Another statistic

Could we have chosen a statistic other than the difference in group means to summarize how different the
two groups' improvement scores were? Yes, for example we could have used the difference in group
medians. Why might we do this? For one reason, the median is less affected by outliers than the mean
(see Section 3.2).

To analyze the difference in group medians, we carry out the 3S Strategy as before, except:
- We calculate the observed value of the difference in *medians* as the statistic.
- After we conduct the re-randomizing, we calculate the difference in medians for the re-randomized
data. Then we repeat this process a large number of times.
- We determine the p-value by counting how many of the simulated statistics are at least as large as the
observed value of the difference in *medians*.

16. Return to the **Multiple Means** applet and use the Statistic pull-down menu (on the left) to select
Difference in Medians.

a. From the Summary Statistics for the original data, record the median improvement score for
each group. Also record the difference between the medians (subtracting unrestricted median
minus deprived median).

Unrestricted median: Deprived median:

Difference (unrestricted – deprived):

b. Enter 1000 for **Number of Shuffles**, and press **Shuffle Responses**. Describe the resulting
    null distribution of difference in group medians. Does this null distribution appear to be
    centered near zero? Does it seem to have a bell-shaped distribution?

c. To calculate a p-value based on the difference in medians, enter the observed value in the
    Count samples box . Then press **Count**. Report both the value that you enter into the applet
    and the resulting p-value.

d. Does this p-value indicate strong evidence that sleep deprivation has a harmful effect on
    improvement score? Explain how you are deciding.

e. With which statistic (difference in *means* or difference in *medians*) do the data provide
    stronger evidence that sleep deprivation has a harmful effect on improvement score? Explain
    how you are deciding.

17. **Step 6: Look back and ahead.**

Looking back: Did anything about the design and conclusions of this study concern you? Issues you may want to critique include:

- Any mismatch between the research question and the study design
- How the experimental units were selected
- How the treatments were assigned to the experimental units
- How the measurements were recorded
- The number of experimental units in the study
- Whether what we observed is of practical value

Looking ahead: What should the researchers' next steps be to fix the limitations or build on this knowledge?

Section 6.2 Summary

The 3S Strategy can be used to assess whether two sample means differ enough to conclude that there is a genuine difference in the population means or long-run means of processes. The reasoning process of statistical significance is the same and the simulation process is very similar to that used for comparing two proportions.

- The null hypothesis can be expressed as no association between the two variables.
  - o Equivalently, the null hypothesis asserts that the two population means are the same, or in the case of processes, the two long-run means are the same.
- The alternative hypothesis says that there is an association between the two variables.
  - o The alternative hypothesis can be one-sided or two-sided, depending on whether the researchers have a particular direction in mind prior to collecting data.
- The difference in sample means between the two groups is often the statistic of interest.
- The key idea in simulating the null distribution of the difference in sample means is to shuffle the values of the response variable and redistribute at random to the explanatory variable groups.
  - o This shuffling generates simulated values of the difference in sample means under the assumption of no association to help us see how much these sample means tend to differ from each other by random chance alone.

- o With a randomized experiment, this shuffling simulates re-randomizing the subjects into the two groups, assuming that the subjects' numerical responses would have been the same regardless of which group they were assigned to.
- As always, the p-value is calculated as the proportion of repetitions of the simulation in which the simulated value of the statistic is at least as extreme as the observed value of the statistic.
  - o Also as always, a small p-value provides evidence against the null hypothesis.

The 2SD Method can be used to produce a confidence interval for estimating the size of the difference between the two population means or two long-run process means.

- As with comparing two proportions, an important question is whether the interval is entirely negative, entirely positive, or includes zero.
  - o If the interval does not include zero, that confirms that our test of significance will reject the null hypothesis that the two population means, or in the case of processes the two long-run means, are not the same.
  - o Whether the interval contains positive or negative values indicates which population (or long-run) mean of the two is estimated to be larger, and by how much.
  - o If the interval contains zero, there is not convincing evidence against the null hypothesis that the two population means or the two long-run process means are equal.

An alternative to comparing the means between two groups is to compare the medians. The 3S Strategy works just as well with comparing medians as with comparing means.

Section 6.3: Comparing Two Means: Theory-Based Approach

Introduction

Earlier we saw that under certain conditions a normal distribution provides a reasonable model for the null distribution of a sample proportion (Chapters 1–3) and for the null distribution of a difference in group proportions (Chapter 5), whereas a *t*-distribution provides a reasonable model for the null distribution of a standardized sample mean (Chapter 2). In these cases we don't have to simulate, we can instead use these theory-based predictions to anticipate what would happen (the distribution of the statistic or standardized statistic) if we were to simulate.

Maybe you already noticed that in Section 6.2 the shape of the distribution of possible values of difference in two group means was somewhat bell-shaped and centered at zero. If that makes you think 'Here we go again…' you're absolutely right. In this section we will, once again, look at a theory-based approach to predict the distribution of our statistic, leading to *t*-procedures: the two-sample *t*-test and the two-sample *t*-interval. It should be no surprise that this approach will work well most of the time, but not all of the time—only when certain validity conditions are met.

Example 6.3: Breastfeeding and Intelligence

An article by researchers Jacobson, Chiodo, and Jacobson published in the journal *Pediatrics* (1999) studied whether and how children who were breastfed during infancy differed from those who weren't breastfed. The study involved 323 white children who were recruited at birth in 1980 through 1981 from four Western Michigan hospitals. After some initial exploration, the researchers deemed that the participants in the study were representative of the community in terms of social class, maternal education, age, marital status, and sex of infant. These children were revisited at age 4 years, and assessed on the McCarthy Scales of Children's Abilities. The General Cognitive Index (GCI) – an overall measure of the child's present level of intellectual functioning – was recorded for each child, as was whether or not the child had been breastfed during infancy.

> **Think about it:** What kind of study is this: experiment or observational? Identify and classify the explanatory and response variables.

This was an observational study because the researchers were not involved at all with determining which children would be breastfed. The explanatory variable is whether or not the baby was breastfed during infancy (categorical), and the response variable is the baby's GCI measure at age 4 (quantitative). Because the study was not a randomized experiment, we cannot consider drawing a *cause-and-effect* conclusion between breast-feeding and higher cognitive functioning.

Because the researchers were interested in any differences in GCI scores between children who were breastfed and those who weren't, they formulated their null and alternative hypotheses as shown here:

Null hypothesis: There is no association between breastfeeding during infancy and GCI at age 4.

Alternative hypothesis: There is an association between breastfeeding during infancy and GCI at age 4.

Alternatively, we can define our parameter of interest to be, $\mu_{breastfed} - \mu_{not}$ where

$\mu_{breastfed}$ = Population mean GCI value at age 4, for the population of children who are breastfed, and

μ_{not} = Population mean GCI value at age 4, for the population of children who are not breastfed.

And using these symbols we can restate our hypotheses to be,

$$H_0: \mu_{breastfed} = \mu_{not} \text{ OR } \mu_{breastfed} - \mu_{not} = 0$$
$$H_a: \mu_{breastfed} \neq \mu_{not} \text{ OR } \mu_{breastfed} - \mu_{not} \neq 0$$

Table 6.4 gives the summary statistics for GCI at age 4 years, separated by whether or not the child was breastfed during infancy.

Table 6.4: Summary statistics for GCI at age 4, separated by whether or not breastfed during infancy

Group	Sample size, n	Sample mean	Sample SD
Breastfed	237	105.3	14.5
Not breastfed	85	100.9	14.0

Based on the summary statistics given in Table 6.4, the distribution of GCI values can be graphed as shown in Figure 6.10.

Figure 6.10: Histograms/boxplots showing the distribution of GCI values, separated by whether or not breastfed

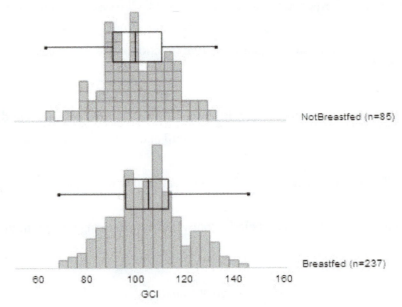

From the summary statistics and histograms/boxplots, we can see that the variability in GCI values was similar in both groups (SD = 14.5 and SD = 14.0), but the mean GCI value for the breastfed group was higher than the mean GCI value for the non-breastfed group (105.3 compared to 100.9). Thus, in the study, the mean GCI values of the two groups differed by 4.4 (= 105.3 – 100.9). *Is this difference unlikely*

to have happened by chance alone, that is, if breastfeeding is not associated with GCI at age 4, and children who are breastfed and those who aren't breastfed have the same population mean GCI at age 4? How do we answer this?

Using the same approach as in Section 6.2, we could use the **Multiple Means** applet to simulate what the difference in sample means could have been by random chance, assuming breastfeeding were not associated with GCI at age 4. As we discussed in Chapter 5, even though this is not a randomized experiment (it is an observational study), we can still use shuffling to redistribute the observed GCI values for the two groups, assuming the null hypothesis of no association, and then approximate a p-value from the simulation results. Figure 6.11 shows the simulated null distribution of the difference in sample means after 1000 shuffles.

Figure 6.11: Distribution of difference in sample mean GCI values, generated assuming breastfeeding is not associated with GCI at age 4

Figure 6.12 circles those shuffles which resulted in values of the difference in sample means that were at least as extreme as that observed in the study, that is, 4.4. Turns out seven out of the 1000 shuffles resulted in values of difference in group means less than or equal to -4.4, and the same number of shuffles resulted in values of difference in group means larger than or equal to 4.4. Adding up the two circled regions as shown in Figure 6.12, we get the approximate two-sided p-value to be .007 + .007 = 0.014. Thus, there is a 0.014 probability of observing a difference in sample mean GCI values for breastfed and non-breastfed children of at least 4.4, just by chance, if breastfeeding were not associated with GCI at age 4.

Figure 6.12: Display of the p-value (circled regions)

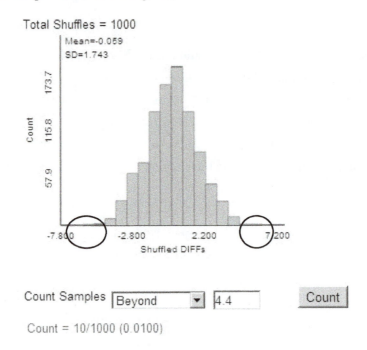

A quick 2SD confidence interval would be $4.4 \pm 2(1.743)$ or (0.914, 7.886). We are about 95% confident that population mean GCI score is .914 to 7.886 points higher for breast-fed babies than babies that aren't breast-fed (about a 5% – 10% increase).

> **Think about it:** What do you notice about the overall shape, center, and variability of the distribution of difference in sample means, as shown in Figure 6.12?

Notice that the shape of the distribution appears to be somewhat like a normal curve, with a mean of approximately 0 (which was difference hypothesized in the null hypothesis!), and standard deviation of 1.743.

> **Think about it:** How many standard deviations away from the hypothesized difference of 0, is the observed difference of 4.4? How are you deciding?

Recall that we use a standardized statistic (e.g., z-statistic or t-statistic) to measure how far away the observed statistic is from the hypothesized value of the parameter as given by the null hypothesis, taking into account the standard deviation of the statistic. When the sample sizes were large, the standardized statistic was also utilized to find a probability based on the area under a normal or a t distribution curve. We can use a similar strategy here. Because the sample sizes in our study are considered large enough (n_1

$= 237$, $n_2 = 85$), we can use the theory-based approach to predict the distribution of the standardized statistic using a mathematical model and hence the p-value from that model.

In the **Multiple Means** applet, if we use the pull-down menu to select ***t*-statistic**, the applet reports the observed *t*-statistic to be 2.46. (This may not match your calculation exactly because above you used the estimated value of the standard error from the simulation rather than the theory-based formula shown at the end of this section.) You should remember *t*-statistics from Chapter 3. Similar to the standardized *z*-statistic for the difference in sample proportions, the *t*-statistic is a standardized statistic for the difference in sample means. Typically the distribution of this *t*-statistic is bell-shaped and symmetric, but it is not exactly normally distributed, being less peaked with fatter tails. As the sample size increases, the distribution of the standardized statistic becomes more normal. At the end of this section we provide a few more details about how this statistic is calculated when comparing two means. As before, the standardized statistic (e.g., $t = 2.46$) is telling us that the observed difference in sample means of 4.4 is 2.46 standard deviations away from the hypothesized difference of 0.

Then the applet allows us to overlay the theoretical *t*-distribution on top of the shuffled *t*-statistics and find the theory-based p-value as shown in Figure 6.13. [Note: You could also use the Theory-Based Inference applet to perform this calculation.] This standardized statistic produces an approximate p-value of 0.0150.

Think about it: Based on the p-value, what conclusion would you draw? Is there statistically significant evidence of a difference between the mean GCI values of the two populations? Can you conclude that breastfeeding improves mean GCI at age 4? To what larger population(s) would you be comfortable generalizing these results?

A p-value of 0.015 is usually considered small enough to provide strong evidence against the null hypothesis. If there were really no association between breastfeeding and GCI values, there would only be a 0.015 chance of obtaining, by random chance alone, sample means as far apart or even farther apart as was found in this study. Thus, we have statistically significant evidence that there is a genuine difference in mean GCI at age 4 between children who are breastfed during infancy and those who are not. Considering that the study was not a randomized experiment, we cannot conclude any cause-and-effect relationship between breastfeeding and GCI at age 4. Also, the participants in this study were all white children born in Western Michigan. This limits the population to whom we can generalize these results. The association between breastfeeding and GCI observed in this study may not hold true for people of other races/ethnicities.

628 Chapter 6: Comparing Two Means

Figure 6.13: Output from the Multiple Means applet, showing the shuffled *t*-statistics and theory-based p-value for the breastfeeding study

theory-based p-value=0.0150, df = 153.01

Notice that the p-value approximated by theory-based method (0.015) very closely matches the p-value approximated by simulation (0.010). The p-values are both small enough, and close enough, to lead you to the same conclusion.

Think about it: So, we were unable to conclude a cause-and-effect relationship between breastfeeding and GCI at age 4. What might be an alternative explanation for the significant difference in mean GCI values of children who were breastfed compared to those who weren't?

Alternative explanations for the difference in mean GCI values of children who were breastfed compared to those who weren't, could be characteristics of the mothers, such as her level of education. Perhaps better-educated mothers are more likely to breastfeed their children and are also more likely to have children of high intellectual functioning. Then this variable, maternal education level, would qualify as a confounding variable, because its effects on intellectual functioning could not be distinguished from the effects of breastfeeding. In that case we cannot say whether the higher GCI values in the breastfeeding group are due to the breastfeeding or due to the mother's education, or perhaps a combination of both, or perhaps neither and some other confounding variable is the real cause. But what the significance test does allow is for us to essentially rule out random chance as a plausible explanation for the large observed difference between the two samples.

> **Think about it:** How could you design a different study that would allow for drawing a cause-and-effect conclusion if a significant difference were found between the groups? Would it be feasible/ethical to conduct such a study?

In order to draw a cause-and-effect conclusion, you would have to perform a randomized comparative experiment. This would mean using random assignment to determine which mothers would breastfeed their babies and which would not. The random assignment would aim to balance out all other variables, such as mother's education level, between the two groups. Then if the group that was breastfed performed significantly better than the other group, we could legitimately conclude that the breastfeeding *caused* the higher intellectual functioning. But conducting such a study would not be feasible or ethical, because such a personal decision as whether to breastfeed a baby cannot be imposed on mothers.

In earlier chapters we have discussed the factors that are associated with the size of the p-value. For the most part, these stay the same. In particular, increasing the sample size still decreases the variability in the null distribution which, in turn, reduces the p-value if nothing else changes. Now that the statistic is the difference in group means, if this difference between the two groups increases then the p-value will decrease. As in Chapter 2, with quantitative data another aspect that also impacts size of the p-value is the underlying data variability, as measured by the standard deviations within each group. In this case the standard deviations are 14.0 and 14.5 in the two groups. If they were larger, the variability in the null distribution would be larger and so the p-value would be larger as well.

Estimation

When you show the *t*-statistic and overlay the theoretical distribution, you can also check a box to display the 95% confidence *t*-interval.

        ☑ Compute 95% confidence interval(s)
        Breastfed - NotBreastfed: (0.8697, 7.9304)*

Notice that this interval is similar to what we found with the 2SD Method earlier.

Think about it: Does the 95% confidence interval contain the value 0? What does that imply?

Recall that a confidence interval is an interval of plausible values for the parameter of interest. In this case, the parameter of interest is $\mu_{breastfed} - \mu_{not}$, where $\mu_{breastfed}$ was defined to be the mean GCI at age 4 for the population of all children who are breastfed during infancy, and μ_{not} to be that for children who aren't breastfed during infancy. The 95% theory-based confidence interval for $\mu_{breastfed} - \mu_{not}$ is calculated by the applet to be (0.8697, 7.9304). Notice that this interval contains only positive values and does not contain 0. Thus, we are 95% confident that the difference in mean GCI values between children who are and are not breastfed is *not* 0. More specifically, we are 95% confident that the mean GCI at age 4 for all children who are breastfed during infancy is 0.847 to 7.953 points higher than the mean GCI for children who aren't breastfed. (Please keep in mind that the same cautions as mentioned above with regard to not drawing a cause-and-effect conclusion and about the relevant population these results apply to, also need to be heeded when interpreting the findings from a confidence interval.) Note that, as we've seen before, increasing the sample size, decreasing the confidence level, or decreasing the data variability (standard deviation) will decrease the width of the confidence interval. The observed value of the statistic (difference in group means) will determine the midpoint of the interval.

The theory-based approach presented above (for both tests of significance and confidence intervals) is valid if *either one* (or both) of the following conditions are met:

Validity Conditions: The quantitative variable should have a symmetric distribution in both groups, or you should have at least 20 observations in each group and the sample distributions should not be strongly skewed.

In the breastfeeding example discussed here, the validity conditions are met (see FAQ 6.3.1 for more details). The sample sizes in each group are larger than 20 (237 and 85) and there is not strong skewness (in fact, the data is fairly symmetric; see Figure 6.10) This is why the p-value and confidence interval obtained from the theory-based approach are so similar to those obtained from simulation.

FAQ 6.3.1 Why is this called Student's t distribution?

Q: Who discovered the *t*-statistic?

A: What really mattered was not the statistic itself, but its distribution. Scientists knew about the statistic and were using it, but they were getting wrong answers because they were wrong about its distribution, which they thought was normal.

Q: I'll revise my question. Who discovered the *t*-distribution?

A: It was your namesake, in a way, but that's a story in itself.

Q: OK, I'll bite. Who was it?

A :   His name was William Seeley Gosset. He worked for the Guinness brewery.

Q: That particular Gosset was not on my radar, but I do know about Guinness.

A: Gosset knew that the distribution of the *t*-statistic was symmetric and bell-shaped, but he discovered that it took on extreme values (far from 0) more often than for a normal distribution. He created tables for the *t*-distribution to use for his work in the brewery, and published his findings in 1908, but didn't use his own name.

Q: Why not?

A :   The Guinness brewery did not allow any of their scientists to publish their results. They were afraid of giving away trade secrets. So at first they refused to let Gosset publish, but he pleaded with them, and tried to convince them that his discovery was too mathematical to be of use to any other brewing company. Finally they gave in and let him publish, but he had to use a made-up name so that no one could recognize his real name and know that he worked for Guinness. The name he chose was "Student." To this day, his test is known as Student's *t*-test.

What are the validity conditions when comparing two means?

Q: Can you clarify this validity conditions thing for me again?

A: When comparing two group means, the theory-based approach (two-sample *t*-test) requires big sample sizes. The bigger the better.

Q: And, by 'better,' you mean what exactly?

A: As the sample sizes increase the theory-based p-value will be closer to the one you would have gotten from simulation. But, it's not just the sample size that matters.

Q: Skewness?

A: Right. If you have fewer than 20 observations in a group, your distributions need to be pretty close to symmetric. With more than 20 in each group, a certain amount of skewness is OK, but not a lot. Of course, 20 isn't a magic number. The bigger the sample size you have, and the less skewness you have the better the theory-based and simulation will match up. These are just some rough guidelines.

Q: OK, but what is 'strong' skewness?

A: Judgment call. If you're not sure, one option is to do the simulation test and see how close your answers are. If they're similar then the skewness wasn't strong enough to matter.

Q: After all that, I'm really feeling like these simulation tests are nice. With none of those pesky validity conditions to worry about, why doesn't everyone just use them all of the time?

A: You know part of the answer to that question.

Q: I guess. You've told me countless times that before computers you couldn't simulate data fast enough for simulation to be practical. The theory-based methods are mathematical predictions of what happens when you simulate. But, c'mon, computers have been on people's desktops for about 20 years now. Are you telling me that change in academia is really that slow?

A: Yes! But, there is more to the story. It's not really accurate to say simulation tests have 'no' validity conditions.

Q: Oh no...

A: For example, in the next chapter you'll learn about pairing and independent groups. The methods you're learning here (and in the previous chapter) need to have independent groups (that is, the people in one group can't be 'connected' in some way to the people in the other group).

Q: OK, but that's more about the way you *get* the sample, not about properties of the data or the sample size, right?

A: Fair point. Sample size is a validity condition only for theory-based methods. Those methods typically include something about a large enough sample size---this isn't an issue for the simulation methods. But, there is more. For example, when you compare two groups using simulation, the null hypothesis is that the distributions are the same. Not just same means, but also same spreads and shapes. If you reject the null, it could be because the means differ, or the spreads differ, or the shapes differ, or some mix of all three. To conclude that the difference is due *only* to the means (or to get confidence intervals for the difference of the means) you have to assume that shapes and spreads are the same.

Q: And the theory based *t*-test and intervals?

A:	The null hypothesis is that the population means are equal. If the SDs are not the same, the theory-based method takes care of it. If the shapes are not the same, but the sample sizes are large enough and the skewness is not too strong, the theory-based method takes care of that, too. So, in this sense the validity conditions for the simulation-based approach are "stronger" (more restrictive) than for the theory-based approach.
Q:	Does it really matter?
A:	Most of the time probably not. But, if you wanted to compare two group means and thought the SDs would be different, the simulation approach we've showed you may not be as good as the theory-based approach.
Q:	So is that it?
A:	One more thing. All of these methods we're showing you are models. All models are wrong---none of them precisely match reality. We've tried to emphasize the most important aspects of the models so you can, in practice, judge whether the models are close enough to reality to be useful. But, there are many subtleties (like the one we just talked about) that go beyond simple rules, and sometimes those subtleties matter. If you really want to become someone who produces statistics, and not just one who reads about them you, should take another course in Statistics and learn more!

Formulas

Earlier you saw how when we applied the theory-based approach we could predict the value of the p-value and the confidence interval for the difference in two means using the standardized t-statistic (see FAQ 6.3.1 for why it's named 't') instead of through simulation. We'll now show a few details of how this works.

First, the value of the *standardized t-statistic* is obtained using:

$$t = \frac{statistic - hypothesized\ value}{SE(statistic)} = \frac{\bar{x}_1 - \bar{x}_2 - 0}{\sqrt{\frac{s_1^2}{n_1} + \frac{s_2^2}{n_2}}}$$

where, $\bar{x}_1$, s_1 and n_1 are the sample mean, sample standard deviation (SD) and sample size of the data in group 1, with the subscript "2" indicating the same quantities in group 2. Notice that there are a lot of similarities between this equation and the equation for the standardized z-statistic shown in Section 5.3—namely, the numerator is the difference in group means (larger difference means larger t-statistic) and the denominator includes the sample sizes (larger sample sizes mean larger t-statistic). The difference here is that the standard deviations of the quantitative variable in each group also impact the value of the t-statistic: namely, larger values of the sample standard deviations decrease the value of the t-statistic. Note that in this case our hypothesized value is 0, but we could use non-zero hypothesized values just as easily.

In many cases sample sizes are large enough that we can use the same approximate guidelines we had for the standardized z-statistic to understand the meaning of the t-statistic. In particular, values of t larger than two generally correspond to p-values less than 0.05 for two-sided tests.

You may be thinking: Why do you call the statistic t instead of z if they are the same?

As you may recall from Chapter 3, the null distributions of the two standardized statistics are not exactly the same. As it turns out, the null distribution of the t-statistic is symmetric, bell-shaped and centered at zero – but it is not normal – its peak in the middle is shorter, and its tails are longer (heavier). The difference between the normal (or z) and t-distributions goes away when the sample size gets larger, and so, most times, interpreting a t-statistic like a z-statistic is fine.

A theory-based confidence interval for $\mu_1 - \mu_2$ is also similar to the approach used in Chapter 5 for estimating $\pi_1 - \pi_2$. In particular, the confidence interval is obtained as:

$$statistic \pm margin\ of\ error$$

$$\text{which is} \qquad (\bar{x}_1 - \bar{x}_2) \pm multiplier \times \sqrt{\frac{s_1^2}{n_1} + \frac{s_2^2}{n_2}}.$$

When the sample size in each group is large, the multiplier is approximately equal to 2 for 95% confidence intervals, 1.645 for 90% confidence intervals and 2.576 for 99% confidence intervals. The actual value of the multiplier from the t distribution depends not only on the confidence level but also on the sample sizes involved.

In the example shown earlier, the *standard error* of $\bar{x}_1 - \bar{x}_2$ is calculated as $\sqrt{\frac{s_1^2}{n_1} + \frac{s_2^2}{n_2}} = 1.7869$. So, the margin-of-error for the 95% confidence interval is approximately equal to $2 \times 1.7869 = 3.57$. Thus, the 95% confidence interval is approximately equal to 4.4 ± 3.57, alternatively written as $(0.83, 7.97)$. Note: The difference between what we show here and the results from the applet are because the applet uses 1.988 as the multiplier instead of 2.

Note: In Chapters 1 and 5 when dealing with categorical data, we discovered when we increased the sample size(s) we increased the evidence against the null if all else remained the same. This is true when dealing with quantitative data as well. However when dealing with quantitative data, we also notice that standard deviations of the sample data play a role in assessing strength of evidence. Let's look at this intuitively. Figure 6.14 shows two pairs of dotplots.

Figure 6.14: These dotplots are designed to show that for the same difference in means, the strength of evidence is increased when standard deviations are smaller

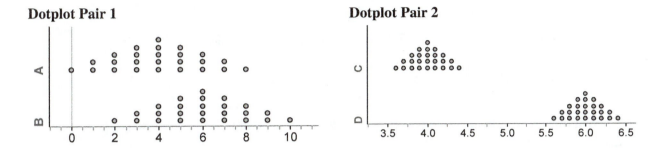

The distributions in Dotplot Pair 1 have group means of 4 and 6, two units apart. The distributions in Dotplot Pair 2 have identical group means as Dotplot Pair 1. This indicates that the variability *between* groups in Dotplot Pair 1 is the same as the variability between groups in Dotplot Pair 2. What is different, however, is the variability *within* the groups. The standard deviations for the samples in Dotplot Pair 1 are much larger than those in Dotplot Pair 2. Now suppose you wanted to show, using tests of significance, that the population means were different in each of the two cases. Which one would have stronger evidence? Hopefully you can see that there is lot of overlap with the two distributions in Dotplot Pair 1 and it could be difficult to convince us that the population means are different, using these data. However, the distributions in Dotplot Pair 2 are completely separated. There is no overlap at all. It would appear that using these data, it would be easy to convince us that the two corresponding population means differ. This all implies that as sample standard deviations decrease (all else the same), the strength of evidence of a difference in population means increases.

Exploration 6.3: Close Friends

How many close friends do you have? You know, the kind of people you like to talk to about important personal matters. Do men and women tend to differ on the number of close friends? And if so, by how much do men and women differ with regard to how many close friends they have? One of the questions asked of a random sample of adult Americans on the 2004 General Social Survey (GSS) was:

> *From time to time, most people discuss important matters with other people. Looking back over the last six months - who are the people with whom you discussed matters important to you? Just tell me their first names or initials.*

The interviewer then recorded how many names each person gave, along with keeping track of the person's gender. The GSS is a survey of a representative sample of U.S. adults who are not institutionalized.

1. Identify the variables recorded. Also classify each as either categorical or quantitative, and identify what each variable's role is: explanatory or response.

2. Did this study make use of random assignment, random sampling, both or neither?

3. Was this an experiment or an observational study? Explain how you are deciding.

4. In words, state the null and the alternative hypotheses to test whether American men and women <u>differ</u> with regard to how many friends they have.

5. Define the parameters of interest, and assign symbols to them.

6. State the null and the alternative hypotheses in symbols.

The survey responses are summarized in the Table 6.5 (and the data can be found in expanded form in the file **CloseFriends**):

Table 6.5: Number of close friends reported, separated by gender

Number of close friends reported	0	1	2	3	4	5	6	Total
Men	196	135	108	100	42	40	33	654
Women	201	146	155	132	86	56	37	813

7. Now you will obtain numerical summaries (statistics, such as, mean and SD) of the survey data, using the **Multiple Means** applet.

   - Open the data file **CloseFriends** to access the raw data. Copy the data (e.g., CTRL-A and CTRL-C).
   - Open the **Multiple Means** applet and press **Clear**. Click inside the Sample data box and paste (e.g., CTRL-V). Then press **Use Data**.

a. Report the sample size, sample mean, and sample SD of the number of close friends reported for each gender.Based on the sample statistics, who tends to have more close friends on average: men or women? How are you deciding?

b. Based on the sample statistics, who tends to have more variability with respect to how many close friends they have: men or women? How are you deciding?

c. Based on the data as presented in Table 6.5, are the distributions of number of close friends symmetric or skewed? If skewed, in which direction?

d. Calculate the observed difference in the mean number of close friends between these two groups (men - women). (Verify your calculation by checking the "Observed diff" output.)

e. Notice that the value recorded in (d) is a negative number. But, number of friends can only be 0 or more. Why is the number recorded in (d) less than 0?

The question we want to answer is, *Is the difference in mean number of close friends between men and women, as seen in the GSS sample data, something that could plausibly have happened by chance, by random sampling alone, if, on average, men and women have the same number of close friends?*

8. Using the same approach as in Section 6.2 with the Multiple Means applet, you can use this applet to generate possible values of the difference in sample means under the null hypothesis by shuffling which response values go with which explanatory variable values. Check the **Show Shuffle Options** box and enter 1000 for the **Number of Shuffles**. Press the **Shuffle Responses** button (there will be a pause, this is a large data set!). The histogram on the right is the simulated null distribution for the difference in sample means.

9.   Enter the observed difference in sample means in the **Count Samples** box and press **Count**.

    a.   Record your estimated p-value.

    b.   Fill in the blanks of the following sentence, to complete the interpretation of the p-value.

        The p-value of _____ is the probability of observing

        _____ assuming _____.

10. If the difference in means was larger (more different than 0), how would this impact the size of the p-value?

11. How would increasing the sample size (all else remaining the same) change the p-value? Why?

12. How would increasing the standard deviation of the number of close friends for both males and females (all else remaining the same) change the p-value? Why?

As we saw earlier, another measure of the strength of evidence against the null hypothesis would be to *standardize* the statistic by subtracting the hypothesized value and dividing by the standard error of the statistic.

13. On the left side of the applet (below the Sample data window) use the Statistic pull-down menu to select the *t*-statistic (again, there will be a pause). Record the value of the *t*-statistic that is computed for your data. Write a one-sentence interpretation of this value.

Notice that the standardized statistic uses the letter *t* (the one you should remember from Chapter 3) instead of the *z* (from Chapters 1 and 5) we saw when testing proportions. This is because the theoretical distribution used is now a *t*-distribution instead of a normal distribution. These *t*-distributions are very similar to normal distributions especially when sample sizes are large. The *t*-statistic, like the *z*-statistic, tells us how many standard deviations our sample difference is above or below the mean and it can be judged in the same manner. More details about the *t*-statistic are given earlier in this section and in the Calculation Details Appendix.

14. In light of the value of the standardized statistic, should you expect the p-value to be large or small? How are you deciding?

15. The histogram on the far right now displays the null distribution of simulated *t*-statistics. Describe the behavior of this distribution.

16. Below the null distribution check the box to **Overlay *t* distribution**. Does the *t*-distribution appear to adequately predict the behavior of the shuffled *t*-statistics? What do you think this suggests about whether the validity conditions will be met for this data?

17. Enter the observed value of the *t*-statistic in the **Count Samples** box and press **Count**. (Remember to use the pull-down menu to specify what direction(s) you want to consider more extreme based on the alternative hypothesis.) How does the p-value from the *t* distribution (in orange) compare to the simulation-based p-value (from 10 to 9 a)?

Validity Conditions

The validity conditions required for this theory-based approach (a "two-sample *t*-test") to be valid are shown in the box.

Validity Conditions: The quantitative variable should have a symmetric distribution in both groups, or you should have at least 20 observations in each group and the sample distributions should not be strongly skewed.

---

18. Do the validity conditions appear to be satisfied for these data? Justify your answer.

19. Based on the above analysis, state your conclusion in the context of the study. Be sure to

comment on the following.

    a.   *Statistical significance:* Do the data provide evidence that the mean number of close friends that men have is different from the mean number of friends women have? How are you deciding?

    b.   *Causation:* Do the data provide evidence that how many friends one has is *caused* by one's gender? How are you deciding?

    c.   *Generalization:* To whom can you apply the results of this study? All people? All adults? How are you deciding?

20. *Estimation:* When you have selected the *t*-statistic and the theory-based overlay, you can also use this applet for estimating the parameter of interest. (You can also go directly to the Theory-Based Inference applet which will also allow you to change the confidence level.)

    •   Check the box next to **Compute 95% CI(s) for difference in means**

    a.   Identify, in words related to the context of this study, the relevant parameter to be estimated here.

    b.   Report the 95% confidence interval for this parameter.

    c.   Does the 95% confidence interval calculated from the GSS sample data contain the value 0? What does that imply? (*Hint:* Recall that a confidence interval is an interval of *plausible* values for the parameter of interest, and the interval is calculated using the sample statistics.)

    d.   Does the 95% confidence interval agree with your conclusion in 16 to 17? How are you deciding?

e. Fill in the blanks of the sentence below to complete the interpretation of the 95% confidence interval:

We are _____% confident that the mean number of close friends men have is _____ than the mean number of close friends that women have, by between _____ and _____ friends.

f. How would the width of the interval change if you increased the confidence level to 99%? Why?

g. How would the width of the interval change if you increased the sample size? Why?

h. How would the width of the interval change if the variability of the number of close friends increased for both males and females? Why?

21. **Step 6. Look back and ahead**

Looking back: Did anything about the design and conclusions of this study concern you?

Issues you may want to critique include:

- The match between the research question and the study design
- How the observational units were selected
- How the measurements were recorded
- The number of observational units in the study
- Whether what we observed is of practical value

Looking ahead: What should the researchers' next steps be to fix the limitations in this study and/or build on this knowledge?

Section 6.3 Summary

Under certain circumstances, essentially when sample sizes are large enough, a theory-based approach can be used to approximate p-values and confidence intervals for the difference between two means.

- The standardized statistic again subtracts the hypothesized value from the observed statistic and then divides by the standard deviation of the statistic.
  - o The statistic in this case is the difference in sample means, that is, $\bar{x}_1 - \bar{x}_2$.

$$t = \frac{\bar{x}_1 - \bar{x}_2 - 0}{\sqrt{\frac{s_1^2}{n_1} + \frac{s_2^2}{n_2}}}$$

  - o The standardized statistic for a hypothesized difference of zero is $\frac{\bar{x}_1 - \bar{x}_2 - 0}{\sqrt{\frac{s_1^2}{n_1} + \frac{s_2^2}{n_2}}}$.
    - ▪ This standardized statistic is called a *t*-statistic rather than a *z*-statistic.

- The confidence interval again takes the form statistic ± margin-of-error.
  - Statistic = $\bar{x}_1 - \bar{x}_2$

  - Margin-of-error = multiplier × standard error of statistic
  - Standard error $$SE = \sqrt{\frac{s_1^2}{n_1} + \frac{s_2^2}{n_2}}$$
- Validity conditions for the theory-based approach to be valid are that both sample sizes are large (generally considered to be at least 20) or the data follow normal distributions.
- It does not matter which group is labeled as group 1 and which as group 2, provided that the labeling is consistent throughout the analysis, including interpretations and conclusions.

Some factors that affect p-values and confidence intervals are also similar to before:

- A larger difference between the sample means provides stronger evidence that the population means differ.
- Larger sample sizes produce stronger evidence that the population means differ.
- Larger within group variability decreases evidence that the population means differ.
- The difference in sample means is the midpoint of the confidence interval.
- Larger sample sizes produce narrower confidence intervals.
- Larger confidence levels produce wider confidence intervals.
- Less within group variability produces narrower confidence intervals.

Type of data	Name of test	Applet	Validity conditions
Two groups compared on a quantitative variable	Simulation test for comparing two groups on quant. variable	Multiple means	---
	Theory-based test or interval for comparing two means (two-sample *t*-test; interval)	Multiple means; Theory-based inference	At least 20 observations in each group without strong skewness or symmetry in distribution of quantitative variable in each group

Chapter 6 Summary

In this chapter we looked at approaches to consider potential differences in the center and/or variability of a quantitative variable between two independent groups. We started by learning how parallel dotplots can be used to examine potential differences in the quantitative variable between the two groups. Comparing centers (e.g., means) and variability (e.g., SDs, IQRs) can tell us a lot, but when the data is skewed or outliers are present, using the median may be more appropriate than the mean, as the median is more resistant to skewness and outliers than is the mean. We also learned that the five-number summary (minimum, lower quartile, median, upper quartile, and maximum) and boxplots can be used to summarize the distribution of a quantitative variable.

We then learned that we can apply the 3S Strategy in a manner very similar to the one we saw in Chapter 5, to assess whether two sample means differ enough to conclude that there is a genuine difference in the population means or long-run means of processes. Similarly, we can use the 2SD Method to generate a 95% confidence interval on the difference in population or long-run means. This approach also can be used to compare the medians of two independent groups. As we've seen before, when certain validity conditions are met, in this case when both sample sizes are large (generally considered to be at least 20) or the data follow normal distributions, a theory-based approach can be used to approximate p-values and confidence intervals for the difference between two means. In this case the standardized statistic is called t.

Chapter 6 Glossary

boxplot

A visual display of the five number summary, summarizing the behavior of a quantitative variable.. 594

causation

Inferring that the explanatory variable is causing the effect seen in the response variable...619

distribution

The characteristics of a variable's behavior... 582

double-blind study

Both the researchers and subjects are blind to which treatment each subject receives.........680

estimation

Using the sample statistic to create a confidence interval to estimate the parameter of interest....629

five-number summary

minimum, lower quartile, median, upper quartile,maximum.............................584, 593

generalization

To extend conclusions from a sample to a population... 619

inter-quartile range (IQR)

the difference between the upper quartile and the lower quartile..........................584, 593

lower quartile

The value for which 25% of the data lie below...584, 593

significance

Are the sample results unlikely to have arisen by chance alone.. 618

strength of evidence

Determining whether the observed statistic provides convincing evidence against the null hypothesis 606

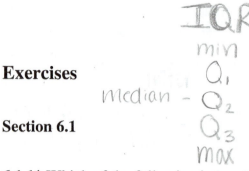

Exercises

Section 6.1

6.1.1* Which of the following is true about the inter-quartile range?

    A. It represents the range of approximately the middle quarter of a data set.

    B. It represents the range of approximately the middle 50% of a data set.

    C. If a value above the upper quartile is increased, the inter-quartile range will also increase.

    D. It a value below the lower quartile is decreased, the inter-quartile range will also decrease.

6.1.2 Which of the following is NOT true about quartiles?

    A. The lower quartile is always smaller than the upper quartile.

    B. Quartiles can be calculated by determining the median of the values above/below the location of the overall median

    C. Approximately 50% of the numbers in a data set lie between the lower and upper quartiles.

    D. The lower quartile is the same as the 25th percentile.

6.1.3* Approximately what does the box portion of a boxplot display?

    A. The middle quarter of the data.

    B. The middle half of the data.

    C. The lower quarter of the data.

    D. The upper quarter of the data.

6.1.4 What are the five components of the five-number summary?

6.1.5* On a data set that is bell-shaped, which will typically be larger, its standard deviation or its inter-quartile range?

Rattlesnakes

6.1.6 The following dotplot gives the ages of 21 male rattlesnakes. Which of the following would be true if one of the rattlesnakes whose age is given as 13 years is actually 15 and that change is made in the data set. (There may be more than one correct answer.)

A. The median would increase.

B. The mean would increase.

C. The inter-quartile range would increase.

D. The standard deviation would increase.

TV Watching*

6.1.7 The following histogram is the distribution of the amount of time, in minutes, that some statistics students said they watched television the previous day. Using this graph to estimate the mean and median age for the amount of television watching time, we can conclude:

mean gets pulled.
By outliers.

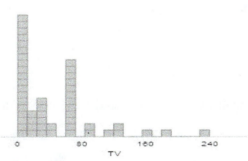

A. The mean is smaller than the median.

B. The mean is larger than the median.

C. The mean and the median have the same value.

D. You can't tell which is larger without the actual data.

Quiz scores

6.1.8 The following set of scores was obtained from a quiz: 4, 5, 8, 9, 11, 13, 15, 18, 18, 18, 20. The teacher computes the usual descriptive measures of central tendency and spread for these data and then discovers that an error was made. One of the 18's should have been a 16. Which of the following measures will NOT need to be changed from the original computations? (There may be more than one correct answer)

    A. Mean
    B. Median
    C. Standard Deviation
    D. IQR
    E. They all will need to be changed.

Changing exam scores*

6.1.9 Answer the following questions about exam scores.
    a) Suppose an instructor decides to add 5 points to every student's exam score in a class. What effect would this have on the five-number summary? On the inter-quartile range? Explain.
    b) Suppose an instructor decides to add 5 points only to the exam score for the highest-scoring student in the class. What effect would this have on the five-number summary? On the inter-quartile range? Explain.
    c) Suppose an instructor decides to add 5 points only to the exam score for the lowest-scoring student in the class. What effect would this have on the five-number summary? On the inter-quartile range? Explain.

Student heights

6.1.10 "A sample of heights (in inches) of 20 female statistics students is as follows":

$$67, 64, 67, 65, 64, 65, 67, 64, 65, 60, 67, 67, 72, 65, 70, 63, 67, 65, 67, 70$$

    a) Find the median height.
    b) Find the first and third quartiles.
    c) Find the inter-quartile range.
    d) Construct a boxplot by hand for the data.

States visited*

Problems 6.1.11 and 6.1.12 refer to the number of U.S. states visited.

6.1.11 An instructor collected data on the number of states that 50 students in her class had visited in the U.S. Here is the dotplot for the data collected.

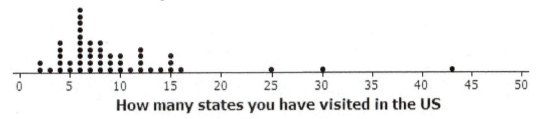

How many states you have visited in the US

a) Use the dotplot to find and report the median value for the number of states visited by the students in this study.

b) Use the dotplot to determine the lower and upper quartile values, and the inter-quartile range, for the number of states visited by the students in this study.

c) Would the mean value for these data be smaller than, larger than, or the same as the median, as reported in (a)? Explain your reasoning.

6.1.12: Reconsider the data in the dotplot in the previous question about the number of states that students had visited. Suppose the observation recorded as 43 states is a typo, and was meant to be 34. If we corrected this entry in the dataset, how would the following numerical statistics change, if at all?

Mean: SMALLER   SAME   LARGER

Median: SMALLER   SAME   LARGER

Standard deviation: SMALLER   SAME   LARGER

Inter-quartile range: SMALLER   SAME   LARGER

Monthly temperatures

6.1.13 The following dotplots show the average monthly temperatures (°F) for San Francisco, CA and Raleigh, NC.

a) What are the observational units here? (*Hint*: The observational units are not "cities.")
b) Identify the response variable, and whether it is categorical or quantitative.
c) Would the mean be a reasonable measure of center for these data? Why or why not?
d) The yearly mean temperature for Raleigh is 59.25°F and the yearly mean temperature for San Francisco is 57.25°F. Because these values are rather close, is it fair to say the cities are similar <u>with respect to average monthly temperatures</u>?
e) Will the standard deviation for Raleigh be larger or smaller than the standard deviation for San Francisco? How are you deciding?

6.1.14 Reconsider the data in the dotplots from the previous question about average monthly temperatures for San Francisco and Raleigh.
a) Determine the median temperature for each city. What do these numbers tell us about the two data sets?
b) Determine the inter-quartile range for each city. What do these numbers tell us about the two data sets?

Study hours*

6.1.15 In a survey of introductory statistics students, an instructor asked students to report how many hours they had spent studying in a typical week during the last term and their gender. Here are dotplots for the data.

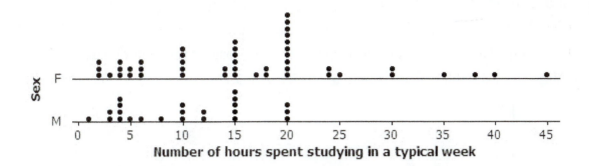

Describe what the graph tells us about any similarities and differences between men and women when it comes to how much time they report spending studying every week. Be sure to compare and contrast the shape, center, and spread for study hours' distributions for males and females.

6.1.16 Reconsider the data in the previous question about number of hours spent studying.

    a)   Find the median number of study hours for both males and females. What do these numbers tell us about the two data sets?

    b)   Find the inter-quartile range for the number of study hours for both males and females. What do these numbers tell us about the two data sets?

    c)   Construct parallel boxplots by hand for the two data sets.

Gettysburg Address

6.1.17 The graph below displays the distribution of word lengths (number of letters) in the Gettysburg Address, which you explored in Exploration 3.1A.

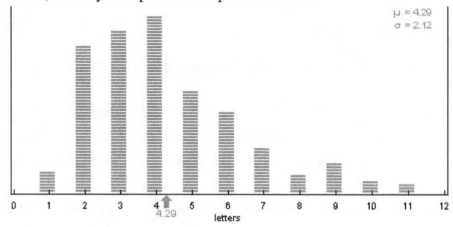

a) Describe the shape of this distribution.
b) Based on this shape, do you expect the median to be less than the mean, greater than the mean, or very close to the mean? Explain.

The following table lists how often each of the word lengths appears for these 268 words.

Word length	1	2	3	4	5	6	7	8	9	10	11
Number of words	7	49	54	59	34	27	15	6	10	4	3

c) Determine the median word length of these 268 words.
d) The mean word length is 4.29 letters per word. Is the median greater than, less than, or very close to the mean? Does this confirm your answer to b)?
e) Calculate the five-number summary of the word lengths.

College student bedtimes*

6.1.18 In a survey, 30 college students were asked what their "usual" bedtime was and the results are shown in the dotplot in terms of hours after midnight. Negative responses are hours before midnight.

Bedtimes

a) Determine the five-number summary for the bed times.

b) What is the inter-quartile range?

c) The earliest bedtime is 11:30pm (represented by -0.5 on the graph). If that person's usual bedtime is actually 9:00pm and that change was made in the dotplot, does that change the inter-quartile range? Would it change the standard deviation?

Candy bars

Problems 6.1.19 and 6.1.20 use the data on weights of candy bars.

6.1.19: Weights of 20 Mounds candy bars and 20 Payday candy bars, in grams, are shown in the dotplots.

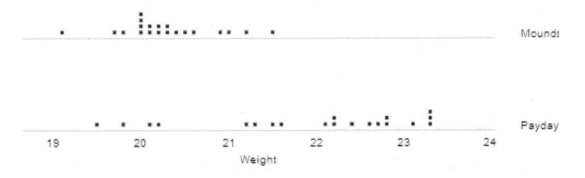

Weight

a) Describe how the distributions of weights of the two types of candy bars differ in both variability and center.

b) Based on your answers to part (a), which set of candy bar weights has the lowest standard deviation? Which has the lowest mean?

c) Would you say there is an association between the type of candy bar and the weight? Why or why not?

6.1.20 Using the candy bar weight data shown in the previous exercise, do the following:

a) Find the five-number summary for both sets of candy bar weights.

b) Construct parallel boxplots for both sets of candy bar weights.

Fast food wait times*

Problems 6.1.21 and 6.1.22 refer to wait times at fast food restaurants.

6.1.21 Student's compared drive-thru times at a couple of fast food restaurants in their town. They kept track of the time (in seconds) for people to go through the drive-thru at Wendy's and Hot N' Now from the time of ordering until they exited the pick-up window during lunch time. The results are shown in the following histograms.

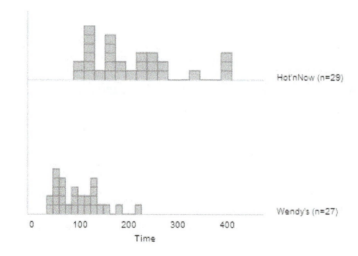

a) The medians for the two sets of times are 85 and 173 seconds. Which belongs to which restaurant?

b) The inter-quartile ranges for the two sets of times are 75 and 116.5 seconds. Which belongs to which restaurant?

c) For each restaurant is the mean larger or smaller than the median?

6.1.22 Reconsider the drive-thru times (in seconds) for a Wendy's and a Hot N' Now restaurant from the previous question.

    a) The means for the two sets of times are 93.7 and 203.0 seconds. Which belongs to which restaurant?

    b) The standard deviations for the two sets of times are 46.7 and 89.6 seconds. Which belongs to which restaurant?

Time in the bathroom

6.1.23 Students collect the time (in seconds) for men and women to use the restroom at a mall in Michigan. The results are shown in the following histograms.

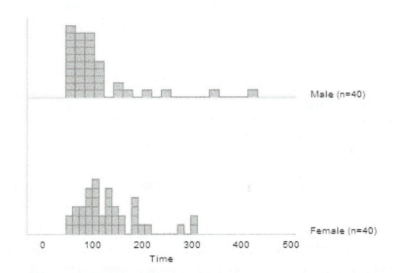

    a) One data set (male or female) has both the larger mean and median. Which one is that?

    b) Do you think the median is larger, smaller, or about the same as the mean for each data set?

    c) One data set has a larger standard deviation, but smaller inter-quartile range than the other. Which one is that? Explain why.

TV Watching*

6.1.24 Statistics students were asked how many minutes of television they watched on the previous day. The results, in minutes, are shown below for the males in top histogram and the females in the bottom histogram.

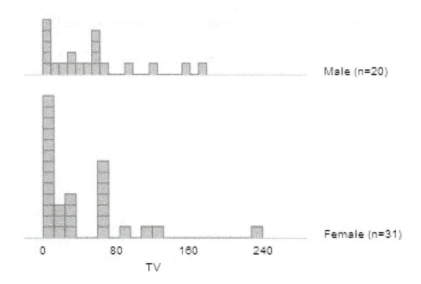

a) In which direction is each distribution skewed?

b) Will the mean be larger, smaller, or about the same as the median for each distribution?

c) Do you think that there is likely an association between TV watching and gender? Explain.

Car crashes and exercise

6.1.25 As part of a survey of statistics students, they were asked if they had ever been in a car crash while they were driving and how many minutes of exercise they had on the previous day. The data file CarCrashExercise.txt has these results. Copy and paste this data set into the **Descriptive Statistics** applet.

a) Based on just how the distributions look, do you think there is an association between being in a car crash while driving and exercise? Is so, in what way? Explain.

b) Find the mean of number of minutes of exercise for those that have been in car crashes while driving and those that have not. With this added information, do you think there is an association between being in a car crash while driving and exercise? Explain.

c) Find the median of number of minutes of exercise for those that have been in car crashes while driving and those that have not. With this added information, do you think there is an association between being in a car crash while driving and exercise? Explain.

6.1.26 Reconsider the previous exercise where students responded to a survey where they were asked if they had ever been in a car crash while they were driving and how many minutes of exercise they had on the previous day. The data file CarCrashExercise.txt has these results. Copy and paste this data set into the **Descriptive Statistics** applet and display boxplots for the data.

a) Is the data skewed to the right or the left?

b) Describe how you can see the skewedness you described in the part (a) when looking at the boxplots.

Sugar in cereals*

6.1.27 Statistics students found the sugar content of breakfast cereals, in grams per serving, for those placed on the high shelves of the store versus the low shelves. The data file CerealSugar.txt has these results. Copy and paste this data set into the **Descriptive Statistics** applet.

a) If you want to compare the sugar content of cereals on high versus low shelves, what would be the explanatory variable and what would be the response.

b) Based on just how the distributions look, do you think there is an association between where the cereal is located and its sugar content? If so, in what way? Explain.

c) Now find the mean grams of sugar per serving for the cereal on the low shelves and the mean for those on the high shelves. With this new information, do you think there is an association between where the cereal is located and its sugar content? Explain.

d) Now find the median grams of sugar per serving for the cereal on the low shelves and the median for those on the high shelves. With this new information, do you think there is an association between where the cereal is located and its sugar content? Explain.

e) Are the median sugar contents between high and low shelved cereal farther apart or closer together than the means are?

Coffee and height

6.1.28 You may have heard someone say that drinking coffee will stunt your growth. To test this, college students were asked their heights and if they were coffee drinkers. The data file CoffeeHeight.txt has the results. Copy and paste this data set into the **Descriptive Statistics** applet.

    a) Based on just how the distributions look, do you think there is an association between being height and drinking coffee? Is so, in what way? Explain.

    b) Notice there are two peaks (or what is called bimodal) in the distribution of heights for the non-coffee drinkers? Why do you think the distribution has this feature?

    c) Find the mean height for the coffee drinkers and the non-coffee drinkers. With this added information, do you think there is an association between drinking coffee and height? Explain.

    d) Find the median height for the coffee drinkers and the non-coffee drinkers. With this added information, do you think there is an association between drinking coffee and height? Explain.

    e) If you could find that coffee drinkers were significantly shorter than non-coffee drinkers in a study like this, would this show there is a tendency for coffee to stunt a person's growth? Explain.

Section 6.2

Haircut prices*

6.2.1: As seen in Exploration 6.1, an instructor collected data on prices (in dollars) paid by students for their most recent haircut, also recording each student's sex, to test whether females in the population paid more on average for their most recent haircut than males. Which of the following are appropriate sets of hypotheses? (Circle all that apply.)

    A. H_0: There is no association between gender and haircut prices.

H_a: There is an association between gender and haircut prices with females tending to pay more than males.

 H_0: Males and females in the population pay the same price on average for their last haircut.

H_a: Males in the population pay less on average than females for their last haircut.

C. H_0: $\mu_{female} - \mu_{male} = 0$

   H_a: $\mu_{female} - \mu_{male} > 0$

D. H_0: $\mu_{female} = \mu_{male}$

   H_a: $\mu_{female} > \mu_{male}$

E. H_0: $\bar{x}_{female} = \bar{x}_{male}$

   H_a: $\bar{x}_{female} > \bar{x}_{male}$

F. H_0: There is no association between gender and haircut prices.

   H_a: There is an association between gender and haircut prices.

G. H_0: $\mu_{female} = \mu_{male}$

   H_a: $\mu_{female} \neq \mu_{male}$

6.2.2: Reconsider the previous exercise about haircut prices. The p-value for the appropriate test turns out to be 0.0022. .What would you conclude? (Circle all that apply.)

A. The sample data provide *strong* evidence that females in the population paid more on average than males for their most recent haircut.

B. The sample data provide *little* evidence that females in the population paid more on average than males for their most recent haircut.

C. The sample data provide strong evidence that females in the population paid *much* more on average than males for their most recent haircut.

6.2.3: Reconsider the previous two exercises about haircut prices. The sample mean prices were $21.85 for males and $54.05 for females, a difference of $32.20. Which are appropriate interpretations of the p-value of 0.0022? (Circle all that apply.)

A. There's a 0.22% chance that women in the population did not pay more on average than males for their most recent haircut.

B. If there were no difference in the population between what females and males paid for their most recent haircut, then there's a 0.22% chance that the sample mean price for females would have been at least $32.20 higher than the sample mean price for males.

C. If there were no difference in the population between what females and males paid for their most recent haircut, then 0.22% of all random shuffles based on the sample data would produce the shuffled mean price for females being at least $32.20 higher than the shuffled mean price for males.

Biking and sleep deprivation

6.2.4: Reconsider the bicycle commuting time study (Example 6.2) and the sleep deprivation study (Exploration 6.2).

a) In what primary way do these studies differ from the dolphin therapy and yawning studies from Chapter 5?
b) Are both of these studies (sleep deprivation and bicycle commuting time) randomized experiments, are both observational studies, or is there one of each type (if so, which is which)?
c) Did we use simulation-based or theory-based methods to analyze these studies?
d) Which, if any, of these two studies uses a two-sided alternative?
e) With which of these two studies did we find strong evidence that there was a cause-and-effect relationship between the variables?

Study hours*

6.2.5: In a survey of introductory statistics students, an instructor asked her students to report how many hours they had spent studying in a typical week during the last term, and their gender. Suppose that the intent is to compare reported study habits of male students to those of female students.

a) Is this an experiment or an observational study? Explain how you know.
b) Identify the observational units.
c) Identify the explanatory and response variable. Also, identify whether each is quantitative or categorical.

d) Identify the appropriate null hypothesis from the following set of five choices.
  A. Male students and female students are equally likely to study.
  B. On average, male students and female students spend an equal amount of time studying.
  C. On average, female students spend a greater amount of time studying compared to male students.
  D. On average, female students spend a lesser amount of time studying compared to male students.
  E. On average, female students differ from male students when it comes to how much time they spend studying.

e) Identify the appropriate alternative hypothesis from the following set of five choices.
  A. Male students and female students are equally likely to study.
  B. On average, male students and female students spend an equal amount of time studying.
  C. On average, female students spend a greater amount of time studying compared to male students.
  D. On average, female students spend a lesser amount of time studying compared to male students.
  E. On average, female students differ from male students when it comes to how much time they spend studying.

How many Facebook friends do you have?

Problems 6.2.6 through 6.2.10 focus on the data pertaining to the number of Facebook friends people have.

6.2.6 In a survey of introductory statistics students, an instructor asked her students to report how many Facebook friends they had. Suppose that the intent is to study whether there is an association between number of Facebook friends and gender.

a) Identify whether the study is an experiment or an observational study. Explain.
b) Identify the observational units.
c) Identify the explanatory and response variables. Also, for each variable identify whether it is categorical or quantitative.
d) Which of the following is the appropriate null hypothesis? **Circle one.**
  A. There is an association between gender and the number of Facebook friends.
  B. There is no association between gender and the number of Facebook friends.
  C. On average, men have more Facebook friends than women.
  D. On average, men have fewer Facebook friends than women.

e) Which of the following is the appropriate alternative hypothesis? **Circle one.**

    Ⓐ. There is an association between gender and the number of Facebook friends.

    B. There is no association between gender and the number of Facebook friends.

    C. On average, men have more Facebook friends than women.

    D. On average, men have fewer Facebook friends than women.

Here are a few summary statistics about the data:

	Sample size	Sample mean	Sample median	Sample SD
Women	35	594.3	532	309.8
Men	13	405.6	485	228.4

f) In this context, which of the following is an (are) appropriate statistic value(s) to compare men to women? **Circle all that apply.**

    A. $35 - 13 = 22$

    B. $594.3 - 405.6 = 188.7$

    Ⓒ. $532 - 485 = 47$

    D. $309.8 - 228.4 = 81.4$

Suppose that we want to use cards to carry out a randomization-test of the appropriate hypotheses.

g) How many cards will we need? **Circle one.**

    A. 35

    B. 13

    C. 48

    D. Doesn't matter

The following is output from carrying out the randomization test using the Multiple Means applet.

h) On the histogram above, shade the region that depicts the p-value. (*Hint:* Is your alternative hypothesis one-sided or two-sided?)

6.2.7: Reconsider the previous exercise about the study of whether there is an association between number of Facebook friends and gender.

    a) Use the information available in the previous exercise to find a standardized statistic to test whether the average number of Facebook friends men have is different from that of women.

    b) Based on the standardized statistic, what is the strength of evidence against the null hypothesis?

6.2.8: Reconsider the previous exercises about the study of whether there is an association between number of Facebook friends and gender. The p-value for the randomization-based test of whether the average number of Facebook friends men have is different from that of women turns out to be 0.0520.

    a) Interpret this p-value in the context of the study.

    b) Based on the p-value, what is the strength of evidence against the null hypothesis?

6.2.9: Reconsider the previous exercises about the study of whether there is an association between number of Facebook friends and gender.

a) Use the 2SD method and the information available in the previous exercises to find a 95% confidence interval for the difference in average number of Facebook friends that men have and that women have.

b) Interpret the confidence interval reported in part a) in the context of the study.

6.2.10: Reconsider the previous exercises about the study of whether there is an association between number of Facebook friends and gender. The following is output from carrying out the randomization test using the Multiple Means applet.

Recall that the summary statistics about the data are as follows:

	Sample size	Sample mean	Sample median	Sample SD
Women	35	594.3	532	309.8
Men	13	405.6	485	228.4

a) On the above histogram, shade the region that depicts the p-value. (*Hint:* Pay close attention to the label on the *x*-axis – what is the statistic of interest here?)

b) Use the information available to find and report the standardized statistic in this context.

How many flip-flops do you have?*

Problems 6.2.11 through 6.2.15 refer to flip-flop data.

6.2.11 In a survey of introductory statistics students, an instructor asked her students how many flip-flops they have. Suppose that the intent is to study whether there is an association between number of flip-flops and gender.

a) Identify whether the study is an experiment or an observational study. Explain.

b) Identify the observational units.

c) Identify the explanatory and response variables. Also, for each variable identify whether it is categorical or quantitative.

d) Which of the following is the appropriate null hypothesis? **Circle one.**

    A. There is an association between gender and the number of flip-flops.

    (B.) There is no association between gender and the number of flip-flops.

    C. On average, men have more flip-flops than women.

    D. On average, men have fewer flip-flops than women.

e) Which of the following is the appropriate alternative hypothesis? **Circle one.**

    (A.) There is an association between gender and the number of flip-flops.

    B. There is no association between gender and the number of flip-flops.

    C. On average, men have more flip-flops than women.

    D. On average, men have fewer flip-flops than women.

Here are a few summary statistics about the data:

	Sample size	Sample mean	Sample median	Sample SD
Women	37	6.22	6.0	3.58
Men	13	2.54	1.0	2.96

f) In this context, which of the following is an (are) appropriate statistic value(s) to compare men to women? **Circle all that apply.**

    A. $37 - 13 = 24$

    B. $6.22 - 2.54 = 3.68$

    C. $6 - 1 = 5$

    D. $3.58 - 2.96 = 0.62$

Suppose that we want to use cards to carry out a randomization-test of the appropriate hypotheses.

g) How many cards will we need? **Circle one.**

    A. 37

    B. 13

    (C.) 50

    D. Doesn't matter

The following is output from carrying out the randomization test using the Multiple Means applet.

h) On the histogram above, shade the region that depicts the p-value. (*Hint:* Is your alternative hypothesis one-sided or two-sided?)

6.2.12: Reconsider the previous exercise about the study of whether there is an association between number of flip-flops and gender.

a) Use the information available in the previous exercise to find a standardized statistic to test whether the average number of flip-flops men have is different from that of women. Report the standardized statistic and interpret it in the context of the study.

b) Based on the standardized statistic, how much evidence do you have against the null hypothesis?

6.2.13: Reconsider the previous exercises about the study of whether there is an association between number of flip-flops and gender. The p-value for the randomization-based test of whether the average number of flip-flops men have is different from that of women turns out to be 0.001.

a) Interpret this p-value in the context of the study.

b) Based on this p-value, how much evidence do you have against the null hypothesis?

6.2.14: Reconsider the previous exercises about the study of whether there is an association between number of flip-flops and gender.

a) Use the 2SD method and the information available in the previous exercises to find a 95% confidence interval for the difference in average number of flip-flops that men have and that women have.

b) Interpret the confidence interval reported in part a) in the context of the study.

c) Do the conclusions from this confidence interval agree with the conclusions from the p-value of 0.001 reported in the previous exercise? Explain your reasoning.

6.2.15: Reconsider the previous exercise about the study of whether there is an association between number of flip-flops and gender. The following is output from carrying out the randomization test using the Multiple Means applet.

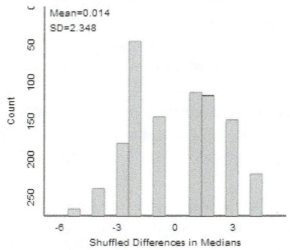

Recall that the summary statistics about the data are as follows:

	Sample size	Sample mean	Sample median	Sample SD
Women	37	6.22	6.0	3.58
Men	13	2.54	1.0	2.96

a) On the above histogram, shade the region that depicts the p-value. (*Hint:* Pay close attention to the label on the *x*-axis – what is the statistic of interest here?)

b) Use the information available to find and report the standardized statistic in this context.

The Anchoring Phenomenon

Problems 6.2.16 through 6.2.19 refer to the data gathered on the anchoring phenomenon.

6.2.16: Anchoring is "the common human tendency to rely too heavily, or "anchor," on one trait or piece of information when making decisions." (Source: *Wikipedia*). A group of students taking an introductory statistics course at a four-year university in California were asked to guess the population of Milwaukee, Wisconsin. Some of the students were randomly chosen to be told that the nearby city of Chicago, Illinois has a population of about 3 million people, while the rest of the students were told that the nearby city of Green Bay, Wisconsin has a population of about 100 thousand. Previous studies have shown that these numbers serve as a psychological anchor, so people told about Chicago tend to guess a higher population for Milwaukee than people told about Green Bay. (For more about this phenomenon, see the book *Nudge: Improving Decisions about Health, Wealth, and Happiness*.) The purpose in analyzing the data is to see whether we find strong evidence of this phenomenon among students like the ones in this study.

   a) Identify the observational units.
   b) Identify the explanatory variable and the response variable in this study. Also classify each as categorical or quantitative.
   c) Is this an observational study or an experiment? Explain.
   d) Were the subjects in this study randomly selected from a population? Explain.
   e) Were the subjects in this study randomly assigned to groups? Explain.

The data set is titled **Milwaukee**. The first column contains the city mentioned to subjects, and the second column contains the guesses for the population of Milwaukee (in thousands, to the nearest thousand).

   f) Using an appropriate applet (for e.g. Multiple Means applet) find and record the following summary statistics for the guesses made about Milwaukee's population, separated by the city which they were told about:

	Sample size	Sample mean	Sample SD
Chicago			
Green Bay			

   g) Find and report the difference in the average guess made by those told about Chicago and those told about Green Bay.
   h) Give two possible explanations for the observed difference in average guesses for the population of Milwaukee reported in part g).

6.2.17 Reconsider the study mentioned in the previous exercise about guessing the size of Milwaukee's population. **Our research question:** *On average, will students like the ones in the study, tend to guess a higher population size for Milwaukee if they are told about Chicago's population size rather than told Green Bay's population size?*
   a) Define (in words) the parameters of interest of this study. Also, assign symbols to the parameters.

b) Express the null and alternative hypotheses for testing whether the class data give strong evidence in support of the anchoring phenomenon described above. Use words only.

c) Express the null and alternative hypotheses for testing whether the class data give strong evidence in support of the anchoring phenomenon described above. Use the symbols defined in part a).

d) Give detailed, step-by-step instructions on how one could conduct a tactile simulation to generate a p-value, keeping in mind that you want to investigate whether there is strong evidence of the anchoring phenomenon among students like the ones in this study. Be sure to include details on the following:

- Would the simulation involve coins, or dice, or index cards?
- How many tosses, or rolls, or cards would be used?
- How many sets of tosses, or rolls, or shuffles would you observe?
- What would you record after every repetition?
- How would you compute the p-value?

6.2.18 Reconsider the study mentioned in the previous exercise about guessing the size of Milwaukee's population. Do the data provide evidence of the anchoring phenomenon among students like the ones in this study? That is, our research question is: *On average, will students like the ones in the study, tend to guess a higher population size for Milwaukee if they are told about Chicago's population size rather than told Green Bay's population size?*
Recall that the data set is titled **Milwaukee**, and that the first column contains the city mentioned to subjects, and the second column contains the guesses for the population of Milwaukee (in thousands, to the nearest thousand).

a) Use an appropriate randomization-based applet to find a p-value to test whether there is strong evidence of the anchoring phenomenon among students like the ones in this study. Report and interpret the p-value, in the context of the study.

b) Use the p-value to indicate the strength of evidence against the null hypothesis.

c) Calculate an appropriate standardized statistic, and interpret the standardized statistic in the context of the study.

d) Use the 2SD Method to find a confidence interval for the parameter of interest. Report this interval, and interpret the interval in the context of the study.

e) Do the results from the 95% confidence interval agree with your conclusion in (b)? How are you deciding?

f) Give a full conclusion including significance, estimation, causation and generalizability.

6.2.19 Reconsider the study mentioned in the previous exercise about guessing the size of Milwaukee's population. Do the data provide evidence of the anchoring phenomenon among students like the ones in this study? Suppose that instead of comparing the average guess for Milwaukee's population (Chicago vs. Green Bay), we wanted to compare the *median* guesses. Recall that the data set is titled **Milwaukee**, and that the first column contains the city mentioned to subjects, and the second column contains the guesses for the population of Milwaukee (in thousands, to the nearest thousand).

   a) Use an appropriate randomization-based applet to find a p-value to test whether there is strong evidence of the anchoring phenomenon among students like the ones in this study. Report <u>and</u> interpret the p-value, in the context of the study.
   b) Use the p-value to indicate the strength of evidence against the null hypothesis.
   c) Calculate an appropriate standardized statistic, and interpret the standardized statistic in the context of the study.
   d) Use the 2SD Method to find a confidence interval for the parameter of interest. Report this interval, and interpret the interval in the context of the study.
   e) Do the results from the 95% confidence interval agree with your conclusion in (b)? How are you deciding?
   f) Give a full conclusion including significance, estimation, causation and generalizability.

Children and lifespan*

Problems 6.2.20 through 6.2.23 use data gathered on the life span of men who have children.

6.2.20 *Do men with children tend to live longer, on average, than those without?* To investigate, a group of Cal Poly students randomly sampled men from the obituaries page on the *San Luis Obispo Tribune*'s webpage between June – November 2012. For each man selected, they noted the age at which person died, and whether or not the person had any children.

   a) Identify the observational/experimental units, explanatory variable, and response variable in this study.
   b) Is this an observational study or an experiment? Explain briefly.

The students' data are in the file **ChildrenandLifespan**.

c) Using an appropriate applet (for e.g. Multiple Means applet) find and record the following summary statistics for the lifespan of men, separated by whether or not the man had children:

	Sample size	Sample mean	Sample SD
Had children			
No children			

d) Find and report the difference in the average lifespan of men with versus without children.

e) Give two possible explanations for the observed difference in average lifespan reported in part d).

6.2.21 Reconsider the study mentioned in the previous exercise about an investigation of whether men with children tend to live longer, *on average,* than men without children.

a) Define (in words) the parameters of interest of this study. <u>Also, assign symbols</u> to the parameters.

b) Express the null and alternative hypotheses for testing whether men with children tend to live longer, *on average,* than men without children. Use words only.

c) Express the null and alternative hypotheses for testing whether men with children tend to live longer, *on average,* than men without children. Use the symbols defined in part a).

d) Give detailed, step-by-step instructions on how one could conduct a tactile simulation to generate a p-value to test the hypotheses stated in parts b) and c). Be sure to include details on the following:

- Would the simulation involve coins, or dice, or index cards?
- How many tosses, or rolls, or cards would be used?
- How many sets of tosses, or rolls, or shuffles would you observe?
- What would you record after every repetition?
- How would you compute the p-value?

6.2.22 Reconsider the study mentioned in the previous exercises about an investigation of whether men with children tend to live longer, *on average,* than men without children. The data are in the file **ChildrenandLifespan**.

a) Use an appropriate randomization-based applet to find a p-value to test whether men with children tend to live longer, *on average,* than men without children. Report <u>and </u>interpret the p-value, in the context of the study.

b) Use the p-value to indicate the strength of evidence against the null hypothesis.

c) Calculate an appropriate standardized statistic, and interpret the standardized statistic in the context of the study.

d) Use the 2SD Method to find a confidence interval for the parameter of interest. Report this interval, and interpret the interval in the context of the study.

e) Do the results from the 95% confidence interval agree with your conclusion in (c)? How are you deciding?

f)   Give a full conclusion including significance, estimation, causation and generalizability.

6.2.23 With randomization techniques we are able to analyze any statistic we desire. What if we wanted to look at whether there the *median* lifespan of men with children was higher than men without children? Using the Multiple Means applet, and the "Median" option on it, answer (a) through (f) of the previous exercise looking at the difference in median mercury levels.

Mercury levels in tuna

6.2.24 Rising mercury levels due to industrial pollution is a recent concern. Many fish have high levels of mercury. Data was collected on random samples of tuna from 1991 to 2010, and are available in the file **Tuna**. Each row in the dataset represents the mercury level of a different fish. *Do fresh Yellowfin Tuna have different levels of mercury on average compared to fresh Albacore Tuna?* Use the fresh tuna data to answer the following questions.
   a)   Is this an experiment or an observational study? Explain how you know.
   b)   Identify the observational units.
   c)   Identify the explanatory and response variable. Also, identify whether each is quantitative or categorical.
   d)   Define (in words) the parameters of interest of this study. Also, assign symbols to the parameters.
   e)   State the null and alternative hypotheses, in words as well as using symbols.
   f)   What statistic will you use?
   g)   Use an appropriate applet to carry out a simulation analysis. Report the p-value and strength of evidence.
   h)   Use the 2SD rule to determine a 95% confidence interval. Interpret the interval, making sure it is clear which parameter (in context) the 95% confidence interval is estimating.
   i)   Give a full conclusion including significance, estimation, causation and generalizability.

6.2.25 With randomization techniques we are able to analyze any test statistic we desire. What if we wanted to look at whether there was a difference in the *median* values of mercury in the Yellowfin tuna compared to the Albacore tuna? Using the Multiple Means applet, and the "Median" option on it, answer (f) through (i) of the previous exercise looking at the difference in median mercury levels.

Bike commuting*

6.2.26 Recall Example 6.2 about the bike commuting times. On each of 56 days (from mid-January to mid-July 2010), researcher Jeremy Groves tossed a £1 coin to decide whether he would be biking the 27 miles to work on his carbon frame (lighter) bicycle that weighed 20.9 lbs. or on his steel frame (heavier) bicycle that weighed 29.75 lbs. He then recorded the commute time (given in minutes) for each trip. The summary statistics were reported to be:

Bike type	Sample size	Sample average	Sample SD
Carbon frame (lighter)	26	108.34 min	6.25 min
Steel frame (heavier)	30	107.81 min	4.89 min

In Example 6.2 you read about conducting a randomization-based test to investigate whether the average commute time differs by bike type.

   a) Suppose we suspect that the lighter carbon frame bike will be less stable, especially in inclement weather conditions, and will have a higher variability in commute times, compared to the heavier steel frame bike. Is there any indication from the data in support of this suspicion? Explain how you are deciding.

To investigate whether the variability in commute times will be higher with the lighter carbon frame bike than the heavier steel frame one, we can examine the ratio of the SDs, such that *Ratio* $= SD_{carbon} \div SD_{steel}$. (Note: $\div$ is "divided by.")

   b) In theory, what numeric values can this *Ratio* take?
   c) For the observed data, what is the numeric value of the statistic *Ratio*?
   d) Describe in detail how you would use simulation (with pennies, dice, spinners, cards, etc.) to find a p-value, to test whether the variability in commute times is *higher* with the lighter carbon frame bike than the heavier steel frame one.

Here is a histogram from an appropriately conducted simulation:

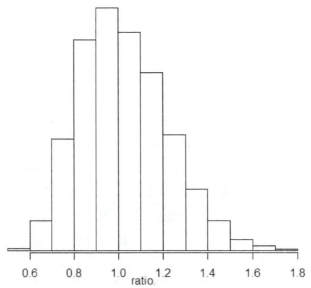

e) As per the above histogram, what numeric values are most common for *Ratio* from the simulation? Why does that make sense?

f) Indicate on the above histogram, the region that depicts the p-value in the context of the study, being sure to include appropriate numbers.

g) Which of the following appears to be the approximate p-value? No explanations needed. **Circle one**.

      A. 1.00   B. 0.10   C. 0.01   D. < 0.01

h) Which of the following is an appropriate statement to make based on the above histogram? **Circle one.**

  A. There is no evidence of the variability in commute times being higher with the lighter carbon frame bike than the heavier steel frame one because the distribution of *Ratio* is not centered at 0.

  B. There is some evidence of the variability in commute times being higher with the lighter carbon frame bike than the heavier steel frame one because the distribution of *Ratio* is not centered at 0, but at 1.

  C. There is no evidence of the variability in commute times being higher with the lighter carbon frame bike than the heavier steel frame one because the distribution of *Ratio* is not normal, but skewed right.

  D. None of the above.

Section 6.3

Time online*

Problems 6.3.1 through 6.3.3 use data on the number of hours spent online each day.

6.3.1: Suppose that randomly sampled college students are asked how many hours they typically spend online each day. You conduct a two-sided test of the null hypothesis that $\mu_{females} - \mu_{males} = 0$, and you also calculate a 95% confidence interval for $\mu_{females} - \mu_{males}$.

a) Describe (in words) what the parameter $\mu_{females} - \mu_{males}$ means here.

b) Now suppose that your friend analyzes the same data but with the order of subtraction reversed (males - females, rather than females – males). Describe the impact (if any) on each of the following. In other words, describe how your friend's findings will compare to yours with regard to each of the following.

    i.   Distribution of simulated statistics under the null hypothesis

    ii.   Standard deviation of the simulated statistics under the null hypothesis

    iii.   Observed value of the statistic (difference in sample means)

    iv.   Approximate p-value from simulation

    v.   Value of t-test statistic

    vi.   p-value from t-test

    vii.   Midpoint of confidence interval

    viii.   Endpoints of confidence interval

    ix.   Width of confidence interval

c) What is the bottom line: Will you and your friend reach the same conclusions even though you disagreed about the order in which to perform the subtraction (males – females or females – males)? Explain.

Haircut prices

6.3.2: Reconsider exercise **6.2.1** in which an instructor collected data on prices (in dollars) paid by students for their most recent haircut, also recording each student's sex. A 95% confidence interval for $\mu_{female} - \mu_{male}$ turned out to be (12.17, 52.23). For each of the following, indicate whether it provides a correct or incorrect interpretation/conclusion from this interval.

a) Between 12.17% and 52.23% of females in the population paid more than males for their haircut.

b) About 95% of all females in the *population* paid between $12.17 and $52.23 more than males for their haircut.

c) About 95% of all females in the *sample* paid between $12.17 and $52.23 more than males for their haircut.

d) We can be 95% confident that the *probability* is between 12.17% and 52.23% that females in the population pay more than males for their haircut.

e) We can be 95% confident that the *average* price paid for a haircut in this population is between $12.17 and $52.23.

f) We can be 95% confident that the average price paid for a haircut by a female in this population is higher than the average price paid by males in this population by between $12.17 and $52.23.

g) We can be 95% confident that *all* females in the population paid more than all males for their haircut by between $12.17 and $52.23.

h) We can be 95% confident that *most* females in the population paid more than most males for their haircut by between $12.17 and $52.23.

i) We can be 95% confident that females in the population paid more *on average* than males for their haircut by between $12.17 and $52.23.

6.3.3: Reconsider the previous exercise about haircut prices. Suppose that the instructor wants to test whether females in the population paid more on average for their most recent haircut than males. Which kind of test might be appropriate? (Circle all that apply.)

a) Simulation-based test comparing proportions
b) Simulation-based test comparing means
c) One-sided test
d) Two-sided test
e) *z*-test for single proportion
f) *z*-test for comparing two proportions
g) *t*-test for comparing means

Too scared to fall asleep?*

Problems 6.3.4 through 6.3.8 deal with data gathered on nightmare frequency.

6.3.4: Selvi et al. (2012) conducted a study on university medical students in Turkey, to investigate whether and how sleep habits were associated with having nightmares. During the study, these students were engaged in lecture-based learning with no hospital duties. All participants were given initial surveys, the results of which were used to determine who among them had "early bird" sleep habits and who "night owl" habits. Next, the students were given the Van Dream Anxiety Scale (VDAS) assessment, which involved questions regarding nightmare frequency and dream anxiety due to frightening dreams during the last 30 days. Here are the summary statistics on nightmare frequency, separated by "early bird" and "night owl." Note that higher scores indicate higher nightmare frequency.

	Sample size	Sample average	Sample SD
"Diagnosed" early bird	67	1.23	0.93
"Diagnosed" night owl	59	2.10	0.99

Analyze the data using a theory-based approach (use the **Theory-based inference** applet) to determine whether early birds differ from night owls, with respect to average nightmare frequency. The data is not strongly skewed within each group. Be sure to include the appropriate hypotheses, a statement about validity conditions, as well as numerical evidence (including p-value) in support of your conclusion with regard to the hypotheses. In your conclusion, include comments about statistical significance, estimation (95% confidence interval (include interpretation)), causation, and generalizability.

6.3.5: Reconsider the previous exercise. How, if at all, would your analysis and conclusions change if the sample sizes had been larger and all else had remained the same? Be sure to refer to the test statistic, p-value, and confidence interval, as well as your conclusion.

6.3.6: Reconsider the previous two exercises. How, if at all, would your analysis and conclusions change if the sample means had been closer together and all else had remained the same as in the original question? Be sure to refer to the test statistic, p-value, and confidence interval, as well as your conclusion.

6.3.7: Reconsider the previous three exercises. How, if at all, would your analysis and conclusions change if the sample standard deviations had been larger and all else had remained the same as in the original question? Be sure to refer to the test statistic, p-value, and confidence interval, as well as your conclusion.

6.3.8: Refer to the previous exercises. The researchers also gave each participant a score for "Fear of sleeping because of anticipated nightmares." Given in the table below are the summary statistics for these scores. Note that higher scores indicate higher fear and that the data is not strongly skewed in either group.

	Sample size	Sample average	Sample SD
"Diagnosed" early bird	67	1.22	0.99
"Diagnosed" night owl	59	1.49	1.19

Analyze the data using a theory-based approach to determine whether early birds differ from night owls, with respect to fear of sleeping in anticipation of nightmares. Be sure to include the appropriate hypotheses, a statement about validity conditions as well as numerical evidence (including p-value) in support of your decision with regard to the hypotheses. In your conclusion, include comments about statistical significance, estimation (95% CI for parameter of interest), causation, and generalizability.

Anchoring

6.3.9: Recall the study about anchoring from exercise **6.2.16**, in which students estimated the population of Milwaukee after some had been told about Chicago and others about Green Bay.

a) Describe in words the parameter(s) of interest, in the context of this study.
b) In the context of the parameter(s) described in (a), express the null and alternative hypotheses for testing whether the class data give strong evidence in support of the anchoring phenomenon described above. If you use symbols, make sure to define them.
c) Check whether the conditions for the validity of the theory-based *t*-test to compare two means are satisfied here.
d) Use appropriate technology (an applet, or a statistical software package) to conduct a *t*-test of significance of the hypotheses that you stated in part (b). Report the test statistic and p-value.
e) State your conclusion in the context of the study, being sure to comment on statistical significance.
f) Determine and interpret the 95% confidence interval using a theory-based approach for the parameter(s) of interest, in the context of the study.
g) Comment on causation and generalizability for this study.

Left-right confusion*

6.3.10: Left-right confusion (LRC) is the term used to describe the difficulty of distinguishing between left and right. In a study related to LRC, researchers Ocklenburg et al. (2011) gave a mental rotation test (MRT) to 91 neurologically healthy men and women. The MRT involved picking two images that were rotated versions of a target figure, from a list of fives images. Each correct answer led to one point. Here are the summary statistics on the scores, separately for the men and women. Note that the data within each group is not strongly skewed.

	Sample size	Sample average	Sample SD
Men	41	14.10	3.78
Women	50	9.56	3.97

Analyze the data to determine whether men differ from women, with respect to long-run average MRT score. Use a theory-based approach. Be sure to include the appropriate hypotheses, as well as numerical evidence in support of your decision with regard to the hypotheses (p-value). Find and interpret a 95% confidence interval. Include comments about causation, and generalizability. Comment on adherence to validity conditions for use of the theory based methods.

Downhill walking

6.3.11: In a study conducted in New Zealand, Parkin et al. randomly assigned volunteers to either wear socks over their shoes (intervention), or wear usual footwear (control), as they walked downhill on an inclined icy path. Researchers standing at the bottom of the inclined path measured the time (in seconds) taken by each participant to walk down the path. Here are the summary statistics. Note that the data is fairly symmetric in both groups.

	n	Mean	SD
Intervention	14	37.7	9.36
Control	15	39.6	11.57

The research question to be addressed through the following questions is: Do the data provide evidence that people wearing socks over their shoes will take a shorter time to walk down an icy, steeply inclined road, on average, compared to those wearing usual footwear?

a) Describe the parameter(s) of interest, in the context of this study.
b) State the null and the alternative hypotheses, in the context of the study.
c) Name the appropriate theory-based method you would use to test the hypotheses stated in (b).
d) State the validity conditions that have to be met to be able to perform the test named in part (c). Are these conditions met? How are you deciding?
e) Regardless of your answer to (d), perform the theory-based test proposed in (c). Be sure to report a test statistic value and a p-value.
f) Determine a 95% confidence interval for comparing the walking time between the two groups.
g) Summarize your conclusions about the research question of the study. Be sure to comment on statistical significance, confidence/estimation, causation, and generalizability.

Pricing pills*

6.3.12 In a randomized, double-blind study reported in the *Journal of American Medical Association*, researchers Waber et al. (2008) administered a pill to each of the 82 healthy paid volunteers from Boston, Massachusetts, but told half of them that the drug had a regular price of $2.50 per pill, whereas the remaining participants were told that the drug has a discounted price of $0.10 per pill, mentioning no reason for the discount. In the recruitment advertisement the pill was described as an FDA approved "opioid analgesic," but in reality both groups were administered placebo pills.

To simulate pain, the participants were administered electric shocks to the wrist. These shocks were calibrated to each participant's pain tolerance, *prior* to administering the pills. In the article, the researchers present the summary statistics on the calibrated maximum tolerance for the participants for the two groups. Researchers wondered if evidence exists for a difference in average pain tolerance between the two groups. The distribution of maximum tolerance scores are not strongly skewed in either group.

	n	Mean	SD
Regular-price pill	41	51.8 volts	18.7 volts
Discount-price pill	41	54.9 volts	23.3 volts

a) Explain why the researchers want to compare the pain tolerance of the treatment groups.
b) State the appropriate null and the alternative hypotheses, in the context of the study.
c) Name the appropriate theory-based method you would use to test the hypotheses stated in (b).
d) State the validity conditions that have to be met to be able to perform the test named in part (c). Are these conditions met? How are you deciding?

e)	Regardless of your answer to (d), perform the theory-based test proposed in (c). Be sure to include a test statistic value and a p-value in your answer.

f)	Interpret your p-value in the context of the study.

g)	Determine a 95% confidence interval for comparing the tolerance levels between the two groups.

h)	Summarize your conclusions about the research question of the study. Be sure to comment on statistical significance, confidence/estimation, causation, and generalizability.

Perceived wealth

6.3.13: Do people tend to spend money differently based on perceived changes in wealth? In a study conducted by Epley et al. (2006), 47 Harvard undergraduates were randomly assigned to receive either a "bonus" check of $50 or a "rebate" check of $50. A week later, each student was contacted and asked whether they had spent any of that money, and if yes, how much. In this exercise we will focus on how much money they recalled spending, when contacted a week later. It turned out that those in the "bonus" group spent an average of about $22, compared to $10 in the "rebate" group.

a)	Identify the observational units.

b)	Identify the explanatory and response variables. Identify each as either categorical or quantitative.

c)	State the appropriate null and alternative hypotheses, in the context of the study.

d)	In the article that appeared in the *Journal of Behavioral Decision Making*, the researchers reported neither the sample size, nor the sample SD of each group. In this exercise you will explore whether and how the strength of evidence is impacted by the sample size and sample SD. Complete the following table by finding the *t*-statistic and a p-value for a theory-based test of significance comparing two means under each of the 4 different scenarios.

Scenario		Sample sizes	Sample means	Sample SDs	*t*-statistic	p-value
1	Bonus	24	22	5		
	Rebate	23	10	5		
2	Bonus	24	22	10		
	Rebate	23	10	10		
3	Bonus	30	22	5		
	Rebate	17	10	5		
4	Bonus	30	22	10		
	Rebate	17	10	10		

e) Summarize what your analysis has revealed about the effects of the sample size breakdown and the sample standard deviations on the values of the *t*-statistic and p-value.

Nostril breathing and cognitive performance*

Problems 6.3.14 through 6.3.16 pertain to data gathered on nostril breathing.

6.3.14: In an article titled "Unilateral Nostril Breathing Influences Lateralized Cognitive Performance" that appeared in *Brain and Cognition* (1989), researchers Block et al. published results from an experiment involving assessments of spatial and verbal cognition when breathing through only the right versus left nostril.

The subjects were 30 male and 30 female right-handed introductory psychology students who volunteered to participate in exchange for course credit. Initial testing on spatial and verbal tests revealed the following summary statistics. Note that the scores on the spatial task can range from 0 to 40, whereas those on the verbal task can go from 0 to 20. The distributions are not strongly skewed on either scale, or for males or females.

	Spatial		Verbal	
Sex	Mean	SD	Mean	SD
Male	10.2	2.7	10.9	3.0
Female	7.8	2.5	15.1	3.4

a) Consider comparing males to females with regard to performance on the spatial assessment task. State the appropriate null and alternative hypotheses, in the context of the study.

b) Explain why it is valid to use the theory-based method for producing a p-value to test the hypotheses stated in (a).

c) Carry out the appropriate test to produce a p-value to test the hypotheses stated in (a), and interpret the p-value.

d) Find a 95% confidence interval for the difference in mean scores of males and females, with regard to performance on spatial assessments. Interpret the interval.

e) Based on your p-value, state a conclusion in the context of the study. Be sure to comment on statistical significance, estimation (confidence interval), causation, and generalizability.

f) Repeat the investigation comparing males and females, this time on verbal performance. Be sure to address the questions asked in (a) – (e).

6.3.15: Reconsider the previous exercise. In this question, we will focus on comparing 10 males who were randomly assigned to breathe only through their left nostril to 10 who were randomly assigned to breathe only through their right nostril while they took the spatial and verbal tests. Note: Ten of the thirty-total males were not assigned to either condition considered here. Here are the summary statistics on the spatial test performance, the data is fairly symmetric in both groups.

	Spatial test	
Condition	Mean	SD
Left nostril breathing	8.9	2.8
Right nostril breathing	11.4	2.0

a) Describe the parameter(s) of interest, in the context of this study.

b) State the null and the alternative hypotheses (two-sided test), in the context of the study.

c) Name the appropriate theory-based method you would use to test the hypotheses stated in (b).

d) State the validity conditions that have to be met to be able to perform the test named in part (c). Are these conditions met? How are you deciding?

e) Regardless of your answer to (d), perform the theory-based test proposed in (c). Be sure to include a test statistic value and a p-value in your answer.

f) Based on your p-value, state your conclusion in the context of the study. Be sure to comment on statistical significance, 95% confidence interval, causation, and generalizability.

g) Without performing any additional calculations, explain how, if at all, the p-value reported in (f) would change if each condition was implemented on 30 males in each group instead of 10 males in each group, but the means and SDs stayed the same.

h) Without performing any additional calculations, explain how, if at all, the p-value reported in (f) would change if the mean for "left" condition was 10.9 and that for "right" was 13.4, but the sample sizes and SDs stayed the same.

i) Without performing any additional calculations, explain how, if at all, the p-value reported in (f) would change if SDs for both conditions were 2.0, but the sample sizes and means stayed the same.

6.3.16: Reconsider the previous two exercises. In this question, we will focus on comparing 10 males who breathe only through their left nostril to 10 males who breathe only through their right nostril while they perform a verbal task. Assume that the verbal scores in each group are fairly symmetric.

a) When the average scores on the verbal task obtained by the males in the study were compared using a two-sided alternative hypothesis and two sample *t*-test, the p-value turned out to be 0.0162. Use this p-value and a 1% level of significance, to state an appropriate conclusion in the context of the study. Explain your reasoning.

b) How, if at all, would your conclusion in (a) change, if you were to use a 5% level of significance?

c) In the context of this study and the related data results, which of the following statements are true?

   (i)   The 95% confidence interval for the difference in the mean scores between males in the "population" who perform a verbal task while breathing through only the left nostril versus only the right nostril, will contain 0, because the p-value is less than 5%.

   (ii)   The 95% confidence interval for the difference in the mean scores between males in the "population" who perform a verbal task while breathing through only the left nostril versus only the right nostril, will *not* contain 0, because the p-value is less than 5%.

   (iii)   The 99% confidence interval for the difference in the mean scores between males in the "population" who perform a verbal task while breathing through only the left nostril versus only the right nostril, will contain 0, because the p-value is greater than 1%.

   (iv)   The 99% confidence interval for the difference in the mean scores between males in the "population" who perform a verbal task while breathing through only the left nostril versus only the right nostril, will *not* contain 0, because the p-value is greater than 1%.

Drive-thru times

Problems 6.3.17 and 6.3.18 use the same data.

6.3.17: Reconsider the data from Exercise 6.1.21 and corresponding statistics from Exercise 6.1.22 about drive-thru times at two fast food restaurants. Conduct a two-sample *t*-test of whether the sample data provide strong evidence that the population mean drive-thru times differ between the two restaurants. Be sure to check whether the validity conditions are satisfied, state hypotheses, report the test statistic and p-value, and summarize your conclusion.

6.3.18: Reconsider the previous exercise. Determine a 95% confidence interval for comparing the population mean drive-thru times between the two fast food restaurants. Also interpret what this interval says. Finally, comment on whether the confidence interval is consistent with the conclusion of your significance test in the previous exercise.

Restroom times*

Problems 6.3.19 and 6.3.20 use the same data.

6.3.19: Reconsider the data from Exercise 6.1.23 about times spent in a public restroom by men and women in a Michigan mall. The average time in the bathroom for the 40 men was 106.65 (sd=79.611), and for the 40 women was 133.5 (sd=62.369). The data is not strongly skewed in either group. Conduct a two-sample t-test of whether the sample data provide strong evidence that the population mean restroom times differ between the two sexes. Be sure to check whether the technical conditions are satisfied, state hypotheses, report the test statistic and p-value, and summarize your conclusion.

6.3.20: Reconsider the previous exercise. Determine a 95% confidence interval for comparing the population mean restroom times between the two sexes. Also interpret what this interval says. Finally, comment on whether the confidence interval is consistent with the conclusion of your significance test in the previous exercise.

Pen color

6.3.21: A psychology study (Rutchick, Slepian, and Ferris, 2010) investigated whether using a red pen causes people to assign lower scores than using a blue pen. A group of 128 students in an introductory undergraduate psychology class were asked to grade an 8th grader's essay on a scale of 0 to 100. Half of the student graders were randomly assigned to use a red pen while grading the essay, and the other half were randomly assigned to use a blue pen. The researchers reported that the group using a red pen gave an average score of 76.2, and the SD of the scores was 12.29. For the group using blue pens, the average score was 80.0, and the SD of the scores was 9.36.

a) State the null and alternative hypotheses to be tested.

b) Without having access to the 128 scores that were assigned to the essays, can you feel comfortable in applying a two-sample *t*-test using the summary statistics? Explain.

c) What additional information is needed to apply a two-sample *t*-test using the summary statistics?

d) Make a reasonable guess for the missing information, and conduct a two-sample *t*-test. Report the values of the test statistic and p-value.

e) Summarize your conclusion from this test.

f) Is it legitimate to draw a cause-and-effect conclusion based on these data and your test result? Explain why nor why not.

g) What would you advise the researchers about how broadly they can generalize the conclusions of this study? For example, would you feel comfortable in generalizing the results of this study to 8th grade teachers, or to college professors? Explain.

h) Do you think the study's conclusions would be stronger if the subjects had been actual teachers or college professors? Why do you think the researchers used introductory students as their subjects?

Got a tip?*

Problems 6.3.22 through 6.3.26 use data gathered on the amount customers tip their server.

6.3.22: Can waitresses increase their tips simply by introducing themselves by name when they greet customers? A waitress collected data on two-person parties that she waited on during Sunday brunch (with a fixed price of $23.21) at a Charley Brown's restaurant in southern California. For each party the waitress used a random mechanism to determine whether to give her name as part of her greeting or not. Then she kept track of how much the party gave for a tip at the end of their meal.

a) Is this an observational study or a randomized experiment? Explain.

b) Identify and classify the explanatory and response variables.

c) State the null and alternative hypotheses, in symbols and in words, for testing the waitress' conjecture.

The sample mean tip amount for the 20 parties to which the waitress gave her name was $5.44, with a standard deviation of $1.75. These statistics were $3.49 and $1.13, respectively, for the 20 parties to which the waitress did not give her name. The data was not strongly skewed in either group.

d) Use appropriate symbols for the six numbers reported in this paragraph.

e) Are the validity conditions met? Explain.

f) Use these summary statistics to calculate the test statistic and p-value using an appropriate theory-based test.

g) What test decision would you reach at the .05 significance level? Explain what this decision means in the context of this study.

6.3.23: Reconsider the previous exercise.

a) Calculate a 95% confidence interval for the difference in population mean tip amounts between the two experimental treatments (giving name or not).

b) Interpret what this interval means in this context.

c) Is this confidence interval consistent with the test decision from the previous exercise? Explain how you know.

d) Summarize your conclusions from this test. Be sure to comment on issues of causation and generalizability, as well as significance and confidence.

6.3.24: Reconsider the previous two exercises.

a) If you change the confidence level to 98%, what impact would you expect this to have on the midpoint and width of the confidence interval?

b) Make this change (from part j), and report the new confidence interval. Was the impact what you expected?

6.3.25: Reconsider the previous exercises.

a) If the waitress had used larger sample sizes (say, 40 in each group), and the sample means and SDs had turned out the same as they did in the study originally, what impact would you expect this to have on the test statistic, p-value, and confidence interval?

b) Make this change, and report the new test statistic, p-value, and confidence interval. Was the impact what you expected?

6.3.26: Reconsider the previous exercises.

a) If the sample responses had shown more variability in each group (say, sd of 2 in each group), and all else had remained the same as originally, what impact would you expect this to have on the test statistic, p-value, and confidence interval?

b) Make this change, and report the new test statistic, p-value, and confidence interval. Was the impact what you expected?

End of Chapter Exercises

Harry Potter **books***

6.CE.1: Suppose that a random sample of teenage boys and girls are asked how many *Harry Potter* books they have read. You conduct a two-sided test of the null hypothesis that $\mu_{girls} - \mu_{boys}$ = 0, and you also calculate a 95% confidence interval for $\mu_{girls} - \mu_{boys}$.

a) Describe (in words) what the parameter $\mu_{girls} - \mu_{boys}$ means here.
b) Now suppose that your friend analyzes the same data but with the order of subtraction reversed (boys – girls, rather than girls – boys). Describe the impact (if any) on each of the following. In other words, describe how your friend's findings will compare to yours with regard to each of the following.
    a. Distribution of simulated statistics under the null hypothesis
    b. Standard deviation of the simulated statistics under the null hypothesis
    c. Observed value of the statistic (difference in sample means)
    d. Approximate p-value from simulation
    e. Value of *t*-test statistic
    f. p-value from *t*-test
    g. Midpoint of confidence interval
    h. Endpoints of confidence interval
    i. Width of confidence interval
c) What is the bottom line: Will you and your friend reach the same conclusions even though you disagreed about the order in which to perform the subtraction (boys – girls, or girls – boys)? Explain.

Flexibility

6.CE.2: Data was gathered on 98 female college students and 81 male college students to see on average which sex was more flexible after participating in a physical wellness class, and are available in the file **FlexibilityPost**.
a) State the research question.
b) How is the study designed? Observational study or randomized experiment.
c) State the hypotheses.
d) Check validity conditions. Can you use the theory based approach, or should you use randomization methods?

e) Regardless of your answer to d) use a simulation analysis to generate a p-value and 95% CI

f) Regardless of your answer to d) use a theory-based approach to generate a p-value and 95% CI

g) Based on your answer to d) compare the p-values and 95% CIs received in e and f.

h) Is zero in the confidence intervals? How does that fit with inference based on the p-values?

NFL Referees*

6.CE.3: Recall from Exercise P.2.5 the data on duration of NFL football in 2012 for the first three weeks of the season, when replacement referees officiated the games, and for the next three weeks when the regular referees returned to officiate the games. Dotplots of these data are reproduced below. The data appear in the file **Referees**.

a) Identify the observational units and variables in this graph. Also identify which variable is explanatory and which is response, and classify the variables as categorical or quantitative.

b) Did the data collection for this study make use of random sampling, random assignment, or neither? Explain.

c) Before you conduct an inferential analysis, what do the dotplots suggest with regard to the question of whether game durations differed for the two types of referees?

d) State the appropriate hypotheses for testing whether the long-run average game duration is different for games played with replacement referees as compared to games played with regular referees.

e) Use an applet to simulate a randomization test for testing the hypotheses in d) and find the p-value.

f) Interpret what this p-value means in this context.

g) Use the simulation results to produce a 95% confidence interval for the relevant parameter.

h) Interpret this confidence interval.

i) Summarize your conclusions from these data address significance, confidence, generalizability and causation.

Elasticity of balsa wood

Problems 6.CE.4 through 6.CE.7 deal with data on the elasticity of balsa wood.

6.CE.4: Student researchers investigated whether balsa wood is less elastic after it has been immersed in water. They took 44 pieces of balsa wood and randomly assigned half to be immersed in water and the other half not to be. They measured the elasticity by seeing how far (in inches) the piece of wood would project a dime into the air.

a) Identify the observational units, explanatory variable, and response variable in this study.
b) Is this an observational study or an experiment? Explain briefly.
c) Express the appropriate null and alternative hypotheses, in the context of the study, for testing the student researchers' conjecture that balsa wood is less elastic after it has been immersed in water.

The students' data are in the file **Balsawood**.

d) Examine graphical displays and summary statistics to compare the elasticity measurements between the two groups. Write a paragraph comparing and contrasting the distributions of elasticity measurements between the two groups

6.CE.5: Reconsider the previous exercise.

a) Give detailed, step-by-step instructions on how one could conduct a tactile simulation to generate a p-value to test the hypotheses stated in (c). Be sure to include details on the following:
   - Would the simulation involve coins, or dice, or index cards?
   - How many tosses, or rolls, or cards would be used?
   - How many sets of tosses, or rolls, or shuffles would you observe?
b) Use an appropriate applet or other technology to find a simulation-based p-value to test the hypotheses stated in the previous exercise. Report the p-value.
c) Interpret the p-value reported in (b), being sure to explain what the number means in the context of how the simulation-based method generated this p-value.
d) Summarize your conclusion from the p-value.

6.CE.6: Reconsider the previous two exercises.

a) Use a theory-based method to determine the test statistic and p-value.
b) Investigate and comment on whether the validity conditions of the theory-based method are satisfied.
c) Does the p-value obtained using the simulation-based method agree with that obtained using the randomization/simulation-based method?
d) Summarize your conclusion from this study, addressing whether the data provide strong evidence to support the students' conjecture.

6.CE.7: Reconsider the previous three exercises.

a) Determine a 95% confidence interval for the difference in mean elasticity (as measured by distance dime is projected) between treated and untreated balsa wood. Use the theory-based approach to find the interval. Be sure to also include an interpretation of the obtained 95% confidence interval.
b) Do the results from the 95% confidence interval agree with your conclusion in the previous exercise? How are you deciding?
c) Address scope of conclusions for this study by discussing generalizability and cause-and-effect in this study.

Sentence lengths*

6.CE.8: A student investigated whether popular novelist John Grisham uses shorter sentences in his books aimed at teenage readers than in his more general books. She took a sample of sentences from two of his books: *The Confession*, aimed at general readers, and *Theodore Boone: Kid Lawyer*, aimed at teenage readers. She then recorded the length (number of words) in each sentence. The sorted data appear below and in the file **GrishamSentences**.

The Confession (55 sentences)

2 2 3 3 4 4 6 7 8 8 8 8 9 9 9

9 9 10 10 10 10 11 11 12 12 12 13 13 13 13

13 14 14 14 15 15 15 15 15 15 16 17 18 19 21

21 21 23 23 24 29 29 32 34 39

Theodore Boone: Kid Lawyer (29 sentences)

3 4 4 7 7 8 11 11 12 12 13 13 14 14 14

15 16 16 17 19 19 20 20 20 22 25 29 30 47

a) Determine the five-number summary for each group.

b) Construct boxplots to compare the two distributions.

c) Comment on what the five-number summaries and boxplots reveal about the question of whether Grisham tends to use shorter sentences for his books aimed at teenage readers than in his books aimed at a general audience.

6.CE.9: Reconsider the previous question.

a) State the appropriate null and alternative hypotheses for testing whether Grisham tends to use shorter sentences for his books aimed at teenage readers than in his books aimed at a general audience.

b) Calculate the sample mean and SD of the sentence lengths for each group.

c) Explain why the calculations in b) indicate that there's no reason to proceed with testing the hypotheses stated in a).

Two tests

Answer the following questions about 2-sample tests and 2-sided tests.

6.CE.10:

a) Explain the difference between a two-*sample* test and a two-*sided* test.

b) Is it legitimate to use a one-sided test with a two-sample test? If so, give an example. If not, why not?

c) Is it legitimate to use a two-sided test with a one-sample test? If so, give an example. If not, why not?

SAT coaching*

6.CE.11: Suppose that 5000 students are randomly assigned to either take an SAT coaching course or not, with the following results in their improvements in SAT scores:

	Sample size	Sample mean	Sample standard deviation
Coaching group	2500	46.2	14.4
Control group	2500	44.4	15.3

    a) Use technology to conduct a test of whether the sample data provide evidence that SAT coaching is helpful. State the hypotheses, and report the test statistic and *p*-value. Draw a conclusion in the context of this study.

    b) Use technology to produce a 99% CI for the difference in population mean improvements between the two groups. Interpret this interval.

    c) Are the test conclusion and CI consistent with each other? Explain.

    d) Do the sample data provide *very* strong evidence that SAT coaching is helpful? Explain whether the *p*-value or the CI helps you to decide.

    e) Do the sample data provide strong evidence that SAT coaching is *very* helpful? Explain whether the *p*-value or the CI helps you to decide.

    f) Comment on what this exercise reveals about the distinction between statistical significance and practical importance.

Runs scored in MLB

6.CE.12: Suppose that a professional baseball fan wants to investigate whether teams in one league (National or American) tend to score more runs on average than teams in the other league. The fan waits until the end of the season, records the number of runs scored by all 30 teams, and classifies the teams as National or American League.

    a) Describe the population of interest in this study.

    b) Explain how you know that this study involves neither random sampling nor random assignment.

    c) Is it legitimate to perform a significance test and calculate a p-value based on the data that this baseball fan collected, even though the fan gathered data on the entire population of all 30 Major League Baseball teams? Explain your thinking.

Investigation: **Memorizing Letters?**

Can people better memorize letters if they are presented in recognizable groupings than if they are not? Students in an introductory statistics class were the subjects in a study that investigated this question. These students were given a sequence of 30 letters to memorize in 20 seconds. Some students (25 of them) were given the letters in recognizable three-letter groupings such as JFK-CIA-FBI-USA-…. The other students (26 of them) were given the exact same letters in the same order, but the groupings varied in size and did not include recognizable chunks, such as JFKC-IAF-BIU-… The instructor decided which students received which grouping by random assignment. After 20 seconds of studying the letters, students recorded as many letters as they could remember. Their score was the number of letters memorized correctly before their first mistake. The instructor conjectured that students in the JFK-CIA-… group would memorize more letters, on average, than students in the JFKC-IAF-… group.

Step 1: Ask a research question

1. State the research question.

Step 2: Design a study and collect data

2. Is this a randomized experiment or an observational study? Explain how you know.

3. What are the observational units?

4. What variables are measured/recorded on each observational unit?

5. Is there an explanatory/response relationship for these variables? Classify the variable in this study as categorical or quantitative.

6. State the null and alternative hypotheses to be investigated with this study. (You can use the symbol μ to represent these parameters if you want).

Go to the textbook website and download the data for this investigation. Paste the data into the **Multiple Means** applet.

Step 3: Explore the data

7. Report the following:

Mean number of letters memorized in recognizable groups_____

SD of number of letters memorized in recognizable groups_____

Mean number of letters memorized when not in recognizable groups_____

SD of number of letters memorized when not in recognizable groups_____

Difference in means_____

Is the difference in the direction conjectured by the instructor (i.e., did the group expected to do better actually do better)?

Step 4: Draw inferences

8. Conduct a simulation analysis with 1000 shuffles. Indicate how to determine the approximate p-value from the simulation results, and report that p-value.

9. Give a conclusion about the strength of evidence against the null, in context, based on the p-value you obtained in 8.

10. Use the 2SD rule to generate a 95% confidence interval on the long-run difference in means between the two groups.

11. Interpret the 95% confidence interval you obtained in #10.

Step 5: Formulate conclusions

12. If the evidence against the null were strong, would you be justified in concluding that you have found strong evidence of a cause-and-effect relationship between recognizable groupings and better memory performance for this study? Why?

13. What broader population are you willing to generalize your findings to, if any? Explain.

An alternative approach: Theory-based inference

14. Explain whether a theoretical approach to computing the p-value and confidence interval would be valid for this data.

15. Now, use the Theory-based inference applet to generate a test statistic, p-value, and confidence interval. Report these, and compare them to what you determined from the simulation analysis. Explain how they are similar or different in light of your answer to #14.

Step 6: Look back and ahead

16. Provide a short paragraph (3-4 sentence) summary of your findings. Make sure to include limitations of the study design. Also indicate at least one possible follow-up study that you now want to conduct.

Research Article: Physical fitness and urbanicity

Read "Does living in urban or rural settings affects aspects of physical fitness in children? An allometric approach." by Tsimeas et al. (2005), 39:671–674 in the *British Journal of Sports Medicine*, and then answer the following questions.

Step 1. Ask a research question

1. In no more than 1-2 sentences state a broad research question (or related questions) that the researchers were trying to investigate in this study.
2. Identify at least two (cited) points of evidence the researchers cite as to why the research question(s) are important. Include the citation in your response.

Step 2. Design a study and collect data

3. How does the article indicate that the children were selected? What do you think this means?

4. For each of the following variables indicate whether the variable is categorical or quantitative and then explain how it was measured

    a. Urban vs. rural

    b. Body Mass Index (BMI)

    c. Sit and Reach (SR)

    d. Basketball throw (BT)

Step 3. Explore the data

5. Provide a 2-3 sentence demographic description of the children in the sample (age, race, etc.).

6. Give the mean and SD of the urban and rural boys, respectively, for BMI, Sit and Reach and Basketball throw.

Step 4. Draw inferences

The researchers conducted a series of two-sided, two-sample t-tests (theory-based approach from Section 6.3) in the paper as shown in Table 1. Although precise p-values are not given, an asterisk indicates comparisons between urban and rural boys that have p-value <0.05.

7. For BMI, state a null and alternative hypothesis, indicate the strength of evidence provided by the p-value ($p < 0.05$ or not) and provide an appropriate (in context) interpretation.

8. Repeat #7 for SR.

9. Repeat #7 for BT.

10. Theory-based approaches require validity conditions to be met. Explain how the validity conditions are met for these tests.

11. Table 1 does not provide confidence intervals on the difference in means, the table only provides confidence intervals on the mean within each group. Indicate whether the corresponding 95% confidence interval on the difference in means (not shown in the paper) would include 0 for each of BMI, SR and ST.

Step 5. Formulate conclusions

12. To whom are you willing to generalize your conclusions from this study? Why?

13. In Table 2, the researchers indicate that they have found significant evidence (p-value < 0.05) between the distance the basketball is thrown for urban and rural girls, does this imply that where the girls were raised (city vs. small village) is *causing* their average basketball throwing distance to change? Why or why not?

Step 6. Look back and ahead

14. Identify at least two things that, if you were running the study and had a reasonable amount of financial resources, you would do differently. You must justify your answer.

15. The researchers conclude that because "only three out of the 14 possible cases (seven fitness parameters for boys and seven for girls) were significantly different between urban and rural children, and (b) these differences were not uniformly distributed in children living in either urban or rural environments, it is concluded that the place of residence has no clear impact on physical fitness as studied herein."

    The researchers are hinting that the three significant p-values may represent a type of statistical error. Which error (Type-1 or Type-2)? In your own words, explain what that means in this context.

16. Explain the authors' statements that (a) the Greek physical education programme and (b) genetics may explain the lack of significant findings.

Chapter 7

Paired Data: One Quantitative Variable

Introduction

The main focus in this chapter is on analyzing paired data. The paired datasets of this chapter have a structure with two features:

---

(1) There is one *pair* of response values for each observational unit.

(2) There is a built-in *comparison*: one number in each pair is for Condition A, the other for Condition B.

Studies leading to paired data are often more efficient – better able to detect differences between conditions, because you have controlled for a source of variation (individual variability among the observational units). In this chapter, we turn to how to formally analyze and draw inferences from paired data. Most of the analysis is not new: You will continue to use the Six-Step Statistical Investigation Method and the 3S Strategy. In particular, the basic idea of the simulation step remains the same as before. You ask, "What if the null hypothesis is true. How will my statistic behave?" Just as before, the answer takes the form of the null ("what if") distribution, and you assess how unlikely the observed result would be if the null hypothesis were true, and you use a p-value to assess the strength of evidence against the null hypothesis.

Just as in Chapter 5, you can think of the null hypothesis in terms of an association between the response and a binary explanatory variable (Condition A or Condition B). To generate the null distribution, you *break the association* and create datasets by shuffling. (Here, because of the pairing, the right way to break the association will be different, but the general idea remains the same.) Section 7.1 focuses on the paired study design. Section 7.2 explains the simulation-based approach. Then Section 7.3 explains the theory-based shortcut.

By the end of the chapter, you will see several examples of the efficiency that can be gained from paired designs.

Section 7.1: Paired Designs

Introduction

So far in this unit all the studies have compared whole *groups* of individuals: mildly depressed patients randomly assigned to swim with dolphins or without dolphins, students randomly assigned to be sleep deprived or get unrestricted amounts of sleep and then see how well they score on a visual learning task. All of these studies used an **independent groups design** – there were no systematic connections relating individuals in one group to individuals in another group. In this section you will learn about a different study design based on pairs of units. For many situations, such **paired designs** allow more focused comparisons. As a result, these designs lead to more powerful tests for differences and narrower confidence intervals (See FAQ 7.1.1 for details).

Your main goal in Section 7.1 should be to learn to tell the difference between an independent subjects design and a paired design. The reason is this: The analysis of the data follows the design of the study that was used. If you are wrong about the design that was used, your analysis will be wrong. In other words, your p-values and confidence intervals may be misleading.

Example 7.1: Can You Study With Music Blaring?

Many students like to study while listening to music. They seem to think that it doesn't hurt their ability to focus on their work. Is that really true? What does the research say? The following excerpt is from "Checking It Out: Does music interfere with studying? (Strauss 2009)"

> *So what happens when your child studies for a science test with Jay-Z pouring from an iPod? According to Stanford University Professor Clifford Nass, the human brain listens to song lyrics with the same part that does word processing, which is the same part that supposedly is being employed for studying, he said. Something has to give, and it is the ability your child has to do the work he or she is supposed to be doing. Instrumental music is another story. For the most part, it is processed on the other side of the brain from the part that is processing language. "So if you are reading and listening to instrumental music you get virtually no interference," Nass said. "The music would not, in fact affect you, unless you are thinking deeply about the music, like, 'I wonder why Chopin chose the F sharp.'"*

Consider the following two experimental designs:

Experiment #1—Random assignment to two separate groups

In the first experiment, 27 students were randomly assigned to one of two groups: a group that listened to music with lyrics and a group that listened to music without lyrics. While students were listening to the music, they were playing a memorization game: They studied a list of 25 common five-letter words for 90 seconds. Then, the students needed to write down as many of the words as they could remember.

Experiment #2—Paired design using repeated measures

In the second experiment, instead of randomly assigning students to each of the two groups, what if students played the memorization game twice: once listening to the song with lyrics and once without. This is called a *paired design*.

> **Think about it:** Identify the explanatory variable and the response variable in this study (regardless of how we design it). Are there any advantages or disadvantages to the two designs?

The response variable in this study is the number of words memorized (quantitative). We would like to compare the outcomes of this variable between the two songs (the explanatory variable). However, there is probably a fair bit of variability in the scores from person to person because different people tend to have different memorization abilities. This variability may make it difficult to see differences between performances on the two songs, especially if those differences are small. For these reasons, the paired design has a couple of obvious advantages. First, instead of focusing on the number of words memorized by each individual, we can instead focus on the *difference* in the number of words memorized between the two songs by each individual. The variable *difference in number of words memorized* may well have less variability than the actual number of words memorized because there will probably be more consistency in how many more words people can remember with the instrumental music than in the number of words remembered among people. Focusing on the differences will provide us with a clearer picture of the "song effect," in particular, increasing the power (see Chapter 2), compared to an independent groups design, to find an effect if there really is one.

> **Definitions:** For a ***paired design***, response values come in pairs, with one response value in the pair for Group 1 and the other for Group 2. Sometimes the pairs come from matching individuals to create groups of two; sometimes the pairs come from measuring the same individual twice, once under each condition. For an ***independent groups design***, each individual in a group is unrelated to all the other individuals in the study. Each individual provides only one response value.

It is important to remember that "randomness" will still play an important role in paired study designs. For example, here we would randomize which song the student listened to first. Otherwise, if everyone listened to the song without lyrics second but tended to perform better, we wouldn't know whether that improvement was due to the absence of lyrics or to having previously heard the song and played the game

Why does the distinction between these two study designs matter? There are two main reasons.

- *When a paired design is possible, you typically get more informative results* because units within a pair tend to be similar to each other. When pairing is effective, differences within each pair on the response variable tend to be due mainly to the explanatory variable.
- *The right way to analyze the data depends on the study design.* The analyses for the two designs are not the same. If you get the design wrong, you get the analysis wrong, and your p-value and confidence interval estimate will be wrong.

You will learn more about the advantages of paired designs and how to use simulation and theory-based methods for inference from paired data in later sections. For now, we focus mainly on differences between the paired design and the independent groups design.

When should we pair?

We should use a paired design when we expect a strong association between repeated observations within the pair (e.g., on the same individual). In the example above, pairing was a good idea because we expect that some students are better memorizers than others, so when students play the game multiple times their memorization scores are strongly associated. FAQ 7.1.2 talks more about this issue.

Do we ever have pairing in observational studies?

Yes, absolutely. There are many cases where this comes up. For example, if you are interested in whether a diet program works you can evaluate the difference pre- and post-diet weights of the same individuals. Did most people lose weight? How much? This is a type of paired design where the response variable focuses on the weight difference. Another example would be if you wanted to compare prices at two different grocery stores. You would one list of products and then both prices for each product at the two stores. On average is there a price difference? By how much?

FAQ 7.1.1 Why do we pair?

Q: What's the main advantage of a paired design?

A: Paired designs are usually more powerful than designs that use independent groups.

Q: Explain, please. What do you mean by powerful?

A: We need an example. Lowley's grocery chain claims to have lower prices than Fresh Goods. How would you test their claim?

Q: Duh. Just compare prices.

A: For all items? That would take too long.

Q: So take a random sample.

A: Do you mean independent groups of items, with a separate random sample from each store?

Q: No, it makes more sense to compare prices for the same items: apples and apples, hot dogs and hot dogs.

A: That's a paired design. Why is it better?

Q: You tell me.

A: Think about *sources of variation*. There are three sources of variability: (1) store-to-store variability, which we want to know about, (2) item-to-item variability, and (3) all other variability. Our standardized statistic compares the size of store-to-store variability with the standard deviation from the design. The SD from the independent groups design has both (2) and (3). The paired design eliminates (2), so its SD has only (3).

Q: We don't care about price differences between items – between apples and hot dogs. The paired design gets rid of this item-to-item variability. With that variation out of the way, the price comparison is more direct – apples to apples, etc.

A: So the price differences for the pairs of items have a much smaller SD than the prices for independent groups of items. The payoff is greater power (less likely to make a type II error – see Chapter 2). The smaller SD means that we'll be more likely to detect a price difference between the two stores if there is one.

Q: Can you say it another way?

A: Think about confidence intervals. Suppose the true difference between store prices is 7 cents. The independent groups design with its much larger SD will give a wider interval. The paired design with its smaller SD will give a narrower interval. The wider interval is more likely to include 0, and fail to reject the null.

Q: So the chance that I reject the null when I should is higher with the paired design.

A: Exactly. That chance is the power. Less formally, the power is the chance of detecting an actual store-to-store difference. The paired design has greater power.

Q: Just like a microsope. Higher power lets you see smaller differences.

A: Good. And for most situations, a paired design is more powerful than a design with independent groups.

Q: Will that always be the case?

A: Well, most of the time the power of the paired design will be greater. But there are times where it doesn't help much, if at all.

Do pairs have to be the same observational unit?

Imagine we want to see whether there is a difference in the average hours spent watching TV by married men and women. One option is to randomly sample some married men and randomly sample some married women and compare the differences in the average hours spent on watching TV between the two groups. However, we might expect a fairly strong association between husband-wife pairs in terms of TV watching, because they might often be watching together. Thus, sampling husband-wife pairs and then testing to see whether the average difference in hours spent watching TV is different than zero will likely be a better study design. In this kind of a study design, the pair becomes the observational unit.

Note: We say here that a pair is an observational unit because we think this is the most intuitive way to understand the difference between and independent groups design and a paired design. However, the standard theory of experimental design says that each pair in a paired design is a *pair of units*. Each response value corresponds to a unit. There is one unit per pair in each treatment group. For now, you can continue to think of the pair as the observational unit, but if you take more statistics, you may need to change your definition.

> **Think about it:** Suppose that we gave each subject a pre-test to see how many words they could remember with no music. How could you use pre-test scores to create pairs of individuals for a paired design?

Experiment #3 — Paired design using matching

So far, we have created the pairs by measuring the same observational unit twice. For example, in the memorization game, each student played the memorization game twice, once under each condition. We got the pairs by "repeated measures." Another way to create pairs could use a pre-test to rank students according to how many words they could memorize, and use the results to create pairs. The top two students would be pair one, the next two would be pair two, and so on. Once we had the pairs, we would randomly assign one person in each pair to each condition.

Both paired designs are quite different from the first design for which the two groups are independent. For this particular context, the second design ("repeated measures") is best because you get two response values from the exact same subject. Sometimes, however, it is not possible, or not a good idea to use repeated measures to create pairs. If you are comparing two different surgeries, for example, you can't expect each patient to have two surgeries. In this case, matching can be useful if you have good criteria (i.e., related to the response variable) on which to base the matching, to try to create pairs where the individuals in each pair are likely to be very similar to each other. As we will see, both paired designs have some distinct advantages over the independent samples design.

FAQ 7.1.2 Should we always use pairs?

Q: So I should always use a paired design?

A: No, mainly for practical reasons.

Q: Practical reasons?

A: Suppose you want to test whether light bulbs burn longer in a room kept at $65°$ or a room kept at $75°$. Tell me how you'd use a paired design.

Q: I can't do it, because each bulb has only one lifetime to give for science.

A: So you can't reuse the bulbs. You get only one response per bulb. Can you create pairs by matching?

Q: Nope. That won't work either. The only matching that makes sense is to pair them according to their length of life, but we run into the same problem. Once matched, the bulbs are dead.

A: More generally, you can't use a paired design if your measurement destroys the observational unit.

Q: Are there other situations where pairing does not work?

A: Clinical trials in medicine. Suppose you want to compare two different treatments for HIV, or two different weight loss diets, or two approaches to quitting cigarettes. If the treatment is successful, then there's no way to get two response values from the same patient. For some studies, it may be possible to created pairs by matching patients according to age, sex, and seriousness of the problem, but for others, not. A lot depends on context.

Q: Context! It's becoming a dirty word.

A: Sorry. But notice that the logic applies not just to medicine but also to situations in education where you want to compare two ways to help students learn something, such as how to read, or how to do long division. Once they've learned, you can't make them unlearn just for the sake of a paired design.

Q: So all these are examples where the combination of treatment and measurement *changes* the observational unit.

A: Nice generalization.

Q: Thanks. Any other kinds of examples?

A: Suppose either the treatment or the measurement process takes a long time. For example, I want to know whether taking a daily zinc supplement helps prevent colds, and my response is number of colds in a year. I could use each subject twice, once taking zinc, and once taking a placebo (in a randomized order, of course). But that paired design would take two years. An independent groups design could be done in half the time.

Q: Are *we* done yet?

A: One last set of examples. If the conditions you want to compare are observational (smoker or not, Catholic or Protestant, fireman or policeman), there's no way to measure your subjects under both conditions. The best you can do to get a paired design is to create pairs by matching. Getting an effective design will depend on context.

Q: That word again. I can see that matching may take some careful planning. Why not just make the effort?

A: Often that's the best approach. Sometimes, though, reality gives you a hard choice. If effective matches are hard to find, you may have to choose between a small number of well-matched pairs, or a much larger set of unmatched subjects.

Q: Is there *anything* in statistics that is simple enough to be reduced to a rule?

A: Only the arithmetic. Arithmetic never depends on context. $1 + 1$ is always 2.

Exploration 7.1: Rounding First Base

Imagine you are at the plate in baseball and have hit a hard line drive. You want to try to stretch your hit from a single to a double. Does the path that you take to "round" first base make much of a difference? For example (see Figure 7.1), is it better to take a "narrow angle" or a "wide angle" around first base? (This exploration is based on an actual study reported in a master's thesis by W. F, Woodward in 1970.)

Figure 7.1: Two methods of rounding first base when a player plans to run to second base

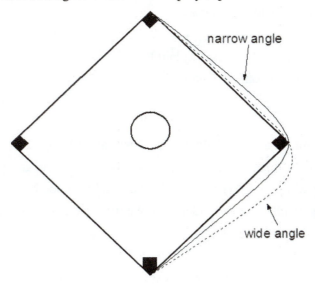

Think about designing a study to investigate this question.

1. Identify the explanatory and response variables in this study.

   Explanatory:

   Response:

2. How would you design an observational study to investigate this question? Explain why an observational study would not allow you to decide which base running angle is better than the other.

3. Suppose 20 baseball players volunteered to participate in an experiment. Suppose you also plan to assign a single angle, either wide or narrow, to each player. How would you decide which player ran with which base running angle?

A reasonable experimental design would be to randomly assign 10 of the 20 players to run with the wide angle and the other 10 to run with the narrow angle.

4.   Some runners are faster than others. Explain how random assignment controls for this, so that speed is not likely to be a confounding variable in this study.

Even though random assignment tends to balance out other variables (such as speed) between the two groups, there's still a chance that most of the fast runners could be in one group and most of the slow runners in the other group. More importantly, there's likely to be a good bit of variability in the runners' speeds, and that variability would make it harder to spot a difference between the base running angles even if one angle really is superior to the other.

5.   Suggest a different way of conducting the experiment, to make sure that speed is completely balanced between the two groups.

Definitions: For a *paired design*, response values come in pairs, with one response value in the pair for Group 1 and the other for Group 2. Sometimes the pairs come from matching individuals to create groups of two; sometimes the pairs come from measuring the same individual twice, once under each condition. For an *independent groups design*, each individual in a group is unrelated to all the other individuals in the study. Each individual provides only one response value.

In this study each runner can use *both* base running angles. That way we can be sure that neither treatment has more of the fast or slow runners, and we can also expect that *differences* in times for each runner will show considerably less variability than individual running times.

6.   What aspect of this experiment should be determined randomly? (*Hint*: The treatment is not determined randomly, because each runner experiences both treatments. But what other factor could still have an effect on the response unless it was randomized?)

7.   What do you suggest using as the variable to be analyzed with this paired design experiment? (*Hint*: Think of a better option than simply analyzing the set of times with the wide angle and the set of times with the narrow angle separately the way you would for an independent groups design of the sort described in #3 and #4.)

With a paired design, we analyze the *differences* in the response between the two treatments. In this case we would calculate the difference in running times between the wide and narrow angles for each player and then analyze the sample of differences. You will learn how to do this in Sections 7.2 and 7.3.

The order in which the players run with the two angles should be determined randomly; otherwise the order could be a confounding variable. Perhaps players would be slower with their second angle because of fatigue, or perhaps they would be faster with their second angle if they were slow to get warmed up. Randomizing the order takes away any worries about an order effect.

8. So far you have seen three designs for this study. The first (#2) was observational. The second (#3 and #4) was a randomized experiment with independent groups. The third (#5 and #6) used a paired design with pairs created by repeated measures. Consider a fourth design: Suppose you have 20 players, as before, and you have the time for each player in a 100-yard dash. Explain how you could use this information to create pairs of runners, and how you would assign one runner in each pair to the narrow angle, and the other to a wide angle.

9. Of the four designs (observational with independent groups, experimental with independent groups, paired design using repeated measures, and paired design using matching) which do you think is best for this context? Explain why. (*Hint*: Pairing works best when the units in a pair are as similar to each other as possible.)

10. As noted in #9, pairing works best when the two units in a pair are as similar as possible. When the units in a pair are not similar, pairing will not be effective. Invent and describe an example where you think pairing will be effective. Give the response and explanatory variable, how you would create the pairs, and why you think pairing will be effective.

11. Invent and describe an example where you think pairing will not be effective. Again, give the response and explanatory variable, how you would create the pairs, and this time why you think pairing will not be effective.

12. For some contexts, the strategy of using repeated measures to create pairs is either impossible or a bad idea. Invent and describe such a context.

13. Pairing can also be used in the design of an observational study. Revisit the observational study in #2, and suppose also that you have times in the hundred yard dash for each of the 20 players. Explain how you could use those times to create pairs. (Suppose that of the 20 players, 12 used the wide angle and only 8 used the narrow angle. This means you can create only 8 pairs.)

Section 7.1 Summary

In **paired designs**, response values come in pairs, with one response value in the pair for each group. Sometimes the pairs come from matching individuals and sometimes the pairs come from measuring the same individual twice. In studies with an **independent groups design**, each individual in a group is unrelated to all the other individuals in the study. Each individual provides only one response value.

- When a paired design is possible, you typically get more informative results because units within a pair tend to be similar to each other. When pairing is effective, subsequent differences within each pair on the response variable tend to be due mainly to the explanatory variable.

- *The right way to analyze the data depends on the study design.* The analyses for the two designs are not the same. If you get the design wrong, you get the simulation wrong, and your p-values and confidence intervals will be wrong.

Section 7.2: Simulation-Based Approach for Analyzing Paired Data

Introduction

In Chapter 6 you learned how to compare two groups on a quantitative response. The data arose either from a randomized experiment or from independent random samples. In Section 7.1, you studied a different experimental design that called for collecting *paired* data. Now you will learn how to analyze data that arise from a matched pairs design. In this section, you will see how to use randomization-based methods to conduct a test of significance for paired data.

Example 7.2: Rounding First Base (continued)

Recall the study from Exploration 7.1 that explored whether taking a "narrow angle" or a "wide angle" while rounding first base would increase your time to second base. (See Figure 7.2.)

Figure 7.2: Two methods of rounding first base when a player plans to run to second base

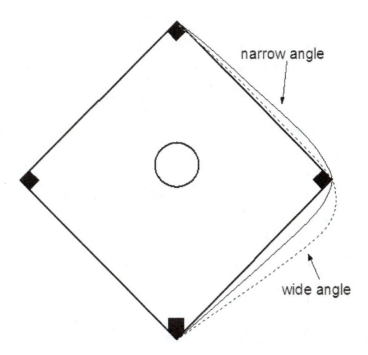

In Woodward's study, he used a stopwatch to time 22 different runners going from a spot 35 feet past home to a spot 15 feet before second. He had each runner use each method, with a rest period in between.

> **Think about it:** Why did he use a stop watch? Why did he use the "middle" of the route instead of the entire distance from home to second? Why did he have each runner use each method? How should he decide which method the runners use first? Why was there a rest period in between?

You should recognize Woodward's study as a ***matched-pairs design*** as discussed in Section 7.1. This design is advantageous because it controls for the runner-to-runner variability: Some runners are faster than others. By having each runner use both methods, we will have a more direct comparison of the effects *of the base running strategies* on the time to second base. The order of the methods used by each runner was determined randomly. The rest period is intended to eliminate fatigue as a confounding variable. If each runner first ran the narrow angle and then the wide angle, we wouldn't be able to separate the fatigue/learning effects from the base running strategy effect.

Key idea: As discussed in Section 7.1, in a matched pairs design, it is important to randomize the order of the treatments.

Although pairing should it make easier to help us separate out any treatment effect from other variables, pairing is not always feasible or helpful. In this study, especially with the rest period in between, pairing is very feasible. Next, we'll investigate whether it was beneficial.

The data in Table 7.1 show the times (seconds) for the first 10 runners, and Figure 7.3 shows dotplots of the times for all 22 runners for the two methods.

Table 7.1: The running times (seconds) for the first ten of the 22 subjects

Subject	1	2	3	4	5	6	7	8	9	10	
narrow angle	5.50	5.70	5.60	5.50	5.85	5.55	5.40	5.50	5.15	5.80	...
wide angle	5.55	5.75	5.50	5.40	5.70	5.60	5.35	5.35	5.00	5.70	...

Figure 7.3: The running times for a sample of 22 runners.

One thing to notice is that there is a lot of overlap in the two distributions (narrow mean: 5.534 sec, wide mean: 5.459 sec) and a fair bit of variability (narrow SD: 0.260 sec, wide SD: 0.273 sec). The slowest times took over six seconds, whereas the fastest times were just under five seconds. As you saw in Chapter 6, it will be difficult to detect a difference between the two methods when there is so much natural variation in the running times.

However, these data are clearly ***paired***. As discussed earlier, such data should <u>not</u> be treated as coming from two ***independent samples***. Instead, we will focus our analysis on the *difference* in times for each runner between the two methods.

Paired response variable: time difference in running to second (narrow angle – wide angle)

Keep in mind that the direction of subtraction here does not make a difference, as long as you are consistent for each pair and throughout the analysis. Table 7.2 shows the differences for the first 10 runners and Figure 7.4 is a dotplot of the differences for all 22 runners.

Table 7.2: Last row is difference in times for each of the first ten runners (narrow – wide)

Subject	1	2	3	4	5	6	7	8	9	10	
narrow angle	5.50	5.70	5.60	5.50	5.85	5.55	5.40	5.50	5.15	5.80	...
wide angle	5.55	5.75	5.50	5.40	5.70	5.60	5.35	5.35	5.00	5.70	...
difference	-0.05	-0.05	0.10	0.10	0.15	-0.05	0.05	0.15	0.15	0.10	...

Figure 7.4: Dotplot of differences in running times (narrow – wide), $\bar{x}_d = 0.075$ sec, $s_d = 0.0883$ sec

The distribution of the differences is slightly skewed left. The mean difference in running times is 0.075 seconds with a standard deviation of 0.0883 seconds. Notice that the mean of the differences (0.075) is equal to the difference in the means between the two methods (5.534 – 5.459). But the standard deviation of the differences (0.0883) is much smaller than the original standard deviations of the running times (0.260 and 0.273). This stems from the fact that although there is considerable runner-to-runner variability, the runners are fairly consistent in how much time they take—typically between 0.05 and 0.20 longer with the narrow method. Whereas we saw differences of up to a second between runners (less than 5 to more than 6 seconds), we did not see any one runner having a difference between the two routes of longer than 0.20 seconds. Figure 7.5 shows the original dotplots but with each observation paired between the two base running strategies.

Figure 7.5: Dotplots of the running times for each method with lines connecting each runner's times

What we notice is that the faster runners with the narrow angle tend to be the faster runners with the wide angle. The slowest runner with the narrow angle is also the slowest runner with the wide angle. So by looking at the differences in running speeds of the runners, we are able to account for a lot of the (runner to runner) variability in the data. Plus we see that most of the connecting lines slant from left to right, indicating a tendency for longer times with the narrow angle. This "leaning" would be difficult to pick up if we just treated them as unrelated dotplots.

So now we can ask whether this mean difference of 0.075 seconds is significantly different from zero.

We can define our parameter of interest to be

μ_d = the long-run mean difference in running times for runners when using the narrow angle instead of the wide angle.

Note that the subscript "*d*" in μ_d is to emphasize that we are looking at an average of *differences* in times. Then, the relevant hypotheses are

H_0: $\mu_d = 0$ (on average, the time to second base does not differ between the two methods)
H_a: $\mu_d \neq 0$ (there is a difference in running times between the two methods, on average)

Key idea: When the parameter of interest is the long-run mean difference or population mean difference, the corresponding statistic is the sample mean difference.

The statistic $\overline{x}_d = 0.075$ is above zero, but we need to ask the same question we've asked before: *Is it surprising to see such a large average difference in running times by chance alone, even if the base running strategy has no genuine effect on the times?*

Think about it: What should you use to measure how surprising the observed results would be, if the null hypothesis were true? Suppose you want to use simulation-based methods to find an approximate p-value, what will you assume to be true in carrying out the simulation? What is the statistic whose could-have-been values you will generate? And then what do you do next to find a p-value?

Another way to think about this question: If the base running strategy doesn't make a difference in general, then the two times for each runner were going to be the same two times regardless of which strategy was being used. Any difference in times for each person was just by chance, perhaps which one they ran first. So it doesn't matter which one we call the "wide angle" time and which we call the "narrow angle" time – the two times are completely interchangeable. In other words, we can consider the pair of measurements for each subject to be "swappable."

To model this using simulation, we will assume the same two times for each runner, but we will randomly decide which time goes with the narrow method and which with the wide method. In other words, for each runner we randomly decide whether to swap the two values or leave them as is. Then we see how much variability this creates in the sample mean difference from repetition to repetition. To decide which time comes first, we can use a coin flip. This is equivalent to using the coin flip to determine which pairs of times to swap and which to leave as is. We will let heads tell us to swap the times and tails tell us to not swap the times. We did this once and the results for the first ten runners are shown in Table 7.3.

Table 7.3: Runners whose responses were swapped are in bold

Subject	1	2	3	4	5	6	7	8	9	10	
narrow angle	5.55	5.70	**5.50**	5.50	**5.70**	**5.60**	5.40	5.50	5.15	**5.70**	...
wide angle	5.50	5.75	**5.60**	5.40	**5.85**	**5.55**	5.35	5.35	5.00	**5.80**	...
Difference	0.05	-0.05	**-0.10**	0.10	**-0.15**	**0.05**	0.05	0.15	0.15	**-0.10**	...

As a result of this one repetition, we ended up with an average difference of $\overline{x}_d = -0.025$ seconds. Note that our observed average difference of 0.075 is larger and further away from the hypothesized long-run mean difference of zero than the average difference found in the simulated results. But we need to repeat this process many times and find many simulated sample mean differences to construct a null distribution. This is how we could really tell whether our original sample mean difference is surprising when there is no connection between the times and which base running strategy was being used. The table below summarizes the key aspects of the simulation:

Null hypothesis = Long-run average difference in times is 0
One repetition = Re-randomizing (potentially swapping) narrow and wide-angle run times within runners
Statistic = Average difference in run times in the sample

To do this we will use the **Matched Pairs** applet. Pressing Randomize, the applet simulates flipping a coin for each runner. If the coin lands "heads" we swap that runner's wide angle times with his narrow angle time; if "tails" no change is made. After this is done for each of the 22 runners, the average of the time differences for each of the 22 runners is recorded. This process is then repeated a large number of times. Figure 7.6 shows the mean differences from 1000 such repetitions.

> **Think about it:** Consider Figure 7.6. The distribution of the simulated sample mean difference appears to be centered at about 0. Does that make sense?

Figure 7.6: Simulated mean differences for 1000 repetitions where for each repetition a coin is flipped for each runner to decide whether the runners' times are swapped

As can be seen in Figure 7.6, the distribution of the simulated sample mean difference appears to be centered around 0 seconds. This makes sense because the simulation is based on the assumption that the null hypothesis is true, and the time to get to second base with the two methods is the same, on average. Figure 7.6 also shows that the SD of the mean difference is 0.024 seconds—an estimate of repetition-to-repetition variability in the mean difference by random chance alone (the coin tossing random assignment).

Think about it: Consider Figure 7.6. With 1000 simulated values of the sample mean difference, is the observed mean difference from the actual study unlikely to have occurred assuming the null hypothesis to be true? How are you deciding?

In the actual study, the mean difference in running times was observed to be 0.075 sec.

We can see from Figure 7.7 that this falls in the upper tail of the distribution. We can use the applet to count whether there are any mean differences beyond 0.075 or -0.075 to estimate a two-sided p-value.

Figure 7.7: The simulated statistics that are at least as extreme as 0.075 (in either direction) are marked

We see just one of the 1000 repetitions of random swappings gave us a $\bar{x}_d$ value at least as extreme as what Woodward found in his study. We can also standardize the 0.075 by dividing by the applet's estimate for the SD of the mean differences (approximately 0.024). [We can estimate this standard deviation theoretically as well by taking the standard deviation of the differences (0.0883) and dividing by the square root of the number of pairs (22) = $0.0883/\sqrt{22} \approx 0.019$.] We note that 0.075 is just over three standard deviations from zero (e.g., 0.075/0.024 is 3.1215), another indication that the observed result would be unlikely to occur by random chance alone.

Think about it: How would you interpret the p-value?

Assuming the running time is the same with a wide angle or a narrow angle, on average, if we were to randomly decide whether to swap which value was recorded for which strategy, we would find a sample mean difference in times of 0.075 seconds or more extreme in about 0.1% of such repetitions.

> **Think about it:** Based on the p-value and standardized statistic, what is your conclusion about the hypotheses? Is there evidence against the null hypothesis? How are you deciding? What are your conclusions about causation and generalization?

This small p-value provides very strong evidence against the null hypothesis and in favor of the alternative hypothesis that runners such as those in this study tend to (on average) take different times getting from home to second when comparing these two running angles. We can draw a cause-and-effect conclusion about the study because the researcher used random assignment between the two methods for each runner (and the small p-value eliminates random chance as a plausible explanation). However, we don't have much information about how these 22 runners were selected to decide whether we can generalize to a larger population.

To estimate how much slower the narrow angle route tends to be, we can approximate a 95% confidence interval for μ_d using the standard deviation of the null distribution:

$$0.075 \pm 2(0.024) = (0.027, 0.124) \text{ seconds.}$$

We are 95% confident that, on average, the narrow angle route takes 0.027 to 0.124 seconds longer than the wide angle route. Because every (fraction of a) second counts, we would recommend that all base runners, at least like those in this study, use a wide angle when rounding the bases.

As you may have already noticed, the strategy we used to find the p-value for this study is the same 3S Strategy that is found in Step 4 of our Statistical Investigation Method, and has been used in analyses involving one or two proportions and two means:

1. **Statistic:** Compute the statistic in the sample. In this case, the statistic we looked at was the observed mean difference (narrow – wide) in running times.

2. **Simulate:** Identify a chance model that reflects the null hypothesis. We tossed a coin for each runner, and if it landed heads we swapped the two running times recorded for that runner – interchanging the wide angle and narrow angle times. If the coin landed tails, we did not flip the times. We repeated this process 1000 times, recording the simulated mean difference in running times each time, and thus obtaining a distribution of the mean differences that could have happened when the null hypothesis is true.

3. **Strength of evidence:** We found that in just one of our 1000 repetitions, the mean difference turned out to be at least as extreme as the observed difference of 0.075 seconds. This gave us a p-value of

0.001, which provided very strong evidence to say that running times for the narrow angle path and wide angle path are different, on average.

Did pairing matter? An incorrect analysis

> **Think about it:** What do you think would happen if we wrongly analyzed the data using a two independent samples procedure? In other words, what if the researcher had treated the 22 observations from the narrow method and the 22 observations from the wide method as coming from two different sets of 22 runners, instead of accounting for the paired nature of the observations? Would the p-value stay the same, increase, or decrease?

We can use the **Multiple Means** applet to simulate what would happen if we were to incorrectly analyze the data using a two independent samples procedure. Figure 7.8 gives the output from the applet, and the approximate p-value is 0.3470, much larger than that obtained from the paired samples analysis.

Figure 7.8: Multiple Means applet output when incorrectly analyzing the base running data using a two independent samples method

Count Samples [Beyond ▼] [.075] [Count]

Count = 347/1000 (0.3470)

This analysis, which assumes no connections between the times within runners, does not detect a significant difference between the two base running strategies. The larger standard deviation of the differences in group means (0.080) comes from not accounting for the runner-to-runner variability we saw back in Figure 7.3. Therefore, the researcher was very wise to use a matched-pairs design for this experiment, because otherwise he likely would not have detected that the wide angle method is significantly faster than the narrow angle.

Exploration 7.2: Exercise and Heart Rate

Raising your heart rate is one of the major goals of exercise. Do some exercises raise your heart rate more than others? Which common exercise, jumping jacks or bicycle kicks, raise your heart rate more? In this exploration, you are going to compare heart rates of you and your classmates doing jumping jacks and bicycle kicks. If you don't want to collect your own data, you may use data that was already collected on a class where jumping jacks were compared to bicycle kicks. The dataset can be found on the textbook website and is titled, **JJvsBicycle**.

Below are links to YouTube vides explaining how to do each of these exercises.
- Jumping jack: http://www.youtube.com/watch?v=dmYwZH_BNd0
- Bicycle kicks: http://www.ehow.com/video_2369096_basic-exercises-bicycle-kicks.html

After you do the exercise for 30 seconds, you will measure your heart rate. For instructions on how to measure your heart rate, visit the National Emergency Medicine Association website (http://www.nemahealth.org/programs/healthcare/heart_rate_pulse.htm) and follow the link for "How to measure your pulse."

It is very important that you follow these protocols carefully so we have high quality data to analyze. Although it is best to have the largest sample size possible, if you feel uncomfortable doing any of these exercises, please don't feel you have to.

1. You will first randomize the order (jumping jacks first or bicycle kick first) in which you measure your heart rate.
   - Flip a coin.
     o If "heads"
       ▪ Do jumping jacks for 30 seconds. Then, measure your **jumping jack heart rate** _____ beats per minute.
       ▪ Sit down and take a break for a minute or so.
       ▪ Now do bicycle kicks for 30 seconds. Then, measure your **bicycle kick heart rate** _____ beats per minute
     o If "tails"
       ▪ Do the bicycle kick exercise for 30 seconds. Then, measure your **bicycle kick heart rate** _____ beats per minute
       ▪ Sit down and take a break for a minute or so.
       ▪ Now do jumping jacks for 30 seconds. Then, measure your **jumping jack heart rate** _____ beats per minute.
   - Find the difference in your two heart rates (jumping jack heart rate – bicycle kick heart rate). _____
   - Your instructor will give you and your classmates instructions on where to record your jumping jack heart rate, bicycle kick heart rate, and difference (jumping jacks – bicycle kick).

2. Explain why it is reasonable to say that the two heart rates you collected should <u>not</u> be treated as *independent* data.

3. Why do you think we randomized the order of jumping jacks and bicycle kick, before measuring heart rates?

As seen in Chapter 6, we can summarize the quantitative data on heart rates using averages or means. But, because the data are *paired*, instead of comparing mean jumping jack heart rate to mean bicycle kick heart rate, we will instead look at the mean *difference* in heart rate between doing jumping jacks and bicycle kick. Thus, we can define our parameter of interest to be

μ_d = long-run mean *difference* in heart rate when doing jumping jacks and bicycle kicks in the population of interest.

Note that the subscript "*d*" in μ_d is used to denote that we are looking at an average of *differences*.

4. State the null and alternative hypotheses (using μ_d), to test whether the mean difference in heart rate between the two exercises is not 0.

Key idea: When the parameter of interest is the long-run mean difference or population mean difference, the corresponding statistic is the sample mean difference.

---

5. Find the average of the differences between the two heart rates for the entire class. This is the statistic we will use to summarize the data.

Your null hypothesis should essentially state that there is no difference in the heart rates between the two exercises, on average. If that is the case, it doesn't really matter if we swap someone's jumping jack heart rate with his or her bicycle kick heart rate. This is how we will model the null to develop a null distribution. To randomly swap some of the values we can just use a coin flip. If the coin lands heads you will swap the two heart rates. If the coin lands tails you won't swap the heart rates.

6. Flip a coin for each pair of heart rates and switch the appropriate ones. Recalculate the difference in heart rates and find the new simulated mean difference. Plot this value on the board in the classroom along with those from the rest of the class. Where does the actual statistic you found in #5 fit in this null distribution? Is it out in the tail?

7. As you know, it would be better to have many more simulations than what your class just did. We will do this by using an applet.

   - Go to the **Matched Pairs** applet.
   - Press **Clear** to erase the default data and then copy and paste the data (both columns – jacks and kicks) into the "Data" window. Then press **Use Data**.
   - Notice that the applet graphs the individual heart rates in each group, along with the means and standard deviations for the two groups.
   - Below that, the applet provides a dotplot of the differences in the heart rates in the sample. Please note that these difference values can be negative numbers because you are looking at *change* or *difference* in heart rates. The "Differences" graph also shows the mean of the differences and the standard deviation of the differences.
   - Write down these values in the following table:

Condition	Sample mean, $\bar{x}$	Sample SD, s
Jumping Jacks	$\bar{x}_{jj} =$	$s_{jj} =$
Bicycle Kicks	$\bar{x}_{bicycle} =$	$s_{bicycle} =$
Diff = JJ – BK	$\bar{x}_d =$	$s_d =$

8. The **Matched Pairs** applet will perform the simulation similar to what you did with flipping a coin.

   - Set the number of times to **Randomize** to 1 and click on **Randomize**.
   - One the coin tosses have determined which time will be in which column, the applet displays the re-randomized data (the colors show you the original column for each observation, so you should see a mix in each group now).
   - The could-have-been value for the mean difference is added to the Average Difference graph.
   - Write down these could-have-been values for the re-randomized data:

Condition	Re-randomized Sample mean, $\bar{x}$	Re-randomized Sample SD, s
Jumping Jacks	$\bar{x}_{jj} =$	$s_{jj} =$
Bicycle Kicks	$\bar{x}_{bicycle} =$	$s_{bicycle} =$
Diff = JJ – BK	$\bar{x}_d =$	$s_d =$

a. How does the actual mean difference compare to your simulated mean difference. Circle one.

     More extreme   Less extreme   Similar

b. How are you deciding?

9. Update the number of times to **Randomize** to 99 (for a total of 100 repetitions), uncheck
    Animate, and press **Randomize**. Consider the graph "Average Difference" that the applet has
    created...

    a. How many dots are in this graph?

    b. What does each dot represent?

The table below summarizes the key aspects of the simulation:

Null hypothesis = Long-run average difference in heart rates is 0
One repetition = Re-randomizing (possibly swapping) exercise heart rates within students
Statistic = Average difference in heart rates in the sample

10. To see many more possible values of mean difference in sample means that could have been, IF
     jumping jacks and bicycle kick rates were *swappable*, update the number of times to **Randomize**
     to 900, and press **Randomize** (for a total of 1000 repetitions). Describe the updated "Average
     Difference" graph with the 1000 samples or repetitions, with regard to the following features.

     a. Shape:

     b. About what number is this graph centered? Explain why you were expecting this.

     c. This graph also reports a value for standard deviation, SD. Report this value, and give a
         simple *interpretation* of this value. As in, what is this value measuring?

11. You now should have generated 1000 possible values of the mean difference in jumping jacks and bicycle kick heart rates that were simulated assuming the null hypothesis was true and these rates were the same, on average. How does the observed mean difference from your data [as reported in #5 and #7] compare to these simulated values? Is an average difference in heart rates like that observed in the actual study unlikely to happen by chance alone, if jumping jacks and bicycle kick heart rates are the same, on average? How are you deciding?

12. To quantify the strength of evidence against the null hypothesis, you can find the p-value.
    - Go back to the **Matched Pairs** applet.
    - In the **Count Samples** box, make an appropriate selection from the drop down menu (*Hint:* In what direction does your alternative hypothesis look), and enter the appropriate number in the box (*Hint:* At least as extreme as what number).
    - Report the approximate p-value.

13. Use the p-value to state a conclusion in the context of the problem. Be sure to comment on statistical significance. Can you conclude that there is strong evidence that jumping jack heart rate and bicycle kick heart rate differ? Why or why not? Can you draw a cause-and-effect conclusion? To what population are you willing to generalize the results?

14. Alternatively, you can summarize the strength of evidence using a standardized statistic. Find the standardized statistic and confirm that the strength of evidence you receive from the p-value is approximately the same as with the standardized statistic.

15. We can again use the 2SD Method to approximate a 95% confidence interval for the mean difference in heart rates between those doing the two exercises. The overall structure of the formula is the same:

$$\text{estimate} \pm 2(SD)$$

where the estimate is the sample mean difference in heart rates for your class and SD is the standard deviation of your null distribution when you did 1000 repetitions in the applet. Use these numbers to find an approximate 95% confidence interval for the mean difference in heart rates for those that do the two exercises your class did.

16. Provide an interpretation of this confidence interval, being sure to explain the parameter in this context.

As you may have already noticed, the strategy we used to find the p-value for this study is the same 3S Strategy that is found in Step 4 of our statistical investigation method, and has been used in analyses involving one or two groups:

1. **Statistic:** Compute the statistic in the sample. In this case, the statistic you looked at was the observed mean difference in heart rates.

2. **Simulate:** Identify a chance model that reflects the null hypothesis. To simulate what could have been if the null hypothesis is true, you can toss a coin for each student, and if it lands heads, swap the two heart rates recorded for that student. If coin lands tails, do not swap the heart rates. Repeat this process 1000 times, recording the mean difference in heart rates each time, and thus obtaining a distribution of these mean differences that were simulated assuming the null hypothesis were true.

3. **Strength of evidence:** If your actual observed statistic falls in the tail of the null distribution, then you have strong evidence that there is a significant difference in the average heart rates between those that do the two exercises.

Note: The heart rates from both exercises were *paired* on the same individuals, and so you used a simulation method that lets you use this information.

Let's check out how things would have worked had we ignored the pairing, and analyzed the data as if the sitting heart rates and standing rates had come from two totally different samples, that were independent of each other.

17. Go to the **Multiple Means** applet, and analyze the data as though we have two independent samples, as you did in Chapter 6. (You will need to create the "stacked data" first with the explanatory variable in one column and the response variable in a second column.)
    - With regard to the graph of the distribution of "Shuffled Differences in Means":
        - Mean = _____
        - SD = _____
    - Find and report the approximate p-value. _____

18. Compare the SD obtained using the *two independent samples* method to that obtained using the *paired samples* method. Which SD is larger?

19. Compare the p-value obtained using the *two independent samples* method to that obtained using the *paired samples* method. Which p-value is smaller and hence stronger evidence against the null hypothesis of no difference?

Note: Using a paired samples method will often give a smaller p-value and hence stronger evidence against the null hypothesis than the two independent samples method. Perhaps this is what you found in Question 18. This would happen if students with high heart rates after doing jumping jacks also had high heart rates after doing bicycles, similarly for students with low heart rates. This would make the variability of the differences in heart rates small. However this may not have happened for students in your class. If the variability within your jumping jack data or within your bicycle data is small, but there is a lot of variability in the differences, you could get a smaller p-value using the two independent samples method.

Section 7.2 Summary

In this section, you learned about analyzing paired samples. You also learned how to use a randomization approach to investigate whether the mean difference in response obtained from paired samples is statistically significant. The null hypothesis said that there was no difference in response, on average. To simulate what could have happened if this null hypothesis were true, we tossed a coin for every observational unit in our sample, and swapped their responses if the coin landed heads, and did not flip the responses if the coin landed tails. After every repetition, we recorded the mean difference that was obtained by chance alone. This process of shuffling and redistributing was repeated many (about 1000) times to get an idea of what could happen in the long run when the null hypothesis is true, and then we compared the long run pattern to what the observed mean difference in the actual study was. When the observed mean difference is unlikely to have happened by random chance, we declare the observed mean difference to be statistically significant. How you decide on what is considered unlikely or unusual is done exactly the same way as you did in the previous chapters. The interpretation of the p-value and what you can conclude from it maintains the same essence it did in earlier chapters. All that has changed is that instead of dealing with one set of responses from one sample, or one set of responses from each of two independent samples, you are now dealing with two sets of responses paired on the same set of observational units.

Section 7.3: Theory-Based Approach to Analyzing Data from Paired Samples

Introduction

In Section 7.2 you probably noticed that the shape of the simulated null distribution tends to be somewhat bell-shaped—something we've seen before! As we've discussed in earlier chapters, when certain validity conditions are met, simulation isn't needed, and we can use theory-based methods to conduct the analysis. We'll look at a theory-based approach, with its validity conditions, for paired data in this section.

Example 7.3: How Many M&Ms Would You Like?

Does your bowl size affect how much you eat? Food psychologist Brian Wansink (www.mindlesseating.org) ran an experiment with a group of undergraduates from the University of Illinois at Urbana-Champaign to investigate whether bowl size would affect how many M&Ms students took. In this example, we will analyze some of the data collected as part of that study. The study was conducted over several days. At one of the sessions, some participants were assigned to receive either a small bowl or a large bowl, and allowed to take as many M&Ms as they planned on eating during the study session. At a following session, the bowl sizes were switched, with those who received small bowls at a previous session now receiving large bowls, and vice versa.

> **Think about it:** What are the observational units in this study? What will you consider the response variable and explanatory variable in this study? Is each variable quantitative or categorical?

We will use the data from these undergraduates (the observational units), to test whether bowl size (the explanatory variable) affects how much we eat, as measured by number of M&Ms taken (the response variable), in this case. Thus, we can state our hypotheses as,

Null hypothesis: There is no association between bowl size and how much we eat, as measured by number of M&Ms taken, in the population.

Alternative hypothesis: When large bowls are presented, people tend to select more M&Ms.

> **Think about it:** Was this observational study or an experiment? What implications might this have with regard to potential conclusions that can be drawn?

You should have noticed that this was a randomized experiment, and the order in which each participant received a small bowl or a large bowl was randomly assigned. This means that if we find a statistically significant difference we will be able to draw a cause-and-effect conclusion between the size of the bowl and the number of M&Ms taken.

Data on the number of M&Ms candies taken by the 17 participants who attended both sessions is shown in Table 7.4. Note that the data are ordered by participant ID; that is, the number of M&Ms taken from the small bowl is matched up with the number of M&Ms taken by that individual from the large bowl, regardless of which bowl the student was given first.

Table 7.4: M&Ms counts separated by whether they used a small bowl or a large bowl

Subject ID	1	2	3	4	5	6	7	8	9	10	11	12	13	14	15	16	17
Small bowl	33	24	35	24	40	33	88	36	65	38	28	50	26	34	51	25	26
Large Bowl	41	92	61	19	21	35	42	50	11	104	97	36	43	62	33	62	32

Because the data are paired and quantitative, we will be looking at the difference in the number of M&Ms taken when a small bowl is used compared to when a large bowl is used.

Thus, we can define our parameter of interest to be

μ_d = Long-run mean difference for the number of M&Ms taken when using a small bowl minus a large bowl.

Using μ_d, we can restate our hypotheses as

$H_0: \mu_d = 0$ (that is, bowl size makes no difference, on average, with regard to number of M&Ms taken)

$H_a: \mu_d < 0$ (that is, bowl size does make a difference, on average, with fewer M&Ms taken from a small bowl)

Table 7.5 and Figure 7.10 show the summary statistics and a dotplot of the distribution of the differences (small bowl – large bowl).

Table 7.5: Summary statistics, including on the difference (small – large), of M&Ms counts

Bowl size	Sample size, n	Sample mean	Sample SD
Small	17	$\bar{x}_s = 38.59$	$s_s = 16.90$
Large	17	$\bar{x}_l = 49.47$	$s_l = 27.21$
Difference = Small - Large	17	$\bar{x}_d = -10.88$	$s_d = 36.30$

Figure 7.9: Dotplot of differences (small – large) in number of M&Ms taken in the study

Difference (= Small - Large)

From Figure 7.9 and Table 7.5, you can see, based on the sample mean and on the bulk of negative differences, that on average, when students used a small bowl, they took about 11 fewer M&Ms than when they used a large bowl. However, there is a fair bit of variability in these differences, with some students taking 40-60 more M&Ms with the small bowl and some taking almost 70 more with the large bowl.

As done in Section 7.2, the Matched Pairs applet can be used to run a simulation of what could have happened when the null hypothesis is true. Figure 7.10 gives a screenshot of the applet output.

Figure 7.10: Distribution of mean difference in M&Ms taken, generated assuming bowl size makes no difference

Figure 7.11 has shaded (to the left of the line) those re-randomizations that resulted in a mean difference less than or equal to -10.88, the mean difference observed in the study. Thus, the approximate p-value is = 122/1000 = 0.122. If bowl size makes no difference to how many M&Ms are taken, on average, there is about a 12.2% chance of seeing a sample mean difference of -10.88 M&Ms or less.

Figure 7.11: Distribution of mean differences in M&Ms taken, generated assuming bowl size makes no difference – area corresponding to p-value shaded red (to left of the line)

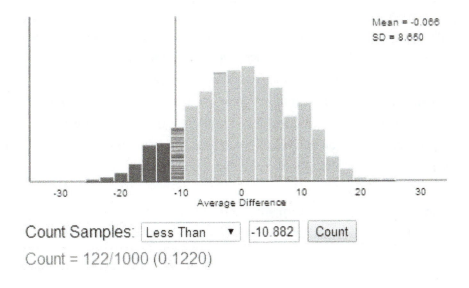

As expected, the null distribution in Figure 7.11 is centered about zero (mean = -0.066). We also note that the null distribution is approximately normal and the standard deviation of the differences is 8.65 candies.

Think about it: How many standard deviations away from the hypothesized difference of 0, is the observed mean difference of -10.88 candies? How are you deciding?

The observed difference, -10.88 candies, is approximately 1.26 standard deviations below 0 (1.26 =(10.88 − 0)/8.65). Thus, the observed mean difference is only 1.26 standard deviations below the mean, which does not put our observed statistic out in the tail of the null distribution—not a rare occurrence. The conclusion from this standardized statistic agrees with the larger sized p-value of 0.122.

Theory-based approach

When the sample size is *large enough* the overall shape of the null distribution of paired differences is approximately normal. With quantitative data, we said we could safely apply the theory-based methods if the response variable was symmetrically distributed, or you had at least 20 observations in (each) sample and the sample distribution and no strong skewness. We will apply these same criteria to our sample of differences: we want at least 20 differences without severe skewness or outliers in the sample distribution, or we want the distribution of the differences to be symmetric.

Validity Conditions for Theory-Based Analysis of paired data: Theory-based methods of inference will work well for paired data if the distribution of differences has a symmetric distribution, or you have at least 20 pairs (i.e., at least 20 differences) and the distribution of the sample differences is not strongly skewed. This test is known as a *paired t-test*.

Thus, if the sample size is large enough or if the differences themselves follow a normal distribution, then the null distribution of the sample mean difference has these properties:
- The shape of the null distribution is approximately normal
- The null distribution is centered at the hypothesized value of the population mean difference.
- The variability of the null distribution depends on the sample size (larger samples will have less variability by chance) and on the variability in the sample differences (the more variability in the differences, the more variability in the null distribution).

As in earlier chapters, we can use this information about the distribution of the sample mean difference in responses (with regard to shape, center, and variability), to estimate p-values and find confidence intervals.

In this case, because there are fewer than 20 pairs (only 17), we look for symmetry. Figure 7.9 illustrates symmetry of the differences and so we will consider the validity conditions to be met for these data.

It is for this reason that the null distribution we obtained when doing the simulation-based method (see Figure 7.11) looked fairly normal. Now, let's run the theory-based test and compare the p-value to the one we obtained from simulation.

Once we have found the differences for each observational unit, the theory-based approach we use is simply the one-sample *t*-procedure (see Chapter 2, Section 2 for more details) applied to the differences. Figure 7.12 shows the results of using the Theory-Based Inference applet to assess the strength of evidence we have against the null hypothesis of no difference.

Figure 7.12: Output from the Theory-Based Inference applet, showing standardized statistic, p-value, and 95% confidence interval for the M&Ms study

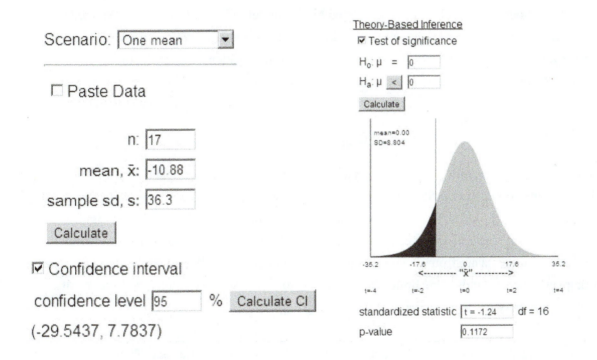

From Figure 7.12, you can see that "One mean" is being compared to a hypothesized value of 0, and that the hypotheses have been adjusted to match our hypotheses (comparing to zero, one-sided). The applet produces a standardized statistic "*t* = -1.24" and a p-value 0.1172. Notice that this standardized statistic and p-value are very close to what we found with the simulated-based analysis.

Think about it: Based on the p-value, what is your conclusion? Do the data provide evidence that bowl size matters with regard to number of M&Ms taken? To what larger population(s) would you be comfortable generalizing these results? Can you draw a cause-and-effect conclusion?

Confidence Interval

The Theory-Based Inference applet output also produces a 95% confidence interval for μ_d (small – large).

Think about it: What parameter is this confidence interval estimating? Does the 95% confidence interval contain 0? What does that imply?

Recall that a confidence interval is an interval of plausible values for the parameter of interest. In this case, the parameter of interest is μ_d, the mean difference in the amount of M&Ms that would be taken from a small bowl versus that taken with a large bowl for the population of students like those in this study. The 95% confidence interval for μ_d is calculated by the applet to be (-29.54, 7.78). We are 95% confident that, on average in the long run, students take up to 29.54 more candies with a large bowl or up to 7.78 fewer candies with a large bowl. (Zero is contained in our confidence interval, confirming what we learned from the test in that we can't declare a clear "winner" here.)

Conclusions

A p-value of 0.1172 is not considered small enough to provide strong evidence against the null hypothesis. Thus, we do not have evidence to conclude that bowl size matters with regard to number of M&Ms taken. Even though the study was a randomized experiment, the lack of statistically significant results prevents us from concluding any cause-and-effect relationship between bowl size and number of M&Ms taken. Also, the data are from undergraduates at a large university. We do not have information on the participants' backgrounds. This limits the population to whom we can generalize these results. These results may or may not be extendable to other age groups.

Think about it: So, we were unable to find evidence to conclude a cause-and-effect relationship between bowl size and number of M&Ms taken. Why might have this happened? What are some possible explanations?

We can think of the following as possible explanations.

- Perhaps, in reality, bowl size does not affect how much we eat.

- Or, maybe in reality, bowl size does affect how much we eat, but the sample size was not big enough to detect the difference, especially considering how much variability there was in the differences of number of M&Ms taken (low power).

- Maybe, bowl size does affect how much we eat of other food, but we looked at data for M&Ms which is considered a snack. Perhaps people don't treat snacks the same way as they would a bowl of soup or pasta.

Formulas

The formula used to calculate the t-statistic, to test the above null hypothesis, is

$$t = \frac{\bar{x}_d - 0}{\frac{s_d}{\sqrt{n}}}$$

where $\bar{x}_d$ is the mean of the paired sample differences, zero is the hypothesized value of the mean difference for the population, s_d is the sample standard deviation of the observed differences, and n is the sample size. The numerator $\bar{x}_d$ for the t-statistic is essentially comparing the observed mean difference to a hypothesized population mean difference of 0. The denominator $s_d/\sqrt{n}$ is a measure of the sample to sample variability of the sample mean difference, and is called the **standard error (SE)** of the sample mean differences. The SE combines information on how much variability there was in the differences in response among observational units (the sample variability in the differences s_d), and how many observational units are in the sample (n). When the sample size is large enough (e.g., at least 20) and the distribution of the sample differences is not strongly skewed or the population distribution of differences follows a normal distribution, then this standardized statistic follows a t-distribution with $n - 1$ degrees of freedom. (Information about the t-distribution was introduced in Section 2.2.)

When the sample sizes are large enough, we can also find an approximate 95% confidence interval for μ_d, as follows:

$$\bar{x}_d \pm multiplier \times \frac{s_d}{\sqrt{n}}.$$

where the multiplier is approximately 2 for 95% confidence intervals (the actual value of the multiplier comes from the t distribution and depends on the confidence level and the number of observational units n).

As before, the formula for confidence interval is *statistic margin-of-error*. In the context of paired data with a quantitative response, the statistic is the observed sample mean difference. The margin-of-error, as before, accounts for (i) confidence level, and (ii) sample to sample variability in the statistic.

Exploration 7.3: Comparing Auction Formats

An economist at Vanderbilt University devised a study to compare different types of online auctions. In one experiment he compared a *Dutch* auction to a *first-price sealed bid* auction. In the Dutch auction the item for sale starts at a very high price and is lowered gradually until someone finds the price low enough to buy. In the first-price sealed bid auction each bidder submits a single sealed bid before a particular deadline. After the deadline, the person with the highest bid wins. The researcher auctioned off collectible trading cards from game *Magic: The Gathering*. He placed pairs of identical cards up for auction; one would go into the Dutch auction and the other to the first-price sealed bid auction. He then looked at the difference in the prices he received on the pair. He repeated this for a total of 88 pairs.

1. Before we look at the data that was collected and start the analysis, let's make sure you understand the study design.

    a. Explain why the price data should be analyzed using paired samples as opposed to two independent samples?

    b. What makes a pair?

    c. What is the explanatory variable? Is it categorical or quantitative?

    d. What is the response variable? Is it categorical or quantitative?

2. State the relevant hypotheses in words. (Use a two-sided alternative.)

    Null Hypothesis:

    Alternative Hypothesis:

3. Define the parameter of interest and give the symbol that should be assigned to it.

4. State the relevant hypotheses in symbols.

   H_0:

   H_a:

5. The data for the auction, **auction**, can be found on the textbook website. The selling prices for the Dutch auction are labeled *Dutch* and the prices for first-price sealed bid auction are labeled *FP*. Paste these data into the **Matched Pairs** applet and answer the following questions.

   a. What is the sample mean price for the Dutch auction?

   b. What is the sample mean price for the first-price sealed bid auction?

   c. What are the mean and standard deviation for the difference in the two prices?

   d. Determine a p-value using the applet. Explain how you did so.

Notice that the simulated null distribution is bell-shaped. This is no coincidence. A theory-based method exists that predicts this to occur when certain validity conditions are met.

Validity Conditions for Theory-Based Analysis of paired data: Theory-based methods of inference will work well for paired data if the distribution of differences has a symmetric distribution, or you have at least 20 pairs (i.e., at least 20 differences) and the distribution of the sample differences is not strongly skewed. This test is known as a ***paired t-test***.

---

6. Are the validity conditions met for these data? Explain.

7. Because the sample is large enough without strong skewness in the distribution of differences, we can use a theory-based approach. We will do this by using the **Theory-Based Inference** applet. To do this:
   - Open the **Theory-Based Inference** applet.
   - Choose **One mean** from the pull-down menu.
   - Enter the sample size, sample mean, and sample standard deviation for the differences as you found in #5.
   - Check the box for **Test of Significance**.
   - Enter the appropriate information for the hypotheses.
     o Make sure the appropriate sign for the alternative hypotheses is chosen.
   - Press **Calculate**.

   a. What is the value of the standardized statistic? What does that number tell you?
   b. In the light of the value of the standardized statistic, do you expect the p-value to be small or not small? How are you deciding?
   c. What is the value of the p-value? Is it similar to the p-value found using the simulation-based method?

8. Recall that the **Theory-Based Inference** applet can also produce confidence intervals for the parameter of interest.

   - Go to the applet, check the **Confidence interval** box and let the **confidence level** be 95%.
   - Press **Calculate CI**.

a. Report the 95% confidence interval here.

b. Interpret the 95% confidence interval in context.

c. Does the 95% confidence interval agree with your conclusion when using the p-value? How are you deciding?

9. Based on the above analysis, state your conclusion in the context of the study. Be sure to comment on:

- *Statistical significance:* Do the data provide evidence that selling cards in a Dutch auction differ than when sold in a first-price sealed bid auction on average? How are you deciding?
- *Estimation:* Find and interpret a 95% confidence interval.
- *Causation:* Can you conclude causation? If yes, what causes what? If not, how are you deciding?
- *Generalization:* Can you extend the results of this study? Other kinds of cards? Other types of items? Anything sold in an auction format on the Internet? How are you deciding?

Section 7.3 Summary

Under certain circumstances, essentially when sample sizes are large enough and/or skewness is minimal, a theory-based approach can be used to approximate p-values and confidence intervals for the population mean difference with a matched pairs design.

- Once again, the standardized statistic subtracts the hypothesized value from the observed statistic, and then divides by the standard deviation of the statistic.
  - o The statistic in this case is the sample mean difference, that is, $\bar{x}_d$.
  - o The standardized statistic is $$t = \frac{\bar{x}_d}{s_d / \sqrt{n}}$$ where s_d is the standard deviation of the differences and n is the number of pairs.
- The confidence interval takes the form: statistic $\pm$ margin-of-error.
  - o Statistic $= \bar{x}_d$
  - o Margin-of-error = multiplier $\times$ standard error of statistic
  - o Standard error $SE = \frac{s_d}{\sqrt{n}}$.
- **Validity conditions** for the theory-based approach for matched pairs data (*paired t-test*) to be valid are that the sample size (of pairs) is large (generally considered to be at least 20) and the distribution of the sample differences is not strongly skewed, or the differences in the population

follow a normal distribution. You can judge this latter possibility by viewing the distribution of the sample differences to check whether it is symmetric.

- The order of subtraction does not matter, provided that the labeling is consistent throughout the analysis, including interpretations and conclusions.

Type of data	Name of test	Applet	Validity conditions
Paired observations of a quantitative variable	Simulation test for paired data	Matched pairs	---
	Theory-based test or interval for paired data (paired *t*-test; paired *t*-interval)	Matched pairs; Theory-based inference	At least 20 pairs without strong skewness or symmetry in distribution of differences

Chapter 7 Summary

In this chapter we saw how a paired design differs from an independent-samples design. A key component of a paired design is to create pairs (or repeated observations on the same observational unit) that should be similar to each other in all respects except for the explanatory variable. This more direct comparison accounts for some of the variability in the data. Ideally the variability of the differences will be smaller than the variability in the response variable making it more likely (greater power) that we will correct detect population differences that exist and giving us a smaller margin-of-error.

- When designing a matched-pairs experiment, randomization is used to determine the order of the treatments.
- After pairing, it is important to adjust the analysis to recognize the paired nature of the data. We do this by computing the differences in the response for each pair. The order of subtraction does not matter, provided that the labeling is consistent throughout the analysis, including interpretations and conclusions.

You then learned how to use a randomization approach to investigate whether the observed mean difference obtained from paired samples is statistically significantly different from zero. This simulation replicated the swapping of the observations within each pair under the null hypothesis that the response would not change depending on the explanatory variable. This allowed us to create a null distribution of mean differences from which we can assess the chance variability in the statistic and approximate a p-value.

The interpretations of the p-value, standardized statistic, confidence interval and what you can conclude from them maintain the same essence as in earlier chapters. All that has changed is that instead of dealing with one set of responses from one sample, or one set of responses from each of two independent samples, you are now dealing with two sets of responses paired on the same set of observational units.

You also saw how one-sample t-procedures that you learned in Chapter 2 (in this context, known as **paired t-tests** and **paired t-intervals**) on the differences can be used as a theory-based approach when the validity conditions are met.

    o   The statistic in this case is the sample mean difference, that is, $\bar{x}_d$.

$$t = \frac{\bar{x}_d}{s_d / \sqrt{n}}$$

    o   The standardized statistic is $\quad$.

    o   The confidence intervals is $\bar{x}_d \pm$ multiplier x $\quad SE = \frac{s_d}{\sqrt{n}}$.

Validity conditions for the theory-based approach (paired t-test and interval) to be valid are that the sample size (of pairs) is large (generally considered to be at least 20) without strong skewness, or that the differences in the population follow a normal distribution. (Check this last case by examining the distribution of differences in the sample.)

Chapter 7 Glossary

μ_d

    Population or long-run parameter for the average of the differences 716

independent groups design

    Each individual in a group is unrelated to all the other individuals in the study. Each individual

    provides one response value. 703

independent samples

    The data recorded on one sample are unrelated to those recorded on the other sample. In other

    words, if the data from the samples can be rearranged without affecting the outcome then the

    samples are independent. 714

matched-pairs design

    Randomize the order in which each subject receives treatment, but each subject receives both

    treatments 713

paired

    Data collected on paired samples consist of two sets of observations on the response variable

    that are recorded on the same set of observational units 715

paired design

    Study design that allows for the comparison of two groups on a response variable but by

    comparing two measurements on each observational unit instead of on completely separate

    groups of individuals. This serves to reduce variability in the response variable. 702

paired t-test

    A theory-based test for paired data 734

standard error (SE)

    The standard deviation of the sample average differences 737

validity conditions for theory-based test of paired data

    A sample size of at least 20 pairs without strong skewness in the distribution of differences, or

    a symmetric distribution of differences. This test is known as a *paired t-test* 734

x-bar$_{d:}$

    Observed sample average of the differences 716

Exercises

Section 7.1

7.1.1* Which of the following statements are true and which are false?

a) In pairs design, each pair of *always* consists of measuring the same individual twice.
b) In pairs design, each pair of *often* consists of measuring the same individual twice.
c) In an independent groups design, each individual in a group is unrelated to all the other individuals in the study.
d) When a paired design is possible, you typically get more informative results because units within a pair tend to be similar to each other.

7.1.2 Which of the following statements are true and which are false?

a) After you collect the data, you can then decide whether or not to use a paired design or independent samples design.
b) The simulations for paired design and independent groups design are the same. Just how the data are collected is different.
c) Paired designs can only be done in experiments and not observational studies.
d) In paired design with repeated measures, you can use random assignment as each individual is measured twice.

7.1.3* For the following scenarios, determine whether or not they represent paired designs.

a) Reaction times for students taken at 8:00 a.m. compared to reaction times for the same students taken at 10:00 p.m.
b) Change in weight for those on the Atkins Diet compared to the change in weight for those on the Paleo Diet.
c) Times for skiers to complete a downhill skiing race course on skis treated with wax A with times for the same skiers to complete the course on skis treated with wax B.

7.1.4 For the following scenarios, determine whether or not they represent paired samples.

Not Paired

Paired

Not Paired

a) Test scores for students in a biology class taught by Professor Quick are being compared to test scores in a different section of the biology class taught by Professor Quack.

b) Pulse rates for students at the beginning of class are being compared to pulse rates for the same students at the end of class.

c) The weights of 10-year olds in 2009 are being compared to the weights of 10-year olds in 1994.

7.1.5* For each of the following, indicate whether or not a *paired* analysis would be appropriate.

a) To investigate the claim that first-year college students tend to gain weight during their first term, you take a random sample of 20 first-year students and weigh them at the beginning and end of their first term.

b) You wonder whether students on your campus tend to drive newer cars than faculty. You take a random sample of 20 students and a random sample of 10 faculty members, and ask each person how old their car is.

7.1.6 For each of the following, indicate whether or not a *paired* analysis would be appropriate.

a) A school cafeteria offers a vegetarian and a non-vegetarian option for lunch every day. For a period of two weeks, you record how many calories are in the vegetarian option and how many calories are in the non-vegetarian option. Your goal is to see whether vegetarian options tend to differ with regard to average number calories from non-vegetarian options.

b) A farmer investigates whether talking to cows by name leads to producing more milk. He randomly selects 30 of his cows and randomly assigns 15 to talk to by name and the other 15 not to.

7.1.7* What is the difference between paired design using repeated measures and paired design using matching?

Testing students

7.1.8 Which of the following is an example of a paired design?

a) A teacher compares the pre-test and post-test scores of a group of students. *Paired*

b) A teacher compares the scores of students using a computer based method of instruction with the scores of other students using a traditional method of instruction. *Not Paired*

c) A teacher compares the scores of students in her class on a standardized test with the national average score. *Not Paired*

d) A teacher calculates the average of scores of students on a pair of tests and tests to see whether this average is significantly larger than 80%. *Not paired.*

The difference first, then at end at the average.

Teaching with technology*

7.1.9 Which of the following is an example of a paired design?

a) A teacher taught one class a lesson using technology and another class without technology. Their scores on a quiz were then compared. *Not Paired*

b) A teacher taught a lesson using technology and then compared the classes score on a quiz to the national average score. *Not paired*

c) A teacher teaches a lesson using technology and then gives her students a pair of quizzes and compares these results to those from her class the previous year. *Not paired*

d) Pairs of students were matched by similar IQs. One of the pair taught a lesson using technology and one without technology. Their scores on a quiz were then compared. *Not Pair*

Football throwing

7.1.10 Suppose that you want to determine whether there is a difference between the distances traveled by an air-filled football and that by a helium-filled football, and have managed to recruit 12 punters of varying abilities to participate in the study.

a) Describe, in detail, the design of an experiment that does not utilize pairing.

b) Next, describe, in detail, the design of an experiment that does utilize pairing.

c) Which study design (a) or (b) do you think is a better study design for the purpose of this study, or does it not matter? Explain your reasoning.

Golf drives*

7.1.11 Suppose that you want to determine whether there is a difference between the distances traveled by two different brands of golf balls. You use the 30 people on your golf league to participate in the study.

a) Describe, in detail, the design of an experiment that does not utilize pairing.
b) Next, describe, in detail, the design of an experiment that does utilize pairing.
c) Which study design (a) or (b) do you think is a better study design for the purpose of this study, or does it not matter? Explain your reasoning.

Caffeine and the long jump

7.1.12 Suppose an instructor wants to study the association between caffeine intake and performance in the long jump. The instructor asks her students to suggest ways to collect data to investigate this question. For each of the suggestions, identify:

- Whether the study is an experiment or an observational study.
- Whether or not the data will be paired.

a) Lara suggests the instructor record everyone's performance on a long jump, and then ask each of them to report their caffeine intake from earlier in the day.
b) Mimi suggests the students be paired by height, and then one student in each pair is randomly assigned to caffeine while the other isn't.
c) Nia suggests all students be given both a caffeine pill and a placebo, but the order be randomized. Students' performances on both these occasions should then be recorded.

Sleep and memorization*

7.1.13 Suppose an instructor wants to see whether there is an association between the amount of sleep a student gets and their ability to memorize words. The instructor asks her students to suggest ways to collect data to investigate this question. For each of the suggestions, identify:
- Whether the study is an experiment or an observational study.
- Whether or not the data will be paired.

a) Julia suggests the instructor have each of the students do two trials. One where they are told to get at least 7 hours of sleep the night before they take the memorization test and one where they are told to limit their sleep to at most 4 hours before they take the memorization test. The trials are a week apart and the students are randomly assigned which they do first.

b) Lindsey suggests half the students (chosen at random) should be told to get at least 7 hours of sleep before the memorization test and the other half should be told to get at most 4 hours of sleep before the test. They all take then take the memorization test on the following day.

c) Mikaela suggests all students are given the memorization test and then are asked how many hours of sleep they got the night before.

Reaction times

7.1.14 To test reaction times, a computer program will display a red circle. When it changes to yellow, the users click the mouse as fast as they can. The program will then display the reaction time. Students want to compare reaction times between people using their dominant hand and non-dominant hand on the mouse.

a) Suppose you had 20 subjects participating in this study. Describe how you would conduct this study with a paired design with repeated measures.

b) If you found a significant difference in reaction times, could you conclude that the hand being used caused the difference?

7.1.15 Reconsider the reaction time study in the previous question.

a) Describe how you would conduct this study with an independent samples design?

b) Which design, paired with repeated measures or independent samples, would you say is more appropriate for this study? Explain why.

7.1.16 Reconsider the reaction time study in the previous questions

a) Describe how you would conduct this study with a paired design using matching?

b) Which design, paired with repeated measures or paired with matching, would you say is more appropriate for this study? Explain why.

Exercise and memorization*

7.1.17 Students wanted to see whether exercising (in particular doing jumping jacks) would help people memorize a list of 10 words that were read to them. They wanted to compare this with the same test of memorizing words while the subjects were sitting down.

a) Suppose you had 20 subjects participating in this study. Describe how you would conduct this study with a paired design with repeated measures.

b) If you found a significant difference in the number of words memorized, could you conclude exercising caused the difference?

7.1.18 Reconsider the memorization and exercise study in the previous question.

a) Describe how you would conduct this study with an independent samples design?

b) Which design, paired with repeated measures or independent samples, would you say is more appropriate for this study? Explain why.

7.1.19 Reconsider the memorization and exercise study in the previous questions

a) Describe how you would conduct this study with a paired design using matching?

b) Which design, paired with repeated measures or paired with matching, would you say is more appropriate for this study? Explain why.

Chewing gum and memorizing

7.1.20 Students wanted to see whether chewing gum would help people memorize words from a given list of 25 words compared to trying to memorize words when not chewing gum.

a) Suppose you had 20 subjects participating in this study. Describe how you would conduct this study with a paired design with repeated measures.

b) If you found a significant difference in the number of words memorized, could you conclude gum chewing caused the difference?

7.1.21 Reconsider the chewing gum and memorization study in the previous question.

a) Describe how you would conduct this study with an independent samples design?

b) Which design, paired with repeated measures or independent samples, would you say is more appropriate for this study? Explain why.

7.1.22 Reconsider the chewing gun and memorization study in the previous questions

a) Describe how you would conduct this study with a paired design using matching?
b) Which design, paired with repeated measures or paired with matching, would you say is more appropriate for this study? Explain why.

Caffeine and running*

7.1.23 Researches wanted to see whether using a caffeine supplement would make runners faster in a five-kilometer run.

a) Suppose you had 30 subjects participating in this study. Describe how you would conduct this study with a paired design with repeated measures.
b) If you found a significant difference in running times, could you conclude caffeine caused the difference?

7.1.24 Reconsider the caffeine and running time study in the previous question.

a) Describe how you would conduct this study with an independent samples design?
b) Which design, paired with repeated measures or independent samples, would you say is more appropriate for this study? Explain why.

7.1.25 Reconsider the caffeine and running time study in the previous questions

a) Describe how you would conduct this study with a paired design using matching?
b) Which design, paired with repeated measures or paired with matching, would you say is more appropriate for this study? Explain why.

Section 7.2

7.2.1* Suppose you are testing the hypotheses $H_0: \mu_d = 0$ and $H_a: \mu_d \neq 0$ in a paired designed test and obtain a p-value of 0.21. Which one of the following could be a possible 95% confidence interval for μ_d?

A. -2.3 to -0.7

B. -1.2 to 0.9

C. 1.5 to 3.8

D. 4.5 to 6.9

7.2.2 Suppose you are testing the hypotheses $H_0: \mu_d = 0$ and $H_a: \mu_d \neq 0$ in a paired designed test and obtain a p-value of 0.02. Also suppose you computed confidence intervals for μ_d. Based on the p-value which of the following is true?

A. Both a 95% confidence interval and 99% confidence interval will contain 0.

B. A 95% confidence interval will contain 0, but a 99% confidence interval will not.

C. A 95% confidence interval will not contain 0, but a 99% confidence interval will.

D. Neither a 95% confidence interval nor a 99% confidence interval will contain 0.

7.2.3* What does one dot in a null distribution made by the Matched Pairs represent?

A. It is a simulated average difference in means under the assumption there is no association between the explanatory and response variable in the study.

B. It is a simulated average difference in means under the assumption there is an association between the explanatory and response variable in the study.

C. It is a simulated difference in proportions under the assumption there is no association between the explanatory and response variable in the study.

D. It is a simulated difference in proportions under the assumption there is an association between the explanatory and response variable in the study.

7.2.4 In the matched pairs tests done in this section:

A. The explanatory variable and response variables are categorical.

B. The explanatory variable and response variables are quantitative.

C. The explanatory variable is categorical and the response variable is quantitative.

D. The explanatory variable is quantitative and the response variable is categorical.

7.2.5* Match the appropriate **test of significance** to each of the following research questions.

1. Testing a single proportion   3. Comparing two proportions
2. Testing a single mean   4. Comparing two means (independent)
                                      5. Testing a mean difference (paired data)

   a) Do a minority of males think cigars smell good?

   b) Are males more likely to donate blood compared to females?

   c) Is the average body temperature of females higher than that for males?

   d) Is there an improvement from the pre-test scores students take before a lesson to their post-test scores after the lesson?

7.2.6 Match the appropriate **test of significance** to each of the following research questions.

1. Testing a single proportion   3. Comparing two proportions
2. Testing a single mean   4. Comparing two means (independent)
                                      5. Testing a mean difference (paired data)

   a) Is the average body temperature higher than 98.6°F?

   b) Will rats go through a maze faster after given caffeine compared to the same rats without caffeine?

   c) Is there an association between gender and GPA?

   d) Is there an association between gender and whether or not a person gets the flu?

7.2.7* A researcher measures body flexibility in college students. She suspects that females will be more flexible than males.

   a) State the null and alternative hypothesis

   b) Identify the explanatory and response variable

   c) Identify whether the study uses an independent groups design or a paired study design.

7.2.8 A researcher takes a sample of sibling-pairs from two-child families. All individuals are given a test of their impatience, hypothesizing that older siblings tend to be have higher levels of impatience.

    a) State the null and alternative hypothesis
    b) Identify the explanatory and response variable
    c) Identify whether the study uses an independent groups design or a paired study design.

7.2.9* The Matched Pairs applet displays parallel dotplots of the response variable for two different conditions of a study. It also displays a dotplot for the differences. Which is the more useful dotplot to look at to help you determine if the test will give you significant results? Explain.

7.2.10 What is the coin flip accomplishing in the Matched Pairs applet and why is it doing it?

Weight gain in college*

7.2.11 Suppose the table below represents statistics about the weights of female college students during their first two years of college. You are interested in trying to show that female college students gain weight during between their first and second years in college. Explain what in this summary tells you should get a smaller p-value when you run your test by doing a matched paired test than if you did an independent samples test.

	Freshmen	Sophomore	Difference
Mean	130 lbs	135 lbs	5 lbs
Standard Deviation	25 lbs	27 lbs	2 lbs

Marriage Ages

7.2.12 To investigate whether or not people tend to marry spouses of similar ages or whether husbands tend to be older than their wives, a student gathered age data from a sample of 24 couples, taken from marriage licenses filed in Cumberland County, PA, in June and July 1993. The data file **MarriageAges** contains the data.

a) Here's a scatterplot of the ages of the husbands and their wives (one dot for each husband-wife pair). What does the scatterplot reveal?

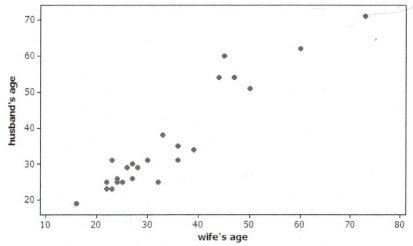

b) Explain why it is reasonable to say that the marriage age of husband and marriage age of wife should <u>not</u> be treated as *independent* observations.

Because the data are *paired*, instead of comparing ages of husbands and wives, we should instead look at the *difference* in age between husband and wife.

c) Define the appropriate parameter of interest, in the context of the study.
d) State the appropriate null and alternative hypotheses, in the context of the study.
e) Use the **Matched Pairs** applet to produce appropriate graphical displays (dotplot) and summary statistics (sample size, mean and SD) for the data. Comment on whether this descriptive analysis of the data suggests whether or not husbands tend to be older than their wives.
f) Use the **Matched Pairs** applet to find an approximate p-value.

g) Interpret the p-value in the context of the study. That is, what is the p-value a probability of?

h) Give a conclusion about the strength of evidence based on the p-value.

7.2.13 In the previous exercise the statistic used was the mean of the difference in ages (husband's age – wife's age). Explain how the following things would change (or not) if you used the mean of the (wife's ages – husband's ages) as your statistic when you conducted this test?

a) Observed statistic

b) Alternative hypothesis

c) p-value

d) Conclusion

Memorizing and listening to music*

7.2.14 Does listening to music effect how many words you can memorize? Student researchers tried to answer this question by having 20 subjects listen to music while trying to memorize words and also had the same 20 subjects try to memorize words when not listening to music. They randomly determine which condition was done first for each of their subjects. Here are their hypotheses:

Null: The average of the difference in number of words memorized (no music – with music) is 0. ($\mu_d = 0$)

Alternative: The average of the difference in number of words memorized (no music – with music) is greater than 0. ($\mu_d > 0$)

The students found the following results in terms of number of words memorized:

	No music	With music	Difference
Mean	12.8	10.5	2.3
Standard Deviation	3.38	3.15	3.76

a) What is the explanatory variable in this study? Is it categorical or quantitative?

b) What is the response variable in this study? Is it categorical or quantitative?

c) What is the observed statistic in this study?

d) Assuming a true null hypothesis, we simulated this study and found 1500 mean differences that are shown in the null distribution below. Based on where the observed statistic lies in this distribution, does the p-value appear to be fairly small?

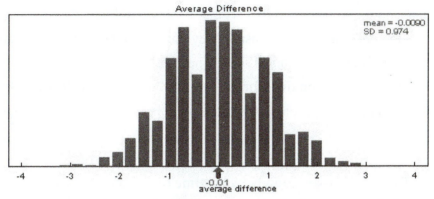

e) The standard deviation of the null distribution is 0.974. Use this to find the standardized statistic for this test.

f) Based on your standardized statistic from part (e), is there strong evidence that people can memorize words better when music is not playing compared to when it is? Explain.

7.2.15 Reconsider the previous exercise about memorization with and without music.

a) The standard deviation of the null distribution shown is 0.974. Use this to find an approximate 95% confidence interval for the mean difference in the number of words memorized (no music – music) for the population of interest.

b) Based on your confidence interval from part (a), is there strong evidence that people can memorize words better when music is not playing compared to when it is? Explain.

7.2.16 Reconsider the previous exercises about memorization with and without music. The statistic used was the mean of the difference in words memorized (no music – music). Explain how the following things would change (or not) if we used the mean of the (music – no music) as the statistic when the test was conducted?

a) Observed statistic

b) Alternative hypothesis

c) p-value

d) The confidence interval

Reaction time of day

7.2.17 Student researchers wanted to see whether a person's reaction time is affected by the time of day. They had a computer program that would display a red circle. When it changed to yellow, the subjects were to click the mouse as fast as they could and the computer would calculate the reaction time. The researchers had their subjects perform this test at 10:00 a.m. and again at 10:00 p.m., recording the reaction time for each student at each time. They randomly determined which time of day was done first for each of their subjects. Here are their hypotheses:

Null: There is no association between time of day and reaction times. ($\mu_d = 0$.)

Alternative: There is an association between time of day an reaction times. ($\mu_d \neq 0$.)

The students found the following results for reaction times in sections:

	Morning	Night	Difference
Mean	0.244	0.306	-0.062
Standard Deviation	0.052	0.082	0.084

a) What is the explanatory variable in this study? Is it categorical or quantitative?

b) What is the response variable in this study? Is it categorical or quantitative?

c) What is the observed statistic in this study?

d) Assuming a true null hypothesis, we simulated this study and found 1000 mean differences that are shown in the null distribution below. Based on where the observed statistic lies in this distribution, does the p-value appear to be fairly small?

e) What is the standardized statistic for this test?

f) Based on your standardized statistic from part (e), is there strong evidence that reaction times differ between morning and night? Explain.

7.2.18 Reconsider the previous exercises about reaction times and time of day.

a) Find an approximate 95% confidence interval for the mean difference in the reaction times (morning – night) for the population of interest.

b) Based on your confidence interval, from part (a), is there strong evidence that reaction times differ between morning and night? Explain.

7.2.19 Reconsider the previous exercises about reaction times and time of day. The statistic used was the mean of the difference in reaction times (morning – night). Explain how the following things would change (or not) if we used the mean of the (night – morning) as the statistic when the test was conducted?

a) Observed statistic

b) Alternative hypothesis

c) p-value

d) The confidence interval

Exercise and memory*

7.2.20 Students wanted to see whether exercising (in particular doing jumping jacks) would help or hinder people's ability to memorize a list of 10 words that were read to them. This would be compared to the same test of memorizing words while the subjects were sitting down. The number of words memorized for the 31 subjects can be found in the data file **ExerciseMemorize.txt**.

a) What is the explanatory and what is the response variable in this study?

b) Write down the hypotheses in either words or symbols.

c) Put the data into the **Matched Pairs** applet. What is the value of the statistic that will be used in the study and what does it represent.

d) Randomize at least 1000 times to develop a null distribution and determine a p-value.

e) Is there strong evidence that exercising while trying to memorize a list of words helps or
    hinders the process?

Memorization delay

7.2.21 Student researchers wanted to see whether a short delay between seeing a list of words
and when people were asked them to recall them would hinder memorization. The subjects were
shown a list of words to memorize for one minute and were then given one minute to recall as
many words as they could. Each subject did this one time with no delay between memorizing
and recall and another with a 30 second wait between memorizing and recall. They were
randomly assigned the order of the two conditions. The number of words memorized under each
condition can be found in the data file **MemorizationDelay**.

a) What is the explanatory and what is the response variable in this study?
b) Write down the hypotheses in either words or symbols.
c) Put the data into the **Matched Pairs** applet. Just by looking at the graphs in the applet,
    does it appear that the delay significantly hinders memorization? Explain.
d) What is the value of the statistic that will be used in the study and what does it represent.
e) Randomize at least 1000 times to develop a null distribution and determine a p-value.
f) Is there strong evidence that a short delay hinders the memorization process?

7.2.22 Reconsider the previous exercise on memorization with and without a delay. If you
haven't already done so, put the data file **MemorizationDelay** into the **Matched Pairs** applet
and do at least 1000 randomizations.

a) What is the standardized statistic in this study?
b) Based on the standardized statistic is there strong evidence that a short delay hinders the
    memorization process? Explain why or why not.

7.2.23 Reconsider the previous exercise on memorization with and without a delay. If you
haven't already done so, put the data file **MemorizationDelay** into the **Matched Pairs** applet
and do at least 1000 randomizations.

a) Determine an approximate 95% 2SD confidence interval for the mean difference in the
    number of words memorized (without delay – delay).

b) Explain what your confidence interval from part (a) means. In other words, we are 95% confident of what?

c) Based on your confidence interval is there strong evidence that a short delay hinders the memorization process? Explain why or why not.

Reaction times and handedness*

7.2.24 Student researchers wanted to see whether a person's reaction time is slower if they use their non-dominant hand to click a computer mouse compared to using their dominant hand. The researchers had a computer program that would display a red circle. When it changed to yellow, the subjects were to click the mouse as fast as they could and the computer would calculate the reaction time. Researchers had their subjects find their reaction time with their dominant hand (e.g. right hand for right-handers) and also with their non-dominant hand. They randomly determined which hand was tested first. The reaction times (in seconds) for the subjects for each condition can be found in the data file **ReactionTimes**.

a) What is the explanatory and what is the response variable in this study?

b) Write down the hypotheses in either words or symbols.

c) Put the data into the **Matched Pairs** applet. Just by looking at the graphs in the applet, does it appear that reaction time is significantly slower with the non-dominant hand? Explain.

d) What is the value of the statistic that will be used in the study and what does it represent.

e) Randomize at least 1000 times to develop a null distribution and determine a p-value.

f) Is there strong evidence that reaction times are slower when people use their non-dominant hand?

7.2.25 Reconsider the previous exercise on handiness and reaction time. If you haven't already done so, put the data file **ReactionTimes** into the **Matched Pairs** applet and do at least 1000 randomizations.

a) What is the standardized statistic in this study?

b) Based on the standardized statistic is there strong evidence that using the non-dominant hand results in slower average reaction times? Explain why or why not.

7.2.26 Reconsider the previous exercise on handiness and reaction time. If you haven't already done so, put the data file **ReactionTimes** into the **Matched Pairs** applet and do at least 1000 randomizations.

a) Determine an approximate 95% confidence interval using the 2SD rule for the mean difference in the reaction times (dominant – non-dominant).
b) Explain what your confidence interval from part (a) means. In other words, we are 95% confident of what?
c) Based on your confidence interval is there strong evidence that using the non-dominant hand results in slower average reaction times? Explain why or why not.

Section 7.3

7.3.1 Design your own experiment: Develop the plan for the investigation of a research question that would be best answered using paired data.

a) State the research question.
b) Explain how pairing will be used
c) Identify the response and explanatory variables.
d) Describe how randomization (random sampling, random assignment, or both) will be used.
e) State the appropriate null and alternative hypotheses, in words.
f) Describe when it will be appropriate to use a theory-based method to test the hypotheses stated above.

Marriage ages*

7.3.2 Refer to Exercise 7.2.12 where a student wanted to investigate whether or not people tend to marry spouses of similar ages, and so gathered age data from a sample of 24 couples, taken from marriage licenses filed in Cumberland County, PA, in June and July 1993.
a) Define, in words, the appropriate parameter(s) of interest in the context of the study. Also, state the appropriate symbol(s) to denote the parameter(s).

b) State, in words, the appropriate null and alternative hypotheses in the context of the study.

c) State, using the symbol(s) described in part a), the appropriate null and alternative hypotheses in the context of the study.

d) Check whether the validity conditions for the theory-based test are satisfied here.

e) Use appropriate technology (such as the Matched Pairs or Theory-Based Inference applet, or a statistical software package) to conduct the theory-based test of significance to find and report the test statistic and p-value.

f) Interpret the p-value from part e) in the context of the study, and then draw a conclusion about statistical significance.

g) Find and interpret the 95% confidence interval for the parameter(s) of interest, in the context of the study.

h) State your conclusion in the context of the study, being sure to comment on statistical significance, estimation, causation, and generalizability.

Perception of child's size

7.3.3 In a study of parents' perceptions of their children's size, researchers Kaufman et al. (*Current Biology,* 2013) asked parents to estimate their youngest child's height. The researchers hypothesized that parents tend to underestimate their youngest child's size because the youngest child is the baby of the family, and everybody else in the family appears bigger compared to the baby. The sample of 39 parents who were surveyed underestimated their youngest child's height by 7.5cm, on average; the standard deviation for the difference in actual heights and estimated heights was 7.2cm and the data is not strongly skewed.

a) Which of the following is an appropriate null hypothesis for this study?
   A. Parents' estimate of their youngest child's height is accurate, on average.
   B. Parents' tend to overestimate their youngest child's height, on average.
   C. Parents' tend to underestimate their youngest child's height, on average.

b) Which of the following is an appropriate alternative hypothesis for this study?
   A. Parents' estimate of their youngest child's height is accurate, on average.
   B. Parents' tend to overestimate their youngest child's height, on average.
   C. Parents' tend to underestimate their youngest child's height, on average.

c) Identify the observational units in this study.
   A. Children

      B. Parents

      C. Actual heights

      D. Estimated heights

  d) Identify the parameter in this study.

      A. The probability that parents will overestimate their youngest child's height, π

      B. The probability that parents will underestimate their youngest child's height, π

      C. The probability that parents will correctly estimate their youngest child's height, π

      (D.) The average amount by which parents' guess of their youngest child's height will differ from the actual height, μ

  e) Explain why the simulation-based method cannot be used to analyze the available information, to investigate whether parents tend to overestimate their youngest child's height.

  f) Explain why a theory-based approach is valid.

7.3.4 Refer to the previous exercise about the study of parents' perceptions of their children's size by researchers Kaufman et al. (*Current Biology,* 2013). Recall that the sample of 39 parents underestimated their youngest child's height by 7.5cm, on average; the standard deviation for the difference in actual heights and estimated heights was 7.2cm.

  a) Is there evidence that youngest children's heights tend to be underestimated by their parents? Carry out a theory-based test using an appropriate applet or statistical software. Find and report a p-value as well as a standardized statistic.

  b) Interpret the p-value in the context of the study.

  c) Using an appropriate applet or statistical software, find a 95% confidence interval for the difference. Interpret the confidence interval in the context of the study.

  d) Summarize your findings, being sure to include a discussion of statistical significance, estimation (confidence), causation, and generalizability in your answer.

  e) What assumption do you have to make about the data in order for the validity conditions of the appropriate theory-based test to be satisfied?

7.3.5 Refer to the previous exercises about the study of parents' perceptions of their children's size. The researchers Kaufman et al. (*Current Biology,* 2013) also surveyed a sample of 38 parents about their eldest-child's height. The parents overestimated their eldest child's height by 0.4cm, on average; the standard deviation for the difference in actual heights and estimated heights was 5.6cm without strong skewness in the data.

a) Is there evidence that parents tend to either over or underestimate eldest children's heights? Carry out a theory-based test using an appropriate applet or statistical software. Find and report a p-value as well as a standardized statistic.

b) Using an appropriate applet or statistical software, find a 95% confidence interval for the difference. Interpret the confidence in the context of the study.

c) Summarize your findings, being sure to include a discussion of statistical significance, causation, and generalizability in your answer.

d) What assumption do you have to make about the data in order for the validity conditions of the appropriate theory-based test to be satisfied?

Infant behavior*

7.3.6 Researchers Hamlin et al. (*Nature, 2007*) conducted a study involving 16 infants, that involved an assessment of infants' expectations of a climber toy's attitude towards a "helper" – a toy figure that helped the climber climb a hill, and a "hinderer" toy – a figure that had hindered the climber's progress up a hill, and whether infants tend to watch the climber's confrontation with the hinderer for a longer time than that with the helper. Each of the sixteen 10-month olds was shown a new display where the climber would alternately approach the helper and the hinderer, of which the latter was considered to be a surprising event, because such an action was perceived as "violation of the expectation paradigm." The lengths of time that each infant spent watching the climber approach the helper and the hinderer was recorded, and it was noted that in the study the infants looked longer at the approach of the climber towards the hinderer than the helper. (Wonder if the infants were expecting the climber to punch the hinderer, and that held their interest longer!) Given below are some summary statistics for the difference (hinderer - helper) in looking times (seconds). Note that the distribution of differences is fairly symmetric.

	n	Mean	SD
Difference = Time$_{Hinderer}$ − Time$_{Helper}$	16	1.140 seconds	1.752 seconds

Do infants tend to look at surprising events or behavior for longer times, on average, compared to uninteresting ones?

a) What are the pairs?

b) What are the two conditions?

c) What is the response variable?

d) State, in words, the appropriate null and alternative hypotheses in the context of the study.

e) Explain why, with the available information, one cannot use a simulation-based method to test the hypotheses.

7.3.7 Refer to the previous exercise about investigating whether infants tend to look at surprising events or behavior for longer times, on average, compared to uninteresting ones?

a) Define, in words, the appropriate parameter(s) of interest in the context of the study. Also, state the appropriate symbol(s) to denote the parameter(s).

b) State, using the symbol(s) described in part a), the appropriate null and alternative hypotheses in the context of the study.

c) State the validity conditions that have to be met to be able to perform a theory-based test to test the hypotheses stated in part b). Are these conditions met? How are you deciding?

d) Regardless of your answer to (c), perform a theory-based test to test the hypotheses stated in part b). Be sure to report a test statistic value and a p-value in your answer.

e) Interpret your p-value in the context of the study.

f) Find and interpret a 95% confidence interval for the parameter of interest, in the context of the study.

g) Give a full conclusion with regards to statistical significance, estimation (confidence), causation, and generalization.

7.3.8 Refer to the previous exercises about investigating whether infants tend to look at surprising events or behavior for longer times, on average, compared to uninteresting ones? Given below are the summary statistics for the difference (hinderer - helper) in looking times (seconds).

	n	Mean	SD
Difference = $\text{Time}_{\text{Hinderer}} - \text{Time}_{\text{Helper}}$	16	1.140 seconds	1.752 seconds

Suppose the summary statistics had been:

	n	Mean	SD
Difference = $\text{Time}_{\text{Hinderer}} - \text{Time}_{\text{Helper}}$	32	1.140 seconds	1.752 seconds

How, if at all, would the values of the following (using the theory-based method) be different?

a) p-value: SAME LARGER

SMALLER

b) Standardized statistic: SAME LARGER

SMALLER

c) Midpoint of the 95% confidence interval: SAME LARGER

SMALLER

d) Width of the 95% confidence interval: SAME LARGER

SMALLER

7.3.9 Refer to the previous exercises about investigating whether infants tend to look at surprising events or behavior for longer times, on average, compared to uninteresting ones? Given below are the summary statistics for the difference (hinderer - helper) in looking times (seconds).

	n	Mean	SD
Difference = Time$_{Hinderer}$ - Time$_{Helper}$	16	1.140 seconds	1.752 seconds

Suppose the summary statistics had been:

	n	Mean	SD
Difference = Time$_{Hinderer}$ - Time$_{Helper}$	16	2.280 seconds	1.752 seconds

How, if at all, would the values of the following (using the theory-based method) be different?

a) p-value: SAME LARGER

SMALLER

b) Standardized statistic: SAME LARGER

SMALLER

c) Midpoint of the 95% confidence interval: SAME LARGER

SMALLER

d) Width of the 95% confidence interval: SAME LARGER

SMALLER

7.3.10 Refer to the previous exercises about investigating whether infants tend to look at surprising events or behavior for longer times, on average, compared to uninteresting ones? Given below are the summary statistics for the difference (hinderer - helper) in looking times (seconds).

	n	Mean	SD
Difference = Time$_{Hinderer}$ - Time$_{Helper}$	16	1.140seconds	1.752seconds

Suppose the summary statistics had been:

	n	Mean	SD
Difference = Time$_{Hinderer}$ - Time$_{Helper}$	16	1.140seconds	0.752seconds

How, if at all, would the values of the following (using the theory-based method) be different?

a) p-value: SAME   LARGER
   SMALLER

b) Standardized statistic: SAME   LARGER
   SMALLER

c) Midpoint of the 95% confidence interval: SAME   LARGER
   SMALLER

d) Width of the 95% confidence interval: SAME   LARGER
   SMALLER

Cholesterol levels

7.3.11 The **Cholesterol Level** data sets give cholesterol levels of heart attack patients. Cholesterol measures are taken 2, 4, and 14 days after a patient has suffered a heart attack.

Is there a significant difference between 2 day cholesterol measures and 4 day cholesterol measures? Carry out a theory-based test and find a 95% confidence interval for the difference. Use the data set **CholesterolLevel_day2-4**.

Summarize your findings, being sure to include a discussion of statistical significance, estimation (confidence interval), causation, and generalization in your answer. Make sure to confirm that validity conditions are met.

7.3.12 Is there a significant difference between 2 day cholesterol measures and 14 day cholesterol measures? Carry out a theory-based test and find a 95% confidence interval for the difference. Use the data set **CholesterolLevel_day2-14**.

Summarize your findings, being sure to include a discussion of statistical significance, estimation (confidence interval), causation, and generalization in your answer. Make sure to confirm that validity conditions are met.

7.3.13 Is there a significant difference between 4 day cholesterol measures and 14 day cholesterol measures? Carry out a theory-based test and find a 95% confidence interval for the difference. Use the data set **CholesterolLevel_day4-14**.

Summarize your findings, being sure to include a discussion of statistical significance, estimation (confidence interval), causation, and generalization in your answer. Make sure to confirm that validity conditions are met.

Estimating energy expenditure*

7.3.14 How many calories do you think you burn or expend with 60 minutes of walking on the treadmill? An article that appeared in the *American Journal of Men's Health* (2010) presented the results of a study conducted by researchers Harris and George, whose objective was to evaluate how accurately men can estimate their energy expenditure (EE) as measured in kilocalories (kcal).

Eighty normal-weight and overweight male participants, ages 21 to 45 years, were recruited from a large, urban university in South Florida. During one session, each participant was made to walk for 60 minutes on a treadmill, and their actual EE was calculated. After they had completed the task, each participant was asked to estimate the number of calories they burned during the task. A summary of the results are shown in the following table.

	Sample size	Sample average (kcal)	Sample SD (kcal)

| Difference = Estimated – Actual | 80 | $\bar{x}_d = 129$ | $s_d = 393.5$ |

The researchers wanted to see whether men tended to overestimate or underestimate their EE.

a) On average, did the men in the study significantly overestimate or underestimate their EE? How do you know?
b) Explain why a matched pairs test would be appropriate to analyze these data.
c) Give the null and alternative hypotheses.
d) What assumption do you need to make about the data in order to use the Theory-based approach to analyze this data?
e) Use the **Theory-Based Inference** Applet to find the standardized statistic and p-value for this test.
f) Use the **Theory-Based Inference** Applet to find a 95% confidence interval. Interpret the 95% confidence interval in the context of the study.
g) Explain how the confidence interval corresponds to the conclusion reported in part d).
h) Provide a full (4-part) conclusion including comments on significance, estimation, causation, and generalization.

7.3.15 Recall the previous exercise about estimating energy expenditure (EE). The previous exercise also reports the summary statistics for the difference in EE (estimate – actual).

$$t = \frac{\bar{x}_d - 0}{\frac{s_d}{\sqrt{n}}}$$

a) Use the formula of the standardized statistic ($\frac{\bar{x}_d - 0}{\frac{s_d}{\sqrt{n}}}$) to calculate its value. Interpret the value of the standardized statistic in the context of the study.

b) Use the formula for the confidence interval ($\bar{x}_d \pm multiplier \times \frac{s_d}{\sqrt{n}}$) to calculate the 95% confidence interval.

7.3.16 Recall the previous exercises about estimating energy expenditure. Previously you carried out a theory-based matched pairs test to investigate whether, on average, men in the study significantly overestimated or underestimated their EE. Here are a few more summary statistics.

	Sample size	Sample average (kcal)	Sample SD (kcal)
Estimated	80	$\bar{x}_e = 411$	$s_e = 505.7$

Actual	80	$\bar{x}_a = 282$	$s_a = 44.7$
Difference = Estimated – Actual	80	$\bar{x}_d = 129$	$s_d = 393.5$

a) Use the **Theory-Based Inference** Applet to carry out a test for two independent means. Report the standardized statistic and p-value for this test.

b) Use the **Theory-Based Inference** Applet to find a 95% confidence interval for the difference between two independent means

c) Explain how the p-value and 95% confidence interval reported in parts a) and b) correspond to those reported in the previous exercise.

d) Although the analysis you conducted in a) and b) is the 'wrong' analysis, explain what it suggests about the effectiveness of pairing in this study.

7.3.17 Recall the previous exercises about estimating energy expenditure. The summary statistics on the difference between estimated EE and actual EE were as follows:

	Sample size	Sample average (kcal)	Sample SD (kcal)
Difference = Estimated – Actual	80	$\bar{x}_d = 129$	$s_d = 393.5$

Suppose the summary statistics had been:

	Sample size	Sample average (kcal)	Sample SD (kcal)
Difference = Estimated – Actual	40	$\bar{x}_d = 129$	$s_d = 393.5$

How, if at all, would the values of the following (using the theory-based method) be different?

a) p-value: SAME   LARGER   SMALLER
b) Standardized statistic: SAME   LARGER   SMALLER
c) Midpoint of the 95% confidence interval: SAME   LARGER   SMALLER
d) Width of the 95% confidence interval: SAME   LARGER   SMALLER

7.3.18 Recall the previous exercises about estimating energy expenditure. The summary statistics on the difference between estimated EE and actual EE were as follows:

	Sample size	Sample average (kcal)	Sample SD (kcal)
Difference = Estimated – Actual	80	$\bar{x}_d = 129$	$s_d = 393.5$

Suppose the summary statistics had been:

	Sample size	Sample average (kcal)	Sample SD (kcal)
Difference = Estimated – Actual	80	$\bar{x}_d = 64.5$	$s_d = 393.5$

How, if at all, would the values of the following (using the theory-based method) be different?

a) p-value: SAME   ⟨LARGER⟩   SMALLER

b) Standardized statistic: SAME   LARGER   ⟨SMALLER⟩

c) Midpoint of the 95% confidence interval: SAME   LARGER   ⟨SMALLER⟩

d) Width of the 95% confidence interval: ⟨SAME⟩   LARGER   SMALLER

7.3.19 Recall the previous exercises about estimating energy expenditure. The summary statistics on the difference between estimated EE and actual EE were as follows:

	Sample size	Sample average (kcal)	Sample SD (kcal)
Difference = Estimated – Actual	80	$\bar{x}_d = 129$	$s_d = 393.5$

Suppose the summary statistics had been:

	Sample size	Sample average (kcal)	Sample SD (kcal)
Difference = Estimated – Actual	80	$\bar{x}_d = 129$	$s_d = 93.5$

How, if at all, would the values of the following (using the theory-based method) be different?

a) p-value: SAME   ⟨LARGER⟩   SMALLER

b) Standardized statistic: SAME   LARGER   ⟨SMALLER⟩

c) Midpoint of the 95% confidence interval: ⟨SAME⟩   LARGER   SMALLER

d) Width of the 95% confidence interval: SAME   ⟨LARGER⟩   SMALLER

Rounding first base

7.3.20 Recall Exploration 7.1 and Example 7.2 about the time needed to round first base, using the narrow angle versus the wide angle. Here are the summary statistics on the difference in times (narrow – wide) for the 22 players:

	Sample size	Sample average (sec)	Sample SD (sec)
Difference = Narrow – Wide	22	0.075	0.088

a) Use an appropriate theory-based method to investigate whether the choice of angle affects the running time. Be sure to find and report a standardized statistic and p-value.
b) Use an appropriate theory-based method to find and report a 95% confidence interval for the parameter of interest.
c) Provide a full (4-part) conclusion including comments on significance, estimation, causation, and generalization.

7.3.21 Recall Exploration 7.1 and Example 7.2 about the time needed to round first base, using the narrow angle versus the wide angle. The previous exercise reports the summary statistics for the difference in times (narrow – wide).

$$t = \frac{\bar{x}_d - 0}{\frac{s_d}{\sqrt{n}}}$$

a) Use the formula of the standardized statistic ($t = \frac{\bar{x}_d - 0}{\frac{s_d}{\sqrt{n}}}$) to calculate its value. Interpret the value of the standardized statistic in the context of the study.
b) Use the formula for the confidence interval ($\bar{x}_d \pm multiplier \times \frac{s_d}{\sqrt{n}}$) to calculate the 95% confidence interval.

M&Ms*

7.3.22 Recall Example 7.3 about the number of M&Ms taken with small versus large bowls. The example uses the following parameter of interest:

μ_d = Long-run mean difference for the number of M&Ms that will be taken when using a small bowl minus a large bowl.

The hypotheses are: H_0: $\mu_d = 0$ versus H_a: $\mu_d < 0$.
The example reports the theory-based p-value to be 0.1172, and the 95% confidence for μ_d to be (-29.54, 7.78).

Suppose we defined the parameter of interest to be
μ_{new} = Long-run mean difference for the number of M&Ms that will be taken when using a *large bowl minus a small bowl*.

a) State the hypotheses in terms of μ_{new}.
b) Without doing any calculations, report what the p-value would be for the hypotheses stated in part a). Explain your reasoning.
c) Without doing any calculations, report what the 95% confidence interval μ_{new} would be. Explain your reasoning.

Tomoxetine

7.3.23 In a study of the effect of tomoxetine as an alternative treatment for attention deficit hyperactivity disorder (ADHD), researchers Spencer et al. (*The American Journal of Psychiatry*, 1998) recruited men and women diagnosed with ADHD. All participants were administered tomoxetine and a placebo – each for three weeks, in a random order and with a week of washout period between the two treatments. The researchers noted that the average increase in standing heart rate when on tomoxetine compared to when on the placebo was 11 beats per minute, with a standard deviation of 15.05 beats per minute, for a sample of 18 individuals. The distribution of differences in standing heart rate is approximately symmetric.

a) Define, in words, the appropriate parameter(s) of interest in the context of the study. Also, state the appropriate symbol(s) to denote the parameter(s).
b) State, using the symbol(s) described in part a), the appropriate null and alternative hypotheses to test whether the use of tomoxetine causes an increase in standing heart rate.
c) State the validity conditions that have to be met to be able to perform a theory-based test to test the hypotheses stated in part b). Are these conditions met? How are you deciding?
d) Regardless of your answer to part c), perform a theory-based test to test the hypotheses

stated in part b). Be sure to report a test statistic value and a p-value in your answer.

e) Interpret your p-value in the context of the study.

f) Find a 95% confidence interval for the parameter of interest.

g) Provide a full (4-part) conclusion including comments on significance, estimation, causation, and generalization

Effect of oxytocin*

7.3.24 The "Reading the Mind in the Eyes Test" (RMET) measures the ability of a person to infer the mental state of another from looking at facial expressions. Researchers Domes et al. (*Biological Psychiatry,* 2007) conducted a study on the effect of oxytocin (a hormone often linked with bonding, trust, and empathy) on people's performance on the RMET. Thirty healthy male volunteers (ages 21 to 30 years) were administered oxytocin and a placebo – each as a nasal spray, in a random order and with a week of washout period between the two treatments. Forty-five minutes after the administering of each treatment, participants were given the RMET. The researchers noted that the average increase in RMET score when on oxytocin compared to when on the placebo was 3, with a standard deviation of 7.54 points. The differences in RMET scores were not strongly skewed.

a) Define, in words, the appropriate parameter(s) of interest in the context of the study. Also, state the appropriate symbol(s) to denote the parameter(s).

b) State, using the symbol(s) described in part a), the appropriate null and alternative hypotheses to test whether the use of oxytocin causes an increase in ability to infer the mental state of others, as measured by the RMET score.

c) State the validity conditions that have to be met to be able to perform a theory-based test to test the hypotheses stated in part b). Are these conditions met? How are you deciding?

d) Regardless of your answer to part c), perform a theory-based test to test the hypotheses stated in part b). Be sure to report a test statistic value and a p-value in your answer.

e) Interpret your p-value in the context of the study.

f) Find a 95% confidence interval for the parameter of interest.

g) Provide a full (4-part) conclusion including comments on significance, estimation, causation, and generalizability.

7.3.25 Refer to the previous exercise about the effect of oxytocin on the ability to infer the mental state of others, as measured by the RMET score. Recall that for the sample of 30 participants, the researchers noted that the average increase in RMET score when on oxytocin compared to when on the placebo was 3, with a standard deviation of 7.54 points.

$$t = \frac{\bar{x}_d - 0}{\frac{s_d}{\sqrt{n}}}$$

a) Use the formula of the standardized statistic ( ) to calculate its value. Interpret the value of the standardized statistic in the context of the study.

b) Use the formula for the confidence interval ($\bar{x}_d \pm multiplier \times \frac{s_d}{\sqrt{n}}$) to calculate the 95% confidence interval.

End of Chapter exercises

7.CE.1* Match each question with the proper significance test (from (A) – (D)) needed to answer the question.

a) _____ Do a majority of students at your college prefer chocolate soft serve ice cream to vanilla?

b) _____ Is the proportion of females at your college who participate in varsity athletics different from the proportion of males who participate in varsity athletics?

c) _____ Is the average study time per week for seniors at your college different from the average study time per week for freshmen?

d) _____ On average, is there a significant difference in one's heart rate before using chewing tobacco and after using chewing tobacco?

A. single proportion test of significance
B. paired samples test of significance
C. two-sample mean test of significance
D. two proportion test of significance

7.CE.2 Does your bowl size influence how much you eat? To investigate, researchers recruited undergraduates from a large university to participate in a study that was conducted over several days. The participants were not told about the goal of the study until the very end of the data collection process.

At the first session, the researchers assigned some participants to receive either a small bowl or a large bowl. After that, each participant was allowed to scoop as many M&Ms into their bowl as they planned on eating that day, and then during the session they counted how many M&Ms their bowl contained. When the data from the first session were analyzed by comparing the average M&Ms count for the "small bowl" group to the average M&Ms count for the "large bowl" group, the p-value was found to be 0.40.

a) Based on the results of the first session, is it okay to say that bowl size influences how much you eat? Why or why not?

b) At the next session, the size of the bowls was switched: those who received small bowls at the first session now received large bowls, and vice versa. So, at the end of the second session, each participant had an M&Ms count for small bowl and an M&Ms count for large bowl. How should the researchers analyze the updated data? Briefly explain.

7.CE.3* Clearly describe what, if anything, is incorrect about each of the following hypotheses.
   a) $H_0: \hat{p}_1 - \hat{p}_2 = 0$ versus $H_A: \hat{p}_1 - \hat{p}_2 \neq 0$
   b) $H_0: \mu = 40$ versus $H_A: \mu < 30$
   c) $H_0: \pi_1 - \pi_2 = 0$ versus $H_A: \pi_1 - \pi_2 \leq 0$

7.CE.4: For each of the following studies, indicate whether the data were collected in a paired manner and therefore require a paired analysis.

a) A farmer wants to investigate whether speaking to cows by name causes them to produce more milk. He randomly selects half of his cows and speaks to them by name while milking them. For the other half of the cows, he does not speak to them by name while milking them.

b) A reporter for a local newspaper wants to compare prices of grocery items at two stores. She selects 30 grocery items that are sold at both stores and records the price of each item at each store.

c) Are some diet programs easier to stick to than other? Researchers randomly assigned overweight people to one of four popular diets, and they recorded how many in each diet successfully completed the program for one year.

d) Does catnip have a negative effect on cats' social interactions with other cats? A student went to the local animal shelter and observed a dozen cats. She counted how many negative interactions each cat had in a fifteen-minute period. Then she exposed each cat to catnip and counted how many negative interactions the cat had in the following fifteen-minute period.

Smoke alarms*

7.CE.5: Some smoke alarms for homes with children now use a recording of the mother's voice calling the child by name, as opposed to the beeping sound of conventional smoke alarms.

a) Describe how you could design a matched-pairs study to investigate whether the new smoke alarms reduce the time needed for a child to wake up and leave the house.

b) Explain why a matched-pairs design is likely to be helpful, as compared to a completely randomized design, in this study.

c) Explain why it's important to randomize the order in which the children receive the treatments.

Music at the restaurant

7.CE.6: Suppose that a restaurant owner want to investigate whether the type of music played in the background affects how much customers spend on their meals. The owner plans to play classical music on some evenings and contemporary music on other evenings for a total of two weeks, comparing the amounts spent by customers on those evenings. The owner is considering three ways to design the study:

- Plan 1: Flip a coin for each evening. If the coin lands heads, play classical music. If the coin lands tails, play contemporary music.
- Plan 2: Write "classical" on 7 index cards and "contemporary" on 7 index cards. Shuffle the cards well. Record the order of "classical" and "contemporary" on the cards, and play the music in that order over the 14 days.
- Plan 3: Flip a coin for each evening of the first week. If the coin lands heads, play classical music. If the coin lands tails, play contemporary music. Then for each day in the second week, play the other kind of music from what was played on that day in the first week.

a) Identify the explanatory and response variables in this study. Also classify the variables as categorical or quantitative.

b) Is this an observational study or an experiment? Explain how you know.

c) Which of these plans is a matched-pairs design?

d) Why might the matched-pairs design be better than a completely randomized design in this context?

Marriage ages*

7.CE.7: Suppose that a student wants to gather data to investigate whether husbands tend to be older, on average, than their wives. The student obtains a sample of 200 marriage licenses that report the ages of both the husband and wife. The student randomly selected 100 of the marriage licenses and recorded the husbands' age from the license, and then the student recorded the wife's age from the other 100 licenses.

a) Explain why a paired analysis would not be appropriate for the student's data.

b) How could the student's data collection method from the 200 marriage licenses be improved so the data would be paired?

c) Why is it reasonable to believe that pairing would be helpful in this situation?

Waterproofing boots

7.CE.8: Suppose that a company has developed a new method of waterproofing boots that it wants to compare in an experiment against a standard method.

a) Describe how to conduct a randomized experiment without pairing to compare the two methods of waterproofing boots.

b) Describe how to conduct a matched-pairs experiment to compare the two methods of waterproofing boots. (*Hint*: Try to think of a very natural way to do the pairing.)

Flexibility*

7.CE.9 During one semester of a healthy living class taught at Hope College, students measured how flexible they were in terms of inches reached while sitting. Flexibility was measured at the beginning of the semester and again at the end of the semester. We will consider this a representative sample of all students taking this class around this time.

a) In chapter 6 we carried out a test to see whether the posttest flexibility scores were different for males and females. We found that they were. Carry out a test to confirm that pretest flexibility scores are significantly different for males and females as well. (Use **FlexibilityPre**.)

b) How would you test to see whether there was a significant gain in flexibility for all students who might take the class? Describe how this test is different than the tests to compare the pretest flexibility and posttest flexibility between males and females.

c) Now, carry out the six step method to test if there is an average gain in flexibility for all students taking the healthy living class. (Use **Flexibility**.)

d) What if you wanted to see whether or not there was a significant difference between males and females as far as their average flexibility gain from this class. Describe what the two variables are that you would use in your analysis and what kind of a test you would use.

e) Carry out the six step method to test whether or not there is a significant difference in the average flexibility gain between males and females. (Use **Flexibility Pre and Post**; Note: You will need to create a new variable in order to do this analysis)

Body fat percentage

7.CE.10 During one semester of a healthy living class taught at Hope College, students were weighed (in kilograms), weighed under water (in kilograms), and had their body fat percentage calculated. These were all done at the beginning of the semester and again at the end of the semester. We will consider this a representative sample of all students taking this class around this time. The data are available in **Weight**, **WaterWeight**, and **BodyFat**.

a) Complete **three** theory-based tests to determine whether or not Hope students weight, underwater weight, and body fat percentage is different at the end of the semester when compared to the beginning. Make sure validity conditions hold before conducting the test. When running the tests, find the mean difference, produce a dotplot of the differences, write out the hypotheses, determine the p-value, and write a conclusion.

b) If the result is significant, determine the direction of the change. Can you explain the differences in these directions?

Chimpanzee collaboration*

7.CE.11 Humans are able to collaborate on projects to accomplish their goals. What about chimpanzees? Do chimpanzees know when they need to recruit help in order to solve a problem and when they are able to solve it on their own? Researchers placed a food platform outside of the chimpanzee's cage. They threaded a rope through metal loops at either end of the platform with the rope ends extending into the cage. If the chimp pulled on only one rope, it would unthread through the platform and the chimp would not be able to get the food. If both ends of the rope were pulled simultaneously, the food platform would move toward the cage and eventually the chimp could reach it. Two different conditions were presented to 8 chimpanzees. In the collaboration condition, the food platform was wide enough that one chimp couldn't reach both ends of the rope at the same time to successfully pull the food platform toward them. They needed to recruit another chimp to pull one end while they pulled the other. In the solo condition the food platform width was such that a single chimp could reach both ends of the rope and unassisted successfully pull the food platform to the cage. For both of these conditions there was an adjacent cage containing another chimp available to help solve the problem. The test chimp had the ability to unlock a sliding door to let the other chimp into their cage to assist them in the retrieval of the food platform. Each of the eight chimps performed several trials of each condition. Data was gathered as the percentage of trials the test chimp opened the door to the adjacent cage for each of the two conditions. Are chimps more likely to recruit a collaborator in the collaboration condition than in the solo condition? (Use data **Chimps**.)

a) State the null and alternative hypotheses.
b) If you compared the average percent of trials the cage was opened in the collaborative condition to the average percent of trials the cage was opened in the solo condition, what piece of information are you ignoring? How would this affect your analysis?
c) Carry out a randomization analysis of these data to see whether chimps are more likely to collaborate in the collaboration condition than in the solo condition. Remember, the same chimp is performing in each of these different conditions.

Running the bases, revisited

7.CE.12: Reconsider the baseball data from Example 7.2. Suppose for now that the data had been from 44 different players rather than paired data on the same 22 players. Test whether one of the methods is significantly different than the other.

a) Perform a simulation of a randomization test to compare the two groups (assuming that the data had been from 44 different players). Report the p-value. What conclusion would you reach?
b) How do the p-value and conclusion change from this analysis, as compared to the paired analysis conducted in Example 7.2?

c) Explain how this exercise demonstrates the value of having conducted a matched-pairs experiment to investigate the question of whether of whether one method of running the bases is better than the other method.

7.CE.13: Reconsider again the baseball data from Example 7.2.

a) Perform the appropriate (theory-based) matched-pairs *t*-test. Report the hypotheses, test statistic, and p-value. Summarize your conclusion.

b) Now suppose that the data had been from 44 different players rather than the same 22 players. Perform a two-sample *t*-test. Report the hypotheses, test statistic, and p-value. Summarize your conclusion.

c) Comment on how different your answers to a) and b) are. What does this indicates about the usefulness of using a matched-pairs design for this study?

7.CE.14: Reconsider again the baseball data from Example 7.2.

a) Repeat the previous exercise, this time producing a 95% confidence interval first based on the actual matched-pairs data, and then based on assuming that the data were from 44 different players rather than the same 22 players.

b) Comment on how the midpoints and widths of the confidence intervals in a) compare. Explain why your answers make sense.

Catnip*

7.CE.15: A student who did volunteer work at an animal shelter wanted to see whether cats really respond aggressively to catnip (Jovan, *Stats Magazine*, 1999). Using a sample of 15 cats, she recorded how many "negative interactions" (such as hissing or clawing) each cat engaged in during a fifteen-minute period prior to being given a teaspoon of catnip and also in the following fifteen-minute period. Her conjecture was that the number of negative interactions would increase after the introduction of the catnip. Her data follow:

Cat name	Negative interactions before catnip	Negative interactions after catnip
Amelia	0	0
Bathsheba	3	6
Boris	3	4
Frank	0	1
Jupiter	0	0
Lupine	4	5
Madonna	1	3
Michelangelo	2	1
Oregano	3	5
Phantom	5	7
Posh	1	0
Sawyer	0	1
Scary	3	5
Slater	0	2
Tucker	2	2

a) Explain how you know that these data came from a matched-pairs design.
b) State the relevant null and alternative hypotheses for testing the student's conjecture.
c) Conduct a simulation analysis to produce a p-value. Report the p-value, and interpret what it means (probability of what, assuming what?).
d) Summarize your conclusion about strength of evidence against the null hypothesis.

7.CE.16: Reconsider the previous exercise.

a) Comment on whether the conditions for applying a theory-based paired-t test are satisfied.
b) Regardless of your answer to a), conduct a theory-based paired-t test. Report the test statistic and p-value.
c) Summarize your conclusion about strength of evidence against the null hypothesis.

7.CE.17: Reconsider the previous two exercises.

a) Use the simulation analysis to produce a 95% confidence interval for the relevant parameter. Interpret what the interval reveals.

b) Determine a 95% theory-based paired *t*-test interval for the relevant parameter. Interpret what the interval reveals.

7.CE.18: Reconsider the previous three exercises.

a) Even though it's not an appropriate analysis, perform a two-sample *t*-test (rather than the appropriate paired *t*-test) on the data. Report the hypotheses, test statistic, and p-value. Summarize the conclusion that you would draw if this were the correct analysis.

b) How has the p-value changed from the paired test?

c) Does this analysis indicate that the pairing was useful in this study? Explain.

Investigation: Filtering Water in Cameroon

Students and professors in the Nursing and Engineering Departments at Hope College went to the Central African country of Cameroon to help improve drinking water quality and community health in rural communities by installing water filters in or near homes in one village. The families living in this village had no electricity, no water distribution system, and got their water from streams. The filters they installed contained a diffuser plate, fine sand, coarse sand, and gravel. Using the new filters, family members were expected to gather their water and filter it before drinking, instead of drinking directly from the stream as was common practice. Students working on this project examined the quality of the filters by looking at many different variables including general observations, filter observations, microbiology observations, household practice observations, user perceptions, and water source observations. It should be noted when making inferences, the water filters in this data set should be treated as a sample of all filters that could be constructed if this pilot project were expanded to other villages. Thus, inference is to an, as yet unbuilt, larger population of filters.

PART 1

Step 1: Ask a research question

There are several research questions we will ask in this investigation. The first one is: On average, is there a significant difference in the E. coli counts between the water that has just been filtered and water that is sitting in the bottom of a filter after it was filtered the previous day?

The data set we will use contains results from 14 water filters each giving E. coli counts (per 100 ml) on the first day and the second day after the water was filtered.

Step 2: Design a study and collect data

1. What are the observational units?

2. What variables are measured/recorded on each observational unit?

3. Is there an explanatory/response relationship for these variables? Classify the variables in this study as categorical or quantitative.

4. Are the samples of the first *e-coli* count independent or dependent of the samples of the second *e-coli* count? Explain. Based on your answer, which test of significance should you perform?

5. Is the sample size of 14 large enough to give a valid p-value from a theory-based test of significance?

6. State the null and alternative hypotheses to be investigated with this study. (You may use the symbol μ to represent these parameters.)

See data file **Ecoli-time**

Step 3: Explore the data

7. What are the average E. coli counts for each day? Did the E. coli in the sample increase or decrease on average from day 1 to day 2? Explain. Also give the standard deviations for the E. coli counts for each day.

8. What is the average of the difference (day 1 – day 2) in E. coli counts? Does the sign of this average correspond to your answer of increasing or decreasing between day 1 and day 2 in the previous question? Explain. Also give the standard deviation for the difference in E. coli counts.

Step 4: Draw inferences

9. Carry out an appropriate test of significance to see whether on average there is a difference between the first *E. coli* count and the second *E. coli* count. Report your p-value and a 95% confidence interval.

Step 5: Formulate conclusions

10. Based on your p-value and confidence interval, what conclusions can you draw from this test?

11. Can generalizations be made to a larger population? Is there a cause-effect relationship between the variables?

Step 6: Look back and ahead

12. Discuss the study design. Why does pairing make sense here? What would you do differently to improve upon this study? What further research would you propose based on your findings from this study?

PART 2

As mentioned earlier, sand is used in the filter. Different filters had different amounts of sand. The students split the filters into two groups; those that had more than two inches of sand and those that had less. We will explore if this difference results in different *E. coli* counts when filtering the water. In this sample, we have results from 20 filters, 14 with more than two inches of sand and 6 with less than two inches of sand. (See data file **Ecoli-sand**.)

13. Is there an explanatory/response relationship for these variables? Classify the variable in this study as categorical or quantitative.

14. What are the average *E. coli* counts for each sand level? Are the counts higher or lower with the higher level of sand in the filter? Also give the standard deviations for the *E. coli* counts for each sand level.

15. Are the samples of the first e-*coli* count independent or dependent of the samples of the second *e-coli* count? Explain. Is the sample large enough to use a theory-based test? Based on your answers, which test of significance should you perform?

16. Carry out an appropriate test of significance to see whether on average there is a difference in the *E. coli* counts between filters with more than two inches of sand and those with less. Report your p-value, a 95% confidence interval, and your conclusion.

PART 3

Let's look at one last research question. As a general rule, the flow rate of a water filter in good working condition should be around 1000 mL/min. Let's test to see whether the water filters have an average flow rate that is significantly different than 1000mL/min. (See data set **Ecoli-flow**.)

17. List some similarities and differences between this research question and our first research question looking at the average difference in first and second *E. coli* counts.

18. Use the **Descriptive Statistics** applet to create a dotplot of the flow rates. Include average and standard deviation for the flow rate. What do you observe?

19. Carry out an appropriate test of significance to see whether the Cameroon water filters flow at a rate different than 1000 mL/min or is an average of 1000 mL/min plausible. State your hypotheses, report your 95% confidence interval, p-value and conclusions.

20. You should have found that 1000 mL/min is a plausible value for average flow rate population of all similar water filters. Does this mean that all the filters have flow rates of about 1000 mL/min? Is testing a single mean here really telling us what we need to know about these filters? Why or why not?

Research article: Lie detection

Read "The eyes don't have it: lie detection and neuro-linguistic programming" by Wiseman et al. (2012), 7(7), e40259 in the journal *PLoS One*, and then answer the following questions.

Step 1. Ask a research question

1. In no more than 1-2 sentences state a broad research question (or related questions) that the researchers were trying to investigate in this study.

2. Identify at least two (cited) points of evidence the researchers cite as to why the research question(s) are important. Include the citation in your response.

Step 2. Design a study and collect data

3. What is the sample size in Experiment #1? How were they recruited?

4. What criteria did all individuals in Experiment #1 have to meet in order to be allowed into the study?

5. Did Experiment #1 involve pairing or was it unpaired?

6. What is the sample size in Experiment #2? How were they recruited?

Step 3. Explore the data

7. Based on Table 1, what was the average number of upper left glances p/second when telling the truth?

8. When telling lies?

9. Upper right gazes when telling the truth?

10. when telling lies?

Step 4. Draw inferences

Table 1 shows the results of four different two-sided paired *t*-tests (called related *t*-tests by these authors) conducted on data in Experiment #1.

11. For UL glances, state the null and alternative hypothesis, indicate the strength of evidence provided by the p-value and provide an appropriate (in context) interpretation.

12. Paired *t*-tests require validity conditions to be met. Explain how the validity conditions are met for all tests in Table 1.

Table 2 shows the results of two different two-sided independent sample *t*-tests conducted on the data in Experiment #2.

13. For Mean correct, state the null and alternative hypothesis, indicate the strength of evidence provided by the p-value and provide an appropriate (in context) interpretation.

14. Independent sample *t*-tests require validity conditions to be met. Explain how the validity conditions are met for both tests in Table 2.

Step 5. Formulate conclusions

15. To which individuals are you willing to generalize your conclusions from this study? Why?

Step 6. Look back and ahead

16. Identify at least two things that, if you were running the study and had a reasonable amount of financial resources, you would do differently. You must justify your answer.

UNIT 3: ANALYZING MORE GENERAL SITUATIONS

So far our datasets have been fairly simple in their structure. That's a good way to start, but life is not always simple. The next three chapters extend what you have seen to more general situations.

Units 1-3 Data Structures

Unit 1: One binary variable

Unit 2: Comparing Two Groups	Response	Explanatory
Chapter 5	Binary	Binary
Chapter 6	Quantitative	Binary
Chapter 7	Quantitative - Paired	

Unit 3: Association Between Variables	Response	Explanatory
Chapter 8	Categorical	Categorical
Chapter 9	Quantitative	Categorical
Chapter 10	Quantitative	Quantitative

Although the data structures in Unit 3 are new, and they vary from one chapter to the next, most of what you have already seen will still apply. We will still use the Six-Step Method to structure a scientific investigation. We will still use the 3S Strategy for testing hypotheses. The statistic will be different for different chapters, but apart from that, the simulation step will be the same. As before, we ask, "What if the null hypothesis was true: How would the statistic behave?" As before, the answer takes the form of a null distribution for the statistic and again we use a p-value to assess the strength of evidence. We will also consider how to follow-up a statistically significant test with confidence intervals to further explore the associations we are analyzing.

Common null and alternative hypotheses

But wait: There's more! For Chapters 8, 9, and 10 there is a single generic way to think about the hypotheses. Abstractly, we will always be testing the same null hypothesis, namely, that there is *no association between the response variable and the explanatory variable* and the alternative is simply that *there is an association*. This way to think about the hypotheses can help you deepen your understanding of how the various chapters and methods are connected. Not only can you state the null hypothesis generically ("no association") but you can think of the 3S Strategy in generic terms: You typically want a statistic that is small when there is no association, but large when the association is strong. To simulate the values of the statistic under the null hypothesis, you break the association by shuffling the response variable (reassigning the values at random to the explanatory variable values) to get the null distribution. As always, a tiny p-value indicates strong evidence of a genuine association, too strong to be attributed just due to random chance.

Issues in Multiple Testing

Chapters 8 and 9 are similar to Chapters 5 and 6. The main difference between 5/8 and 6/9 being that now (Chapters 8 and 9) the categorical explanatory variables might have multiple categories instead of being binary (as in Chapters 5 and 6). Thus, Chapter 8 deals with multiple proportions and Chapter 9 deals with multiple means.

You might think that to compare multiple proportions (or multiple means), one simply conducts all the possible two-sample comparisons between the groups using the methods you learned in Chapters 5 and 6. For example, with three groups, we could test all the different ways we could pair them up. If our groups are labeled A, B, C, and D our comparisons are A vs. B, A vs. C, A vs. D, B vs. C, B vs. D, and C vs. D. Suppose each of these tests is tested at the 5% significance level (i.e., rejecting the null hypothesis if the p-value is smaller than 0.05). Now suppose the null hypotheses in these tests are all true (all four of the group parameters are actually the same). This means that for each test there is a 5% chance that we will make mistake and reject the null hypothesis even though the statistic actually did happen by chance alone. In Section 2.3 this was defined as a **Type I Error**, or false alarm. Each time we do a test, we have a chance of making a Type I Error, which we control through the level of significance. The problem is, these Type I Errors "accumulate" when we do more and more tests on the same data. (For an analogy, suppose that every time you ski down a certain run, there's a 5% chance you will fall. If you ski down the run 15 times, what's the chance you will fall at least once?) If we conduct four tests each at the 5% significance level, our overall Type I Error rate (the probability of rejecting at least one of the null hypotheses when it was really true) jumps to almost 19%. An alternative approach is to use one test that compares all three proportions at once. If we fail to reject the null hypothesis, say at the 5% level, we are "done," in that we will conclude that we don't have evidence that the long-run parameters differ. By constructing a statistic that compares all sample proportions or sample means at once, we can perform just one test and thus we can keep the probability of a Type I Error as small as we want. But, to do this, we need to find a single statistic that measures the differences in all our proportions (or means) at once so we can run one overall test. We'll use the Mean Absolute Difference (MAD) statistic in both Chapters 8 (differences in proportions) and 9 (differences in means) to do this. We'll also see the commonly used theory-based alternatives: the chi-square test and ANOVA. In these cases we'll also explore a follow-up method that that allows us to begin to construct a more detailed picture of the nature of the relationship, when evidence exists that there is a relationship.

Chapter 8

Comparing More Than Two Proportions

Chapter Overview

In Chapter 5 you learned how to compare two groups using proportions. In this chapter, we will show you how to compare more than two groups. In Chapter 5 you worked with data in the form of x tables. Here we'll work first with x tables, and eventually with larger tables as well. Recall before Chapter 1, in the Preliminaries, we looked at a study that explored the best way to recruit people to become organ donors. The data from that study could have been summarized in a x table like that shown here.

		Default Option		
		Opt In	**Opt Out**	**Neutral**
Organ Donor	**Yes**	23	41	44
	No	32	9	12

Just as in Chapter 5, the response variable is binary. However in Chapter 5, the explanatory variable was always binary and here our explanatory variable (Which group?) has three categories. With three or more groups, we can often state the null hypothesis to be that all groups have the same probabilities of success. If any one group probability differs significantly from the others, there is evidence against the null. Another equivalent way to state the null hypothesis is that there is no association between the explanatory variable (default option) and the response variable (organ donor). This kind of test is called an overall test because it tests all three long-run probabilities at once.

Many things will be the same as in Chapter 5. As always, inference will take place as part of the Six-Step Method. You will use segmented bar graphs to show proportions, just as before. You will use the 3S Strategy to compute p-values by simulation. When the overall test tells you that group differences are statistically significant, you can follow up using pairwise tests and intervals to pin down where the population differences are. For these pairwise comparisons, you'll use the same methods as in Chapter 5. (But you can think of the overall test as preventing us from doing all these comparisons when it is not necessary; see the Unit 3 Introduction for some discussion about this.)

Chapter 8 has only two sections. Section 8.1 takes a simulation-based approach—but we need to think carefully about what to use as our statistic. Section 8.2 shows how one choice of statistic leads to a theory-based shortcut. This latter statistic also easily extends to any number of groups for both the explanatory variable and/or the response variable.

Section 8.1: Simulation-Based Approach to Compare Multiple Proportions

Introduction

In Chapter 5 we saw how to use simulation-based methods to investigate the statistical significance of the difference between two conditional proportions. This was accomplished by taking the difference in the two conditional proportions as our statistic and shuffling (re-randomizing) the response outcomes to the explanatory variable groups to simulate the null distribution under the null hypothesis of no association between the response and explanatory variables. This allowed us to compare the observed difference in the sample proportions to these simulated differences in conditional proportions in order to draw a conclusion about strength of evidence.

Would that approach work when comparing categorical variables that have more than two categories? Yes, but the key question is first defining a reasonable statistic to measure the degree of the association between two categorical variables with multiple categories.

Example 8.1: Coming to a Stop

Students at Virginia Tech studied which vehicles come to a complete stop at an intersection with four-way stop signs. They looked at several factors to see which (if any) were associated with coming to a complete stop. (They defined a complete stop as "the speed of the vehicle will become zero at least for an [instant]"). Some of these variables included the age of the driver, how many passengers were in the vehicle, and type of vehicle. The variable we are going to investigate is the arrival position of vehicles approaching an intersection all traveling in the same direction. They classified this arrival pattern into three groups: whether the vehicle arrives alone, is the lead in a group of vehicles, or is a follower in a group of vehicles. The students studied one specific intersection in Northern Virginia at a variety of different times. Because random assignment was not used, this is an observational study. Also note that no vehicle from one group is paired with a vehicle from another group. In other words, there is independence between the different groups of vehicles.

You should have seen in Chapters 5 and 6 that we can often state our null and alternative hypotheses in terms of no association (null) and association (alternative). In this case, the null and alternative hypotheses are:

Null hypothesis: There is no association between the arrival position of the vehicle and whether or not it comes to a complete stop.

Alternative hypothesis: There is an association between the arrival position of the vehicle and whether or not it comes to a complete stop.

Another way to write the hypothesis

> **Think about it:** What does no association mean in this context? Specifically, what does no association imply about the similarity or differences in the conditional proportions that stop for the three stopping positions? What should be true about the segmented bar graph if there is no association?

As we discussed in Chapter 5, no association implies that the long-run probabilities for each of the explanatory variable conditions are the same. The difference in this example compared to what we saw in Chapter 5 is that now there are three long-run probabilities of stopping (single, lead, or following vehicle). If the null hypothesis (no association) is true, then these three long-run probabilities are all actually equal. In other words, the three bars of a segmented bar graph corresponding to these long-run probabilities will be shaded similarly. Note: This doesn't mean that the shaded areas within each bar will be the equally sized (e.g., 50/50). Here is another way to write the null hypothesis that reflects this:

Null Hypothesis: The long-run probability that a single vehicle will stop is *the same as* the long-run probability a lead vehicle will stop, which is *the same as* the long-run probability that a following vehicle will stop. In other words all three long-run probabilities are actually the same.

Here is the null hypothesis written in symbols:

$$H_0: \pi_{Single} = \pi_{Lead} = \pi_{Follow}$$ where π is the long-run probability a vehicle will stop.

One more way to think about this is that the null hypothesis means that regardless of what the arrival position is, the chance the vehicle will stop is the same.

We've already stated that the alternative hypothesis says: "There is an association between arrival position and whether or not the vehicle stops". Now let's translate that into what it means for the conditional proportions.

Think about it: What does there is an association mean in this context? Specifically, what does an association imply about the similarity or differences in long-run probabilities of stopping across the arrival position categories?

Saying there is an association is a very generic, all-purpose kind of statement as an association can look many different ways. In short, it means that the null hypothesis is not true—at least one of the pairs of long-run probabilities of stopping is not equal. Here is the alternative hypothesis in words:

H_a: At least one of the three long-run probabilities of stopping is different from the others.

A common student mistake is to think that the alternative hypothesis means that *all* of the long-run probabilities must be different from each other. But, even if just one of the long-run probabilities is different from the others, that's an association between the two variables.

Explore the data

The results for this study are shown in Table 8.1.

Table 8.1: A two-way table and segmented bar graph showing the number (and conditional percentage) of vehicles that came to a complete stop for three different arrival positions to the intersection

	Single Vehicle	**Lead Vehicle**	**Following Vehicle**	**Total**
Complete Stop	151 (85.8%)	38 (90.5%)	76 (77.6%)	**265**
Not Complete Stop	25 (14.2%)	4 (9.5%)	22 (22.4%)	**51**
Total	**176**	**42**	**98**	**316**

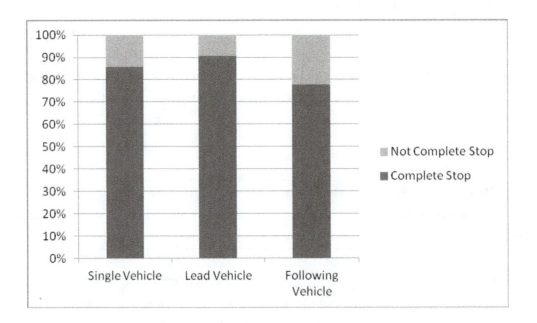

We see from Table 8.1 that a vehicle was most likely to stop (90.5%) if it was leading a group of vehicles as it approached the intersection, followed by single vehicles (85.8%), and then by following vehicles (77.6%). Although these percentages are all different, are they different enough to be statistically significant? Are they different enough to convince us that the differences didn't arise just by random chance? In other words, do these sample proportions provide convincing evidence that any of the long-run probabilities of stopping differ for the different arrival positions of vehicles at this intersection?

Applying the 3S Strategy

The first step in the 3S Strategy is to specify a statistic. The statistic should be a single number that summarizes all the evidence for the alternative hypothesis based on the sample data.

> **Think about it:** Propose a formula to determine a single number (statistic) that summarizes how different the conditional proportions are in our sample. That is, how might we measure the discrepancies in these conditional proportions across the three groups with one numerical value? What types of values for that statistic (e.g., large or small, positive or negative) would you consider evidence against the null hypothesis?

The alternative hypothesis simply says at least two of the probabilities differ from each other. So if we have any pair of proportions with a large difference that is evidence against the null hypothesis and in favor of the alternative hypothesis. One approach is to examine each pair of differences. Let's start by computing the three differences in proportions.

$\hat{p}_s$ proportion of single vehicles that stopped $= 0.858$
$\hat{p}_L$ proportion of lead vehicles that stopped $= 0.905$
    Difference in conditional proportions $= 0.858 - 0.905 = -0.047$

$\hat{p}_F$ proportion of following vehicles that stopped $= 0.776$
$\hat{p}_s$ proportion of single vehicles that stopped $= 0.858$
    Difference in proportions $= 0.776 - 0.858 = -0.082$

$\hat{p}_L$ proportion of lead vehicles that stopped $= 0.905$
$\hat{p}_F$ proportion of following vehicles that stopped $= 0.776$
    Difference in proportions $= 0.905 - 0.776 = 0.129$

We now have three numbers (-0.047, -0.082 and 0.129) measuring how different the pairs of sample proportions are. Whereas we could evaluate each of these differences separately, this has the potential to inflate the overall Type I error rate by doing multiple tests (see Unit 3 introduction for more about this, as well as FAQ 8.1.1). The null hypothesis we want to test is an overall test of no association, so the statistic must be a single number to test this single overall hypothesis.

> **Think about it:** How can we combine the three separate differences in proportions into a single number that summarizes the entire strength of evidence for the alternative hypothesis?

One option is to subtract the smallest group proportion from the largest: 0.129. This represents the largest of the differences. However, this doesn't tell us much about the average difference.

Another option is to simply average these three numbers: (-0.047 + (-0.082) + 0.129)/3 = 0. Uh-oh. That's not good. We certainly wouldn't say there are no differences in the conditional proportions in this sample. The differences in proportions cancelled out because some were negative and some were positive. Also if we subtracted the proportions in different orders we would come up with a different statistic, so this method does not work well. To get around this, one option is to make all the differences positive by taking the absolute value before averaging them. You could also square the numbers to make them positive, or as we will discuss in the next section standardize and square to yield a chi-square statistic.

1. Statistic – Mean Absolute Difference

A single number summarizing the sample evidence for the alternative hypothesis is: (0.047 + 0.082 + 0.129)/3 = 0.258/3 = 0.086. This is the average (or mean) of the (absolute value of the) differences in the conditional proportions, which we will refer to as the **MAD** statistic. It provides a measure of how far apart these sample proportions are on average.

> **Think about it:** What would have to be true for the average of the absolute differences to equal 0? Could the average of absolute differences ever be negative? What types of values of this statistic (e.g., large or small) would provide evidence in favor of the alternative hypothesis?

The average of the absolute differences is equal to 0 when each pairwise difference is zero. This only happens when all of the sample proportions are the same. So, small values of the MAD offer weak evidence against the null. Conversely, large values of the MAD offer strong evidence against the null hypothesis and in support of the alternative that at least one of those long-run probabilities is different from the others. The absolute value of a difference will always be non-negative. If we average several non-negative numbers we will always get a non-negative number. So the MAD statistic is always going to be non-negative (zero or positive), with larger and larger values indicating stronger and stronger evidence against the null hypothesis.

2. Simulate

We know that the next step in the 3S Strategy is to simulate the null distribution of the statistic.

Think about it: How can we simulate MAD values assuming the null hypothesis of no association? Could we do this with playing cards?

We've seen already that the null hypothesis of no association means that the value of the explanatory variable (the arrival position of the vehicle) has no impact on or no relationship with the outcome of the response variable (whether or not the car comes to a complete stop).

To simulate this with cards, we could let black cards represent the vehicles that stopped in our sample and red represent vehicles that did not stop. Specifically we would have 265 black cards to represent the stoppers and 51 red cards to represent the non-stoppers. Simulating the null hypothesis of no association means that we can shuffle all the cards and no matter which group they are dealt into (lead, follow, single) they will always stop if they are a stopper (black card), and always not stop if they are a non-stopper (red card). So then we would:

1. Deal 176 of the cards to represent the single group, 42 to represent the lead group, and 98 to represent the following group.

2. Determine the proportion of black cards in each group to calculate the simulated proportions of stoppers under the assumption that the null is true. From these simulated proportions we can now calculate a simulated MAD statistic.

We can also think of this simulation done not with cards, but the raw data. To simulate no association using the data table, we can shuffle the values of the response variable (stopping or not) and reassign each to an existing outcome of the explanatory variable (vehicle arrival position). This accomplishes the same thing as the simulation done with cards. We will do this with the Multiple Proportions applet. Figure 8.1 shows part of the spreadsheet for the observed data from the study, the conditional proportions for the vehicles that stopped, a segmented bar graph for the sample data, and the observed value of our statistic, 0.086.

The table below summarizes the key aspects of the simulation:

Null hypothesis = Long-run proportions are the same in each population
One repetition = Re-randomizing cars to the order they arrived without changing whether or not they stopped
Statistic = mean absolute difference in proportions between the three groups

---

FAQ 8.1.1 Multiple testing

Q: I'm still not clear about what you called "issues of multiple testing" back at the beginning of the section.

A: It's not an easy issue. Remember the example about skiing?

Q: If I have a 5% chance of a fall on each downhill run and I do 15 runs, I'm pretty likely to fall at least once.

A: Right. Your chance of at least one fall is about 54%.

Q: And for testing?

A: Just as with skiing, there are two different error rates. The usual false alarm rate for a single test is like the chance of a fall in one downhill run. Let's say 5%.

Q: That 5% is arbitrary.

A: Absolutely. I'm using 5% just for illustration.

Q: And a bunch of tests is like the 15 downhill runs?

A: Yes. In stat-speak, a set of parallel tests is called a "family." The chance of at least one false alarm is called the "family-wise" error rate.

Q: So for the skiing example, the individual false alarm rate was 5%, but the family-wise error rate was 54%.

A. Right.

Q: An error rate of 54% seems really high. What should I do?

A: Apologies in advance: it depends. This time it's not so bad. I can boil it down to three strategies.

Q: First one?

A: Its official name is "adjusting" for multiple tests or intervals. This is what most books teach. Your goal is to keep the family-wise error rate down to some small number, like 5%. To do this, you have to "adjust" the individual error rates – that is, use a much lower false alarm rate for the individual tests.

Q: Example please.

A: If you've got four groups there are six pairs of groups to compare. Suppose you want to test all six at once. That set of six pairwise comparisons is your "family." To keep the family-wise error rate at 5% you'd have to adjust the false alarm rate for each individual test from 5% down to about 0.85%.

Q: Ouch! That really jacks up the chance of a missed opportunity.

A: Good thinking. As always, there's a tradeoff between false alarms and missed opportunities.

Q: Suppose I care more about missed opportunities?

A: That's the second strategy. Ignore the family-wise error rate. Keep the individual rates at 5% or whatever. That way you won't increase the chance of a missed opportunity.

Q: Say more about when this would be a good strategy.

A: Just as you said – when you care more about missed opportunities. Suppose your first study is just prospecting, exploring to find possible differences. You plan to follow up with a second study to re-test any differences you find. Then a false alarm is not a big deal, but a missed opportunity could be really bad. The first strategy (adjusting) would be a bad choice because it increases the chance of missed opportunities.

Q: Does this happen in real life?

A: All the time. Medical screening tests are allowed to have high false positives because it's worse to fail to detect a disease.

Q: When would I want to adjust the individual error rates?

A: Suppose you are the editor of a scientific journal. Your inbox overflows with research reports. You want to publish the best that science has to offer, but you do not want to publish claims that later turn out to be false. For you, it's the family-wise error rate that matters.

Q: Got it. You said there's a third strategy?

A: It's a compromise, due to Sir Ronald Fisher, who was knighted for his work in statistics and genetics. Fisher's compromise is the one we use in this chapter.

Q: Is there a quick summary?

A: First do a single test for the whole family at once, sometimes called an omnibus or "overall" test. The tests based on $\hat{p}_{max}$ − $\hat{p}_{min}$, MAD, and chi-square are omnibus tests of the null that all of the probabilities are equal. If you get strong evidence of at least one difference, then follow up with unadjusted individual tests.

Figure 8.1: The sample data are pasted into the applet and the display shows a segmented bar graph of the sample as well as our observed value of the statistic, the mean of the absolute values of all pairwise differences (MAD)

First we will shuffle the response variable one time. The results from the applet are shown in Figure 8.2.

Figure 8.2: The results after one shuffling of the response variable. Part of the shuffled data is shown on the right along (using the Data option) with a segmented bar graph (using the Plot option) for these shuffled data and the mean of absolute differences for the shuffled data.

When viewing the **Most Recent Shuffle** window, you can see the outcomes for the explanatory variable did not change whereas the outcomes for the response variable were reassigned to those outcomes. The segmented bar graph and the conditional proportions also reflect the results from the shuffled data. One last thing to note is the sum of the absolute values for the shuffled data is 0.080. This outcome shows that for this shuffle we obtained three sample proportions that were a bit closer than they were for the original data, because this shuffled MAD value is lower than the observed MAD of 0.086. Will this always be the case? Let's shuffle some more and find out.

We used the applet to conduct 1000 shuffles of the data, which yielded 1000 values of the simulated values of the MAD statistic as shown in Figure 8.3.

Figure 8.3: Distribution of the simulated values of the MAD statistic when there is no association between arrival position of the vehicle and whether or not it comes to a complete stop at the intersection

Think about it: Why is the simulated distribution of the statistic not a bell-shaped curve centered at zero like we are used to seeing?

By making all of the differences in the proportions non-negative (by taking the absolute value), we cannot get a negative value of the statistic. The result is a right-skewed distribution of statistics that could happen when the null hypothesis of no association is true. We have fewer and fewer large values for the statistic, as that would indicate a larger difference among the groups, which shouldn't happen very often when the null hypothesis is true.

3. Strength of evidence

Now that we have a distribution of simulated statistics for the null hypothesis and the value of the observed statistic from our data, we can compute a p-value as a measure of the strength of evidence against the null hypothesis. In the past we have used the way we've phrased our alternative hypothesis to determine whether the test is one-sided or two-sided and, thus, how to calculate the p-value (one tail or both tails). But in this case, values of the MAD statistic that are more extreme than what we observed will always be larger (farther from zero), so we only care about values that are equal to or even larger than the observed value of the statistic.

As always, the p-value will be the probability, assuming the null is true, that we obtain a statistic value as extreme or more extreme than what we observed in our sample. In this case the p-value turns out to be 0.080 because we saw a value of 0.086 or greater occur 80 times (out of 1000) when simulating the null hypothesis. To see this, look back at Figure 8.3.

With a p-value of 0.080 we have moderate evidence against the null hypothesis, but this evidence is not extremely strong. So, these sample data provide only moderate evidence that the probability of coming to a complete stop is related to the arrival position of the vehicle. It is probably unwise to generalize these results to intersections beyond the one in Northern Virginia that the students studied. Each intersection probably has many different factors that would influence someone to come to a complete stop or not. We also would not want to draw any cause-and-effect conclusions between arrival position and coming to a complete stop, because this is an observational study and not a randomized experiment.

If we had found more evidence of a difference in the groups, we might then proceed to examine confidence intervals to help us see which arrival position's stopping probabilities are significantly different from each other. This idea will be discussed in more detail with theory-based methods in the next section.

Exploration 8.1: Recruiting Organ Donors

In Example P.1 we considered a study that investigated how to encourage people to be more likely to become organ donors. At the time, we did not formally analyze the data from the study. Now we are studying methods for comparing three groups, so we can further analyze the data and draw inferential conclusions from this study.

Recall that researchers asked volunteer subjects to imagine that they had moved to a new state, applied for a driver's license, and needed to decide whether or not to become an organ donor. Researchers created three different versions of the survey, each with a different donor recruiting strategy. Subjects were randomly assigned to one of the three types of recruiting strategies:

- The default option is to be an organ donor, and individuals have to "opt-out" of organ donation (Opt-out group)
- The default option is to not be an organ donor, and individuals have to "opt-in" to organ donation (Opt-in group)
- There is no default option, and individuals have to choose whether or not they will be an organ donor (Neutral group)

1. What are the observational units? What are the variables? Are the variables categorical or quantitative? If categorical, how many categories do they have? Also, identify the roles (explanatory or response) of the variables.

   Observational units:

   Explanatory variable: Type:

   Response variable: Type:

2. Did this study make use of random sampling, random assignment, both, or neither? Also describe the implication of your answer in terms of scope of conclusions that can be drawn from this study.

3. Did this study pair subjects in one group with subjects in another group, or are the subjects in the different recruiting groups independent of each other?

4.
    a. Write the appropriate null and alternative hypotheses (in words) using the language of association between the explanatory and response variables.

    b. Express this null hypothesis in symbols. (*Hint*: Remember that hypotheses are always about parameters, so use appropriate symbols for parameters.)

5. Researchers found that 23 of 55 subjects in the opt-in group chose to be organ donors, 41 of 50 in the opt-out group chose to be organ donors, and 44 of 56 in the neutral group chose to be organ donors.
    a. Based on these counts, produce a two-way table of counts, putting the explanatory variable in columns.

    b. For each of the three groups, calculate the conditional proportion who agreed to become an organ donor.

    c. Produce a segmented bar graph (by hand or with the **Multiple Proportions** applet) to compare these conditional proportions across the three groups.

    d. Comment on what your calculations and graph reveal about whether the default option used in the question appears to affect how likely the person is to become an organ donor.

We see some differences in the sample proportions of individuals who chose to be organ donors across the three treatments, but are these differences large enough to be statistically significant? In other words, do these data provide strong evidence of an association between donor recruiting strategy (opt-in, opt-out, neutral) and choosing to become an organ donor in the population?

6. Suggest two possible explanations for how these observed proportions could have turned out to be as different from each other as they are. (*Hint*: Think about what our null and alternative hypotheses are.)

Applying the 3S Strategy

To investigate we will see how we can apply the 3S Strategy to these data. Because researchers are attempting to minimize (or at least control) the probability of Type I errors, in cases with multiple groups they first do an overall test (see the Unit 3 introduction for more discussion on this). However, this requires computing a single statistic and considering the null distribution of that statistic. Let's look at this approach now.

1. Statistic

7.

a. Propose a formula for a statistic that could be used to measure how different the three sample proportions are. (*Hint*: The key here is that the statistic needs to provide a *single* number that somehow focuses on differences between/among the observed sample proportions.)

b. For the statistic you propose, would a large value or small value give evidence for the alternative hypothesis that at least one proportion is different in the population?

When we compared *two* proportions in Chapter 5, we used the difference in two proportions as the statistic. In this chapter, though, there are multiple proportions. There are actually several reasonable ways of summarizing the differences among the groups in one number. Once you have settled on a statistic, you apply the 3S strategy as before – simulate the distribution of the statistic under the null hypothesis and then see where the observed value of the statistic falls in that distribution.

8. One possible statistic is *max $\hat{p}$ – min $\hat{p}$*. Review the conditional proportions you determined in #5b and calculate the observed value of this statistic for these data. Why might this be a less than satisfactory choice?

However, this statistic ignores some of the groups. Another reasonable statistic to calculate is the mean of the absolute values of differences for each pair. (We could call this statistic "*MAD*" for mean of absolute differences.)

9. Let's construct the MAD statistic for these data by going through the following steps.

    a. Review the conditional proportions of organ donors you determined in #5b and calculate the differences in these proportions for each pair:

        Opt-out minus opt-in:

        Opt-out minus neutral:

        Opt-in minus neutral:

    b. Calculate the mean of the absolute values of these differences.

So which statistic should you use? And how can you use the statistic to estimate a p-value? We'll consider those questions in the following sections.

2. Simulation

In Chapter 5 we first saw how we can simulate a null hypothesis of no association by shuffling the response variable outcomes across the explanatory variable groups. This models the random assignment process used in the data collection process, and, assuming the null hypothesis is true, breaks any potential association between the response and explanatory variables. We modeled this with playing cards in Chapter 5.

10. Describe how you would model a simulation of the null hypothesis with red and black playing cards now that there are three explanatory variable groups instead of two. Make sure to tell how many black cards you would need and how many red ones. Also tell what each color of cards would represent. How many cards would you deal out to each group? What would you calculate after dealing them out?

11. Now let's use the **Multiple Proportions** applet to simulate a null distribution of these MAD statistics. This null distribution is again used to measure the strength of evidence provided by the observed statistic against the null hypothesis. Open the **OrganDonor** data file and paste the two columns (with column names) into the applet and press **Use Data**. (Or enter/paste in the two-way table of counts in the Sample data window and press **Use Table**.) Check the **Show table** box and verify that the sample proportions of organ donors match your answers from #5b and the observed MAD statistic shown matches your answer from #9b. If you cannot see the whole table, click on the edge and drag it to make it larger.

a. Now check **Show Shuffle Options**, and leave **Number of Shuffles** at 1. Press **Shuffle Response** to perform one shuffle of the response variable values. You should see the shuffled response variable, the new two-way table for the simulated data, and the value of the simulated statistic (Most Recent Shuffled MAD) which is also placed in blue on the graph on the right. Select the **Plot** radio button to see the shuffled segmented bar graph. How does the distribution across the groups for the shuffled data compare to the original data? Is the simulated MAD statistic value closer to zero than the observed value of the MAD statistic?

b. Now enter 999 for the **Number of Shuffles**, and press **Shuffle Responses**. This repeats the shuffling of the response variable 999 more times for a total of 1000 repetitions. You should see a graph of a null distribution of the MAD statistics. What is the shape of this distribution? Why is it not a bell-shaped curve centered at zero?

3. Strength of Evidence

12. To estimate a p-value, determine how often the observed value of the statistic, or something even larger, occurred in the null distribution. (*Hint*: Enter the observed value for the MAD statistic from the research study in the **Count Samples** box and press **Count**.)

13. Based on the p-value, summarize the strength of evidence that the sample data provide against the null hypothesis.

Generalization

14. Comment on how the sample was obtained and to which population your conclusion can be generalized.

Causation

15. Comment on how the study was designed and whether a cause-and-effect conclusion is warranted from this study.

Look Back and Ahead

16. Summarize your conclusion for the researchers. Do you have any concerns about the study design, any comments on sample size? Are there other limitations that you feel need to be addressed? Are there specific improvements you would suggest to make this study better? Based on what you concluded from this study, can you suggest one or two follow up studies that could be carried out?

Section 8.1 Summary

The 3S strategy for simulation-based inference can be applied to comparing proportions across multiple groups. Neither the explanatory nor response variable has to be binary. The key is choosing a formula for the statistic that reasonably summarizes the overall evidence against the null hypothesis.

- This approach provides an overall test of the multiple groups simultaneously, as opposed to testing each pair of groups separately, in order to guard against inflating the Type I error rate.
- The null hypothesis is that there is no association between the two variables.
    o This is equivalent to saying that the long-run probabilities of the response variable are identical for all categories of the explanatory variable in the population.
- The alternative hypothesis is that there is an association between the two variables in the population.
    o No direction is specified for the association in the alternative hypothesis.
    o The alternative hypothesis can be expressed as saying that at least one population proportion (or long-run probability) differs from the others.

The choice of a statistic to measure how different the groups are is not straightforward. Many reasonable statistics can be used.

- One reasonable statistic is to calculate the difference in sample proportions for each pair of groups, and then take the **mean of the absolute differences (MAD)**.

o Larger values of the MAD statistic indicate a greater difference in sample proportions across the groups.

- Another statistic is the largest sample proportion minus the smallest sample proportion, however this statistic may be less powerful than MAD in many cases.
- Other statistics also exist. The key thing to remember is that a single number is needed and that the general approach of the 3S strategy works regardless.

-

The simulation method is to shuffle the values (outcome categories) of the response variable and redistribute to the explanatory variable outcomes.

- This shuffling method simulates the results of the study under the assumption of no association between the variables (the response variable outcomes were going to be the same regardless of which group they are assigned to).
- As we've seen before, the p-value is calculated as the proportion of repetitions in which the simulated value of the statistic is at least as extreme as the observed value of the statistic.
  - o **As always,** a small p-value provides evidence against the null hypothesis.
- As before, we will often use this shuffling method to obtain a p-value even with random sampling and studies with no randomness in the study design.

Section 8.2: Theory-Based Approach to Compare Multiple

Proportions

Introduction

As we've seen numerous times already, simulation methods are quite flexible because they work in many situations, but they do require a computer to perform enough repetitions to understand the long-run pattern of the null distribution. Theory-based approaches avoid the need for simulation, but only work under certain conditions—which must be met in order to accurately predict the null distribution that would have occurred if you were to simulate.

In the past three chapters, our simulated null distributions were bell-shaped and centered at zero. Because of the way we constructed our statistics in the previous section, we saw that our null distribution was no longer bell-shaped nor centered at zero. Unfortunately, the theory to accurately predict the shape of the null distribution for the MAD statistic is challenging.

As it turns out, using a slightly different statistic (the *chi-square statistic*), instead of the MAD statistic, we can accurately predict the behavior of the null distribution by a mathematical model, called a chi-square distribution. In this section we will see how to use the chi-square statistic and chi-square distribution instead of simulation to conduct a theory-based test in order to evaluate the strength of evidence for an association between two categorical variables with multiple (two or more) categories.

Example 8.2: Sham Acupuncture

Increasingly, health insurers, federal agencies, medical organizations, and others are insisting on the use of evidence-based medicine in making health-care decisions. Evidence-based medicine means only using medical treatments where evidence (measured using Statistics, most typically based on randomized experiments) exists that a particular treatment is both effective and better than other available treatments.

Recently, a randomized experiment was conducted exploring the evidence for the effectiveness of acupuncture in the treatment of chronic lower back pain (Haake et al., 2007). Acupuncture involves the insertion of needles into the skin of the patient at specific locations (acupuncture points) around the body to treat a variety of ailments.

In the experiment, there were three treatment groups of interest: (1) Verum acupuncture practiced according to traditional Chinese medicine principles; (2) Sham acupuncture where needles were inserted into the skin, but not at acupuncture points and not very deep (this is often used as a control in experiments involving acupuncture); and (3) Traditional (non-acupuncture) therapy consisting of a combination of drugs, physical therapy, and exercise. A total of 1162 patients were randomly assigned to each of the three treatment groups, yielding 387 patients in groups 1 and 2, and 388 patients in group 3.

Think about it: What are the explanatory and response variables in this study? Is there any pairing in the design of the study? If we find an effect, are we able to draw cause/effect conclusions?

The explanatory variable is the type of treatment. It is a categorical variable with three categories: verum acupuncture, sham acupuncture, and traditional therapy. The response variable is also categorical. It has two categories: whether or not there was significant relief from lower back pain. There wasn't any pairing of subjects between the treatment groups; rather the subjects in the three different treatment groups are independent of each other.

Think about it: What are the null and alternative hypotheses being tested in this study?

Hypotheses

The null hypothesis is that there is no association between the type of treatment received (real, sham, or no acupuncture) and whether someone shows substantial reduction in back pain or not in this population. In other words, the long-run probability of improvement is the same across the three treatments. The alternative hypothesis is that there is an association between these two variables in the population. In other words, at least one of the long-run probabilities of improvement is different from the others.

The data gathered on the 1162 subjects is presented in Table 8.2 along with the conditional proportions comparing the treatment groups to whether or not patients showed a substantial reduction in lower back pain.

Table 8.2: Two-way table of pain reduction by acupuncture group

	Real acupuncture	Sham acupuncture	Non-acupuncture	Total
Substantial reduction in pain	184 (0.475)	171 (0.442)	106 (0.273)	461
Not a substantial reduction in pain	203 (0.525)	216 (0.558)	282 (0.727)	701
Total	387	387	388	1162

We also put the two-way table into the Multiple Proportions applet (pressing **Use Table**) and produced the conditional proportions of those that had substantial reduction in pain (better) or not, as well as a segmented bar graph. (See Figure 8.4.)

Figure 8.4: Segmented bar graph of pain reduction by acupuncture group and the conditional proportions for those that had substantial pain reduced (better) or not

	real	sham	none
better	184	171	106
not	203	216	282
Prop	0.475	0.442	0.273

Both Table 8.2 and Figure 8.4 illustrate that real acupuncture showed the highest percentage of patients with reduction in pain symptoms—but is it significantly better than the other treatments?

Simulation-based approach

In Section 8.1 we learned how to take a simulation-based approach to the analysis by applying the 3S Strategy to test the statistical significance of an association between two categorical variables. Initially in that section, our choice of statistic was the mean of the absolute values of the differences in the conditional proportions between the groups (the MAD statistic).

1. Statistic: The absolute value of the differences in proportions of improvement comparing all three groups is:

Real vs. Sham: $0.475 - 0.442 = 0.034$

Real vs. None: $0.475 - 0.273 = 0.202$

Sham vs. None: $0.442 - 0.273 = 0.167$

Thus, the observed statistic is MAD $= (0.034 + 0.202 + 0.167) = 0.403/3 = 0.135$

2. Simulate: We can use the Multiple Proportions applet to shuffle the response variable (equivalent to re-randomizing the subjects to treatments, assuming their responses will not change) and recompute the value of the statistic for each of the shuffled datasets. We repeated this process 1000 times to obtain the null distribution shown in Figure 8.5.

Figure 8.5: Null distribution for acupuncture study

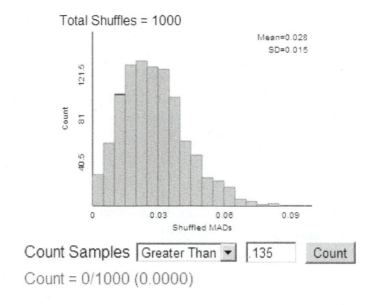

3. Strength of Evidence: The p-value in this study is approximately 0, because a statistic of 0.135 or larger never occurred in this simulation that assumes the null hypothesis to be true. This gives us very strong evidence against the null hypothesis. We don't think it's plausible that the observed differences in conditional proportions that we saw in this study arose solely from the random chance in the random assignment process. We have very strong evidence that the treatment (type of acupuncture) used is truly associated with pain reduction.

Alternate choice of statistic – chi-square

The MAD statistic focuses on how all pairs of proportions compare to each other. Another option is to consider how the proportions compare to some average proportion. As it turns out, this new statistic, called a chi-square statistic, has the advantage of being well predicted by a mathematical model (theory-based approach) and, so is commonly used in practice for these types of problems. It also can be used on larger sized tables (two categorical variables, with any number of categories each; see Exploration 8.2b).

> **Think about it:** Reconsider Table 8.2. What is the overall proportion in the sample of 1162 subjects that experienced substantial reduction in pain? What would we expect to be true about each of the group proportions if there really is no difference among the treatments?

If we ignored the type of treatment, then 461/1162 or about 39.7% of the patients experience substantial reduction in pain. If the different treatments really don't differ in their effectiveness, then we would expect this same percentage in each of the three categories.

> **Think about it:** How could we compare each group proportion to this overall proportion? How could we standardize each comparison?

When we standardize, we take the observed statistic, subtract the typical (average) value, and divide by the standard deviation. So, similar to standardized z-statistics that we have calculated before, we could standardize each group proportion ($\hat{p}_i$) by comparing it to this overall proportion ($\hat{p}$) and dividing by the standard deviation of the group proportion for that sample size assuming the null is true ($\sqrt{\hat{p}(1-\hat{p})/n_i}$ —where n_i is each individual group's sample size).

Real	**Sham**	**None**
$\dfrac{.475-.397}{\sqrt{.397(1-.397)/387}}=3.14$	$\dfrac{.442-.397}{\sqrt{.397(1-.397)/387}}=1.81$	$\dfrac{.273-.397}{\sqrt{.397(1-.397)/388}}=-4.99$

giving us standardized statistics which tell us how different each group proportion is from the average (overall) proportion. In particular, notice that both the Real group and the None group are quite a bit different than the overall proportion because their standardized statistics are greater or less than 2.

So how do we combine these into one overall standardized statistic? We could just add them together, but again we don't want the positive and negative values to cancel each other out. This time we will square them before we add them together: $(3.14^2) + (1.81^2) + (-4.99)^2 \approx 38.03$.

$$\chi^2 = \left[\left(\frac{0.475 - 0.397}{\sqrt{.397(1-.397)/387}}\right)^2 + \left(\frac{0.442 - 0.397}{\sqrt{.397(1-.397)/387}}\right)^2 + \left(\frac{0.273 - 0.397}{\sqrt{.397(1-.397)/388}}\right)^2 = 38.03\right] \text{(Se}$$

e the Formulas section for an alternate (and more generalizable) way to do this calculation.)

Think about it: What value would this statistic have if the group proportions had all been the same? What types of values will this statistic have if there is a strong association between the explanatory variable and the response variable?

The calculation just performed is a special case of the chi-square statistic. The smallest value this statistic can be is zero, which would only happen if all the group proportions had been exactly the same. Larger values of the chi-square statistic indicate stronger evidence against the null hypothesis. To decide whether this chi-square statistic is larger than we would expect by chance, we need to see how it performs in repeated shuffles. Figure 8.6 shows a simulated null distribution for the chi-square statistic for these data. (Like with the MAD statistic, we only need to worry about values larger than our observed statistic to be considered more extreme for this always positive statistic.)

Figure 8.6: Null distribution of chi-square statistic

Not surprisingly, the results are similar to when we used the mean of the absolute differences in proportions as the statistic. The null distribution is skewed to the right and the observed value of the statistic (38.05) is very far in the tail of the distribution. Similar to what we found with the MAD statistic, our observed statistic is not even close to the largest simulated values giving us very strong evidence against the null hypothesis, in favor of the alternative hypothesis.

How do we decide which statistic to use?

See FAQ 8.2.1 for more details, but the chi-square statistic does have some advantages. It is a standardized statistic in that it takes into account the sample sizes in our groups. In fact looking at the individual components of the chi-square sum (think of them as mini z-statistics) helps us identify which cells have the largest discrepancy from what the null hypothesis would predict. Furthermore, chi-square statistics can more directly be compared to other chi-square statistics from different studies (as long as the tables are similar in size).

Another huge advantage is that the chi-square statistic is often well predicted by a mathematical model--- that is, it can be used to obtain a theory-based p-value. Yet another advantage to the chi-square statistic is that it easily generalizes to a categorical response variable with more than two categories (see Exploration 8.2b for details).

Theory-based approach: The chi-square test

Instead of actually simulating data (shuffling) to generate a null distribution, can we use mathematical theory to predict the results that would occur by simulation? For the Max-Min and MAD statistics, the answer is no. However, if we use the chi-square statistic, the answer is yes. Figure 8.7 shows the theoretical *chi-square distribution* overlaid on the simulated chi-square statistics. This appears to be a close match to the simulated distribution.

Figure 8.7: A chi-square distribution used to predict our simulated null distribution for the acupuncture study

This also predicts the p-value for the test of no-association to be 0.0000, which is comparable to what we obtained through simulation.

The applet can also display more details about the chi-square statistic calculations and output for the theory-based test if you check the **Show X² output** box (Figure 8.8).

Figure 8.8: Chi-square Test from Multiple Proportions applet

```
             Chisq Cell Contributions
                    real    sham     none
             better  6.05    1.99    14.92
             not     3.98    1.31     9.82
             Sum=38.05
             df=2
Show χ² output: ☑  p-value= 0.0000
```

Note: The "***Chisq Cell Contributions***" show the components being summed together for the other version of the formula as shown in the Formulas section. The cell contribution values are useful for helping you detect where larger differences arise (e.g., much of the relationship here comes from the smaller than expected number of participants who felt better without acupuncture and the larger than expected number that did not feel better without acupuncture).

FAQ 8.2.1 Which test statistic?

Q: We've looked at $\hat{p}_{max} - \hat{p}_{min}$, MAD and chi-square as possible test statistics for comparing group proportions. Which one is best?

A: We start with $\hat{p}_{max} - \hat{p}_{min}$ and MAD because they are more intuitive and easier to understand.

Q: Why bother with chi-square, then?

A: The chi-square test has a theory-based approximation for its null distribution. The theory for $\hat{p}_{max} - \hat{p}_{min}$ and MAD is very messy. For those test statistics, simulation is pretty much the only option.

Q: Any other reasons to prefer chi-square?

A: Definitely. The chi-square test generalizes easily to tables with three or more rows and three or more columns. Tests using $\hat{p}_{max} - \hat{p}_{min}$ and MAD do not generalize in a single simple way to tables 3x3 and larger.

Q: So far we have two advantages for chi-square: approximation and generalization. Anything else?

A: Standardization. The chi-square statistic adjusts for (divides by) the standard error. The other two test statistics don't do that. (For 3x3 tables and larger, the standardization is trickier, but it's there nonetheless.)

Q: Approximation, generalization, standardization. Is there still more?

A: One last thing: power. Tests using chi-square are more powerful than tests based on $\hat{p}_{max}$ $\hat{p}_{min}$. Here's an example of a two-way table. What do you think?

	Group 1	Group 2	Group 3	Group 4
Success	1	3	0	0
Failure	0	0	3	1

Q: That looks like pretty strong evidence of association. Groups 1 and 2 have 100% success; Groups 3 and 4 have 0% success.

A: Agreed. So we want a test statistic that gives us a low p-value.

Q: Right.

A: The chi-square test gives a p-value of 0.03.

Q: And the test based on $\hat{p}_{max} - \hat{p}_{min}$?

A: It gives a p-value of 0.745.

Q: Not good. What about MAD?

A: For this data set it gives the same p-value as the chi-square test. But for bigger tables, there's no single right way to do a test based on MAD, and so the power is also not clear.

Q: What if I only had two groups?

A: Well, the MAD and the Max-Min would be the same. In fact, if you think back to chapter 6, the only difference is that now your 'difference in proportions' has to be positive. Furthermore, if you did a chi-square test comparing two groups you'd get the exact same p-value as if you did a two-sided, theory-based test comparing two proportions .

Q: Are you kidding? Exactly? This is statistics…

A: In this case, if you square your z-statistic when comparing two groups you get the chi-square statistic and the p-values from both will be identical.

Validity Conditions

Of course, this theory-based approach only works well when certain validity conditions are met. When the conditions are not met, the simulation-based method still gives accurate estimates of the p-value, whereas the theory-based approach does not. When the validity conditions are met, then the results will be similar regardless of the method (simulation- or theory-based) you choose.

Like we have seen with other theory-based tests, the validity condition for a chi-square test is that we need large sample sizes. When we were comparing two proportions, we said we had large sample sizes if the number of successes and failures were both at least 10 in each explanatory variable group. We will use the same guideline here. If each cell of the two-way table contains at least 10 observations, then we will consider our sample sizes to be large enough for the theory-based chi-square test to work well. (Note: This is a rather conservative guideline and if your data do not satisfy it, you might still consider the theory-based approach. See the Calculation Details Appendix for more details and alternative validity conditions.)

Did we have large enough sample sizes in our acupuncture example? Yes. The smallest number of observations back in Table 8.2 was 106. So our overall sample size was plenty big for the chi-square model to provide a good approximation of the null distribution of the chi-square statistic.

> **You might be thinking:** So, I found evidence of an association—what now? How do I follow up a chi-square test? How do I decide which groups differ significantly from which others, and by how much?

Although there is no conventional method for following up a chi-square test, one approach is to simply follow up a statistically significant chi-square test (based on a specific level of significance) by looking at all the pairwise comparisons between the groups. We can do that by producing theory-based confidence intervals for the difference in proportions for each pair of groups. We can calculate these intervals by checking the **Compute 95% confidence interval(s)** box (when the chi-square statistic has been chosen).

☑ Compute 95% confidence interval(s)
real - sham: (-0.0366, 0.1038)
real - none: (0.1356, 0.2689)*
sham - none: (0.1022, 0.2351)*

These results tell us that we have no evidence of a difference between the sham and real acupuncture (because that interval contains 0); it is plausible that there is no additional effect of real acupuncture as compared to sham acupuncture. However, because the other two intervals do not contain 0 there is a significant difference for pain reduction between real acupuncture and none as well as between sham acupuncture and none.

Exploration 8.2: Conserving Hotel Towels

If you have stayed at a hotel recently, you may have noticed how hotels are increasingly encouraging their guests to practice conservation habits by not having their towels washed every day—instead hanging them back on the rack to be reused. A recent study (Goldstein et al., 2008) conducted a randomized experiment to investigate how different phrasings on signs placed on bathroom towel racks impacted guests towel reuse behavior. In particular, the researchers were interested in evaluating how messages that communicated different types of social norms impacted towel reuse. One week prior to a guest staying in the room, rooms at a particular hotel were randomly assigned to receive one of the following five messages on a sign hung on the towel bar in the room:

Message 1: HELP SAVE THE ENVIRONMENT. You can show your respect for nature and help save the environment by reusing your towels during your stay. (No social norm)

Message 2: JOIN YOUR FELLOW GUESTS IN HELPING TO SAVE THE ENVIRONMENT. In a study conducted in fall 2003, 75% of the guests participated in our new resource savings program by using their towels more than once. You can join your fellow guests in this program to help save the environment by reusing your towels during your stay (Guest identity norm)

Message 3: JOIN YOUR FELLOW GUESTS IN HELPING TO SAVE THE ENVIRONMENT. In a study conducted in fall 2003, 75% of the guests who stayed in this room (#xxx) participated in our new resource savings program by using their towels more than once. You can join your fellow guests in this program to help save the environment by reusing your towels during your stay. (Same room norm)

Message 4: The citizen identity descriptive norm message stated "JOIN YOUR FELLOW CITIZENS IN HELPING TO SAVE THE ENVIRONMENT. In a study conducted in fall 2003, 75% of the guests participated in our new resource savings program by using their towels more than once. You can join your fellow citizens in this program to help save the environment by reusing your towels during your stay." (Citizens norm)

Message 5: JOIN THE MEN AND WOMEN WHO ARE HELPING TO SAVE THE ENVIRONMENT. In a study conducted in fall 2003, 76% of the women and 74% of the men participated in our new resource savings program by using their towels more than once. You can join the other men and women in this program to help save the environment by reusing your towels during your stay." (Gender identity norm)

Data were collected on 1595 instances of potential towel reuse. For each of the 1595 instances, room attendants recorded whether or not the hotel guest reused their towels.

1. In your own words, state the research question of interest in this study.

2. Identify the explanatory and response variables in this study, along with their types (categorical or quantitative). For any explanatory variables, indicate how many categories it involves.

    Explanatory: Type:

    Response: Type:

3. Did the study design make use of any pairing of subjects, or are the subjects who receive the different signs independent of each other?

4. State the appropriate null and alternative hypotheses for testing the researchers' conjecture, using both words and symbols.

5. Paste the data from the **Towels** file into the Multiple Proportions applet (and press **Use Table**).

   a. Check the box next to **Show table** in order to present the two-way table of counts and calculate conditional proportions of those who reused towels. (You can grab the bottom right corner of the table in the applet to extend it.) Report these five proportions below.

   b. Looking at the segmented bar graph, as well as the conditional proportions, conjecture on whether you think there will turn out to be a significant difference among the five groups with regard to towel reuse.

Applying the 3S Strategy

6.
   a. *Statistic*: Report the observed value of the MAD (mean absolute difference) statistic for this study, as given by the applet.

b. *Simulate*: Use the applet to conduct 1000 shuffles of the response variable. (First check **Show Shuffle Options**, and then enter 1000 for the **Number of Shuffles**. Press **Shuffle Responses**.) Then enter the observed value of the MAD statistic in the **Count Samples** box and press **Count**. Report the p-value.

c. *Strength of evidence:* Based on the p-value, evaluate the strength of evidence provided by the data against the null hypothesis. Summarize your conclusion in the context of this study.

Another choice of statistic: chi-square

The MAD statistic is fairly easy to understand but is not widely used in part because it cannot be easily predicted theoretically. A more commonly used statistic is called a chi-square (X^2) statistic. You can think of this statistic as measuring how far the group proportions of success are from the overall proportion of success. These deviations from the overall proportion are then squared and summed together across all of the explanatory variable groups. We'll do this calculation now.

7. Determine the overall proportion, $\hat{p}$, of rooms that reused towels.

8. We will calculate the standard error of a group proportion by computing $\sqrt{\hat{p}(1-\hat{p})/n_i}$ where $\hat{p}$ is the overall proportion found in #7 and n_i is the sample size for the ith group. Find the standard error of the group proportion for each of the five groups.

9. Create a standardized z-statistic for each conditional proportion compared to this overall proportion. To do this take each of the five group proportions, subtract the overall proportion (#7) and then divide the result by the standard deviation you found for that group in #8.

10. Which of the five standardized statistics are larger than 2 or smaller than −2, suggesting strong evidence that the group proportion is significantly different than the overall proportion?

11. Square each of these values and sum them to calculate the chi-square statistic.

12. What kinds of values (larger, small, positive, negative) will the chi-square statistic have if the null hypothesis is true? What kinds of values will provide evidence against the null hypothesis?

From the **Statistic** pull-down menu, select X^2. What does the applet report for the observed value of the chi-square statistic? Confirm that this matches your answer to the previous questions (Note: You may be a bit different due to rounding).

13.   The applet will also now display the null distribution for the x^2 statistic rather than the MAD statistic. Determine the p-value based on the simulated x^2 statistics. (*Hint*: Change the value in the **Count Samples** box to the observed value of the x^2 statistic.)

14.   How does the p-value based on the x^2 statistic compare to the one based on the MAD statistic? Are they similar? Is your conclusion about strength of evidence provided by the data against the null hypothesis similar?

Theory-Based Approach

The primary advantage of using the x^2 statistic is that its null distribution can be predicted well by a theoretical distribution. So, once again a theory-based approach could be used without conducting a simulation in the first place.

15.   Below the graph of the simulated chi-square statistics, check the box to **Overlay Chi-square distribution**. Does the theoretical distribution match the distribution of simulated statistics reasonably well? How does the theory-based p-value compare to what you found in #13? Does it lead you to a similar conclusion about strength of evidence?

Validity Conditions

Like all theory-based tests, this one also comes with the condition of having a large sample size. We will (very conservatively) consider a sample size large if the sample data include at least 10 observations in each cell of the two-way table.

16.   Go back to the applet, and make sure the **Show Table** box is checked.

a. How many cells are in this table? In other words, how many counts need to be checked that they are at least 10?

b. Is the validity condition for a theory-based chi-square test satisfied for these data? Justify your answer.

Follow-up Analysis

When a chi-square test produces a significant result, we conclude that at least one long-run probability or population proportion differs from at least one of the others. A sensible next step is to try to identify *which* group proportion differs significantly from which others. We can do this by producing confidence intervals for pairwise differences in population proportions.

17. Below the p-value output, check the box to **Compute 95% confidence interval(s)**.

a. How many intervals are produced?

b.   Which conditions are significantly different from each other? How are you deciding?

c.   For one of the intervals you identified in b, write a one-sentence interpretation of the interval being very clear what is supposed to be captured inside the interval and which condition has a more positive effect on towel use and by how much.

Scope of Conclusions

18.   Is this an observational study or a randomized experiment?

19.   Explain how the conclusions of this study are limited by conducting the study at a single hotel.

Look Back and Ahead

20. Summarize your conclusion for your analysis of the data from this study. Include a recommendation for the hotel manager about whether it matters what message is used, which message should be used, and how large of an impact the best message would be expected to have. Think about the study and its design. Were there any limitations you feel need to be addressed? Did the design match the research question? Can you suggest a better sampling method of hotel guests, a better way to collect data on towel use? Share some ideas for a logical next study.

Exploration 8.2b: Near-sightedness and Nightlights revisited

Recall Example 4.1, which described a study investigating whether there is a relationship between use of night lights in a child's room before age 2 and the child's eyesight condition a few years later. In Chapter 4, we presented a two-way table of counts from the study examining whether or not the child was near-sighted as the response variable:

	Darkness	Night light	Room light
Near-sighted	18	78	41
Not near-sighted	154	154	34
Total	172	232	75

The conditional proportions of near-sightedness in the three lighting groups are 0.105, 0.336, and 0.547, respectively, which **suggests** that the more light used in the child's room, the more likely the child is to become near-sighted. Now we know how to calculate a standardized statistic and estimate a p-value to measure the strength of evidence provided by these sample data for the conjecture that eyesight is associated with the type of lighting used in the child's room.

1. State null and alternative hypotheses in terms of population proportions.

2. State null and alternative hypotheses in terms of association between variables.

3. Are the validity conditions necessary for a chi-square test met for these data? Explain.

Alternative formula for chi-square statistic

Another popular and more general way to formulate the chi-square statistic is in terms of observed and expected cell counts. The *observed counts* are what you actually observed in the study. The *expected cell counts* are what you would expect for the count in a cell if the null hypothesis **were** true.

4. To find out how many children we expect to be in the Darkness/Near-sighted cell of the table do the following:

a. Find the overall proportion of children who were near-sighted in the sample.

b. How many children in the sample slept in darkness?

c. Multiply these two numbers together to find the expected number of children who would be near-sighted in the slept in darkness group if the null hypothesis were true (the probability of near-sightedness is the same in each condition).

d. How different is this expected cell count from what was observed for that cell? What does this suggest about the strength of evidence against the null hypothesis?

A similar calculation is performed to find the expected cell count for each of the six cells of the table. The chi-square statistic is obtained by squaring the differences in the observed and expected counts, dividing by the expected count (a form of standardizing) and then summing the six values. This can be thought of as $(O - E)^2/E$, where $O =$ observed and $E =$ Expected. See the Formulas section at the end of this exploration for the complete details.

5. Enter the two-way table of counts into the **Multiple Proportions** applet (remember to use one-word variable names and to not include the totals). Press **Use Table** and check the box to **Show X^2 output**. Report the chi-square statistic and theory-based p-value. What conclusion do you draw about the statistical significance of the sample results?

6. Examine the **Chisq Cell Contributions** (remember you can expand that table by pulling out the lower right corner). Verify that the square of the number you calculated in #4c is the first term. These six values are added together to yield the chi-square statistic; in other words, they are the six $(O - E)^2/E$ values. Which cell contributions are largest and so are contributing the most to the chi-square statistic value? According to the segmented bar graph, how are the observed counts in these cells deviating from what the null hypothesis would predict? (In other words, which is larger, the observed count or the expected count? How does that relate to our research question?)

The original research study actually classified the response variable into three categories of eyesight, as shown in the following table:

	Darkness	Night light	Room light
Near-sighted	18	78	41
Normal refraction	114	115	22
Far-sighted	40	39	12

One advantage to the more general chi-square statistic formula using observed and expected counts is that it works for any sized table (any number of categories for the explanatory variable and the response variable).

7. Create a segmented bar graph to summarize this table. Discuss what the graph reveals, including what this graph tells you that we did not learn from the previous table showing only near-sighted vs. not near-sighted.

8. With this new table, do you need to change your null and alternative hypotheses from #2?

9. Are the validity conditions for the chi-square test met for this table?

10. Enter this table into the Multiple Proportions applet and carry out the chi-square test. Report the test statistic and p-value. Is the strength of evidence of a genuine association between the variables in the population stronger, weaker, or essentially the same as before? Explain your answer.

11. Carry out a randomization test (shuffling) for this table using the chi-square statistic. Overlay the chi-square distribution on the simulated null distribution of chi-square statistic values. Does it appear to be a reasonable model for the null distribution?

12. Examine the cell contributions. Is the pattern of deviations from the null similar to before?

13. Does this study suggest that use of night lights and room lights *causes* an increase to the chance that a child is near-sighted? Why or why not?

Formulas: Details of the chi-square statistic and distribution

In calculating the chi-square statistic when you had a binary response variable (Example 8.2 and Exploration 8.2), you used the formula

$$\sum \frac{(\hat{p}_i - \hat{p})^2}{\hat{p}(1-\hat{p})/n_i}$$

where the overall proportion $\hat{p}$ came from assuming the null hypothesis was true and we expect the same probability of success in each group. The goal was to compare each group proportion to an estimate of this common probability. In Exploration 8.2b we saw an alternate approach that used observed and expected cell counts. Observed counts were what we actually observed in the study, and expected counts were what we would get, on average, if the null hypothesis were true.

For example, in the night light study:

Observed counts	Darkness	Night light	Room light	Total
Near-sighted	18	78	41	137
Normal refraction	114	115	22	251
Far-sighted	40	39	12	91
Total	172	232	75	479

The null hypothesis says we expect a similar proportion to be near-sighted in each lighting group. We can estimate the proportion by $137/479 \approx .286$. Similarly, we expect a bout $251/479$ or 52.4% to have normal refraction and $91/479$ or about 19% to be far sighted.

We need to work with the proportions instead of the counts because we have different samples sizes in our explanatory variable groups. By multiplying these proportions by the respective samples sizes we get a table of *expected counts*—the expected count of children in the cell if the null hypothesis were true. Verify that these expected counts would produce the same segmented bar graph for each lighting category.

Expected counts	Darkness	Night light	Room light	Total
Near-sighted	$172 \times 0.286 =$ 49.192	$232 \times 0.286 =$ 66.352	$75 \times 0.286 =$ 21.45	137
Normal refraction	$172 \times 0.524 =$ 90.128	$232 \times 0.524 =$ 121.568	$75 \times 0.524 =$ 39.3	251
Far-sighted	$172 \times 0.19 =$ 32.68	$232 \times 0.19 =$ 44.08	$75 \times 0.19 =$ 14.25	91

Notice that the *expected* counts need not be integers, but that each column will still sum to the column total and each row will sum to the row total.

By comparing the observed counts to the expected counts we get a sense of how much evidence there is against the null hypothesis. To do this, each expected count is subtracted from each observed count and then squared. This squared difference is then divided by the expected count to standardize the value, ensuring that bigger differences count more if the cell count wasn't expected to be very large. All of the standardized, squared differences are added up to yield the chi-square statistic as shown below.

$$\chi^2 = \overset{\substack{\textit{Total number of cells} \\ \textit{in the two-way table}}}{\underset{i=1}{\sum}} \frac{(observed - expected)^2}{expected} = \overset{\substack{\textit{Total number of cells} \\ \textit{in the two-way table}}}{\underset{i=1}{\sum}} \frac{(O_i - E_i)^2}{E}$$

$$+ \frac{(114 - 90.18)^2}{90.18} + \frac{(115 - 121.57)^2}{121.57} + \frac{(22 - 39.3)^2}{39.3}$$

$$+ \frac{(40 - 32.68)^2}{32.68} + \frac{(39 - 44.08)^2}{44.08} + \frac{(12 - 14.29)^2}{14.29}$$

$$= 19.76 + 2.04 + 17.82 + 6.32 + 0.36 + 7.62 + 1.64 + 0.58 + 0.35 = 56.51$$

The terms being summed together, one from each cell, are often called the *cell contributions*. They are often useful in identifying where cells have the largest discrepancies between the observed and expected counts.

In order to use the chi-square statistic to estimate a p-value you also need to know the *degrees of freedom* of the test. The degrees of freedom are computed by multiplying the number of categories in the explanatory variable minus 1 by the number of categories in the response variable minus 1. It turns out that the mean of this distribution equals the degrees of freedom. We expect larger values in the distribution as we are summing the results for more cells together.

In our case, that is $(3 - 1) \times (3 - 1) = 4$. Thus, we have a 4 degree of freedom test.

In general, the chi-square distribution changes as the degrees of freedom increases. Notice in Figure 8.9 that the distribution is often right skewed.

Figure 8.9. Chi-square distributions with various degrees of freedom (k)

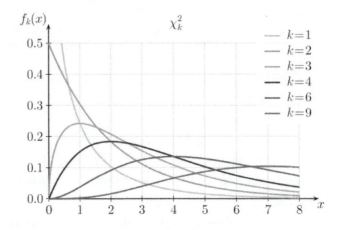

Section 8.2 Summary

A theory-based approach can be used to test association between two categorical variables under certain conditions.

- This procedure is called a **chi-square test**.
- The statistic for theory-based inference is called a **chi-square statistic**.

- With a binary response variable, this statistic is: $\sum\limits_{i\,groups}\left(\dfrac{\hat{p}_i - \hat{p}}{\sqrt{\hat{p}(1-\hat{p})/n_i}}\right)^2$

  More generally, this statistic is $\sum\limits_{all\,cells}\dfrac{(observed - expected)^2}{expected}$, where 'observed' refers to the observed cell counts and 'expected' refers to the expected cell counts, computed assuming the null hypothesis to be true. To find the expected cell counts multiply the row total times the column total and divide by the column total.

    o This chi-square statistic becomes larger as the discrepancies in the sample proportions become larger.
    o Larger values of the chi-square statistic provide stronger evidence against the null hypothesis.
    o Software produces the contribution to the chi-square statistic for each cell of the table.
      - Cells with larger contributions indicate the greatest discrepancy between what was observed and what would have been expected if there were no association between variables.
- The validity condition for applying the chi-square test is that all cells of the table should each have at least 10 observations.

When a chi-square test reveals strong evidence that at least one of the population proportions differs from the others, confidence intervals for the difference in population proportions can be calculated for pairs of groups.
- Confidence intervals that do not include zero indicate which pairs of sample proportions differ significantly.

Chapter 8 Summary

In this chapter we saw that the 3S strategy for simulation-based inference can be applied to comparing proportions across multiple groups. Neither the explanatory nor response variable has to be binary. This approach provides an overall test of the multiple groups simultaneously, as opposed to testing each pair of groups separately. (Something you will see again in Chapter 9 when we compare multiple means.)

Because comparing more than two groups is a bit more complicated than just two, the choice of a statistic to measure how different the groups are is not straightforward. We saw that a reasonable statistic to use is the MAD or the mean of the absolute differences. Using this statistic, it is easy to see that the larger values of the MAD statistic indicate a greater difference in sample proportions across the groups. (Again, something you will see being repeated in Chapter 9 with multiple means.)

The Max – Min and MAD statistics are fairly easy to understand but are not widely used. A more commonly used statistic is called a chi-square (χ^2) statistic. We also saw that the chi-square statistic has some similar properties to that of the MAD statistic. This statistic is used as the basis for the theory-based test, appropriately called the chi-square test.

When a chi-square test reveals strong evidence that at least one of the population proportions or long-run probabilities differs from the others, confidence intervals for these differences can be calculated for pairs of groups. Confidence intervals that do not include zero indicate which pairs of sample proportions differ significantly. (This again, is something we will see repeated in Chapter 9 when we compare multiple means.)

Type of data	Name of test	Applet	Validity conditions
A categorical explanatory variable with 2 or more categories and a categorical response variable with 2 or more categories	Simulation test for multiple proportions	Multiple proportions	---
	Theory-based test for multiple proportions (chi-square test/statistic)	Multiple proportions	At least 10 successes and failures in each of the groups

Chapter 8 Glossary

cell contributions

Contribution of cell in 2-way table to the chi-square statistic. Helpful in determining where large differences from *observed data* to what would be expected if the null hypothesis were true 836

chi-square distribution

A non-negative, right-skewed distribution used in theory-based test for an association between two categorical variables 819

chi-square statistic

A standardized statistic for summarizing the relationship between two categorical variables which has a predictable null distribution, making it a popular theory-based test 796

expected counts

The predicted outcomes for the cells of the two-way table assuming the null hypothesis is true. Found by applying the row proportions to the column totals 839

MAD

A for testing an association between two categorical variables of more than two categories. (M)ean of the (A)bsolute values of the (D)ifferences in the conditional proportions 796

observed counts

The values in the cells of the two-way table from the study 835

validity conditions for chi-square test

Each cell of the two-way table must have at least 10 observations 824

Exercises

Section 8.1

8.1.1* Which of the following are true about the MAD statistic? There may be more than one statement that is true.

    A. The MAD statistic can never be negative.

    B. The MAD statistic is the average distance sample proportions are from each other.

    C. The MAD statistic is the average of a group of sample proportions.

    D. A MAD statistic of zero means that all the proportions are the same distance apart.

    E. As the MAD statistic increases, the p-value decreases.

8.1.2 As observed proportions move farther apart:

    A. Both the MAD statistic and the p-value will decrease.

    B. Both the MAD statistic and the p-value will increase.

    C. The MAD statistic will increase and the p-value will decrease.

    D. The MAD statistic will decrease and the p-value will increase.

8.1.3* Null distributions using the MAD statistic will be:

    A. Symmetric and centered at zero.

    B. Symmetric and centered at the MAD statistic.

    C. Skewed right.

    D. Skewed left.

8.1.4 What has to be true about the observed proportions if the MAD statistic is zero?

8.1.5* Given the three sample proportions 0.25, 0.30, and 0.35, determine the MAD statistic.

8.1.6 Given the four sample proportions 0.66, 0.45, 0.73, and 0.44, determine the MAD statistic.

8.1.7* Consider the MAD statistic. What would happen if you did not take absolute values before calculating the sum of differences between group proportions? Why would this not be a useful calculation without taking absolute values?

New law

8.1.8 Suppose a poll is done to see if voters agree with a new law passed in Congress. The results are summarized in the following table along with the voters' party affiliation. You plan to test to see if there is an association between party affiliation and agreement with the new law.

		Republican	Democrat	Independent	Total
Agree with a	**Yes**	125	201	178	504
new law?	**No**	233	156	184	573
	Total	358	357	362	1077

    a) Determine the proportion of each political group that agrees with the new law.
    b) Determine the MAD statistic.

Giving blood and political party*

8.1.9 Suppose that you want to compare proportions of registered Democrats, Independents, and Republicans who have donated blood within the past year.

a) Which would be appropriate for this study: random sampling, random assignment, both, or neither? Explain.

b) If you find that the sample proportions who have donated blood are 0.20, 0.15, and 0.12, what more information would you need to determine if these proportions differ significantly?

Giving blood and state

8.1.10 Suppose you want to compare proportions of adults who have donated blood within the past year among California, Iowa, Massachusetts, and Michigan residents. You take a random sample of 200 adults from each state.

a) Which of the following sets of sample proportions (A, B, or C) would produce the *smallest* p-value? Explain.

b) Which of the following sets of sample proportions (A, B, or C) would produce the *largest* p-value? Explain.

	California roportion	Iowa roportion	Massachusetts roportion	Michigan oportion
A	0.30	0.25	0.28	0.22
B	0.40	0.15	0.18	0.37
C	0.35	0.20	0.23	0.30

Gender and phone type*

8.1.11 Suppose 100 students are asked if the cellphone they use is a smartphone or a basic phone and the following two-way table summarizes the results.

	Male	Female	Total
Smartphone	25	30	45
Basic Phone	25	20	55
Total	50	50	100

a) Find the proportion of males that own smartphones minus the proportion of females that own smartphones.

b) Determine the MAD statistic.

c) When comparing two proportions, you could use the difference in the proportion or the MAD as the statistic. How so these two statistics compare in this example? Will they always have the same relationship as in this example? Explain.

d) Put the data from the table, in the **Multiple Proportions** applet and make null distributions using both the difference in proportions and the MAD statistic. Describe how these distributions are different.

e) Based on your answer from part (c), explain why a distribution of MAD statistics is skewed right when the distribution of differences in proportions is bell-shaped.

Generation and marriage views

8.1.12 Do different generations view marriage differently? A 2010 survey of a random sample of adult Americans conducted by the Pew Research Center asked the following question of each participant, "*Is marriage becoming obsolete?*" Here is how the data from Pew Research Center's study turned out:

		Generation				
		Millennial (ages 18-29)	Gen X (ages 30-45)	Boomers (ages 46-65)	Age 65+	Total
Marriage obsolete?	Yes	236	313	401	68	1018
	No	300	416	745	143	1604
Total		536	729	1146	211	2622

a) Find and report appropriate numerical statistics to describe the association, if any, between generation and opinion about marriage.

b) Calculate the MAD statistic for this data.

c) Create a well-labeled graph to display the above data. Also, describe in 1-2 sentences what this graph tells you about the association between generations and whether marriage is viewed as becoming obsolete. Does it appear that generations tend to view marriage differently, and if so, how?

8.1.13 Reconsider the previous question on how different generations view marriage. The results are shown in the table in the previous exercise and can also found in the file **MarriageViews.**

		Generation				
		Millennial (ages 18-29)	Gen X (ages 30-45)	Boomers (ages 46-65)	Age 65+	Total
Marriage obsolete?	Yes	236	313	401	68	1018
	No	300	416	745	143	1604
Total		536	729	1146	211	2622

a) Identify the observational units.
b) Identify the variables recorded. For each variable, identify the role (explanatory or response), and the type (categorical or quantitative).
c) Give two ways in which this study is different from the "*Is yawning contagious?*" study which was introduced in **Exploration 5.2**.
d) Describe the parameter(s) of interest, and assign appropriate symbol(s) to the parameter(s).
e) State in words the appropriate null and alternative hypotheses in the context of the study.

f) State in symbols the appropriate null and alternative hypotheses in the context of the study.

g) Use a simulation-based approach to test the hypotheses stated in part (e) and (f). Use the **Multiple Proportions** applet and paste the table located in the file **MarriageViews** and press **Use My Table**.

Be sure to report the following:

- The name and the observed value of the statistic (from the study)
- The p-value
- Your conclusion in the context of the study. (Be sure to comment on significance, causation, and generalization.)

8.1.14 Reconsider the previous exercises. If the different generations *did not* view marriage differently what would you expect the cell counts in the following table to be? (I.e. There is no association between the age groups and whether or not someone thinks marriage is obsolete.) Complete the table.

		Generation				Total
		Millennial (ages 18-29)	**Gen X (ages 30-45)**	**Boomers (ages 46-65)**	**Age 65+**	
Marriage obsolete?	**Yes**					1018
	No					1604
	Total	536	729	1146	211	2622

Heart disease and baldness*

8.1.15 Many studies have been done to look at the relationship between heart disease and baldness. In one study, researchers selected a sample of 663 heart disease male patients and a control group of 772 male patients not suffering from heart disease from hospitals in eastern Massachusetts and Rhode Island. Each was asked to classify their degree of baldness. The results are given in the following table. Use these data to answer the following questions.

		Baldness				
		None	**Little**	**Some**	**Much**	**Total**
Heart	**Yes**	251	165	195	52	663
Disease	**No**	331	221	185	35	772
	Total	582	386	380	87	1435

Notice the degree of baldness uses ordered categories. This makes it possible to reduce the 2x4 table to a simpler one:

I. Of those in the control group, what percent claimed to have:
   a) Little or no baldness?
   b) Some or much baldness?

II. Of those with heart disease, what percent claimed to have:
   a) Little or no baldness?
   b) Some or much baldness?

At this stage in the investigation, state whether you think there is a relationship between heart disease and baldness. Explain your answer.

We are now going to run a test of significance on these data using all four degrees of baldness. Use the data set **HeartDiseaseAndBaldness**.

III. State the null and alternative hypotheses you would use to test whether there is an association between heart disease and baldness.

IV. Use a simulation-based approach to test the hypotheses stated in (c).
   Be sure to report the following:
   - The name and the observed value of the statistic (from the study)
   - The shape of the null distribution and the p-value
   - Your conclusion in the context of the study. (Be sure to comment on significance, causation, and generalization.)

Note: Save your results. You will need them for exercise 8.CE.20

Night lights and vision

8.1.16 In **Exploration 8.2b**, we looked at a study that was done to determine if there is an association between the amount of light in infants' bedrooms at night (darkness, nightlight, room light) and whether or not the child was near-sighted when they were older. A summary of the data from that study is shown in the following table. We are interested in determining if there is an association between the type of light in the room and whether or not the child is near-sighted.

	Darkness	Nightlight	Room light	Total
Near-sighted	18	78	41	137
Normal vision	154	154	34	342
Total	172	232	75	479

a) What is the explanatory variable in this study? Is it categorical or quantitative?

b) What is the response variable in this study? Is it categorical or quantitative?

c) Write out the null and alternative hypotheses for this study.

d) Compute the observed proportions of children that are near-sighted for each lighting condition.

e) Compute the MAD statistic.

f) A null distribution was generated for this data set and is shown. Based on this distribution along with the MAD statistic, what can you say about the value of the p-value?

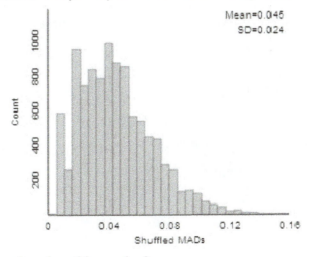

g) What is your conclusion based on this p-value?

Political jokes*

8.1.17 Researchers at the University of Pennsylvania studied the humor of late-night comedians Jon Stewart, Jay Leno, and David Letterman. Between July 15 and September 16, 2004, they performed a content analysis of the jokes made by Jon Stewart during the "headlines" segment of *The Daily Show* and by Jay Leno and David Letterman during the monologue segments of their shows. (We will consider this a random sample of all their jokes.) We want to determine if there is an association between the comedian and the political nature of their jokes. A summary of the data they gathered was:

- Leno: 315 of the 1313 jokes were of a political nature
- Letterman: 136 of the 648 jokes were of a political nature
- Stewart: 83 of the 252 jokes were of a political nature

a) Organize the data into a two-way table, with the explanatory variable as the columns and the response variable as the rows.
b) What is the explanatory variable in this study? Is it categorical or quantitative?
c) What is the response variable in this study? Is it categorical or quantitative?
d) Write out the null and alternative hypotheses for this study.
e) For each comedian, calculate the conditional proportion of his jokes that were political in nature.
f) Compute the MAD statistic.
g) A null distribution was generated for this data set and is shown. Based on this distribution along with the MAD statistic, what can you say about the value of the p-value?

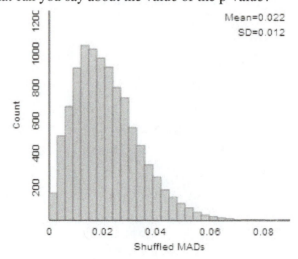

h) What is your conclusion based on this p-value?

Contagious yawning

8.1.18 Student researchers wanted to see if it there was an association between the time of day and whether or not someone would yawn right after seeing another person yawn. To gather their data, they walked around campus at various times of day yawning at people and watched to see if the subjects yawned. The results are shown in the following table.

		Time of Day				
		8-11a	**11a-2p**	**2-5p**	**5-8p**	**Total**
Subject	**Yes**	25	38	15	29	107
Yawned?	**No**	14	28	18	15	75
	Total	39	66	33	44	182

a)   What is the explanatory variable in this study? Is it categorical or quantitative?

b)   What is the response variable in this study? Is it categorical or quantitative?

c)   Write out the null and alternative hypotheses for this study.

d)   What time period did the highest proportion of the subjects yawn?

e)   What time period did the lowest proportion of subjects yawn?

f)   The table for these results can be found in the file **YawnTime.** Put that table in the **Multiple Proportions** applet (make sure you click on **Use Table**). What is the value of the MAD statistic?

g)   Do at least 1000 repetitions and find a p-value.

h)   Based on your p-value, do you have strong evidence that there is an association between the time of day and whether or not someone will yawn after they see a yawn?

8.1.19 Reconsider the previous exercise about contagious yawning and time of day. The results of this test are not significant. If this study were redone, what suggestions do you have that would help make the results significant?

Stop and frisk*

8.1.20 In a February 2013 Quinnipiac University poll the following question about a typical practice of the New York Police Department, was asked to a random sample of New York City voters. *"As you may know, there is a police practice known as stop and frisk, where police stop and question a person they suspect of wrongdoing and, if necessary search that person. Do you approve or disapprove of this police practice?"* The results from the poll are shown in the following table by age groups.

	Age			
	18-34	**35-54**	**55+**	**Total**
Approve	68	128	173	369
Disapprove	164	204	165	533
Total	232	332	338	902

a) What proportion in each age category approves of stop and frisk?

b) Does there seem to be an association with age and whether or not someone approves of stop and frisk? If so, describe the association.

c) Write out the null and alternative hypotheses for this study.

d) The table for these results can be found in the file **FriskAge.** Put that table in the **Multiple Proportions** applet (make sure you click on **Use Table**). What is the value of the MAD statistic?

e) Do at least 1000 repetitions and find a p-value.

f) Based on your p-value, do you have strong evidence that there is an association between age and whether or not someone approves of the stop and frisk practice?

8.1.21 Reconsider the previous problem on approval of stop and frisk. Complete a test of significance to determine if there is an association between age category and whether or not someone approves or disapproves of the police practice known as stop and frisk using the maximum proportion – minimum proportion as the statistic. Make sure you given the hypotheses, your p-value, and a conclusion.

Health care reform

8.1.22 In a January 2014 Quinnipiac University poll the following question was asked to a random sample of U.S. voters. *"Do you support or oppose the health care law passed by Barack Obama and Congress in 2010?"* The results from the poll are shown in the following table by age groups.

	Age				
	18-29	**30-49**	**50-64**	**65+**	**Total**
Support	128	236	213	153	730
Oppose	122	367	331	247	1067
Total	250	603	544	400	1797

a) What proportion in each age category supports the health care law?

b) Which age category is most supportive of the health care law? Which age category is the least supportive?

c) Write out the null and alternative hypotheses for this study.

d) The table for these results can be found in the file **HealthcareAge.** Put that table in the **Multiple Proportions** applet (make sure you click on **Use Table**). What is the value of the MAD statistic?

e) Do at least 1000 repetitions and find a p-value.

f) Based on your p-value, how much evidence is there against the null hypothesis?

g) Comment on the ability to generalize and draw cause-and-effect conclusions from this data.

Cellphone ownership*

8.1.23 In a September 2013 poll conducted by the Pew Research Center, they asked a random sample of U.S. adults if they owned a cellphone. The results are shown in the following table by race.

		White	**Black**	**Hispanic**	**Total**
Own a	**Yes**	3801	611	614	5026
cellphone?	**No**	422	53	68	543
	Total	4223	664	682	5569

a) What proportion in each race category owns a cellphone?

b) Does there seem to be an association between race and whether or not someone owns a cellphone?

c) Write out the null and alternative hypotheses for this study.

d) The table for these results can be found in the file **CellphoneRace.** Put that table in the **Multiple Proportions** applet (make sure you click on **Use Table**). What is the value of the MAD statistic?

e) Do at least 1000 repetitions and find a p-value.

f) Based on your p-value, do you have strong evidence that there is an association between race and whether or not someone owns a cellphone?

8.1.24 Reconsider the previous problem on cellphone use. Complete a test of significance to determine if there is an association between race and whether or not someone owns a cellphone using the maximum proportion – minimum proportion as the statistic. Make sure you given the hypotheses, your p-value, and a conclusion.

8.1.25 In a September 2013 poll conducted by the Pew Research Center, they asked a random sample of U.S. adults if they owned a smart phone. The results are shown in the following table by race.

		White	Black	Hispanic	Total
Own a	**Yes**	2238	372	382	2992
smartphone?	**No**	1985	292	300	2577
	Total	4223	664	682	5569

a) What proportion in each race category owns a smartphone?

b) Does there seem to be an association between race and whether or not someone owns a smartphone?

c) Write out the null and alternative hypotheses for this study.

d) The table for these results can be found in the file **SmartphoneRace.** Put that table in the **Multiple Proportions** applet (make sure you click on **Use Table**). What is the value of the MAD statistic?

e) Do at least 1000 repetitions and find a p-value.

f) Based on your p-value, write a complete conclusion. Be sure to include to whom you can generalize the result.

8.1.26 Reconsider **EXERCISES 8.1.23** and **8.1.25** on cellphone and smartphone ownership. We considered our population to be that of all adults in the US. Now suppose we want the population to be all the cellphone users in the US.

    a) Combine the results in the tables from these two exercises so you have a new two-way table where race as the explanatory variable and the response variable is whether or not someone owns a smartphone given that they own a cellphone.

    b) What proportion of the cellphone owners of each race own smartphones?

Political party survey

8.1.27 Suppose two surveys are done to see if Democrats, Independents, and Republicans differ on their agreement with a certain issue and the results are shown in the following two tables.

Survey A

	Dem	Ind	Rep	Total
Agree	30	25	20	75
Disagree	70	75	80	225
	100	100	100	300

Survey B

	Dem	Ind	Rep	Total
Agree	300	250	200	750
Disagree	700	750	800	2250
	1000	1000	1000	3000

    a) Input the table for Survey A in the **Multiple Proportions** applet and report the MAD statistic.

    b) Do at least 1000 shuffles to find a null distribution for Survey A. Report your p-value and standard deviation of the null distribution.

    c) Input the table for Survey B in the **Multiple Proportions** applet and report the MAD statistic.

    d) Do at least 1000 shuffles to find a null distribution for Survey B. Report your p-value and standard deviation of the null distribution.

    e) You should have found the MAD statistics are the same for both surveys. Explain why they are the same.

    f) You should have also found the p-value from Survey B was much smaller than that for Survey A. Explain why this makes sense intuitively.

    g) Even though the MAD statistics are the same, explain why the p-value for Survey B is smaller based on the two null distributions.

Coming to a stop*

8.1.28 In **Example 8.1** you used the MAD statistic to evaluate the strength of evidence for association between the order in which a car arrived at an intersection and whether or not it came to a complete stop.

   a) Complete a test of significance to determine if there is an association between order of car arrival and whether or not it came to a complete stop using the maximum proportion – minimum proportion as the statistic. Make sure you given the hypotheses, your p-value, and a conclusion.
   b) Compare and contrast the null distribution you found with the max-min statistic vs. the MAD statistic.
   c) Compare and contrast the strength of evidence using Max-Min vs. MAD on this dataset

Recruiting organ donors

8.1.29 In **Exploration 8.1** you used the MAD statistic to evaluate the strength of evidence for association between the way a question was asked and whether or not someone chose to be an organ donor.

   a) Complete a test of significance to determine if there is an association between survey type arrival and whether or not someone chose to become an organ donor using the maximum proportion – minimum proportion as the statistic. Make sure you given the hypotheses, your p-value, and a conclusion.
   b) Compare and contrast the null distribution you found with the max-min statistic vs. the MAD statistic.
   c) Compare and contrast the strength of evidence using Max-Min vs. MAD on this dataset

Section 8.2

8.2.1* Consider the chi-square statistic.
   a) Can the value of this statistic be negative? Why or why not?
   b) When will the chi-square statistic be 0?
   c) Which of (A) – (C) describes the shape of the distribution of the chi-square statistic:
      (A) Symmetric
      (B) Skewed toward high values (long right tail)

(C) Skewed toward low values (long left tail)

d) Give an intuitive explanation of why you expect the shape of distribution of the chi-square statistic to be what you described in (c).

HIV intervention programs

Problems 8.2.2 through 8.2.5 use data from the following study on HIV.

8.2.2 Researchers Jemmott III et al. (*Journal of American Medical Association*, 1998) reported a study in which a group of inner-city African American adolescents from Philadelphia were randomly divided into three groups as part of an experiment to assess the effectiveness of different HIV prevention programs. One group was given "abstinence HIV intervention," the next group was given "safer-sex HIV intervention," and the third group was given "health promotion intervention," which was to serve as a control. One outcome that was measured was whether subjects who were sexually active during a three-month period reported consistent condom use. Here is the table of the summarized data:

		Abstinence Intervention	Safer-Sex Intervention	Control	Total
Consistent	**Yes**	14	20	21	**55**
Condom Use?	**No**	20	12	20	**52**
	Total	**34**	**32**	**41**	**107**

a) Is this an experiment or an observational study? How are you deciding?
b) Identify the observational units.
c) Identify the explanatory and response variable. Also, for each variable, identify whether categorical or quantitative.
d) Define the parameters of interest. Assign symbols to these parameters.
e) State the appropriate null and the alternative hypotheses.
f) Find the table of expected cell counts; that is, complete the table below, assuming the null hypothesis is true.

		Abstinence Intervention	Safer-Sex Intervention	Control	Total
Consistent	Yes				55
Condom Use?	No				52
	Total	34	32	41	107

g) Describe how one might use everyday items (for example, coins, dice, cards, etc.) to conduct a tactile simulation-based test of the hypotheses. Be sure to clearly describe how the p-value will be computed from the simulation.

8.2.3 Refer to the previous exercise about the study of different HIV intervention programs.
   a) State the appropriate null and the alternative hypotheses *in words*.
   b) Explain why it would be okay to use the theory-based method to find a p-value for this study.
   c) Use an appropriate applet (for example, the **Multiple Proportions** applet) to find and report the following from the data.
      •   The name of the statistic and its numeric value.
      •   The theory-based p-value.
   d) Comment on the strength of evidence based on the p-value.
   e) Comment on Causation and **Generalization**. Be sure to explain how you are arriving at your conclusion.
   f) Explain why it is not particularly useful to look at confidence intervals for the parameter(s) of interest in this study.

8.2.4 Refer to the previous exercise about the study of different HIV intervention programs. Show how to calculate the chi-square statistic by setting up the formula with appropriate numbers.

8.2.5 Refer to the previous exercise about the study of different HIV intervention programs. Suppose that the observed cell counts were actually twice as big as those reported in the original study.

For parts (a) – (c): how, if at all, would the values of the following change?
   a) The expected cell counts: INCREASE   DECREASE   STAY THE SAME
   b) The chi-square statistic: INCREASE   DECREASE   STAY THE SAME
   c) The p-value: INCREASE   DECREASE   STAY THE SAME

   d) Would the evidence in support of an association between type of HIV intervention program and consistent condom usage among adolescents such as the ones in the study be stronger or weaker compared to the original study? Explain how you are deciding.

Views of marriage*

8.2.6 Recall exercise 8.1.12 about whether different generations view marriage differently, and that the corresponding data are available in the file **MarriageViews**.

a) State the appropriate null and the alternative hypotheses *in words*.

b) Explain why it would be okay to use the theory-based method to find a p-value for this study.

c) Use an appropriate applet (for example, the **Multiple Proportions** applet) to find and report the following from the data.
   - The name of the statistic and its numeric value.
   - The theory-based p-value.

d) Comment on the strength of evidence based on the p-value.

e) Comment on Causation and Generalization. Be sure to explain how you are arriving at your conclusion.

f) Use an applet to find and report a set of confidence intervals appropriate for the context. Also, interpret the confidence intervals in the context of the study.

Sleep apnea and hypertension

Problems 8.2.7 through 8.2.10 use the the following data on sleep apnea.

8.2.7 In a study of the relationship between sleep apnea and hypertension (*New England Journal of Medicine*, 2000), researchers Peppard et al. categorized the participants by the severity of sleep apnea at the beginning of the study, and four years later recorded whether or not the participant had hypertension. Here is the table of the summarized data:

		\multicolumn{4}{c}{Severity of sleep apnea}			
		Not	**Somewhat**	**Severe**	**Very**	**Total**
Severe	**Yes**	32	142	63	40	**277**
Hypertension?	**No**	155	365	69	27	**616**
	Total	**187**	**507**	**132**	**67**	**893**

a) Is this an experiment or an observational study? How are you deciding?

b) Identify the observational units.

c) Identify the explanatory and response variable. Also, for each variable, identify whether categorical or quantitative.

d) Define the parameters of interest. Assign symbols to these parameters.
e) State the appropriate null and the alternative hypotheses in symbols.
f) Find the table of expected cell counts; that is, complete the table below, assuming the null hypothesis were true.

		Severity of sleep apnea				
		Not	Somewhat	Severe	Very	Total
Severe	**Yes**					**277**
Hypertension?	**No**					**616**
	Total	**187**	**507**	**132**	**67**	**893**

g) Describe how one might use everyday items (for example, coins, dice, cards, etc.) to conduct a tactile simulation-based test of the hypotheses. Be sure to clearly describe how the p-value will be computed from the simulation.

8.2.8 Refer to the previous exercise about the study of the relationship between sleep apnea and hypertension.
a) State the appropriate null and the alternative hypotheses *in words*.
b) Explain why it would be okay to use the theory-based method to find a p-value for this study.
c) Use an appropriate applet (for example, the **Multiple Proportions** applet) to find and report the following from the data.
    i.   The name of the statistic and its numeric value.
    ii. The theory-based p-value.
d) Interpret the p-value in the context of the study.
e) State a complete conclusion about this study including Significance, Causation and Generalization. Be sure to explain how you are arriving at your conclusion.
f) If relevant, use an applet to find and report a set of confidence intervals appropriate for the context. Also, interpret the confidence intervals in the context of the study. If not relevant, explain why it is not particularly useful to look at confidence intervals for the parameter(s) of interest in the context of this study.

8.2.9 Refer to the previous exercise about the study of the relationship between sleep apnea and hypertension. Show how to calculate the chi-square statistic by setting up the formula with appropriate numbers.

8.2.10 Refer to the previous exercise about the study of the relationship between sleep apnea and hypertension. Suppose that the observed cell counts were actually three times as big as those reported in the original study.

For parts (a) – (c): how, if at all, would the values of the following change?
  a) The expected cell counts: INCREASE   DECREASE   STAY THE SAME
  b) The p-value: INCREASE   DECREASE   STAY THE SAME
  c) The chi-square statistic: INCREASE   DECREASE   STAY THE SAME

  d) Would the evidence in support of an association between sleep apnea and hypertension for people such as the ones in the study be stronger or weaker compared to the original study? Explain how you are deciding.

Heart Disease and Baldness?*

Problems 8.2.11 and 8.2.12 use the heart disease and baldness data from Exercise 8.1.15.

  a) Identify the observational units.
  b) Describe the parameter(s) of interest, and assign appropriate symbol(s) to the parameter(s).
  c) State in words the appropriate null and alternative hypotheses in the context of the study.
  d) State in symbols the appropriate null and alternative hypotheses in the context of the study.
  e) Use a theory-based approach to test your hypotheses from the previous question. Report the p-value, and state your complete conclusion. Discuss your scope of conclusions: whether your results necessarily mean that heart disease is caused by baldness or baldness is caused by heart disease & whether you can extend your conclusions to a broader population.

8.2.12 Refer to the previous exercise about the relationship between heart disease and baldness. The data are provided in the file **HeartDiseaseAndBaldness**.
  a) Compute the conditional proportions of heart disease given each degree of baldness.
  b) There are six pairwise differences. Which two are the largest? Which one is the smallest?
  c) Use the applet for Theory-Based Inference to compute p-values and confidence intervals for the two largest differences and the smallest difference.
  d) Are the results from the theory-based approach valid for the data in this study? State the validity conditions, and explain how you are deciding whether or not these conditions are met.

Physical activity and aging

Problems 8.2.13 through 8.2.16 refer to the following study on physical activity and aging.

8.2.13 An eight-year cohort study of 3,454 initially disease-free men and women (ages around 60 years or more) recorded the extent of physical activity the participants engaged and whether or not they were showing signs of "health aging" (*British Journal of Sports Medicine,* 2013). Here is the table of summarized counts at the conclusion of the study:

		Activity level			
		Inactive	Moderate	Vigorous	Total
Healthy aging?	Yes	55	345	265	665
	No	598	1347	844	2789
	Total	653	1692	1109	3454

a) Is this an experiment or an observational study? How are you deciding?

b) Identify the observational units.

c) Identify the explanatory and response variable. Also, for each variable, identify whether categorical or quantitative.

d) State the appropriate null and the alternative hypotheses.

e) Describe how one might use everyday items (for example, coins, dice, cards, etc.) to conduct a tactile simulation-based test of the hypotheses. Be sure to clearly describe how the p-value will be computed from the simulation.

8.2.14 Refer to the previous exercise about the study of the relationship between level of physical activity and aging.

a) Explain why it would be okay to use the theory-based method to find a p-value for this study.

b) Use an appropriate applet (for example, the **Multiple Proportions** applet) to find and report the following from the data: (i) The name of the statistic and its numeric value (ii) The theory-based p-value.

c) Interpret the p-value in the context of the study.

d) State a complete conclusion about this study including Significance, Causation and Generalization. Be sure to explain how you are arriving at your conclusion.

e) If relevant, use an applet to find and report a set of confidence intervals appropriate for the context. Also, interpret the confidence intervals in the context of the study. If not relevant, explain why it is not particularly useful to look at confidence intervals for the parameter(s) of interest in the context of this study.

8.2.15 Refer to the previous exercise about the study of the relationship between level of physical activity and aging. Show how to calculate the chi-square statistic by setting up the formula with appropriate numbers.

8.2.16 Refer to the previous exercise about the study of the relationship between level of physical activity and aging. Suppose that the observed cell counts were actually half as small as those reported in the original study.

    a) Without carrying out any extensive calculations, say whether the p-value would increase, decrease, or stay the same. Explain your reasoning.

    b) Without carrying out any extensive calculations, say what the value of the chi-square statistic would be. Explain your reasoning.

    c) Would the evidence in support of an association between physical activity and healthy aging for people such as the ones in the study be stronger or weaker compared to the original study? Explain how you are deciding.

Politics and taxes*

8.2.17 Given below is a table of summarized data from a 2013 study of a nationally representative sample of U.S. adults by the Gallup organization, where participants were asked about their political leaning as well as their opinion on whether there should be a tax on unhealthy food/soda.

		Political leaning			
		Republican	**Democrat**	**Independent**	**Total**
Favor tax on unhealthy	**Yes**	127	273	246	**646**
food/soda?	**No**	403	334	500	**1237**
	Total	**530**	**607**	**746**	**1883**

    a) Is this an experiment or an observational study? How are you deciding?

    b) Identify the observational units.

    c) Identify the explanatory and response variable. Also, for each variable, identify whether categorical or quantitative.

    d) Define the parameters of interest. Assign symbols to these parameters.

    e) State the appropriate null and the alternative hypotheses in symbols.

    f) State the appropriate null and the alternative hypotheses *in words*.

    g) Explain why it would be okay to use the theory-based method to find a p-value for this study.

    h) Use an appropriate applet (for example, the **Multiple Proportions** applet) to find and report the following from the data. (i) The name of the statistic and its numeric value. (ii) The theory-based p-value.

    i) Interpret the p-value in the context of the study.

    j) State a complete conclusion about this study including Significance, Causation and Generalization. Be sure to explain how you are arriving at your conclusion.

k) If relevant, use an applet to find and report a set of confidence intervals appropriate for the context. Also, interpret the confidence intervals in the context of the study. If not relevant, explain why it is not particularly useful to look at confidence intervals for the parameter(s) of interest in the context of this study.

Fish and cancer

8.2.18 In an article published in the Lancet (2001), researchers shared their findings from a study where they followed 6272 Swedish men for 30 years to see whether there was an association between the amount of fish in the diet and likelihood of prostate cancer. The results are presented in the following two-way table:

		Large	**Moderate**	**Small**	**None**	**Total**
Prostate	**Yes**	42	209	201	14	466
Cancer?	**No**	507	2769	2420	110	5806
	Total	549	2978	2621	124	6272

a) Is this an experiment or an observational study? How are you deciding?
b) Identify the observational units.
c) Identify the explanatory and response variable. Also, for each variable, identify whether categorical or quantitative.
d) Define the parameters of interest. Assign symbols to these parameters.
e) State the appropriate null and the alternative hypotheses in symbols.
f) State the appropriate null and the alternative hypotheses *in words*.
g) Explain why it would be okay to use the theory-based method to find a p-value for this study.
h) Use an appropriate applet (for example, the **Multiple Proportions** applet) to find and report the following from the data. (i) The name of the statistic and its numeric value. (ii) The theory-based p-value.
i) Interpret the p-value in the context of the study.
j) State a complete conclusion about this study including Significance, Causation and Generalization. Be sure to explain how you are arriving at your conclusion.
k) If relevant, use an applet to find and report a set of confidence intervals appropriate for the context. Also, interpret the confidence intervals in the context of the study. If not relevant, explain why it is not particularly useful to look at confidence intervals for the parameter(s) of interest in the context of this study.

Gender and breakfast*

8.2.19 In a survey of 50 students in an introductory statistics students, data were collected on gender and breakfast eating habits. Here is a table of the summarized data:

		Females	**Males**	**Total**
Eat breakfast?	**Every morning**	21	5	26
	Most mornings	7	4	11
	Some mornings	7	3	10
	Never	2	1	3
	Total	37	13	50

a) Is this an experiment or an observational study? How are you deciding?

b) Identify the observational units.

c) Identify the explanatory and response variable. Also, for each variable, identify whether categorical or quantitative.

d) State the appropriate null and the alternative hypotheses *in words*.

e) Explain why it would NOT be okay to use the theory-based method to find a p-value for this study.

f) Describe how one might use everyday items (for example, coins, dice, cards, etc.) to conduct a tactile simulation-based test of the hypotheses. Be sure to clearly describe how the p-value will be computed from the simulation.

g) Use an appropriate applet (for example, the **Multiple Proportions** applet) to find and report the following from the data: The name of a statistic and its numeric value and the simulation-based p-value.

h) Interpret the p-value in the context of the study.

i) State a complete conclusion about this study including Significance, Causation and Generalization. Be sure to explain how you are arriving at your conclusion.

Design your own study

8.2.20 Design your own study: Develop the plan for the investigation of a research question that can be answered using a "chi-square test."

a) State the research question.

b) Will you conduct an experiment or an observational study? How are you deciding? Also, describe how the randomization (random sampling, random assignment, or both) will be carried out.

c) Identify the observational units.

d) Identify the response and explanatory variables. Also, identify whether each is categorical or quantitative.

e) State the appropriate null and alternative hypotheses, in words.

f) Describe when it will be appropriate to use a theory-based method to test the hypotheses stated above.

Cocaine treatment*

8.2.21 Researchers Gawin et al. (*Journal of American Medical Association Psychiatry,* 1989) reported a study where each of 72 cocaine abusers were randomly assigned to receive one of three treatments for six-weeks: 24 subjects received an antidepressant drug desipramine hydrochloride, 24 subjects received lithium carbonate (the usual treatment), and the remaining 24 received a placebo. Forty-one percent of the desipramine group had a relapse during the six-week period, compared to 75% for the lithium carbonate and 83% for the placebo.

a) Use the above information to create a well-labeled table of observed cell counts.

b) Is this an experiment or an observational study? How are you deciding?

c) Identify the observational units.

d) Identify the explanatory and response variable. Also, for each variable, identify whether categorical or quantitative.

e) State the appropriate null and the alternative hypotheses *in words*.

f) Explain why it would NOT be okay to use the theory-based method to find a p-value for this study.

g) Describe how one might use everyday items (for example, coins, dice, cards, etc.) to conduct a tactile simulation-based test of the hypotheses. Be sure to clearly describe how the p-value will be computed from the simulation.

h) Use an appropriate applet (for example, the **Multiple Proportions** applet) to find and report the following from the data: The name of a statistic and its numeric value and the simulation-based p-value.

i) Interpret the p-value in the context of the study.

j) State a complete conclusion about this study including Significance, Causation and Generalization. Be sure to explain how you are arriving at your conclusion.

Happiness and income

8.2.22 To investigate whether there is an association between happiness and income level, we will use data from the 2002 General Social Survey (GSS), cross-classifying a person's perceived happiness with their family income level. The GSS is a survey of randomly selected U.S. adults who are not institutionalized.

		Not too happy	Pretty happy	Very happy	Total
Income?	**Above average**	21	159	110	290
	Average	53	372	221	646
	Below average	94	249	83	426
	Total	168	780	414	1362

a) State the appropriate null and the alternative hypotheses *in words*.
b) Explain why it would be okay to use the theory-based method to find a p-value for this study.
c) Use an appropriate applet (for example, the **Multiple Proportions** applet) to find and report the following from the data: The name of the statistic and its numeric value and the theory-based p-value.
d) Interpret the p-value in the context of the study.
e) State a complete conclusion about this study including Significance, Causation and Generalization. Be sure to explain how you are arriving at your conclusion.

Demographics of treatment groups*

8.2.23 Often research articles include tables of comparisons for demographics of experimental groups. In one such article (*Psycho-Oncology,* 2006) investigating the effect of art therapy on women diagnosed with cancer, researchers Monti et al. report the counts for race for the control group and the treatment group.

		Control group	Treatment group	Total
Race?	**Caucasian**	38	45	83
	African-American	13	10	23
	Asian	1	1	2
	Hispanic	2	0	2
	Other	1	0	1
	Total	55	56	111

a) Is this an experiment or an observational study? How are you deciding?

b) State the appropriate null and the alternative hypotheses *in words*.

c) Explain why it would NOT be okay to use the theory-based method to find a p-value for this study.

d) Use an appropriate applet (for example, the **Multiple Proportions** applet) to find and report the following from the data: The name of a statistic and its numeric value and the simulation-based p-value.

e) Explain why providing such a table would be in the best interest of the researchers. What does the p-value tell you about the distribution of race in the control group compared to the treatment group? Be sure to explain how you are arriving at your conclusion.

Estimating energy expenditure and energy intake

Problems 8.2.24 and 8.2.25 both use the energy data described here.

8.2.24 How good are you at estimating how many calories you expend? Researchers Harris et al. (*American Journal of Men's Health,* 2010) conducted a study where 80 males (ages 21 to 45 years) were first given a test to identify whether had high or low dietary restraint. Dietary restraint measures the conscious control over food intake. After a session of exercising, each participant was asked how many calories they thought they had expended, and according to their response they were classified into one of the three categories: underestimation, within 50% estimation, or overestimation. Here are the summarized data:

	High restraint	**Low restraint**	**Total**
Underestimation	7	10	17
Within 50% estimation	29	10	39
Overestimation	12	12	24
Total	48	32	80

a) State the appropriate null and the alternative hypotheses *in words*.

b) Explain why it would NOT be okay to use the theory-based method to find a p-value for this study.

c) Use an appropriate applet (for example, the **Multiple Proportions** applet) to find and report the following from the data: The name of the statistic and its numeric value, and the simulation-based p-value.

d) Interpret the p-value in the context of the study.

e) State a complete conclusion about this study including Significance, Causation and Generalization. Be sure to explain how you are arriving at your conclusion.

8.2.25 Refer to the previous exercise about the study of the relationship between dietary restraint and energy expenditure. Another response variable the researchers were interested in the estimation of energy intake. After a session of exercising, each participant was given a mean and then asked how many calories they thought they had consumed, and according to their response they were classified into one of the three categories: underestimation, within 50% estimation, or overestimation. Here are the summarized data:

	High restraint	**Low restraint**	**Total**
Underestimation	18	17	35
Within 50% estimation	28	13	41
Overestimation	2	2	4
Total	48	32	80

a) State the appropriate null and the alternative hypotheses *in words*.

b) Explain why it would NOT be okay to use the theory-based method to find a p-value for this study.

c) Use an appropriate applet (for example, the **Multiple Proportions** applet) to find and report the following from the data: The name of a statistic and its numeric value and the simulation-based p-value.

d) State a complete conclusion about this study including Significance, Causation and Generalization. Be sure to explain how you are arriving at your conclusion.

Rolling dice*

8.2.26 Suppose you roll a fair six-sided die 600 times. The die has sides numbered 1, 2, …, 6. For each of the following, find the probability, π , and the expected number of:

a) Ones

b) Even numbers

c) Values greater than 2

d) Values greater than 1

Cards

8.2.27 An ordinary deck of 52 cards has four sets of 13 cards each, labeled clubs, diamonds, hearts and spades. Clubs and spades are black, diamonds and hearts are red. Each set of 13 has a second label: Ace, 2, 3, …, 10, Jack, Queen, King. State n, π , and find the expected number for each of the following situations.

a) The expected number of hearts in a set of 10 randomly chosen cards.

b) The expected number of kings in a set of 10 randomly chosen card.

c) In the game of bridge, a hand is a set of 13 cards dealt at random from the deck of 52. Find the expected number of hearts in a bridge hand.

d) Find the expected number of face cards in a bridge hand. (A face card is a Jack, Queen, King, or Ace.)

e) In some versions of poker, deuces (2s) are particularly valuable. If a poker hand is five cards dealt at random from a deck, find the expected number of deuces in a hand.

f) For other versions of poker, you get 7 cards dealt at random. Find the expected number of deuces.

Finding expected values*

8.2.28 Find n, $\hat{p}$, and expected number for each of the following situations.

Marine. In Example 2.1 the dog Marine had 33 tries at identifying the one person with cancer among a set of five patients. (Each set of five involved different people.) There are two expected values to find.

a) Murderous Nurse (Exploration 5.1). Nurse Kristen Gilbert was present for 257 of 1641 consecutive shifts at the intensive care unit. Of those shifts 74 experienced a death. (There are four expected values to find.)

b) Organ Donors (Exploration 8.1). In this study there were three conditions: Opt In (54 subjects), Opt Out (54 subjects), and Neutral (53 subjects.) Of the 161 subjects, 109 agreed to become a donor. (There are six expected values to find.)

Harvesting alligators

Problems 8.2.29 through 8.2.36 use the data on lakes and alligators in Florida.

8.2.29 Alligators have become a big concern in Florida. Here is some "Gator Data."

OBS	2000	2005	2010	Total
Pierce	12	8	16	36
Marian	30	50	90	170
Hatchineha	36	71	135	242
Blue Cypress	31	66	128	225
Seminole	16	25	16	57
Total	125	220	385	730

The rows are for five small Florida lakes. They stretch along a line from the west coast near Tampa to the east coast. The columns are for years, spaced half a decade apart. The counts tell the numbers of alligators "harvested." (Data Source: *myfwc.com/media/1357388/Alligator_Annual_Summaries.pdf*)

Look first at the column totals: Is there a trend over time? Then look at row totals: How do lake sizes compare? Finally, think about association. Is the pattern over time pretty much the same for all five lakes?

8.2.30 Referring to the previous exercise, two possible data tables are shown below. Which one has the observational units right? Why?

ID	Lake	Year	Harvest
1	Pierce	2000	12
2	Pierce	2005	8
3	Pierce	2010	16
4	Marian	2000	30
5	Marian	2005	50
6	Marian	2010	90
7	Hatch	2000	36
8	Hatch	2005	71
9	Hatch	2010	135
10	Blue	2000	31
11	Blue	2005	66
12	Blue	2010	128
13	Seminole	2000	16
14	Seminole	2005	25
15	Seminole	2010	16

ID	Year	Lake
1	2000	Pierce
.	.	.
.	.	.
.	.	.
12	2000	Pierce
13	2005	Pierce
.	.	.
.	.	.
20	2005	Pierce
21	2010	Pierce
.	.	.
.	.	.
36	2010	Pierce
37	2000	Marian
.	.	.
.	.	.
.	.	.
.	.	.
730	2010	Seminole

8.2.31 Referring to the previous exercises, If both variables are categorical, you have two choices for segmented bar graphs. Either one of the two variables can go on the x-axis, with percentages for the other on the y-axis. Here are both segmented bar graphs for the Gator Data.

(a) Tell what you learn from both taken together.

(b) Is one more useful than the other? (If so, why? If not, why not?)

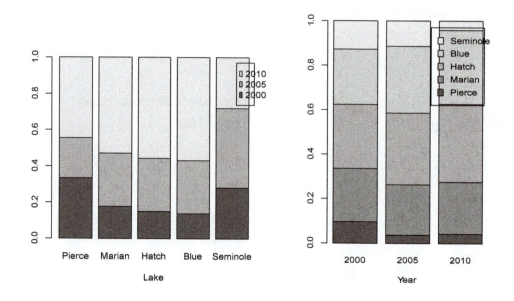

Segmented bar graphs for the Gator Data

8.2.32 Referring to the previous exercises, sometimes a categorical variable comes from a quantitative variable. For the Gator Data, Year can be regarded as quantitative. Here's a plot that takes advantage of that quantitative underpinning. What do you see in this plot that was harder to see in the two segmented bar graphs? Try to name at least two or three systematic patterns. Do keep in mind that small changes are likely due to chance. Large patterns are the ones that matter.

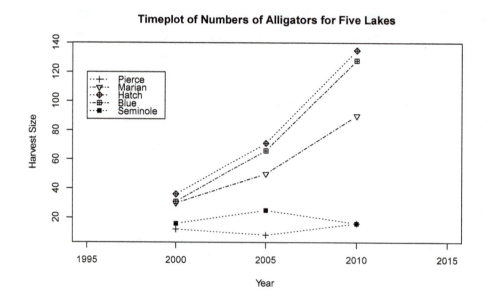

8.2.33 Referring to the previous exercises, lakes Pierce and Seminole have smaller harvests that show little or no evidence of a time trend.

OBS	2000	2005	2010	Total
Pierce	12	8	16	36
Seminole	16	25	16	57
Total	28	33	32	93

Is it reasonable to treat them as a group? Use the applet to compare the two lakes using segmented bar graphs, and to conduct an appropriate test of whether the pattern over time differs for the two lakes.

8.2.34 Referring to the previous exercises, the other three lakes have larger harvests that show steep increases over time:

OBS	2000	2005	2010	Total
Marian	30	50	90	170
Hatchineha	36	71	135	242
Blue Cypress	31	66	128	225
Total	97	187	353	637

Here are values of $(O-E)^2/E$, the contribution to chi-square, for each cell.

$(O-E)2/E$	2000	2005	2010	Total
Marian	0.65	0.00	0.19	0.84
Hatchineha	0.02	0.00	0.01	0.03
Blue Cypress	0.31	0.00	0.09	0.40
Total	0.98	0.00	0.28	1.27

Based on these values, is it reasonable to regard the time trend for these three lakes as essentially the same?

8.2.35 As we hope you agree, there seem to be two kinds of lakes. Two (Pierce and Seminole) have small harvests that don't show a time trend. The other three lakes have much bigger harvests that increase steeply over time. Use the applet to compare the two groups over time, and to test the strength of evidence that the two groups differ.

Lake Group	Year			TOTAL
	2000	2005	2010	
Pierce and Seminole	28	33	32	93
The other three	97	187	353	637
TOTAL	125	220	385	730

8.2.36 Referring to the previous exercise, what can you say about cause-and-effect and generalization based the Gator Data?

End of chapter exercises

Coffee consumption*

Problems 8.CE.1 through 8.CE.5 deal with data on sex and coffee consumption.

8.CE.1 Suppose that you want to investigate whether men and women differ with regard to coffee consumption. You plan to ask people to indicate their sex and also to say whether they drink coffee every day, sometimes, or almost never.
   a) Identify the two variables in this study. Also classify each variable as categorical or quantitative. For the categorical variable(s), also report how many categories are involved.
   b) Could this study involve random sampling, random assignment, or both? Explain.

Suppose that you collect survey responses on these questions for a sample of 70 students at your school. You then give the responses to your friend to ask him to tally them. He reports the following results back to you:

> Sex: 38 men, 32 women
>
> Coffee consumption: 22 every day, 22 sometimes, 26 almost never

   c) Does this provide enough information for you to investigate whether the sample data provide strong evidence that men and women differ with regard to coffee consumption? If not, describe what additional information is needed.

8.CE.2 Reconsider the previous exercise about sex and coffee consumption. The sample data could be organized in a two-way table as follows:

	Men	Women	Total
Every day			22
Sometimes			22
Almost never			26
Total	38	32	70

   a) Fill in the table with counts (which you make up) in such a way that there's almost *no association* between sex and coffee consumption. Also draw the appropriate graph to display what your made-up data reveal.

b) Now fill in the table with counts (which you make up) in such a way that there's a *strong* association between sex and coffee consumption. Again draw the appropriate graph to display what your made-up data reveal.

8.CE.3 Reconsider the previous two exercises. Students at Cal Poly collected data from their classmates and produced the following two-way table of counts:

	Men	Women	Total
Every day	7	15	22
Sometimes	12	10	22
Almost never	19	7	26
Total	38	32	70

a) Produce a segmented bar graph to display these data. Comment on what the graph reveals.
b) Perform a simulation-based analysis using the chi-square statistic. Submit a screen capture of the simulated null distribution, and report the approximate p-value.
c) Based on the approximate p-value from the simulation analysis, would you conclude (at the .05 significance level) that there is an association between sex and coffee consumption?
d) Summarize your conclusions from this analysis.

8.CE.4 Reconsider the data from the previous exercise. Check whether the validity conditions for a theory-based analysis are met. Regardless of your answer, conduct such a theory-based analysis. Report the test statistic and p-value. Summarize your conclusions.

8.CE.5 Reconsider the data from the previous two exercises. Suppose that the sample size had been twice as large, with the counts for every cell in the table also being twice as large.
a) What effect would this change have on the segmented bar graph?
b) What effect would this change have on the p-value? (Do not bother to perform any analysis to answer this question.)
c) What effect would this change have on the strength of evidence of an association between sex and coffee consumption?

Giving Blood

Problems 2.CE.6 through 2.CE.8 look at data on donating blood.

8.CE.6 The General Social Survey asked a random sample of adult Americans about their political viewpoint and also about whether they had given blood in the previous year. The sample results for the 2012 survey are organized in the following two-way table of counts:

	Liberal	Moderate	Conservative
Gave blood	41	33	48
Did not give blood	317	411	368

a) For each political viewpoint, determine the sample proportion who claimed to have given blood in the past year.

b) Calculate the value of the MAD statistic for these sample data.

c) Express the null hypothesis to be tested with these data.

d) Conduct a simulation-based analysis of these data, based on the MAD statistic. Submit a screen capture of the null distribution, and shade the area in the null distribution that corresponds to the p-value.

e) Report the p-value from the simulation analysis. Also write a sentence interpreting what this p-value represents (probability of what, assuming what?).

f) At the .05 significance level, do the sample proportions who gave blood differ enough among the three political groups to consider the observed association to be statistically significant? Explain how you determine the answer.

8.CE.7 Reconsider the previous exercise. Conduct a theory-based analysis, including a check of validity conditions, and summarize your conclusion.

8.CE.8 Reconsider the data from the previous two exercises. Use what you learned in the first two chapters to address the following questions. Provide a complete analysis and summary of conclusions for both questions.

a) Do the sample data provide convincing evidence that more than 10% of adult Americans gave blood in the year preceding the survey?

b) Estimate with 95% confidence the proportion of adult Americans who gave blood in the year preceding the survey.

Trusting people*

Problems 8.CE.9 through 8.CE.11 refer to the data on trusting people and highest level of education.

8.CE.9 The 2010 General Social Survey asked a random sample of Americans about whether they generally trust people. The following two-way table classifies responses by the highest educational degree achieved by the person:

	Less than high school	High school	Junior college	Bachelor's	Graduate
Trusts people	29	183	28	121	97
Does not trust people	181	490	63	124	56
Total	210	673	91	245	153

a) Identify the explanatory and response variables in this study.
b) For each category of educational attainment, calculate the proportion indicating they generally trust people. Also produce a graph to display these proportions. Comment on what the calculations and graph reveal.
c) State the null hypothesis to be tested from these data.
d) Calculate the value of the MAD statistic for these sample data.
e) Conduct a simulation-based analysis, using the MAD statistic. Submit a screen capture of the null distribution, and report the approximate p-value.
f) Interpret what the p-value says in this context.
g) Would you reject the null hypothesis at any reasonable significance level?
h) State your conclusion about the strength of evidence

8.CE.10 Reconsider the previous exercise. Conduct a theory-based analysis. Report the test statistic and p-value, and summarize your conclusions. Also, compute 95% confidence intervals as a follow-up, including these in your summary.

8.CE.11 Reconsider the previous two exercises. There were actually three categories of the response variable. The response "it depends" had been classified as not trusting people in the table above but is separated out into its own category in the table below:

	Less than high school	High school	Junior college	Bachelor's	Graduate
Trusts people	29	183	28	121	97
"It depends"	9	32	11	12	11
Does not trust people	172	458	52	112	45
Total	210	673	91	245	153

Perform a chi-square test on these data. Report the test statistic and p-value, and summarize your conclusion. Also comment on whether the analysis is very different from the previous exercise when two of the response categories had been combined.

Drinking at college

8.CE.12 A study of drinking habits of college students at a particular college produced the following table of counts::

	Reside on campus	Reside off campus, not with parents	Reside off campus, with parents	Total
Abstain from drinking	46	17	43	106
Light or moderate drinking	126	72	68	266
Heavy drinking	130	52	32	214
Total	302	141	143	586

Analyze these data with appropriate descriptive and inferential methods. Write a short report summarizing and justifying your conclusions.

School Goals*

Problems 8.CE.13 and 8.CE.14 refer to the data on goals for boys and girls grades 4 through 6.

8.CE.13 Students (grades 4-6) in the Lansing, Michigan area were asked, "What is the most important goal in school: getting good grades, being popular, or excelling in sports." The data are tabled below. (The raw data is in the **Goals** file.)

	Boy	Girl
Grades	117	130
Popular	50	91
Sports	60	30

a) What proportion of boys believes that getting good grades is the most important goal?

b) What proportion of girls believes that getting good grades is the most important goal?

c) What proportion of boys believes that being popular is the most important goal?

d) What proportion of girls believes that being popular is the most important goal?

e) What proportion of boys believes that excelling in sports is the most important goal?

f) What proportion of girls believes that excelling in sports is the most important goal?

g) What is the (marginal) proportion of students who believe that getting good grades is the most important goal?

h) What is the (marginal) proportion of students who believe that being popular is the most important goal?

i) What is the (marginal) proportion of students who believe that excelling in sports is the most important goal?

j) Based on your answers for (a) – (i), do you think that boys and girls in 4th to 6th grade have different goals? Explain your reasoning.

8.CE.14 Reconsider the previous exercise. Carry out a theory-based test to investigate whether the sample data provide strong evidence of an association between gender and goals of 4th to 6th graders. Follow up, if applicable, with confidence intervals for any differences you may find. Be sure to include the following in your report:

- The appropriate null and alternative hypotheses
- The statistic you are using, as well its observed value.
- The p-value
- An appropriate conclusion that comments on significance, estimation, causation, and generalization, in the context of the study.

Social Norms

8.CE.15 What happens when you break social norms? Pedestrians walking on sidewalks typically follow the same traffic patterns as cars driving in the streets. What happens if you walk on the wrong side of the sidewalk? Do gender, location and speed of walker affect the number of people in the oncoming traffic that run into you?

Data were collected using both a male experimenter and a female experimenter, and are given in the data file **SocialNormsSex**.

a) Is there an association between whether you collide with oncoming pedestrians and the sex of the experimenter? Calculate appropriate descriptive statistics (*HINT:* percent collisions by the gender of confederate) and create segmented bar graphs. Write a few sentences (2 – 3) about what these numerical and graphical summaries reveal to you. Explain how you are deciding.

b) Carry out a simulation-based test of significance to test whether there is evidence of an association between whether you collide with oncoming pedestrians and the gender of the confederate. State the hypotheses, an appropriate conclusion, and be sure to include your reasoning for your conclusion.

8.CE.16 Recall the previous exercise about social norms. Data were also collected in a hallway inside a building and on a sidewalk outside in a park, and are given in the data file **SocialNormsLocation**. Is there an association between whether you collide with oncoming pedestrians and location?

a) State the appropriate null and alternative hypotheses, in the context of the study.

b) Choose either a simulation-based approach or a theory-based approach to test your hypotheses from (a).
- If you have chosen to use a theory-based approach, be sure to explain why you think this is a valid approach. That is, state the validity conditions, and explain how you are deciding whether or not these conditions are met.
- Analyze the data using technology (either an applet or statistical software). Be sure to report the following: the name of the statistic you are using, and its value (from the study), and the p-value.
- State your conclusion in the context of the study.

8.CE.17 Recall the previous exercises about social norms. Data were also collected on whether the experimenter was walking at a slow speed or was walking at a fast speed, and are given in the data file **SocialNormsSpeed**. Is there an association between whether you collide with oncoming pedestrians and your speed, as measured by the speed of experimenter and whether the experimenter was in a collision?
a) State the appropriate null and alternative hypotheses, in the context of the study.
b) Choose either a simulation-based approach or a theory-based approach to test your hypotheses from (a).
- If you have chosen to use a theory-based approach, be sure to explain why you think this is a valid approach. That is, state the validity conditions, and explain how you are deciding whether or not these conditions are met.
- Analyze the data using technology (either an applet or statistical software). Be sure to report the following: the name of the statistic you are using, and its value (from the study), and the p-value.
- State your conclusion in the context of the study.

Bonus or Rebate?*

Problems 8.CE.18 and 8.CE.19 refer to the data gathered on rebates versus bonuses.

8.CE.18 Do people tend to spend money differently based on perceived changes in wealth? In a study conducted by researchers Epley et al. (2006), 47 Harvard undergraduates were randomly assigned to receive either a bonus check of $50 or a rebate check of $50. A week later, each student was contacted and asked whether they had spent any of that money, and if yes, how much. In this exercise we will focus on whether or not they had spent any of that money as the response variable of interest. It turned out that 36% of those in the bonus group spent none of the money, compared to 73% in the rebate group.

a) Identify the observational units.
b) Identify the explanatory variable. Is it categorical or quantitative?

c) Is the response variable categorical or quantitative?

d) State the appropriate null and alternative hypotheses, in the context of the study.

8.CE.19 Reconsider the previous exercise. In the article that appeared in the *Journal of Behavioral Decision Making*, the researchers did not report the sample size of each group. The following 2x2 table gives a set of observed counts that are consistent with the rest of the study data. This table can be found in the file **Rebate**.

	Bonus	Rebate	Total
Saved	9	16	25
Spent	16	6	22
Total	25	22	47

a) Use the **Two Proportions** applet to test the hypotheses stated in the previous question. Be sure to specify the statistic you used, and its value (from the study). Also, find and report the p-value.

b) Next, use the **Multiple Proportions** applet to test the hypotheses stated in the previous question. Be sure to specify the statistic you used, and its value (from the study). Also, find and report the p-value.

c) How do the p-values from (a) and (b) compare? Are they about the same or very different? Explain why.

Heart Disease and Baldness

8.CE.20 The study used for exercises in sections 8.1 and 8.2 actually included a fifth degree of baldness, Extreme. Here is a table that summarizes the full data set:

COUNTS	None	Little	Some	Much	Extreme	Total
Yes	251	165	195	50	2	663
No	331	221	185	34	1	772
Sum	582	386	380	84	3	1435

What stands out about the new category is the tiny sample size: only 3 patients with extreme baldness. This exercise explores what can happen when a category has so few observational units. Use the data set **HeartDiseaseAndBaldness2**.

Simulation-based test using MAD.

a) Use the **Multiple Proportions** applet to find the MAD, its simulation-based null distribution, and the p-value.

b) Compare the null distribution with the null distribution you obtained in 8.1.15 (recreate it if you need to). What are the main differences? How do you account for these differences?

c) Compare the conclusions from the two simulation-based analyses using the MAD. Which conclusion do you consider more trustworthy? Why?

Simulation-based test using chi-square.

d) Use the **Multiple Proportions** applet to find the chi-square statistic, its simulation-based null distribution, and the p-value.

e) Compare the shape of the distribution for chi-square with the shape of the simulation-based distribution in for the MAD (a).

f) Compare the conclusions from the simulation-based analysis using MAD in (c) above and using chi-square. Which conclusion do you consider more trustworthy? Why?

Theory-based test using chi-square.

g) Use the applet to find the p-value and contributions to chi-square for a theory-based test.

h) Is the theory-based test trustworthy? Why, or why not?

The bottom line.

i) Which analysis do you consider more persuasive: the one based on four degrees of baldness, or the one based on all five? Justify your answer.

j) A third approach to the data combines the two categories "much" and "extreme." What results do you expect from this analysis? Explain your reasoning.

Which test?

k) What lesson should you take away from your comparison of the three tests?

Investigation: Who Yields to Pedestrians?

Researchers from the University of California Berkley wondered whether upper-class individuals behave more unethically than lower-class individuals. To further investigate this question, they studied behaviors of drivers in different makes and models of cars. The cars were a surrogate measure for the drivers' class status. According to California Vehicle Code, cars must yield to a pedestrian if said pedestrian is crossing within any marked crosswalk. Data was collected from an intersection in the San Francisco Bay area on three weekdays between 2:00 pm and 5:00 pm in June of 2011. A coder, positioned near the crosswalk, rated the approaching car on a scale from 1 to 5 based on make, age and physical appearance of the car. The lowest status received a code of 1 and the highest status a code of 5. The coder also recorded whether or not the approaching car yielded to the pedestrian (a confederate of the study) waiting to cross. There were 152 drivers who were scored.

Step 1: Ask a research question

1. State the research question.

Step 2: Design a study and collect data

2. What are the observational/experimental units?

3. Is this a randomized experiment or an observational study? Explain how you know.

4. State the two variables measured on each unit.

5. Is there an explanatory/response relationship for these variables? Classify the variables in this study as categorical or quantitative.

6. State the null and alternative hypotheses to be investigated with this study. (You may use the symbol π to represent these parameters.)

The following table provides descriptive statistics on the 152 drivers:

	Status				
	1	2	3	4	5
Did yield	5	20	41	25	7
Didn't yield	0	8	20	20	6
Total	5	28	61	45	13

Step 3: Explore the data

7. Calculate the conditional proportions of those who didn't yield to the pedestrian for each level of status.

8. Calculate the MAD (mean absolute value of the differences) for these data.

Step 4: Draw inferences

9. Use the appropriate applet to construct a simulated null distribution using the MAD statistic. Mark the observed MAD statistic calculated in the previous question on this graph.

   a) Paste a screen shot or draw your null distribution below with the observed MAD marked and the approximate p-value shaded in.

b) Is the observed statistic out in the tail of this null distribution, or is it a fairly typical result?

c) What is the p-value from your simulation? Based on this simulation analysis, would you conclude that the experiment provides *strong* evidence against the null hypothesis and conclude that there is a significant difference in at least one of the long-run probabilities of not yielding for the 5 different status groups? Explain your reasoning.

10. Are the validity conditions met to complete a theory-based Chi-square test for these data? Explain. Whether the conditions are met or not, use a theory-based Chi-square test to find a p-value for this test. How does this p-value compare to the p-value your found using the randomization method?

11. Are the validity conditions met so that we can perform follow-up analyses using theory-based techniques? Explain.

12. Find confidence intervals for each pair of differences in the long-run probabilities of not yielding. Comment on how this shows where differences can be established and where it is plausible there are no differences between long-run probabilities.

Step 5: Formulate conclusions

13. Are you able to conclude that the cause of any differences was due to the class status of the driver? Explain.

14. What generalizations are you willing to make? Explain.

Step 6: Look back and ahead

15. Summarize the findings from the study. What worked well in this study design? What would you change? What are some follow-up research questions you would like to explore based on the findings from this study?

http://www.pnas.org/content/109/11/4086.full?sid=6d9e4159-8d2c-4979-a63c-62a204f42158

Research article: Music problems

Read "Early adolescent music preferences and minor deliquency." by ter Bogt, Keijsers and Meeus et al. (2013), 131(2), 1-10 in the journal *Pediatrics*, and then answer the following questions.

Step 1. Ask a research question

1. In no more than 1-2 sentences state a broad research question (or related questions) that the researchers were trying to investigate in this study.
2. Identify at least two (cited) points of evidence the researchers cite as to why the research question(s) are important. Include the citation in your response.

Step 2. Design a study and collect data

3. How many different adolescents were in the study?
4. What age were the adolescents when the study started?

Step 3. Explore the data

5. Based on Table 1, at age 12, on average, what were the two most liked genres of music? What were their average values?
6. What were the two genres that were least liked at age 12? What were their average values?
7. Answer question #5 for the sample at age 16.
8. Answer question #6 for the sample at age 16.

Step 4. Draw inferences

Table 2 shows the results of fitting regression lines or quadratic regression lines (curves) to the musical preference scores over the four measurements (age 12, 14, 15, 16) in the study.

9. The chart pop musical preference regression line is given as Preference Score = $4.12 - 0.12*$(Age $- 12$), where Age $- 12$ is the age of the participant minus 12 (so, when the participant is age 12 the value of the variable is 0. **Ignore the quadratic slope coefficient for this question.**
   a. Interpret the intercept.
   b. Interpret the slope
   c. Notice that the slope is significant ($p < 0.05$, as indicated by an asterisk). State the null and alternative hypothesis, indicate the strength of evidence provided by the p-value and provide an appropriate (in context) interpretation.
10. Answer the same questions as in #9 for the hip-hop genre using the regression line Preference Score = $3.27 + 0.10*$(Age $- 12$)

Table 3 shows the correlations between numerous variables in the study. For each of the following pairs of variables State the null and alternative hypothesis, indicate the strength of evidence provided by the p-value ($p > 0.05$ (no asterisk), $p < 0.05$ (*) or $p < 0.01$ (**) and provide an appropriate (in context) interpretation.

11. Chart pop preference and R&B preference at age 12 (correlation = 0.44**)

12. Classic preference and Hip-hop preference at age 12 (correlation = -0.22**)

13. Metal preference at age 12 and Delinquency at age 16 (correlation = 0.29**)

14. Chart pop preference at age 12 and Delinquency at age 16 (correlation = -0.07)

Step 5. Formulate conclusions

15. To which adolescents are you willing to generalize your conclusions from this study? Why?

16. Does the significant result in question #13 mean that liking the Metal genre at age 12 causes adolescents to be more delinquent at age 16?

Step 6. Look back and ahead

17. Identify at least two things that, if you were running the study and had a reasonable amount of financial resources, you would do differently. You must justify your answer.

Chapter 9

Comparing More Than Two Means

Chapter Overview

In Chapter 6 you learned how to compare two groups using means. Chapter 9 will show you how to compare more than two groups on a quantitative response. This chapter very much parallels Chapter 8 and also follows from Chapter 6. In Chapter 6 you learned simulation-based and theory-based methods for determining p-values and confidence intervals for comparing two means from independent samples. In this chapter you will extend those approaches to compare more than two groups. Many of the details (statement of hypotheses, interpretation of p-values and confidence intervals, scope of conclusions) are virtually unchanged. What does need to change is determining a statistic we can calculate from our sample data that summarizes the information we have about how the group means differ. Once we have such a formula, we can easily assess its statistical significance using the same 3S Strategy. As described in the Unit 3 introduction, we will do a single test instead of multiple comparisons of two groups to minimize the impact of Type I errors on our analysis.

Chapter 9 has two sections. Section 9.1 explores the simulation-based approach and then Section 9.2 explores the theory-based approach for a particular choice of statistic.

Section 9.1: Simulation-Based Approach for Comparing More than Two Groups with a Quantitative Response
Introduction

The focus of this chapter is on comparing more than two groups on a quantitative response. Your first question might be, why don't we simply conduct all the possible two-sample comparisons between the groups? For example, with three samples, we could conduct three separate independent samples *t*-tests for the three different comparisons between the groups. (That is, if our groups are A, B, and C, our comparisons are A vs. B, A vs. C, and B vs. C.). As we began discussing in the introduction to Unit 3, this approach can increase the overall chance of making at least one false alarm because each comparison has a chance of being a false alarm. So, instead we will construct a statistic that compares all sample means at once, letting us perform just one test and keeping the probability of a Type I error as small as we want. The following example shows how to construct such a statistic.

Example 9.1: Comprehending Ambiguous Prose

Sometimes when reading esoteric prose we have a hard time comprehending what the author is trying to convey. A student project group decided to partially replicate part of a seminal 1972 study by Bransford and Johnson on memory encoding ("Contextual prerequisites for understanding: Some investigations of comprehension and recall," *Journal of Verbal Learning and Verbal Behavior*, 11, pp. 717-726). The study examined college students' comprehension of following ambiguous prose passage.

> *If the balloons popped, the sound wouldn't be able to carry since everything would be too far away from the correct floor. A closed window would also prevent the sound from carrying, since most buildings tend to be well insulated. Since the whole operation depends on a steady flow of electricity, a break in the middle of the wire would also cause problems. Of course, the fellow could shout, but the human voice is not loud enough to carry that far. An additional problem is that a string could break on the instrument. Then there could be no accompaniment to the message. It is clear that the best situation would involve less distance. Then there would be fewer potential problems. With face to face contact, the least number of things could go wrong. (p. 719)*

Did you understand what the passage was describing? Would it help to have a picture? The picture that goes along with the passage is shown in Figure 9.1.

Before the college students were tested to see whether they understood the passage, they were randomly assigned to one of three groups, and then each group was read the passage under one of the following cue conditions:

- Students were shown the picture in Figure 9.1 before they heard the passage.
- Students were shown the picture in Figure 9.1 after they heard the passage.
- Students were not shown any picture before or after hearing the passage.

Step 1: Ask a research question. Is student comprehension of an ambiguous prose passage affected by viewing a picture designed to aid them in their understanding either before or after or not at all? So our research conjecture might be: The long-run mean comprehension score differs among the three treatments.

Step 2: Design a study and collect data.

Design a study: Fifty-seven students were randomly assigned to be in one of the three groups with nineteen in each group. After hearing the passage under the assigned cue condition, they were given a test and their comprehension of the passage was graded on a scale of 1 to 7 with 7 being the highest level of comprehension. Note that the data collected from this study design will not be paired, but will be independent, with the outcomes in one treatment group not affecting the outcomes in the other treatment groups.

Figure 9.1: Does this picture help explain the previous passage?

When comparing means across multiple samples, the null hypothesis is that all the long-run or population means are the same. The alternative hypothesis is that not all the long-run or population means are the same. Notice that this alternative hypothesis does not imply that all the means are different. It only implies that at least one population mean is different from at least one of the others. So for our experiment we have the following hypotheses.

Null Hypothesis: The long-run mean comprehension scores are the same under all three cue conditions.

Alternative Hypothesis: At least one of the long-run mean comprehension scores is different.

We can write these hypotheses using symbols as well as we have done in previous chapters. Remember that the symbols in our hypotheses represent process or population parameters.

H_0: $\mu_{\text{no picture}} = \mu_{\text{picture before}} = \mu_{\text{picture after}}$

H_a: At least one μ is different from the others

where $\mu_{\text{no picture}}$ is the long-run mean comprehension score after seeing no picture; $\mu_{\text{picture before}}$ and $\mu_{\text{picture after}}$ are defined similarly.

We can also (again) state these hypotheses in terms of association:

H_0: In the population, there is no association between when/whether picture is shown and student comprehension of this passage

H_a: There is an association between the variables in the population.

Notice in each of these phrasings of the alternative hypothesis we are only saying there is a difference somewhere, we are not specifying anything about where that difference is or in what direction or how many differences there might be.

Collect data: We are not given much information about the characteristics of the students in the study other than they were college students at a small Midwestern college.

Step 3: Explore the data. Graphical and numerical summaries for this study are displayed in Figure 9.2 and Table 9.1. Notice that Figure 9.2 graphically displays the data in both dotplots and boxplots. (Recall from Chapter 6 that boxplots display the five number summary: min, lower quartile, median, upper quartile, and max.)

Figure 9.2: Comparative dotplots and boxplots for comprehension scores from an ambiguous prose passage for 57 students under three different cue conditions

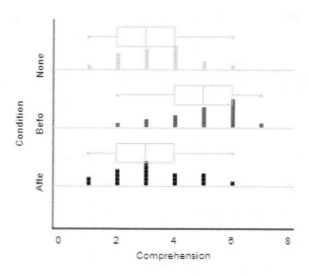

From our plots in Figure 9.2, it seems that there is a tendency for the "after" and the "none" groups to have fairly similar scores, and have lower scores than the "before" group. We also note that the variability and shapes of the three distributions are similar. The three distributions are reasonably symmetric though the "before" sample distribution is slightly skewed to the left.

Table 9.1: Numerical summary statistics (including the five-number summary) for comprehension scores of an ambiguous prose passage for 57 students under three different cue conditions

Group	Sample size	Mean	SD	Minimum	Lower quartile	Median	Upper quartile	Maximum
None	19	3.37	1.26	1.00	2.00	3.00	4.00	6.00
Before	19	4.95	1.31	2.00	4.00	5.00	6.00	7.00
After	19	3.21	1.40	1.00	2.00	3.00	4.00	6.00

In Table 9.1, note that from the sample means, medians, and quartiles, it seems that the "Picture after Reading" and the "No Picture" groups have fairly similar scores. Showing the students the picture before the reading does appear to produce comprehension results that are higher on average (4.95) than the other two samples. We also note the three distributions have similar SDs (1.40, 1.31, and 1.26).

Step 4: Draw inferences beyond the data. But maybe more of the students that had higher comprehension abilities just happened to randomly end up in the "Picture Before" sample? What we want to know is: *If there were no underlying difference in comprehension levels among the three treatments, how unlikely is it that we would get sample means as different as we have observed in this study just by chance alone?*

Let's envision how we might do a randomization test.

Think about it: What statistic would we use? Remember what we used when we compared two groups?

For just two groups we took the difference in means. Now, we are comparing three groups. We still want to measure, overall, how different the means are. There are many choices for what statistic you could use to measure the overall difference. For example, you could find the difference in the largest and smallest values: $\max \overline{x} - \min \overline{x}$. But this doesn't tell us as much about the average difference between the groups. Another statistic is the Mean Absolute Difference (MAD):

$$MAD = \frac{[|avg1 - avg2| + |avg1 - avg3| + |avg2 - avg3|]}{3}$$

where | | means "take the absolute value of." (This ensures that positive and negative differences don't cancel each other out when we sum them up.) Let's apply this to our study. The mean absolute difference would be

MAD-

observed

$= ([|3.21 - 4.95| + |3.21 - 3.37| + |4.95 - 3.37|])/3 = (1.74 + 0.16 + 1.58)/3 = 3.48/3 = 1.16$

So, on average, the difference between the group means (averages) is 1.16. This is our observed statistic. Is that overall difference consistent with random chance or is this overall difference convincing evidence that the alternative hypothesis is true?

> **Think about it:** What types of values (e.g., large, small, positive, negative) will this statistic have when the differences among the groups are small? What types of values will you consider evidence against the null hypothesis?

When all the sample means are the same, the value of the MAD is zero. As the sample means get farther and farther apart, the MAD gets larger and larger. A MAD of zero is what we would expect if the null were true. Larger values of the MAD offer support of the alternative hypothesis.

Significance: Here is where we employ our 3S Strategy. We have our observed *statistic*: $MAD_{observed} = 1.16$. Next we *simulate* the null hypothesis by breaking the association between our explanatory variable (experimental condition) and response (comprehension score). You know from earlier chapters that we do this by shuffling the response variable and reassigning the values back to the explanatory variable groups and then recomputing the statistic. We repeat this process many, many times to see how extreme our observed statistic of 1.16 is compared to the simulated statistics in the null distribution.

> **Think about it:** If you had to do a tactile simulation, what items would you need to conduct the simulation: coins, dice, spinners, or cards? How many? What would you do on each repetition of the simulation?

As was done in Chapter 6, using index cards would be the best in this scenario, as well. You would need 57 index cards, with one comprehension score written on each card. Then you would shuffle and deal them into 3 piles, with 19 in each pile, to mimic what happened in the random assignment used in the original study. Next, for these simulated data, you would calculate and record the MAD statistic value. Then, repeat this "shuffle and deal" many times, each time recording the MAD statistic value. Instead of doing this tactile simulation, we can use the **Multiple Means** applet. Our result from repeating the simulation 1000 times is shown in Figure 9.3.

The table below summarizes the key aspects of the simulation:

Null hypothesis = Long-run means are the same for each treatment.
One repetition = Re-randomizing students to when (if at all) they saw the picture, without changing their comprehension score
Statistic = mean absolute difference in means between the three groups

Think about it: Describe the shape, center, and variability of this null distribution. Why is the simulated null distribution not centered at zero?

Figure 9.3: Simulated null distribution of the mean of the absolute values of the difference in means

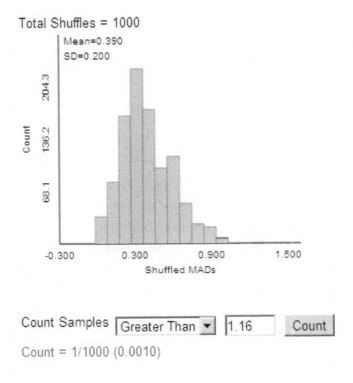

What do you notice about the shape of the simulated null distribution for the mean absolute difference? The distribution is NOT bell-shaped and centered at zero. This occurs because the numerator is the sum of non-negative numbers (due to the absolute value). Keep this in mind as you reach the next section on the theory-based test. Because of this non-symmetric shape, we will *not* use the "within two standard deviations" criteria to measure strength of evidence with this null distribution.

From Figure 9.3, we can see that the value of the MAD statistic, 1.16, occurred only once out of 1000 repetitions in our simulation. For assessing the *strength of evidence* this yields an approximate p-value of 0.001. Thus, we have strong evidence against the null hypothesis, and conclude that there is convincing evidence of an association between experimental condition and comprehension levels in the population. Stated another way, we have evidence that at least one of the long-run mean comprehension scores is different than the others.

Estimation: So, does one treatment work better than the others? By looking at the sample data, we can see that the mean score for the picture before is higher than the other two. So it would appear that viewing a picture to help interpret a prose passage before that passage is read does cause improved comprehension of that passage. In the next section, we will show how to do (theory-based) follow-up analysis to confirm this sort of conclusion using confidence intervals.

Step 5: Formulate conclusions.

Causation: Because the 57 students were randomly assigned to the three different treatment groups, we can conclude that the statistically significant difference in comprehension scores was *caused* by the treatments imposed.

Generalization: We are not told whether the 57 students are a random sample of college students, thus it is risky to extend our conclusions to a broader population. We should at least restrict our conclusions to college students like those in the study.

Step 6: Look back and ahead. To answer the research question: "Is a student's comprehension of an ambiguous prose passage affected by viewing a picture before, after, or not at all?" data were gathered from 57 college students in a randomized experiment. It was hypothesized that the mean comprehension score would be different in at least one of the three treatment populations. The sample mean test scores for the students were 3.21 for the picture after reading group, 4.95 for the picture before reading group, and 3.37 for the no picture group. The observed mean (absolute) difference between the sample means was 1.16. A randomization test was conducted and the resulting p-value was 0.001. This provides extremely strong evidence against the null hypothesis and we are able to conclude that there is a least one long-run mean comprehension score that is different. It appears from our data the long-run mean comprehension for the "Picture Before" treatment is higher than the other two treatments. Because this was a randomized experiment (and statistically significant) we can conclude a causal relationship between the presentation of the picture before the prose passage and improved comprehension of that passage.

However, the 57 students were not randomly selected so we can only cautiously extend our conclusions to the population of students similar to those in this study.

In a future study, the researchers could explore the option of selecting the participants in a more random fashion, so that they would be able to generalize the results to different age-groups, educational backgrounds, and so on. The researchers could also try to ask different kinds of questions to see whether comprehension is better on certain types of questions, and differs between the treatments.

Exploration 9.1: Exercise and Brain Volume

Brain size typically shrinks as people age past adulthood, and such shrinkage may be linked to dementia. Therefore, any intervention that can protect against brain shrinkage could help to protect the elderly against dementia and Alzheimer's disease. Researchers in China recently investigated whether different kinds of exercise/activity might help to prevent brain shrinkage or perhaps even lead to an increase in brain size (Mortimer et al., 2012).

The researchers randomly assigned elderly adult volunteers into four activity groups: tai chi, walking, social interaction, and no intervention. Except for the group with no intervention, each group met for about an hour three times a week for 40 weeks to participate in their assigned activity. The tai chi group was led by a tai chi master and an assistant, the walking group walked around a track, the social interaction group met at a community center and discussed topics that interested them, and the no intervention group just received four phone calls during the study period. A total of 120 participants started the study, and 13 dropped out along the way, so 107 completed the study

Each participant had an MRI to determine brain size before the study began and again at its end. The researchers measured the percentage increase or decrease in brain size during that time. They thought that physical activity would help increase brain size, hence they anticipated that the tai chi and walking groups would tend to show larger increases in brain size during the study than the other activity groups.

Step 1: Ask a research question.

1. What is the main research question that the researchers hoped to answer?

Step 2: Design a study and collect data.

2. Identify the observational units in this study.

3. Identify the explanatory and response variables in this study. Indicate whether they are quantitative or categorical. If they are categorical, indicate how many categories.

   Explanatory: Type:

   Response: Type:

4. What type of study was this: an experiment or an observational study? Are the responses of the participants in one treatment group paired with or independent of the responses of the participants in other treatment groups?

We need a procedure that considers information from across all four groups to allow for comparing all four groups simultaneously. Just as in Chapter 6, when we compared two groups, the null hypothesis can be written in terms of no association between the explanatory and response variable, whereas the alternative asserts that there is an association between the explanatory and response variables. More specifically, the null hypothesis asserts that all four long-run means are equal, and the alternative states that at least one long-run mean differs from the others.

5. State the null and the alternative hypotheses, in symbols and in words, in the context of this study.

The researchers calculated the percentage change in size of each participant's brain. If a person's brain size increased, then this percentage change was positive; if brain size decreased then this percentage change was negative. These data can be found in the file **Brain** on the textbook website.

Step 3: Explore the data.

6. Paste the data into the **Multiple Means** applet and press **Use Data**. The applet will produce dotplots and descriptive statistics. Check the **Boxplots** box to overlay those as well.

    a. Record the means and standard deviations in the following table, and provide estimates of the five number summaries of each group from the boxplots

Group	Sample size	Mean	SD	Minimum	Lower quartile	Median	Upper quartile	Maximum
Tai chi								
Walking								
Social								
No interaction								

    b. Which activity group tended to have the largest increase in brain size percentage change? Which tended to have the smallest increase (i.e., largest decrease).

         Largest: Smallest:

7. Describe what the graphs and statistics reveal about whether the four activities appear to differ with regard to percentage change in brain size for the participants in this study.

8. Suppose for the moment that there really is *no effect* of the different activities on percentage change in brain size.

    a. Would it be *possible* to obtain sample means as far apart as these four are in this sample?

b. Describe a general process for assessing how unlikely it would be to obtain sample means as far apart as these four are.

Step 4: Draw inferences (3S Strategy).

1. Statistic

To minimize the chance of making a Type I error, we won't run six different tests comparing each group to each other (see Unit 3 introduction). So, instead we need a single statistic to help of test the overall null hypothesis of no association.

9. The first step in applying the 3S Strategy is to decide on a statistic that measures how far apart the sample means are. We could focus on the max $\bar{x}$ – min $\bar{x}$ that you identified in #6b. Why might this not be the best statistic to use?

10. Propose an alternative statistic to use for this purpose. In other words, come up with a formula based on the four sample means that will result in a single number that summarizes how far apart the four sample means are. (*Hint*: You should be able to calculate the value of your statistic from the observed sample data.)

When we compared means between *two* groups in Chapter 6 using simulation-based approaches, we used the *difference* between the two means for the statistic.

But in this study involving four groups, there are six different pairs of groups for which we could take the differences in means.

11. Suggest how you might combine those six differences into one statistic.

One suggestion for this statistic is to calculate the mean of the absolute values of the differences in means (called **MAD**, for mean absolute difference):

$$MAD = [$$

$$\frac{|Avg1 - Avg2| + |Avg1 - Avg3| + |Avg1 - Avg4| + |Avg2 - Avg3| + |Avg2 - Avg4| + |Avg3 - Avg4|}{6}$$

$$]$$

12. Explain why it's important to take the absolute values before averaging the six differences.

13. What is the value of the MAD statistic shown in the applet?

14. What types of values will the MAD statistic have if the null hypothesis is true (large, small, positive negative)? If there is strong evidence against the null hypothesis?

2. Simulate

15. Suppose that you were to conduct a tactile simulation to generate many possible values of the MAD statistic that could have happened if the null hypothesis were true, that is, by random chance alone. As you did in Chapter 6, which was a simpler case with comparing just two means, you can use index cards here. Answer the following questions about what else needs to be kept in mind:

    a. How many cards will you need?
    b. Do the cards just need to be certain colors (like in Chapter 5), or do they need something written on them (like in Chapter 6), and if so, what?
    c. When you shuffle and deal, how many piles will you make?
    d. How many cards will you place in each pile?
    e. What should you be recording after the completion of each shuffle and deal?
    f. What else do you need to do to find the p-value?

Instead of doing the tactile simulation (which would be fun of course, but time consuming) let us use the **Multiple Means** applet to run the simulation. Check the **Show Shuffle Options** box in the applet. Enter 1 for **Number of Shuffles**, and press **Shuffle Responses**. Notice in the Most Recent Shuffle data window, the Treatment column has not changed, but the BrainChange column has mixed up all its values. If you select the **Plot** radio button, you will see the new dotplots.

16. Notice that the new dotplots and statistics are produced for the shuffled data. The individual with the smallest percentage change, originally in the Walking group is now in which group? Report the value of the MAD statistic for the shuffled data. Is this value more extreme than the observed value of the MAD statistic from the experimental data?

17. Press **Shuffle Responses** again, with the Plot option selected. Notice the applet pooling all the responses together, and then randomly reassigning them to their new groups. What did you get for the MAD value for these shuffled responses?

18. Press **Shuffle Responses** three more times. Record the three new values of the MAD statistic for the shuffled data. Are these all the same? Are any of them more extreme than the observed value of the MAD statistic from the experimental data?

19. Now enter 995 for the **Number of Shuffles**, producing a total of 1000 shuffles.
    a. Describe the shape of the null distribution of the MAD statistic. Is it symmetric or skewed? If it is skewed, in which direction is it skewed?
    b. Is the null distribution centered at zero? Explain why your answer makes sense.

3. Strength of Evidence

20. Now you will calculate a p-value, as you have done from simulated null distributions many times before, in order to assess the strength of evidence that the experimental data provide against the null hypothesis that activity has no effect on change in brain size.

    a.   To calculate the p-value, you will count how many of the simulated MAD statistics are equal to _____ or <u>larger or smaller</u>. (Review your answer to #16 to consider what values are "more extreme" that then observed.)

    b.   Check the **Count Samples** box in the applet, and then enter the observed value of the MAD statistic. Report the p-value.

    c.   You should find that the p-value is fairly close to a common cutoff value for assessing statistical significance. So, go ahead and produce 9000 more shuffles to produce a more accurate estimate of the p-value (there will be a pause). Report this p-value.

    d.   Interpret this p-value: It is the probability of obtaining a MAD statistic of _____ or _____, assuming that _____.

    e.   Based on this p-value, evaluate the strength of evidence: The experimental data provide _____ evidence against the null hypothesis that activity group has no effect on percentage change in brain size.

21. **Significance**:

    a.   Summarize your conclusion, based on the p-value.

    b.   Does this analysis allow you to determine *which* activities differ significantly from which others with regard to percentage change in brain size? Explain why or why not.

22. **Estimation**: Based on what you learned in earlier chapters, suggest how you might go about estimating the magnitude of the difference in the long-run average percentage change in brain size between the tai chi and no intervention groups. (You will see a method for doing this later in this chapter.)

Step 5: Formulate conclusions.

23. **Generalization**: To what population would you feel comfortable generalizing these results?

24. **Causation**: Would you feel comfortable with drawing a cause-and-effect conclusion from this study? Between what two variables? Explain why or why not

Step 6. Look back and ahead.

25. **Looking back:** Did anything about the design and conclusions of this study concern you? Issues you may want to critique include:
    - The match between the research question and the study design
    - How the experimental units were selected
    - How the treatments were assigned to the experimental units
    - How the measurements were recorded
    - The number of experimental units in the study
    - Whether what we observed is of practical value

26. **Looking ahead**: What should the researchers' next steps be to fix the limitations or build on this knowledge?

Section 9.1 Summary

The 3S strategy for simulation-based inference can be applied to comparing multiple groups with a quantitative response variable.
- This approach provides an overall test of the multiple groups simultaneously, as opposed to testing each pair of groups separately.
- The null hypothesis is again that there is no association between the two variables.
    - o   If there is no association, then the population means or long-run means of the response variable are identical for all categories of the explanatory variable in the population.
- The alternative hypothesis is that there is an association between the two variables.
    - o   No direction is specified for the association in the alternative hypothesis.
    - o   The alternative hypothesis can be expressed as saying that at least one population mean or long-run mean differs from the others.

The choice of a statistic to measure how different the groups are is not straightforward. Several reasonable statistics can be used.
- One reasonable statistic is to calculate the difference in sample means for each pair of groups, and then take the **mean of the absolute differences (MAD)**.
- Larger values of the MAD statistic indicate a greater difference in sample means across the groups.

The simulation method is again to shuffle the values of the response variable.

- This shuffling method simulates the results of the study under the assumption of no association between the variables.
- The p-value is again calculated as the proportion of repetitions in which the simulated value of the statistic is at least as extreme as the observed value of the statistic.
  - o   Also as always, a small p-value provides evidence against the null hypothesis.

Section 9.2: Theory-based Approach to Comparing More than Two Groups with a Quantitative Response

Introduction

The MAD statistic in Section 9.1 is a simple generalization of what we used in Section 6.2 to test for a difference between two sample means. Unfortunately, however, the MAD statistic does not lead easily to a simple shortcut based on theory. It is also not a standardized statistic in that it does not convey information about the sample sizes involved. (Though it gives you a bit more information about the number of pairs involved than max $\bar{x}$ – min $\bar{x}$ does.) The usual theory-based test uses a different statistic, the ratio of two measures of variability. The numerator of the ratio measures variability *between* groups and the denominator measures variability *within* groups. This ratio is large when the group means are spread out and the response values within each group are close together. The null distribution for this statistic, called the *F*-distribution in honor of statistician Ronald Fisher, has a theory-based approximation that works well provided certain validity conditions are satisfied.

Example 9.2: Recalling Ambiguous Prose

In Example 9.1 we looked at comprehension scores for an ambiguous prose passage. You might recall the students were read a passage that was quite ambiguous and thus difficult to understand. However, a picture went along with the passage that helped make things much more understandable. To test how the picture would help in understanding, the students were randomly assigned to one of three groups.

- Students were shown the picture before they heard the passage.
- Students were shown the picture after they heard the passage.

- Students were not given any picture before or after hearing the passage.

In Example 9.1, we saw that the students were tested on their comprehension of the passage and those that saw the picture before they were read the passage scored higher, on average, than the other two groups. The experiment did not stop there. A few hours later they took the comprehension test again to see how well they could recall the ambiguous passage. Let's take a look at the *recall scores* to see whether any of the three treatment groups tended to have different recall scores from the others.

Let's first repeat what we did in Example 9.1 with this new set of scores. Our hypotheses can be written in words as follows.

Null Hypothesis: All three of the long-run mean recall scores for students under the different cue conditions are the same ($\mu_{\text{no picture}} = \mu_{\text{picture before}} = \mu_{\text{picture after}}$).

Alternative Hypothesis: At least one of the long-run mean recall scores for students under the different cue conditions is different.

where $\mu_{\text{no picture}}$ the long-run mean recall score when no picture shown; $\mu_{\text{picture before}} = \mu_{\text{picture after}}$ defined similarly.

We put the results from the tests into the Multiple Means applet and the results are shown in Figure 9.4. You can again see that showing the picture before gave test scores with a higher sample mean (8.26) than the other two cue conditions (5.37, 6.63)

Figure 9.4: Summary of the **recall scores** from showing the students the picture after, picture before, and not showing a picture at all

Summary Statistics:

	n	Mean	SD
None	19	6.63	2.01
Befo	19	8.26	1.82
Afte	19	5.37	1.46
pooled	57	6.75	1.78

Observed MAD=1.930

When we run the test using the sum of the absolute values as our statistic we get an approximate p-value of 0 as shown in Figure 9.5. We again see the right skewed shape, but it turns out there is actually no simple mathematical model that can be used to predict this null distribution.

Figure 9.5: The null distribution and p-value when comparing recall scores under the three different cue conditions using the MAD statistic

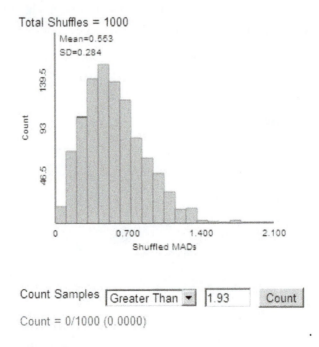

Another Statistic – *F*-statistic (ANOVA)

Notice that the value of the MAD statistic for recall scores (1.93) is larger than when we compared comprehension scores (1.16). But can we directly compare these two numbers? Does the larger MAD necessarily imply stronger evidence against the null hypothesis? To assess that, we really need to look at the p-values. Both p-values were less than 0.001, but are the null distributions equivalent? Figure 9.6 shows the two null distributions.

Figure 9.6: Null distributions of MAD statistic for recall scores and for comprehension scores

(a) Recall   (b) Comprehension

Think about it: Based on these null distributions, which statistic (1.93 for recall or 1.16 for comprehension) provides strong evidence against the null? How are you deciding?

Notice that a MAD statistics of 1.16 gives a p-value of < 0.001 on the Comprehension null distribution, but a p-value of 0.02 on the Recall null distribution. This means that you can't directly compare MAD statistics for different studies. A MAD statistic of 1.16 may be very strong evidence in one study and less strong or not strong evidence at all in another study. This is not a nice characteristic of the statistic.

The way we've solved this issue before is by standardizing. So next we will look at a different statistic that does combine all of this information together into one number.

Now examine the distributions shown in Figure 9.7. The graphs on the left are the original recall scores and the graphs on the right are hypothetical recall scores. Notice that the means for each cue condition are similar across the two studies (around 5 for "after," around 6 for "none" and around 8 for "before").

Think about it: Which study shows stronger evidence of a difference in the long-run mean recall scores among the three cue conditions? Why?

Even though the means are similar, the hypothetical results in 9.7(b) would provide even stronger evidence that at least one of the treatment cue conditions causes an increase in scores. This is because there is an even clearer distinction between the three groups. This happens because there is less "within group" variability and less overlap between the distributions. With less within group variability, the same differences in sample means now seem larger.

Key idea: Instead of comparing only the differences in means, a statistic could standardize those differences by comparing to the amount of within group or natural variability in the response variable, while also taking sample sizes into account.

Figure 9.7: (a) Actual and (b) hypothetical recall scores, with group means displayed

(a) Actual   (b) Hypothetical

An alternative to the MAD statistic is the **F-statistic** (see **Formula** section at the end of this section for details; see FAQ 9.2.1 for more on why it's needed), but, put simply, this is a ratio of "between-group variability" and "within-group variability" and can be expressed as:

$$F = \frac{\text{between-group} \quad \text{variability}}{\text{within-group} \quad \text{variability}}$$

FAQ 9.2.1 Why bother with z and t? Why not just use chi-square and F?

Q: Back in Chapter 5 you showed me how to use a z-statistic to compare two proportions. Now you tell me I can use chi-square instead. You also tell me that with three or more proportions, I can do an overall test for them all using chi-square, but not with z.

A: Guilty as charged.

Q: So why did you make me learn about z? Why can't I just use chi-square all the time?

A: While you're at it, why don't you ask the same sort of question about t and F?

Q: OK. Why did we bother learning about t back in Chapter 6? Chapter 9 has given us F for an overall test to compare two or more means. Why don't we just use F when we want to compare two means?

A: Three reasons. First, most people find z and t easier to learn than chi-square and F. It's easier to start with just two groups, and the standardizing is straightforward. The chi-square and F statistics are less intuitive.

Q: Second?

A: The z and t tests can be either one-sided or two-sided. They tell you which proportion or mean is larger. They also give you confidence intervals for the difference. Chi-square and F statistics don't do either one.

Q: Third?

A: Suppose you've got three or more groups and your chi-square or F test shows strong evidence of at least one pairwise difference.

Q: OK.

A: You still need z or t intervals for follow-up to pin down where the differences are and how big they are.

What does this formula tell us? The numerator is a measure of how much the group means deviate from the overall mean of the dataset (averaging all the values together). The denominator is a combined estimate of the variability within the groups, giving more weight to the groups that have larger sample sizes. So if the differences between the group means are large relative to the differences within the groups (as is happening with the hypothetical data in Figure 9.7), the numerator is relatively larger than the denominator, and the *F-statistic* is large, relatively speaking. If the variability in the data within the groups is large relative to the variability between the group means, then the F-statistic will be relatively small, and the evidence against the null hypothesis will be weak, relatively speaking. So, larger values of the F-statistic are stronger evidence against the null hypothesis. See **FAQ 9.2.2** for more details.

The value of this statistic for these data turns out to be 12.67. Sounds large, but keep in mind that we can't interpret this as "number of standard deviations" anymore. So how do we decide whether this is larger than we would expect by chance alone? We can again run a simulation, scrambling the response variable, recomputing this F-statistic each time, and then see where 12.67 falls in this null distribution (Figure 9.8).

Figure 9.8: Simulated null distribution of the *F*-statistic

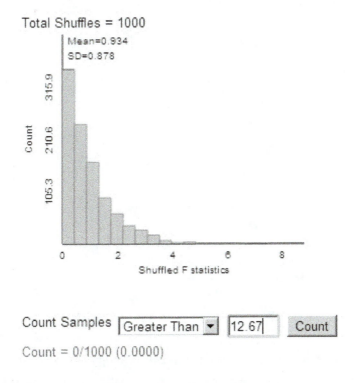

As you can see in Figure 9.8, the *F*-statistic behaves pretty similarly to the one you saw in Figure 9.5 for the MAD and will be nonnegative and right skewed (even more so than before). The scaling is different from the MAD statistic but again our observed statistic, $F_{observed} = 12.67$, happened less than 1 in 1000 repetitions, giving us very strong evidence against the null hypothesis.

But this time, this distribution has a known theoretical model – the ***F-distribution*** (named after famous statistician R.A. Fisher). Below we overlay the *F*-curve on our histogram of our simulated data.

Figure 9.9: Simulated *F*-statistics with *F*-distribution overlaid

The advantage of this probability model is we can have technology compute the probability of interest, the area under the curve at 12.67 and above, from the mathematical model rather than running the simulation. As our simulation foretold, it is very unlikely to get an *F*-statistic as large as 12.67 by chance alone!

Most statistical packages will do all of this for you in one step, called an ***ANOVA test*** or Analysis of Variance. We can check the **Show ANOVA Table** box in the applet (Figure 9.10).

Figure 9.10: Applet ANOVA output for testing the difference in recall scores

Source	df	SS	MS	F	p-value
Treatment	2	80.04	40.02	12.67	0.0000
Error		54	170.53	3.16	
Total		56	250.56		

The main thing to note in the output in Figure 9.10 is that the p-value is zero to at least four decimal places. Thus we have strong evidence against the null hypothesis and thus can conclude that at least one of the long-run mean recall scores for the students under the three cue conditions will be different. Also note that this p-value is similar to what we found in the simulation-based approach.

Before we look more closely at which mean might be different (although we think you know which it is) let's try to understand what some of the other numbers mean in the output in Figure 9.10. Though the p-value is the most important number to understand, if we know a little bit more about some of the other numbers we will understand what an ANOVA test is doing.

The *mean squares for treatment* (40.02 above) is measuring the variation between the three groups. In fact, it is similar to the numerator of the chi-square statistic from 8.2 which compares each group proportion to an overall proportion as a measure of how far apart the group proportions were. In this case, we will use means to find the sum of the squared deviations between each group mean and the overall mean of the data. The *mean squares for error* (3.16 above) is a measure of the within group variation. It is similar to averaging the sample standard deviations across the three groups. The *F*-statistic is a ratio of these two measures of variation; that is, $F = 40.02/3.16 = 12.67$. As we saw above, when the variation between groups is large compared to the variation within groups, we have stronger evidence against the null hypothesis. This is the case we have here. Remember that this is an overall test and only tells us at least one difference exists. Our next step is to find out where the significant difference(s) is (are).

Just like the other theory-based approaches of conducting tests of significance, there are validity conditions that must be met in order to run an ANOVA test.

Validity conditions for the ANOVA test

The *F*-distribution is a good approximation to the null distribution of the *F*-statistic as long as

- Either the sample size is at least 20 for all the groups, or the distribution of the response variable is approximately normal in all the samples (examine the dotplots for strong skewness or outliers).

> • Also, the standard deviations of the samples are approximately equal to each other (largest standard deviation is not more than twice the value of the smallest standard deviation).

Let's take a look at our dotplots again to assess whether our F-distribution is predicted to be a good fit to the simulated null distribution we created in our simulation-based test (Figure 9.11).

Figure 9.11: Scores for the recall of an ambiguous prose passage for 57 students under three different cue conditions

Because the sample sizes are not large enough (barely less than 20) for all the three samples, we look at the dotplots and find that we don't have sample distributions that are highly skewed or have outliers; so we will consider the condition of normality to be met, even though the sample sizes are not particularly large. The standard deviations for our three groups are 2.01, 1.82, and 1.46. The largest (2.01) is not more than twice the smallest (1.46) so the equal standard deviations condition is considered to be met. Therefore it is okay for us to run an ANOVA test like we did. In fact, statisticians have found that this F-distribution works pretty often, even with skewed data, especially with larger sample sizes.

For this *Ambiguous Prose* study, the choice of statistic (max – min or MAD or F) is not that important, all three choices here give very similar results here: We have strong evidence against the null hypothesis and thus can conclude that at least one of the long-run mean recall scores for the three cue conditions is different. Because the 57 students were randomly assigned to the three different treatment groups, we can conclude that the difference in recall scores was *caused* by the treatment. See **FAQ 9.2.2** for more on choice of statistic.

Follow-up Analysis

The ANOVA test is an overall test. A statistically significant ANOVA test tells us that at least one long-run or population mean is different from the others, but does not tell us which one or ones. (Though looking back at the sample means and dotplots we had our suspicions.) To determine which mean or means differ *significantly* can conduct follow up analyses to the ANOVA. There are many different follow up analyses that can be done. We will use a "pooled" version of the two-sample t confidence intervals for the difference in means from Chapter 6. (Basically, we use the same estimate of the pooled standard deviation for each pair.) We can do this by checking the Compute 95% confidence interval(s) box in the applet. Doing so we get the following intervals for the differences in long-run means:

☑ Compute 95% confidence interval(s)
After - Before: (-4.05, -1.74)*
After - None: (-2.42, -0.11)*
Before - None: (0.4756, 2.7875)*

These tell us that we have convincing evidence of a difference in long-run mean recall scores between all three conditions (because none of our intervals contain 0). For example, we have a significant difference when we show the picture before compared to either after (all negative values) or no picture (all positive values). This means that we have strong evidence that showing the picture before reading the passage will result in higher recall scores than showing no picture (on average 0.476 to 2.879 higher), and higher recall scores than showing the picture after reading the passage (on average 1.74 to 4.05 higher).

> **Key idea:** If an ANOVA test finds that there is at least one long-run or population mean that is different, there are many different follow up tests designed to help pinpoint where difference(s) occur. One option is to look at all the pairwise theory-based confidence intervals for the difference in long-run or population means.

Formulas

In this section we will show you the details of how to compute the *F*-statistic. Software packages can do this automatically, but knowing how to do the computation by hand can be useful to ensure you really understand what the ratio means.

The *F*-statistic (or "*F*-ratio") is given as

$$F = \frac{\text{between} - \text{group} \quad \text{variability}}{\text{within} - \text{group} \quad \text{variability}} = \frac{\dfrac{\sum\limits_{i=1}^{I} n_i (\bar{x}_i - \bar{x})^2}{I-1}}{\dfrac{\sum\limits_{i=1}^{I} (n_i - 1)s_i^2}{n-I}}$$

There is quite a bit of new notation there! Briefly, stated, there are *I* groups being compared, and *n* is the overall sample size, and *i* tells us what group we're in. So, $\bar{x}_i$ and s_i represent the sample mean and standard deviation in group *i*, respectively. Finally, $\bar{x}$ is the overall mean, combining all the observations together. The numerator is our measure of how much the group means deviate from the overall mean and the denominator is our measure of how much variability there is in the measurements themselves, pooled across all of the groups.

Recall the following summary statistics for recall score in this study.

```
Summary Statistics:

              n       Mean    SD
None         19       6.63    2.01
Befo         19       8.26    1.82
Afte         19       5.37    1.46
pooled       57       6.75    1.78
```

To find the observed value of the *F*-statistic, let us first calculate the numerator:

- *between-group variability* =

$$\frac{\sum n_i (\bar{x}_i - \bar{x})^2}{I - 1} = \frac{19\,(6.63 - 6.75)^2 + 19\,(8.26 - 6.75)^2 + 19\,(5.37 - 6.75)^2}{3 - 1} =$$

$$\frac{80.04}{2} = 40.02 \quad \text{(also known as the Mean Squares Treatment; MST)}$$

- *within-group variability =*

$$\frac{\Sigma(n_i - 1)s_i^2}{n - I} = \frac{(19 - 1)(2.01)^2 + (19 - 1)(1.82)^2 + (19 - 1)(1.46)^2}{57 - 3} = \frac{170.53}{54} = \approx 3.16$$

(also known as the Mean Squares Error; MSE)

- Thus, the observed value of the F-statistic $= 40.02/3.16 = 12.67$.

Note how these values correspond to output received from the applet when the checkbox for ANOVA is clicked.

Source	df	SS	MS	F	p-value
Treatment	2	80.04	40.02	12.67	0.0000
Error	54	170.53	3.16		
Total	56	250.56			

In particular, the mean-squares treatment is the between groups variability (40.02), the mean-squares error is the within groups variability (3.16). The treatment df (degrees of freedom) is $I - 2$, and the error degrees of freedom is $n - I$.

Recall Figure 9.6 where we saw that MAD statistics were not comparable across studies. Here we show the distribution of simulated F-statistics across the two studies. Notice that the two distributions are nearly identical.

Figure 9.12: Null distributions of *F*-statistic for recall scores and for comprehension scores
(a) Recall   (b) Comprehension

Thus, the *F*-statistic is a standardized statistic and is fairly comparable across studies. With this in mind, because the comprehension scores *F*-statistic is 10.01 compared to 12.67 for the recall scores, there is more evidence of an association when comparing recall scores. This is also reflected in the p-values (0.0002 for comprehension scores and <0.0001 for recall scores). In this case the statistics are directly comparable because the sample sizes in the groups are identical. If that is not the case, comparing F-statistics requires more effort.

FAQ 9.2.2 The *F* statistic

Q: Remember back in Chapter 8 I asked which statistic was best: $[(\hat{p})_{max} - \hat{p}_{min}]$, MAD or chi-square?

A: Sure. FAQ 8.2.1.

Q: I suppose the answer is similar for $\bar{x}_{Max} - \bar{x}_{min}$, MAD, or *F*: Which is best?

A: The logic is almost exactly the same. For some situations, it doesn't make much difference, for others, it does. I won't repeat all the prose, but here instead is a summary table:

	Max - Min	MAD	chi-square and F
Intuitive?	Yes	Yes	No
Theory-based approximation?	No	No	Yes
Straightforward extension to larger tables?	No	No	Yes
Standardized?	No	No	Yes
Power	Lower	Depends	Higher

Q: That fits with what I thought. But how did anyone ever come up with the F statistic? The other two, $(\bar{x}_{Max} - \bar{x}_{min})$ and MAD, seem natural enough, but F is less intuitive.

A: I agree. The main reason it's harder is that it's based on a ratio. You've seen that before, with t and z, but for F it's less clear because we work with squares of SDs. In all there are three things to understand: (1) the numerator, (2) the denominator, and (3) what the ratio tells.

Q: OK. What's the numerator?

A: It's the SD^2 of the group means—how spread out the group means are from each other– multiplied by the group size. Like the MAD, the numerator is zero when those means are all equal, it's small when they are close to each other, and it's large when at least two means are far apart.

Q: So the numerator is like the MAD times group size except it uses squares instead of absolute values?

A: That's a good way to think about it.

Q: And the denominator?

A: If group sizes are equal the denominator is the average of the SD^2s for the individual groups. So, the denominator tells you whether response values in the same group are close together or far apart. The closer they are, the smaller the denominator, just as with any SD. So the denominator measures within-group variability.

Q: And if the group sizes aren't equal?

A: There are minor adjustments for group sizes but the main idea is still the same. The denominator summarizes the variability (average SD^2) within groups.

Q: I'm ready for the ratio.

A: There are two related ways to think about it. One is that the ratio compares the variability *between* groups (numerator) and the variability *within* groups.

Q: And the other way?

A: Remember how when comparing two group proportions z^2 = chi-square? Well, if the variability within the two groups being compared is similar than the t-statistic from a two-sample t-test can be squared to get the corresponding F statistic. In other words, $F = t^2$.

Q: Right.

A: And the t-statistic comes from standardizing: we divide $(\bar{x}_1 - \bar{x}_2)$ by its SE.

Q: And since $F = t^2$ ...

A: ... the F statistic also divides in order to standardize, although everything is in the squared scale.

Q: Are we done?

A: Not quite. Before we quit, it's worth considering how the F-ratio behaves when group means are far apart and when they are close together. First, suppose the means are far apart compared to the within-group SDs.

Q: Then the numerator is large, the denominator is small, and so the F-ratio is large.

A: Exactly. That's why the p-value comes from the right tail of the null distribution. We consider values to be more extreme if they produce larger F-statistics.

Q: And if the group means are all the same the numerator is 0 and so F is 0.

A: That's true, but we care more about how F behaves when the null hypothesis is true.

Q: "H_0 true" means there's no difference between groups. All the observational units belong to one big group.

A: Right. If that's true, the numerator is just another way to estimate the within-groups SD. Numerator and denominator are both estimating the same number.

Q: So the F-ratio should be near 1.

A: Yes, you've got it. To end with, here's a summary of the F-ratio: Apart from a multiplier based on the group sample sizes, the F-ratio equals the variability between groups divided by the variability within groups.

Q: And the multiplier?

A: Basically, the size of a group.

Q: Can you make this more concrete?

A: Sure, here's a quick example. Consider this dataset.

Group	A	B	C
Group size	5	5	5
Group mean	2	5	8

Group SD	4	5	3	

What is the overall mean? Ignore group.

Q: I guess since the groups are equally sized, the overall mean is 5 since the means for group A and C kind of 'cancel out.'

A: Exactly. Now, what's the standard deviation of the three group means: 2, 5 and 8.
Q: Can I use an applet? Or is that cheating?

A: Knowing when and how to use technology isn't cheating!

Q: OK, the SD of the three group means is 3.

A: So, the numerator is the group size (5) times the SD of the group means (9).

Q: So, 45. Why does the equation in the Formulas section look harder?

A: Well, the example I'm showing you is nice because the group sizes are the same. That makes the math easier. The formula in the book covers the more complicated cases also.

Q: OK, so the denominator…I see that the SDs within each group are 4, 5 and 3. So, I square them first. That gives me 16, 25 and 9. Now what?

A: Average those three numbers together.

Q: OK, 16 + 25 + 9 = 50. Dividing by three gives 16.7. This means that the F-statistic is 45/16.7, meaning 2.7.

A: Nice work. Now a question for you. What happens if I double the sample size from 5 to 10 in each group… but everything else stays the same.

Q: Well. The denominator won't change. Just the numerator. So, my F-statistic doubles, meaning more evidence against the null hypothesis.

A: Great job. You've got it now.

Exploration 9.2: Comparing Popular Diets

Because about two-thirds of Americans are considered overweight, weight loss is big business. There many different types of diets, but do some work better than others? Is low fat better than low carb, or is some combination best? Researchers (Garnder er al. 2007) conducted a study involving four popular diets: Atkins (very low carb), Zone (40:30:30 ratio of carbs, protein, fat), LEARN (high carbohydrate, low fat) and Ornish (low fat). They randomly assigned women aged 25 to 50 with a body mass index (BMI) of 27 to 40 (overweight and obese) to one of the three diets. The 311 women who volunteered for the program were educated on their assigned diet and were observed periodically as they stayed on the diet for a year. At the end of the year, the researchers calculated the amount of weight loss for each woman and compared the results across the four diets.

Step 1. Ask a research question.

    1.   What is the overarching research question the researchers hoped to answer?

Step 2. Design a study and collect data.

    2.

        a.   What are the observational units in this study?

        b.   Identify the explanatory and response variables. Classify them as categorical or quantitative. For categorical variables, indicate how many categories are used.

            Explanatory: Type:

            Response: Type:

    3.

        a.   Does this study make use of random sampling, random assignment, both, or neither? What are the implications of your answer with regard to scope of inference?

        b.   Did the researchers collect the data as paired data or as independent samples? As in, according to the study design, are the responses from one treatment group paired with or independent of the responses from other treatment groups?

4. State the null and alternative hypotheses, both in words and symbols, for testing the research conjecture. (Recall that the response variable is weight loss; positive values indicate that amount of weight lost between the beginning of the study and the end of the study; negative values indicate the woman gained weight during the study.)

    Null hypothesis:

    Alternative hypothesis:

Step 3. Explore the data.

5. The data for this investigation appear in a file called **Diets**, available on the textbook website. Copy and paste the data into the **Multiple Means** applet. (Recall that the response variable is *weight loss*.) Press **Use Data**, and the applet will produce dotplots and summary statistics. You can also check the **Boxplots** box to overlay those as well.

    a. Report the means, standard deviations, and sample sizes for the weight loss amounts for each diet.

    b. Describe what the graphs and statistics reveal about whether the diets appear to differ with regard to weight loss and, if so, which diet appears to be best and which worst.

Step 4. Draw inferences beyond the data.

6. Conduct a simulation-based test, as in the last section, to compare the four group means using the mean of the absolute values of the differences (MAD) as the statistic. Report the observed value of this statistic. Then determine and report a p-value based on at least 1000 shuffles.

7. Based on this p-value, evaluate the strength of evidence against the null hypothesis provided by the experimental data. In other words, do the data provide strong evidence that the four diets differ with regard to average weight loss?

Although the MAD statistic is fairly easy to understand and calculate, it is not commonly used, in part because there is no theory-based model for the null distribution. Another downside is that the MAD statistic is not standardized so it is not comparable across studies (e.g., a MAD statistic of 1 in one study might be strong evidence, but in another study little to no evidence).

> **Key idea:** Instead of comparing only the differences in means, a statistic could standardize those differences by comparing to the amount of within group or natural variability in the response variable, while also taking sample sizes into account.

A much more commonly used statistic for comparing multiple groups on a quantitative response, which does have a theory-based distribution, is called an **F-statistic**. As with the MAD statistic, the *F*-statistic equals zero only when the group means are all identical. Otherwise the *F*-statistic is positive, with larger values indicating larger differences across the group means.

The *F*-statistic (see Formula section at the end of Example 9.2 for details) is a ratio of "between-group" and "within-group" variability. Thus,

$$F = \frac{\text{between-group} \quad \text{variability}}{\text{within-group} \quad \text{variability}}$$

The numerator is a measure of how much the group means differ from each other, and the denominator is a measure of how much variation there is within the groups (related to the SD within the groups).

8. Select the **Statistic** pull-down menu (on the left) to select **F-statistic**, and the null distribution changes from the MAD statistic to the *F*-statistic.

   a. Report the observed value of the *F*-statistic, which appears on the left side of the applet.

   b. Use the simulation results, and the observed value of the *F*-statistic, to determine the p-value.

   c. Is this p-value similar to the one based on the MAD statistic? Is your conclusion about strength of evidence the same with both statistics?

The *F*-statistic has a known theoretical distribution when the null hypothesis is true, called (surprisingly enough) an **F-distribution**.

9.

    a.   Check the box to **Overlay *F* distribution**. Would you say there is good agreement between theoretical prediction and your simulated null distribution?

    b.   How does the theory-based p-value compare to what you found in #8b?

But what does $F = 3.80$ really mean? Check the box by **Show ANOVA Table**, to see more output related to the calculation of the value of the *F*-statistic.

10. Confirm that the *F*-statistic is the ratio of the Mean Squares (MS) for treatment to Mean Squares (MS) for error.

11. To explore this statistic further, switch to the **Descriptive Statistics** applet.
    - Type in the four group means for the original data. (Clear the existing data first and uncheck the **Includes header** box.)
    - Have the applet calculate the standard deviation of these four group means.
    Write a one sentence interpretation of this number.

12. To estimate the natural variation in these BMI values, we want to basically average the "within group" variability across the four groups. Take the four group standard deviations and average them. (You can use the same applet). Write a one sentence interpretation of this number.

13. Now compare your answer to #11 to your answer to #12 by finding the ratio of the **squares** of these values (variance = standard deviation squared), multiplying the numerator by 77, roughly the sample size of each group.

        77 X Variability between the group means / Variability within the groups

Note: There is some more discussion about the *F*-statistic in Example 9.2 and FAQ 9.2.2. If you haven't already done so, we encourage you to read more about the *F*-statistic there. The formula for the *F*-statistic is given at the end of the section.

14. You have now calculated a p-value in three different ways, two using a simulation-based approach (but with different statistics, MAD and *F*) and one using a theory-based approach (using the *F* statistic). Are they all similar? What is your conclusion about whether the diets differ with regard to average BMI changes?

Follow-up Analysis

The ANOVA *F*-test is an overall test. A significant test result tells us that we have strong evidence that at least one population mean is different from the others, but does not tell us which one(s) differ from which other one(s). To determine this we need to apply a follow-up procedure to ANOVA. There are many different follow-up tests that can be done. We will use pairwise confidence intervals for the difference in two means as we did in Chapter 6.

Key idea: If an ANOVA test finds that there is at least one long-run or population mean that is different, there are many different follow up tests designed to help pinpoint where difference(s) occur. One option is to look at all the pairwise theory-based confidence intervals for the difference in long-run or population means.

---

15. When you have the *F*-statistic selected and the *F*-distribution overlaid (or are showing the ANOVA table) you can check the box to **Compute 95% CI(s) for difference in means.** Report the intervals below.

    a. Atkins Diet vs. Zone Diet.

    b. Atkins Diet vs. Ornish Diet.

    c. Atkins Diet vs. LEARN Diet.

    d.   Zone Diet vs. Ornish Diet.

    e.   Zone Diet vs. LEARN Diet.

    f.   Ornish Diet vs. LEARN Diet

16. Which of the confidence intervals calculated in #15 indicate a significant difference between groups? How are you deciding?

17. Based on your analysis, would you conclude that there is one diet that works significantly better than the other three? If so, which one?

Validity Conditions

As with other theory-based approaches of conducting tests of significance, certain validity conditions must be met in order to conduct an ANOVA F-test.

> **Validity conditions for the ANOVA F-test**
>
> The F-distribution is a good approximation to the null distribution of the F-statistic as long as
> - Either the sample size is at least 20 for all the groups without strong skewness our outliers, or if the sample size is less than 20, then the distribution of the response variable is approximately symmetric in all the samples (examine the dotplots for strong skewness or outliers).
> - Also, the standard deviations of the samples are approximately equal to each other (largest standard deviation is not more than twice the value of the smallest standard deviation).

18. Do the sample data for the diet study appear to satisfy the validity conditions for conducting an ANOVA F-test? How are you deciding?

Step 5. Formulate conclusions.

19. Are you comfortable with concluding from this study that the diet used causes a difference in BMI change? Justify your answer.

20.   To what population are you comfortable generalizing the results of this study? Justify your answer.

Step 6. Look back and ahead.

21.   **Looking back:** Did anything about the design and conclusions of this study concern you? Issues you may want to critique include:
   - The match between the research question and the study design
   - How the experimental units were selected
   - How the treatments was assigned to the experimental units
   - How the measurements were recorded
   - The number of experimental units in the study
   - Whether what we observed is of practical value

22.   **Looking ahead**: What should the researchers' next steps be to fix the limitations or build on this knowledge?

Section 9.2 Summary

A theory-based approach can be used to test equality of several group means under certain conditions.
   - This procedure is called **analysis of variance (ANOVA)**.
   - The statistic is called an *F*-statistic.
     - The *F*-statistic is calculated as a ratio of between-group variability to within-group variability.
     - Larger values of the *F*-statistic indicate greater discrepancies among the sample means.
     - Larger values of the *F*-statistic provide stronger evidence that the long-run or population means are not all the same.
   - The validity conditions for the ANOVA *F*-test are

     - That all the sample sizes are at least 20 for all the groups, or that the quantitative response variable has an approximately normal distribution in each of the populations.
     - That the populations all have the same variability as measured by SD. We will consider the condition met if no sample standard deviation is more than twice as large as another sample standard deviation.

When an *F*-test reveals strong evidence that at least one of the population means differs from the others, confidence intervals for the difference in population means can be calculated for pairs of groups.

- Confidence intervals that do not include zero indicate which pairs of groups differ significantly.

Type of data	Name of test	Applet	Validity conditions
A categorical explanatory variable with 2 or more categories and a quantitative response variable	Simulation test for multiple means	Multiple means	---
	Theory-based test for multiple means (*F*-test test/statistic; ANOVA)	Multiple means	At least 20 in each group, or mostly symmetric distribution of response variable in each group; standard deviations of response variable within each group are within factor of 2 each other

Chapter 9 Summary

This chapter highlighted that when comparing multiple means there are actually two kinds of variability that need to be considered. The variation between the groups is what we are most interested in because we are trying to determine whether the group means are significantly different from each other. When the group means are farther apart from each other (larger variation between the groups), the evidence against the null hypothesis is stronger. However, this difference can be clouded if there is also a lot of variability within the groups. An *F*-statistic is a convenient way to examine the ratio of these two sources of variability. Also, the *F*-statistic has a nice mathematical model for the null distribution.

We also realized that there are two steps to finding differences between more than two groups.

First we need an overall test to determine whether any differences exist. If so, then we can use follow-up pairwise confidence intervals to ascertain where those differences are.

Chapter 9 Glossary

ANOVA test

Analysis of variance test, is an overall test of multiple means that explores the variation between groups compared to the variation within groups 9-26

F-distribution

Theory-based approximation for simulated null distribution of F-statistic is non-negative and skewed right 9-25, 9-38

Follow-up Analysis

A second step in the analysis process that follows a significant ANOVA test. A follow-up test tells where significant differences between pairs of groups are found. This is usually presented as confidence intervals for the difference in each pair of means 9-28, 9-40

F-statistic

Ratio of variation between the groups to the variation within the groups 9-24, 9-38, 9-43

mean squares error

Denominator of the F-statistic. Measures the within group variation. It is similar to averaging the standard deviations across the groups being compared. 9-27

mean squares for treatment

Numerator of the F-statistic. Measures the variation between the groups. 9-27

validity conditions for ANOVA

For the F-statistic to follow an F distribution, need each sample distribution to be approximately normal or large sample sizes and the sample standard deviations should be similar 9-27

Exercises

Section 9.1

9.1.1* Suppose you were conducting a simulation with cards to compare the means of three groups of test scores. The groups consisted of samples of size 8, 10, and 12. Which of the following best describes one repetition of the simulation?

   A. Take 30 cards and write down each test score on a card, shuffle all the cards and make 3 piles of 10 cards. Find the means for all three piles and from these compute a MAD statistic.
   B. Take 8 red, 10 blue, and 12 green cards, shuffle all the cards and make a pile of 8 cards, a pile of 10 cards and a pile of 12 cards. Find the proportion of red cards in each group and from these compute the MAD statistic.
   C. Take 30 cards and write down each test score on a card, shuffle all the cards and make a pile of 8 cards, a pile of 10 cards and a pile of 12 cards. Find the means for all three piles and then find the average of all these means.
   D. Take 30 cards and write down each test score on a card, shuffle all the cards and make a pile of 8 cards, a pile of 10 cards and a pile of 12 cards. Find the means for all three piles and from these compute a MAD statistic.

9.1.2 When finding determining the p-value using the MAD statistic:

   A. We always count the values in the null distribution that are greater than or equal to our observed MAD statistic.
   B. We always count the values in the null distribution that are less than or equal to our observed MAD statistic.
   C. We either count the values in the null distribution that are greater than or equal to our observed MAD statistic or the values less than or equal to the negative of our observed MAD statistic. It depends on the direction of the alternative hypothesis.
   D. We always count the values in the null distribution that are greater than or equal to our observed MAD statistic as well as the values that are less than or equal to the negative value of our observed MAD statistic because these are all two-sided tests.

9.1.3* When testing for an association between class level (Fresh, Soph, Junior, Senior) and IQ score, which of the following is NOT an appropriate way to write the alternative hypothesis?

    A. There is an association between class level and IQ.

    B. At least one of the population IQ means will be different.

    C. Not all the population IQ means are the same.

    D. $\mu_F \neq \mu_S \neq \mu_J \neq \mu_S$

9.1.4 Suppose you are comparing just two means. Among the possible statistics you could use is the difference in means, the MAD, or the max − min (the difference between the largest mean and the smallest mean).

    a) Will all three of these statistics always be the same? Will any pair of these statistics always be the same?

    b) If one of the statistics is different from the other two, how is it related to them?

9.1.5* Suppose there is no association between class level (Fr, So, Jr, Sr) and the amount of sleep student get per night at a certain college. What does no association imply about the mean sleep hours for each class level?

9.1.6 Consider the following four boxplots constructed from the four data sets A, B, C, and D.

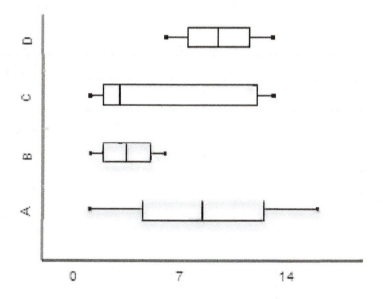

a) Which data set has the largest interquartile range? Which has the smallest?

b) Which data set has the largest median? Which has the smallest?

c) Which data set looks like it could be skewed? In which direction could it be skewed?

d) Which data set has the largest lower quartile?

e) Which data set has the smallest sample size?

9.1.7* Consider the MAD statistic. What would happen if you did not take absolute values before calculating the sum of differences between group means? Why would this not be a useful calculation without taking absolute values?

9.1.8 For its numerator, the MAD statistic uses the sum of absolute pairwise differences. The absolute values get rid of possible negative differences. Instead of taking absolute value, what is another way to avoid summing negative differences?

9.1.9* Suppose three group means are 4, 5 and 10. Compute the value of the MAD statistic.

9.1.10 Suppose four group means are 2, 5, 7, and 8. Compute the value of the MAD statistic.

Baseball payrolls

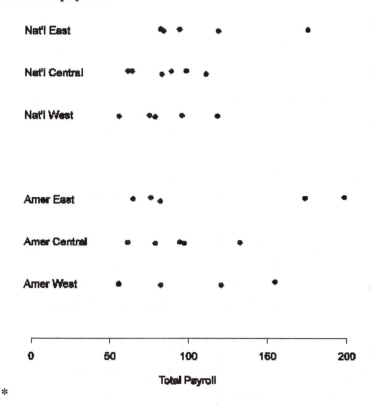

9.1.11: The plot below shows total payroll in millions of dollars, for the 30 major league teams of professional baseball, sorted by league (National or American) and division (East, Central, West).

a) Identify the observational units, the response, and the explanatory variable.

b) For plots like the one above:
  i. Each point represents a _____ (unit, response value, value of the explanatory variable)
  ii. Each horizontal cluster, taken as a whole, corresponds to a value of the _____ (response, explanatory) variable.
  iii. Values along the horizontal axis represent values of the _____ (response, explanatory) variable.

c) The overall distribution of the response is (Choose one):

    A. Symmetric

    B. Skewed right (a long tail points to the larger values)

    C. Skewed left (a long tail points to the lower values)

    D. No way to tell – need to see a histogram.

d) Comparing groups by eye suggests that (Choose one):

    A. Group means are roughly equal. A p-value will show weak evidence of differences between group means.

    B. Group means show substantial differences, but total payroll varies a lot within divisions. A p-value will not show strong evidence of group differences in means.

    C. Group means are roughly equal, but within-group variability is small enough that with several response values per group, even the small differences in means will register as significant.

    D. Group means show substantial differences, and overall variability within divisions is small enough by comparison to differences in group means that a p-value will show evidence, possibly strong evidence, of differences.

    E. There is no way to tell without getting a computer to find the p-values.

Statistics student survey

9.1.12: Statistics students were surveyed and two of the questions they were asked was what type of cellphone they use (Basic, iPhone, Smart Phone that is not an iPhone) and how much sleep they typically get on a school night in hours to see if there is an association between sleep and type of phone. For the purposes of this exercise, we will consider this sample to be representative of all students at the school. The summary statistics are shown in the following table and the data are shown in the dotplots.

Type of Phone	Sample Size	Mean	Standard Deviation
Smart Phone	18	7.47	1.00
iPhone	61	7.27	1.07
Basic	15	6.40	1.08

a) Identify the explanatory variable and the response variable in this study.
b) State, in words or symbols, the null and alternative hypotheses.
c) Using the summary statistics given, compute the MAD statistic.
d) A null distribution for this study is shown. Using the MAD statistic you calculated in part (c), will the p-value for this test be large or small? Explain.

e) Does it appear there is an association between type of cell phone used and the amount of sleep students get? If so, can you conclude the type of cellphone being used is causing the difference in sleep times?

Extracurricular activities*

9.1.13: Hope College students were surveyed and were asked their class level and how many hours of extracurricular activities they participated in during the past month. For the purposes of this exercise, we will consider this sample to be representative of all students at the school. The summary statistics are shown in the following table and the data are shown in the dotplots.

Class	Sample Size	Mean	Standard Deviation
Senior	13	43.54	33.18
Junior	29	33.83	34.33
Sophomore	19	28.63	26.13
Freshman	25	26.24	24.47

a) Identify the explanatory variable and the response variable in this study.
b) State, in words or symbols, the null and alternative hypotheses.
c) Using the summary statistics given, compute the MAD statistic.
d) A null distribution for this study is shown. Using the MAD statistic you calculated in part (c), will the p-value for this test be large or small? Explain.

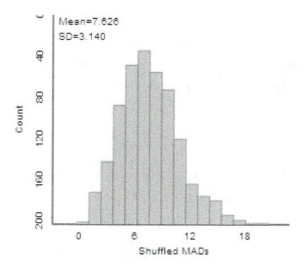

e) Does it appear there is an association between class level and the amount of extracurricular activities students participate in?

Height and phone type

9.1.14: Statistics students were surveyed and two of the questions they were asked was what type of cellphone do they use (Basic, iPhone, Smart Phone that is not an iPhone) and how tall they are, in inches to investigate a possible association between type of phone and student height. For the purposes of this exercise, we will consider this sample to be representative of all students at the school. The summary statistics are shown in the following table and the data are shown in the dotplots.

Type of Phone	Sample Size	Mean	Standard Deviation
Smart Phone	18	70.58	2.55
iPhone	61	66.90	4.65
Basic	15	67.10	2.93

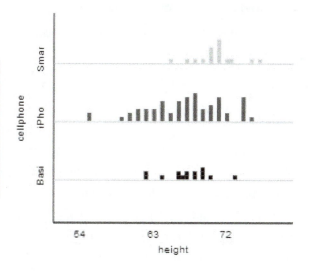

a) State, in words or symbols, the null and alternative hypotheses.

b) Using the summary statistics given, compute the MAD statistic.

c) A null distribution for this study is shown. Using the MAD statistic you calculated part (b), will the p-value for this test be large or small? Explain.

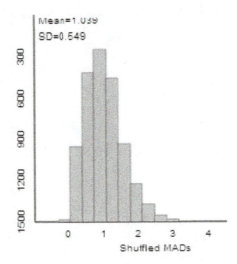

d) Does it appear there is an association between type of cell phone used and height? If so, can you conclude the type of cellphone being used is causing the difference in sleep times? What is a possible confounding variable in this?

Video watching and emotions*

Problems 9.1.15 through 9.1.17 refer to the data gathered on emotional states and video watching.

9.1.15: Student researchers at Hope College studied the effects that watching certain types of videos on a person's emotional state. The subjects were randomly shown a short happy video (babies laughing), a short sad video (abused dogs in a shelter), or no video. After this, the subjects were given an assessment of their emotional state with high numbers representing 'good emotional state.' The results can be found in the dataset **VideoEmotion**.

a) Identify the explanatory variable and the response variable in this study.

b) State, in words or symbols, the null and alternative hypotheses.

c) Put the data in the **Multiple Means** applet and describe what the dotplots tell us about the association between type of video and emotional state.

d) Find the MAD statistic. What is that number describing?

e) Use a simulation-based approach to test whether video type is affects a person's emotional state. Be sure to report a p-value.

f) Based on the p-value, state an appropriate conclusion in the context of the study. Be sure to comment on statistical significance, causation, and generalization, along with how you are making your decision.

9.1.16: Reconsider **Exercise 9.1.15** on video type and emotion. Besides assessing their subjects' emotional state after watching the videos, they also assessed their mood. The results can be found in the dataset **VideoMood** with the higher the mood score, the better the subject's mood.

a) Identify the explanatory variable and the response variable in this study.

b) State, in words or symbols, the null and alternative hypotheses.

c) Put the data in the **Multiple Means** applet and describe what the dotplots tell us about the association between type of video and a person's mood.

d) Find the MAD statistic. What is that number describing?

e) Use a simulation-based approach to test whether video type is affects a person's mood. Be sure to report a p-value.

f) Based on the p-value, state an appropriate conclusion in the context of the study. Be sure to comment on statistical significance, causation, and generalization, along with how you are making your decision.

9.1.17: Reconsider **Exercise 9.1.15** on video type and emotion. Besides assessing their subjects' emotional state after watching the videos, they also assessed their stress. The results can be found in the dataset **VideoStess** with the higher the stress score, the lower the stress level; so high numbers are good.

a) Identify the explanatory variable and the response variable in this study.

b) State, in words or symbols, the null and alternative hypotheses.

c) Put the data in the **Multiple Means** applet and describe what the dotplots tell us about the association between type of video and a person's stress level.

d) Find the MAD statistic. What is that number describing?

e) Use a simulation-based approach to test whether video type is affects a person's stress level. Be sure to report a p-value.

f) Based on the p-value, state an appropriate conclusion in the context of the study. Be sure to comment on statistical significance, causation, and generalization, along with how you are making your decision.

Diet and weight gain

9.1.18: Which diet is best? A fairly recent article in the *Journal of the American Medical Association* (Dansinger, Griffith, Gleason et al., 2005) reported on a randomized, comparative experiment in which 160 subjects were randomly assigned to one of four popular diet plans: Atkins, Ornish, Weight Watchers, and Zone (40 subjects per diet). These subjects were recruited through newspaper and television advertisements in the greater Boston area; all were overweight or obese with body mass index values between 27 and 42. Data for the 93 subjects who completed the 12-month study are in datafile **Diets2**. The file contains which diet the subject was on and the weight loss in kilograms (positive values indicate weight loss and negative values indicate weight gain).

a) Identify the explanatory variable and the response variable in this study.
b) What are the observational units in this study?
c) State, in words or symbols, the null and alternative hypotheses.
d) Put the data in the **Multiple Means** applet and describe what the dotplots tell us about the effectiveness of the four diet plans in aiding weight loss.
e) Using the **Multiple Means** applet find and report the value of a statistic that summarizes the discrepancies between the four diet plans. What is this statistic called?
f) Using a simulation-based approach, test whether there is evidence of an association between weight loss and diet. Be sure to report a p-value.
g) Based on the p-value reported in (f), state an appropriate conclusion in the context of the study. Be sure to comment on statistical significance, causation, and generalization, along with how you are making your decision.

Halo*

Problems 9.1.19 through 9.1.22 refer to data gathered on Halo video games.

9.1.19: Halo is a popular science fiction video game. It can be played on different levels. Statistics students at Hope College wanted to determine whether or not the level on which the game was played affects how well a person plays. Data were collected from ninety different games of Halo; thirty from each of the three different levels. Consider the number of Kills to be the response variable. The data are given in **HaloKills**.

a) State, in words, the null and alternative hypotheses to test whether number of Kills differs by group level.

b) Describe in words the parameter(s) of interest, and assign symbol(s) to the parameter(s).

c) State, in symbols, the null and alternative hypotheses.

d) *Pretend, for the purposes of this study, that each of the 90 games was played by a different person.* Explain why this matters.

e) Using the **Multiple Means** applet find and report the value of a statistic that summarizes the discrepancies between the three levels. What is this statistic called?

f) Using the **Multiple Means** applet implement a simulation-based approach to test the hypotheses stated in (a). Be sure to report a p-value.

g) Based on the p-value reported in part (f), state an appropriate conclusion in the context of the study. Be sure to comment on statistical significance, causation, and generalization, along with how you are making your decision.

9.1.20: Reconsider **Exercises 9.1.19** on the game Halo. Besides the number of kills on each level, the students also recorded the number of deaths on each level. Use the simulation-based approach to run a test for the number of deaths to determine whether at least one of the group (or level) means is different. Give the p-value and state a conclusion. (The data file is **HaloDeaths**.)

9.1.21: Reconsider **Exercises 9.1.19** on the game Halo. Besides the number of kills on each level, the students also recorded the number of assists on each level. Use the simulation-based approach to run a test for the number of assists to determine whether at least one of the group (or level) means is different. Give the p-value and state a conclusion. (The data file is **HaloAssists**.)

9.1.22: Reconsider **Exercises 9.1.19** on the game Halo. Besides the number of kills on each level, the students also recorded the number of medals on each level. Use the simulation-based approach to run a test for the number of medals to determine whether at least one of the group (or level) means is different. Give the p-value and state a conclusion. (The data file is **HaloMedals**.)

Major and sudoku

Problems 9.1.23 through 9.1.25 refer to data on student majors and mastery of sudoku puzzles.

9.1.23: A group of Hope College statistics students wanted to see if there was an association between students' major and the time (in seconds) it takes them to complete a small Sudoku-like puzzle. They grouped majors into four categories: applied science (as), natural science (ns), social science (ss), and arts/humanities (ah). Their results can be found in the data file **MajorPuzzle.**

    a) Identify the explanatory variable and the response variable in this study.

    b) State, in words or symbols, the null and alternative hypotheses.

    c) Put the data in the **Multiple Means** applet and describe what the dotplots and means tell us about the association between major and time to complete the puzzle.

    d) What is the value of the MAD statistic?

    e) Use a simulation-based approach to determine if there is an association between a student's majors and the time it takes them to complete the puzzle. Be sure to report a p-value. Is there strong evidence of an association?

9.1.24: Reconsider **Exercise 9.1.23** on major and time needed to complete a puzzle. There was one outlier in the applied science group. The researchers reported that this subject kept getting distracted while completing the puzzle. Suppose the first step in computing the MAD statistic was to find the differences in medians instead of the differences in means. How do you think this would change the value of the MAD statistic in this example? Explain.

9.1.25: Reconsider **Exercise 9.1.23** on major and time needed to complete a puzzle. There was one outlier in the applied science group. The researchers reported that this subject kept getting distracted while completing the puzzle. Put the data **MajorPuzzle** into the **Multiple Means** applet and run a simulation-based test using the MAD statistic.

    a) Remove the outlier (it should be at the top of the list) and run the test again. What is the new MAD statistic and p-value?

b) After removing the outlier, you should have seen the MAD statistic decrease compared to the MAD statistics with the outlier in (see your answer to 9.1.23, part d). Explain why this makes sense.

c) After removing the outlier, and getting a smaller MAD statistic, you might think that the p-value should increase. The p-value, however, decreased slightly compared to your answer to 9.1.23, part e). What else changed that caused this to happen?

Comparing two studies*

9.1.26: Suppose two studies are to compare three means. The results are found in the data sets **Study1** and **Study2**.

a) Copy and paste the data from Study 1 into the **Multiple Means** applet and report the three means, the three standard deviations, and the MAD statistic.

b) Do at least 1000 shuffles to find a null distribution for Study 1. Report your p-value and the mean/standard deviation of the null distribution.

c) Copy and paste the data from Study 2 into the **Multiple Means** applet and report the three means, the three standard deviations, and the MAD statistic.

d) Do at least 1000 shuffles to find a null distribution for Study 2. Report your p-value and standard deviation of the null distribution.

e) You should have found the MAD statistics are the same for both surveys. Explain why they are the same.

f) Even though the MAD statistics are the same, explain why the p-value for Study 2 is smaller based on the two null distributions.

g) Does this suggest it is wise to compare MAD statistics across different studies?

Section 9.2

9.2.1* What is an ANOVA test used for?

9.2.2 If we are comparing means from three different groups, why would we use ANOVA rather than doing three independent samples t-tests?

9.2.3* Suppose you are conducting an ANOVA test in a situation where the alternative hypothesis is true. Are you more likely or less likely to be able to conclude the alternative hypothesis if:

    a) The differences in the population means increase?

    b) The population standard deviations increase?

    c) Your significance level increases?

    d) Sample size increases.

9.2.4 Suppose you are conducting an ANOVA test with four groups and each group has a sample size of 20.

    a) Suppose the *variability between groups increases.* Does the F-statistic increase or decrease? Does the p-value increase or decrease?

    b) Suppose the variability within groups increases. Does the F-statistic increase or decrease? Does the p-value increase or decrease?

9.2.5* The pair of plots below shows two artificial data sets, A and B.

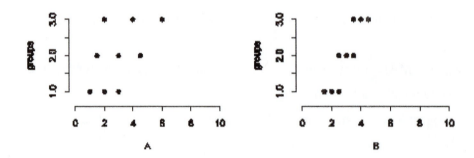

    a)   Which data set will have the larger mean absolute pairwise difference:

        A. Larger for A than for B

        B. Larger for B than for A

        C. Same size for both

        D. Can't tell

b) Suppose your null hypothesis is that there is no association between response value and group number. You use simulation to find a p-value for each data set based on its mean absolute pairwise difference. Which data set will have the larger p-value:

A. Larger for A than for B

B. Larger for B than for A

C. Same size for both

D. Can't tell

c) Suppose instead you use the *F*-statistic and theory-based test. Which data set will have the larger mean square for groups (between groups):

A. Larger for A than for B

B. Larger for B than for A

C. Same size for both

D. Can't tell

d) Which data set will have the larger mean square error (within groups):

A. Larger for A than for B

B. Larger for B than for A

C. Same size for both

D. Can't tell

e) Based on the *F* statistic, which data set will have the larger p-value?

A. Larger for A than for B

B. Larger for B than for A

C. Same size for both

D. Can't tell

Diets and weight loss

9.2.6 Recall Exercise 9.1.18 about the article in the *Journal of the American Medical Association* (Dansinger, Griffith, Gleason et al., 2005) that reported on a randomized, comparative experiment in which 160 subjects were randomly assigned to one of four popular diet plans: Atkins, Ornish, Weight Watchers, and Zone (40 subjects per diet). Data for the 93 subjects who completed the

12-month study are in datafile **Diets2**. The file contains which diet the subject was on and the weight loss in kilograms (positive values indicate weight loss and negative values indicate weight gain)

  a) State the null and alternative hypotheses in the context of the study.
  b) Use technology (either an applet or statistical software) to find and report the value of the F-statistic for these data.
  c) Recall that the theory-based test of significance in this scenario is called the ANOVA test or F-test. Use technology (either an applet or statistical software) to implement the ANOVA test to test whether weight loss differs significantly for any of the diets. Be sure to report a p-value.
  d) Based on the p-value reported in (c), state an appropriate conclusion in the context of the study. Be sure to comment on statistical significance, causation, and generalization, along with how you are making your decision.
  e) State the conditions under which the results from the theory-based ANOVA test would be valid. In the context of this study, is it valid to use the ANOVA test? Explain how you are deciding.
  f) If appropriate, perform follow up analysis. If you do not think follow-up analysis is appropriate, explain why not.
  g) How do your results compare from (c) and Exercise 9.1.18?

Heart rates and gender*

9.2.7 *This question explores a dataset considered in chapter 6 using an independent samples t-test to make your conclusion. We will compare those results with those using ANOVA.* Do heart rates of men and women differ? To answer this question, we will look at a data set found in "What's Normal? - Temperature, Gender, and Heart Rate" by Allen L. Shoemaker of Calvin College found in the *Journal of Statistics Education* v.4, n.2 (1996). Use this data set (**HeartRates**) to answer the question, "Do heart rates, on average, differ between males and females?"

  a) Give the null and alternative hypotheses for this study.
  b) Use a theory-based approach to compare two independent samples using the t-statistic on these data and give your t-statistic value, p-value, and state your complete conclusion, along with reasoning.
  c) Complete an ANOVA test on these data and give the values of your F-statistic value and p-value.
  d) Compare your conclusions in parts (b) and (c).
  e) What additional validity condition needs to be met to conduct an ANOVA test that is not needed for an independent samples (theory-based) t-test?

Mercury in fish

9.2.8 Alarming levels of mercury are starting to creep into our food chain: our fish are showing up with levels of mercury in them that used to be absent. Are the mercury levels same in all fish? The dataset **Mercury** gives mercury levels in part per million for four types of fish.

   a) What is the explanatory variable? What is the response variable?
   b) What are the observational units for this data set?
   c) Are the validity conditions met to carry out a theory-based test? Explain.
   d) If the result is significant, follow up with 95% confidence intervals to find where differences are.
   e) Write, as if for a cnn.com health blog, a paragraph summarizing your findings.

Halo*

9.2.9 Problems 9.2.9 through 9.2.13 recall data from exercises 9.1.19 through 9.1.22 about the popular science fiction video game Halo. Statistics students at Hope College, Holland, MI wanted to determine whether or not the level on which the game was played affects how well a person plays, and so data were collected from ninety different games of Halo; thirty from each of the three different levels.

   a) Use a theory-based approach to run an ANOVA test to investigate whether the average number of Kills is different for at least one of the groups (or levels). Report the value of the F-statistic and the corresponding p-value. (Note: As done earlier, pretend for purposes of this study that each of the 90 games was played by a different player.)
   The data file is **HaloKills**.
   b) Based on the p-value, state your conclusion in the context of the study.

9.2.10 Reconsider Exercises 9.1.19 through 9.1.22 as well as the previous exercise about the popular science fiction video game Halo. Statistics students at Hope College, Holland, MI wanted to determine whether or not the level on which the game was played affects how well a person plays, and so data were collected from ninety different games of Halo; thirty from each of the three different levels.

a) Use a theory-based approach to run an ANOVA test to investigate whether the average number of Deaths is different for at least one of the groups (or levels). Report the value of the F-statistic and the corresponding p-value. (Note: As done earlier, pretend for purposes of this study that each of the 90 games was played by a different player.)
The data file is **HaloDeaths**.

c) Based on the p-value, state your conclusion in the context of the study.

9.2.11 Reconsider Exercises 9.1.19 through 9.1.22 as well as the previous exercise about the popular science fiction video game Halo. Statistics students at Hope College, Holland, MI wanted to determine whether or not the level on which the game was played affects how well a person plays, and so data were collected from ninety different games of Halo; thirty from each of the three different levels.

a) Use a theory-based approach to run an ANOVA test to investigate whether the average number of Assists is different for at least one of the groups (or levels). Report the value of the F-statistic and the corresponding p-value. (Note: As done earlier, pretend for purposes of this study that each of the 90 games was played by a different player.)
The data file is **HaloAssists**.

d) Based on the p-value, state your conclusion in the context of the study.

9.2.12 Reconsider Exercises 9.1.19 through 9.1.22 as well as the previous exercise about the popular science fiction video game Halo. Statistics students at Hope College, Holland, MI wanted to determine whether or not the level on which the game was played affects how well a person plays, and so data were collected from ninety different games of Halo; thirty from each of the three different levels.

a) Use a theory-based approach to run an ANOVA test to investigate whether the average number of Medals is different for at least one of the groups (or levels). Report the value of the F-statistic and the corresponding p-value. (Note: As done earlier, pretend for purposes of this study that each of the 90 games was played by a different player.)
The data file is **HaloMedals**.

b) Based on the p-value, state your conclusion in the context of the study.

9.2.13 Reconsider Exercises 9.1.19 through 9.1.22 as well as the previous exercise about the popular science fiction video game Halo. Statistics students at Hope College, Holland, MI wanted to determine whether or not the level on which the game was played affects how well a person plays, and so data were collected from ninety different games of Halo; thirty from each of the three different levels.

Perform an appropriate test of significance to determine whether there is a relationship between the level the game is played and if a player wins (1) or loses (0). (The data file is **HaloWins.** *Hint:* You do not want to use an ANOVA for this test.)

Social Media and Self Esteem

Problems 9.2.14 through 9.2.16 refer to the following study on social media and self esteem.

9.2.14 A group of statistics students wanted to investigate the effect of social media on a person's self-esteem. For this purpose, they conducted a study where participants were randomly assigned to one of three groups: browsing *ENews* for 2 minutes and then taking a survey, browsing *Facebook* for 2 minutes and then taking the survey, or a control group where the participant just took the survey. The survey, based on the Rosenberg self-esteem scale, was the tool used to measure each participant's self-esteem. Each participant was then given a score based on their responses to the survey, and higher scores are considered indicative of higher self-esteem.

a) Is this an experiment or an observational study? How are you deciding?
b) Identify the observational units.
c) Identify the explanatory and response variable. Also, for each variable, identify whether categorical or quantitative.
d) State the appropriate null and the alternative hypotheses *in words*.
e) State the appropriate null and the alternative hypotheses *in symbols*. Be sure to define the symbols used.
   The data file is **SocialMediaandSelfEsteem**.
f) Use an appropriate applet to use a simulation-based approach to find a p-value to test the hypotheses stated.
g) Recall that the theory-based test of significance in this scenario is called the ANOVA test or *F*-test. Use technology (either an applet or statistical software) to implement the ANOVA test to test whether self-esteem scores differ significantly for any of the treatments. Be sure to report the *F*-statistic and the corresponding p-value.
h) Based on the p-value reported in (g), state an appropriate conclusion in the context of the study. Be sure to comment on statistical significance, causation, and generalization, along with how you are making your decision.
i) State the conditions under which the results from the theory-based ANOVA test would be valid. In the context of this study, is it valid to use the ANOVA test? Explain how you are deciding.

j) If appropriate, perform follow up analysis. If you do not think follow-up analysis is appropriate, explain why not.

9.2.15 Reconsider the previous exercise about the effect of social media on a person's self-esteem. In the study conducted by the statistics students, there were 27 people assigned to each of the three groups, for a total of 81 subjects. Suppose that instead of 27, each group had 30 people and yet had resulted in the same statistics (means and SDs) as before. How, if at all, would the following be different from before?

a) The F-statistic: BIGGER   SMALLER   SAME
b) The p-value: BIGGER   SMALLER   SAME
c) The strength of evidence: STRONGER   WEAKER   SAME
d) Explain your reasoning for your choices in (a) – (c).

9.2.16 Reconsider the previous exercise about the effect of social media on a person's self-esteem. In the study conducted by the statistics students, participants were given a score that measured their self-esteem. Suppose that instead of recording a numeric score, participants were categorized as having high self-esteem or low self-esteem.

Would ANOVA still have been a valid approach to analyzing the data? If yes, explain why. If no, explain why not and suggest an alternative approach to analyze the data.

Donations*

Problems 9.2.17 through 9.2.19 refer to donations taken after occurrences of different disasters.

9.2.17 Students in a statistics class in a four-year university in California were randomly assigned to be asked one of the following three questions:

- Suppose that a part of the state of <u>California</u> was affected by a devastating earthquake, and you were contacted via email by the American Red Cross, requesting you to make a monetary donation. How much ($) would you be willing to donate?
- Suppose that a part of the state of <u>Kansas</u> was affected by a devastating tornado, and you were contacted via email by the American Red Cross, requesting you to make a monetary donation. How much ($) would you be willing to donate?

- Suppose that a part of the state of <u>Maine</u> was affected by devastating floods, and you were contacted via email by the American Red Cross, requesting you to make a monetary donation. How much ($) would you be willing to donate?

Research question: Does being farther away from a disaster site affect our giving spirit?

a) Identify the observational units.

b) Identify the variable(s) measured. Also identify the type (categorical or quantitative) and the role (explanatory or response) for each variable.

c) Define the parameter(s) of interest in the context of the study, and assign symbol(s) to the parameter(s).

d) State your hypotheses, in symbols.

The data are available in the file **Donation**.

e) Run an appropriate simulation-based analysis to test the hypotheses stated in (d). Report the test statistic and p-value.

f) Use technology (either an applet or statistical software) to implement the ANOVA test to test whether self-esteem scores differ significantly for any of the treatments. Be sure to report the *F*-statistic and the corresponding p-value.

g) Based on the p-value reported in (f), state an appropriate conclusion in the context of the study. Be sure to comment on statistical significance, causation, and generalization, along with how you are making your decision.

h) State the conditions under which the results from the theory-based ANOVA test would be valid. In the context of this study, is it valid to use the ANOVA test? Explain how you are deciding.

i) If appropriate, perform follow up analysis. If you do not think follow-up analysis is appropriate, explain why not.

j) Even though this study was a randomized experiment, there is at least one possible confounding variable. Identify the confounding variable. (*Hint:* There is more than one difference between the three questions. So, any observed difference in responses might have been attributable to either the state or …)

9.2.18 Reconsider the previous exercise about the effect of distance on size of donation. In the study conducted on the statistics students in California, there were 14 people assigned to "California," 14 to "Kansas" and 16 to "Maine" for a total of 44 subjects. Suppose that instead each group had 35 people and yet had resulted in the same statistics (means and SDs) as before. How, if at all, would the following be different from before?

a) The F-statistic: BIGGER   SMALLER   SAME
b) The p-value: BIGGER   SMALLER   SAME
c) The strength of evidence: STRONGER   WEAKER   SAME
d) Explain your reasoning for your choices in (a) – (c).

9.2.19 Reconsider the previous exercise about the effect of distance on size of donation. In the study conducted on the statistics students in California, participants were asked for an amount that they would be willing to donate. Suppose that instead the participants were asked whether or not they would be willing to make a donation, and their responses were recorded as either a "yes" or a "no."

Would ANOVA still have been a valid approach to analyzing the data? If yes, explain why. If no, explain why not and suggest an alternative approach to analyze the data.

Flight Times

Problems 9.2.20 and 9.2.21 refer to data on flight times.

9.2.20 In a study of the effect of wing length on the flight time of paper helicopters, two wing lengths were of interest: 2 cm and 3cm. Two paper helicopters were made—one with a wing length of 2cm and the other with 3cm. All other aspects of the helicopters were made to be as identical as possible. Each helicopter was repeatedly dropped from a height of 10 feet and flight time was measured in seconds taken for the helicopter to hit ground.

a) Identify the observational units.
b) Is this an experiment or an observational study? How are you deciding?
c) Identify the explanatory and response variable. Also, for each variable, identify whether categorical or quantitative.
d) State the appropriate null and the alternative hypotheses *in words*.
e) State the appropriate null and the alternative hypotheses *in symbols*. Be sure to define the symbols used.
    The data file is **FlightTimes**.
f) Use technology (either an applet or statistical software) to implement a two sample *t*-test to test whether the flight times differ significantly for the two wing lengths. Report the mean, SD and

sample size of each group. Also, report the t-statistic as well as a p-value. (Note: Recall that the t-test is the theory based test for comparing two groups on a quantitative response.)

g) Now, use technology (either an applet or statistical software) to implement the ANOVA test to test whether the flight times differ significantly for the two wing lengths. Be sure to report the F-statistic and the corresponding p-value.

h) Compare the two p-values reported in (f) and (g). Are they about the same or different?

i) State an appropriate conclusion in the context of the study. Be sure to comment on statistical significance, causation, and generalization, along with how you are making your decision.

j) State the conditions under which the results from the theory-based ANOVA test would be valid. In the context of this study, is it valid to use the ANOVA test? Explain how you are deciding.

k) If appropriate, perform follow up analysis. If you do not think follow-up analysis is appropriate, explain why not.

l) Why is important to state that "All other aspects of the helicopters were made to be as identical as possible"?

9.2.21 Reconsider the previous exercise about the study of the effect of wing length on the flight time of paper helicopters. Recall that there were two wing lengths of interest: 2 cm and 3cm, and that flight time was measured in seconds taken for a paper helicopter to hit ground after being dropped from a height of 10 feet. Suppose that instead of two wing lengths, there were four wing lengths of interest: 2cm, 2.5cm, 3cm, and 3.5cm.

a) Would a two sample t-test still be a valid approach to analyzing the data? If yes, explain why. If no, explain why not.

b) Would ANOVA still be a valid approach to analyzing the data? If yes, explain why. If no, explain why not.

Effects of yoga*

Problems 9.2.22 through 9.2.25 refer to the study on yoga.

9.2.22 Researchers Oken et al. (*Altern Ther Health Med.*, 2006) conducted a study where 135 generally healthy men and women aged 65-85 years were randomly assigned to either 6 months of Hatha yoga class (44 people), walking exercise (47 people), or wait-list control (44 people). One of the outcomes of interest was change in "chair sit and reach" – a measure of how far the subject can reach out while sitting on a chair, without losing balance. The researchers reported the corresponding p-value to be 0.049. (*Note:* We have presented a simplified version of the study here.)

a) State the appropriate null and the alternative hypotheses *in words*.

b) State the appropriate null and the alternative hypotheses *in symbols*. Be sure to define the symbols used.

c) Based on the p-value reported by the researchers state an appropriate conclusion in the context of the study. Be sure to comment on statistical significance, causation, and generalization, along with how you are making your decision.

d) State the conditions under which the results from the theory-based ANOVA test would be valid.

9.2.23 Reconsider the previous exercise of the study of the Hatha yoga, walking exercise and wait-list control. Another outcome of interest was change in simple reaction time (msec). The researchers reported the corresponding p-value to be 0.760. (*Note:* We have presented a simplified version of the study here.)

a) State the appropriate null and the alternative hypotheses *in words*.

b) State the appropriate null and the alternative hypotheses *in symbols*. Be sure to define the symbols used.

c) Based on the p-value reported by the researchers state an appropriate conclusion in the context of the study. Be sure to comment on statistical significance, causation, and generalization, along with how you are making your decision.

9.2.24 Reconsider the previous exercise of the study of the Hatha yoga, walking exercise and wait-list control. Another outcome of interest was change in number of words correctly recalled from a list, after a time delay. The researchers reported the corresponding p-value to be 0.38. (*Note:* We have presented a simplified version of the study here.)

a) State the appropriate null and the alternative hypotheses *in words*.

b) State the appropriate null and the alternative hypotheses *in symbols*. Be sure to define the symbols used.

c) Based on the p-value reported by the researchers state an appropriate conclusion in the context of the study. Be sure to comment on statistical significance, causation, and generalization, along with how you are making your decision.

9.2.25 Reconsider the previous exercise of the study of the Hatha yoga, walking exercise and wait-list control. The researchers were also interested in change in subjects' perception of their own health. An outcome of interest was change in perception of one's vitality. The researchers reported the corresponding p-value to be 0.006. (*Note:* We have presented a simplified version of the study here.)

a) State the appropriate null and the alternative hypotheses *in words*.
b) State the appropriate null and the alternative hypotheses *in symbols*. Be sure to define the symbols used.
c) Based on the p-value reported by the researchers state an appropriate conclusion in the context of the study. Be sure to comment on statistical significance, causation, and generalization, along with how you are making your decision.

End of Chapter Exercises

9.CE.1*

Effects of p-values.

a) If all other things remain the same, then larger differences between group means have what effect on the p-value (smaller, larger, or no effect)?
b) If all other things remain the same, then larger variation within groups has what effect on the p-value (smaller, larger, or no effect)?
c) If all other things remain the same, then larger sample sizes have what effect on the p-value (smaller, larger, or no effect)?

9.CE.2 Suppose that you collect data on heights of singers in order to investigate whether the four different singing parts in a chorus tend to have different heights, on average. Your friend Josephine says that there are four explanatory variables in this study (soprano, alto, tenor, bass), and your friend Geraldine says that there is only one explanatory variable (singing part). Identify which friend is right, and explain the mistake that the other friend is making.

9.CE.3* The symbols μ and $\bar{x}$ both stand for means/averages but represent different quantities. What's the difference between μ and $\bar{x}$?

9.CE.4 Why do the μ and $\bar{x}$ symbols in this chapter have subscripts? What do the subscripts represent?

9.CE.5* Is the average of group averages always equal to the overall average? Let's make this question more precise: Suppose that an instructor teaches 4 sections of a course, and the average scores on an exam for the four sections are 77, 80, 83, and 84. Is it necessarily true that the average exam score among all of the instructors' students is equal to $(77+80+83+84)/4$, which is 81? If not, under what circumstances would the overall average equal the average of the group averages?

9.CE.6 Give your own example of a research question that you could use methods of this chapter to investigate. Include the following components:

   a) the research question itself
   b) the observational units
   c) the explanatory variable
   d) the number of categories/groups of the explanatory variable
   e) the response variable
   f) whether the study would be observational or experimental
   g) whether the data collection would use random sampling, random assignment, both, or neither
   h) the null hypothesis
   i) the alternative hypothesis

9.CE.7* Suppose that two different studies (A and B) have the same sample sizes in each of four groups, with similar standard deviations in the four groups. Furthermore, those sample sizes and sample SDs are also the same, or very similar, in Study A and Study B. Consider the following group means:

	Group 1 mean	Group 2 mean	Group 3 mean	Group 4 mean
STUDY A	40	40	20	20
STUDY B	40	30	30	20

   a) Calculate the value of the MAD statistic for each study.
   b) Which study provides stronger evidence of a significant difference among the group means? Explain your thinking, without performing any calculations or simulations.
   c) Which study would you expect to have the larger value of the F-statistic? Explain your thinking.

9.CE.8 Consider summary statistics from three different studies (A, B, C) presented below:

STUDY A	Sample size	Sample mean	Sample SD
Group 1	10	40	10
Group 2	10	30	10
Group 3	10	20	10
STUDY B	Sample size	Sample mean	Sample SD
Group 1	100	40	10
Group 2	100	30	10
Group 3	100	20	10
STUDY C	Sample size	Sample mean	Sample SD
Group 1	10	40	12
Group 2	10	30	12
Group 3	10	20	12

a) Which study provides the *least strong* evidence of a significant difference among the group means? Explain your thinking, without performing any calculations or simulations.

b) Which study provides the *strongest* evidence of a significant difference among the group means? Explain your thinking, without performing any calculations or simulations.

Earnings by degree*

9.CE.9 The file **EarningsByDegree** contains data from random samples of 50 American adults taken in the year 2000 in each of five categories of higher educational achievement:

- some higher education
- associate degree
- bachelor's degree
- master's degree
- doctorate

The response variable is the individual's yearly earnings, in thousands of dollars.

a) Produce graphical displays for comparing yearly earnings across the five education levels.

b) Calculate numerical summaries of the yearly earnings values for each of the five higher education levels.

c) Write a paragraph or two comparing and contrasting the distributions of yearly earnings values across the five higher education levels.

d) Perform a simulation-based analysis of whether the sample data provide strong evidence to believe that average earnings differ across the five education levels in the population.

e) Perform a theory-based analysis of whether the sample data provide strong evidence to believe that average earnings differ across the five education levels in the population.

f) If you find strong evidence that the population means are not all the same, produce confidence intervals to determine which groups differ significantly from which other groups.

g) Write a summary paragraph specifically responding to these three questions:
   - Is more higher education associated with greater earnings?
   - If so, how much do earnings increase with higher levels of education?
   - Which degree is associated with the largest increase in earnings over the preceding degree level?

Heights of singers

9.CE.10 Each singer in the New York Choral Society in 1979 self-reported his or her height to the nearest inch. Their voice parts in order from highest pitch to lowest pitch are Soprano, Alto, Tenor, Bass. The first two are typically sung by female voices and the last two by male voices. The data can be found in the datafile **SingerHeights**.

a) Produce comparative boxplots of the distributions of heights by singing part. Comment on similarities and differences in the distributions of heights across the four parts. Use these boxplots to answer:
   - Which part has the highest median height? What is the value of that median?
   - Which part has the tallest singer? How tall is that singer?
   - Which part has the smallest inter-quartile range of heights? What is the value of that IQR?
   - Which part has the smallest lower quartile of heights? What is the value of that lower quartile?

b) Calculate the sample size, sample mean, and sample standard deviation of the heights within each group (singing part). Use these statistics to answer:
   - Which part has the most singers? What is the value of that sample size?
   - Which part has the largest sample mean height? What is the value of that mean?
   - Which part has the largest standard deviation of heights? What is the value of that standard deviation?

c)   Treat these data as a random sample of all singers. Conduct an ANOVA analysis to test whether the sample data provide evidence that the mean heights differ significantly among the four parts. Report the ANOVA table, hypotheses, test statistic, and p-value. Also summarize your conclusion.

d)   Check and comment on whether the validity conditions for ANOVA appear to be satisfied here.

e)   Produce confidence intervals to determine which parts' mean heights differ significantly from which others. Summarize your findings.

Car crash dummies*

Problems 9.CE.11 and 9.CE.12 refer to data gathered on car crashes.

9.CE.11 The datafile **CarCrashDummies** contains data on automobile crash test results. The explanatory variable is the number of doors on the vehicle, and the response variable is a measurement of the extent of injury on the dummy's head.

a)   Produce graphical displays of the head injury measurements by the number of doors. Describe the distributions, paying particular attention to the question of whether the head injury measurements appear to differ significantly among the three groups.

b)   Does it appear that the validity conditions of the ANOVA procedure and F-test are satisfied with these data? Explain.

c)   Perform a simulation-based analysis of these data. Report the p-value, and summarize your conclusion.

9.CE.12 Reconsider the previous exercise.

a)   The datafile **CarCrashDummiesLog** contains the same data, but with the response variable being the logarithm (base 10) of the original head injury measurements. Re-examine the distributions of these (transformed) measurements by the number of doors. Do the validity conditions appear to be satisfied now? Explain.

b)   Determine the ANOVA table and interpret the results. Do the data provide strong evidence that head injury measurements differ based on the number of doors on the vehicle? Explain.

c)   Produce confidence intervals to determine which groups differ significantly from which others. Summarize your findings.

Liar, Liar

9.CE.13 A study was conducted to investigate whether how well you know a person is associated with your ability to detect a lie told by that person. Ten statements were devised about the researcher (e.g., My toothbrush is blue)—five of which were true, and five of which were false—but for which no one but the researcher would know the true answer---not even their close friends. The researcher told these statements to 86 participants: some of whom were close friends of the researcher, others acquaintances and others total strangers. The 10 statements were read by the researcher to the participant. The participant then had to identify whether each statement was true or false. The number of correctly identified questions out of 10 was recorded for each participant. The data is found in the **LiarLiar** dataset.

a) Create dotplots (stacked by relationship groups) for the number of questions correct, and find the average number correct for each of the three relationship groups. Do you think any of the averages differ significantly from others? If so, which one or ones? Explain how you are deciding.

b) Run an ANOVA test to determine whether at least one of the group averages is different. State the null and alternative hypotheses; find the p-value, and tell whether the null hypothesis should be rejected. (Because the study design is complicated the scope of inference is also complicated. You are not expected to discuss the scope of inference for this study.).

c) If your result from part (b) was significant, complete the follow-up analysis by generating confidence intervals for pairwise differences to determine which mean is different from which other(s), how and by how much. Write a paragraph summarizing your findings.

Noise in NYC*

9.CE.14 Here are recorded noise levels in decibels for a sample of locations in four different parts of New York City.

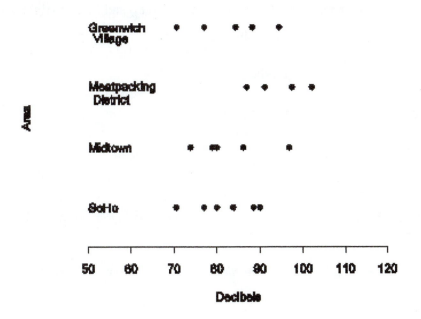

a) What is the observational unit? What is the response variable? What is the explanatory variable?

b) Which of the following best describes the pattern in the plot? (Choose one.)

  A. Within-group variability is roughly constant (roughly the same in in each group) and group means are roughly equal. A theory-based test is appropriate, but the p-value will not be significant. (The evidence of a difference in means is weak.)

  B. Within-group variability is roughly constant (roughly the same in in each group) but at least one group mean differs substantially from the others.   A theory-based test is appropriate, and the p-value may be significant.

  C. The pattern in the plot indicates that the theory-based *F*-test is not appropriate. A simulation test should be done instead.

  D. None of the above.

c) Discuss the scope of inference in the context of the following facts about the study: All six of the SoHo locations were inside stores. None of the other 14 locations were. All of the locations in the Meatpacking District, and four of the five locations in the Village were in restaurants. Three of the five Midtown locations were in restaurants. In the study, restaurants tended to be noisy, especially later in the evening. Stores tended to be quiet.

Rattlesnake ages

9.CE.15. The two plots below show distributions of Age (left panel) and Length (right panel) for male and female rattlesnakes caught at two different sites. Here are the four groups:

1: Site 1, male   3: Site 2, male

2: Site 1, female   4: Site 2, female

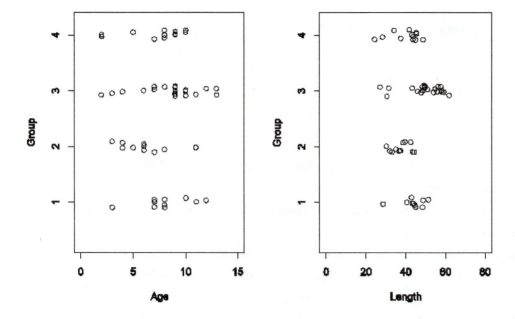

a)   Classify each of the following statements as True or False:
   A. The observational unit is the same for both plots.

   B. The response variable is the same for both plots.

C. The explanatory variable is the same for both plots.

Compare the two plots for age (left) and length (right). For each plot, eyeball the differences between groups, and the variability within groups. Which plot – Age or Length --shows the stronger evidence of differences between groups?

b)   Now focus on Age (left plot). Here are the group averages:

Ave Age	Sex m	Sex f
Site 1	6.45	8.10
Site 2	7.25	8.57

Consider two comparisons, Site and Sex:

Site: Site 1 (Groups 1 & 2) versus Site 2 (Groups 3 & 4)

Sex: Male (Groups 1 & 3) versus Female (Groups 2 & 4)

Which average difference is larger?

c)   Now compare all four groups at once. *Without* computing all the pairwise differences, estimate the mean absolute pairwise difference in age to the nearest 5 years. Choose from 0, 5, 10, 15, ... (*Hint*: Estimate the largest absolute difference and the smallest to the nearest whole number. You know the mean difference must be in between these extremes.)

d)   Consider the p-values 0.023 and 0.230. They differ by a factor of ten. One is approximately correct, based on a simulation using 10,000 repetitions, for comparing mean age across the four groups. The other is badly wrong. Which one is correct: 0.023 or 0.23?

e)   Now change your focus from Age to Length (right panel). Here are the group averages ordered from smallest to largest: 37.5, 40.1, 43.7, 49.5. Use the plot to match each average to its group.

f)   On average, which has the longer snakes, Site 1 or Site 2? On average, which snakes are longer, Males or Females?

g) Now compare all four groups at once. *Without* computing all pairwise differences, estimate the mean absolute pairwise difference in length to the nearest half foot (6 inches). (Choose from 0, 6, 12, 18, …) See the hint in (d) above.

h) Consider the p-values 0.0005, 0.2 and 0.5. One is approximately correct, using a theory-based F-test. The other two are wrong. Which one is correct?

i) Suppose we had measured the Age of the rattlesnakes in months rather than years: 1 year = 12 months, 2 years = 24 months, etc. Several quantities are listed below. For each one, choose one of

A: The quantity won't change[1]

B: The quantity will be multiplied by 12

C: The quantity will change, but the multiplier will not be 12

D: Impossible to tell.

i)   The mean age for each group.
ii) The mean absolute pairwise difference.
iii) The p-value based on the mean absolute pairwise difference.
iv) The mean square for groups (between groups)
v)   The mean square within groups
vi) The F-statistic
vii) The p-value based on the F-statistic

Background music*

9.CE.16 A British study (North, Shilcock, and Hargreaves, 2003) examined whether the type of background music playing in a restaurant affected the amount of money that diners spent on their meals. The researchers asked a restaurant to alternate silence, popular music, and classical music on successive nights over eighteen days. Each type of music was played for 6 nights (the order was randomly determined to guard against confounding). The data file **RestaurantMusic** gives the type of music played and the amount spent on food and drinks (in British pounds) for each customer.

---

[1] For the purpose of this question, ignore differences due solely to the simulation.

a) Report the sample means and standard deviations for the amount spend on food and drinks for each music type.

b) What is the response variable and explanatory variable? What are the experimental units?

c) Construct dotplots, stacked by the three different treatments of music, of amounts spent on food and drink.

d) Use a simulation-based approach to analyze the data. Be sure to state the relevant hypotheses, a p-value, and an appropriate conclusion.

e) State and check validity conditions to see whether you can run a theory-based ANOVA test. Be sure to explain how you are checking.

f) If appropriate, carry out the theory-based test of significance. Be sure to report the p-value.

g) If a significant result was found, construct additional follow-up confidence intervals to see where differences between the groups are. Report and interpret 95% confidence intervals.

h) How do your results compare from parts (d) and (f)?

i) In both cases what are your scope of conclusions? As in, to whom can the results of the study be generalized?

Regions of the US

Problems 9.CE.17 through 9.CE.19 use the same explanatory variable and observational units, but look at different response variables.

9.CE.17 The US Census Bureau classifies the 50 US states into four regions:

Northeast (9 states, New England and Middle Atlantic):

CT, ME, MA, NH, NJ, NY, PA, RI, VT

South (16 states, South Atlantic, East South Central, and West South Central):

AL, AR, DE, FL, GA, KY, LA, MD, MS, NC, OK, SC, TN, TX, VA, WV

Midwest (12 states, East North Central and West North Central):

IL, IN, IA, KS, MI, MN, MO, NE, ND, OH, SD, WI

West (12 states, Mountain and Pacific):

AK, AZ, CA, CO, HI, ID, MT, NV, MN, OR, UT, WA

(When the District of Columbia is included, it is part of the South.)

a)   The graph above shows dot plots by region for four different response variables.   For these four data sets, the response differs, but the observational unit and the explanatory variable are always the same. What is the observational unit? What is the explanatory variable?

b)   Two of the overall means (all 50 states) are between 8 and 9, and two are between 13 and 14. Which is which? Note that the horizontal scales are the same for each plot.

c)   Here are four pairs of group means for the Northeast and the West: (8.5, 9.3), (11.9, 8.9), (11.2, 13.8), (11.0, 16.8). Which pair goes with which response variable?

d) Eyeball each plot and think about the size of the mean absolute pairwise deviation. One value is clearly the largest, one is clearly the smallest, and two are in between. Which is the largest? ... the smallest?

e) (Hard) Discuss the scope of inference for these data sets. In particular, what does it tell you if you reject the null hypothesis of no association?

f) The four plots below show histograms for the simulated null distributions of the mean absolute pairwise difference for the first set of four response variables. The observed values of the MAD statistic are 2.8 (Poverty), 0.8 (Unemployment), 2.0 (Attorneys) and 3.8 (Not Covered). For which variables, if any, do you conclude that there is strong evidence of an

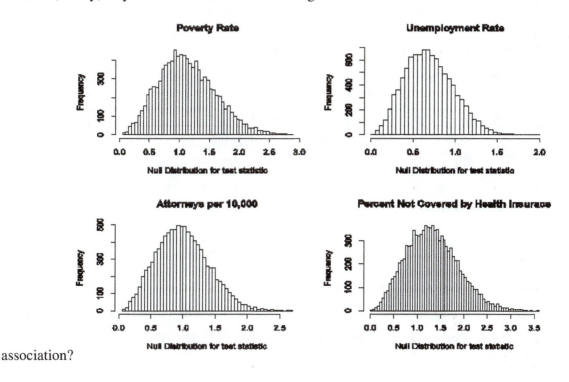

association?

g) Use the observed values and histograms from the previous question -- to match each response variable to its p-value, choosing from 0.00002, 0.0004, 0.03, and 0.33.

9.CE.18 (See 9.CE.17) Keep the observational unit and explanatory variable the same as in exercise 9.CE.17. Consider another four response variables: Year that a state joined the union, number of Macdonald's restaurants in the state, number of payday lenders in the state, and number of hazardous waste sites in the state. For each of these, imagine testing the null hypothesis that on average there are no differences from one region to another. The plots below show simulated null distributions for the standardized statistic mean absolute pairwise difference (MAD).

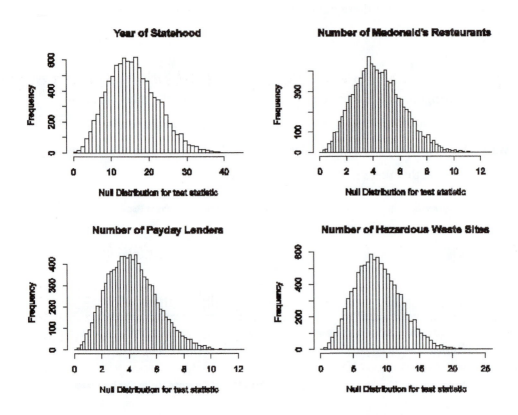

a) The observed values of the MAD statistic are: 56.7 (Year), 2.7 (Macdonald's), 5.6 (Lenders), and 13.3 (Waste Sites). Use these values, together with the histograms, to assess the strength of evidence of an association. For which variables, if any, do you conclude that there is strong evidence of an association?

b) Use the information in the last question to rank order the four variables according to the size of the p-value, from smallest to largest.

9.CE.19 (See 9.CE.17.) Keep the observational unit and explanatory variable the same as in exercise 9.CE.17. The dot plots below are for a third set of response variables: Precipitation (a state's annual average), Temperature (again, a state's annual average), a state's High School Graduation rate, and a state's Cremation rate. (Note that all four plots use the same horizontal scale, from 0 to 100.)

a) Use the dot plots to choose a label according to the value of the mean absolute pairwise difference (MAD), choosing from (a) one of the two smallest, (b) the largest, and (c) in

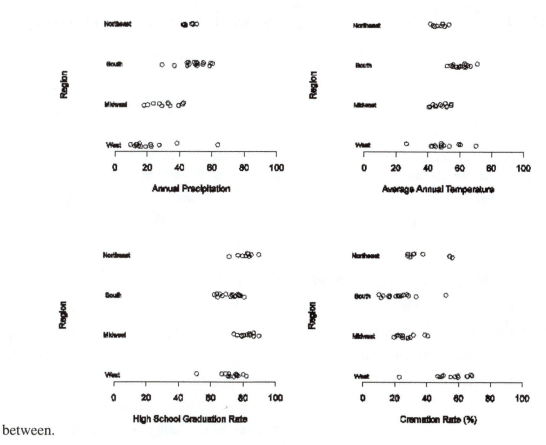

between.

b) Choose a label for each plot based on the size of the variability within groups, choosing from (a) one of the two smallest and (b) one of the two largest.

c) All four p-values are miniscule: the largest of the four is 0.00002. Clearly, the evidence of an association is extremely strong. Discuss the scope of inference for these data sets.

The next set of five questions illustrates what an outlier can do. (We don't expect you to know in advance about the effect of an outlier. Rather, these questions ask about your conceptual understanding of the ideas from the chapter.) The key thing to know is that when DC is included, the Census Bureau puts it in the South.

d) The two panels in the plot below show the number of attorneys per 10,000 people for the US states. The panel on the left is for the 50 states only, excluding DC. The plot on the right includes DC. (Note that the horizontal scales differ.) What is it about the District of Columbia that makes it an outlier?

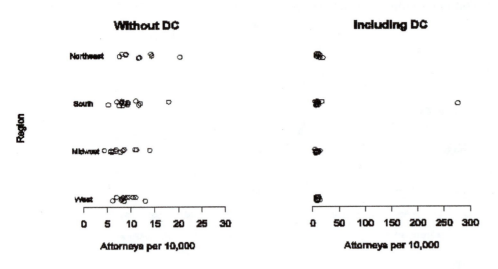

e) If we use the MAD statistic, what will including DC do to its value (Choose one):

    A. Cause it to get a lot larger.

    B. Cause it to get a lot smaller.

    C. Leave it pretty much the same.

    D. No way to tell.

f) If we use the MAD statistic, what will including DC do to the variability of the null distribution (Choose one):

    A. Cause it to get a lot larger.

    B. Cause it to get a lot smaller.

    C. Leave it pretty much the same.

    D. No way to tell.

g) Can you tell from just the size of the MAD statistic which analysis will give the stronger evidence of a difference (smaller p-value)? Why or why not?

h) (Hard.) The two histograms below show null distributions for the MAD without DC (left panel) and with DC included (right panel). Both null distributions come from scrambling to break the association. On the left (no DC) the shape is familiar. On the right (outlier included) the distribution has three peaks, and looks like nothing you've seen before. Explain where the peaks come from, and why scrambling gives only three peaks instead of four when you have four groups.

Investigation: Aggression

In an article titled, "Are People More Aggressive When They Are Worse Off or Better Off Than Others?" published in *Social Psychological and Personality Science* (November 2012), reported on an experiment where researchers told the subjects in the study that they were interested in impression formation. The 72 French female college students who agreed to participate were asked to fill out a food preference survey on different foods such as dairy, spicy, sweet, salty, fruits, etc. rating each 1 to 21 with 1 being strongly dislike and 21 being strongly like. Each subject also wrote an essay on a particularly nice day in their life. Finally they all participated in an illusory conjunction test where they were shown patters of simple shapes for 70 milliseconds and they had to decide whether the dollar sign symbol was present or absent.

The female college students were told that all of these instruments were going to be used with a female partner to form impressions of them. The partner is fictitious in the study. The students were then all told they received 65% correct on their illusory conjunction test. Each was then randomly told one of three things about their partner's score: downward feedback group was told their partner scored 50% correct, upward feedback group was told their partner scored 80% correct and the control group was told they couldn't provide feedback about their partner's score due to a computer error. Then they received their own essay back which had been negatively rated by their partner, including a hand-written comment of "Poor writing, middle school level." They also received their partner's food likes/dislikes with salty and spicy both rated as a 3. Next the students were told they had been randomly assigned to taste a sweet drink prepared by their partner and they were to prepare a tomato juice drink for their partner. They were told they could add salt and Tabasco sauce to the drink if they wished. They tasted a small sample of the drink before sending it to their partner. Then they tasted a drink of water sweetened with glucose which was prepared by their partner.

A statistic was constructed based on the amount of Tabasco sauce added and salt added to the tomato drink as a measurement of aggression. Researchers wondered if there would be any differences in the level of aggression between the downward feedback group (those that thought their partners scored less than them), the upward feedback group (those that thought their partners scored more than them), and the control group.

Step 1: Ask a research question

1. State the research question.

Step 2: Design a study and collect data

2. Is this a randomized experiment or an observational study? Explain how you know.

3. What are the observational/experimental units?

4. Although there are many variables recorded on each subject, state the two variables that will help to answer the research question.

5. Is there an explanatory/response relationship for these variables? Classify the variables in this study as categorical or quantitative.

6. State the null and alternative hypotheses to be investigated with this study. (You may use the symbol μ to represent these parameters.)

The following table provides descriptive statistics on the measurements for aggression:

	Experimental group		
	Downward	Control	Upward
Average	0.78	-0.52	-0.27
SD	1.89	1.72	1.56

Step 3: Explore the data

7. Calculate the MAD (mean absolute value of the differences) for these data.

Step 4: Draw inferences

8. Use the appropriate applet to construct a simulated null distribution using the MAD statistic. Mark the observed MAD statistic calculated in the previous question on this graph.

   a) Paste a screen shot or draw your null distribution below with the observed MAD marked and the approximate p-value shaded in.

   b) Is the observed statistic out in the tail of this null distribution, or is it a fairly typical result?

   c) What is the p-value from your simulation? Based on this simulation analysis, would you conclude that the experiment provides *strong* evidence against the null hypothesis and conclude that there is a significant difference in at least one of the mean aggression scores for the three experimental groups? Explain your reasoning.

9. Are the validity conditions met to complete a theory-based ANOVA test for these data? Explain. If the conditions are met, use a theory-based ANOVA test to find a p-value for this test.

10. Are the validity conditions met so that we can perform follow-up analyses using theory-based techniques? Explain.

11. Find confidence intervals for each pair of differences in population means. Comment on how this shows where differences can be established and where it is plausible there are no differences between population means.

Step 5: Formulate conclusions

12. Because this is a randomized experiment, can you conclude that you have found strong evidence of a cause and effect relationship between the three groups that received different kinds of feedback and their aggressiveness toward their partner? Explain.

13. To what broader population are you willing to generalize your findings? Explain.

Step 6: Look back and ahead

14. Summarize the findings from the study. What worked well in this study design? What would you change? What are some follow-up research questions you would like to explore based on the findings from this study?

Research article

Read "Awe expands people's perception of time, alters decision making and enhances well-being," by Rudd et al. (2013, forthcoming), *Psychological Science*, and then answer the following questions.

Step 1. Ask a research question

1. In no more than 1-2 sentences state a broad research question (or related questions) that the researchers were trying to investigate in this study.

2. Identify at least two (cited) points of evidence the researchers cite as to why the research question(s) are important. Include the citation in your response.

Step 2. Design a study and collect data

3. What is the sample size in Experiment #1?
4. What is the sample size in Experiment #2?
5. What is the sample size in Experiment #3?

Step 3. Explore the data

6. What was the average amount of awe experienced by awe condition participants in experiment #1? For happiness condition participants?
7. What was the average amount of happiness experienced by awe condition participants in experiment #1? For happiness condition participants?

Step 4. Draw inferences

8. In discussing Experiment #1, the authors state that they "conducted an analysis of variance (ANOVA) on the perceived time availability index, which revealed the predicted effect of emotion condition ($p = 0.01$). Awe condition participants (Mean = 3.38, SD = 1.49) perceived time as more plentiful than did happiness condition participants (Mean = 2.44, SD = 1.20). State the null and alternative hypothesis, indicate the strength of evidence provided by the p-value and provide an appropriate (in context) interpretation.
9. Confirm that the validity conditions for the ANOVA are met for the data in the previous question.
10. In discussing Experiment #2, the authors state that "As predicted, awe condition participants (Mean = 3.90, SD = 1.57) reported feel less impatient than did happiness condition participants (Mean = 4.64, SD = 1.48), $p = 0.03$)." The p-value is from an ANOVA comparing the two group means. State the null and alternative hypothesis, indicate the strength of evidence provided by the p-value and provide an appropriate (in context) interpretation.
11. Confirm that the validity conditions for the ANOVA are met for the data in the previous question.
12. In discussing Experiment #3, the authors state that "Testing whether awe expands time perceptions, an ANOVA on the perceived time availability index revealed the predicted condition effect ($p < 0.01$). Awe condition participants (Mean = 3.38, sd = 1.38) perceived time as more plentiful than did neutral condition participants (Mean = 3.09, sd = 1.22)." State the null and

alternative hypothesis, indicate the strength of evidence provided by the p-value and provide an appropriate (in context) interpretation.

13. Confirm that the validity conditions for the ANOVA are met for the data in the previous question.

14. In this study the researchers repeatedly used ANOVA, but they could have used a different approach as the researchers typically compared only two groups. What is the name of the other theory-based approach they could have used?

Step 5. Formulate conclusions

15. Do the significant results in Experiments 1, 2 and 3 suggest that awe is causing the change in people time perceptions, and other variables? Why or why not?

Step 6. Look back and ahead

16. Identify at least two things that, if you were running the study and had a reasonable amount of financial resources, you would do differently. You must justify your answer.

Chapter 10

Two Quantitative Variables

Chapter Overview

In Chapters 5–9 the focus was on comparing two or more groups. Chapter 10 is different. Until now the explanatory variable has been categorical. In Chapter 10 both variables are quantitative. This allows us to address questions about an association between two quantitative variables and to predict the value of one variable based on the other variable.

Graphical summaries will use a new kind of plot called a *scatterplot*. These are like the dotplots we've been using throughout the book, except that they now tell you information about two (quantitative) variables at once. In describing these graphs, we will focus on the ***Form***, ***Direction***, and ***Strength*** of the relationship. Here are some quick examples: (1) You measure the heights of classmates, first in centimeters, then in inches. Graphing the height in centimeters on the *y*-axis vs. the height in inches along the *x*-axis, the dots will fall along a straight line. The direction of the relationship is positive: *y* goes up as *x* goes up. The relationship cannot be stronger, because the points lie exactly on the line and we can perfectly predict one variable from the other. (2) You measure shoe size and height of your classmates. Taller people tend to have bigger feet (the relationship is positive), but there is a lot of variability (the relationship is not as strong). (3) You plot age and height for one person each year on her birthday, from year 1 to year 60. This relationship is *not* linear – people stop getting taller, and the plot flattens out. (4) You plot age and height for a sample of adults. There may be no systematic relationship. Once we stop getting taller, age and height for a sample of adults are unrelated.

Chapter 10 has five sections. The first section uses scatterplots to explore form, direction, and strength of a relationship between two quantitative variables, and introduces a new statistic, the correlation coefficient, which measures direction and strength of a linear association on a scale from -1 to 1, with 0 indicating no linear relationship. (Heads up: The strength of a relationship is not the same as the strength of evidence that a relationship exists.) Section 10.2 illustrates how to conduct simulation-based tests based on the correlation coefficient.

Section 10.3 indicates how to fit a line to give a quantitative description of a linear relationship. The slope of the fitted line might tell us, for example, that between the ages of 8 and 13, kids tend to get taller by about 2 inches per year. Section 10.4 uses simulation to test a null hypothesis of no association using the slope as the statistic. Finally, Section 10.5 presents a theory-based approach for both the regression slope and the correlation coefficient.

It is important to keep in mind throughout this chapter that *association does not imply causation.* Beware of confounding variables. For a sample of children in grades K-12, kids with larger feet tend to drink more diet soft drinks, but drinking diet soda won't make your feet grow. State by state, there is a strong positive relationship between the number of McDonald's restaurants and the number of cases of lung cancer, but Big Macs do not cause lung cancer.

Section 10.1: Summarizing the Relationship Between Two Quantitative Variables Using the Correlation Coefficient

Introduction

Do you think there is an association between a student's grade point average and the number of classes they skip per semester? How about the fuel efficiency of cars and their weight? These questions are different from what we have studied so far, because both variables are *quantitative*. But we will follow the same general strategy as before – first producing graphical summaries that can illuminate the relationship exhibited in the sample and then numerical summaries or statistics that measure the strength of the association. In the next section (10.2), we will draw inferences, again using simulation to compare the observed statistic of association to what we might expect to see by chance alone. These inferential techniques employ the same reasoning process as techniques we used in previous chapters. We start, however, by spending some time with scatterplots and the correlation coefficient before moving into simulation-based inference techniques.

Example 10.1: Exam Times and Exam Scores

Is there an association between the time it takes a student to take an exam and their score on the exam? One of the authors wondered about this question and decided to collect data from one of his mathematics classes.

Think about it: Which variables do you consider to be explanatory and response?

Even though this was not an experiment, it makes sense to consider the time spent on the test as the explanatory variable, helping to explain why some scores were higher and some were lower. It also makes sense to think of predicting test score (response) from time spent on test (explanatory). Note: With this goal, we often call the explanatory variable a *predictor* variable.

Exploring the Data: Graphical Summary

To examine an association between these variables graphically, it would not be helpful to make separate dotplots of time spent on test and score on test. Doing so would do nothing to tell us about the *relationship* between the two variables. Instead, a *scatterplot* shows the two values for each observational unit. In constructing a scatterplot we put the explanatory variable on the horizontal axis and the response variable on the vertical axis. (If you don't have a clear explanatory/response variable distinction, then this choice is more arbitrary.) Then each dot on the scatterplot shows the values for both variables on that observational unit. The results for 22 students are shown in Figure 10.1.

Figure 10.1: A scatterplot showing the association between students' scores on a mathematics test and the time it took them to take the test

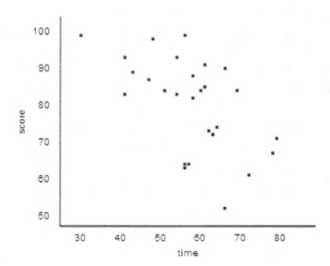

Think about it: Based on the graph, approximate the score and time for the student who was the first to turn in her/his exam. Also describe what you learn about these data from the scatterplot.

When describing a scatterplot we look for three aspects of association: direction, form, and strength. The *direction* of the association in Figure 10.1 is negative because larger values on one variable correspond to smaller values on the other variable. This shows that the longer it takes students to finish the test, the lower their scores tend to be. (This is not a hard and fast rule of course, some student who took longer scored higher; but overall, there is this tendency.)

In describing the *form* of an association, we describe whether it follows a linear pattern or some more complicated curve. For the test time and score data, it appears that a line would do a reasonable job of summarizing the overall pattern in the data, so we say that the association between score and time is linear. We will see lots of examples in which the form is fairly linear. Occasionally, however, the form is clearly not linear and that will be very important to note.

In describing the *strength* of association revealed in a scatterplot, we see how closely the points follow a pattern. If you imagine a line through the test time and score scatterplot, the data points all fall pretty close to that line, so we say that the association is reasonably strong.

It is also important to consider any *unusual observations* that do not follow the overall pattern. In this example, the fastest student finished in about 30 minutes and scored 100 on the exam. Even though this time is smaller than all the rest, and might be considered an outlier if looking at the time variable alone (no one else finished in less than 40 minutes!), this (time, score) combination appears consistent with the overall pattern of the sample. However, if we saw a student that finished in 30 minutes and only received a score of 60, we might try to follow-up on the data and make sure there wasn't an error recording their score, their time, or some other issue.

Key idea: Association between quantitative variables can be described with direction, form, and strength.

- If above average values of one variable tend to correspond to above average values of the other variable, the *direction* is positive. If, however, above average values of one variable are associated with below average values of the other, the *direction* is negative.
- The *form* of the association is whether the data follow some linear pattern or some more complicated pattern.
- The *strength* of an association refers to how closely the data follow a particular pattern. When an association is strong, the value of one variable can be accurately predicted from the other.
- Also look for and investigate further any *unusual observations* that do not fit the pattern of the rest.

Exploring the Data: Numerical Summary

We can also summarize the association between two quantitative variables numerically. The *correlation coefficient* measures the strength and the direction of a *linear* association between two quantitative variables. The correlation coefficient, usually denoted by the letter r, has the property that $-1 \leq r \leq 1$. If the correlation coefficient is close to -1 or 1, then the variables are highly correlated, revealing a strong linear association. In fact, if $r = 1$ or $r = -1$, the data all fall exactly in a straight line. If the correlation coefficient is close to 0, then the data points do not have a linear association. If the correlation coefficient is positive, then there is a positive linear association between the two variables. If the correlation coefficient is negative, then there is a negative linear association between the two variables.

Key Idea: The correlation coefficient uses a rather complex formula (see the **Appendix Calculation Details** for details) that is rarely computed by hand; instead, people almost always use a calculator or computer to calculate the value of the correlation coefficient. But you should be able to apply the above properties to interpret the correlation coefficient that is found.

The correlation coefficient between test time and score is approximately $r = -0.56$. The negative value indicates that students with longer times tend to have lower test scores. The magnitude of the -0.56 correlation appears to indicate a fairly strong linear association between time and test score, so you could predict a student's test score reasonably well knowing only how long he/she took on the test. Be aware that evaluating the strength of the association will be subjective, especially across different research areas. A correlation of -0.56 might be considered extremely strong for a psychologist trying to predict human behavior but very weak for a physicist trying to predict how far a particle will travel. (To help develop your judgment for how a scatterplot looks for various correlation values, you can play the Correlation Guessing Game applet.)

Key Idea: The correlation coefficient is only applicable for data which has a linear form; non-linear data is not summarized well by the correlation coefficient. In fact, we could say that the correlation coefficient is a numerical summary of the strength and direction of a linear association between two quantitative variables.

Caution: Influential Observations

We haven't been completely honest in presenting our test time and score dataset. There were three additional students in the class, the last three students to turn in their tests. Their times were 93, 96, and 100 minutes. These three students also scored very well on the exam: 93, 93, and 97 respectively. We added these scores to the scatterplot shown in Figure 10.2.

Figure 10.2: A scatterplot of the score on a math test and the time (in minutes) it took to complete the test after the last three tests turned in were added for all 25 students

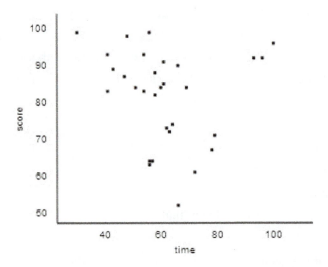

The three added scores can be seen on the far right side of the graph. These three students took a long time and had high scores, so they go against the general tendency that students who took longer tended to score lower. Without these three students we would view the relationship differently than we do with these students. For this reason, these unusual observations are called ***influential observations***. They are influential because the addition of just these three points changes the correlation coefficient dramatically (and our perception of the strength of the relationship), from -0.56 (a fairly strong negative correlation) to -0.12 (a weak negative correlation). Influential points tend to be ones that are far to the right or left on the graph, having extreme values for the explanatory variable like these three. The amount of influence of even just one observation can sometimes be dramatic. For example, if the person that finished the test in 30 minutes would have scored 50 instead of 100, the correlation coefficient would switch from negative to positive (0.13).

Note: There are two kinds of unusual observations seen in scatterplots: influential observations and outliers. An observation is *influential* if removing it from the data set dramatically changes our perception of the association. Typically influential observations are extreme in the explanatory variable. Outliers are observations that don't follow the overall pattern of the relationship. They may or may not be influential and they may or may not be extreme in either variable individually, but are unusual in terms of the combination of values.

Exploration 10.1: Are Dinner Plates Getting Larger?

Figure 10.3: Are the squares the same size? Or is one larger than the other?

1. Look at the two pictures in Figure 10.3. What do you think: Are the squares the same size, or is one larger than the other?

For many people, the square on the right looks smaller, but in fact they are about the same size. You have probably seen many articles and television reports about the growing obesity problem in America. Many reasons are given for this problem, from fast food to lack of exercise. One reason that some have given is simply that the size of dinner plates are increasing. Figure 10.3 illustrates that that people may tend to put more food on larger dinner plates without even knowing they are doing so.

Step 1: Ask a research question

An interesting research question is whether dinner plates are getting larger over time.

Step 2: Design a study and collect data.

To investigate the claim that dinner plates are getting larger, some researchers gathered data for American dinner plates being sold on ebay.com on March 30, 2010 (Van Ittersum and Wansink, 2011). They recorded the size and the year of manufacture for each distinct plate in a sample of 20 American-made dinner plates.

2. What is the explanatory variable? Is it categorical or quantitative?

3. What is the response variable? Is it categorical or quantitative?

4. In your own words, explain what it would mean for there to be an association between the explanatory and response variables in this study.

5. What are the observational units? How many are there?

6. Is this study an experiment or an observational study? Why?

This study is different than studies we've looked at before because both of the variables of interest are quantitative. For this reason, the graphical and numerical techniques we will use to summarize the data are different than what we've seen before.

Step 3: Explore the data,

The data in Table 10.1 represent a subset of the data values reported in the research paper.

Table 10.1: Data for size (diameter, in inches) and year of manufacture for 20 American-made dinner plates

Year	1950	1956	1957	1958	1963	1964	1969	1974	1975	1978
size	10	10.75	10.125	10	10.625	10.75	10.625	10	10.5	10.125

Year	1980	1986	1990	1995	2004	2004	2007	2008	2008	2009
size	10.375	10.75	10.375	11	10.75	10.125	11.5	11	11.125	11

Graphical summary of two-quantitative variables: Scatterplots

7. Paste the **PlateSize** data set into the **Corr/Regression** applet. Which variable is on the x (horizontal) axis? Which variable is on the y-(vertical) axis?

A *scatterplot* is a graph showing a dot for each observational unit, where the location of the dot indicates the values of the observational unit for both the explanatory and response variables. Typically, the explanatory variable is placed on the *x*-axis and the response variable is placed on the *y*-axis.

8. Create a rough sketch of the scatterplot from the applet here. Be sure that the axes are labeled well. Show that you understand how a scatterplot is created by circling the dot on the scatterplot corresponding to a plate made in 1958 that is 10 inches in diameter.

When describing a scatterplot we look for three aspects of association: direction, form, and strength.

The *direction* of association between two quantitative variables is either positive or negative, depending on whether or not as the explanatory variable (year) increases the response variable (size) tends to increase (positive association) or decrease (negative association).

9. **Direction.** Is the association between year and plate size positive or negative? Interpret what this means in context.

The *form* of association between two quantitative variables is described by indicating whether a line would do a reasonable job summarizing the overall pattern in the data or if a curve would be better. It is important to note that, especially when the sample size is small, you don't want to let one or two points on the scatterplot to change your interpretation of whether or not the form of association is linear. In general, assume that the form is linear unless there is compelling (strong) evidence in the scatterplot that the form is not linear.

10. **Form.** Does the association between year and size appear to be linear, or is there strong evidence (many observational units) suggesting the relationship is non-linear?

In describing the *strength* of association revealed in a scatterplot, we see how closely the points follow the form: that is how closely do the points follow a straight line or curve. If all of the points fall pretty close to this straight line or curve, we say the association is strong. Weak associations will show little pattern in the scatterplot, and moderate associations will be somewhere in the middle.

11. In your opinion, would you say that the association between plate size and year appears to be strong, moderate or weak?

It is also important to consider any *unusual observations* that do not follow the overall pattern.

12. Are there any observational units (dots on the scatterplot, representing individual plates) that seem to fall outside of the overall pattern?

In this example, the largest plate diameter was 11.5 inches. Compared to other plate diameters in the sample, this is fairly large. But based on the scatterplot we see that the plate doesn't appear to be unusual. On the other hand, if the 11.5 inch plate had been manufactured in 1950, then the scatterplot would indicate that to be unusual.

Note: There are two kinds of unusual observations seen in scatterplots: influential observations and outliers. An observation is *influential* if removing it from the data set dramatically changes our perception of the association. Typically influential observations are extreme in the explanatory variable. *Outliers* are observations that don't follow the overall pattern of the relationship. They may or may not be influential and they may or may not be extreme in either variable individually, but are unusual in terms of the combination of values.

> **Key idea:** Association between quantitative variables can be described with direction, form, and strength.
> - If above average values of one variable tend to correspond to above average values of the other variable, the ***direction*** is positive. If, however, above average values of one variable are associated with below average values of the other, the ***direction*** is negative.
> - The ***form*** of the association is whether the data follow some linear pattern or some more complicated pattern.
> - The ***strength*** of an association refers to how closely the data follow a particular pattern. When an association is strong, the value of one variable can be accurately predicted from the other.
> - Also look for and investigate further any ***unusual observations*** that do not fit the pattern of the rest.

Numerical Summaries

Describing the direction, form, and strength of association based on a scatterplot, along with investigating unusual observations, is an important first step in summarizing the relationship between two quantitative variables. Another approach is to use a statistic. One of the statistics most commonly used for this purpose is the correlation coefficient.

The ***correlation coefficient***, often denoted by the symbol r, is a single number that takes a value between -1 and 1, inclusive. Negative values of r indicate a negative association, whereas positive values of r indicate a positive association.

13. Will the value of the correlation coefficient for the year-plate size data be negative or positive? Why?

> **Key Idea:** The correlation coefficient uses a rather complex formula (see the **Appendix Calculation Details** for details) that is rarely computed by hand; instead, people almost always use a calculator or computer to calculate the value of the correlation coefficient. But you should be able to apply the above properties to interpret the correlation coefficient that is found.

The stronger the linear association is between the two variables, the closer the value of the correlation coefficient will be to either -1 or 1, whereas linear associations will have correlation coefficient values closer to 0. Moderate linear associations will typically have correlation coefficients in the range of 0.3 to 0.7, or -0.3 to -0.7.

14. Without using the applet, give an estimated range for the value of the correlation coefficient between plate size and year based on the scatterplot:

15. Now, check the **Correlation coefficient** box in the applet to reveal the actual value of the correlation coefficient. Report the value here.

Key Idea: The correlation coefficient is only applicable for data which has a linear form; non-linear data is not summarized well by the correlation coefficient. In fact, we could say that the correlation coefficient is a numerical summary of the strength and direction of a linear association between two quantitative variables.

Another point about correlation is that it is sensitive to unusual observations called influential points. Influential points are unusual observations which substantially impact the value of the correlation based on whether or not they are present in the dataset.

16. In the **Add/remove observations** section of the applet, enter a year (x) of 1950 and plate size (y) of 11.5 and press **Add**. Note how this is an unusual observation on the scatterplot. How did the correlation change as a result?

We have now seen a graphical technique (scatterplot) and numerical summary (correlation coefficient) to summarize the relationship between two quantitative variables. Using techniques you will learn in the next section, this correlation coefficient is significantly different than zero (p-value ≈ 0.01), meaning that there is strong evidence of an association between year and plate size, though we need to be cautious about generalizing too far because this is only a random sample of plates for sale on a single day on eBay.

Section 10.1 Summary

Examining the **association** between two quantitative variables involves looking at scatterplots.

- The **direction** of the association indicates whether large values of one variable tend to go with large values of the other (**positive** association) or with small values of the other (**negative** association).

- The **form** of association refers to whether the relationship between the variables can be summarized well with a straight line or some more complicated pattern.

- The **strength** of association entails how closely the points fall to a recognizable form such as a line.

- As always, **unusual observations** that do not fit the pattern of the rest of the observations are worth examining closely or have an undue impact on our perception of the association.

  o Influential observations are observations whose removal would dramatically impact the overall direction, form, or strength of the association. (Often you find them by looking for observations that are extreme in the x-direction.)

  o Outliers are unusual observations that do not follow the overall pattern of the association. They may or may not be influential.

A numerical measure of the association between two quantitative variables is the **correlation coefficient, r**.

- A correlation coefficient is between -1 and 1, inclusive.
- The closer the correlation coefficient to -1 or 1, the stronger the linear association between the variables.
- The closer the correlation coefficient to 0, the weaker the linear association between the variables.
- Correlation coefficient is not resistant to extreme values.

Section 10.2: Inference for the Correlation Coefficient: A Simulation-based Approach

Introduction

In the previous section we learned about how the correlation coefficient and scatterplots are used to describe the association between two quantitative variables for the sample. We will now learn how to test whether there is convincing evidence of any association between the two quantitative variables in the larger population or process.

Example 10.2: Exercise Intensity and Mood Changes

You have probably heard that exercise helps relieve stress. Exercising has been shown to decrease tension and elevate one's mood. However, is this change in mood related to the intensity of the exercise? Szabo (2003) investigated this question with a sample of female college students who jogged or ran on a treadmill for 20 minutes. Participants were allowed to go at any speed they desired as long as they didn't walk. Using heart rate monitors, researchers calculated their exercise intensity as a percentage of their maximal heart rate. (The subjects' maximal heart rates were estimated from their base heart rates. As a result a couple of these percentages are over 100.)

Before exercising, the subjects also took a quick test about their mood. The test measured things like anger, confusion, and depression. This test was given again after exercising, and the researchers determined a numerical score for the change in mood of the subjects before and after exercising. Negative numbers indicate a decrease in measures of bad mood like anger, and so forth. Although the researchers found significant results in the participants' change of mood when comparing before and after exercise, we are interested to see whether there is a significant association between the *change of mood* and *exercise intensity*. Let's look at this study using the six step statistical investigation method.

Step 1: Ask a research question. Is there an association between how much the mood of a person changes during exercise and the intensity of the exercise?

Step 2: Design a study and collect data. The study design was detailed above. What we haven't done yet, however, is to convert our research question into hypotheses. Writing our hypotheses out in words should look very familiar to the way we have done this in the past few chapters.

> Null Hypothesis: There is no association between exercise intensity and changes in mood in the population.

> Alternative Hypothesis: There is an association between exercise intensity and changes in mood in the population.

We are going to use the correlation coefficient to measure the association between these two variables. See FAQ 10.2.1 for further discussion.

The researchers used 32 volunteer female college students as their subjects. Their results are shown in Table 10.2.

Table 10.2: Exercise intensity as a percentage of maximal heart rate and the change in mood test scores

Ex. Intensity	41.8	56.1	56.8	59.1	59.5	60.9	61.6	62.6	64.9	65.6	66.4
MoodChange	-12	-4	-5	-6	-24	1	-24	-3	-5	-12	-16

Ex. Intensity	66.5	67.1	67.8	68.3	68.4	69.7	69.9	70.1	70.3	72	72.2
MoodChange	-3	-8	-2	-24	-10	-12	-4	-12	-18	-13	-31

Ex. Intensity	73.9	76.1	77	77.3	79.9	82.5	83.3	87	106	109.6
MoodChange	-20	-26	-3	8	2	3	-11	-3	5	-13

Step 3: Explore the data. A scatterplot of the change in mood and exercise intensity is shown in Figure 10.4.

Figure 10.4: The data were put in the Corr/Regression applet and a scatterplot of change in mood and exercise intensity was produced

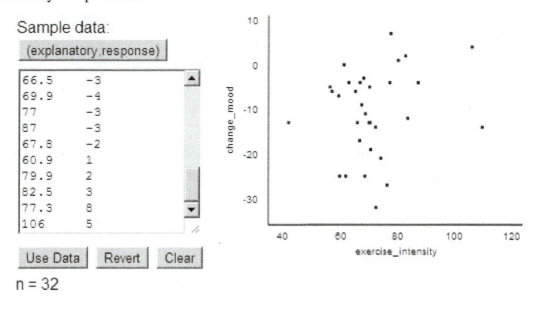

Think about it: Do these data support the research conjecture?

The scatterplot shows a weak positive association between change in mood and intensity of exercise, with a sample correlation of $r = 0.187$.

Now suppose there is no association between these two variables in the population. Would it be *possible* to get a sample correlation coefficient as large as $r = 0.187$ just by random chance? The answer to this question, as always, is yes, it's possible. The real question though, is whether it is *unlikely*. Answering this question will take more investigation.

One way to explore this question is to investigate how much sample to sample variability we can expect to see in the correlation coefficient when the two variables are not related; that is, if the change in mood numbers were essentially assigned at random to the exercise intensity numbers, how strong is the typical "by chance" association?

Think about it: Suggest a procedure for evaluating the strength of evidence against the null hypothesis.

Step 4: Draw inferences. We will proceed just as before with our simulation-based approach, starting with the 3S Strategy.

1. **Statistic:** In this case, we found the observed correlation coefficient between exercise intensity and mood change for our sample to be $r = 0.187$.

2. **Simulate:** We will assume there is no relationship between change in mood and exercise intensity in the population (this is our null hypothesis). If this is the case, it does not matter which change in mood value paired up with which exercise intensity value. Therefore we will shuffle the response variable, change in mood, and find the correlation coefficient for our new shuffled data. We will repeat this random shuffling a large number of times to obtain a null distribution that models the behavior of the correlation coefficient under the null hypothesis.

   To better understand the simulation approach, consider the following copy of the original data table where we have removed the mood change values from the table.

Table 10.3. Data table of exercise intensity and mood change

Ex. Intensity	41.8	56.1	56.8	59.1	59.5	60.9	61.6	62.6	64.9	65.6	66.4
Mood Change											

Ex. Intensity	66.5	67.1	67.8	68.3	68.4	69.7	69.9	70.1	70.3	72	72.2
Mood Change											

Ex. Intensity	73.9	76.1	77	77.3	79.9	82.5	83.3	87	106	109.6
Mood Change										

	-12	-4	-5	-6	-24	1	-24	-3	-5	-12	-16
Mood Change	-3	-8	-2	-24	-10	-12	-4	-12	-18	-13	-31
	-20	-26	-3	8	2	3	-11	-3	5	-13	

Imagine that the actual mood change scores were now randomly matched to the different exercise intensity values. You can envision this random matching process by imagining that the 32 different mood change scores were each written on a slip of paper, so you could shuffle all the mood change scores and randomly pair up each mood change score with an exercise intensity score. Thus, this simulation means that for this process there really is no underlying association between exercise intensity and mood change, any association observed in the sample is simply due to chance.

3. **Strength of evidence:** If the observed statistic (correlation coefficient of our sample, $r = 0.187$) is in the tail of the null distribution created by our simulation, then we say we have found evidence against the null hypothesis in favor of the alternative hypothesis.

To complete steps 2 and 3 of our 3S strategy, we used the Corr/Regression applet to shuffle the mood change values 1000 times, and after each shuffle calculated the correlation coefficient. These 1000 shuffled correlation coefficients are shown in a histogram in Figure 10.5. We counted that 328 of these 1000, or 32.8%, were as large as or larger than our sample correlation coefficient of 0.187 or as small or smaller than -0.187 (two-sided), giving us an approximate p-value of 0.328. Therefore, our sample correlation coefficient is not very unlikely to occur by chance alone when the null hypothesis is true. Hence these data do not provide strong evidence against the null hypothesis and do not support the alternative hypothesis that change in mood is associated with intensity of exercise in the population.

Figure 10.5: The null distribution of correlation coefficients for 1000 shuffles of mood change assuming no association between mood change and exercise intensity

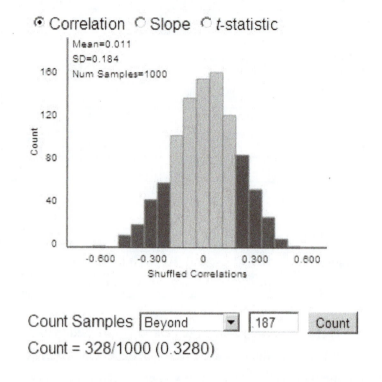

Here is a summary of the simulation model:

Null hypothesis = No association between x and y variables	
One repetition = Re-randomizing the response outcomes to the explanatory variable outcomes	
Statistic = correlation coefficient	

Step 5: Formulate conclusions. We do not have strong evidence that there is a genuine association between change in mood and exercise intensity in the population. Hence, we cannot really talk about cause-and-effect. Even if we found a significant association, we would still not be able to draw a cause-and-effect conclusion because this was an observational study in which subjects chose for themselves the level of exercise intensity. Also, this study does not appear to have selected a random sample, so we should be cautious in generalizing these results to a larger population. At best, they apply to female college students. The study did report that they had significant results when comparing mood before and after exercise, however the intensity of the exercise does not seem to significantly influence the amount of the change in mood for students like those in this study.

Step 6: Look back and ahead. The researchers in this study now have some important decisions to make. The results of this analysis did not provide compelling evidence that exercise intensity and mood change were associated in the population. But is this an example of a missed opportunity to establish that such a hypothesis is true? Or, are the researchers incorrect in what they've hypothesized? Although we'll never know for sure from a single study, there are a number of possible improvements if the researchers do a follow-up study: including potentially

(a) gathering a larger sample size in a follow-up study,

(b) considering whether that sample should be randomly chosen from a population of interest, (c) randomly assigning subjects to different exercise intensity levels,

(d) improving the way that "mood" and "mood change" were measured (e.g., using reliable and valid instruments),

(e) considering starting only with subjects who had a negative mood before exercising as those with positive moods before exercising may show little change afterwards.

FAQ 10.2.1 Where's the parameter? Why no interval?

Q: I've got a question about the draft lottery (Exploration 10.2).

A: OK.

Q: What's the parameter?

A: That's a complicated question. The short answer is there isn't one, but to explain why, I want to go back and get some background. Are we asking about a process or a population?

Q: A process. Mix the numbered capsules, then take them out one at a time.

A: Right. And the null hypothesis?

Q: There's no association between sequential date and draft number.

A: Right. Which number goes with which birthday is purely a matter of random chance. What's the alternative hypothesis?

Q: There's some sort of association. The pattern in the scatter plot is not due just to random chance.

A: Right.

Q: I'm still waiting. What's the parameter?

A: One more question. Suppose the null is false and there is some sort of association. What form will the association take?

Q: How should I know? There could be all sorts of patterns hidden in that scatterplot. Possibly linear, possibly curved, possibly up and down. Who knows.

A: Exactly. Until we can say more about the structure of the alternative – about the form of the pattern – we can't specify a parameter.

Q: You took long enough to say you don't know.

A: Yes, but there's an important lesson: *You have to know (or assume) some important features of your data in order have a meaningful parameter. In early chapters we assumed the probability of success or the difference in treatment means was all that mattered, so those were our parameters.*

Q: But you can measure strength of evidence against the null even without a parameter? Seems like magic.

A: No magic. As always, there's no free lunch. You don't need a parameter, but your test may not be very powerful, because it has to work for all possible alternatives.

Q: I thought I asked a simple question.

A: An important question, even if the answer isn't simple.

Q: One last question. Why don't we do simulation-based confidence intervals?

A: The short answer is that you can't have confidence intervals if you don't have a parameter. But we'll come back to all of this in Section 5.

Q: I can hardly wait. Be still my heart!

Exploration 10.2: Draft Lottery

In 1970, the United States Selective Service conducted a lottery to decide which young men would be drafted into the armed forces (Fienberg, 1971). Each of the 366 birthdays in a year (including February 29) was assigned a draft number. Young men born on days assigned low draft numbers were drafted.

We will regard the 366 dates of the year as observational units. We will consider two variables recorded on each date: *draft number* assigned to the date, and *sequential date* in the year (so January 31 is sequential date 31, February 1 is sequential date 32, and so on).

1. In a perfectly fair, random lottery, what should be the value of the correlation coefficient between *draft number* and *sequential date of birthday*?

Figure 10.6 displays a *scatterplot* of the assigned draft numbers and the sequential dates. There are 366 dots, one for each day of the (leap) year.

Figure 10.6: A scatterplot of the draft numbers and sequential dates

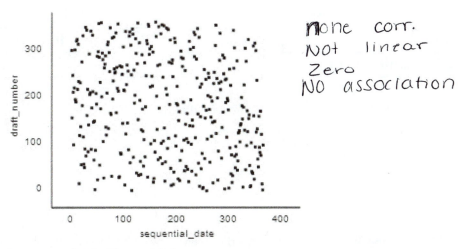

none corr.
Not linear
Zero
NO association

2. Let's take a look at the scatterplot in Figure 10.6.
    a. Does the scatterplot reveal much of an association between draft number and sequential date?

    b. Based on the scatterplot, guess the value of the correlation coefficient.

c.   Does it appear that this was a fair, random lottery?

It's difficult to see much of a pattern or association in the scatterplot, so it seems reasonable to conclude that this was a fair, random lottery with a correlation coefficient near zero.

But let's dig a little deeper …

3.   The 1970 Draft Lottery data sheet (at the end of the exploration) shows the draft number assigned to each of the 366 birthdays.
   a.   Find and report the draft number assigned to *your* birthday.

   b.   Is your draft number in the bottom third (1-122), middle third (123-244), or top third (245-366)?

4.   The second table at the end of the exploration has ordered the draft numbers within each month.
   a.   Use this table to determine the *median* draft number for your birth month.

   b.   Collaborate with your classmates to determine and report the median draft number for all twelve months.

Month	Jan	Feb	Mar	Apr	May	June	July	Aug	Sep	Oct	Nov	Dec
Median draft number												

   c.   Do you notice any pattern or trend in the median draft numbers over the course of the year? (*Hint*: If you do not see a trend, compare the six medians from Jan – June with the 6 medians from July – Dec.)

5.   The correlation coefficient for these data is $r = -0.226$. What does this number reveal? Is it consistent with the scatterplot?

6.
   a.   Suggest two possible explanations (hypotheses) which could have generated the value of the non-zero correlation coefficient.

b. In your own words, how could we go about determining whether random chance is a plausible explanation for the observed correlation value between sequential date and draft number? Explain how the 3S Strategy could be applied here, in particular identifying a simulation strategy you could conduct "by hand." Note: You do not need to actually carry out a simulation analysis.

   i. What is the statistic?

   ii. How would you simulate?

   iii. How would you evaluate the strength of evidence?

The null hypothesis to be tested is that the lottery was conducted with a fair, random process. The null hypothesis would therefore mean that there is *no* association between sequential date and draft number for this process. The alternative hypothesis is that this lottery was *not* conducted with a fair, random process, so there *is* an association between sequential date and draft number. See FAQ 10.2.1 for further discussion.

How can we assess whether the observed correlation coefficient of $r = -0.226$ is far enough from zero to provide convincing evidence that the lottery process was not random? Like always, we ask how unlikely it would be for a fair, random lottery to produce a correlation value as far from zero as -0.226. Also like always, we answer that question by simulating a large number of fair random lotteries, calculating the correlation coefficient for each one, and seeing how often we obtain a correlation coefficient as or more extreme (as far from zero) as -0.226.

7. Open the **Corr/Regression** applet. Copy the data from **DraftLottery** into the applet (remember to include the column titles).
   a. Check the **Correlation Coefficient** box and confirm that the correlation coefficient is -0.226.

   b. Check the **Show Shuffle Options** box and select the **Correlation** radio button to the right to keep track of that statistic. Then press **Shuffle Y-values** to simulate one fair, random lottery. Record the value of the correlation coefficient between the shuffled draft numbers and sequential date.

c.   Press **Shuffle Y-values** four more times to generate results of four more fair, random lotteries. Record the values of the shuffled correlation coefficients in the table below.

Repetition	1	2	3	4	5
Correlation coefficient					

d.   Did any of these simulated statistics from fair, random lotteries produce a correlation coefficient as extreme (far from zero) as the observed -0.226?

e.   Change the **Number of Shuffles** from 1 to 995 and press **Shuffle Y-values** to simulate 995 more fair, random lotteries. Look at the null distribution of these 1000 correlation coefficients. Where is this distribution centered? Why does this make sense?

f.   Next to the **Count Samples** box choose **Beyond** from the pull-down menu. Specify the observed correlation coefficient (-0.226) and press **Count**. What proportion, of the 1000 simulated random lotteries produced a correlation coefficient as extreme (as far from zero in either direction) as -0.226? Report the approximate p-value.

g.   Interpret this p-value: This is the probability of what, assuming what?

h.   What conclusion would you draw from this p-value? Do you have strong evidence that the 1970 draft lottery was not conducted with a fair, random process? Explain the reasoning behind your conclusion.

Once they saw these results, statisticians were quick to point out that something fishy happened with the 1970 draft lottery. The irregularity can be attributed to improper mixing of the balls used in the lottery drawing process. (Balls with birthdays early in the year were placed in the bin first, and balls with birthdays late in the year were placed in the bin last. Without thorough mixing, balls with birthdays late in the year settled near the top of the bin and so tended to be selected earlier.) The mixing process was changed for the 1971 draft lottery (e.g., two bins, one for the draft numbers and one for the birthdays), for which the correlation coefficient turned out to be $r = 0.014$.

8.   Use your simulation results to approximate the p-value for the 1971 draft lottery. Is there any reason to suspect that this 1971 draft lottery was not conducted with a fair, random process? Explain the reasoning behind your conclusion. Also explain why you don't need to paste in the data from the 1971 lottery first.

1970 Draft Lottery Data

date	Jan	Feb	Mar	Apr	May	Jun	Jul	Aug	Sep	Oct	Nov	Dec
1	305	86	108	32	330	249	93	111	225	359	19	129
2	159	144	29	271	298	228	350	45	161	125	34	328
3	251	297	267	83	40	301	115	261	49	244	348	157
4	215	210	275	81	276	20	279	145	232	202	266	165
5	101	214	293	269	364	28	188	54	82	24	310	56
6	224	347	139	253	155	110	327	114	6	87	76	10
7	306	91	122	147	35	85	50	168	8	234	51	12
8	199	181	213	312	321	366	13	48	184	283	97	105
9	194	338	317	219	197	335	277	106	263	342	80	43
10	325	216	323	218	65	206	284	21	71	220	282	41
11	329	150	136	14	37	134	248	324	158	237	46	39
12	221	68	300	346	133	272	15	142	242	72	66	314
13	318	152	259	124	295	69	42	307	175	138	126	163
14	238	4	354	231	178	356	331	198	1	294	127	26
15	17	89	169	273	130	180	322	102	113	171	131	320
16	121	212	166	148	55	274	120	44	207	254	107	96
17	235	189	33	260	112	73	98	154	255	288	143	304
18	140	292	332	90	278	341	190	141	246	5	146	128
19	58	25	200	336	75	104	227	311	177	241	203	240
20	280	302	239	345	183	360	187	344	63	192	185	135
21	186	363	334	62	250	60	27	291	204	243	156	70
22	337	290	265	316	326	247	153	339	160	117	9	53
23	118	57	256	252	319	109	172	116	119	201	182	162
24	59	236	258	2	31	358	23	36	195	196	230	95
25	52	179	343	351	361	137	67	286	149	176	132	84
26	92	365	170	340	357	22	303	245	18	7	309	173
27	355	205	268	74	296	64	289	352	233	264	47	78
28	77	299	223	262	308	222	88	167	257	94	281	123
29	349	285	362	191	226	353	270	61	151	229	99	16
30	164		217	208	103	209	287	333	315	38	174	3
31	211		30		313		193	11		79		100

The following table arranges the draft numbers in order for each month:

rank	Jan	Feb	Mar	Apr	May	Jun	Jul	Aug	Sep	Oct	Nov	Dec
1	17	4	29	2	31	20	13	11	1	5	9	3
2	52	25	30	14	35	22	15	21	6	7	19	10
3	58	57	33	32	37	28	23	36	8	24	34	12
4	59	68	108	62	40	60	27	44	18	38	46	16
5	77	86	122	74	55	64	42	45	49	72	47	26
6	92	89	136	81	65	69	50	48	63	79	51	39
7	101	91	139	83	75	73	67	54	71	87	66	41
8	118	144	166	90	103	85	88	61	82	94	76	43
9	121	150	169	124	112	104	93	102	113	117	80	53
10	140	152	170	147	130	109	98	106	119	125	97	56
11	159	179	200	148	133	110	115	111	149	138	99	70
12	164	181	213	191	155	134	120	114	151	171	107	78
13	186	189	217	208	178	137	153	116	158	176	126	84
14	194	205	223	218	183	180	172	141	160	192	127	95
15	199	210	239	219	197	206	187	142	161	196	131	96
16	211	212	256	231	226	209	188	145	175	201	132	100
17	215	214	258	252	250	222	190	154	177	202	143	105
18	221	216	259	253	276	228	193	167	184	220	146	123
19	224	236	265	260	278	247	227	168	195	229	156	128
20	235	285	267	262	295	249	248	198	204	234	174	129
21	238	290	268	269	296	272	270	245	207	237	182	135
22	251	292	275	271	298	274	277	261	225	241	185	157
23	280	297	293	273	308	301	279	286	232	243	203	162
24	305	299	300	312	313	335	284	291	233	244	230	163
25	306	302	317	316	319	341	287	307	242	254	266	165
26	318	338	323	336	321	353	289	311	246	264	281	173
27	325	347	332	340	326	356	303	324	255	283	282	240
28	329	363	334	345	330	358	322	333	257	288	309	304
29	337	365	343	346	357	360	327	339	263	294	310	314
30	349		354	351	361	366	331	344	315	342	348	320
31	355		362		364		350	352		359		328

Section 10.2 Summary

The 3S Strategy can be used to assess whether a sample correlation coefficient is extreme enough to provide strong evidence that the variables are associated in the population.

- The null hypothesis is that there is no association between the two variables in the population.
- The statistic used to measure the (linear) association is the correlation coefficient.
- Shuffling the values of the response variable is used to produce simulated values of the statistic.
  - o This shuffling simulates values of the statistic under the assumption of no underlying association between the two variables.
- As always, the p-value is calculated as the proportion of repetitions in which the simulated value of the statistic is at least as extreme as the observed value of the statistic.
  - o Also, as always, a small p-value provides evidence against the null hypothesis in favor of the alternative hypothesis.
  - o We can test for positive or negative linear associations (one-sided p-values) or for any association (two-sided p-value).

Section 10.3: Least Squares Regression

In the last two sections we used graphical and numerical methods to describe the observed association between two quantitative variables, and then looked at a simulation method for drawing inferences based on the correlation coefficient. If we decide the association is linear, it is helpful to develop a mathematical model of that association. This is especially helpful in allowing us to make predictions about the response variable from the explanatory variable. The most common way to do this is with a ***least squares regression line*** (often called the best fit line, regression line or, simply, regression) which is the line that gets as "close" as possible to all of the data points.

Example 10.3: Are Dinner Plates Getting Larger (revisited)?

In Exploration 10.1 we looked at a recent study investigating a potential association between the year of manufacture and dinner plate size, finding strong evidence of a positive association, meaning that, based on a random sample of plates for sale on eBay on a given day, plate sizes appear to be increasing over time. A scatterplot of the data is presented again in Figure 10.7.

This scatterplot reveals a moderately strong positive association, as there does appear to be a trend of increasing plate sizes in more recent years. This observation is supported by the correlation coefficient ($r = 0.604$). The association also appears to be roughly linear, indicating that a line would do a reasonable job of summarizing the overall relationship. We have added such a line, specifically the least squares regression line, to the scatterplot in Figure 10.7.

Figure 10.7: A scatterplot and least squares regression line for the size of American dinner plates and the year of manufacture

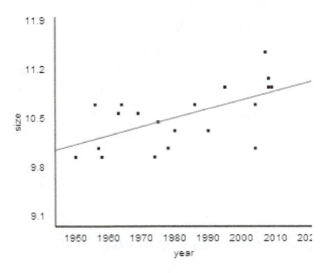

As you may recall, lines can be summarized by an equation which indicates the slope and y-intercept of the line. The line shown above has a slope of 0.0128 and a y-intercept of -14.80.

Notation: The equation of the best fit line is written as $\hat{y} = a + b(x)$ where

- a is the **y-intercept** and b is the **slope**
- x is a value of the explanatory variable
- $\hat{y}$ is the predicted value for the response variable

The least squares line for the dinner plate data turns out to be $\hat{y} = -14.8 + 0.0128\,x$, or better written as

$$\text{diameter} = -14.8 + 0.0128(\text{year}).$$

Such an equation allows us to predict the plate diameter in a particular year. If we substitute 2000 in for the year, we predict a plate diameter of $-14.8 + 0.0128(2000) = 10.8$ inches.

Think about it: What do you predict for a plate diameter for a plate manufactured in the year 2001? How does this compare to our prediction for the year 2000?

The slope of this equation is $b = 0.0128$. This means that, according to our line, dinner plate diameters are predicted to increase by 0.0128 inches per year on average. Your prediction for 2001 should have been roughly 0.0128 inches larger than your prediction for 2000. This is true no matter what two years you use, as long as they are one year part. If you wanted to predict the change in average plate size associated with a 10 year difference in manufacturing dates, you could simply multiply the slope by 10.

Key idea: The slope of a least squares regression line is interpreted as the predicted change in the average response variable for a one-unit change in the explanatory variable.

Note that the slope of the regression line for the dinner plate data is positive, as was the correlation coefficient. These two statistics will always have the same sign.

Key idea: For a given dataset, the signs (positive, negative, or zero) for the correlation coefficient and the slope of the regression line must be the same.

You can think of the y-intercept in a couple of different ways. Graphically, it is where the regression line crosses the y-axis. Because the x-value of the y-axis is 0, the y-intercept can also be thought of as the $\hat{y}$-value (or predicted response) when the x-value (or explanatory variable) equals 0. We had a y-intercept of -14.8 in the dinner plate equation. This tells us that, according to the line, the predicted size of American dinner plates in year 0 was -14.8 inches. Now this doesn't make much sense, does it? For one, we can't have a negative value for the size of dinner plates. Our equation might work reasonably well within the range of values of the explanatory variable with our given data, but may fail miserably outside that range. In fact, our equation should only be used to predict the size of dinner plates from about 1950 to 2010. We don't know whether the same relationship between year and plate size holds outside this range, so we should not make predictions outside this range. Going way back to year 0, for example, gives us an answer that makes no sense at all.

Key idea: Predicting values for the response variable for values of the explanatory variable that are outside of the range of the original data is known as *extrapolation*, and can give very misleading predictions.

We should also make sure the interpretation is meaningful in the study context. In many situations the intercept can have a meaningful interpretation, such as the start-up (time $= 0$) cost of a manufacturing process.

Key idea: The y-intercept of a regression line is interpreted as the predicted value of the response variable when the explanatory variable has a value of zero (though be wary of extrapolation in interpreting the intercept or other values outside the original data range).

What makes this line the "best"?

We've talked about how the line above is the "best" line. But what makes it the best? Intuitively, we said that the best fitting line gets as close as possible to all of the points, but we can be more precise by better understanding the idea of a *residual*. The vertical distance from a point to the line is called a residual. So the residuals measure the difference between the predicted and observed value of the response variable (in this case, plate size) for each x value. Figure 10.8 shows the residuals for the plate example.

Key idea: A *residual* is the difference between an observed response and the corresponding prediction made by the least squares regression line (*residual = observed – predicted*). Thus, negative residuals occur when points are below the best fit line and positive residuals occur when points are above the best fit line.

Figure 10.8: Residuals for best-fit line of year vs. plate size

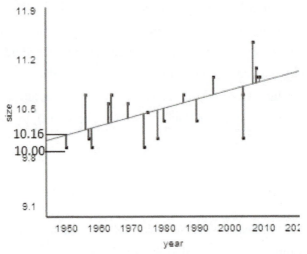

Each observational unit (plate) has a residual which can be found by subtracting the predicted plate size from the observed (real) plate size (in other words finding the value of $y - \hat{y}$). For example, the plate manufactured in 1950 has a predicted plate size of 10.16 inches, but is actually 10 inches, and so the residual is $10 - 10.16 = -0.16$ inches.

There are actually a number of ways to get "as close" to all of the points as possible, one way would be to add up all of the values of the residuals (making them positive first, so they didn't cancel out), and finding the line that minimized the sum of the (absolute) residuals. In practice, however, finding such a line is very difficult. Instead, we will find the line that minimizes the sum of the squared residuals.

> **Key Idea:** The least squares regression line minimizes the sum of squared residuals.

Although the reasons for this change are due to calculus (and not something we really want you to worry about!), the implications aren't too different than if we were minimizing the sum of the absolute residuals. The only subtle difference is that because we are squaring, unusual observations (extreme in the x direction) have more "pull" as the squared value of the residual is larger relative to more typical points. Thus, these kinds of points can be particularly influential — just like they were when we explored the correlation coefficient. See FAQ 10.3.1 for more.

> **Key idea:** An observation or set of observations is considered *influential* if removing the observation from the dataset substantially changes the values of the correlation coefficient and/or the least squares regression equation. Typically, observations that have extreme explanatory variable values (far below or far above $\bar{x}$) are potentially influential. They may not have large residuals, having pulled the line close to them.

One final statistic that is of interest in a linear regression is called the *coefficient of determination* (r^2). This statistic measures the percentage of total variation in the response variable that is explained by the linear relationship with the explanatory variable. The value is literally the square of the correlation coefficient. In this example, the correlation coefficient is 0.604, and so the coefficient of determination is $(0.604)^2 \approx 0.365$, meaning that 36.5% of the observed variation in plate sizes is explained by knowing the year of manufacture.

Another way to compute the coefficient of determination is as follows:

$$r^2 = 100\% \times \frac{SSE(\bar{y}) - SSE\ (\text{regression line})}{SSE(\bar{y})}$$

where *SSE(regression line)* is the sum of the squared residuals for the least squares regression line and $SSE(\bar{y})$ is the sum of squared residuals for the horizontal line at the average value of the response variable. We can think of $SSE(\bar{y})$ as a measure of the overall variability in the response, and we are seeing how much less variability there is about the regression line compared to this "no association" line.

There is a third way to think about the coefficient of determination. If we look at the plate sizes alone, variability of the plate sizes, as measured by the standard deviation is 0.430. We can also use the squared standard deviation or *variance* to reflect this, $0.430^2 \approx 0.185$. But if we look at the residuals, the "unexplained variability" is down to 0.343. That is, the standard deviation of the residuals is 0.343, meaning that that the variance of the residuals is $0.343^2 \approx 0.118$, a reduction of $0.185 - 0.118 = 0.067$. So we've explained about 0.067/.430 or 36% of the original variability in the plate sizes.

Figure 10.9a. Variability in plate sizes

Figure 10.9b. Variability in residuals

Key idea: The coefficient of determination (r^2) is the percentage of total observed variation in the response variable that is accounted for by changes (variability) in the explanatory variable

---

FAQ 10.3.1 Why least squares?

Q: Why least squares? Why do we square the deviations? Why not absolute values?

A: Not only is your question a thoughtful one, it's also a question that invites us back in history. In the 1700s, scientists tried to do what you suggest – to minimize the sum of absolute deviations: they wanted to use least MAD(deviation) instead of least sum of squared deviations. The scientific logic was good, but the math turned out to be too messy. Sometimes it worked, and there was a single "best" line that minimized the MAD statistic. But sometimes there would be infinitely many "best" lines, with no way to choose from among them. "Best" and "infinitely many" don't go together. Even when there was a single best line, it was devilishly hard to find. Overall, the math was just too hard.

Q: So least squares is pretty much just the fallback Plan B after least absolute deviations failed?

A: Yes. The *justification* for least absolute deviations makes good sense, but the *theory* is much simpler for least squares. As Fred Mosteller, one of the greatest statisticians of the last century, used to tell his students at Harvard, "Mathematicians love calculus, and calculus loves squares."

Q: Does the choice – absolute values or squares -- make much difference in the line you get?

A: Another good question. Sometimes not, but sometimes it does.

Q: Say more.

A: Remember SD and IQR and outliers?

Q: You mean SD is sensitive to outliers and IQR is resistant?

A: Exactly. Remember why?

Q: Because the SD squares the deviations, and the IQR does not. Squaring gives extra weight to large deviations.

A: Apply the same logic to the least squares line.

Q: Outliers have a big effect because their large deviations get extra weight from the squaring.

A: Bravo! This is why some observations can become influential points – the least squares line won't let their residuals get too large. If this were second grade in the 1950s you'd get to stay after class and dust the erasers. If this were the 1970s and you were a statistician, you'd be able to get a big research grant from the National Science Foundation. If this was Dr. Bastian's lab and you were a dolphin, you'd have just earned a fish.

Exploration 10.3: Predicting Height from Footprints

Can a footprint taken at the scene of a crime help to predict the height of the criminal? In other words, is there an association between height and foot length? To investigate this question, a sample of 20 statistics students measured their height (in inches) and their foot length (in centimeters).

1. Let's think about this study.
    a.   What are the observational units in this study?

b. Identify the two variables recorded for each observational unit. Which is the explanatory variable, and which is the response? Also classify these variables as categorical or quantitative.

        Explanatory: Type:

        Response: Type:

c. Is this an observational study or a randomized experiment?

The sample data for the 20 students appear in the **Corr/Regression** applet.

2. Open the applet, and look at the scatterplot for the data already there. Describe the association between the variables as revealed in the scatterplot. (*Hint*: Remember to comment on direction, strength, and form of association as well as unusual observations.)

3. Would you say that a straight line could summarize the relationship between height and foot length reasonably well?

4. Check the **Show Movable Line** box to add a blue line to the scatterplot. If you place your mouse over one of the green squares at the ends of the line and drag, you can change the slope of the line and move it. You can also use the mouse to move the green dot up and down vertically to change the intercept of the line.

    a. Move the line until you believe your line "best" summarizes the relationship between height and foot length for these data. Write down the resulting equation for your line (using traditional statistical notation).

    b. Why do you believe that your line is "best?"

    c. Did all students in your class obtain the same line/equation?

    d. How can we decide whether your line provides a better fit to the data than other students' lines? Suggest a criterion for deciding which line "best" summarizes the relationship.

One way to draw the best fit line is to minimize the distance of the points to the line (these distances are called *residuals*).

Key idea: A *residual* is the difference between an observed response and the corresponding prediction made by the least squares regression line (*residual = observed – predicted*). Thus, negative residuals occur when points are below the best fit line and positive residuals occur when points are above the best fit line.

---

5. Would points above the line have a positive or negative residual, or is it impossible to tell? What about points below the line?

    Above line: Below line:

6. Check the **Show Residuals** box to visually represent these residuals for your line on the scatterplot. The applet also reports the sum of the values of the residuals (**SAE**). SAE stands for "Sum of the Absolute Errors." The acronym indicates that we need to make residuals positive before we add them up and that sometimes people call residuals "errors."

Record the SAE value for your line _____

What is the best (lowest) SAE in the class?_____

7. It turns out that a more common criterion for determining the "best" line is to instead look at the sum of the *squared* residuals (**SSE**). This approach is similar to simply adding up the residuals, but is even more strict in not letting individual residuals get too large.
Check the **Show Squared Residuals** box to visually represent the squared residual for each observation. Note that we can visually represent the squared residual as the area of a square where each side of the square has length equal to the residual.

    a. What is the SSE (sum of squared residuals) for your line?_____

        What is the best (lowest) SSE in the class?_____

b. Now continue to adjust your line until you think you have minimized the sum of the squared residuals.

Report your new equation_____

Report your new SSE value_____

What is the best SSE in the class?_____

Key Idea: The least squares regression line minimizes the sum of squared residuals.

---

8. Now check the **Show Regression Line** box to determine and display the equation for the line that actually does minimize (as can be shown using calculus) the sum of the squared residuals.

a. Record the equation of the least squares regression line by indicating the appropriate slope and intercept of the line. Note that we've used variable names in the equation, not generic *x* and *y*. And put a carat ("hat") over the *y* variable name to emphasize that the line gives predicted values of the *y* (response) variable.

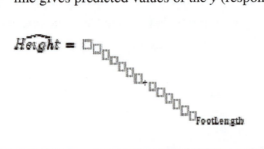

Notation: The equation of the best fit line is written as $\widehat{y} = a + b(x)$ where

- a is the **y-intercept**
- b is the **slope**
- x is a value of the explanatory variable
- $\widehat{y}$ is the predicted value for the response variable

---

b. Did everyone in your class obtain the same equation?

c. Is the slope positive or negative? Explain how the sign of the slope tells you about whether your data display a positive or a negative association.

Key idea: For a given dataset, the signs (positive or negative) for the correlation coefficient and the slope of the regression line must be the same.

9.

   a.   Use the least squares regression line to predict the height of someone whose foot length is 28 cm. (Simply plug in the value of 28 cm for foot length in the equation of the line.)

   b.   Use the least squares regression line to predict the height of someone whose foot length is 29 cm.

   c.   By how much do your predictions in (a) and (b) differ? Does this number look familiar? Explain.

Key idea: The *slope* coefficient of a least squares regression model is interpreted as the predicted change in the response (y) variable for a one-unit change in the explanatory (x) variable.

---

   d.   Interpret the slope in context:

   The slope of the regression line predicting height based on foot length is

   _____, meaning that for every additional _____ cm increase in foot

   length, their predicted height increases by _____ inches.

10.

   a.   Use the least squares regression line to predict the height of someone who has a foot length of 0 cm.

   b. Your answer to (a) should look familiar. What is this value?

> **Key idea:** The *y*-intercept of a regression line is interpreted as the predicted value of the response variable when the explanatory variable has a value of zero (though be wary of extrapolation in interpreting the intercept or other values outside the original data range).

> **Key idea:** Predicting values for the response variable for values of the explanatory variable that are outside of the range of the original data is known as *extrapolation* and can give very misleading predictions.

      c. Explain how your prediction of the height of someone whose foot length is 0 cm is an example of extrapolation.

11. Earlier, we explored the notion of a residual as the distance of a point to a line.

      a. Using the equation for the best fit line that you reported in #8, find the residual for the person whose foot length is 32 centimeters and height is 74 inches. Note: Find this value by taking the actual height and subtracting the person's predicted height from the equation.

      b. Is this person's dot on the scatterplot above or below the line? How does the residual tell you this?

Residuals are helpful for identifying unusual observations. But not all observations of interest have large residuals.

12. Uncheck the **Show Movable Line** box to remove it from the display and check the **Move observations** box.

a. Now click on one student's point in the scatterplot and drag the point up and down (The original regression line will remain in grey.) Does the regression line change much as you change this student's height?

b. Repeat the previous question, using a student with a very small x (foot size) value and then a point with an x value near the middle and then a point with a very large x value. Which of these seem(s) to have the most *influence* on the least squares regression line? Explain.

Key idea: An observation or set of observations is considered ***influential*** if removing the observation from the dataset substantially changes the values of the correlation coefficient and/or the least squares regression equation. Typically, observations that have extreme explanatory variable values (far below or far above $\bar{x}$) are potentially influential. They may not have large residuals, having pulled the line close to them.

Residuals also help us measure how accurate our predictions are from using the regression line. In particular, we can compare the "prediction errors" from the regression line to the prediction errors if we made no use of the explanatory variable.

13. Press the **Revert** button to reload the original data set. Recheck the **Show Movable Line** box to redisplay the blue line. Notice that this line is flat at the mean of the y (height) values.

a. Check the **Show Squared Residuals** box (under the Movable Line information) to determine the SSE if we were to use the average height ($\bar{y}$) as the predicted value for every x (foot size). Record this value.

b. What is the slope of this line?

c. If the slope of the best fit line is zero, our data shows _____ (positive/negative/no) association between the explanatory and response variables.

14. Now compare this to the SSE value from the regression line. Which is smaller? Why does that make sense?

Coefficient of Determination (r^2)

15. A quantity related to the correlation coefficient is called the coefficient of determination, or r-squared (r^2).

    a. To measure how much the regression line decreases the "unexplained variability" in the response variable, we can calculate the percentage reduction in SSE.

$$r^2 = 100\% \times \frac{SSE(\bar{y}) - SSE\ (\text{regression line})}{SSE(\bar{y})}$$

> **Key Idea:** The *coefficient of determination* indicates the percentage observed in the response variable (height) that is explained by changes in the explanatory variable (foot length).

    b. Complete the following statement: The coefficient of determination is

        _____ percent, this means that _____ percent of the

        variation in people's _____ is attributable to

        changes in their _____.

        If we think of $\hat{y} = \bar{y}$ as the "worst" line (no association), then we are seeing how much better the actual line is than this worst line.

    c. Find the value of the correlation coefficient using the applet, and confirm that when you square the correlation coefficient you get the same number as the applet reports for the coefficient of determination.

For more on the coefficient of determination see Example 10.3.

Section 10.3 Summary

When the association between two quantitative variables has a linear form, a line can be used as a mathematical model for summarizing the relationship and for predicting values of the response variable from the explanatory variable.

- The **least squares regression line** selects the intercept and slope coefficients to minimize the sum of the squared vertical deviations between the observations (*y* values) and the line.
  - o These vertical deviations between the observed *y* values and the line are called **residuals**.
  - o A primary use of the least squares regression line is to make predictions for the response variable based on the explanatory variable.
    - ▪ Making predictions for values of the explanatory variable not included in the range of our data values is called **extrapolation** and is often unwise.
  - o The **slope coefficient** is interpreted as the predicted change in the response variable associated with a one-unit change in the explanatory variable.
  - o The **intercept coefficient** is interpreted as the predicted value of the response variable when the explanatory variable equals zero.
    - ▪ In many contexts the interpretation of the intercept is not sensible and/or may be extrapolation.
  - o The square of the correlation coefficient (r^2) indicates the proportion of variability in the response variable that is explained by the least squares line with the explanatory variable.
    - ▪ This is sometimes called the **coefficient of determination**.

- The sign of the slope (positive, negative, or zero) matches the sign of the correlation coefficient (positive, negative, or zero).

- **Influential observations** affect the regression line substantially.
  - o Influential observations typically have extreme values of the explanatory variable.
  - o Outliers in a regression setting have large residuals.
  - o Influential observations are typically not outliers because they pull the line toward themselves.

Section 10.4: Inference for Regression Slope: Simulation-Based Approach

In the previous section, we looked at how least squares linear regression can be used to describe a linear relationship between two quantitative variables. In that section, we mentioned that the sign on the slope of the regression equation and the sign on the correlation coefficient are always the same. For example, when there is a positive association in the data, both the correlation coefficient and the slope of the regression equation are positive. In Section 10.2, we saw how to use the sample correlation coefficient in a simulated-based test about a null hypothesis of no association. In this section, we will see how we can do the same type of inference but now with the population slope as the parameter of interest.

Example 10.4: Do students who spend more time in non-academic activities, tend to have lower GPAs?

College students have lots of opportunities to be involved in different activities. Some also hold jobs or participate in athletics. Both of these can be very time consuming. Time just socializing with friends also takes up lots of time for some. Does time spent on these types of activities have a negative association with students' GPAs? Many researchers have conducted studies to answer this question. We will look at results from a study carried out by a couple of undergraduate students at the University of Minnesota (Ock, 2008). They surveyed 42 students (34 from a research methods class and 8 from a business management class). They asked questions about time spent per week on various nonacademic activities like work, watching TV, exercising, and socializing. They calculated the total time spent on all nonacademic activities and compared that with the students' GPAs. Those data, gleaned from a scatterplot in their paper, are presented in Figure 10.10, along with the least squares regression line.

Figure 10.10: A scatterplot of students' GPAs and the total number of hours spent on nonacademic activities for 42 students

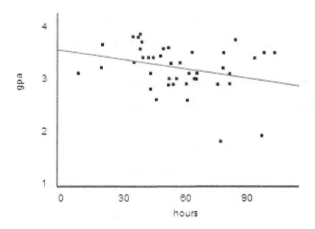

Think about it: Describe the nature of the association between GPA and hours spent on nonacademic activities revealed by this scatterplot.

We discussed outliers briefly earlier. In the regression setting, an outlier is defined as an observation with a large residual. It is often useful to identify the outliers separately, if possible, and then describe the resulting association as if the outliers were removed from the dataset. Apart from the two low outliers (two identified students with GPAs below 2.0), these data exhibit a moderate to weak negative association. The correlation coefficient for the data in Figure 10.9 is $r = -0.290$ and the regression equation is:

$$\widehat{GPA} = 3.60 - 0.0059(\text{nonacademic hours}).$$

As you can see, for the sample of 42 students both the correlation coefficient and the slope of the regression equation are negative.

Think about it: Interpret the slope and intercept of this regression line in the context of non-academic hours and GPA.

The slope of -0.0057 indicates that the predicted GPA decreases by 0.0057 grade points for each additional hour spent on nonacademic activities per week. The intercept reveals that the predicted GPA for a student who spends absolutely no time on nonacademic activities is 3.60.

What can we infer about the larger population represented by this sample? Are we convinced that the slope in the population is negative, as we suspected in advance? Or is it plausible that the GPA values we observed are just arbitrarily paired up with the hours spent on nonacademic activities, and that the association we observed in this sample arose just by chance? Our hypotheses are as follows.

Null Hypothesis: There is no association between the number of hours students spend on nonacademic activities and student GPA in the population.

Alternative Hypothesis: There is a negative association between the number of hours students spend on nonacademic activities and student GPA in the population.

In Section 10.2 we tested the statistical significance of an association between two quantitative variables using the correlation coefficient as our statistic. In this example we will use the slope of the sample regression equation as our statistic. We previously said that a regression line can be represented by the equation $\hat{y} = a + b(x)$ where a is the y-intercept and b is the slope. This is the form of the equation that represents a sample. For this reason, both the y-intercept (a) and slope (b) are statistics.

We will now use the sample slope as the statistic in test of no association. To do this we will use the Corr/Regression applet we used for correlation coefficients. We do this exactly the same way as was done with correlation except select the slope radio button so that slope is used as our statistic instead of the correlation coefficient.

In particular, we shuffle the values of the response variable in the data table, which is equivalent to "dealing" the response variable values out randomly to the values of the explanatory variable in the data table. We did 1000 shuffles and the results are shown in Figure 10.11. See FAQ 10.4.1. for more on the unique behavior of the 1000 shuffled lines. We look for the lower-tail p-value to coincide with our one sided alternative hypothesis.

Figure 10.11: Assuming there was no genuine association between number of hours of nonacademic activities and GPA, we did 1000 shuffles of the response variable and calculated the slope of the resulting regression equation each time

Here is a summary of our simulation model:

Null hypothesis = No association between x and y variables	
One repetition = Re-randomizing the response outcomes to the explanatory variable outcomes	
Statistic = slope coefficient	

From our 1000 shuffles we found 33 simulated slopes that were as small or smaller as our observed sample slope of -0.0057. Thus our simulated p-value is 0.033, and we can conclude there is strong evidence of a negative association between the number of hours students spend on nonacademic activities and their GPA for this population. If we had used the correlation coefficient for our statistic instead of the slope, our results would have been exactly the same (unlike what we saw in Chapters 8 and 9 when we used the MAD statistic as an alternative to a chi-squared or ANOVA F-statistic). We will revisit this idea in the next section.

Key idea: For a given dataset, a test of association based on a slope is equivalent to a test of association based on a correlation coefficient.

We must ask to what population we can make this inference. The researchers used a convenience sample of mostly Research Methods students. They also added a few business management students in order for the sample to more closely match the racial demographics of the undergraduates at the University of

Minnesota. It would seem they did this because their population of interest was all undergraduates at the university. Do you think the sample collected is representative of all students at the university in terms of the association that was being studied? If we had a simple random sample of all students, then we could easily make this connection. A convenience sample, however, brings with it some concerns. When we are talking about student behavior, it is probably not safe to assume that a convenience sample will produce representative results. We should be very cautious as to exactly what population we can make this inference. Perhaps all social science students (the type that would take a research methods class) might be okay, but beyond that we would have a pretty weak argument.

Finally, we should note that because this study is observational, we cannot conclude that time spent on nonacademic activities *causes* a decrease in GPA.

Exploration 10.4: Perceptions of Heaviness

Researchers believe a person's body is used as a perceptual ruler and people will judge the size of an object based on its relationship to parts of their body. Specifically, some researchers thought people with smaller hands will perceive objects to be bigger and hence heavier than those with larger hands. Linkenauger, Mohler, and Proffitt (2011) collected data on 46 participants, recording their hand width and estimated weight of bean bags. The results are shown in Table 10.4.

Table 10.4: Hand widths (in centimeters) and estimated weight of bean bags (in grams) for 46 participants

Hand Width	Estimated Weight	Hand Width	Estimated Weight	Hand Width	Estimated Weight	Hand Width	Estimated Weight
7.4	75.2	7.7	89.2	8.6	96.8	9.2	92.1
7.4	83.6	7.7	83.6	8.6	103.1	9.2	95.9
7.4	86.2	8.0	97.0	8.6	111.9	9.2	103.0
7.4	92.6	8.0	97.0	8.6	112.5	9.5	88.5
7.4	98.1	8.0	97.0	8.9	110.5	9.8	99.3
7.4	99.2	8.0	103.3	8.9	103.2	9.8	80.9
7.4	103.1	8.0	112.5	8.9	95.4	9.8	79.3
7.0	107.2	8.1	108.6	8.9	92.8	9.8	79.0
7.7	104.3	8.3	105.1	8.9	91.9	10.1	72.8
7.7	100.6	8.3	89.9	8.9	85.6	10.1	87.0
7.7	96.8	8.6	71.6	9.2	86.2		
7.7	96.4	8.6	93.0	9.2	89.1		

1. Write the null and alternative hypotheses for this study in words (use the term "association").

In Section 10.2 we used the Corr/Regression applet to perform a simulation-based test for a correlation coefficient. Doing the simulation-based test for the slope of the regression line is extremely similar. The only difference is we use the slope as our statistic instead of the correlation coefficient. The applet has a radio button to select to switch from one statistic to the other.

2. Paste the data, **handwidth**, in the **Corr/Regression** applet and make a scatterplot. Describe the direction, form, and strength of the scatterplot.

3. Use the applet to determine the least squares regression line for predicting estimated weight based on hand width. What is the value of the slope of the regression line? What does this number imply in terms of hand width and estimated weight?

4. You should have found a negative association between hand width and estimated weight of the bean bag in the sample. The question, however, is if there were no association between hand width and the weight of an object in the population, how likely is it that we would get a slope as small as we did. Let's apply the 3S Strategy.

   a. Statistic: What is the value of the slope in the sample?

   b. Simulate: To simulate you can use the same general approach we used for correlation in Section 10.2. Explain how you would conduct the approach by hand. Assume you have 46 slips of paper with the sample hand widths written on them, and 46 slips of paper with the perceived weight values written on them.

   c. Strength of evidence: Explain how you will calculate the strength of evidence in support of the conjecture that people with larger hands tend to perceive more weight in the bean bags.

5.
   a. Let's shuffle the response variable (estimated weight) and see what we get for sample slope values when we break the association between the data pairs. Use the applet to do this five times and write down the five simulated slopes you get. Are any of them as small or smaller than the value of the actual sample slope?

b. Now do at least 1000 shuffles with the applet and find the p-value.

c. Click on **Plot** in the Shuffle Options section of the applet. This will now display each of the 1000 simulated regression lines (in blue). Describe the behavior of the 1000 regression lines across the different shuffles (see FAQ 10.4.1. for some discussion about this).

d. Explain what your p-value measures in context of the study.

e. Can we conclude that there is strong evidence of a genuine negative association between hand width and estimated weight of the object in the population?

6. As you did in Section 10.2, find the p-value corresponding to the correlation coefficient. How does this p-value compare with the p-value corresponding to the slope? (Note: You should have the p-value for the slope in #5d.)

Key idea: For a given dataset, the test for slope is equivalent to the test for correlation coefficient.

---

7. The sample used here was not a random sample. The article just said that the researchers obtained data on 46 participants.

a. Describe the population to which we could legitimately make our inference. Explain your reasoning.

b. Can we conclude that having smaller hands causes a person to estimate the weight of the bean bag to be smaller? Explain your answer.

FAQ 10.4.1 The point of averages

Q: I've got a question about the **capitalized regression** applet. When I shuffle to simulate the null hypothesis, all of the resulting 'shuffled' lines go through the same point. How come?

A: That point is sometimes called the "point of averages" because its coordinates are $(\bar{x}, \bar{y})$. There's more.

Q: Yes?

A: The point of averages is the center of gravity or balance point of the scatterplot. Imagine trying to find the point where the scatterplot balances. That point is always $(\bar{x}, \bar{y})$.

Q: But that doesn't answer my question. Why do the lines always go through that point?

A: When you shuffle, the x's and y's don't change, so $\bar{x}$ and $\bar{y}$ don't change.

Q: More please!

A: Here's the key idea: Regardless of how the x's and y's are paired up, the least squares regression line will go through the point of balance $(\bar{x}, \bar{y})$. Because shuffling doesn't change the x and the y values the 'shuffled' lines all go through the same point.

Q: Why does the least-squares regression line go through there in the first place?

A: There's some pretty complicated math in the full answer. The short answer is that the best way to minimize the squared residuals is to incorporate the balance point in the regression equation. Mathematically you can show that's always the case.

Q: Does this have anything to do with using the mean as a measure of center?

A: Yes! The mean is the balance point when you have just one variable, and minimizes the sum of the squared deviations of the observations in your sample.

Section 10.4 Summary

The 3S Strategy can be used with the sample slope as the statistic of interest and applying the same simulation strategy we have used before.

- We can use the shuffling of the values of the response variable, to produce simulated values of the statistic that could have been if there was no genuine association between the two variables.
  - o   As always, the p-value is calculated as the proportion of repetitions in which the simulated value of the statistic is at least as extreme as the observed value of the statistic; and, a small p-value provides evidence against the null hypothesis.
- For a given dataset, the test of significance for the slope is equivalent to the test of significance for the correlation coefficient.

Section 10.5: Inference for the Regression Slope: Theory-Based Approach

Introduction

You may have noticed a very familiar looking pattern to the null distributions you generated of sample correlation coefficients in Section 10.2 and of sample slopes in Section 10.4. For example, two null distributions from Examples 10.2 and 10.4 are shown again in Figure 10.12.

Figure 10.12: Null distributions for sample correlation (left) and sample slope (right) from previous examples

Example 10.2: Exercise and Mood Intensity

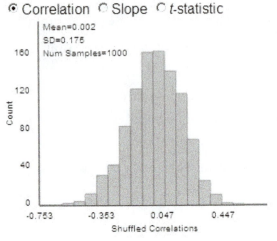

Example 10.4: GPA and nonacademic hrs.

The null distribution on the left in Figure 10.11 is of sample correlations from Example 10.2 (change in mood and exercise intensity), and the null distribution on the right is of sample slopes from Example 10.4 (GPA and hours spent on nonacademic activities). Both are centered around zero and fairly bell-shaped.

Although we won't precisely predict the correlation or slope null distributions we obtained through simulation, under certain conditions, the null distributions of a standardized version of these statistics are well modeled by t-distributions. Thus, theory-based inference can be conducted. This theory-based test has some additional differences from the simulation version which we will also explore.

Example 10.5A: Predicting Heart Rate from Body Temperature

We will use data on the heart rates and body temperature of healthy adults to see whether there is an association between body temperature and heart rate. The dataset consists of body temperatures and heart rates from 65 females and 65 males.

Let's start by looking at a scatterplot (see Figure 10.13) of the heart rate (beats per minute) and body temperature (degrees Fahrenheit) for the 130 subjects.

> **Think about it:** What does the scatterplot reveal about the relationship between heart rate and body temperature for this sample?

Figure 10.13: A scatterplot of the body temperature and heart rate of 130 individuals

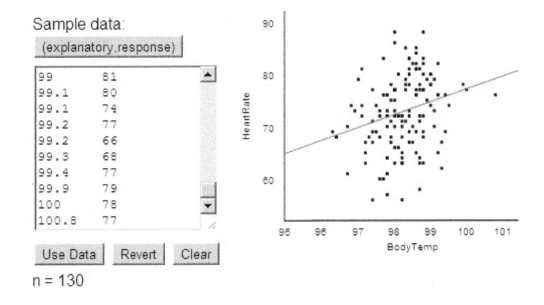

From the scatterplot it appears there is a fairly weak positive linear association between these two variables. We also found the correlation coefficient and regression equation for predicting heart rate from body temperature by using the applet. The correlation coefficient is 0.257 and the regression equation is

Predicted Heart Rate = −166.3 + 2.44(Body Temp).

Think about it: Interpret the slope value from this equation.

From the regression equation, we can see the slope is 2.44. This means that we predict heart rate to increase 2.44 beats per minute for every 1 degree rise in body temperature.

To evaluate whether this is convincing evidence that there is in fact a relationship between heart rate and body temperature in the population, we will first use a simulation-based approach. Our hypotheses are as follows:

Null Hypothesis: There is no association between heart rate and body temperature in the population.

Alternative Hypothesis: There is an association between heart rate and body temperature in the population.

The response variable (heart rate) in the dataset of 130 individuals was shuffled and redistributed to the body temperature values 10,000 times, and each time a simulated value for the slope of the regression line was computed. A graph of those simulated slopes is shown in Figure 10.14. We see that our observed sample slope of 2.44 is very unlikely in the null distribution. Only 36 out of 10,000 shuffles yielded a slope that was as extreme as 2.44 (either greater than or equal to 2.44 or less than or equal to -2.44). Thus our p-value of 0.0036 gives us strong evidence of a relationship between heart rate a body temperature in the population.

Figure 10.14: A null distribution for slope between heart rate and body temperature along with other output from the Corr/Regression applet

As you can see from Figure 10.14, the null distribution has the nice, familiar bell shape. However, as has been the case before, we need a standardized version of the test statistic in order to obtain an accurate theory-based prediction of the null distribution.

Formulas

The theory-based approach computes a standardized statistic (*t*-statistic) using either of the following equations:

$$ t = \frac{r}{\sqrt{\frac{1-r^2}{n-2}}} = \frac{b}{SE(b)}, $$

More details are provided in the Calculation details appendix, but notice that the *t*-statistic can be computed based on either the correlation coefficient or the slope, and it yields the same standardized value-further underscoring that tests for the correlation and for the slope are essentially identical.

Figure 10.15 shows the results of simulating when using the *t*-statistic.

Figure 10.15: Simulated standardized statistics with *t*-distribution overlaid

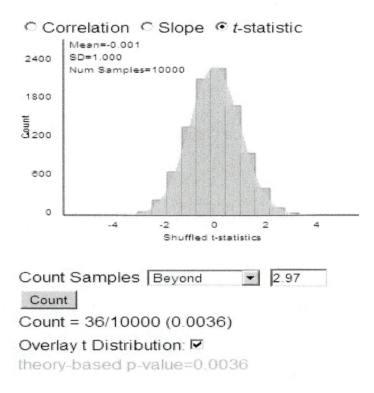

Count = 36/10000 (0.0036)

Overlay t Distribution: ☑

theory-based p-value=0.0036

Validity Conditions

As has been the case before, theory-based tests have validity conditions which are needed in order to obtain an accurate prediction of the simulated distribution.

Validity conditions for a theory-based test for regression

1. The general pattern of the points in the scatterplot should follow a linear trend; the pattern should not show curved or other non-linear patterns

2. There should be approximately the same distribution of points above the regression line as below the regression line (symmetry about the regression line) and,

3. The variability of the points around the regression line should be similar regardless of the value of the explanatory variable; the variability (spread) of the points around the regression line should not differ as you slide along the *x*-axis (equal variance/standard deviation).

To check the validity conditions for using the theory-based approach, we examine the scatterplot in more detail. First, the pattern of the points in the scatterplot doesn't looked curved; a linear trend seems reasonable. Second, we note that the points above the regression line appear similar in spread and shape as the points below the regression line (like a mirror image to the pattern). Finally, the variability in heart rates is approximately the same for different body temperatures (e.g., we don't see that low body temperatures have low variability in heart rates, and high body have high variability in heart rates, or vice versa). When checking validity conditions, we use logic much like we do when testing hypotheses–we will assume the condition is true unless we see strong evidence that it is not.

You know by now that theory-based tests, when valid, predict the behavior of the null distribution (of the standardized statistic) you would have gotten from simulation, if you had simulated. In this case, we know that the theory-based approach will accurately predict the shape of the null distribution because the validity conditions are met. This is confirmed by overlaying the theoretical t-distribution on our simulated slopes (Figure 10.15). But, it's actually a bit more complicated than that as FAQ 10.5.1 and the discussion in the next sections point out.

Figure 10.16 shows a more formal presentation of the output for a regression analysis illustrating some of the calculation details.

Figure 10.16: Formal Regression Table output for theory-based approach

Regression Table: ☑

Term	Coeff	SE	t-stat	p-value
Intercept	-166.28	80.91	-2.06	0.0419
BodyTemp	2.44	0.82	2.97	0.0036

The last line of the regression output shows the theory-based p-value for our test. This p-value of 0.0036 is the same as our simulated p-value of 0.0036. The theory-based p-value will always be given as a two-sided p-value in the Regression Table (see Figure 10.16). In this case, we are conducting a two-sided test. If we had suspected beforehand that heart rate and body temperatures have a positive association, then our p-value would simply be half of the theory-based p-value of 0.0036, namely 0.0018.

Also note that the output reports the standardized statistic of $t = 2.97$ for the slope. This tells us that the observed sample slope (2.44) is almost 3 standard errors (0.82) above the hypothesized slope of zero. Note that the standard error here is telling us about the shuffle-to-shuffle variation in the sample slopes. This value is smaller than what we saw in the simulation for the standard deviation of the simulated slopes (there SD of simulated statistics was 0.842 (see Figure 10.14)) because (as discussed in FAQ 10.5.1) the simulation is for a more general 'no association' hypothesis, whereas the theory-based p-value is for a more specific set of hypotheses about linearity. In particular we can now specify that we are testing for linearity,

Null Hypothesis: There is no *linear* association between heart rate and body temperature in the population.

Alternative Hypothesis: There is a *linear* association between heart rate and body temperature in the population.

This gives us the ability to state hypotheses using population parameters. We use the symbol β (beta) to indicate the population slope and ρ (rho) to indicate the population correlation. We can write the hypotheses as shown here:

Null Hypothesis: $\beta = 0$ (the population slope $= 0$)
Alternative Hypothesis: $\beta \neq 0$ (the population slope is not 0)

Or, equivalently,

Null Hypothesis: $\rho = 0$ (the population correlation $= 0$)
Alternative Hypothesis: $\rho \neq 0$ (the population correlation is not 0)

Confidence interval

The theory-based confidence interval (Figure 10.17) for the slope tells us that we are 95% confident that a one degree increase in body temperature is associated with an average increase of 0.8137 to 4.0728 heart beats per minute in the population represented by this sample. Note that this can be approximated using the 2SD method using the more precise estimate of the standard deviation from the regression table (0.82), yielding $2.44 \pm 1.64 = (0.80, 4.08)$. See FAQ 10.5.1 for more details.

Figure 10.17: Confidence interval for regression slope

95% Confidence interval for Slope: ☑
(0.8137 , 4.0728), df = 128

FAQ 10.5.1 Simulation, parameters, and intervals

Q: I haven't forgotten about the draft lottery. You still owe me more about parameters and intervals.

A: We're now in a better position to take a closer look. How would you ask your question based on what you've seen since Section 10.2?

Q: Back then we did a simulation of the null hypothesis of no association by shuffling the response variable values. We had no parameter, and without a parameter, there was no way to get a confidence interval.

A: Right.

Q: Then in the very next section parameters popped up like mushrooms after a rain. What changed?

A: That's definitely a thoughtful question. You've put your finger on a very important idea. Remember that with the lottery you said you couldn't tell what the pattern of possible association might look like, and I agreed.

Q: OK.

A: Now look back and compare the scatterplot for the lottery (Figure 10.6) and the scatterplot for the dinner plates (Figure 10.7).

Q: What's your point?

A: For the dinner plates, the form of the plot is roughly linear, and we are most interested in the underlying linear relationship.

Q: OK again.

A: We can make that linearity part of our framework for inference.

Q: Framework?

A: We make linearity one of our validity conditions.

Q: Why would I want to do that?

A: *If the relationship is linear, we get parameters: the slope of the line, the intercept, and the correlation coefficient.* These parameters don't make sense unless the relationship is linear.

Q: If I understand you, we have two kinds of situations. For the first kind, like the draft lottery, we know the null hypothesis – no association –but we don't specify anything more specific than "is an association" for the alternative hypothesis. We can test the statistical significance of the observed association by shuffling, but we don't have a parameter, so we don't have intervals.

A: Excellent! And the other kind?

Q: We assume that the underlying relationship is linear. We have parameters, and we can find theory-based tests and intervals.

A: Well said.

Q: But I've still got a question.

A: Sure.

Q: Why can't we do a simulation-based significance test for the dinner plates? Doesn't it work when the alternative is a linear relationship?

A: You *could* do it, and it *does* work, but the theory-based test is more powerful.

Q: How come?

A: If you assume linearity, you automatically rule out any and all non-linear forms of association. You narrow your question to whether the population slope is zero. Nothing else matters.

Q: That makes sense.

A: One way you can see that it's more powerful is by looking at the variability in the null distribution. The standard deviation of the null distribution of slopes when we simulate using shuffling is larger than the standard deviation of the sample slopes using the theory-based test.

Q: …because of the extra assumption of being linear. That makes sense.

A: Here's another way to look at it. You go to the doctor. The null hypothesis is that don't have a disease. The alternative hypothesis is that do have some disease. What's the parameter?

Q: I get it. No way to define a parameter without additional assumptions.

A: Now suppose the question is much more narrow: whether you have diabetes. The doctor knows to do a test based on the level of your fasting blood sugar.

Q: The parameter.

A: Exactly.

Q: One last question. Are you telling me that we can't do simulation-based confidence intervals for the slope?

A: I'm glad you asked that, because you actually can get simulation-based intervals if you have a linear relationship. It's a little more complicated, so we've saved it for your next course in statistics.

Example 10.5B: Smoking and Drinking

The scatterplot in Figure 10.18 shows the relationship between number of alcoholic drinks consumed per week (*x*-axis) and number of cigarettes smoked per week (*y*-axis) for a random sample of students at a small college.

Note: The scatterplot will only show one "dot" for people who have the same values of the explanatory/response variable. So, although we only see one "dot" on the point (0, 0) there are actually 524 students at that point!

Figure 10.18: A scatterplot of alcoholic drinks and cigarettes smoked per week for a sample of college students

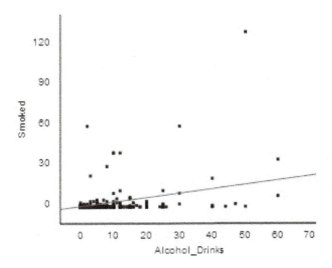

Think about it: Does the data in the scatterplot appear to meet the validity conditions to conduct a theory-based test?

Examining the scatterplot of the *sample* data should help reveal any strong violations of the validity conditions described in Example 10.5A. Both conditions appear violated. In particular, the scatterplot in Figure 10.18 reveals that most dots lying below the regression line (along the *x*-axis) are much closer to the line on average than the values lying above the regression line. For example, at any value of number of drinks per week, there are very many people who smoked zero cigarettes, a few with a small number of cigarettes smoked, and then even fewer with very few large values for number of cigarettes smoked. In other words, the distribution of cigarettes smoked per week for each different value of alcoholic drinks per week is highly right skewed. The other condition is that there is similar variability of the dots around the regression line for different values of the explanatory variable (alcoholic drinks per week). The scatterplot suggests that there is evidence that this condition is also violated because the distributions of the number of cigarettes smoked seem to be more variable for larger values of alcoholic drinks per week than for smaller values.

Because these two conditions are violated, we should not apply theory-based inference procedures to these data. In Figure 10.19 we see both the theory-based prediction for the null distribution of the standardized correlation coefficient and simulation results for 1000 shuffles. Notice that the theory-based prediction of the null distribution is mound-shaped (a *t* distribution) with a mean of zero, but the simulated null distribution for the standardized correlation coefficient is highly right-skewed. In this case, the theory-based prediction for the null distribution is not a good match to the actual null distribution, so p-values and confidence intervals from the theory-based approach could be quite misleading.

Figure 10.19: Theory-based and simulated null distribution for *t*-statistics by standardizing shuffled slopes of alcohol drinks and cigarettes smoked per week

When the validity conditions are not met for theory-based inference for a population correlation coefficient or slope, you can rely on the simulated-based approach. Another common strategy is to re-express ("**_transform_**") the data on a different scale, such as a logarithmic scale, in such a way that the conditions would be met. See Exercise 10.5.20 for an example.

Caution: Outliers and Influential Observations

Reconsidering Figure 10.18, we also note that there may be some influential points in this sample. Specifically, the person who says they smoke 130 cigarettes a week is also one of largest alcohol consumers (50 drinks per week). If we remove this person from the dataset, the correlation coefficient drops from 0.37 to 0.30, a noticeable decrease. This is the best way to check whether extreme observations unduly affect the analysis: Remove them from the dataset and see how the relationship changes. It is a judgment call about how big of a change is large, but any point that substantially changes the correlation coefficient or regression slope is considered influential. Depending on the context, influential points should be considered for removal from the analysis in order to get a more robust estimate of the correlation coefficient and regression line, but should only be done so with proper justification.

Key idea: Remove possible influential points and outliers and recompute the correlation coefficient and/or regression slope. If the statistic has changed a lot after removing the point, the point is considered influential and you may want to keep that point out of the analysis (or report your conclusions both with and without that observation).

Exploration 10.5: Predicting Brain Density from Number of Facebook Friends

How often did you check your Facebook account today? How many Facebook friends do you have? Does everyone have the same number? Might the number of facebook friends that a person has be associated in some way with the person's brain structure? Researchers used MRI's to examine areas of the brain that are involved with social interaction, memory, and emotional responses (Kanai, Bahrami, Roylance, & Rees, 2011). Their subjects were 40 university students in London. They examined five areas of the brain that have been previously linked to social perception and associative memory, and they compared each to the number of Facebook friends the subject reported. The results from each brain area were quite similar. We will look at one set of measurements focusing on the left MTG (middle temporal gyrus which has been linked in other studies to face recognition).These results are shown in Table 10.5. Note that number of friends is given in units of 100 friends, so that 0.30=30 friends, 1.09=109 friends, and so on.

Table 10.5: The number of Facebook friends (in 100s of friends) and the brain density (in arbitrary units) for 40 university students

friends	density	friends	density	friends	Density	friends	density
0.30	-2.14	3.41	-0.72	5.08	-0.17	2.93	1.24
1.09	-1.09	3.97	-0.74	5.15	-0.09	3.09	0.93
0.39	-0.72	4.10	-0.85	5.01	-0.03	3.65	1.02
1.20	-0.53	4.39	-0.69	3.92	0.26	4.73	0.75
0.80	-0.43	4.60	-0.71	3.13	0.42	3.95	1.34
1.71	-0.26	5.07	-1.07	2.53	0.46	6.14	0.96
2.24	-0.34	5.08	-0.49	1.51	0.23	7.32	1.05
2.40	-1.51	5.83	-0.59	1.73	0.27	6.17	1.6
3.23	-1.45	6.34	-0.41	1.52	0.55	5.47	1.97
3.81	-1.48	5.63	0.21	2.71	1.09	5.70	2.08

The researchers wanted to explore whether there is a positive relationship between these two variables, and we will do the same. In particular, the researchers tried to predict brain density based on number of facebook friends, asking: Does brain density tend to increase as a person's number of facebook friends increases?

1. What are the null and alternative hypotheses for this study?

2. Paste the data (**Facebook**) in the **Corr/Regression** applet and make a scatterplot where brain density is the response variable and number of Facebook friends is the explanatory variable. (Note that the number of facebook friends is reported in terms of hundreds of friends.)

   Describe the direction, form, and strength of association between the variables as revealed in the scatterplot. Are there any unusual observations?

3. What is the slope of the regression equation for predicting brain density based on number of Facebook friends? How does this value support your answer to the previous question?

4. You should have found a positive slope between brain density and number of Facebook friends. The question, however, is if there were no association between brain density and number of Facebook friends, how likely is it that we would get a slope as high as we did.

   a. In your own words, explain how you could use slips of paper to model the null hypothesis of no association between brain density and number of Facebook friends.

   b. Let's shuffle the response variable (brain density) to see what we get for the slope if we break the association between the data pairs. Use the applet to do this five times and write down the five simulated slopes you get. Are any of them as large as or larger than the actual value of the sample slope?

c.   Now do at least 1000 shuffles with the applet and find your p-value.

d.   Interpret what your p-value measures in context of the study.

e.   Can we conclude that there is a positive association between the density of the left MTG and the number of Facebook friends a person has in the population? Explain your reasoning.

5.   What is the shape of the null distribution of shuffled slopes?

By now, you've seen many times that when the null distribution of statistics takes a familiar, mound-shaped curve, we can often use theory-based methods to predict the null distribution of related standardized statistics---as long as certain validity conditions are true. This is no different for regression (and correlation!).

The theory-based approach computes a standardized statistic (*t*-statistic) using one of the following equations:

$$t = \frac{r}{\sqrt{\frac{1-r^2}{n-2}}} = \frac{b_0}{SE(b)}$$

More details are provided in the Calculation details appendix, but notice that the *t*-statistic can be computed based on either the correlation coefficient or the slope, and it yields the same value-further underscoring that tests for the correlation coefficient and for the slope are essentially identical.

6.   Use the applet to find the correlation coefficient.

7. Use the correlation coefficient to find the t-statistic using the equation shown above.

There are three validity conditions for regression which are needed in order to use the theory-based approach to yield a p-value

---

Validity conditions for a theory-based test for regression

1. The general pattern of the points in the scatterplot should follow a linear trend; the pattern should not show curved or other non-linear patterns,

2. There should be approximately the same distribution of points above the regression line as below the regression line (symmetry about the regression line) and,

3. The variability of the points around the regression line should be similar regardless of the value of the explanatory variable; the variability (spread) of the points around the regression line should not differ as you slide along the x-axis (equal variance/standard deviation).

---

8. Based on the scatterplot you used in #2, is Validity Condition 1 met? Namely, is the general pattern of the scatter plot linear?

9. Based on the scatterplot you used in #2, is Validity Condition 2 met? Namely, is the distribution of points above the regression line the same as below the line?

10. Based on the scatterplot you used in #2, is Validity Condition 3 met? Namely, is the variability of the points around the line similar regardless of the value of the explanatory variable?

11. Select the **t-statistic** radio button, then check the **Overlay t distribution** button. Does the t-distribution appear to do a good job of predicting the distribution of the simulated t-statistics? Why?

12. In the applet, click the box for **Regression Table**. This table provides the observed t-statistic for the data, as well as the two-sided, theory-based test p-value. Use the theory-based approach to test whether there is a significant association between the variables in the population. Report the standardized statistic and p-value. Summarize your conclusion.

When using the theory-based approach for regression you can write your hypotheses in terms of population parameters. The relevant population parameters are the population slope (indicated by the Greek letter β) and the population correlation (indicated by the Greek letter ρ). See FAQ 10.5.1 for more discussion about parameters.

13. Write the null and alternative hypotheses in terms of the population slope

14. Write the null and alternative hypotheses in terms of the population correlation.

15. Produce and interpret a 95% confidence interval for the population slope coefficient. (*Hint*: Make sure that your interpretation refers to how to interpret the slope coefficient in the population and remember that the units of the explanatory variable is '100s of facebook friends'.) To get the confidence interval, check the box below the regression table for 95% Confidence interval for Slope.

16. Imagine if we added someone to this dataset whose brain density was -10 and had 5000 Facebook friends. Uncheck the Show Shuffle Options and enter this observation into the dataset using **Add/remove observations** or typing this new row into the Sample data window. Conduct a two-sided test. Caution: Make sure you enter 5000 facebook friends in terms of the units used in this study. What would 5000 facebook friends be in terms of '100s of facebook friends'?

    a. Find the p-value using a simulation-based approach with shuffled t-statistics.

    b. Find the p-value using the theory-based approach.

    c. Why are the p-values from the two approaches not very similar?

17. Remember that the sample used here was not randomly selected, but rather a group of 40 volunteer university students.

    a.   Describe a population in which you would be comfortable drawing inferences. Explain your reasoning.

    b.   Can we conclude that the acquiring more facebook friends would lead to an increase in a person's brain density? Explain your answer.

    c.   What about the opposite direction: Can we conclude that have a larger brain density causes a person to acquire more facebook friends? Explain.

18. The researchers originally searched for positive associations between many areas of the brain and the number of Facebook friends for 125 subjects. They found significantly positive associations for a number of brain areas. The second part of the study is with the 40 subjects that we looked at in this exploration.

Why do you think this method of "look at a bunch of regions to see which are significant" is often criticized? (*Hint*: If you run lots of tests of significance using a 5% level of significance, how often will you reject the null hypothesis when you really shouldn't have?)

Section 10.5 Summary

A theory-based method can be used to conduct inference for the population slope coefficient or population correlation coefficient. We consider the two methods identical.

- The test of significance and the confidence interval are based on the *t*-**distribution**.
- The validity conditions for the slope coefficient are that the response variable has approximately the same distribution of points above the regression line as below the regression line (symmetry) and, and the variability of the points around the regression line should be similar regardless of the value of the explanatory variable (equal variance).

Chapter 10 Summary

In this chapter we looked at relationships between two quantitative variables. We saw that scatterplots are used to explore the relationship between two quantitative variables. In creating a scatterplot, we put the explanatory variable on the horizontal axis and the response variable on the vertical axis. Which variable is which is sometimes straightforward (especially if you are doing an experiment) and sometimes not so straightforward.

We saw that the correlation coefficient is a measure of the strength and direction of a *linear* relationship between two quantitative variables.

A regression line has the form $\hat{y} = a + bx$ where a is the y-intercept, b is the slope, x is a value of the explanatory variable, and $\hat{y}$ is the predicted value for the response variable. You should always put the variables in context when writing out the equation and be able to give contextual interpretations of the slope and intercept (but watch for extrapolation). For a specific value of x, the corresponding $y - \hat{y}$ is a residual. The least squares regression line is found by minimizing the sum of the squared residuals.

The simulation-based technique that we learned in the previous chapters was adapted to work with looking for an association in quantitative data. We again used the 3S Strategy:

1. **Statistic:** In this chapter our statistic was either the correlation coefficient or the regression slope.

2. **Simulate:** Shuffle the response variable, recompute the statistic, and repeat this process many times.

3. **Strength of evidence:** We find evidence against the null hypothesis if the original sample statistic is in the tail of the distribution used to model the null hypothesis.

Similar to previous chapters, we saw that the p-value is the probability of getting a value of the (standardized) statistic equal to or more extreme than the one from our original sample data when the null hypothesis is true. We also used the standard deviation from the null distribution and the two standard deviation rule to approximate a 95% confidence interval for the parameter.

The theory-based test for a regression slope was discussed and compared to the simulation-based test. We saw how the predicted null distribution was similar to all the others we have seen so far: symmetric, bell-shaped, and centered at zero. This allowed us (the computer) to formally standardized the statistic and use the t-distribution (with $n - 2$ degrees of freedom) to approximate the p-value and estimate the confidence interval.

Chapter 10 Glossary

correlation coefficient
   Statistic that measures direction and strength of a **linear** relationship between two quantitative variables 10-4, 10-11

extrapolation
   Predicting values for the response variable for given values of the explanatory variable that are outside of the range of the original data..10-31, 10-40

influential observations
   An observational is considered influential if removing it from the data set dramatically changes the correlation coefficient or regression line. Often have extreme *x* values. ... 10-6, 10-32, 10-41

predictor
   Another word for explanatory variable, often used in correlation/regression settings 10-2

r
   Symbol for correlation coefficient, values range from -1 to 1 and are unit-less. Values close to -1 and 1 denote a strong linear relationship whereas values close to 0 denote a weak or no linear relationship.. 10-4

residuals
   The vertical distances between a point and the least squares regression line 10-31, 10-37

slope
   Change in predicted response variable divided by change in explanatory variable 10-30, 10-38, 10-39

SSE
   Sum squared error. It is the sum of all the squared residuals. 10-37

transform
   Express data on a different scale, such as logarithmic, often used to meet validity conditions 10-62

unusual observations
   Any observations that do not fit in with the pattern of the rest of the observations ... 10-4, 10-11

validity conditions for regression theory-based test
   Symmetry of data about the regression line and equal variability in response variable across the *x*-values 10-56, 10-66

Exercises

Section 10.1

10.1.1* Suppose that a study measures how far each student sits from the front of the classroom and also records the student's final exam score. If better students tend to sit closer to the front, would the correlation between distance and exam score be positive, negative, or close to zero? Explain briefly.

10.1.2 Suppose that you record the daily high temperature and the daily amount of ice cream sold by an ice cream vendor at your favorite beach this summer, starting on Memorial Day weekend and ending on Labor Day weekend. Would you expect to find a positive or negative correlation between these variables? Explain briefly.

10.1.3* Suppose that every student in this class scored 5 points lower on the second exam than on the first exam. Consider the correlation coefficient between *first exam score* and *second exam score*. What would the value of this correlation coefficient be? Explain briefly. (Hint: You might draw a scatterplot of hypothetical data that fit the description.)

10.1.4 Six statements about what you can learn from the correlation coefficient are listed below. Label each statement, choosing from A (can always tell), N (can never tell), or S (can sometimes tell. For each statement you label S for sometimes, explain briefly when you can tell and when you cannot.
  a) Whether there is a systematic relationship between *x* and *y*.
  b) Whether the relationship between *x* and *y* is linear or curved.
  c) Whether the least squares line has a positive or negative slope.
  d) How closely the points fall near a line.
  e) Whether there are outliers or influential points.
  f) How steep the least squares line is.

10.1.5* Fill in the blanks, choosing from: values of the response, values of the explanatory variable, cases:

In a scatterplot, the points are _____ ; the *x*-axis represents _____ ;
and the *y*-axis represents _____ .

10.1.6 The data shown in the following scatterplot shows a very nice relationship between the
two variables. However, the correlation here is 0.03, very close to zero. Explain why we can
have a nice relationship between two quantitative variables and yet have a correlation of 0.

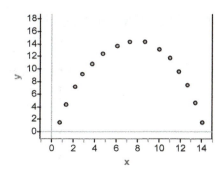

10.1.7* Which of the following is **not** a property of correlation, *r*?

A. $-1 \le r \le 1$
B. Correlation measures the strength of a linear relationship between two quantitative
   variables.
C. The sign on *r* tells the direction of the linear relationship between two quantitative
   variables.
D. If the correlation between two quantitative variables is zero, then there is no relationship
   between these two variables.

10.1.8 Which of the following statements is correct?

A. Changing the units of measurements of the explanatory or response variable does not
   change the value of the correlation.
B. A negative value for the correlation indicates that there is no relationship between the two
   variables.
C. The correlation has the same units (e.g., feet or minutes) as the explanatory variable.
D. Correlation between *y* and *x* has the same number but opposite sign as the correlation
   between *x* and *y*.

10.1.9* If two variables are negatively associated, then we know that

    A. Above average values in one variable correspond to below average values in the other variable.

    B. Above average values in one variable correspond to above average values in the other variable.

    C. Below average values in one variable correspond to below average values in the other variable.

    D. Below average values in one variable correspond either above average or below average values in the other variable.

10.1.10 For each of the following statements, say what, if anything is wrong.

    a) Because the correlation coefficient between test time and test score is -0.56, the correlation coefficient between *test score and test time* must be 0.56 (the positive value).

    b) There is a strong positive correlation between a person's yard size and whether or not they have a dog.

    c) For a sample of 50 students, the correlation coefficient between weight (kg) and height (inches) was found to be 0.78kg/inches.

    d) A correlation coefficient of $r = 0.84$ denotes just as strong a relationship as a correlation coefficient of $r = -0.84$.

House Prices*

10.1.11 The data file **HousePrices** contains data on prices ($) and sizes (in square feet) for a random sample of houses that sold in the year 2006 in Arroyo Grande, California

    a) Identify the type of study randomized experiment or observational study.

    b) Identify the experimental/observational units in this study.

    c) Identify the two variables of interest, and whether each is categorical or quantitative. Which variable do you think makes more sense to use as the explanatory variable, and which as the response variable?

    d) Enter the data into the **Corr/Regression** applet. Produce a scatterplot, and comment on the association between the variables as revealed in the scatterplot. (Remember to comment on form, direction, strength, and unusual observations.)

e) Report the value of the correlation coefficient. Does the value of the correlation coefficient support your answer in part d)? Explain how you are deciding.

Roller coasters

Data set

10.1.12 The data file **RollerCoasters** contains data (collected from www.rcdb.com) on the maximum speed (miles per hour) and the maximum height (feet) for each of a sample of roller coasters in the U.S.

a) Identify the type of study: randomized experiment or observational study.
b) Identify the experimental/observational units in this study.
c) Identify the two variables of interest, and whether each is categorical or quantitative. Which variable do you think makes more sense to use as the explanatory variable, and which as the response variable?
d) Enter the data into the **Corr/Regression** applet. Produce a scatterplot, and comment on the association between the variables as revealed in the scatterplot. (Remember to comment on form, direction, strength, and unusual observations.)
e) Report the value of the correlation coefficient. Does the value of the correlation coefficient support your answer in part d)? Explain how you are deciding.

Height and finger length*

10.1.13 An article titled "Giving the Finger to Dating Services" appeared in *Chance* magazine (2008) that investigated the association between length of index finger and height. This research was spurred by a report that online dating sites were asking members for the length of their index fingers. Perhaps this information could be used to verify whether the member was reporting his/her correct height on the profile? The data file **HeightAndFingerLength** contains data on the height (inches) and the length of index finger of dominant hand (cm) for students in a statistics class.

a) Identify the type of study: randomized experiment or observational study.
b) Identify the experimental/observational units in this study.

c) Identify the two variables of interest, and whether each is categorical or quantitative. Which variable do you think makes more sense to use as the explanatory variable, and which as the response variable?

d) Enter the data into the **Corr/Regression** applet. Produce a scatterplot, and comment on the association between the variables as revealed in the scatterplot. (Remember to comment on form, direction, strength, and unusual observations.)

e) Report the value of the correlation coefficient. Does the value of the correlation coefficient support your answer in part d)? Explain how you are deciding.

10.1.14 Refer to the previous exercise that investigated the association between length of index finger and height. The data file **HeightAndFingerLength** contains data on the height (inches) and the length of index finger of dominant hand (cm) for students in a statistics class.

a) Using appropriate technology (applet or statistical software) create a scatterplot of the data. On the scatterplot, locate the individual who reported a finger length of 5 cm and height of 70 inches. Does this individual appear to be an unusual observation? Explain you reasoning.

b) Next, report the correlation coefficient for these data.

c) Now, remove this individual (finger length of 5 cm and height of 70 inches). Recalculate the correlation coefficient for the data without this individual.

d) Did the correlation coefficient increase (move farther from zero) or decrease (move closer to zero) on the removal of the individual who reported a finger length of 5 cm and height of 70 inches? Explain why that makes sense.

10.1.15: Refer to the previous exercises that investigated the association between length of index finger and height. The data file **HeightAndFingerLength** contains data on the height (inches) and the length of index finger of dominant hand (cm) for students in a statistics class.

a) Once again, using appropriate technology (applet or statistical software) create a scatterplot of the data. On the scatterplot, locate the individual who reported a finger length of 10.16 cm and height of 78 inches. Does this individual appear to be an unusual observation? Explain you reasoning.

b) Next, report the correlation coefficient for these data.

c) Now, remove this individual (finger length of 10.16 cm and height of 78 inches). Recalculate the correlation coefficient for the data without this individual.

d) Did the correlation coefficient increase (move farther from zero) or decrease (move closer to zero) on the removal of the individual who reported a finger length of 10.16 cm and height of 78 inches? Explain why that makes sense.

Relationship between parents' and their children's heights

10.1.16 In 1885, Sir Francis Galton first used regression to explore the association between children's heights and (biological) parents' heights. The data file **MomandChildHeights** contains data for students in a statistics class on the following two variables: student's height (inches) and biological mother's height (inches).

a) Identify the type of study: randomized experiment or observational study.
b) Identify the experimental/observational units in this study.
c) Identify the two variables of interest, and whether each is categorical or quantitative. Which variable do you think makes more sense to use as the explanatory variable, and which as the response variable?
d) Enter the data into the **Corr/Regression** applet. Produce a scatterplot, and comment on the association between the variables as revealed in the scatterplot. (Remember to comment on form, direction, strength, and unusual observations.)
e) Report the value of the correlation coefficient. Does the value of the correlation coefficient support your answer in part d)? Explain how you are deciding.

10.1.17 Refer back to the previous exercise that explores the association between children's heights and (biological) mothers' heights. The data file **DadandChildHeights** contains data for students in a statistics class on the following two variables: student's height (inches) and biological father's height (inches).

a) Identify the type of study: randomized experiment or observational study.
b) Identify the experimental/observational units in this study.
c) Identify the two variables of interest, and whether each is categorical or quantitative. Which variable do you think makes more sense to use as the explanatory variable, and which as the response variable?

d) Enter the data into the **Corr/Regression** applet. Produce a scatterplot, and comment on the association between the variables as revealed in the scatterplot. (Remember to comment on form, direction, strength, and unusual observations.)

e) Report the value of the correlation coefficient. Does the value of the correlation coefficient support your answer in part d)? Explain how you are deciding.

10.1.18 Refer back to the previous two exercises that explore the association between children's heights and (biological) parents' heights. Recall that the data file **DadandChildHeights** contains data for students in a statistics class on the following two variables: student's height (inches) and biological father's height (inches), whereas the data file **MomandChildHeights** contains data for students in a statistics class on the following two variables: student's height (inches) and biological mother's height (inches).

a) Report the value of the correlation coefficient for children and fathers' heights.

b) Report the value of the correlation coefficient for children and mothers' heights.

c) In which case does there appear to be a stronger association between child and parent's heights: with father's height or with mother's height? Explain your reasoning.

Scrabble names*

10.1.19 Have you ever played the game *Scrabble*? It is a word building game where certain letters earn you more points than others when you use them in your word. Here is the table of letters and the corresponding Scrabble points:

A	B	C	D	E	F	G	H	I	J	K	L	M
1	3	3	2	1	4	2	4	1	8	5	1	3
N	O	P	Q	R	S	T	U	V	W	X	Y	Z
1	1	3	10	1	1	1	1	4	4	8	4	10

In a statistics class, students were asked to calculate the number of Scrabble points their names would earn. For example, the Scrabble score for the name Tom Sawyer would be $1 + 1 + 3 + 1 + 1 + 4 + 4 + 1 + 1 = 17$ points. The data file **ScrabbleNames** contains data on the following two variables: number of letters in a student's name and the corresponding Scrabble score.

a) Identify the type of study: randomized experiment or observational study.

b) Identify the experimental/observational units in this study.

c) Identify the two variables of interest, and whether each is categorical or quantitative. Which variable do you think makes more sense to use as the explanatory variable, and which as the response variable?

d) Enter the data into the **Corr/Regression** applet. Produce a scatterplot, and comment on the association between the variables as revealed in the scatterplot. (Remember to comment on form, direction, strength, and unusual observations.)

e) Report the value of the correlation coefficient. Does the value of the correlation coefficient support your answer in part d)? Explain how you are deciding.

10.1.20 Refer back to the previous exercise about exploring the relationship between the number of letters in a name and the corresponding Scrabble score. Recall that students were asked to calculate the number of Scrabble points their names would earn. Students were also asked to calculate the points per letter, that is, *ratio* = (Scrabble score)/(number of letters in name). The data file **ScrabbleRatio** contains Scrabble points for a student's name and the corresponding *ratio*.

a) Enter the data into the **Corr/Regression** applet. Produce a scatterplot, and comment on the association between the variables as revealed in the scatterplot. (Remember to comment on form, direction, strength, and unusual observations.)

b) Report the value of the correlation coefficient. Does the value of the correlation coefficient support your answer in part a)? Explain how you are deciding.

10.1.21 Refer back to the previous two exercises that explore the association between number of letters in a name and the corresponding Scrabble score. Recall that the data file **ScrabbleNames** contains data on number of letters in students' names and the corresponding Scrabble points, whereas the data file **ScrabbleRatio** contains Scrabble points for a student's name and the corresponding ratio of Scrabble points to number of letters.

a) In which case is the linear association stronger: between the number of letters in a student's name and the corresponding Scrabble points, or between Scrabble points for a student's name and the corresponding ratio? How are you deciding?

b) With regard to your finding in part a), explain why that makes sense.

TV and life expectancy

10.1.22 The data file **TVLife** provides information on life expectancy and number of televisions per thousand people in a sample of 22 countries, as reported by the *2006 World Almanac and Book of Facts.*

a) Identify the type of study: randomized experiment or observational study.

b) Identify the experimental/observational units in this study. (*Hint:* "People" is not the answer.)

c) Enter the data into the **Corr/Regression** applet. Produce a scatterplot using life expectancy as the response variable and number of TVs per 1000 as the explanatory variable. Comment on the association between the variables as

d) revealed in the scatterplot. (Remember to comment on form, direction, strength, and unusual observations.)

e) Report the value of the correlation coefficient. Does the value of the correlation coefficient support your answer in part c)? Explain how you are deciding.

f) As you (should) have discovered in parts c) and d), the association between life expectancy and number of televisions per thousand people appears pretty strong. Based on this finding, is it okay to conclude that simply sending televisions to the countries with lower life expectancies would cause their inhabitants to live longer? Explain why or why not.

Section 10.2

10.2.1* An instructor wanted to investigate whether there was an association between height (inches) and hand span (cm). She collected data from 10 students, and after analyzing the data found the p-value for to be 0.022. For each of the following statements, indicate whether or not the statement is VALID or INVALID.

a) The p-value says that there is a 2.2% probability that there is no association between height and hand span.

b) The p-value says that there is a 2.2% probability that there is an association between height and hand span.

c) If there were no association between height and hand span, the probability of observing the association observed in the sample data of 10 students is 0.022.

d) If there were no association between height and hand span, the probability of observing the association observed in the sample data or an even stronger association in a sample of 10 students is 0.022.

e) If there were an association between height and hand span, the probability of observing the association observed in the sample data or an even stronger association in a sample of 10 students is 0.022.

10.2.2 A researcher wants to investigate whether there is a relationship between annual company profit ($) and median annual salary paid by the company ($). The researcher collects data on a random sample of companies, and after analyzing the data finds the p-value to be 0.56. Which of the following is an appropriate conclusion based on this p-value?

1. There is no relationship between annual company profit ($) and median annual salary paid by the company ($).
2. There is a relationship between annual company profit ($) and median annual salary paid by the company ($), and corresponding value of correlation coefficient is $r = 0.56$.
3. There is no evidence of a relationship between annual company profit ($) and median annual salary paid by the company ($).
4. There is a 56% chance that there is no relationship between annual company profit ($) and median annual salary paid by the company ($).

10.2.3* Two researchers want to investigate whether there is a relationship between annual company profit ($) and median annual salary paid by the company ($). Researcher Bart collects data on a random sample of 40 companies, and researcher Lisa collects data on a random sample of 140 companies. After analyzing their respective data sets, each finds a correlation coefficient of $r = 0.601$.

a) Who will have a smaller p-value?

A.	Bart
B.	Lisa
C.	Both will find the same p-value
D.	More information is needed to answer this question

b) Explain the reasoning your choice in part a).

Height and finger length

10.2.4 Recall from Exercise 10.1.13 that the data file **HeightAnd FingerLength** contains data on the height (inches) and the length of index finger of dominant hand (cm) for students in a statistics class. State in words the appropriate null and alternative hypotheses to test whether there is an association between height and length of index finger.

10.2.5 Refer to the previous exercise about the association between height and length of index finger. Recall that the data file **HeightAnd FingerLength** contains data on the height (inches) and the length of index finger of dominant hand (cm) for students in a statistics class. Describe how one might use everyday items (for example, coins, dice, cards, etc.) to conduct a tactile simulation-based test of the hypotheses. Be sure to clearly describe how the p-value will be computed from the simulation.

10.2.6 Refer to the previous exercise about the association between height and length of index finger. Recall that the data file **HeightAnd FingerLength** contains data on the height (inches) and the length of index finger of dominant hand (cm) for students in a statistics class.

The correlation coefficient for the sample data is 0.474. A simulated null distribution is shown to test whether the sample data provide evidence of a linear association between height and length of index finger.

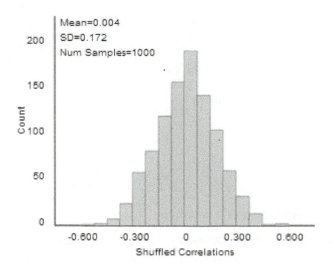

On the null distribution indicate the region that denotes the p-value, being sure to include any number that you are using to decide on the region. Also, make sure that you answer is consistent with the hypotheses chosen in part a).

Used Honda Civics*

10.2.7 The data in the file **UsedHondaCivics** come from a sample of used Honda Civics listed for sale online in July 2006. The variables recorded in this data file are the car's age (calculated as 2006 minus year of manufacture) and price. Consider conducting a simulation analysis to test whether the sample data provide strong evidence of a linear association between a car's price and age in the population.

a) Identify the type of study: randomized experiment or observational study.
b) Identify the experimental/observational units in this study.
c) Identify the two variables of interest, and whether each is categorical or quantitative. Which variable do you think makes more sense to use as the explanatory variable, and which as the response variable?
d) State the appropriate null hypothesis *in words*.
e) State the relevant alternative hypothesis *in words*.
f) Enter the data into the **Corr/Regression** applet. Produce a scatterplot, and comment on the association between the variables as revealed in the scatterplot. (Remember to comment on form, direction, strength, and unusual observations.)
g) Report the value of the correlation coefficient. Does the value of the correlation coefficient support your answer in part f)? Explain how you are deciding.
h) Describe how one might use everyday items (for example, coins, dice, cards, etc.) to conduct a tactile simulation-based test of the hypotheses. Be sure to clearly describe how the p-value will be computed from the simulation.

10.2.8 Refer to the previous exercise about the association between the age and price of used Honda Civics. Consider conducting a test of significance to investigate whether the sample data provide strong evidence of an association between a car's price and age in the population of used Honda Civics. The data are in the file **UsedHondaCivics**

a) Enter the data into the **Corr/Regression** applet, and conduct a simulation-based test of significance using the correlation coefficient as the statistic. Report the approximate p-value.

b) Interpret the p-value reported in (a).

c) Summarize your conclusion (significance, causation, generalization) from this simulation analysis. Also describe the reasoning process by which your conclusion follows from your simulation results.

House prices

10.2.9 Recall from Exercise 10.1.1 that the data file **HousePrices** contains data on prices ($) and sizes (in square feet) for a random sample of houses that sold in the year 2006 in Arroyo Grande, California.

a) State in words the appropriate null and alternative hypotheses to test whether there is a linear association between prices and sizes of houses.

b) Describe how one might use everyday items (for example, coins, dice, cards, etc.) to conduct a tactile simulation-based test of the hypotheses. Be sure to clearly describe how the p-value will be computed from the simulation.

10.2.10 Refer to the previous exercise about the association between prices and sizes of houses. Recall that the data file **HousePrices** contains data on prices ($) and sizes (in square feet) for a random sample of houses that sold in the year 2006 in Arroyo Grande, California.

a) Enter the data into the **Corr/Regression** applet, and conduct a simulation-based test of significance. Report the approximate p-value.

b) Interpret the p-value reported in part (a).

c) Summarize your conclusion (significance, causation, generalization) from this simulation analysis. Also describe the reasoning process by which your conclusion follows from your simulation results.

Roller coasters*

10.2.11 Recall from Exercise 10.1.12 that the data file **RollerCoasters** contains data (collected from www.rcdb.com) on the maximum speed (miles per hour) and the maximum height (feet) for each of a sample of roller coasters in the U.S.

a) State in words the appropriate null and alternative hypotheses to test whether is a linear association between maximum speed and maximum height of such roller coasters.

b) Describe how one might use everyday items (for example, coins, dice, cards, etc.) to conduct a tactile simulation-based test of the hypotheses. Be sure to clearly describe how the p-value will be computed from the simulation.

10.2.12 Refer to the previous exercise about the association between maximum speed and maximum height of roller coasters. Recall that the data file **Rollercoasters** contains data on the maximum speed (miles per hour) and the maximum height (feet) for each of a sample of roller coasters in the U.S.

a) Enter the data into the **Corr/Regression** applet, and conduct a simulation-based test of significance. Report the approximate p-value.

b) Interpret the p-value reported in part (a).

c) Summarize your conclusion (significance, causation, generalization) from this simulation analysis. Also describe the reasoning process by which your conclusion follows from your simulation results.

Relationship between parents' and their children's heights

10.2.13 Recall from Exercise 10.1.16 that the data file **MomandChildHeights** contains data for students in a statistics class on the following two variables: student's height (inches) and biological mother's height (inches).

a) State in words the appropriate null and alternative hypotheses to test whether is a linear association between a person's height and their biological mother's height.

b) Describe how one might use everyday items (for example, coins, dice, cards, etc.) to conduct a tactile simulation-based test of the hypotheses. Be sure to clearly describe how the p-value will be computed from the simulation.

10.2.14 Refer to the previous exercise about the association between a person's height and their biological mother's height. Recall that the data file **MomandChildHeights** contains data for students in a statistics class on the following two variables: student's height (inches) and biological mother's height (inches).

a) Enter the data into the **Corr/Regression** applet, and conduct a simulation-based test of significance. Report the approximate p-value.
b) Interpret the p-value reported in part (a).
c) Summarize your conclusion (significance, causation, generalization) from this simulation analysis. Also describe the reasoning process by which your conclusion follows from your simulation results.

10.2.15 Refer back to the previous exercises that explores the association between children's heights and (biological) mothers' heights. The data file **DadandChildHeights** contains data for students in a statistics class on the following two variables: student's height (inches) and biological father's height (inches).

a) State in words the appropriate null and alternative hypotheses to test whether is a linear association between a person's height and their biological father's height.
b) Describe how one might use everyday items (for example, coins, dice, cards, etc.) to conduct a tactile simulation-based test of the hypotheses. Be sure to clearly describe how the p-value will be computed from the simulation.

10.2.16 Refer to the previous exercise about the association between a person's height and their biological father's height. Recall that the data file **DadandChildHeights** contains data for students in a statistics class on the following two variables: student's height (inches) and biological father's height (inches).

a) Enter the data into the **Corr/Regression** applet, and conduct a simulation-based test of significance. Report the approximate p-value.

b) Interpret the p-value reported in part (a).

c) Summarize your conclusion (significance, causation, generalization) from this simulation analysis. Also describe the reasoning process by which your conclusion follows from your simulation results.

Scrabble names*

10.2.17 Recall from Exercise 10.1.19 that in a statistics class, students were asked to calculate the number of Scrabble points their names would earn. For example, the Scrabble score for the name Tom Sawyer would be 17 points. The data file **ScrabbleNames** contains data on the following two variables: number of letters in a student's name and the corresponding Scrabble score.

a) State in words the appropriate null and alternative hypotheses to test whether is a linear association between the number of letters in a person's name and the Scrabble points their name earns.

b) Describe how one might use everyday items (for example, coins, dice, cards, etc.) to conduct a tactile simulation-based test of the hypotheses. Be sure to clearly describe how the p-value will be computed from the simulation.

10.2.18 Refer back to the previous exercise about exploring the relationship between the number of letters in a name and the corresponding Scrabble score. Recall that the data file **ScrabbleNames** contains data on the following two variables: number of letters in a student's name and the corresponding Scrabble score.

a) Enter the data into the **Corr/Regression** applet, and conduct a simulation-based test of significance. Report the approximate p-value.

b) Interpret the p-value reported in part (a)

c) Summarize your conclusion (significance, causation, generalization) from this simulation analysis. Also describe the reasoning process by which your conclusion follows from your simulation results.

10.2.19 Refer back to the previous exercise about exploring the relationship between the number of letters in a name and the corresponding Scrabble score. Students were also asked to calculate the points per letter, that is, *ratio* = (Scrabble score)/(number of letters in name). The data file **ScrabbleRatio** contains Scrabble points for a student's name and the corresponding *ratio*.

a) State in words the appropriate null and alternative hypotheses to test whether is a linear association between the Scrabble points for a student's name and the corresponding *ratio*.

b) Describe how one might use everyday items (for example, coins, dice, cards, etc.) to conduct a tactile simulation-based test of the hypotheses. Be sure to clearly describe how the p-value will be computed from the simulation.

10.2.20 Refer back to the previous exercise. Recall that the data file **ScrabbleRatio** contains Scrabble points for a student's name and the corresponding ratio of Scrabble points to number of letters.

a) Enter the data into the **Corr/Regression** applet, and conduct a simulation-based test of significance. Report the approximate p-value.

b) Interpret the p-value reported in part (a).

c) Summarize your conclusion (significance, causation, generalization) from this simulation analysis. Also describe the reasoning process by which your conclusion follows from your simulation results.

TV and life expectancy

10.2.21 Recall from Exercise 10.1.22 that the data file **TVLife** provides information on life expectancy and number of televisions per thousand people in a sample of 22 countries, as reported by the *2006 World Almanac and Book of Facts*.

a) State in words the appropriate null and alternative hypotheses to test whether is a linear association between the life expectancy and number of televisions per thousand people in such countries.

b) Describe how one might use everyday items (for example, coins, dice, cards, etc.) to conduct a tactile simulation-based test of the hypotheses. Be sure to clearly describe how the p-value will be computed from the simulation.

10.2.22 Refer back to the previous exercise. Recall that the data file **TVLife** provides information on life expectancy and number of televisions per thousand people in a sample of 22 countries, as reported by the *2006 World Almanac and Book of Facts*.

a) Enter the data into the **Corr/Regression** applet, and conduct a simulation-based test of significance. Report the approximate p-value.

b) Interpret the p-value reported in part (a).

c) Summarize your conclusion (significance, causation, generalization) from this simulation analysis. Also describe the reasoning process by which your conclusion follows from your simulation results.

Section 10.3

10.3.1* Which of the following does the method of least squares minimize?

A. Sum of vertical distances between observations and the line

B. Sum of squared vertical distances between observations and the line

C. Sum of perpendicular distances between observations and the line

D. Sum of squared perpendicular distances between observations and the line

E. Sum of horizontal distances between observations and the line

F. Sum of squared horizontal distances between observations and the line

10.3.2 Explain what is wrong with the following statements about correlation and regression.

a) The correlation between the number of years of education and yearly income is 1.23.

b) The regression equation that describes the relationship between the age of a used Ford Mustang and its value is $\hat{y} = -1000x + 10,000$ where the explanatory variable is the age of the automobile (in years) and the response is its value (in dollars). The correlation describing this relationship is 0.83.

10.3.3* It can be shown that the *sum* of residuals from a least squares line always equals zero.

a) Does it follow from this result that the *mean* of the residuals from a least squares line always equals zero? Explain briefly.

b) Does it follow from this result that the *median* of the residuals from a least squares line always equals zero? Explain briefly.

10.3.4 Answer the following questions about slope coefficients.

a) Can a slope coefficient ever be negative?

b) Can a slope coefficient ever be greater than one?

c) Can a slope coefficient ever equal zero?

10.3.5* Reconsider the previous exercise. For each of the descriptions presented that is indeed possible, produce a scatterplot of hypothetical data to show that the description is indeed possible.

10.3.6 Reconsider the previous two exercises. Answer the questions with regard to the *intercept* coefficient rather than the slope coefficient.

Legos*

10.3.7 A statistics professor went to the lego.com website in February 2014 and recorded the number of pieces and the sales price for 157 Lego products listed there. The data appear in the datafile **legos**.
   a) Which variable do you think makes more sense to use as the explanatory variable, and which as the response variable?
   b) Enter the data into the **Corr/Regression** applet. Produce a scatterplot, and comment on the association between the variables as revealed in the scatterplot. (Remember to comment on form, direction, strength, and unusual observations.)
   c) Report the value of the correlation coefficient.
   d) Calculate the value of r^2, and interpret what the value means in this context.

10.3.8 Reconsider the previous exercise and the datafile **legos**.
   a) Determine the equation of the least squares line for predicting price from number of pieces. Report this equation using good statistical notation.
   b) Interpret the value of the slope coefficient in this context.
   c) Interpret the value of the intercept coefficient. Is this a context for which the value of the intercept provides relevant information? Explain.

10.3.9 Reconsider the previous two exercises and the datafile **legos**. The equation of the least squares line turns out to be $\widehat{price} = 4.86 + 0.105 \times pieces$, and the value of $r^2 = 0.949$. Identify which of the following interpretations are correct, which are incorrect, and which are correct but incomplete or poorly worded.
   a) The Legos cost 10.5 cents per piece.
   b) For each additional dollar in price, the predicted number of pieces in the set increases by about 10.5.
   c) For each additional piece in the set, the predicted price increases by about 10.5 cents.

d) The price goes up 10.5 cents for each additional piece in the set.

e) The predicted price for a set with 0 pieces is $4.86.

f) The predicted price increases by $10.5 for each additional 100 pieces in the set.

10.3.10 Reconsider the previous three exercises and the datafile **legos**. The equation of the least squares line turns out to be $\widehat{price} = 4.86 + 0.105 \times pieces$, and the value of $r^2 = 0.949$. Identify which of the following interpretations are correct, which are incorrect, and which are correct but incomplete or poorly worded.

a) 10.5% of the observed variability in prices is explained by the least squares line with number of pieces.

b) 94.9% of the observed variability in prices is explained by the least squares line with number of pieces.

c) 94.9% of the prices fall on the least squares line based on number of pieces.

d) The least squares line correctly predicts the price for 94.9% of the Lego products.

10.3.11 Reconsider the previous four exercises and the datafile **legos**.

a) Determine the predicted price for a product with 500 pieces.

b) Determine the predicted price for a product with 1500 pieces.

c) Determine the difference between the two predictions in parts a) and b). Also indicate how you could have determined this value directly from the equation of the least squares line.

d) Determine the predicted price for a product with 5000 pieces.

e) Do you feel very confident with this prediction in part e)? Explain why or why not.

10.3.12 Reconsider the previous five exercises and the datafile **legos**. The last product listed in the datafile has 415 pieces and a price of $49.99.

a) Determine the predicted price for such a product.

b) Determine the residual value for this product.

c) Interpret what this residual value means.

d) Does the product fall above or below the least squares line in the graph? Explain how you can tell, based on its residual value.

10.3.13 Reconsider the previous six exercises and the datafile **legos**. This is very unrealistic, but suppose that one of the products were to be offered at a price of $0.

a) Would you expect this change to affect the least squares line very much? Explain.

b) For which one product would you expect this change to have the greatest impact on the least squares line? Explain how you choose this product.

c) Change the price to $0 for the product that you identified in part b). Report the (new) equation of the least squares line and the (new) value of r^2. Have these values changed considerably?

Crickets

10.3.14 Consider the following two scatterplots based on data gathered in a study of 30 crickets, with temperature measured in degrees Fahrenheit and chirp frequency measured in chirps per minute:

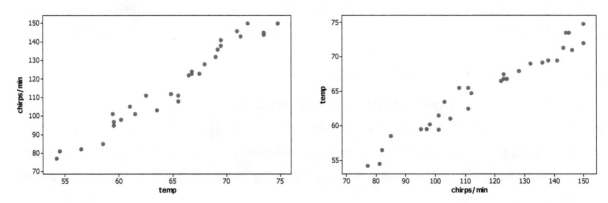

a) If the goal is to predict temperature based on a cricket's chirps per minute, which is the appropriate scatterplot to examine- the one on the left or the one on the right? Explain briefly.

One of the following is the correct equation of the least squares line for predicting temperature from chirps per minute:

     A. predicted temperature = 35.78 + 0.25 chirps per minute

     B. predicted temperature = -131.23 + 3.81 chirps per minute

     C. predicted temperature = 83.54 – 0.25 chirps per minute

b) Which is the correct equation? Circle your answer, and explain briefly.

c) Use the correct equation to predict the temperature when the cricket is chirping at 100 chirps per minute.

d) Interpret the value of the slope coefficient, in this context, for whichever equation you think is the correct one.

Cat jumping*

10.3.15 Harris and Steudel (2002) studied factors that might be associated with the jumping performance of domestic cats. They studied 18 cats, using takeoff velocity (in centimeters per second) as the response variable. They used body mass (in grams), hind limb length (in centimeters), muscle mass (in grams), and percent body fat, in addition to sex, as potential explanatory variables. The data can be found in the **CatJumping**. A scatterplot of takeoff velocity vs. body mass is shown here:

a) Describe the association between these variables.
b) Use the **Corr/Regression** applet to determine the equation of the least squares line for predicting a cat's takeoff velocity from its mass.
c) Interpret the value of the slope coefficient in this context.
d) Interpret the value of the intercept coefficient. Is this a context in which the intercept coefficient is meaningful?
e) Determine the proportion of variability in takeoff velocity that is explained by the least squares line with mass.

10.3.16 Reconsider the previous exercise and the **CatJumping** datafile.
a) Determine the predicted takeoff velocity for a cat with a mass of 5000 grams (which is about 11 pounds).
b) Determine the predicted takeoff velocity for a cat with a mass of 10,000 grams. Also explain why it's not advisable to have much confidence in this prediction.

c) Determine the predicted takeoff velocity and the residual value for the cat with the largest mass. Also interpret what this residual value means.

10.3.17 Reconsider the previous two exercises. Answer the following based on the scatterplot presented above. Do not bother to perform any calculations.

    a) Which cat has the *largest predicted* value for its takeoff velocity?

    b) Which cat has the *smallest predicted* value for its takeoff velocity?

    c) Which cat has the *largest residual* value?

    d) Which cat has the *smallest residual* value?

10.3.18 Reconsider the previous three exercises and the **CatJumping** datafile. Investigate the association between the response variable (takeoff velocity) and the other explanatory variables (hind limb length, muscle mass, percent body fat).

a) Select the explanatory variable that has the strongest association with the response. Describe this association.

b) Report the equation of the least squares line for predicting the cat's takeoff velocity using this explanatory variable.

c) Interpret the value of the slope coefficient.

d) Determine and interpret the value of r^2.

Honda civic pricing

10.3.19 The data in the file **UsedHondaCivics** come from a sample of used Honda Civics listed for sale online in July 2006. The variables recorded are the car's year of manufacture, age (calculated as 2006 minus year of manufacture), mileage, and price.

    a) Identify the observational units.

    b) Produce a scatterplot of price vs. age. Describe the association revealed in the graph.

    c) Determine the least squares line for predicting price from age, and produce a scatterplot with the least squares line superimposed.

    d) Report and interpret the value of the slope coefficient.

    e) What percentage of the variability in car prices is explained by knowing the car's age?

Textbook prices*

10.3.20 Two Cal Poly freshmen gathered data on a random sample of textbooks from the campus bookstore in November of 2006. Two of the variables recorded were the price of the book and the number of pages that it contained. These data are in the file **TextbookPrices**.

a) Identify the explanatory and response variables in this study.
b) Determine the equation of the least squares line for predicting price from number of pages. Report the equation, being sure to use good statistical notation.
c) Use the least squares line to predict the price of a 500-page textbook. Then do the same for a 1500-page textbook. Which prediction would you have more confidence in? Explain.
d) Interpret what the slope coefficient means in this context.
e) Determine the proportion of variability in textbook prices that is explained by knowing the number of pages in the book.

Day hikes

10.3.21 The book *Day Hikes in San Luis Obispo County* lists information about 72 hikes, including the distance of the hike (in miles), the elevation gain of the hike (in feet), and the time that the hike is expected to take (in minutes). Consider the scatterplot below, with least squares regression lines superimposed:

a) Report the value of the slope coefficient for predicting time from distance.
b) Write a sentence interpreting the value of the slope coefficient for predicting time from distance.

c) Use the line to predict how long a 4-mile hike will take.

d) Would you feel more comfortable using the line predict the time for a 4-mile hike or for a 12-mile hike? Explain your choice.

e) The value of the correlation coefficient between time and distance is 0.916, and the value of r^2 = 0.839. Complete this sentence to interpret what this value means:

      83.9% of _____ is explained by _____.

10.3.22 Reconsider the previous exercise. The following scatterplot displays hiking time vs. elevation gain, with the least squares line superimposed:

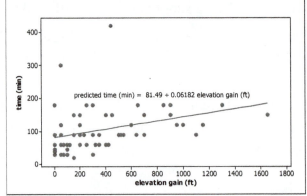

a) Report the value of the slope coefficient for predicting time from elevation gain.

b) Write a sentence interpreting the value of the slope coefficient for predicting time from elevation gain.

c) Use the line to predict how long a hike with an 800-feet elevation gain will take.

d) Would you feel more comfortable using the line predict the time for a hike with an 800-feet elevation gain or a hike with a 2800-feet elevation gain? Explain your choice.

e) The value of the correlation coefficient between time and distance is 0.344, and the value of r^2 = .119. Complete this sentence to interpret what this value means:

      11.9% of _____ is explained by _____.

10.3.23 The slope b and intercept a of a least squares line can be calculated from the means and standard deviations of the two variables, along with the correlation coefficient, as follows:

$b = r \dfrac{s_y}{s_x}$ and $a = \bar{y} - b\bar{x}$. These summary statistics for the day hikes described in the previous two exercises are:

	Mean	SD	Correlation with time
Distance	3.283	1.856	0.916
Elevation gain	333.2	355.5	0.344

Time	102.08	63.79	1.000

a) Use these statistics to determine the slope b and intercept a for predicting time from *distance*.

b) Use these statistics to determine the slope b and intercept a for predicting time from *elevation gain*.

c) Use the least squares line to predict the time needed for a hike with distance equal to the mean of the distances.

d) Use the least squares line to predict the time needed for a hike with elevation gain equal to the mean of the elevation gains.

e) What do you notice about your predictions in parts c) and d)?

Introductory Statistics Tests*

10.3.24 The following scatterplot represents scores on test 2 and test 3 in Introductory Statistics for a random sample of students.

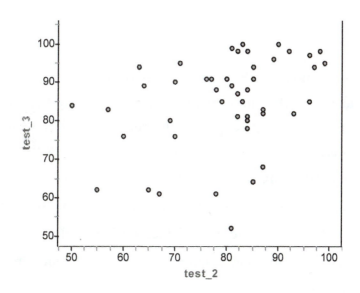

a) In the figure above, there appears to be a positive relationship between scores on test 2 and test 3. Does this mean that a person scores above the mean on test 2 is expected to also score above the mean on test 3? What about those who score below the mean on test 2, are they expected to score below the mean on test 3?

The regression equation for these data is $\hat{y} = 55.2 + 0.3708x$, where x is the score on test 2 and $\hat{y}$ is the expected score on test 3. The mean for test 2 was 79.8 and the mean for test 3 was 84.8.

b) If a student scores a 79.8 on test 2, what is their predicted score for test 3? If that student then actually gets a 90 on test 3, what is the residual?

c) If someone scores 10 points *above* the mean on test 2, how many points above the mean on test 3 is their expected score?

d) If someone scores 10 points *below* the mean on test 2, how many points below the mean on test 3 is their expected score?

10.3.25 Reconsider the previous exercise.

a) Determine the predicted score on test 3 for a student who scores 55 on test 2.

b) Determine the predicted score on test 3 for a student who scores 95 on test 2.

c) Of the two students in parts a) and b), which student is predicted to achieve a *higher* score on test 3 than he/she scored on test 2? Which student is predicted to achieve a *lower* score on test 3 than he/she scored on test 2? (This phenomenon is known as the **regression effect**.)

Section 10.4

10.4.1* When testing the hypothesis that there is no association (null) vs. an association (alternative) you can use either the sample correlation coefficient or the sample slope as the statistic. How will the p-values compare when using both approaches on the same data set?

A. The result from using slope as our statistic is equivalent to using the correlation coefficient as our statistic in a test; therefore our p-values will be identical.

B. The result from using slope as our statistic is similar to using the correlation coefficient as our statistic in a test; therefore our p-values will be similar.

C. As slope and the correlation coefficient measure two different things, our p-values will, most likely, not be similar.

D. There is no way to tell without running the tests.

Foot length and height

10.4.2 Suppose a sample of 10 people had their foot lengths measured (in cm) and heights (in inches) and the results, along with the regression line are shown in the following scatterplot. The equation of the regression line, correlation coefficient and r^2 are also shown.

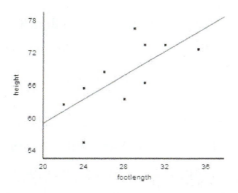

Show Regression Line: ☑
height^ = 37.48 + 1.10 x footlength
Show Residuals: ☐
Show Squared Residuals: ☐

Correlation coefficient ☑ $r = 0.700$
R-squared ☑ $r^2 = 48.9\%$

Which one of the following applet screen shots is the appropriate way to test to determine if there was a positive relationship between people's foot length and height?

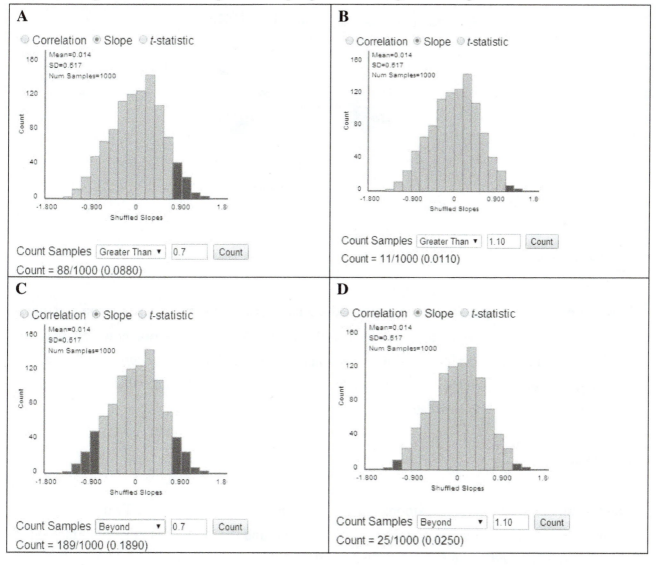

10.4.3* Referring to the previous exercise, describe how you would construct a null distribution for this situation by hand using the slope of the regression line as your statistic. Assume you have 20 slips of paper.

10.4.4 Using foot length and height data, we shuffled the *y*-values 100 times and each time plotted the resulting regression line in the graph below. The regression line for the observed data is the one being pointed at with the dashed arrow. Based on the plot of these lines, what can you say about the p-value if we are testing to see whether there is a positive association between height and foot length? Explain.

Height and leap*

10.4.5 Suppose you are testing to see whether there is an association between a person's height and their vertical leap. You've collected data on this from 20 people. Describe how you would construct a null distribution for this situation by hand using the slope of the regression line as your statistic. Assume you have 20 slips of paper containing the 20 heights and 20 slips of paper containing the 20 vertical leap distances.

Textbook prices

10.4.6 A random sample of 15 textbooks in a campus bookstore was selected. Two of the variables recorded were the price of the book (in dollars) and the number of pages it contained. The data are displayed in the following scatterplot.

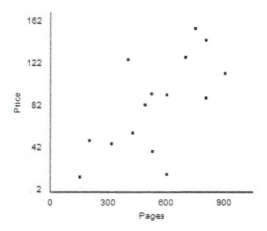

a) Is this an experiment or an observational study?

b) What are the observational units?

c) The equation of the least squares regression line for predicting price from number of pages is: $\widehat{price} = 14.11 + 0.13(pages)$

    i.   Use the regression equation to predict the price of a 500-page textbook.

    ii.   Use the regression equation to predict the price of a 1500-page textbook.

    iii.   Which prediction, the one for the 500 page book or 1500 page book, should you have more confidence in? Explain.

10.4.7 Refer to the information in the previous exercise to answer the following.

  a) State the null and alternative hypotheses for a test of possible association   between pages and price.

  b) The following null distribution was created to test the hypotheses stated in the previous question using slope as the statistic.

   i. Based on information shown in the null distribution, how many standard deviations is our observed statistic above the mean of the null distribution? (i.e. What is the standardized statistic?)

   ii. Based on your standardized statistic, do you have strong evidence of an association between number of pages and price of textbooks? Explain.

10.4.8 Reconsider the previous two exercises about textbook prices. The equation of the least squares regression line for predicting price from number of pages is:

$$\widehat{price} = 14.11 + 0.13(pages)$$

   a) Interpret what the slope coefficient means in the context of pages and price.

   b) Interpret the intercept. Is this an example of extrapolation? Why or why not?

Sleep and maze performance*

10.4.9 Student researchers asked their subjects how much sleep they had the previous night (in hours) and then timed how long it took them (in seconds) to complete a paper and pencil maze. The results are shown in the scatterplot along with the regression line.

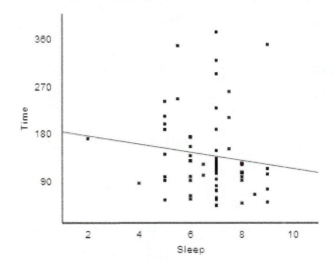

   a) What is the explanatory variable?

   b) What is the response variable?

   c) The equation of the least squares regression line for predicting price from number of pages is: $\widehat{time} = 190.33 - 7.76(sleep)$. The following null distribution was created to test the association between number of time and the amount of sleep using slope as the statistic. The null and alternative hypotheses for this test can be written as: Null: No association between sleep and time in the population, Alt: Association between sleep and time in the population.

i. Based on information shown in the null distribution, how many standard deviations is our observed statistic above the mean of the null distribution? (i.e. What is the standardized statistic?)

ii. Based on your standardized statistic, do you have strong evidence of an association between the time it takes someone to complete the maze and how much sleep they got the night before? Explain.

10.4.10 Reconsider the previous exercise about the amount of sleep (in hours) obtained in the previous night and time to complete a paper and pencil maze (in seconds). The equation of the least squares regression line for predicting price from number of pages is:

$\widehat{time} = 190.33 - 7.76(sleep)$.

a) Interpret what the slope coefficient means in the context of sleep and time to complete the maze.

b) Interpret the intercept. Is this an example of extrapolation? Why or why not?

Weight loss and protein

10.4.11 In a research study was looking to see if there was an association between weight loss and the amount of a certain protein in a person's body fat, the researchers measured a number of different attributes in their 39 subjects at the beginning of the study. The article reported, "These subjects were clinically and ethnically heterogeneous." Two of the variables they measured were body mass index (BMI) and total cholesterol. The results are shown in the scatterplot along with the regression line.

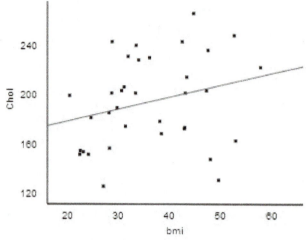

a) What are the observational units in the study?

b) The equation of the least squares regression line for predicting total cholesterol from BMI is: $\widehat{cholesterol} = 162.56 - 0.9658(BMI)$. The following null distribution was created to test the association between people's total cholesterol number and their BMI using slope as the statistic. The null and alternative hypotheses for this test can be written as: Null: No association between cholesterol and BMI in the population, Alt: Association between cholesterol and BMI in the population.

   i.   Based on information shown in the null distribution, how many standard deviations is our observed statistic below the mean of the null distribution? (i.e. What is the standardized statistic?)

   ii. Based on your standardized statistic, do you have strong evidence of an association between a people's total cholesterol and their BMI? Explain.

10.4.12 Reconsider the previous exercise about the cholesterol and BMI. The equation of the least squares regression line obtained was $\widehat{cholesterol} = 162.56 - 0.9658(BMI)$.

    a) Interpret what the slope coefficient means in the context of cholesterol and BMI.

    b) Interpret the intercept. Is this an example of extrapolation? Why or why not?

    c) Based on your confidence interval do you have strong evidence of an association between cholesterol and BMI? Explain.

Honda Civic prices*

10.4.13 The data in the file **UsedHondaCivics** come from a sample of used Honda Civics listed for sale online in July 2006. The variables recorded are the car's year of manufacture, age (calculated as 2006 minus year of manufacture) and price. Consider conducting a simulation analysis to test whether the sample data provide strong evidence of an association between a car's price and age in the population, in terms of the population slope.

    a) State the appropriate null and alternative hypotheses.

    b) Conduct a simulation analysis with 1000 repetitions. Describe how to find your p-value from your simulation results, and report this p-value.

    c) Summarize your conclusion from this simulation analysis. Also describe the reasoning process by which your conclusion follows from your simulation results.

10.4.14 Reconsider the previous exercise on prices of Honda Civics.

    a) Find the regression equation that predicts the price of the car given its age.

    b) Interpret the slope and intercept of the regression line.

Weight and haircut price

10.4.15 In a survey of statistics students at Hope College, two of the questions asked were their weight (in pounds) and the cost of their last haircut, including any hair treatments (in dollars). The data can be found in the file **WeightHaircut**. Let's explore whether or not there is evidence of a strong association between weight and cost of haircuts.

    a) State the appropriate null and alternative hypotheses.

    b) Conduct a simulation analysis with 1000 repetitions. What is your p-value?

c) Summarize your conclusion from this simulation analysis. If there is an association, describe if it is positive or negative in the context of the study.

d) Can you conclude a cause and effect between weight and cost of a haircut? Why or why not?

10.4.16 Reconsider the previous exercise on weight and haircut price.

a) Find the regression equation that predicts the haircut price based on weight.

b) Interpret the slope and intercept of the regression line.

Height and haircut price*

10.4.17 In a survey of statistics students at Hope College, two of the questions asked were their height (in inches) and the cost of their last haircut, including any hair treatments (in dollars). The data can be found in the file **HeightHaircut**. Let's explore whether or not there is evidence of a strong association between height and cost of haircuts.

a) State the appropriate null and alternative hypotheses

b) Conduct a simulation analysis with 1000 repetitions. What is your p-value?

c) Summarize your conclusion from this simulation analysis. If there is an association, describe if it is positive or negative in the context of the study.

d) Can you conclude a cause and effect between height and cost of a haircut? Why or why not?

10.4.18 Reconsider the previous exercise on height and haircut price.

a) Find the regression equation that predicts the haircut price based on height.

b) Interpret the slope and intercept of the regression line.

Age and BMI

10.4.19 Researchers in a clinical study collected information from their subjects at the beginning of the study. Two of the variables were body mass index (BMI) and age. We are interested in seeing whether there is an association between BMI and age. The data from the study can be found in the file **AgeBMI.**

a) State the appropriate null and alternative hypotheses.

b) Conduct a simulation analysis with 1000 repetitions. What is your p-value?

c) Summarize your conclusion from this simulation analysis. If there is an association, describe if it is positive or negative in the context of the study.

10.4.20 Reconsider the previous exercise on age and BMI. We are interested in seeing whether there is an association between BMI and age. The data from the study can be found in the file **AgeBMI.**

a) Conduct a new simulation analysis with 1000 repetitions using slope as the statistic. What is your p-value?

b) Using the same null distribution as in part (a), use correlation as the statistic. What is the p-value?

c) Are you two p-values exactly the same? (The two p-values should be the same; they may be just slightly different due to rounding of the statistics.)

Missing class and GPA*

10.4.21 In a survey of statistics students at Hope College, two of the questions asked were their current grade point average (GPA) and how many classes they failed to attend during the past three weeks at the college. The data can be found in the file **MissClassGPA**. Let's explore whether or not there is evidence of a strong association between these two variables.

a) State the appropriate null and alternative hypotheses

b) Conduct a simulation analysis with 1000 repetitions. What is your p-value?

c) Summarize your conclusion from this simulation analysis. If there is an association, describe if it is positive or negative in the context of the study.

d) Can you conclude a cause and effect between missing class and GPA? Why or why not?

Sleep and GPA

10.4.22 In a survey of Hope College students, two of the questions asked were their current grade point average (GPA) and how much sleep the got (in hours) on the previous school night. The data can be found in the file **SleepGPA**. Let's explore whether or not there is evidence of a strong association between these two variables.

a) What is the value of the correlation? Based on this number, does it appear there is a strong association between our two variables?

b) What is the value of the slope of the regression line? What does this number mean in terms of sleep and GPA?

c) State the appropriate null and alternative hypotheses.

d) Conduct a simulation analysis with 1000 repetitions. What is your p-value?

e) Based on your p-value, is there a strong association between sleep and GPA for Hope College students?

f) There should be some conflict between your estimate of the strength of association in part (a) and your answer to part (e). Why did that occur with this data set?

Stroop effect*

10.4.23 You may have seen the test or puzzle where names of colors are written out in colors not denoted by the name. For example the word GREEN might be written with red ink. The job of the person completing the test is to say the word and not the color. This interference in the reaction time of saying the words is called the Stroop effect. Student researchers at Hope College wanted to see if there was an association between age and the time it takes people to read a list of 20 of these colored words. Their results can be found in the data file **StroopAgeTime.** The age is given in years and the time to complete the task is given in seconds. Put the data in the **Corr/Regression** applet and complete the following.

a) Write out the null and alternative hypotheses for this study using a two-sided alternative.

b) What is the slope of the regression line? What does that number mean in the context of age and time to complete the task?

c) Conduct at least 1000 shuffles of the data, find and report the resulting p-value.

d) Summarize your conclusion from this analysis.

10.4.24 Reconsider the previous exercise on the Stroop effect.

a) There was one point in the scatterplot that was a clear outlier. One of the researchers' great aunt, who was 87 years old, took 100.6 seconds to complete the task. This was almost three times as long as the next longest time. Remove that point from the data set and find a new p-value. Are your results significant?

b) After the great aunt's point is removed from the scatterplot, there now appears as a couple of other points representing an 8 year-old and a 10 year-old might be outliers. Perhaps an association between age and time won't be demonstrated when the subjects are too young. For this reason, delete all the points representing anyone 10 and under. There should be 5 of these. Find a new p-value. Are your results significant?

c) With the 87 year-old included, there is a significant association. When the result from the 87 year-old is deleted from the data, there is not a significant association. When results from everyone 10 and under are deleted from the study, there is a significant association. So, how would you answer the question, is there an association between age that the time needed to complete this task of reading of color names? Explain.

Section 10.5

10.5.1* Which one of the following is a validity condition for theory-based tests for regression?
A. The variability of the points around the regression line should not differ as you slide along the x-axis.
B. There should be at least 10 success and 10 failures in the data.
C. The variability of the explanatory variable should be the same as the variability of the response variable.
D. The scatterplot should show a positive association between the explanatory and response variables.

10.5.2 For a given dataset, a test of association based on a slope is equivalent to a test of association based on a correlation coefficient. Being equivalent means which of the following is true?
A. The confidence intervals for the population correlation and population slope will be the same.
B. The observed correlation will be the same as the observed slope of the regression line.
C. The p-value will be the same whether you use correlation as the statistic or the slope of the regression line as the statistic.
D. All the above.

10.5.3* The theory-based p-value given in the regression table in the **Corr/Regression** applet is:
A. A two-sided p-value.
B. A one-sided p-value.
C. Either a one-sided or two-sided p-value depending how count your samples (greater than, less than, or beyond) in the simulation part of the applet.

   D. A one-sided p-value when you using correlation as the statistic, but a two-sided p-value when you are using slope as the statistic.

10.5.4 To explore the relationship between the runtime and profit made by movies, data were collected on a random sample of 15 movies released in the last 4 years. The p-value corresponding to the two-sided test for the slope turned out to be 0.06.

   a) Based on the p-value, is it okay to conclude that *"there is evidence that there is no association between runtime and profit made by movies"*? How are you deciding?
   b) Suppose that you were to create a 90% confidence interval for the population slope, using the data from this study. Would this confidence interval contain 0? How are you deciding?
   c) How, if at all, would the p-value change if you wanted to test whether there is a positive association between runtime and profit made by movies? Explain.

10.5.5* If there are influential points or outliers in your data, why is it a good idea to run the analysis with those points removed?

10.5.6 For the data in each of the following scatterplots, are the validity conditions met to run a theory-based test for regression? If the answer is no on any of them, explain why not.

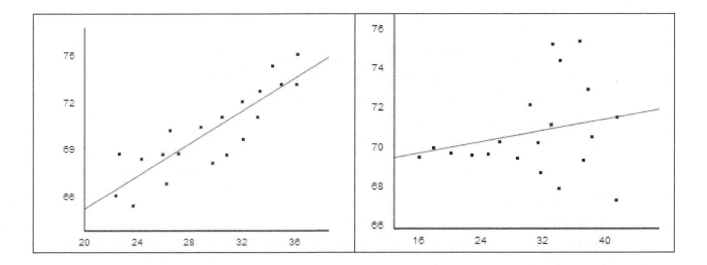

Sleep and maze completion*

10.5.7 A regression table is shown based on data used to test an association between the amount of sleep someone had the previous night (in hours) and the time needed to complete a paper and pencil maze (in seconds). Sleep is the explanatory variable and time need to complete the maze is the response.

Term	Coeff	SE	t-stat	p-value
Intercept	198.33	51.57	3.85	0.0003
Sleep	-7.76	7.50	-1.03	0.3052

a) What is the regression equation where time to complete the maze is predicted from the amount of sleep?
b) If we were testing to the alternative hypothesis $\beta \neq 0$, what is the p-value?
c) If we were testing to the alternative hypothesis $\beta < 0$, what is the p-value?

BMI and glucose levels

10.5.8 Problems 10.5.8 and 10.5.9 refer to data on BMI and glucose levels.

In a research study that was looking to see if there was an association between weight loss and the amount of a certain protein in a person's body fat, the researchers measured a number of different attributes in their 39 subjects at the beginning of the study. The article reported, "These subjects were clinically and ethnically heterogeneous." Two of the variables they measured were body mass index (BMI) and blood glucose levels and the results are in the file **BMIGlucose**. We are interested in seeing whether there is an association between BMI and glucose levels.

a) Describe the population slope in the context of the study, and assign a symbol to it.

b) State the appropriate null and alternative hypotheses in terms of the population slope, using the symbol used in (a).

c) Is a theory-based test appropriate? Why or why not?

d) Regardless of your answer to the previous question, use a theory-based approach to find the p-value to test the hypotheses stated in (b). Report and interpret this p-value, in the context of the study.

e) Summarize your conclusion, based on the p-value reported in (d).

f) If you were testing to determine if there was a *positive* association between BMI and glucose levels, what is the theory-based p-value?

10.5.9 Reconsider the previous exercise on BMI and blood glucose levels.

a) Determine a 95% confidence interval for the population slope coefficient, and interpret what this interval represents.

b) Based on your confidence interval, is there a positive association between BMI and blood glucose levels? Explain.

Height and Weight*

10.5.10 At the beginning of the semester, students in a statistics class were asked to give their estimates of how much their professor weighed (in pounds) and his height (in inches). We want to see if students who give large estimates for height will they also give large estimates for weight. In other words, is there are positive association between these two variables. The data from the study can be found in the file **HeightWeight.** (The professor actually weighed about 160 pounds and was 72 inches tall.)

a) Describe the population slope in the context of the study, and assign a symbol to it.

b) State the appropriate null and alternative hypotheses in terms of the population slope, using the symbol used in (a).

c) Is a theory-based test appropriate? Why or why not?

d) Regardless of your answer to the previous question, use a theory-based approach to find the p-value to test the hypotheses stated in (b). Report and interpret this p-value, in the context of the study.

Age and BMI

10.5.11 Problems 10.5.11 and 10.5.12 refer to data on age and BMI. Researchers in a clinical study collected information from their subjects at the beginning of the study. Two of the variables were body mass index (BMI) and age. We are interested in seeing whether there is an association between these two variables. The data from the study can be found in the file **AgeBMI.**

a) Describe the population slope in the context of the study, and assign a symbol to it.

b) State the appropriate null and alternative hypotheses in terms of the population slope, using the symbol used in (a).

c) Is a theory-based test appropriate? Why or why not?

d) Use a theory-based approach to find p-value.

e) Summarize your conclusion, based on the p-value reported in (d).

f) If you were testing to determine if there was a *positive* association between age and BMI, what is theory-based p-value?

10.5.12 Reconsider the previous exericse on age and BMI.

a) Determine a 95% confidence interval for the population slope and interpret what this interval represents.

b) Based on your confidence interval, is there strong evidence for a positive association between age and BMI? Explain.

c) Based on a one-sided p-value, is there strong evidence for a positive association between age and BMI? Explain.

d) Your answers to parts (b) and (c) might contradict each other. Explain why that is.

Honda Civic prices*

10.5.13 Problems 10.5.13 and 10.5.14 refer to data gathered on Honda Civic prices. The data in the file **UsedHondaCivics** come from a sample of used Honda Civics listed for sale online in July 2006. The variables recorded are the car's year of manufacture, age (calculated as 2006 minus year of manufacture) and price. Consider conducting an analysis to test whether the sample data provide strong evidence of an association between a car's price and age in the population, in terms of the population slope.

a) Describe the population slope in the context of the study, and assign a symbol to it.

b) State the appropriate null and alternative hypotheses in terms of the population slope, using the symbol used in (a).

c) Is a theory-based test appropriate? Why or why not?

d) Use a theory-based approach to find a p-value to test the hypotheses stated in (b).

e) Summarize your conclusion, based on the p-value reported in (d).

10.5.14 Reconsider the previous exericise on Used Honda Civics.

a) Determine a 95% confidence interval for the population slope coefficient, and interpret what this interval represents.

b) Based on your confidence interval, is there a negative association between age and price for used Honda Civics? Explain.

Textbook prices

10.5.15 Two Cal Poly freshmen gathered data on a random sample of textbooks from the campus bookstore in November of 2006. Two of the variables recorded were the price of the book and the number of pages that it contained. These data are in the file **TextbookPrices**.

a) Describe the population slope in the context of the study, and assign a symbol to it.

b) Suppose that you want to test whether there is a positive association between price and number of pages in the population of all textbooks for sale in the bookstore in November of 2006. State the appropriate null and alternative hypotheses in terms of the population slope, using the symbol used in (a).

c) Is a theory-based test appropriate? Why or why not?

d) Determine and report the value of the t statistic and p-value for testing the hypotheses stated in (b).

e) Do the sample data provide strong evidence that number of pages is a significant predictor of price? Explain how you can tell.

f) Determine and interpret a 95% confidence interval for the value of the population slope.

Height and BMI*

10.5.16 Problems 10.5.16 and 10.5.17 refer to data gathered on height and BMI. The data in the file **HeightBMI** was obtained from a survey of statistics students. The variables recorded are the student's heights (in inches) and their body mass index (BMI). Consider conducting an analysis to test whether the sample data provide strong evidence of an association between height and BMI, using slope as the statistic.

a) Describe the population slope in the context of the study, and assign a symbol to it.

b) State the appropriate null and alternative hypotheses in terms of the population slope, using the symbol used in (a).

c) Is a theory-based test appropriate? Why or why not?

d) Use a theory-based approach to find a p-value.

e) Summarize your conclusion, based on the p-value reported in (d).

10.5.17 Reconsider the previous exercise on height and BMI.

a) Determine a 95% confidence interval for the population slope coefficient, and interpret what this interval represents.

b) Based on your confidence interval, is there an association between height and BMI? Explain.

Gestation and Life Expectancy

10.5.18 The data in the file **GestationLifeExpectancy** gives the gestation period (in day) and the life expectancy (in years) for a sample of mammals. We will consider this sample to be representative of all mammals for these two variables. Consider conducting an analysis to test whether the sample data provide strong evidence of an association between gestation period and life expectancy.

a) What is the value of the slope of the regression line and what does it mean in terms of gestation period and life expectancy?

b) State the appropriate null and alternative hypotheses in terms of the population slope.

c) Use a theory-based approach to find a p-value. Do you have strong evidence of an association between the two variables?

d) Remember that a TYPE II error is not rejecting a false null hypothesis (or a missed opportunity). Do you think you made a TYPE II error for this exercise? Why or why not?

Sleep and height*

10.5.19 The data in the file **SleepHeight.txt** was obtained from a survey of statistics students. The variables recorded are the student's heights (in inches) and the amount of sleep they typically get on a school night (in hours). Consider conducting an analysis to test whether the sample data provide strong evidence of a negative association between sleep and height.

a) Describe the population slope in the context of the study, and assign a symbol to it.

b) State the appropriate null and alternative hypotheses in terms of the population slope, using the symbol used in (a).

c) Use a theory-based approach to find a p-value. Do you have strong evidence of a negative association between sleep and height?

d) There are a couple of errors in the data set. It appears as though two students reported their heights as 55 inches and they sleep 10 hours a night. Because all of the answers on the survey were identical for these two entries, it would appear they came from the same person. It is also obvious to the instructor that there was nobody in class as short as 55 inches, so we know the height is an error. Eliminate these two points (they are the last ones on the list) and find a new p-value. Do you now have strong evidence of a negative association?

GDP and infant mortality

10.5.20 The following scatterplot gives data from a random sample of 27 countries where the explanatory variable is infant mortality (deaths per 1000 live births) and the response is the gross domestic product (GDP) per capita (in thousands of dollars). The gross domestic product per capita is a representation of a country's standard of living.

a) Describe the relationship between infant mortality and GDP per capita. Is this relationship linear?

b) The regression line is also included in the scatterplot. Will this regression line be a good prediction of infant mortality given GDP per capita? Why or why not?

Sometimes we can transform data that don't fit a linear pattern in such a way that they do fit a linear pattern. We calculated the logarithm of all the values of both the infant mortality and GDP per capita and plotted the results in the scatterplot below.

c) Describe the relationship between logarithm of infant mortality and logarithm of GDP per capita. Is this relationship linear?

d) The regression line is also included in the scatterplot. Will this regression line be a good prediction of the logarithm of infant mortality given the logarithm of GDP per capita? Why or why not?

End-of-Chapter Exercises

10.CE.1* Answer the following questions on correlation.

a) Suppose that every student in your class scores 10 points *higher* on the final exam than on the midterm exam. What would the value of the correlation coefficient between the two exam scores equal? Explain. (*Hint*: Draw a scatterplot for some hypothetical data.)

b) Suppose that every student in your class scores 10 points *lower* on the final exam than on the midterm exam. What would the value of the correlation coefficient between the two exam scores equal? Explain. (*Hint*: Draw a scatterplot for some hypothetical data.)

c) Suppose that every student in your class scores *twice* as many points on the final exam than on the midterm exam. What would the value of the correlation coefficient between the two exam scores equal? Explain. (*Hint*: Draw a scatterplot for some hypothetical data.)

10.CE.2 Answer the following questions on correlation and regression.

a) When calculating a correlation coefficient between two quantitative variables, does it matter which is considered explanatory and which response?

b) When calculating a least squares line between two quantitative variables, does it matter which is considered explanatory and which response?

c) When calculating a residual value from a least squares line, does it matter whether you calculate the vertical horizontal, or perpendicular distance to the line?

Teacher's age and distance to school*

10.CE.3 Is there an association between a teacher's age and how far he/she lives from school? For each of the following descriptions, draw a scatterplot of hypothetical data for a sample of 10 teachers that reveal what's described. (Be sure to label and put a reasonable scale on both axes.)

a) No association between age and distance

b) Strong, positive, linear association between age and distance

c) Moderate, positive, non-linear association between age and distance

d) Moderate evidence that older teachers tend to live closer to school than younger teachers

e) Very little association between age and distance, except for a severe outlier

Major league baseball

10.CE.4 Suppose that you record the following information about 15 Major League Baseball games played next Saturday:

        A. Total number of runs scored in the game

        B. Whether or not the home team wins the game

        C. Time (in minutes) required to play the game

        D. Attendance (number of people) at the game

        E. Whether the game is played in the afternoon or evening

        F. Temperature when the game begins

a) Identify all *pairs* of variables for which it does make sense to calculate the correlation coefficient between the variables.

b) Identify all *pairs* of variables for which it does *not* make sense to calculate the correlation coefficient between the variables.

(*Hint*: There should be a total of 15 pairs of variables between your answers to parts a) and b).

Test-Taking Time*

10.CE.5 Problems 10.CE.5 through 10.CE.7 refer to data on test-taking times. The datafile **TestTimes** contains data on the time (in minutes) taken by 33 students to complete a multiple choice exam, and the students' scores (as a percentage out of 100) on the exam. Investigate whether the data provide evidence of an association between the time to take the test and the score achieved on the test.

    a) Produce a scatterplot, and describe the association between the variables.

    b) Report the value of the correlation coefficient between these variables.

    c) Conduct a simulation-based analysis of whether the correlation coefficient differs significantly from zero. Report the p-value, along with a screen capture of the simulated null distribution of correlation values.

    d) Conduct a *t*-test based on the correlation coefficient. Report the value of the test statistic and p-value.

    e) Summarize your conclusion.

10.CE.6 Reconsider the previous exercise and the data in **TestTimes**.

    a) Report the equation of the least squares line for reporting test score from time to take the test.

    b) Interpret the value of the slope coefficient.

    c) Interpret the value of the intercept coefficient. Does this make sense in this context?

    d) Report and interpret the value of r^2.

10.CE.7 Reconsider the previous two exercises and the data in **TestTimes**.

    a) Conduct a *t*-test of whether the slope coefficient differs significantly from zero. Report the value of the test statistic and p-value.

    b) Produce a 95% confidence interval for the population slope coefficient.

    c) Interpret this confidence interval.

    d) Are the test result and confidence interval consistent with each other? Explain.

Planets

10.CE.8 Problems 10.CE.8 and 10.CE.9 refer to data gathered on the planets. The following data on the eight planets in our solar system were obtained from Wikipedia:

	Mercury	Venus	Earth	Mars	Jupiter	Saturn	Uranus	Neptune
Mean distance from sun (in million km)	57.9	108.2	149.6	227.9	778.4	1426.7	2871.0	4498.3
Orbital period (in Earth years)	0.24	0.62	1.00	1.88	11.86	29.45	84.02	164.79

a) Describe the association between these variables as described in the graph on the scatterplot below. Comment on direction, strength, and form of the association.

b) The correlation coefficient between these variables equals 0.988. Would you conclude that the relationship between these variables is linear? Explain.

10.CE.9 Reconsider the previous exercise. The following graph displays the same data, with the least squares line superimposed:

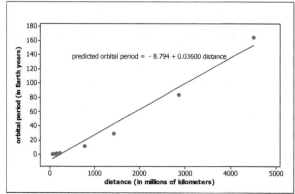

Would you feel comfortable using this line to make predictions? Explain.

Day hikes*

10.CE.10 Problems 10.CE.10 and 10.CE.11 reconsider the data on day hikes in San Luis Obispo county from Exercises 10.3.21 – 10.3.23. The data on hiking time (in minutes) and distance (in miles) can be found in the datafile **HikeDistances**.

a) Conduct a *t*-test to determine whether the data provide strong evidence of a positive association between hiking time and distance. Report the hypotheses, test statistic, and p-value. Summarize your conclusion.

b) Determine a 95% confidence interval for the population slope coefficient. Interpret this interval, including an interpretation of what slope means in this context.

10.CE.11 Repeat the previous exercise, replacing distance with elevation gain as the explanatory variable. The data on hiking time (in minutes) and elevation gain (in feet) can be found in the datafile **HikeElevations**.

Walking straight

10.CE.12 While a high school student in Texas, Andrea Axtell conducted a project in which she investigated how well blindfolded students can walk in a straight line. She recruited 30 subjects by randomly selecting students at her school. She put the subjects on the center hash mark of a football field's goal line, blindfolded them, and asked them to walk in a straight line toward the opposite goal line. She then recorded the yard line at which the subject crossed the sideline, so larger values indicate that the subject walked farther before veering off course. The data, including the heights of the 30 subjects, appear in the file **WalkingStraight**.

Analyze the data to investigate whether taller people tend to walk farther before veering off course. Write a paragraph summarizing your findings. Include an appropriate graph and numerical summaries. Also conduct inference (p-values, confidence intervals) using both simulation- and theory-based methods. Summarize your conclusions.

Late to class*

10.CE.13 Does having a professor show up late for class, influence students to be late for class? Or when professors are early, do students tend to show up early? In the spring of 2008, some statistics students randomly chose 31 different classes and recorded how many minutes a professor was early or late for class and the average number of minutes the students in the class were early or late. They wanted to see if there was a positive linear relationship between the professor's and the students' arrival times. We input these data into statistical software and found the correlation was 0.227. We then shuffled the data 1000 times and found the corresponding 1000 correlations. The simulated correlations are shown in the following graph.

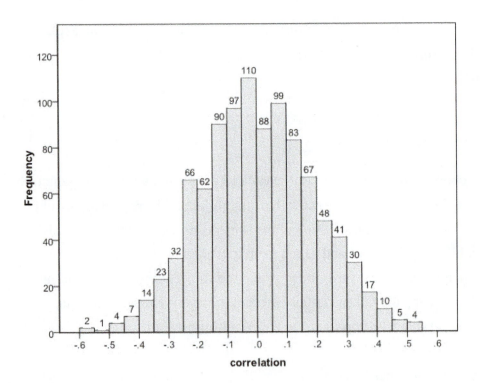

a) What is the null hypothesis in this study?
b) What is the alternative hypothesis in this study?
c) Which variable would it seem is the explanatory variable and which is the response?
d) Based on the histogram, what is the approximate p-value for this study?

e) What would be your complete conclusion for the one-sided study?

f) If this were a two-sided test, how would the alternative hypothesis change and what would be the new approximate p-value?

g) Even if you conclude that the association between these variables is statistically significant, can you legitimately conclude that the instructor's being late caused students to be late? Explain.

Least squares

10.CE.14 Consider a miniature data set:

x	0	1	2
y	1	0	2

Four copies of the scatterplot are shown below, each with a fitted line. Find the residuals for each line, and the sum of squares. Match the sum of squared residuals to the plots, choosing from 3/2, 2, 2, and 9/4.

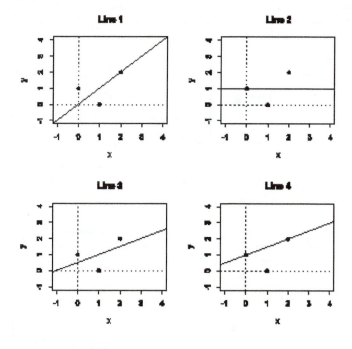

Noise in New York*

10.CE.15 Problems 10.CE.15 through 10.CE.20 refer to the data on decibel levels in New York. On July 20, 2012 *The New York Times* ran an article about the noise level at 33 different locations in New York City. The scatterplots below are all based on that study. Each shows noise level in decibels (dB) versus time of day (24 hour clock).

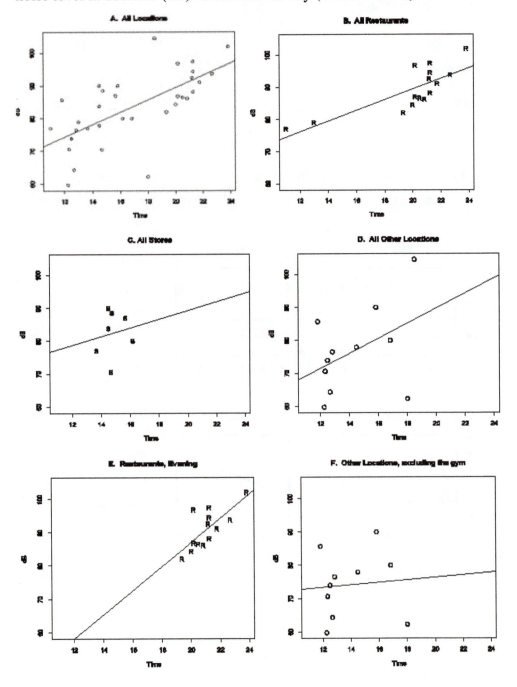

10.CE.15 Match each scatterplot with its coefficient and regression slope, choosing from (a) – (f):

	(a)	(b)	(c)	(d)	(e)	(f)
Correlations:	0.08	0.15	0.43	0.66	0.73	0.79.
Slopes:	0.38	1.31	2.29	1.83	3.62	1.64

10.CE.16 Match each scatterplot with its verbal description (a) – (e). You may need to use some descriptions more than once, and others may not be needed.

a) No big outliers, no influential points; strong linear relationship.

b) No big outliers, no influential points; no evidence of a linear relationship.

c) No big outliers, but two influential points. Removing these points will decrease the correlation and increase the fitted slope.

10.CE.17 Fill in the blanks based on the patterns in the data.

a) For restaurants, the correlation between average noise level and time becomes _____ (stronger, weaker) if you exclude the two lunch-time measurements.

b) For stores, the noise level _____ (is, is not) related to time of day.

c) For the "other" locations, the suggestion of a possible relationship between noise level and time of day becomes _____ (stronger, weaker) if you exclude the gym measured at 6:30 pm.

d) On average, restaurants are _____ (more, less) noisy than stores.

e) On average, decibel levels measured later in the day tend to be _____ (higher, lower) than those measured earlier in the day.

f) Time of day _____ (is, is not) confounded with type of location.

g) For _____ (restaurants, stores, other locations) the fitted noise level increases by more than 10 decibels during the three hours from 7 pm to 10 pm.

h) If you go by the fitted line for all restaurants, the predicted decibel level at 4 pm is closest to _____ (round to the nearest 10 dB).

10.CE. 18 Refer to description in 10.CE.15.

a) Why is it not possible to conclude from the data that decibel levels increase as it gets later in the day?

b) How could the design of the study be improved in order to allow better conclusions about the effect of time of day on noise level?

10.CE. 19 Refer to plots in 10.CE.15.

a) Refer to plot A, the scatterplot for all locations in the survey. Note that between 12 noon and midnight (24) the fitted decibel level goes from about 75 to about 95. Use this fact to estimate the decibel level at the y-intercept. (For comparison, a measurement taken in the Evergreens Cemetery was 60 dB.)

b) What time of day does the y-intercept correspond to? Explain why the fitted decibel level for time 0 cannot possibly be right.

10.CE. 20 Three histograms are shown below. Each shows simulated values of r, obtained by breaking the association on one of the scatterplots (A) – (C) in (10.CE.15). Which histogram is for each plot? How can you tell?

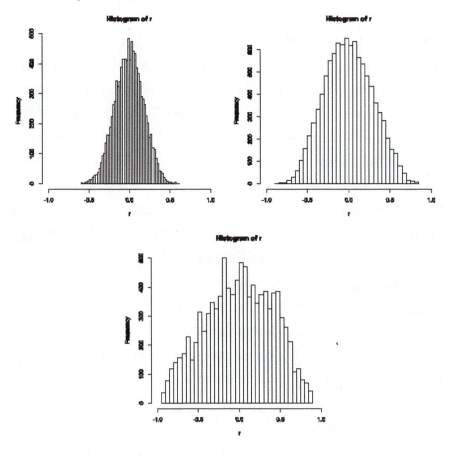

Coffee and height

10.CE.21 Suppose a friend of yours is telling you about a study reported in a popular magazine. The friend said that the study reported a correlation of zero between the amount of coffee a person drank as a child and his or her height as an adult. Your friend interprets this correlation to mean that the more coffee a child drinks, the shorter that person will be as an adult. In other words, drinking coffee stunts one's growth. Explain what is wrong with your friend's interpretation and explain what this type of correlation means.

US States*

10.CE.22 This question is based on the fifty US states. Each state is an observational unit, and the variables listed are measured separately for each of the 50 states. For each of the following pairs of variables, would you expect the correlation to be: strong and positive, moderate and positive, weak or very weak, moderate and negative, or strong and negative.

a)   2012 popular vote for Romney and 2012 popular vote for Obama
b)   Total 2012 popular vote for president and 2012 number of electoral votes
c)   Number of McDonald's restaurants and number of attorneys
d)   High school graduation rate and percentage of people over 25 with a four-year college degree
e)   Year of statehood and number of years as a state
f)   Year of statehood and longitude (east/west)
g)   Longitude and latitude (north/south)
h)   Latitude and average annual temperature

Investigation: Association between hand span and candy?

Is hand span a good predictor of how much candy you can grab? Using 45 college students as subjects, researchers set out to explore whether a linear relationship exists between hand span (cm) and the number of tootsie rolls each subject could grab.

Step 1: Ask a research question

1.   State the research question.

Step 2: Design a study and collect data

2. What are the observational/experimental units?

3. Is this a randomized experiment or an observational study? Explain how you know.

4. State the two variables measured on each unit.

5. Is there an explanatory/response relationship for these variables? Classify the variables in this study as categorical or quantitative.

6. State the null and alternative hypotheses to be investigated with this study.

Step 3: Explore the data

7. Use the data set **HandSpan** to calculate means, standard deviations and the correlation coefficient.

8. Create a scatterplot of the data with hand span on the x axis and number of tootsie rolls on the y axis.

9. Does there appear to be an association between hand span and number of tootsie rolls? Describe the direction, form and strength of the association: is it positive, negative, weak, strong, linear?

10. Are there any unusual observations?

11. Find the least squares regression equation that predicts number of tootsie rolls based on hand span.

12. Interpret the slope of the least squares regression equation in terms of hand span and predicted number of tootsie rolls grabbed.

Step 4: Draw inferences

13. Use the appropriate applet to construct a simulated null distribution using the slope of the least squares regression line as the statistic. Mark the observed least squares slope on this graph.

   a) Paste a screen shot or draw your null distribution below with the observed slope marked and the approximate p-value shaded in.

   b) Is the observed statistic out in the tail of this null distribution, or is it a fairly typical result?

   c) What is the p-value from your simulation? Based on this simulation analysis, would you conclude that the data provides *strong* evidence against the null hypothesis and conclude that there is a significant association between hand span and number of tootsie rolls grabbed? Explain your reasoning.

14. Are the validity conditions met to complete a theory-based *t*-test on these data? Explain. Whether the conditions are met or not, use a theory-based *t*-test to find a p-value. How does this p-value compare to the p-value your found using the simulation-based method?

15. Find a confidence interval for the slope that would describe the association between all hand spans and number of tootsie rolls grabbed. Is zero in this interval? Does this make sense based on your p-value from your test of significance?

Step 5: Formulate conclusions

16. Are you able to conclude that the association between hand span and number of tootsie rolls grabbed is causal? Explain.

17. What generalizations are you willing to make? Explain.

Step 6: Look back and ahead

18. Summarize the findings from the study. What worked well in this study design? What would you change? What are some follow-up research questions you would like to explore based on the findings from this study?

Research Article: Music preferences and delinquency in youth

Read "Early adolescent music preferences and minor deliquency." by ter Bogt, Keijsers, and Meeus et al. (2013), 131(2), 1-10 in the journal *Pediatrics*, and then answer the following questions.

Step 1. Ask a research question

1. In no more than 1-2 sentences state a broad research question (or related questions) that the researchers were trying to investigate in this study.

2. Identify at least two (cited) points of evidence the researchers cite as to why the research question(s) are important. Include the citation in your response.

Step 2. Design a study and collect data

3. How many different adolescents were in the study?

4. What age were the adolescents when the study started?

Step 3. Explore the data

5. Based on Table 1, at age 12, on average, what were the two most liked genres of music? What were their average values?

6. What were the two genres that were least liked at age 12? What were their average values?

7. Answer question #5 for the sample at age 16.

8. Answer question #6 for the sample at age 16.

Step 4. Draw inferences

Table 2 shows the results of fitting regression lines or quadratic regression lines (curves) to the musical preference scores over the four measurements (age 12, 14, 15, 16) in the study.

9. The chart pop musical preference regression line is given as Preference Score = 4.12 - 0.12*(Age-12), where "Age-12" is the age of the participant minus 12 (so, when the

participant is age 12 the value of the variable is 0. **Ignore the quadratic slope coefficient for this question.**

    a. Interpret the intercept.

    b. Interpret the slope

    c. Notice that the slope is significant ($p < 0.05$, as indicated by an asterisk). State the null and alternative hypothesis, indicate the strength of evidence provided by the p-value and provide an appropriate (in context) interpretation.

10. Answer the same questions as in #9 for the hip-hop genre using the regression line Preference Score $= 3.27 + 0.10*(Age-12)$

Table 3 shows the correlations between numerous variables in the study. For each of the following pairs of variables State the null and alternative hypothesis, indicate the strength of evidence provided by the p-value ($p > 0.05$ (no asterisk), $p < 0.05$ (*) or $p < 0.01$ (**) and provide an appropriate (in context) interpretation.

11. Chart pop preference and R&B preference at age 12 (correlation $= 0.44$**)

12. Classic preference and Hip-hop preference at age 12 (correlation $= -0.22$**)

13. Metal preference at age 12 and Delinquency at age 16 (correlation $= 0.29$**)

14. Chart pop preference at age 12 and Delinquency at age 16 (correlation $= -0.07$)

Step 5. Formulate conclusions

15. To which adolescents are you willing to generalize your conclusions from this study? Why?

16. Does the significant result in question #13 mean that liking the Metal genre at age 12 causes adolescents to be more delinquent at age 16?

Step 6. Look back and ahead

17. Identify at least two things that, if you were running the study and had a reasonable amount of financial resources, you would do differently. You must justify your answer.

APPENDIX: CALCULATION DETAILS

PRELIMINARIES: INTRODUCTION TO STATISTICAL INVESTIGATIONS

Notation

- x_i represents the value of the ith observation for a variable of interest

- n represents the sample size

- $\bar{x}$ represents the sample mean

- s represents the sample standard deviation

- Σ represents summation

A common measure of location is the (arithmetic) **mean**, sometimes called the average.

$$1. \quad \text{Sample } \textbf{mean} = \bar{x} = \frac{\Sigma x_i}{n} = \frac{x_1 + x_2 + \ldots + x_n}{n} = \frac{\text{sum of observations from the sample}}{n}$$

Example: Consider the following data on how much students paid for their last haircut (including tip) in dollars:

$$\$18, \$55, \$23, \$75, \$36.$$

$$x_1 = 18, x_2 = 55, x_3 = 23, x_4 = 75, x_5 = 36$$

$$\text{Sample mean} = \bar{x} = \frac{18 + 55 + 23 + 75 + 36}{5} = \$41.40.$$

People in this sample paid $41.40 for their last haircut, on average.

A common measure of data variability is the **standard deviation (SD).** The standard deviation can be loosely interpreted as the average deviation of the data values from their mean.

$$2. \quad \text{Sample } \textbf{SD} = s = \sqrt{\frac{\Sigma(x_i - \bar{x})^2}{n-1}} = \sqrt{\frac{(x_1 - \bar{x})^2 + (x_2 - \bar{x})^2 + \ldots + (x_n - \bar{x})^2}{n-1}}$$

$$= \sqrt{\frac{\text{sum of observed squared differences from the sample mean}}{n-1}}$$

Example: Recall that the observed sample mean haircut price, $\bar{x} = \$41.40$.

I	1	2	3	4	5	Sum
x_i	$x_1 = 18$	$x_2 = 55$	$x_3 = 23$	$x_4 = 75$	$x_5 = 36$	
$x_i - \bar{x}$	$18 - 41.40 =$ -23.4	$55 - 41.40 =$ 13.6	$23 - 41.40$ $= -18.4$	$75 - 41.40$ $= 33.6$	$36 - 41.40$ $= -5.4$	$\Sigma (x_i - \bar{x})$ $= 0$
$(x_i - \bar{x})^2$	$(-23.4)^2 =$ 547.56	$(13.6)^2 =$ 184.96	$(-18.4)^2 =$ 338.56	$(33.6)^2 =$ 1128.96	$(-5.4)^2 =$ 29.16	$\Sigma (x_i - \bar{x})^2$ $= 2229.2$

Thus, sample SD $= s = \sqrt{\dfrac{2229.2}{5-1}} = \sqrt{\dfrac{2229.2}{4}} = \sqrt{557.3} = \$23.61$. Thus, the typical variation for a haircut

cost from the sample mean ($41.4) was $23.61, for this sample of 5 people.

$(n=5)$
mean = 41.400
SD = 23.607

Other Measures of Location:

3. **Minimum** = smallest number in the (sorted) dataset

4. **Maximum** = largest number in the (sorted) dataset.

Example: Reconsider the data on how much each of 5 students paid for their last haircut (including tips) in dollars:

18, 55, 23, 75, 36

Sorted data:

18, 23, 36, 55, 75

Then, observed Minimum = \$18.00, and observed Maximum = \$75.00.

CHAPTER 1: SIGNIFICANCE: HOW STRONG IS THE EVIDENCE?

Notation

- π represents the underlying process probability of success or population proportion of successes

- π_0 represents the hypothesized value of the underlying process probability or population proportion, if the null hypothesis is true.

- $\hat{p}$ represents the sample proportion of "successes"

- n represents the sample size

- z represents the standardized statistic for a single proportion

1. Sample proportion $= \hat{p} = \dfrac{\text{number of successes in the sample}}{n}$

2. In general a standardized statistic is

$$\text{standardized statistic} = \frac{\text{statistic} - \text{mean of null distribution}}{\text{SD of null distribution}}$$

$$z = \frac{\text{sample proportion} - \text{mean of null distribution}}{\text{SD of null distribution}}$$

3. Let, the null hypothesis be $H_0: \pi = \pi_0$

Then, the theory-based formula for the standardized statistic, for a single proportion is given by

$$z = \frac{\hat{p} - \pi_0}{\sqrt{\dfrac{\pi_0(1 - \pi_0)}{n}}}$$

Example:

Suppose that in 32 attempts, the dolphin Buzz had pushed the correct button, 30 times. Does this provide evidence that Buzz was doing better than guessing which button to push?

Then, the

Null hypothesis, H_0: Buzz is just guessing, so his long-run probability of choosing the correct button is 0.5.

Alternative hypothesis, H_a: Buzz is doing better than guessing, so his long-run probability of choosing the correct button is greater than 0.5.

- Let, π represent Buzz's (unknown) long-run probability of pushing the correct button.

- Then, H_0: $\pi = 0.5$ versus H_a: $\pi > 0.50$

  where $\pi_0 = 0.5$.

- From the sample data, $n = 32$, and $\hat{p} = 30/32 = 0.9375$.

- Then, $z = \dfrac{0.9375 - 0.5}{\sqrt{\dfrac{0.5(1-0.5)}{32}}} = \dfrac{0.4375}{0.088} = 4.95$

- The observed proportion of successes (0.9375) that Buzz had is 4.95 SDs above the hypothesized proportion 0.5, which is what Buzz's long-run probability of success would have been if he had been guessing.

- Note: Using the standardized statistic for a single proportion along with the theory-based approach to find a p-value is valid if the sample size is large enough, that is, at least 10 "successes" and at least 10 "failures." In this example, the theory-based p-value here is less than 0.0001.

CHAPTER 2: GENERALIZATION: HOW BROADLY DO THE RESULTS APPLY?

Notation

- μ represents the population mean

- σ represents the population standard deviation (SD)

- n represents the sample size

- $\bar{x}$ represents the sample mean

- s represents the sample SD

- μ_0 represents the hypothesized value of the population mean, if the null hypothesis is true.

- t represents the standardized statistic for a mean

1) **Median:** Another common measure of location is the **median** divides the dataset into two equally sized pieces. When there is an *odd* number of observations, the median equals the middle value in a sorted dataset. When there is an even number of observations, we report the average of the two middle values in the sorted dataset as the median.

Another way to think about the median is that its location is the $(n+1)/2^{th}$ position in the sorted dataset.

Example: Consider the weights (lbs) of 20 male basketball:

160, 180, 185, 185, 190, 190, 194, 210, 220, 224, 225, 225, 230, 230, 235, 237, 240, 240, 245, 263

Then, observed Median = (224 + 225)/2 = 224.5lbs.

2) In general a standardized statistic is

$$\text{standardized statistic} = \frac{statistic - mean\ of\ null\ distribution}{SD\ of\ null\ distribution}$$

3) With quantitative data, let the null hypothesis be H_0: $\mu = \mu_0$.

Then, the standardized statistic for a single mean is given by

$$t = \frac{\bar{x} - \mu_0}{s/\sqrt{n}}$$

Example:

To test whether students at her school use sunscreens with an SPF different than 30, on average, a student surveyed a random sample of 48 students, who reported an average SPF of 35.29 and a SD of 17.19.

Then, the

Null hypothesis, H_0: $\mu = 30$

Alternative hypothesis, H_a: $\mu \neq 30$

where μ represents the average SPF of sunscreens used by all students at this school, and $\mu_0 = 30$.

- From the sample data, $n = 48$, and $\bar{x} = 35.29$ and $s = 17.19$.

- Then, $t = \dfrac{35.29 - 30}{17.19/\sqrt{48}} = \dfrac{5.29}{2.481} = 2.132$

- The observed value of sample average SPF (35.29) from the sample of 48 students is 2.132 SDs above the hypothesized average of 30, which is what the average SPF for sunscreens used by all students at this school is assumed to be by the null hypothesis.

- Note: Using the standardized statistic for a single mean along with the theory-based approach to find a p-value is valid if the population distribution is normally distributed (make sure the sample is roughly symmetric) or the sample size is large (at least 20 or quite large if the sample is more heavily skewed). In this example, the theory-based (two-sided) p-value equals 0.0383.

CHAPTER 3: ESTIMATION: HOW LARGE IS THE EFFECT?

A. Theory-based confidence intervals for a single proportion

Notation

- π represents the underlying process probability of success or population proportion of "successes"

- $\hat{p}$ represents the sample proportion of "successes"

- n represents the sample size

1. Sample proportion $= \hat{p} = \dfrac{\text{number of successes in the sample}}{n}$

2. In general, a confidence interval is given by: *statistic   margin of error*

   Alternatively we can write this as: *statistic   multiplier   SE*

3. **Theory-based confidence intervals for π**, the underlying process probability or population proportion, is

   given by $\hat{p} \pm$ multiplier $\times \sqrt{\dfrac{\hat{p}(1-\hat{p})}{n}}$, where SE $= \sqrt{\dfrac{\hat{p}(1-\hat{p})}{n}}$.

4. For 90%, 95%, and 99% confidence levels specifically, the intervals are as given below:

Confidence level	Margin of error	Confidence interval
90%	$1.645 \times \sqrt{\dfrac{\hat{p}(1-\hat{p})}{n}}$	$\hat{p} \pm 1.645 \times \sqrt{\dfrac{\hat{p}(1-\hat{p})}{n}}$
95%	$1.96 \times \sqrt{\dfrac{\hat{p}(1-\hat{p})}{n}}$	$\hat{p} \pm 1.96 \times \sqrt{\dfrac{\hat{p}(1-\hat{p})}{n}}$
99%	$2.576 \times \sqrt{\dfrac{\hat{p}(1-\hat{p})}{n}}$	$\hat{p} \pm 2.576 \times \sqrt{\dfrac{\hat{p}(1-\hat{p})}{n}}$

- For a different confidence level, change the multiplier used in the margin of error formula. The multiplier comes from the standard normal distribution.

- These theory-based confidence intervals are valid when the sample size is large enough, that is, there are at least 10 "successes" and at least 10 "failures" in the sample data.

Example:

In a July 2012 Gallup survey of 1,014 randomly selected U.S. adults, 5% said that they consider themselves to be vegetarians. Find a 95% confidence interval of the proportion of all U.S. adults who consider themselves to be vegetarians.

- Let, π represent the proportion of all U.S. adults who consider themselves to be vegetarians.

- From the sample data, $n = 1014$, and $\hat{p} = 0.05$. Thus, there are $(1014) \times (0.05) = 50.7 \rightarrow 51$ "successes" and $(1014) \times (1\text{-}0.05) = 963.3 \rightarrow 963$ "failures" in the sample. Both these values exceed 10.

- Then, the theory-based 95% confidence interval is given by

$$\hat{p} \pm 1.96 \sqrt{\frac{\hat{p}(1-\hat{p})}{n}} = 0.05 \pm 1.96 \sqrt{\frac{0.05(1-0.05)}{1014}} = 0.05 \pm 1.96\,(0.0068) = 0.05 \pm 0.0134 = (0.037, 0.063)$$

- So, we are 95% confident that the proportion of all U.S. adults who consider themselves to be vegetarians is somewhere between 0.037 and 0.063.

B. Theory-based confidence intervals for a single mean

Notation

- μ represents the population mean

- n represents the sample size

- $\bar{x}$ represents the sample mean

- s represents the sample SD

1. **A theory-based confidence interval for μ**, the population mean or underlying process mean is given by

$$\bar{x} \pm \text{multiplier} \times \frac{s}{\sqrt{n}}$$

- For 95% confidence, the multiplier will be roughly in the ballpark of 2.

- For a different confidence level, change the multiplier used in the margin of error formula; the multiplier comes from the t distribution with $n - 1$ degrees of freedom.

- These theory-based confidence intervals are valid when the sample size is large enough; that is, if the population distribution is normally distributed (make sure the sample is roughly symmetric) or the sample size is large (at least 20 or quite large if the sample is more heavily skewed).

Example:

A student conducts a survey of a random sample of 48 students at her school, asking each participant the SPF of the sunscreen they use. She finds the average SPF in the sample to be 35.29, with a standard deviation of 17.19. Find a 95% confidence interval for the average SPF of sunscreens used by all students in this school.

Let μ represent the average SPF of sunscreens used by all students at this school.

- From the sample data, $n = 48$, and $\bar{x} = 35.29$ and $s = 17.19$.

  Then, the 2SD confidence interval for the average SPF of sunscreens used by all students in this school is

  given by: $\bar{x} \pm \text{multiplier} \times \frac{s}{\sqrt{n}} = 35.29 \pm 2 \times \frac{17.19}{\sqrt{40}} = 35.29 \pm 4.96 = (30.33, 40.25)$

1119

- We are 95% confident that the average SPF of sunscreens used by all students at this school is between 30.33 and 40.25.

- Note: Using theory-based confidence intervals along with the theory-based approach to find a confidence interval is valid if the population distribution is normally distributed (make sure the sample is roughly symmetric) or the sample size is large (at least 20 or quite large if the sample is more heavily skewed). In this case, a sample size of 48 seems reasonably large. The 95% one-sample t-interval is (30.30, 40.28).

CHAPTER 5: COMPARING TWO PROPORTIONS[1]

Notation

- π_1 represents the underlying probability for the 1st process or population proportion for 1st population (of "successes")

- π_2 represents the underlying probability for the 2nd process or population proportion for 2nd population (of "successes")

- $\hat{p}_1$ represents the sample proportion of successes in the 1st sample

- $\hat{p}_2$ represents the sample proportion of successes in the 2nd sample

- $\hat{p}_1$ and $\hat{p}_2$ are also called the conditional proportions of successes

- $\hat{p}$ represents the overall proportion of successes across the two samples $= \dfrac{n_1\hat{p}_1 + n_1\hat{p}_2}{n_1 + n_2}$

- n_1 represents the sample size of 1st sample

- n_2 represents the sample size of 2nd sample

- z represents the standardized statistic for a difference in sample proportions

---

In general a standardized statistic is

$$\text{standardized statistic} = \frac{statistic - mean\ of\ null\ distribution}{SD\ of\ null\ distribution}$$

Let, H_0: $\pi_1 - \pi_2 = 0$

- Then, the corresponding **standardized z- statistic** is given by $z = \dfrac{\hat{p}_1 - \hat{p}_2 - 0}{\sqrt{\hat{p}(1 - \hat{p})\left(\frac{1}{n_1} + \frac{1}{n_2}\right)}}$

- Using the standardized z- statistic along with the theory-based approach to find a p-value is valid if the sample size is large enough, that is, there are at least 10 observations in each of the four cells, when the data are written out as a 2×2 table.

---

[1] Note: There are no calculation details for Chapter 4

And, the **theory-based confidence interval for** $\pi_1 - \pi_2$ is given by:

$$\hat{p}_1 - \hat{p}_2 \pm multiplier \times \sqrt{\frac{\hat{p}_1\left(1 - \hat{p}_1\right)}{n_1} + \frac{\hat{p}_2\left(1 - \hat{p}_2\right)}{n_2}}$$

- The *multiplier* changes depending on the confidence level. For example, for a 95% confidence interval the *multiplier* is 1.96.
- The theory-based confidence interval is valid provided the sample sizes are large enough. That is, there are at least 10 observations in each of the four cells, when the data are written out as a 2×2 table.

Example: Let $\hat{p}_1 = 0.548$, $\hat{p}_2 = 0.451$, $n_1 = 3602$, $n_2 = 565$.

Then, $\hat{p} = (3602*0.548 + 565*0.451)/(3602+565) = 0.535$.

- The standardized z- statistic for a difference in proportions can be computed as

$$z = \frac{\hat{p}_1 - \hat{p}_2}{\sqrt{\hat{p}(1 - \hat{p})\left(\frac{1}{n_1} + \frac{1}{n_2}\right)}} = \frac{0.548 - 0.451}{\sqrt{0.535(1 - 0.535)\left(\frac{1}{3602} + \frac{1}{565}\right)}} = 4.30$$

- Thus, the observed difference in the sample proportions of 0.097 is 4.3 SDs above the hypothesized difference of 0. Recall that the null hypothesis says $\pi_1 - \pi_2$ is equal to 0.

- Because the sample sizes are large enough, the theory-based 95% confidence interval for $\pi_1 - \pi_2$ can be computed as

$$0.548 - 0.451 \pm 1.96 \sqrt{\frac{0.548\,(1 - 0.548)}{3602} + \frac{0.451(1 - 0.451)}{565}} = (0.053, 0.141)$$

- We are 95% confident that the π_1 is higher than π_2 by somewhere between 0.053 and 0.141.

☑ Test of significance

H_0: $\pi_1 - \pi_2$ = [0]

H_a: $\pi_1 - \pi_2$ [≠] [0]

[Calculate]

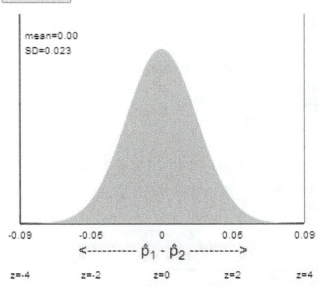

standardized statistic [z = 4.30]

p-value [0.0000]

☑ Confidence interval

confidence level [95] % [Calculate CI]

(0.0529, 0.1411)

CHAPTER 6: COMPARING TWO MEANS

A. Exploring quantitative data: quartiles, inter-quartile range, five number summary

- The value for which 25% of the data lie below that value is called the *lower quartile* (or 25th percentile).

- Similarly, the value for which 25% of the data lie above that value is called the *upper quartile* (or 75th percentile).

- Quartiles can be calculated by determining the median of the values (strictly) above/below the location of the location of the sample median. (Remember to sort the data first.)

- The difference between the quartiles is called the *inter-quartile range* (IQR), another measure of variability along with standard deviation.

- The *five-number summary* for the distribution of a quantitative variable consists of the minimum, lower quartile, median, upper quartile, and maximum.

Example: Weights (lbs) of 20 male basketball players (contd.):

160, 180, 185, 185, (190)(190) 194, 210, 220, 224| 225, 225, 230, 230,(235)(237) 240, 240, 245, 263

- Note that the data has been sorted from smallest to largest values.

- $n = 20$

- The observed median is $(224 + 225) = 224.5$lbs.

- Then, the observed lower quartile = $(190+190)/2 = 190$lbs, and observed upper quartile = $(235+237)/2 = 236$lbs.

- The inter-quartile range is IQR = $236 - 190 = 46$lbs.

  160, 180, 185, 185, (190)(190) 194, 210, 220, 224| 225, 225, 230, 230,(235)(237) 240, 240, 245, 263

This tells us the "length" of the middle 50% of the (sorted) data values.

- Thus, the five-number summary is:

  Min = 160lbs

  Lower quartile = 190lbs

  Median = 224.5lbs

  Upper quartile = 236lbs

  Maximum = 263lbs

B. Calculations involved in inference for comparing two means

Notation

- μ_1 represents the "population" average for population 1

- μ_2 represents the "population" average for population 2

- $\bar{x}_1$ represents the sample mean for sample 1

- $\bar{x}_2$ represents the sample mean for sample 2

- s_1 represents the sample SD for sample 1

- s_2 represents the sample SD for sample 2

- n_1 represents the sample size for sample 1

- n_2 represents the sample size for sample 2

- t represents the standardized statistic for a difference in means

In general a standardized statistic is

$$\text{standardized statistic} = \frac{statistic - mean\ of\ null\ distribution}{SD\ of\ null\ distribution}$$

-

Let, $H_0: \mu_1 - \mu_2 = 0$

The formula used to calculate the **standardized t-statistic** to compare two groups on a quantitative response is

$$t = \frac{\bar{x}_1 - \bar{x}_2 - 0}{\sqrt{\frac{s_1^2}{n_1} + \frac{s_2^2}{n_2}}}$$

A **theory-based confidence interval for $\mu_1 - \mu_2$**, can be written as follows:

$$(\bar{x}_1 - \bar{x}_2) \pm \text{multiplier} \times \sqrt{\frac{s_1^2}{n_1} + \frac{s_2^2}{n_2}}$$

- For 95% confidence, the multiplier will be roughly in the ballpark of 2.

- For a different confidence level, change the multiplier used in the margin of error formula; the multiplier comes from the t distribution with the smaller of $n_1 - 1$ and $n_2 - 1$ degrees of freedom.

- The theory-based p-value and confidence interval are valid when both population distributions are normally distributed (make sure that both samples' distributions are roughly symmetric) or the sample sizes are large (at least 20 or quite large if the sample distributions are more heavily skewed).

Example: Consider data on BMI of women participating in a randomized experiment of lifestyle change programs.

Group	Sample size	Sample mean	Sample SD
Intervention	$n_{intervention} = 60$	$\bar{x}_{intervention} = 30 \text{ kg/m}^2$	$s_{intervention} = 2.1 \text{ kg/m}^2$
Control	$n_{control} = 60$	$\bar{x}_{control} = 34 \text{ kg/m}^2$	$s_{intervention} = 2.4 \text{ kg/m}^2$

- We can define our parameters of interest to be,
  $\mu_{intervention}$ = Average BMI after 2 years of being enrolled in an individualized lifestyle change program, for all (obese) women like those in the study, and
  $\mu_{control}$ = Average BMI after 2 years of being enrolled in a "one size fits all" lifestyle change program, for all (obese) women like those in the study

- And using the symbols μ_1 and μ_2 we can restate our hypotheses to be,
  Null hypothesis, H_0: $\mu_{intervention} - \mu_{control} = 0$
  Alternative hypothesis, H_a: $\mu_{intervention} - \mu_{control} \neq 0$

- The **standardized t-statistic** can be computed as follows:

$$t = \frac{30 - 34}{\sqrt{\frac{2.1^2}{60} + \frac{2.4^2}{60}}} = \frac{-4}{\sqrt{0.0735 + 0.096}} = \frac{-4}{0.4117} = -9.72$$

- This tells us that the observed difference in sample means (-4) is 9.72 SDs below the hypothesized difference in population means of 0.

- Using the data from the study, we can see that each sample had 60 observational units, which is larger than 20. Thus, the **2SD confidence interval for** $\mu_{Intervention} - \mu_{control}$, is:

$$(30 - 34) \pm 2 \sqrt{\frac{2.1^2}{60} + \frac{2.4^2}{60}} = -4 \pm 2\,(0.4117) = -4 \pm 0.8234 = (-4.8234, -3.1776)$$

- Thus, we are 95% confident that enrollment in individualized lifestyle change programs (compared to "one size fits all" programs) decreases the average BMI of women like the obese Italian women in our study, by somewhere between 3.18 kg/m^2 to 4.82 kg/m^2.

☑ Test of significance

$H_0: \mu_1 - \mu_2 = $ 0

$H_a: \mu_1 - \mu_2 \neq $ 0

[Calculate]

mean=0.00
SD=0.412

standardized statistic t = -9.72   df = 115.96
p-value   0.0000

☑ Confidence interval

confidence level 95 % [Calculate CI]

(-4.8155, -3.1845)

CHAPTER 7: PAIRED DATA: ONE QUANTITATIVE VARIABLE

Notation

- μ_d represents the mean difference for the "population"
- $\overline{x}_d$ represents the sample mean difference,
- s_d represents the SD of the sample differences, and
- n represents the sample size (number of differences).
- t represents the standardized statistic for a mean difference

Let, H_0: $\mu_d = 0$.

The standardized *t*-statistic for paired data on a quantitative response is $t = \dfrac{\overline{x}_d - 0}{s_d / \sqrt{n}}$.

We can also find a **theory-based confidence interval for μ_d**, as follows:

$$\overline{x}_d \pm multiplier \times \dfrac{s_d}{\sqrt{n}}.$$

- For 95% confidence, the multiplier will be roughly in the ballpark of 2.

- For a different confidence level, change the multiplier used in the margin of error formula; the multiplier comes from the *t* distribution with $n - 1$ degrees of freedom.

- Using the standardized *t*-statistic along with the theory-based approach to find a p-value, and/or finding the theory-based confidence interval is valid if (i) the sample size (that is, the number of pairs observed) is large enough (at least 20) OR (ii) the sample of differences comes from a normally distributed population.

Example: Suppose that we have the following data on the estimated versus actual number of calories burned, by a sample of 80 men.

	Sample size	Mean of sample differences (kcal)	SD of sample differences (kcal)
Difference = Estimated – Actual	80	$\overline{x}_d = -435$	$s_d = 527.7$

- Let, μ_d represent the average difference in estimated and actual energy intake, by all men like those in the study.

- Then, H_0: $\mu_d = 0$, versus H_a: $\mu_d \neq 0$.

- The standardized t-statistic can be found as
$$t = \frac{435}{527.7 / \sqrt{80}} = \frac{435}{58.999} = -7.37$$

- Thus, the observed average difference between estimated and actual energy intake, 435kcal, is 7.37 SDs below 0.

- Using the data from the study, because there are 80 observational units (that is, 80 pairs of responses), the 2SD confidence interval for μ_d, is:

$$(-435) \pm 2 \, \frac{393.5}{\sqrt{80}} = 129 \pm 2 \, (43.99) = -435 \pm 67.99 = (-523, -347) kcal$$

- Thus, we are 95% confident that, on average, men such as the ones in this study, underestimate their energy intake by somewhere between 347kcal to 523kcal.

☑ Test of significance

H_0: μ = [0]

H_a: μ [≠] [0]

[Calculate]

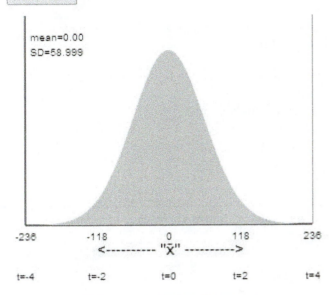

mean=0.00
SD=58.999

-236   -118   0   118   236

<---------- "x̄" ---------->

t=-4   t=-2   t=0   t=2   t=4

standardized statistic [t = -7.37]   df = 79

p-value   [0.0000]

☑ Confidence interval

confidence level [95] % [Calculate CI]

(-552.4390, -317.5610)

CHAPTER 8: COMPARING MORE THAN TWO PROPORTIONS

Notation

- π_i represents the underlying probability of success for the ith process or population proportion of success for i^{th} population.

- $\hat{p}_i$ represents the sample proportion of successes in i^{th} sample

- n_i represents the sample size of ith sample

- $\hat{p}$ represents the overall proportion of successes in the entire study

- χ^2 represents the chi-square statistic

- Σ represents summation

The **chi-square statistic for comparing multiple groups on a binary response variable** is given by

$$\chi^2 = \sum \left(\frac{(\hat{p}_i - \hat{p})}{\sqrt{\frac{\hat{p}(1-\hat{p})}{n}}} \right)^2$$

Example: Consider the following data from Chapter 8

observed counts	Real acupuncture	Sham acupuncture	Non-acupuncture	Total
Substantial reduction in pain	184	171	106	461
Not a substantial reduction in pain	203	216	282	701
Total	387	387	388	1162

- The hypotheses can be written as

H_0: There is no association between type of treatment and whether or not person experiences substantial reduction in pain,

H_a: There is an association between type of treatment and whether or not person experiences substantial reduction in pain.

- $\hat{p}_1$ = sample proportion of subjects experiencing substantial reduction in pain among those who received the real acupuncture = 184/387 = 0.495

- Similarly, $\hat{p}_2$ = 171/387 = 0.442, and $\hat{p}_3$ = 106/388 = 0.273

- $n_1 = 387$, $n_2 = 387$, $n_3 = 388$

- $\hat{p}$ = overall proportion of successes in the entire study = $(184 + 171 + 106)/(1162) = 0.397$

- Then, the observed value of the chi-square statistic can be calculated as

$$\chi^2_{observed} = \left[\left(\frac{0.475 - 0.397}{\sqrt{.397(1-.397)/387}} \right)^2 + \left(\frac{0.442 - 0.397}{\sqrt{.397(1-.397)/387}} \right)^2 + \left(\frac{0.273 - 0.397}{\sqrt{.397(1-.397)/388}} \right)^2 = 38.03 \right]$$

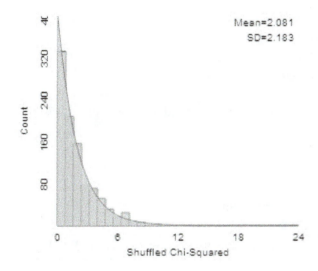

Mean=2.081
SD=2.183

Count Samples [Greater Than ▼] [38.05]

[Count]

☑ Overlay Chi-squared distribution
theory-based p-value=0.0000

Another form of the chi-square statistic:

The **chi-square statistic for comparing multiple groups on a categorical response variable** can also be written

out as

$$\chi^2 = \sum \frac{(observed_i - expected_i)^2}{expected_i}$$

where

- $observed_i$ represents the observed number of observational units in the ith cell (O_i)

- *expected$_i$* represents the number of observational units we would expect in the ith cell if the null hypothesis were true, and there were no association between the explanatory variable and the response variable; these *expected$_i$* are often referred to as the expected cell counts. (E_i)
- Σ represents summation

To calculate the expected counts, we determine the overall proportion in each response category, and then apply this proportion to each explanatory variable group. One way of implementing this is by calculating the expected cell counts (E_i) using the following formula

$$E_i = \frac{(row\ total) \times (column\ total)}{total\ number\ of\ observational\ units\ in\ the\ study}$$

where the row and column totals correspond to the row and column in which the ith cell appears.

Note: Using the chi-square statistic along with the theory-based approach to find a p-value is valid provided the sample sizes are large enough. That is, there are at least 10 observations in each of the cells, when the data are written out as a two-way table.

Example: Let us use the data from Chapter 8 again, where the observed counts are as follows

Observed counts (O_i)	Real acupuncture	Sham acupuncture	Non-acupuncture	Total
Substantial reduction in pain	184	171	106	461
Not a substantial reduction in pain	203	216	282	701
Total	387	387	388	1162

Then, **expected cell counts, E_i,** can be calculated as

Expected counts (E_i)	Real acupuncture	Sham acupuncture	Non-acupuncture	Total
Substantial reduction in pain	= (461 × 387)/1162 = 153.5	= (461 × 387)/1162 = 153.5	= (461 × 388)/1162 = 153.9	461
Not a substantial reduction in pain	= (701 × 387) /1162 = 233.5	= (701 × 387)/1162 = 233.5	= (701 × 388)/1162 = 233.1	701
Total	387	387	388	1162

And, the **observed value of the chi-square statistic** is then

$$\chi^2 = \overset{\text{Total number of cells in the two-way table}}{\underset{i=1}{\sum}} \frac{(O_i - E_i)^2}{E_i} = \frac{(184 - 153.5)^2}{153.5} + \frac{(171 - 153.5)^2}{153.5} + \frac{(106 - 153.9)^2}{153.9}$$

$$+ \frac{(203 - 233.5)^2}{233.5} + \frac{(216 - 233.5)^2}{233.5} + \frac{(282 - 233.1)^2}{233.1}$$

$$= 6.05 + 1.99 + 14.92 + 3.98 + 1.31 + 9.82 = 38.08$$

Each calculated value of $\dfrac{(O_i - E_i)^2}{E_i}$ is called a chi-square cell contribution. For example, the chi-square cell contribution for the cell corresponding to "real acupuncture" and "substantial reduction in pain" is given by

$$\frac{(184 - 153.5)^2}{153.5} = 6.05.$$

CHAPTER 9: COMPARING MORE THAN TWO MEANS

Notation

- μ_i represents the "population" average for the ith population

- $\overline{x}_i$ represents the sample mean for the ith sample

- s_i represents the sample standard deviation (SD) for the ith sample

- n_i represents the sample size for the ith sample

- $\overline{x}$ represents the overall sample mean

- I = number of samples/groups being compared

- $N = n_1 + n_2 + \ldots + n_I$

- F represents the F-statistic

- Σ represents summation

- The hypotheses can be written as

H_0: $\mu_1 = \mu_2 = \ldots = \mu_I$, versus H_a: At least one of the μ_i is different

Or as,

H_0: There is no association between the two variables, versus

H_a: There is an association between the two variables

Then, the **F-statistic for comparing more than two groups on a quantitative response** is given by

$$F = \frac{between - group\ variability}{within - group\ variability} = \frac{\dfrac{\sum\limits_{i=1}^{I} n_i(\overline{x}_i - \overline{x})^2}{I-1}}{\dfrac{\sum\limits_{i=1}^{I}(n_i - 1)s_i^2}{N-I}}$$

Example: Consider the following data from Chapter 9 on the comprehension scores of an ambiguous prose passage. Using this notation:

	Sample size	Sample mean	Sample SD
No picture	$n_1 = 19$	$\overline{x}_1 = 3.37$	$s_1 = 1.26$
Picture shown before	$n_2 = 19$	$\overline{x}_2 = 4.95$	$s_2 = 1.31$
Picture shown after	$n_3 = 19$	$\overline{x}_3 = 3.21$	$s_3 = 1.40$
Overall	$N = 57$	$\overline{x} = 3.84$	

and $I = 3$.

To find the observed value of the F-statistic, let us first calculate the numerator:

- *between-group variability =*

$$\frac{\sum n_i (\overline{x}_i - \overline{x})^2}{I - 1} = \frac{19\,(3.37 - 3.84)^2 + 19\,(4.95 - 3.84)^2 + 19\,(3.21 - 3.84)^2}{3 - 1} = \frac{4.20 + 23.41 + 7.54}{2}$$

$$= \frac{35.15}{2} = 17.57$$

Next we calculate the denominator:

- *within-group variability =*

$$\frac{\sum (n_i - 1)s_i^2}{N - I} = \frac{(19 - 1)(1.26)^2 + (19 - 1)(1.31)^2 + (19 - 1)(1.4)^2}{57 - 3} = \frac{28.58 + 30.89 + 35.28}{54} =$$

$$\frac{94.75}{54} = 1.75$$

- The denominator ("within-group variability") is also called the *Mean Square Error*. In this study, Mean Square Error = 1.75.

- Thus, the observed value of the F-statistic = 17.57 /1.75 = 10.02.

Count Samples [Greater Than ▾] [10.01|]

[Count]

☑ Overlay *F* distribution

theory-based p-value=0.0002

Note: Using the *F*-statistic along with the theory-based approach to find a p-value is valid as long as

(i) There are at least 20 observations in each sample,

OR, the samples each come from normally distributed populations.

(ii) The variability (SD) in each of the populations is the same. (The ratio of the largest sample standard deviation to the smallest is not larger than two.)

CHAPTER 10: TWO QUANTITATIVE VARIABLES

Notation

- ρ ('rho') represents the population correlation coefficient

- r represents the sample correlation coefficient

- β ('beta') represents the population slope

- b represents the sample slope

- a represents the sample intercept

- x_i represents the observed value of the explanatory variable for the ith observational unit

- y_i represents the observed value of the response variable for the ith observational unit

- n represents the sample size (number of x, y pairs)

- $\bar{x}$ represents the sample mean of the observed values of the explanatory variable

- $\bar{y}$ represents the sample mean of the observed values of the response variable

- s_x represents the sample SD of the observed values of the explanatory variable

- s_y represents the sample SD of the observed values of the response variable

- SE_b is the standard error of the sample slope.

- t represents the standardized t-statistic

- Σ represents summation

The **sample correlation coefficient** for data on a quantitative explanatory variable and a quantitative response variable, is given by

$$r = \frac{1}{n-1}\sum \frac{(x_i - \bar{x})}{s_x}\frac{(y_i - \bar{y})}{s_y}$$

The **least squares regression line** for data on a quantitative explanatory variable and a quantitative response is given by

$\hat{y} = a + bx$

where, the **sample slope** is given by

$$b = r \frac{s_y}{s_x}.$$

And, the **sample intercept** is given by

$$a = \bar{y} - b\bar{x}$$

Let, $H_0: \rho = 0$ versus $H_a: \rho \neq 0$

Then, the **corresponding statistic for the hypotheses about the population correlation coefficient (ρ)** is given as

$$t = \frac{r}{\sqrt{\frac{1 - r^2}{n - 2}}}$$

The theory-based confidence interval for the slope, r, is valid if both the explanatory variable and the response variable follow a normal distribution in the population.

Let, $H_0: \beta = 0$ versus $H_a: \beta \neq 0$

Then, the **corresponding statistic for the hypotheses about the population slope (β)** is given as

$$t = \frac{b}{SE_b}$$, where SE_b can be read off applet or software output.

And, an **approximate theory-based 95% confidence interval for β** is given by

$$b \quad 2SE_b$$

The theory-based confidence interval for the slope, β, is valid if
(i) The values of the response variable at each possible value of the explanatory variable have a normal distribution in the population from which the sample was drawn.
(ii) Each of these normal distributions has the same SD.

Example: Let us use the data on height (inches) and handspan (cm) for a sample of 10 college students.

	Mean	SD

Height (inches), y_i	64	67	65	72	71	70	66	62	73	65	67.50	3.75
Handspan (cm), x_i	17	21	20.3	26	24	22	21	19	20	19	20.93	2.59

Then,

- $\bar{x}$ = 20.93cm
- $\bar{y}$ = 67.50 inches
- s_x = 2.59 cm
- s_y = 3.75 inches
- n = 10

The observed value of the **sample correlation coefficient** can be calculated to be

$$r = \frac{1}{(10-1)(2.59)(3.75)} [(17-20.93)(64-67.5) + (21-20.93)(67-67.5) + (20.3-20.93)(65-67.5) \ldots + (19-20.$$

To find the least squares regression line, we can calculate the sample slope and sample intercept as

$$b = (0.706)\frac{(3.75)}{2.59} = 1.02$$

$$a = 67.5 - (1.02)(20.93) = 46.12$$

Thus, the calculated least squares regression line for the sample data is found to be

$$\widehat{height} = 46.12 + 1.02\, handspan$$

The standardized t-statistic for H_0: $\rho = 0$ versus H_a: $\rho \neq 0$, can be calculated to be

$$t =$$

$$\frac{0.706}{\sqrt{\dfrac{1-0.706^2}{10-2}}} = \frac{0.706}{\sqrt{\dfrac{0.4959}{8}}} = \frac{0.706}{\sqrt{0.062}} = \frac{0.706}{0.25} = 2.82$$

Thus, the observed (sample) correlation coefficient 0.706 between height and handspan is about 2.82 SDs above the hypothesized value 0 of the population correlation coefficient, the value ρ would be if the null hypothesis were true.

Here is statistical software output with b, SE_b, and the corresponding value of the standardized t-statistic circled.

```
Predictor    Coef   SE Coef     T       P
Constant    46.117    7.632    6.04   0.000
Handspan    (1.0217) (0.3621) (2.82)  0.022
```

Notice that the standardized t-statistic for H_0: $\beta = 0$ versus H_a: $\beta \neq 0$, can be calculated to be

$$t = \frac{b}{SE_b} = \frac{1.02}{0.36} = 2.82$$

And, the **2SD 95% confidence interval for β** can be calculated to be

$$1.02 \pm 2 \, (0.36) = 1.02 \pm 0.72 = (0.3, 1.74)$$

We are 95% confident that the increase in average height associated with an increase of 1cm in hand span is somewhere between 0.30 inches and 1.74 inches for this population.

Appendix: Stratified and Cluster Samples

There are several ways to get random samples from a population. Simple random sampling (SRS) is often used, but two other methods are also common: stratified random sampling and cluster sampling. This appendix describes each of these two methods in turn:

> How the method works,
>
> An example, and
>
> Why the method is used.

We then conclude with a brief comparison. We start with stratified sampling.

Stratified random samples

1. Divide or classify your entire population into non-overlapping groups (called strata).

2. Choose a sample size for each stratum; then choose a simple random sample from each one.

Example A.1. Medicare Fraud

Auditing claims from health care providers for payment by Medicare is expensive. It takes the time of a highly trained professional to evaluate the paperwork and decide whether a claim is justified. Because of the high cost, auditors rely on random samples rather than reviewing all existing claims. They evaluate the claims in the sample, then apply the results to the whole population of claims from the provider. Here is a summary from an actual case. (The numbers are real; the names are not.)

TrueMed, an auditing firm, demanded summary information for all claims submitted by Family Home Health during 2009. In all, there were 11,059 claims. TrueMed classified these into three strata according to the amount paid:

			Dollar Amount of claim	
	Dollar			
	Range of	Number		
Stratum	Claim	of Claims	Average	SD
1	< $89	9646	$58.41	$18.08
2	$89 - $400	1116	$71.36	$87.55

3	> $400	297	$646.83	$237.30

Several points are worth noting:

1. Every claim belongs to a stratum, and no claim belongs to more than one. (This is a requirement for every stratified sampling plan.)

2. For this particular plan, the strata have very different sizes. Stratum 1 is almost nine times as large as Stratum 2, and more than thirty times as large as Stratum 3.

3. The strata are chosen so that
   - *within* each stratum, claim amounts are similar, and
   - *between* strata, amounts are different.

   You can see this by comparing the claim averages of the three strata, and noticing that the differences between averages are large compared to the standard deviations of the claim amounts.

4. In the actual case, TrueMed took a simple random sample of size 30 from each stratum, for a total of 90 claims.

> **Think about it:** Suppose that instead of stratifying, TrueMed had taken one simple random sample of size 90 from the population as a whole. Based on the numbers of claims in the strata, roughly how many claims from Stratum 1 would you expect in a SRS of size 90? How many from Stratum 3?

It's likely that the SRS would contain only about 2 or 3 claims from Stratum 3, whereas TrueMed's stratified sample has 30 claims.

> **Think about it:** Why does it make sense to "over-represent" Stratum 3?

Stratum 3 contains the largest claim amounts, with an average that is more than ten times as large as the average for Stratum 1. These large claims are more important to the audit because they contribute more to the overall total, so it makes sense to ensure that we have many claims from that stratum, rather than just 2 or 3.

Stratum 3 also shows the largest variability, with SD more than 10 times as large as the SD for Stratum 1. This means that it takes more than 30 observational units from Stratum 3 to learn as much (similar precision) as you get from just 3 observational units from Stratum 1. So, although TrueMed used equal sample sizes of 30, 30, and 30, there may be other ways to allocate 90 units among the three strata depending on the goal of the analysis.

Advantages of Stratified random samples

1. You can make sure you get enough observational units to learn about smaller strata.
2. You can get better (less variable) estimates using fewer observational units.

Finally, notice that stratified random sampling is not possible unless you have a complete list of all the units in the population. If such a list is not possible, or is too expensive to get, you can't stratify, but you may be able to use another method, cluster sampling. For this method, you don't need a list of all the units in the population as long as you can define and list a set of non-overlapping groups that include all the units.

Cluster samples

Step 1. Divide your entire population into non-overlapping groups called clusters. List the clusters and choose a simple random sample *of clusters*.
Step 2. For each of the clusters in your sample, list all the units in the cluster, then either
(a) record the value of your variable(s) for every unit in the cluster, or
(b) take a simple random sample of units from each cluster.

Here are two examples:

Example A.2 Sizes of Trees.

Suppose you want to know the distribution of diameters for trees in a one-square mile area of forest. You have a map of the area, but you don't have a list of all the trees, and because you estimate there are over a million trees in the square mile, you don't really want to try to list them all. Without such a list, you can't take a simple or stratified random sample. You can, however, take a cluster sample.

Step 1. Divide your one-square-mile area into a 100-by-100 grid of squares of equal size, 52.8 feet on a side. Each square is a cluster, and you have 10,000 of them in all. Number your clusters, and choose a simple random sample of 40 clusters. (The 40 is arbitrary. More clusters would cost more and take more time, but would give more information.)

Step 2. List all the trees in each of your 40 clusters. Then either (a) measure the diameter of every tree in each square, or (b) take a random sample of 15 trees in each square and measure the diameters of those trees.

The main advantage of cluster samples is that instead of listing all the units in your population, you only need to list a small fraction of units – the ones in the randomly selected clusters. In the tree example, you only needed to list the trees in 40 of the 10,000 little squares.

Here's another example in the same spirit. In the last example, you had to *create* the clusters, and the size of a square was somewhat arbitrary. In this next example, the structure of the population offers natural clusters.

Example A.3. First graders in a city.

For a nutrition study, you want to measure several quantities for a random sample of first grade pupils in a major city. Although it would be possible to compile a list of all first graders, it would be much quicker to use a cluster sample.

Step 1. List all the elementary schools in the city, and choose a simple random sample of schools.

Step 2. For each school in the sample, list all the first graders, and then choose a simple random sample of first graders within each selected school and record your measurements on those individual students

As these two examples of cluster sampling illustrate, there is one main advantage of using clusters for choosing a random sample.

Advantage of cluster samples

Cluster sampling lets you take random samples even when you don't have and can't easily get a list of all the units in your population.

You may have noticed that strata and clusters are defined in the same way: you divide or classify your entire population into non-overlapping groups. What's the difference between a stratum and a cluster? Although the definitions of the two kinds of groups are abstractly the same, the uses of the groups are quite different. With strata you typically have only a small number of large groups, you have a list of all the units in the whole population, and you take a SRS from every stratum. With clusters, you may have a very large number of clusters, you don't have a listing of all the units in the whole population, and you don't sample from all clusters, just from a SRS of clusters. Stratifying is useful as a way to reduce the SD of the overall (population) estimate. Cluster sampling is not. Cluster sampling is useful when you don't have a list of all the units in your population. Stratifying is not. Both methods should produce a representative sample, but our estimates tend to be more precise with stratified sampling.

Here's a chart that summarizes comparisons.

	Strata	Clusters
Is a listing of the whole population required	Yes	No
Number of groups	Comparatively few strata	Typically a large number of clusters
Sizes of groups	Often large	Smaller
Measure units from	Every stratum	Only a simple random sample of clusters
Reason(s) for using	To ensure small strata are represented and to reduce the SD of estimates	To get a random sample without having to list all units in the population

Preliminaries

P.1.1

a. Observational units: 47 students, Variables: (1) How much each student spent ($), (2) what the student was told (rebate or bonus)

b. Observational units: Typical American consumers, Variable: (1) How much each consumer spent ($)

c. Observational units: College students, Variables: (1) GPA of each student, (2) Whether or not each student pulls all-nighters (yes/no)

d. Observational units: College students, Variables: (1) Alcohol consumption of each student (e.g., typical drinks p/week), (2) Residence situation of each student (on-campus, off-campus with parents, off-campus without parents)

e. Observational units: Cats, Variables: (1) How far the cat can jump (inches), (2) How long the cat is (inches)

P.1.3

a. Variables: (1) How much each student spent ($, quantitative), (2) what the student was told (rebate or bonus, categorical-binary)

b. Variable: (1) How much each consumer spent ($, quantitative)

c. Variables: (1) GPA of each student (quantitative) (2) Whether or not each student pulls all-nighters (yes/no, categorical-binary)

d. Variables: (1) Alcohol consumption of each student (e.g., typical drinks p/week, quantitative), (2) Residence situation of each student (on-campus, off-campus with parents, off-campus without parents, categorical–not binary)

e. Variables: (1) How far the cat can jump (inches, quantitative), (2) How long the cat is (inches, quantitative)

P.1.5

Answers will vary.

P.1.7

a. Violin students

b. How much time spent practicing

c. Which of the three groups (international soloist, good violinists, teachers) the student was in

P.1.9

a. Quantitative variable

b. Research question

c. Categorical variable

d. Categorical variable

e. Research question

P.1.11

a. Do novice skydivers tend to have higher levels of self-reported anxiety prior to a skydive than experienced skydivers?

b. i. The 24 skydivers ii. From a parachute center in Northern England iii. Novice or expert skydiver (Categorical) and anxiety score (quantitative)

c. i. 43 among novice skydivers and 27 among experienced skydivers. ii. There are 11 first time skydivers and 13 experienced skydivers

d. Self-reported anxiety levels are substantially higher among novice skydivers

e. That these 24 skydivers have similar anxiety levels to most skydivers. In other words that these 24 'represent' (look-like) most skydivers

f. If older people tend to have lower anxiety levels, this could explain the difference. In a future study, researchers could make sure the ages of the novice and experienced sky divers are similar.

P.1.13

Answers will vary.

P.1.15

Answers will vary

P.1.17

Answers will vary

P.2.1

We would expect Nellie to have a larger standard deviation because the high temperatures in New York City should vary more throughout the year (high temperatures may not get above 30 in the winter but may exceed 100 in the summer), whereas in San Diego there is much less variation in daily temperatures, almost all values falling rather close to the overall average high temperature.

P.2.3

Quiz B has the smallest standard deviation (zero) because all the quiz scores are the same.

Quiz D has the next smallest standard deviation because most of the scores are equal to the mean, with only four students deviating from the mean by 2 points.

Quiz C has the second largest standard deviation because four students different from the mean by 2 points as with Quiz D but four other students also deviate by one point, so there is less consistency in the scores than in Quiz D.

Quiz A has the largest standard deviation. All quiz scores deviate by 5 points (as much as possible) from the mean score.

P.2.5

a. Replacement $3/43 \approx 0.07$; Regular $8/48 \approx 0.167$

b. Replacement $5/48 \approx 0.104$; Regular $12/43 \approx 0.279$

c. The replacement referees tended to have longer games (about 195 minutes on average compared to 185 minutes).

d. The replacement referees also tended to have more variability in the game lengths.

e. Neither distribution of game durations is symmetric, because both have a few games that took a very long time compared to the others. The games with replacement referees tended to take a bit longer than those with regular referees, roughly 10 minutes longer on average. Games with replacement referees also displayed more variability in game durations, as compared to the slightly greater consistency in game durations for regular referees. The two distributions show considerable overlap: Both types of referees saw most games take between roughly 180 and 210 minutes (3 – 3.5 hours).

P.2.7

a. The distribution of predicted high temperatures in San Luis Obispo county on July 8, 2012 shows a two clustered distribution, with many predictions between 63 and 73 degrees, but with another cluster of predictions between 85 and 96 96 degrees. The center of the overall distribution is near 70 to 75.

b. The bi-modal distribution is likely due to widely varying geography in San Luis Obispo County, including locations nearer to the ocean (cooler) and farther away from the ocean (warmer)

P.2.9

a. US, CA, SLO-County, World (highest to lowest average high temperature)

b. World, US, CA, SLO-County (most to least variability)

c. Anchorage or Juneau, Alaska.

d. Northern and southern hemisphere cities.

Section P.3

P.3.1

About 60% of all new businesses (in the long run) close or change owners within the first three years.

P.3.3

B, C and E cannot be probabilities, because probabilities must be numbers between 0 and 1.

P.3.5

a. If these two teams play each other many many times under identical conditions, team A will win 2/3 of the games in the long run.

b. No, this is a long-run proportion, not the proportion for every set of 3 games.

c. No, although we expect the proportion to tend to be closer to 2/3, this is still not a guarantee.

d. We expect the proportion of games that team A wins to be close to 2/3, but with so many different possibilities that are also close to two-thirds (such as 18, 19, 21, 22 wins), there could still be a low probability that A wins exactly 2/3 of the 30 games.

P.3.7

a. Let heads represent a boy and tails a girl. Flip the coin four times and record the number of boys (heads) in those four children (tosses). Repeat this a large number of times (say 1,000) and look at what proportion of those 1,000 repetitions resulted in 2 boys and 2 girls.

b. In a very large number of couples with four children, roughly 37.5% of the couples will have 2 boys and 2 girls (assuming each birth is equally likely to be a boy or a girl).

c. Having exactly two of each gender is pretty specific. Even though a specific 3-1 or 4-0 split is less likely than a 2-2 split, there are more ways to obtain a result other than a 2-2 split.

Chapter 1

1.1.1

a) The true proportion of times the racquet lands face up

b) parameter

c) 50%

d) 48 out of 100 does not constitute strong evidence that the spinning process is not fair, because if the spinning process was fair (50% chance of racquet landing face up) getting 48 out of 100 spins landing face up is a typical result.

e) Plausible that the spinning process is fair.

1.1.2

a) 24 out of 100 does constitute strong evidence that the spinning process is not fair, because if the spinning process was fair (50% chance of racquet landing face up) getting 24 out of 100 spins landing face up is an atypical result.

b) Statistically significant evidence that spinning is not fair

1.1.3

A (100, 1000)

1.1.4

C

1.1.8

a) Parameter

b) Statistic

c) The correct matches are shown below

Column A	Column B
Coin flip	Author plays a game of *Minesweeper*
Heads	Author wins a game
Tails	Author loses a game
Chance of Heads	Long-run probability that author wins
One repetition	One set of 20 *Minesweeper* games played by author

1.1.11

a) Statistic

b) Parameter

c) Yes, it is possible to get 17 out of 20 first serves in if Mark was just as likely to make his first serve as to miss it

d) Getting 17 out of 20 first serves if Mark was just as likely to make the serve as to miss it is like flipping a coin 20 times and getting heads 17 times, this is fairly unlikely, so 17 out of 20 first serves in is not a very plausible outcome if Mark is just as likely to make his first serve as to miss it

1.1.15

a) The long-run probability that Zwerg chooses the correct object.

b) Zwerg is just guessing, or Zwerg is choosing the correct cup because she understands the cue.

c) 37 out of 48 attempts seems fairly unlikely to happen by chance, since 24 out of 48 is what we would expect to happen in the long-run.

d) 50%

1.1.16

a) 37 times out of 48 attempts

b) Applet input: Probability of success is 0.5, Sample size is 48, Number of samples is1000

← Number of successes →

c) Yes, it appears as if the chance model is wrong, as it is highly unlikely to obtain a value as large of 37 when there is a 50% of picking the correct object

d) We have strong evidence that Zwerg can correctly follow this type of direction more than 50% of the time.

e) The results are statistically significant because we have strong evidence that the chance model is incorrect.

1.1.17

a) Zwerg is just guessing, or Zwerg is picking up on the experimenter cue to make a choice.

b) 26 out of 48 seems like the kind of thing that could happen just by chance since 24 out of 48 is what we would expect on average in the long run

c) 50%

1.1.18

a) 26 times out of 48 attempts.

b) Applet input: Probability of success is 0.5, Sample size is 48, Number of samples is 1000

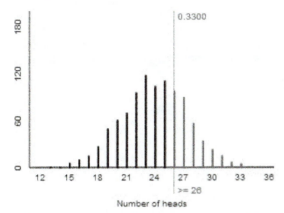

c) We cannot conclude the chance model is wrong since a value as large or larger than 26 is fairly likely.

d) We do not have strong evidence that Zwerg can correctly follow this type of direction more than 50% of the time.

e) The chance model (Zwerg guessing) is a plausible explanation for the observed data (26 out of 48)

f) Less convincing evidence that Zwerg can correctly follow this type of direction more than 50% of the time. We could have anticipated this since 26 out of 48 is closer to 24 out of 48 than is 37 out of 48.

g) This does not prove that Zwerg is just guessing. Guessing is just one plausible explanation for Zwerg's performance in this experiment. We cannot rule out guessing as an explanation for Zwerg getting 26 out of 48 correct.

1.1.23

a) 0.5

b) 20

c) 1000 (or some large number)

d) 12 out 20 is a fairly likely value because it occurred frequently in the simulated data.

1.1.24

a) 0.5

b) 100

c) 1000 (or some large number)

d) 60 out of 100 is somewhat unlikely because it occurred somewhat infrequently in the simulated data

e) The sample size was different (20 serves vs. 100 serves).

1.2.1

D

1.2.3

A

1.2.5

A

1.2.7

C

1.2.9

a). 0.25

b) 25 (since 0.25*100=25)

1.2.11

$\hat{p}$ is the value of the observed statistic, while the p-value is the probability that the observed statistic or more extreme occurs if the null hypothesis is true; p-value is a measure of strength of evidence

1.2.13

a) Null: The long run probability Hope will go to the correct object is 0.50, Alt: The long-run probability that Hope will go to the correct object is more than 0.50

b) H_0: $\pi=0.5$, H_0: $\pi>0.5$

c) 0.23 (23 dots are 0.60 or larger)

d) Since 3 of the 100 simulated outcomes gave a result of 7 or more, the p-value is 0.03.

e) 0.70

f) i.

1.2.18

a) Obs units: each of the 40 monkeys. Variable: correct choice or not (categorical)

b) The long-run probability that a monkey will make the correct choice

c) 30/40=0.75. Statistic.we use the symbol $\hat{p}$ to denote this quantity

d) Null hypothesis: The long-run probability that rhesus monkeys make the correct choice when observing the researcher jerk their head is 50% (just guessing). Alt hypothesis: The long-run probability that rhesus monkeys make the correct choice is more than 50%

H_0: $\pi=0.5$, H_0: $\pi>0.5$

e) Flip a coin 40 times, record the number of heads. Repeat this process 999 more times, yielding a set of 1000 values of the number of heads received in 40 coin tosses. Compute the p-value as the proportion of times 30 or larger was obtained by chance in the 1000 sets of 40 coin tosses. If the p-value is small (indicating 30 or larger rarely occurs by chance), then this is convincing evidence that rhesus monkeys can interpret human gestures better than by random chance.

f) The approximate p-value from the applet (using $\pi=0.5$, n=40, Number of samples=1000) is 0.001 (probability of 30 or greater). This small p-value means that 30 out of 40 is strong evidence that the rhesus monkeys are not guessing, which may lead us to believe that rhesus monkeys may be able to understand a head jerk to indicate which box to choose.

1.2.19

a) Obs units: each of the 40 monkeys. Variable: correct choice or not (categorical)

b) The long-run probability that a monkey will make the correct choice

c) 31/40=0.775. Statistic. We use the symbol $\hat{p}$ to denote this quantity

d) Null hypothesis: The long-run probability that rhesus monkeys make the correct choice when the researcher looks towards the correct box is 50% (just guessing). Alt hypothesis: The long-run probability that rhesus monkeys make the correct choice is more then 50%

H_0: $\pi=0.5$, H_0: $\pi>0.5$

e) Flip a coin 40 times, record the number of heads. Repeat this process 999 more times, yielding a set of 1000 values of the number of heads received in 40 coin tosses. Compute the p-value as the proportion of times 31 or larger was obtained by chance in the 1000 sets of 40 coin tosses. If the p-value is small (indicating 30 or larger rarely occurs by chance), then this is convincing evidence that rhesus monkeys can interpret human looking better than by random chance.

f) The approximate p-value from the applet (using $\pi = 0.5$, $n = 40$, Number of samples $= 1000$) is 0.001 (probability of 31 or greater). This small p-value means that 31 out of 40 is strong evidence that the rhesus monkeys are not guessing, which may lead us to believe that rhesus monkeys can interpret looking to indicate which box to choose.

1.2.20

p-value is approximately 0.25. We don't have strong evidence that the author's long-run probability of winning Minesweeper is greater than 50%. The null hypothesis (long-run probability is 50%) is a plausible explanation for her winning 12 out of 20 games.

1.2.22

a) The long run probability of a spun penny landing heads

b) p-value= 0.16. There is little to no evidence that a spun penny lands heads less than 50% of the time.

c) Null would be the same, Alternative would be >0.5. To calculate the p-value find the probability that 29 or larger (58% or larger) occurred.

1.2.24

A simulation analysis using a null hypothesis probability of 0.75 yields a p-value of 0.10, meaning that the set of 20 free throws by your friend (and making 12/20 of them) provides little to no evidence that your friends long-run probability of making free-throws is worse than the NBA average.

1.2.25

A simulation analysis using a null hypothesis probability of 0.75 yields a p-value of 0.02, meaning that the set of 40 free throws by your friend (and making 24/40 of them) provides strong evidence that your friend's long-run probability of making free throws is worse than the NBA average.

1.3.1

C

1.3.3

A, because the standard deviation is smaller

1.3.5

a) FALSE, b. TRUE, c. FALSE, d. FALSE

1.3.7

a) -3.47 (100 out of 400; 25%), -3.80 (20 out of 120; 16.7%), -4.17 (65 out of 300; 21.7%)

b) 65 out of 300 is the strongest evidence, 100 out of 400 is the least strong evidence

1.3.8

a) Friend D because they played more games.

b) Friend D because this is more evidence against the null hypothesis.

c) Friend D because this is more evidence against the null hypothesis.

d) Friend D because a smaller standard deviation leaves to a larger standardized statistic.

1.3.9

Friend F because the value of their statistic (30 out of 40) is larger than Friend G (15 out of 40) and thus will be farther in the tail of the distribution.

1.3.11

Simulation yields a standard deviation of the null distribution of 0.125, and a standardized statistic of approximately 2.86, which provides very strong evidence that the long run probability of Buzz pushing the correct button is higher than 50%.

1.3.12

Simulation yields a standard deviation of the null distribution of approximately 0.094, and a standardized statistic of approximately 0.76, which provides little to no evidence that the long run probability of Buzz pushing the correct button is higher than 50%.

1.3.15

a). Null: The long-run proportion of all couples that have the male say 'I love you' first is 50%. Alt: The long run proportion is more than 50%.

b) $z = (0.7 - 0.5)/0.079 = 2.53$.

c) The observed proportion of couples leaning their heads to the right while kissing is 2.53 standard deviations away from the null hypothesized parameter value of 0.5.

d) We have strong evidence that the proportion of couples for which the male says 'I love you' first is more than 50%.

1.3.17

a) The long-run probability the lady correctly identifies which was poured first, π

b) Null: $\pi = 0.5$, Alt: $\pi > 0.5$

c) $8/8 = 1 = \hat{p}$

d) 0.5, because that is the value of the parameter if the null hypothesis is true. The standard deviation will be positive because the standard deviation must be at least 0, and is only equal to zero if there is no variability in the values (there will be variability in the simulated statistics).

e) $(1 - 0.5)/0.177 = 2.82$.

f) The observed proportion of times the lady correctly identified which was poured first is 2.82 standard deviations away from the null hypothesized parameter value of 0.5.

g) We have strong evidence that the long-run probability that lady makes the correct identification is greater than 50%.

1.3.20

a) The long-run probability that 10-month olds choose the helper toy, π

b) Null: $\pi=0.5$, Alt: $\pi>0.5$

c) $14/16=0.875=\hat{p}$

d) $(0.875-0.5)/0.125=3$.

e) The observed proportion of times the 10-month old babies chose the helper toy is 3 standard deviations away from the null hypothesized parameter value of 0.5.

f) We have strong evidence that the long-run probability that 10-month old babies choose the helper toy is greater than 50%.

1.3.21

a) Yes, the p-value will be small because the standardized statistic is large

b) p-value of approximately 0.002. The p-value is the probability observing 14/16 or larger assuming the null hypothesis is true.

c) We have strong evidence that the long-run probability that 10-month old babies choose the helper toy is greater than 50%.

d) Yes, because both the p-value and the standardized statistic are measuring the strength of evidence (how far out in the tail is the observed data), and so should lead to the same conclusion.

1.4.1

D

1.4.2

A

1.4.3

D

1.4.7

a) Stronger. The statistic (18/20=90%) is much farther away from the null hypothesized value (50%) than before (12/20=60%)

b) Stronger, the statistic is the same (60%) but the sample size is much larger (100 vs. 20)

c) No, 12/30=40% is less than the null hypothesis value of 50%, thus this is not evidence that the long-run probability of winning Minesweeper is more than 50%

1.4.13

a) Sample size, Distance

b) Stronger

1.4.14

a) Distance, Sample size

b) Weaker

1.4.15

a) Double it. The alternative would now be two-sided

b) Weaker

1.4.18

a) The long run probability that Krieger chooses the correct object, π.

b) 50%

c) Null: The long run probability that Krieger chooses the correct object is 50%; Alt: The long run probability that Krieger chooses the correct object is more than 50%

1.4.19

a) 6 out of 10 for a proportion of 0.6

b) Mean=0.50, sd=0.16

c) The p-value (The probability of obtaining 0.6 or larger when the true chance Krieger chooses the correct object is 0.5) is approximately 0.38

d) We do not have strong evidence that Krieger will choose the correct object more than 50% of the time; It is plausible that Krieger will choose the correct object 50% of the time

e) Stronger

1.4.20

a) Values of the long-run probability less than 0.5

b)Increase

c) The two-sided p-value will approximately double to 0.75.

d) Stronger

1.4.21

a) Decrease

b) Yes

c) Stronger

1.4.22

a) The long run proportion of times Krieger makes the correct choice when the experimenter leans towards the object

b) 50%

c) Null: The long run proportion of times Krieger makes the correct choice when the experimenter leans towards the object is 50%; Alt: The long run proportion of times Krieger makes the correct choice when the experimenter leans towards the object is more than 50%

d) Decrease, observed statistic is farther from the hypothesized values under the null.

1.4.23

a) 9 out 10 (0.9)

b) Mean=0.50, SD=0.16

c) 0.01

d) Approximately 0.7

1.4.24

a) (0.9-0.5)/0.16=2.5

b) We have strong evidence that the long run proportion of times that Krieger makes the correct choice is more than 50%

c) Yes, the p-value got smaller (evidence got stronger).

1.4.25

a) Analysis method	Sample size n	Null value π_0	Value of $\hat{p}$
A: 1.4.8 – 1.4.12	_938,223_	_0.50_____	.516276
B: 1.4.25	__88_____	_0.50___	1.0_

b) Analysis method A provides strong evidence, but method B is overwhelming.

1.4.27

a) one-sided

b) 6 out of 8 seems fairly likely to occur by chance

c) P-value = 0.1094+0.0313+0.0039=0.1446

1.4.30

D=Sample size is $n = 25$, alternative hypothesis is right-sided: $\square > \frac{1}{2}$.

A=Sample size is $n = 225$, alternative hypothesis is right-sided: $\square > \frac{1}{2}$.

C=Sample size is $n = 25$, alternative hypothesis is left-sided: $\square < \frac{1}{2}$.

B=Sample size is $n = 225$, alternative hypothesis is left-sided: $\square < \frac{1}{2}$.

E=Sample size is $n = 25$, alternative hypothesis is two-sided: $\square \neq \frac{1}{2}$.

F=Sample size is $n = 225$, alternative hypothesis is two-sided: $\square \neq \frac{1}{2}$.

1.4.31

a) Increasing: B,C
b) Decreasing: A,D
c) Up-down: E,F

1.4.32

a) 100%
b) 50%
c) 0%

1.4.33

a) Smaller, decreasing, A,D
b) Larger, increasing, B,C
c) Up-down,E,F
d)

Alternative hypothesis	Strongest evidence	Shape of curve
$\square > 0.5$	$\hat{p} = 1$	Decreasing
$\square < 0.5$	$\hat{p} = 0$	Increasing
$\square > 0.5$	$\hat{p} = 0$ or $\hat{p} = 1$	Up-down

1.4.34

a) Can't tell—it depends on the alternative
b) Can't tell—it depends on the alternative
c) Can't tell—it depends on the alternative
d) Less steep

1.4.35

	Increasing	Decreasing	Up-Down
Steeper	B	A	F
Flatter	C	D	E

1.5.1

C

1.5.3

D, 60. Smaller standard deviations occur when the sample size is larger, for the same value of the proportion, π.

1.5.5

The value of the standardized statistic, z.

1.5.7

The standardized statistic; a measurement of how many standard deviations from the mean the observed statistic is on the null distribution

1.5.9

a) Null: The long-run probability that a penny lands heads when spun is 0.2
   Alt: The long-run probability that a penny lands heads when spun is >0.2

b) The simulation based p-value of 0.077, because the validity conditions are not met. There are not at least 10 times where the penny landed heads and at least 10 times where the penny landed tails in the sample.

c) No, we do not have strong evidence that a penny will land heads more than 20% of the time in the long run, since the p-value is only 0.077.

1.5.11

a) 0.152, this is very close to the hypothesized parameter value of 0.15

b) The validity conditions are not met. There are not at least 10 successes and 10 failures in the data (only 8 and 2)

1.5.12

a) 0.15, this is the hypothesized parameter value of 0.15

b) 0.019, $\sqrt{\dfrac{\pi(1-\pi)}{n}} = \sqrt{\dfrac{.15(1-.15)}{361}} = 0.019$

c) Yes

d) Because the validity conditions are met for this data set (larger sample size)

1.5.15

a) Null: The long-run probability that the male says I love you first is 50%

Alt: The long-run probability that the male says I love you first is more than 50%

b) 0.0057

c) We have very strong evidence that the long-run probability that the male says I love you first is more than 50%

d) Z=2.53. That the proportion of males that say I love you first is 2.53 standard deviations above the mean of the null distribution.

e) 0.0057 x 2 = 0.0114

1.5.17

a) Null: The long-run probability that a player starts with scissors is 33%

Alt: The long-run probability that a player starts with scissors is different than 33%

Null: $\pi = 33\%$

Alt: $\pi \neq 33\%$

b) p-value=0

c) We have very strong evidence that the long-run probability that a player starts with scissors is different than 33%

1.5.18

a) Null: The long-run probability that a player starts with rock is 33%

Alt: The long-run probability that a player starts with rock is different than 33%

Null: $\pi=33\%$

Alt: $\pi\neq33\%$

b) p-value=0

c) We have very strong evidence that the long-run probability that a player starts with rock is different than 33%

1.5.20

a) Null: The long-run probability that the most competent looking candidate wins is 50%

Alt: The long-run probability that the most competent looking candidate wins is more 50%

b) n=279, $\hat{p}=67.7\%$

c) 0

d) We have very strong evidence that the most competent looking candidate wins more than 50% of the time

e) Z=1.46. That the proportion of the sample that chose Tim is 1.46 standard deviations above the mean of the null distribution.

1.5.23

a) The long-run probability of getting a six

b) Null: The long-run probability of getting a six is 16.7%

Alt: The long-run probability of getting a six is more than 16.7%

c) 0.1492

d) We have little to no evidence that the long-run probability of getting a six is more than 16.7%

1.5.24

a) The long-run probability of getting a one

b) Null: The long-run probability of getting a one is 16.7%
Alt: The long-run probability of getting a less than 16.7%

c) 0.0896

d) We have little to no evidence that the long-run probability of getting a one is less than 16.7%

End of chapter exercises

1.CE.1

Null

1.CE.3

a) Null: The probability the statistics professor wins is 0.2, Alternative; The probability the statistics professor wins is larger than 0.2

b) Start with 5 playing cards—one red, and four black. Shuffle the cards. Randomly choose one card, record if it is red or not and then place the card back in the deck. Shuffle and randomly choose cards until 12 cards have been selected. Record the number of red cards selected out of the 12 selections. Repeat this entire process 999 more times to generate a distribution of counts of red cards. If 7 out of 12 red cards chosen rarely happened in the 1000 simulations, then this would be convincing evidence that 7 out of 12 was unlikely to have occurred by chance.

c)

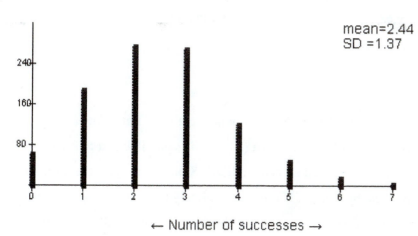

mean=2.44
SD =1.37

← Number of successes →

The observed data (7 wins out of 12 attempts) provide convincing evidence that the statistics professor's probability of winning in one week was larger than would be expected if the 5 competitor's were equally likely to win because 7 out of 12 rarely happens by chance, when everyone is equally likely to win; in particular the p-value is 0.005.

d) The p-value is the probability of observing 7 or more successes out of 12 attempts when each attempt has a 20% chance of being correct.

e) The theory based approach is not appropriate here because the resulting simulated distribution of statistics is not normal---this is because the sample size is not large enough---in particular, there are only 7 successes, and 5 failures---instead of at least 10 of each.

1.CE.5

a) Even though 34.6% (the % of players suspended for PED use who were from the US) is less than the percent of all baseball players born in the US (57.3%), it's possible that this could have happened by chance (just like it's possible to flip a coin and get heads 8 times out of 8), the question is how likely this (34.6%) would happen just by chance. If it is quite unlikely then we say that the result is statistically significant, meaning that there is something about US baseball players which make them less likely to be suspended for PED use

b) The likelihood of having only 34.6% of 595 suspended baseball players be from the US when the proportion of all baseball players who are from the US is 57.3% is extremely unlikely; in other words, 34.6% is (statistically) significantly less than 57.3%

c) z= -11.19. This tells us that the observed proportion (34.6%) is 11.19 sd's less than the mean of the null distribution, confirming that we have extremely strong evidence that the null hypothesized value of the parameter is incorrect.

1.CE.6

a) Even though 61.8% (the % of players suspended for PED use who were from Latin America) is more than the percent of all baseball players born in Latin America (34.6%), it's possible that this could have happened by chance (just like it's possible to flip a coin and get heads 8 times out of 8), the question is how likely this (61.8%) would happen just by chance. If it is quite unlikely then we say that the result is statistically significant, meaning that there is something about Latin American baseball players which make them more likely to be suspended for PED use

b) The likelihood of having 61.8% of 595 suspended baseball players be from Latin America when the proportion of all baseball players who are from the US is 34.6% is extremely unlikely; in other words, 61.8% is (statistically) significantly more than 34.6%

c) z= 13.95. This tells us that the observed proportion (61.8%) is more than 13.95 sd's less than the mean of the null distribution, confirming that we have extremely strong evidence that the null hypothesized value of the parameter is incorrect.

1.CE.9

Not necessarily, a larger sample size yields a smaller p-value if the value of the statistic is the same, there is no guarantee the value of the statistic (proportion of heads) will be the same in Jose and Roberto's separate samples

1.CE.10

A is the only correct answer

1.CE.13

a) The long-run probability that Rick makes a free throw underhanded

b) a) Null: The long –run probability that Rick makes a free throw underhanded is 90%
Alt: The long –run probability that Rick makes a free throw underhanded is more than 90%

Null: $\pi = 90\%$

Alt: $\pi > 90\%\%$

1.CE.14

a) The long-run probability that Lorena makes a 10 foot putt

b) a) Null: The long –run probability that Lorena makes a 10-foot putt is 60%

Alt: The long –run probability that Lorena makes a 10-foot putt is more than 60%

Null: $\pi = 60\%$

Alt: $\pi > 60\%$

1.CE.17

a) We don't have strong evidence that the probability a spun tennis racquet lands with the label up is different from 0.5

b) If the probability that a spun tennis racquet lands with the label is actually 0.5, then it is quite likely to get 46 spins out of 100 with the label up; thus, a reasonable (plausible) explanation for the author's data (46 out 100) is that the tennis racquet is fair (0.5 chance of label landing up)

1.CE.18

Null hypothesis probability = 0.5

Statistic = 0.46

P-value = 0.484

Chapter 2

2.1.1. A

2.1.3. B

2.1.5. A. FALSE, B. FALSE, C. FALSE

2.1.7. B

2.1.9. While some of these could be argued the other way, all of them could likely not be representative. If you have food that finches like and other birds didn't, you would overestimate the proportion of finches. You could have food that finches don't like and you would rarely see more than one eating at a time. The proportion of male birds could be species dependent and depending on the type of food you have could affect the type of species and hence affect the proportion of male birds that come to your feeder. This proportion could then be different than the proportion of males in your area as well as those that typically visit feeders.

2.1.18. a) The population is all the likely voters in the city. b) Since it is a random sample, I would think the proportion that favor the incumbent in the sample is similar to that for the population.

2.1.19. a) The proportion of all city voters that plan to vote for the incumbent. B) The proportion of those in the sample that plan to vote for the incumbent (0.65).

2.1.20. a) The variable is who they plan to vote for or whether or not they plan to vote for the incumbent. B) Categorical. C) Proportion. D) Bar graph.

2.1.21. A) Null: 50% of the all the city voters plan to vote for the incumbent. Alternative: A majority of all the city voters plan to vote for the incumbent. B) The proportion of all city voters that plan to vote for the incumbent. C) 0.65.

2.1.22. A) P-value = 0. B) We have strong evidence that a majority of all the city voters plan to vote for the incumbent. C) We can infer these results to all likely city voters since it came from a random sample. D) Using theory-based methods is appropriate and we also obtain a p-value of 0.

2.1.23. A) How long people send reading or learning about local politics. B) Quantitative. C) Mean or Median. D) Dotplot.

2.1.24.

Voter	Time spent reading/learning about local politics	Voting for incumbent?
#1	0	Yes
#2	0	Yes
#3	60	No
…	…	…

2.1.25. The study was done using a random sample. It could have been done by obtaining a list of all the voters in the city and then assign every voter a number. Then have a random number generator give 267 random numbers. The voter's names that match the numbers chosen will be the random sample.

2.1.34. A) The population is all the sharks at the zoo. B) The proportion should be similar to the population since it came from a random sample.

2.1.35. A) The parameter is the proportion of sharks in the zoo that have the disease. B) The statistic is the proportion of sharks in the sample that have the disease.

2.1.36. A) The variable measured is whether or not they have the disease. B) Categorical. C) Proportion. D) Bar graph.

2.1.37. Null: The proportion of sharks at the zoo that have the disease is 0.25. Alternative: The proportion of sharks at the zoo that have the disease is less than 0.25. B) The parameter is the proportion of sharks in the zoo that have the disease. C) $3/15 = 0.2$.

2.1.38 A) P-value = 0.46. B) We do not have any evidence that the proportion of sharks in the zoo that have the disease is less than 0.25. C) The sharks at the zoo. D) A theory-based approach using the normal distribution is not reasonable to use since there were only 3 sharks with the disease. We need at least 10.

2.1.39. A) The shark's blood oxygen content. B) Quantitative. C) Mean or median. D) Dotplot.

2.1.40.

Shark	Has Disease?	Blood Oxygen Content
#1	Yes	1.2%
#2	No	5.6%

#3	No	6.2%
...		...

2.1.41. The study was done using a random sample. It could have been done by obtaining a list of all the sharks at the zoo and assign each a number. Then have a random number generator give 15 random numbers. The sharks that match the numbers chosen will be the random sample.

2.1.44. A) All the customers of the store. B) The 100 people asked to fill out the survey. C) The proportion of all customers that visit the store because of the sale on coats. D) The proportion of the sample that said they visited the store because of the sale on coats.

2.1.45. It may not be representative since it was not a random sample.

2.2.1. a) Since the distribution is skewed to the left, the mean will be to the left of the median, hence 65.86°F is the mean and 67.50°F is the median. b) Mean: Larger, Median: Larger, Standard Deviation: Smaller.

2.2.6. a) The mean and median would increase by 5 points, the standard deviation would stay the same since the entire distribution would move up but the variability stays the same. b) The mean would increase since 5 would be added to the total of the scores. The median would stay the same since the high score remains the high score. The standard deviation would increase since there would be more variability with the larger number. c) The mean would increase since 5 would be added to the total of the scores. We can't really say for certain how or if the median or standard deviation would change. It depends on the original distribution of scores. For example if adding the 5 points to the lowest score could still keep it the lowest score or it could change it to the highest score.

2.2.8. B)

2.2.10. a) The distribution is skewed to the right. b) Since the distribution is skewed to the right we should expect the mean to be higher than the media. c) The median is $35 and the mean is $45.68. The mean is higher, as expected. d) If a $150 haircut is changed to $300 we should expect the median to stay the same (since $150 or $300 is just another larger value), but the mean should increase (since the total of all haircut costs will increase). When the change is made, we can see the median does stay the same and the mean increases to $48.68.

2.2.16. a) The SPF value for students at a certain school. b) μ = mean SPF of sunscreens used by all students at this school. c) H_0: μ = 15 versus H_a: μ > 15. d) n = 48, $\bar{x} = 3.01$, s = 17.19. e) No, it just came from students in her class. f) They probably don't differ much from the students as a whole on this issue. g) It may not be representative for students at a Midwestern college where it is very cloudy.

2.2.17. a) You would have to fabricate a large data set to represent the population of SPF numbers for all students at the school with the variability similar to that of the sample data and a mean of 30. From that data you would take a sample of 48 and find its mean. Repeat this at least 1000 times to develop a null distribution. To determine the p-value, determine the proportion of simulated statistics that are at more than 35.29. b)

Simulation		Real study
One repetition	=	A sample of 48 students
Null model	=	Population mean is 30
Statistic	=	Average SPF number in the sample

2.2.18. a) Yes since the sample size is larger than 20. b) The standardized statistic: t = 8.18 and the p-value = 0. (applet output is shown below) c) If the mean SPF for all students at the school is 30, the probability we would get a sample mean as large or larger than 35.29 from a random sample of 30 is 0. d) We have strong evidence that the average SPF used by students at the school is more than 30.

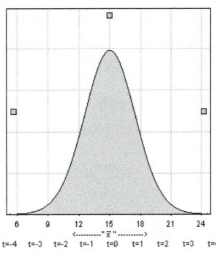

2.2.22. a) The distribution is fairly symmetric. b) The mean and the median will be about the same. c) The mean is 98.105 and the median is 98.100. Yes they are very close. d) The actual standard deviation is 0.699.

2.2.23. a) Null: The average body temperature for males is 98.6 degrees ($\mu = 98.6$), Alt: The average body temperature for males is not 98.6 ($\mu \neq 98$). b) The standardized statistic is -5.71 and the p-value is 0. c) Since the p-value is less than 0.05 we have strong evidence that the average male body temperature is not 98.6. d) Any generalization should be done with caution, but we can probably generalize it to healthy male adults similar to those that were in the study.

2.2.24. a) Null hypothesis: The average body temperature for females is 98.6 degrees ($\mu=98.6$) Alt hypothesis: the average body temperature for females is not 98.6 degrees ($\mu\neq98.6$). b) Hard to tell, there is a lot of variability in the data. c) The standardized statistic is -2.24 and the p-value = 0.0289. d) Since the p-value is less than 0.05 we have strong evidence that the average body temperature for females is different than 98.6 degrees.

2.3.1 a) 5%, b) Larger, because each test itself has a 5% chance, so if we do multiple tests, the overall chance of at least one making an error will be larger than 5%. This is like saying the chance of getting at least one head when flipping a coin 10 times is larger than 50%.

2.3.3. a) Fail to reject, b) Fail to reject, c) Reject, d) Fail to reject

2.3.5. a) Reject, since the p-value is less than 0.01, it must be less than 0.10, b) Reject, c) Fail to reject, d) Reject

2.3.6. a) P-values less than or equal to 0.05, b) p-values larger than 0.05

2.3.8. a) 0.001, b) Type II error rate will decrease

2.3.10. a) We find strong evidence that Buzz is not guessing, but he is guessing b) We do not have evidence that Buzz is not guessing (guessing is plausible), when Buzz is actually not guessing

2.3.11. A Type I error is possible here, which means that we conclude there is strong evidence Buzz is not guessing, when in fact Buzz is guessing.

2.3.14 Because this would increase the type II error rate, meaning we'd frequently conclude "not enough evidence to reject the null hypothesis" when the null hypothesis is false, leading to many "missed opportunities"

2.3.17 a) The proportion of all students at the school who state they would likely read a campus newspaper b) Null: The proportion of all students at the school who state they would likely read a campus newspaper is 10%, Alt: The proportion of all students at the school who state they would likely read a campus newspaper is more than 10% c) We find strong evidence the proportion is more than 10%, when it is not. The consequence is that readership of the newspaper may be lower than anticipated. d) We don't have evidence the proportion of student reader's is more than 10%, when it actually is more than 10%. You won't start the newspaper even though you would have enough readership.

2.3.19 a) We have strong evidence the subject is not telling the truth b) It's plausible the subject is telling the truth (we can't rule out the fact that they are telling the truth) c) We have strong evidence the subject is not telling the truth, when in fact they are telling the truth d) We don't have strong evidence the subject is not telling the truth (it's plausible they are telling the truth), when in fact they are lying

2.3.21 a) We have strong evidence the new treatment is better, when it actually isn't, b) We don't have evidence the new treatment is better, when it actually is

2.3.23 a) We have strong evidence that the proportion of organ donors between the three default options are different between the groups, but the proportions actually are not different b) We don't have evidence that the three default options are different, when they actually are

2.3.25 a) We have strong evidence that trick or treaters prefer candy over toys, when actually there is no preference (long-run probability of choosing candy is 50%), b) We don't have strong evidence that trick or treaters prefer candy over toys, when actually trick or treaters do prefer candy over toys c) No d) Type II error is possible

2.CE.1 a) No, all games during a certain period were selected, instead of randomly choosing some b) Number of runs is likely more representative of the population (all games) since attendance fluctuates dramatically during the year due to weather, opponent and timing of the games during the season

2.CE.3 a) No, the instructor did not list all games in the 2010 season and randomly choose some. b) Null hypothesis: 75% of all Major league baseball games have a 'big bang' ($\pi=0.75$), Alt hypothesis: Less than 75% of all games have a 'big bang' ($\pi<0.75$) c) $\hat{p} = \frac{21}{45} = 0.467$ d) Using applet with 0.75=prob of success, n=45, and number of samples=1000, calculate probability of 0.467 or less. The p-value is approximately 0. e) The data provide strong evidence that the true proportion of all Major league baseball games with a big bang is less than 75%, since it's extremely unlikely that in a sample of 45 games only 21 would have big bang if the true proportion of all games with a big bang was 75%.

2.CE.5 a) Obs unit=Each of the brides, variable = whether or not kept own name b) Pop=US Brides, Sample = 600 brides with wedding announcement in NY Times c) The proportion of all US Brides that keep their own names d) Yes, concerned because people with wedding in NY Times are likely to be different than 'typical' US adults. Probably reasonable to generalize to other brides who could/would have their wedding announcement in the NY Times

2.CE.6 a) p-value = 0.0451. Strong enough evidence at 0.05 significance level since 0.0451<0.05, b) Strong enough evidence at 0.10 significance level since 0.0451<0.10, c) No enough evidence at 0.01 significance level since 0.0451>0.01

2.CE.9 a) The sample size is large, but it would be good to know that the data was not strongly skewed to feel better about running a theory-based test on the data b) Null: The average adult body temperature is 98.6 degrees, Alt: The average adult body temperature is not 98.6 degrees, p-value is 0.0000, we have strong evidence that average body temperature is different than 98.8, c) t-statistic of 6.32 d) Yes, the t-statistic is very different than 0 (in particular >3) which means there is very strong evidence against the null hypothesis

2.CE.10 a) Type I b) The p-value is >0.05, c) Change significance level, increase sample size

Chapter 3

3.1.1

B

3.1.3

A

3.1.5

D

3.1.7

B

3.1.9

A

3.1.11

A

3.1.13

    a) B
    b) B

3.1.15 0.01

3.1.17 The confidence interval is estimating the *proportion of all American adults* that thought math was the most valuable subject they studied in school.

3.1.19

    a) (0.48, 0.56)
    b) (0.46, 0.59)

3.1.21 p-values will differ for each students' simulation, one possible ex. follows

Null	p-value	Null	p-value
Proportion = 0.45	0.001	Proportion = 0.53	0.640
Proportion = 0.46	0.006	Proportion = 0.54	0.316
Proportion = 0.47	0.010	Proportion = 0.55	0.130
Proportion = 0.48	0.053	Proportion = 0.56	0.058
Proportion = 0.49	0.140	Proportion = 0.57	0.014
Proportion = 0.50	0.342	Proportion = 0.58	0.007
Proportion = 0.51	0.667	Proportion = 0.59	0.001
Proportion = 0.52	1.000	Proportion = 0.60	0.000

95% confidence interval is (0.48, 0.56)

3.1.24

a) The symbol π represents the proportion of all American adults that think same-sex marriages should be legal and have the same rights as traditional marriages.

b) The p-value is about 0.028.

c) No, 0.5 does not appear to be a plausible value for π. Since we have a small p-value, we have evidence against $\pi = 0.5$, so 0.5 is NOT a plausible value.

d) A 95% confidence interval for π is 0.505 to 0.555.

e) We are 95% confident that the value for π (the proportion of all American adults that think same-sex marriages should be legal and have the same rights as traditional marriages) is between 0.505 and 0.555.

3.1.25 A 99% confidence interval is (0.495, 0.561) (answers will vary due to simulation)

3.2.1
B

3.2.3
B

3.2.5
D

3.2.7

0.38 ± 0.11

3.2.9

a) We need to know the sample size.

b) As the sample size gets larger, the variability will be smaller and hence the confidence intervals will be narrower. Therefore the confidence interval based on the sample size of 250 will be widest and the one based on the sample size of 4000 will be the narrowest.

c) For a sample size of 250, we got a standard deviation of 0.032 and a confidence interval of 0.182 $\pm$ 0.064. For a sample size of 1000, we got a standard deviation of 0.016 and a confidence interval of 0.182 $\pm$ 0.032. For a sample size of 4000, we got a standard deviation of 0.008 and a confidence interval of 0.182 $\pm$ 0.016.

d) The midpoints of the intervals are all the same.

e) As the sample size increases, the widths of the intervals decrease.

f) Four times as much.

g) Yes

3.2.11

a) $0.60 \pm 2(0.070)$ or $(0.46, 0.74)$

b) No, since values less than 0.50 are contained in the confidence interval of plausible values.

3.2.12

a) $(0.4642, 0.7358)$

b) $(0.2642, 0.5358)$

3.2.15

a) 95% $(0.5503, 0.7831)$

b) 99% $(0.5137, 0.8197)$

c) The p-value for the test is less than 0.01

3.2.16 $0.667 +/- 2(0.063) = 0.667 +/- 0.126$, $(0.541, 0.793)$. Similar because validity conditions are met.

3.2.18

a) Ho:$\square\square = 0.33$ Ha: :$\square\square \neq 0.33$ p-value $=0.9030$ which offers weak evidence against the null so we can say that it is plausible that the long-run probability of correctly identifying the cup with Pepsi is 0.33.

b) $0.344 \pm 2(0.058)$ or $(0.228, 0.460)$

c) Yes. Since our p-value was larger than 0.05 the value under the null was found to be plausible, thus it would be contained in an approximate 95% confidence interval of plausible values.

3.2.19

a) $(0.2274, 0.4602)$

b) 0.1164

3.2.22 $(0.4267, 0.6071)$

3.2.23 $(0.1036, 0.2528)$

3.2.27 0.65 ± 0.0418 or $(0.6082, 0.6918)$

3.3.1

FALSE

3.3.3

FALSE

3.3.6

a) Male rattle snakes, all male rattle snakes, 21 male rattle snakes

b) Sample size is greater than 20 and data is not strongly skewed.

c) Average age of all male rattle snakes at this single site

d) False, the parameter is fixed and is either in the interval or not.

3.3.8

a) Students

b) Hrs/week studying Statistics outside of class, quantitative

c) $(7.1113, 9.2887)$

d) Yes, sample size is greater than 20 and the data was not strongly skewed.

e) We are 95% confident that the average hours/week all statistics students will spend studying is between 7.11 and 9.29.confidence interval is completely above 30

3.3.11

a) Students
b) Number of U.S. states visited, quantitative
c) Probably, since statistics students wouldn't be different on the number of states visited compared to all students at the school.
d) Ho $\mu = 16$ and Ha: $\mu \neq 16$
e) The midpoint of the interval is 9.48. The endpoints of the interval are found by subtracting 2* the SD of the simulated null from 9.48 and adding 2*SD of the simulated null to 9.48.

3.3.12

a) (7.4536, 11.5064)
b) Yes, 16 is not in the confidence interval
c) Yes, sample size was greater than 20. And the data was not strongly skewed

3.3.13

a) (7.8844, 11.0756)
b) 2SD method is a rough approximation for a 95% confidence interval. It will not work for other levels of confidence.

3.3.14

a) INVALID
b) VALID
c) INVALID
d) VALID
e) INVALID
f) INVALID

3.3.19

1181

a) the 49 students
b) number of Facebook friends
c) sample size is greater than 20 and data are not strongly skewed
d) (453.60, 624.80) We are 95% confident that the population average number of Facebook friends is between 453.60 and 624.80.

3.3.20

a) INVALID
b) VALID
c) INVALID

3.3.22

a) Albacore Tunas
b) Mercury level in parts per million, quantitative
c) The sample size is greater than 20 and the data is not strongly skewed.
d) (0.3155, 0.4005) We are 95% confident that the population average ppm of mercury in Albacore Tunas is between 0.3155 and 0.400513.
e) (0.3168, 0.3992) it is close

3.3.23

a) The sample size is greater than 20 and the data is not strongly skewed
b) (0.3241, 0.3839) We are 95% confident that the population average ppm of mercury in Yellowfin Tunas is between 0.3155 and 0.400513.

3.3.25

a) The sample standard deviation is very large compared to the mean.
b) The sample size is greater than 20 and the data is not strongly skewed
c) (1.8963, 2.0777) We are 95% confident that the population average number of close friends of U.S. adults is between.
d) (1.8964, 2.0776) it is very close

3.4.1

a) $\hat{p}$ is the sample proportion (statistic) and $\square\square$ is the population proportion (parameter)

b) $\hat{p}$

c) Standard deviation of $\hat{p}$ (or the standard error)

d) Margin of error

3.4.3 The sample size, the level of confidence and the value of $\hat{p}$. The value of $\hat{p}$ is hardest to control.

3.4.5 a) 90%: 0.09993, 0.1327; 95%: 0.0962, 0.1358; 99%: 0.0899, 0.1421, b) Midpoints are all the same, c) 90% is narrowest, 99% is widest; as confidence level increases width of interval increases d) No, because all of the intervals include 10% e) Yes, because there are at least 10 people who said 'yes' (116) and at least 10 who said 'no' (884)

3.4.6 a) Narrower, b) Wider

3.4.8 E

3.4.10 decreases, slowly

3.4.13

a)

n	SD of $\hat{p}$
10	0.158
20	0.112
40	0.079
100	0.050
500	0.022
1000	0.016

b)

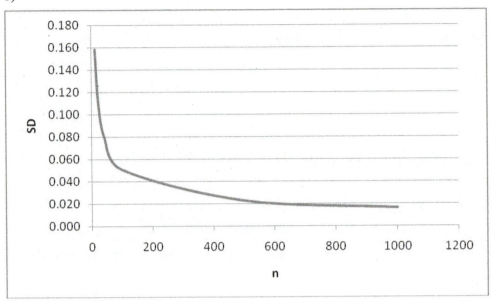

c) decreases

d) less

3.4.14

a) Rogers, the benefit of the additional 25 flips decreases as the sample size increases

b) The will go down similar amounts because both Rogers and Hammerstein are doubling the number of flips they are doing

3.4.23 A

3.4.24 D

3.4.25 decrease

3.4.26 increase

3.4.27 a) remain same b) increase

3.4.28 increase

3.4.29 decrease

3.4.30 95%

3.4.31 99%

3.4.33 B

Section 3.5

3.5.1

a) If one coffee bar opens earlier than the other, the lines may be different at the bars when they first open (if one opens before first class starts, may have a longer line than one that opens after first class starts.)

b) A more representative sample of students might be observed at different times of the day and on different days of the week.

3.5.3 The question is awkwardly phrased. "Does it seem possible or impossible…it never happened?" The latter part ('impossible it never happened') involves a double negative.

3.5.4 This question is more clearly phrased, they got rid of the double negative

3.5.5. a)

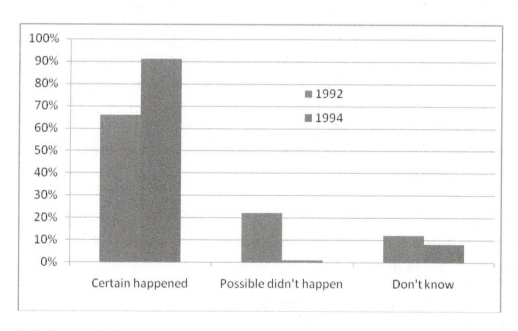

b) Substantially fewer individuals reported that it was 'possible it didn't happen' in 1994 as compared to 1992

c) for 1992: (0.6305, 0.6895) for 1994: (0.8922, 0.9278)

d) The confidence intervals are quite different, they do not overlap.

e) Wording of the question can dramatically impact survey results

3.5.7 Probably the question which did not give two options will yield a higher percentage of affirmative responses.

3.5.9 Affordable care act since Obamacare links the program to the president which may prove more divisive than the program itself.

3.5.11 Whether the respondent supports social/welfare programs may differ by race

3.5.12 Self reporting of sexually transmitted diseases might differ based on gender of interviewer

3.5.13 If the clothing worn by the interviewer suggests that interviewer is wealthy, the respondent may be less likely to report support of increased taxes of the wealthy

3.5.14 If the interviewer is smoking, respondents may be less likely to report anti-smoking opinions

3.5.17 a) Maybe overstate, since students may tend to think they are getting more sleep than they are b) Maybe overstate, since students may tend to think they are volunteering more than they are c) Maybe overstate, since students may tend to think they are attending church more often than they actually are, d) Maybe overstate since students may tend to think they are studying more than they actually are e) Maybe overstate since students may tend to think they are wearing a seat belt more often than they actually do

3.5.18 a) p-value b) confidence interval c) Statistical significance tells you about the strength of evidence against a hypothesized value while practical importance tells you about where we believe the parameter exists

3.5.19 It matters for large sample sizes because things can be statistically significant without being practically important. For small sample sizes differences can be not statistically significant, but the true difference may be practically important (further studies would be needed to determine that).

3.5.23 a) (0.4242, 0.4274) b) Yes, significantly less than 50%. The difference may be only modestly practically important since it's not that much lower than 50%. c) this sample was taken in such a way that we should have major concerns about the ability to infer that the 350,847 participating in the survey are representative of all US citizens

3.CE.1

a) 0.115 to 0.229

b) We can be 95% confident that between 11.5% and 22.9% of all water specimens from aircraft carrying domestic and international passengers will test positive for the presence of bacteria, thus failing to meet federal safety standards.

3.CE.3

C. Same midpoint, Frank has wider interval

3.CE.4

B. Same midpoint, Elsa has wider interval

3.CE.6

a) It is unlikely that our sample proportion will be the same as the population proportion, but if the sample was taken randomly, we can be fairly sure that it is close and that closeness can be represented through an interval estimate.

b) Country B will have the widest interval since it had the smallest sample size. Country C will have the narrowest since it had the largest sample size.

c) The interval for country B will contain 0.2.

d) The p-value for this test for country B will be large (above 0.05) since the interval contained 0.2. The p-value for this test on the other two countries will be small (less than 0.05) since their intervals did not contain 0.2 and so 0.2 is not plausible.

3.CE.7

a) A: $z=8.16$, B: -1.44, C: 3.95, b) Researcher A has strongest evidence, Researcher B has least evidence

3.CE.10

a) 0.121 to 0.159.

b) Values between 0.121 and 0.159 should not be rejected at the 0.01 significance level since they are plausible values for the population proportion.

3.CE.11

a) Null Hypothesis: One-third of all American households include a pet cat. Alternative Hypothesis: A proportion different than one-third of all American households include a pet cat $z = -4.29$ and p-value < 0.0001. We can conclude that the proportion of American households that include a cat is not one-third. Since our sample proportion is less than one-third, we can also conclude that the population proportion is also lsess than one-third.

b) The p-value turned out to be so small because the sample size is huge.

c) 0.318 to 0.330.

d) Yes, one third (or about 0.3333) is not included in the interval.

e) The interval is very close to one-third.

3.CE.13 Because they will often be extremely wide and uninformative

3.CE.14 Because the difference in width between an interval based on a sample of 1000 Americans vs. 1,000,000 Americans may not be that practically different, but the sample of 1,000,000 Americans will cost substantially more

3.CE.15 Because 70% is not a very high confidence level, researchers often want to be more confident in their conclusion than just 70% confident

3.CE.16 a) Increase b) Decrease

3.CE.19 a) Statistics because they are based on samples b) (0.2863, 0.3057). We are 95% confident that the proportion of Americans who reported that the state of the US economy would affect their Halloween spending is between 28.63% and 30.57% c) The sample standard deviation

3.CE.20 a) $55.67, $56.95, We are 95% sure that the average amount Americans will spend on Halloween is between $55.67 and $56.95. b) ($55.25, $57.37). The interval is wider. c) The distribution of expected Halloween spending amounts should not be strongly skewed

3.CE.23 a) The distribution of difference in ages is fairly symmetric and centered slightly above zero, indicating a typical difference of approximately 1 to 2 years in male and female ages (males slightly older), though the overall distribution ranged from -7 to 18, with a slight tail to the right indicating a handful of marriages where the male is more than 10 years older than the female. b) Yes, data is not strongly skewed and sample size is more than 20. c) 0.5943, 3.2457, d) We are 99% confident that the true difference in male and female ages of soon to be married couples in Cumberland County is between 0.59 and 3.25 e) Yes, 0 is not in the interval f) soon to be married couples in Cumberland County

3.CE.24 a) 27 out 100 (0.27) b) Yes, because there are at least 10 couples where wife is older than husband (27), and at least 10 couples where husband is at least as old as wife (73). c) (0.197, 0.343) d) Yes, 50% is not in the interval.

3.CE.25 a) INVALID, b) INVALID, c) INVALID, d) VALID

3.CE.27 a) The random sample because the 4.5% of women who returned the questionnaire are likely quite different than the 95.5% of women who didn't b) (0.3938, 0.4862) c) We are 95% confident that proportion of all American women who believe that they give more emotional support than they receive is between 39.38% and 48.62%. d) The Hite survey would have a smaller margin of error because the sample size is larger.

Chapter 4

4.1.1 a) B, b) A, c) A, d) Categorical, e) Quantitative

4.1.2 B

4.1.3

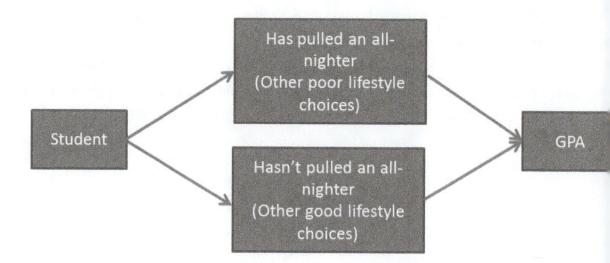

In short, students who pull all-nighters are more likely to make other poor lifestyle choices (e.g., poor diet, lack of exercise, etc.) and those lifestyle choices may have a negative impact on GPA

4.1.5 a) 2622 adults in the survey b) Explanatory = Generation, Response: Marriage obsolete c) Yes, the proportions are different between the groups suggesting the different generations tend to view marriage differently

4.1.7 A confounding variable may explain this relationship: Wealth of the country (e.g., GDP). In particular, wealthier countries have better healthcare and longer life expectancy, and also can afford to buy more televisions

4.1.8 a) C, b) B, c) Quantitative, d) D, e) Quantitative

4.1.10 a) B, b) C, c) Yes, the proportions are quite a bit different

4.1.12 a) Exp: Eat breakfast, Resp: Maintain weight loss, b) Getting enough sleep

4.1.14 a) Pregnant women b) Smoking status (Yes/no), categorical c) Weight of baby (quantitative), d) people of lower socio-economic status are more likely to smoke, and also more likely to have a poorer diet (and other factors) that may lead to lower weight babies

4.1.16 Since 46% of people who are overweight have overweight friends, compared to only 30% of people who are not overweight there is an association; the proportions are different

4.1.18 a) Income, categorical, b) Happiness, categorical, c) The proportions are different between the groups showing evidence of association. d) can't conclude cause and effect. A potential confounding variable is intelligence

4.1.20 a) Exp: Political party (categorical), Resp: Humans evolved or not (categorical), b) Yes, the proportions are different

4.1.22 a) Attend religious service, categorical b) Blood pressure, quantitative c) Diet could be associated with church attendance and blood pressure

4.1.24 a) Political party, categorical b) Ignore federal gun laws? Categorical c) Yes, the proportions are different between the different political parties

4.1.26 a) Countries b) Exp: Chocolate consumption, Resp: Number of nobel prizes c) Wealth of nation.

Section 4.2

4.2.1 A

4.2.3 A

4.2.5   Random sampling

4.2.7   Random assignment

4.2.9 Random assignment

4.2.11   Yes, a study can have both random sampling and random assignment. In this case researchers can generalize to the population and infer cause-and-effect

4.1.13   No, not always. On average random assignment balances variables between two groups, so there is a tendency for balance, but confounding variables won't be exactly balanced every time due to the randomness in the assignment

4.2.15   a) Yes, cause-and-effect; type of video changed quiz score b) No cause and effect c) No cause and effect

4.2.17   a) Observational studies b) No c) babies d) mother smoked while pregnant (yes/no) e) Weight of baby at birth

4.2.18   a) To conduct a randomized experiment you would have to randomly assign some mothers to smoke and others not to. This would be unethical, especially if we feared potential negative health outcomes for the baby b)Diet, other healthy living behaviors of the mother

4.2.20   a) Observational b) No c) children d) classical music e) Intelligence

4.2.22   a) Experiment because of random assignment to receive friend request from male or female student b) Students who received friend requests on face book c) Sex of person giving friend request (categorical; explanatory), Accepted friend request (categorical, response) d) No, all students from one class, so can't generalize to all students at the college, or elsewhere e) Yes, individuals were randomly assigned to receive the friend request from a male or female f) Yes, because of the random assignment

4.2.24   a) Experiment, researchers determined who would be in which pose b) Participants (42 of them) c) Pose (categorical; explanatory), Took bet? (categorical; response) d) No random sampling— participants were volunteers; Yes, random assignment---researchers random assigned who would be in which pose e) Yes, because the random assignment suggests cause-and-effect is possible

4.CE.1A

4.CE.3a) Type of visit b) Heart rate, blood pressure, anxiety levels, etc. c) Randomized experiment because patients were randomly assigned to one of the three types of experimental groups d) To potentially draw a cause-and-effect conclusion e) No, they were hopefully controlled for by the random assignment, eliminating them as potential confounding variables

4.CE.5a) Explanatory: Price of pill (categorical), Response: Reduction in pain? Yes or no (categorical), b) Randomized means subjects were randomly assigned which price group they were in c) Double-blind means neither the participant nor the researchers interacting with the participants knew which group the participants were in d) The results are statistically significant, meaning that there is evidence of a potential cause-and-effect relationship between the price of the pill and perceived reduction in symptoms. Because there is not random sampling this result should be generalized, with caution, only to individuals with similar characteristics of the volunteers

4.CE.7 a) Explanatory: Gender, Response: Satisfaction with attractiveness, b) Observational study c) Random sampling, to generalize to all US adults, but not random assignment because you can't random assign people's gender to them

4.CE.9a) Experiment because visitors to google were assigned to see one add or the other b) Explanatory: Name of book on ad, Response: Click through? Yes or no, c) Yes, assuming google visitors were randomly assigned which ad to see

4.CE.11 a) Explanatory: US or UK (categorical), Response: Number of Harry Potter books read (quantitative) b) Random sampling because you want to generalize to US and UK teenage populations, random assignment makes less sense because you would have to randomly assign teenagers to live in the US or UK

4.CE.13Randomly assign student essays to be graded by a professor with a blue or red pen, and then keep track of the scores given on the essays

4.CE.15Randomized experiment is not really feasible because you would have to randomly assign some people to go to church regularly and others not

Chapter 5

5.1.1 a) Tables 1 and 4 have the same pair of conditional proportions. Namely, 0.5 of A's are Yes, and 0.5 of B's are Yes in both tables b) The difference in conditional proportions is largest in Table 3, where 0.667 of A's are Yes and 0.333 of B's are Yes. c) The difference in conditional proportions is largest in Tables 1, 2, and 4, where the difference equals zero because the proportion of A's that are Yes equals the proportion of B's that are Yes.

5.1.3

		Male	Female	Total
Played sports in high school	Yes	83	117	200
	No	67	93	160
Total		150	210	360

5.1.5 a) The observational units are the college students b) The response variable is whether or not the student is wearing clothing that displays the college name or logo on a particular day. Categorical c) Yes, random sampling can be used. Obtain a numbered list of the students at each college, and use software to randomly select the number that you want for the sample size. d) No, random assignment cannot be used. You cannot randomly assign students to attend one college or the other.

e) A 2×2 table suitable for summarizing the data would be:

	ESU	MSU	Total
Wearing school name/logo			
Not wearing school name/logo			
Total			

5.1.7 a) More rattlesnakes were caught at site G than at site B, so looking at conditional proportions (rather than counts) takes this into account. b) The explanatory variable is site, so we calculate the conditional proportions of males/females at each site c) The proportion of rattlesnakes that are female are $11/21 \approx 0.524$ at site b and $12/33 \approx 0.364$ at site g. Reasoning informally, these proportions appear to be different enough to provide evidence of an association between site and sex of rattlesnakes.

5.1.12 a) Observational study because researchers did not assign respondents to attend church in differing amounts b) Explanatory: Amount of church attendance, Response: Smoke (yes/no), c)

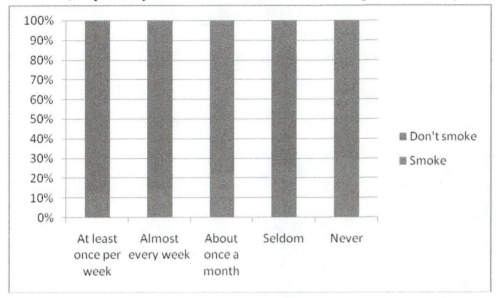

d) Because the percentages are different, this is evidence of an association between regularity of church attendance and smoking e) No, because this is an observational study

5.1.13 a) $0.30/0.12 = 2.5$. People who never attend church are 2.5 times as likely as those who attend church weekly to be smokers b) Seldom vs. At least once: 2.08, About once p/month vs. At least once: 1.83, Almost every week vs. At least once: 1.17

5.1.16 a) Experiment, b) Explanatory: Receive AZT yes/no, c)

	AZT	No AZT	Total
HIV pos	13	40	53
HIV neg	151	120	271
Total	164	160	324

d) 0.079 (AZT) vs. 0.25 (no AZT), e) 3.16, f)

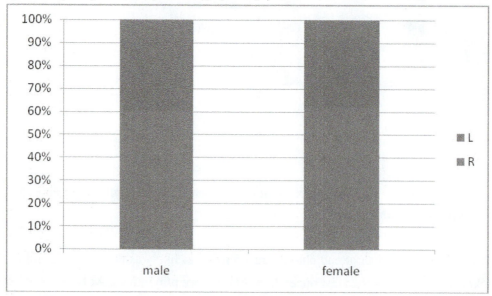

g) Due to the difference in conditional proportions there appears to be an association between taking AZT and a reduction in the chance that the child of an HIV positive woman will be HIV positive

5.1.19

a)

	Male	Female	Total
Coffee	5	6	11
No Coffee	5	10	15
Total	10	16	26

b) Males: 0.5, Females: 0.375

c)

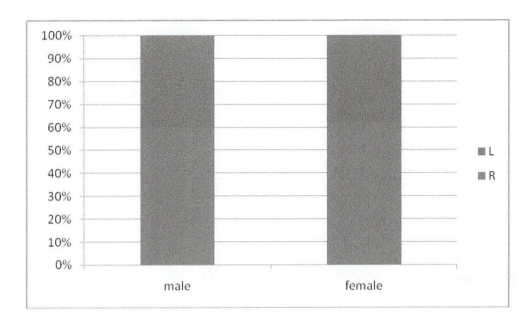

d) Because the proportion of coffee drinking men is different than coffee drinking women this is some evidence of an association between gender and coffee drinking

5.1.20

a)

	Male	Female	Total
R	6	8	14
L	4	5	9
Total	10	13	23

b) Proportion of people choosing R, Males: 0.6, Females: 0.62

c)

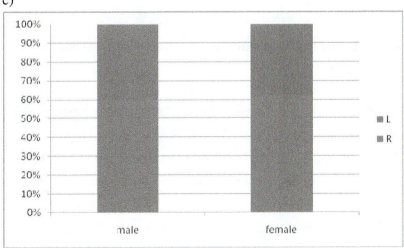

d) The proportions of men and women choosing R are quite similar suggesting little association between gender and letter choice

5.1.21

a)

	Coffee	No Coffee	Total
R	9	5	14
L	2	7	9
Total	11	12	23

b) Proportion of people who chose R, Coffee: 0.82, No coffee: 0.42

c)

d) There is an association between letter choice and coffee drinking because the proportions are quite different

5.1.24 a) Male acceptance rate: 533/1198=0.445, Female acceptance rate: 113/449=0.252. Males have a higher acceptance rate overall b) Males: 511/825=0.619, Females: 89/108=0.824. Females have higher acceptance rate. c) Males: 22/373=0.059, Females: 24/341=0.07, d) Yes, even though the overall proportion of Males accepted is higher, Females have a higher acceptance rate within each group e) Program F is highly competitive, but Program A is not. However, the majority of female applicants received applied to Program F, while the majority of males applied to program A. Thus, the overall proportion of males accepted was higher.

5.2.1 a) C, b) D, c) B, D, d) C, e) C, D, f) 30 in each pile, representing 30 medium claims and 30 small claims (explanatory variable), g) Difference in conditional proportions or relative risk h) Find how often the observed statistic value or more extreme occurred in the simulation (both tails since two-sided)

5.2.2 a) 0, because if the null hypothesis is true there is no difference in the two group proportions b)

c) There is a 0.184 chance of getting 0.20 (the value of the statistic) or larger, or -0.20 or smaller, if the null hypothesis is true

d) There is little to no evidence against the null hypothesis.

e) (0.20-0)/0.125=1.6. The statistic is approximately 1.6 standard deviations from the mean of the null distribution

f) They give similar indications about evidence---there is not much evidence against the null hypothesis. Note: The standardized statistic (1.6) is near the 'cutoff' of moderate evidence, but this modest difference isn't a problem, neither gives 'strong' evidence

g) $0.20 \pm 0.125 \times 2$ (-0.05, 0.45).

h) We are 95% confident that the true difference in proportions of overpaid claims between the two groups is between -5% and 45%.

i) Yes, 0 is in the interval meaning that it's plausible that there is no difference in the proportions

j) We have little to no evidence that the proportion of overpaid claims is different between the two groups. We're 95% sure that true difference in the proportion of overpaid claims is between -5% and 45%. This is a random sample, and so this conclusion can be generalized to other Medicare/Medicaid claims at the company being audited, however this is not a randomized experiment and so cause-and-effect conclusions cannot be drawn.

5.2.7 a) C, b) C, c) A, d) B, e) Categorical f) Categorical g) A, D h) The difference happened by chance because no difference in survival rate between CC and CPR techniques or there is an actual difference in the survival rate from the two different techniques

5.2.8 a) A, b) B, c) 3031, d) Two different colors, 389 of one color and 2642 of the other color, e) Two piles, 1500 in one pile, 1531 in the other pile because that's how many people receive the two techniques in the actual study f) The difference in the proportion of one of the colors of cards between the two groups subtracting the group with 1531 from the group with 1500 (or use ratio of proportions) g) Find the proportion of times that the actual difference in the study (0.024) or larger occurred in the simulated statistics h) C

5.2.9 a) p-value is approximately 0.024. We obtained statistics 0.024 or larger in 2.4% of simulations b) We have strong evidence against the null hypothesis c) 0.024/0.012=2. The observed statistic is 2 SDs from the mean of the null distribution. d) Yes, both suggest strong evidence against the null hypothesis e) 0.024 ±- 2×0.012 = (0,0.048). We are 95% sure that the difference in survival rates when comparing CC to CPR is between 0 and 4.8% better with CC f) Yes, more or less. 0 is (barely) in the interval and the p-value rejects 0. Note that the test is one-sided and the interval is two sided, and so the strength of evidence will change a bit depending on which approach you take g) We have strong evidence that the true difference in survival percentages is greater for CC than for traditional CPR (p=0.024; 95% CI: 0, 0.048). Because this is a randomized experiment, this suggests that the use of CC is causing the survival rate to increase. There is no information provided as to how the sample was gathered, but it was likely not a random sample of calls for help, but instead all calls during certain time periods in one or more cities. Thus, this result may not generalize to other locations with different types of people who may respond differently to different types of CPR.

5.2.10 a) 0.042 b) 0.024 c) The Hallstrom et al. study had a larger difference in proportions d) The one-sided p-value is 0.10. This is the probability of obtaining 0.042 or larger in the simulated data e) We have little to no evidence against the null hypothesis f) The Hallstrom et al. study provides less evidence than the meta-analysis g) The sample size is smaller h) The 95% CI for the Hallstrom study is approximately -0.016 to 0.099. The Hallstrom study has a wider interval than the meta-analysis (0, 0.042) because the sample size is larger. The Hallstrom study CI is centered on 0.042, while the meta-analysis CI is centered on 0.024. CIs are centered on the statistic.

5.2.15 a) 3/20 − 18/22 = -0.668 b) There is an actual difference in the likelihood of taking the double or nothing bet based on pose or this result occurred merely by chance

5.2.16 a) The parameter is the difference in probabilities of selecting the high-risk option between those who held a high-power pose and those who held a low-power pose. This parameter can be denoted as $\pi_{high} - \pi_{low}$

b) The null hypothesis is that the probability of selecting the high-risk option is the same, regardless of whether the person holds a high-power pose or a low-power pose. The alternative hypothesis is that the probability of selecting the high-risk option is higher for a person who holds a high-power pose than for a person who holds a low-power pose.

c) $H_0: \pi_{high} = \pi_{low}$, $H_a: \pi_{high} > \pi_{low}$

d) Use 21 red cards to represent the people who chose the risky (double-or-nothing) option, and use 21 black cards to represent the people who chose the safe option. Shuffle the 42 cards thoroughly. Deal out 20 cards to represent the people assigned to the low-power pose, with the remaining 22 cards then representing the people assigned to the high-power pose. Count how many red cards (representing the high-risk choice) are in each group. Calculate the proportion of red cards in each group, and subtract those proportions, taking the high-power pose group's proportion minus the low-power pose group's proportion. Repeat this process a large number of times, say 1000. The p-value will be approximated by the proportion of repetitions in which this difference is ($18/22 - 3/20 \approx 0.668$) or larger.

5.2.17a) The approximate p-value is 0.000, which means that a difference in group proportions who choose the risky option of 0.668 or larger would almost never occur by random assignment alone if there were really no difference in the probabilities of selecting the risky option between the two groups.

b) The very small p-value indicates that the experimental data provide extremely strong evidence that people who hold a high-power pose are more likely to choose the risky, double-or-nothing option than people who hold a low-power pose.

c) $-0.668/0.157 = 4.25$. The observed statistic is 4.25 SDs from the mean of the null distribution.

d) Yes, both give very strong evidence against the null hypothesis

e) $-0.668 \pm 2 \times 0.157 = (-0.982, -0.354)$. We are 95% confident that the group in the low power pose takes the double or nothing between 0.982 and 0.354 less often than the group in the high power pose.

f) Yes, 0 is not in the interval.

g) We have very strong evidence that the pose yields different proportions of people taking the bet ($p<0.001$). We are 95% confident that the difference in actual probability of choosing the risky option is between 0.360 and 0.976 higher for people who hold a high-power pose than for people who hold a low-power pose. Because this was a randomized experiment, we conclude that the high-power pose causes an increased probability of selecting the risky option. We are not told how the participants were chosen, so it's not clear how broadly we can reasonably generalize this conclusion to other people beyond those involved in this experiment.

5.2.18 a) 5.46 (or 0.183) b) Null: The relative risk is 1, Alt: The relative risk is not 1, c) The p-value is <0.001. Meaning that the chances of getting a relative risk as or more extreme than the one observed (5.46) is <0.001 d) The conclusions are the same because the data is not changing, just our way of summarizing the data.

5.2.22 a)

		College degree	No college degree	Total
Knows Census required by law	Yes	195	271	466
	No	331	702	1033
Total		526	973	1499

b) 271/973 - 195/526 = -0.092, c) The proportion of people who know that participating in the Census is required by law is the same for those with a college degree as for those without, the observed difference is due to chance alone or there is actually a difference in the two proportions in the population d) Observational study, the researchers have not assigned the values of either variable to the observational units

5.2.23 a) Null: There is no association between education level and knowledge that the census is required by law, Alt: There is an association between education level and knowledge that the census is required by law b) Null: $\pi_{College} = \pi_{NoCollege}$, $\pi_{College}$ = Proportion of people with college degree in population of all US adults who know participating in the census is required by law, Alt: $\pi_{College} \neq \pi_{NoCollege}$ c) Use 466 red cards to represent the people who know census is required and 1033 black cards to represent the people who don't. Shuffle the 1499 cards thoroughly. Deal out 526 cards to represent the college group, with the remaining 973 cards representing the no college group. Count how many red cards are in each group. Calculate the proportion of red cards in each group, and subtract those proportions, taking the no college proportion minus the college proportion. Repeat this process a large number of times, say 1000. The p-value will be approximated by the proportion of repetitions in which this difference is -0.092 or larger or 0.092 or smaller.

5.2.24 a) The p-value is approximately 0, meaning that we did not observe any simulated differences in proportions larger than 0.092 or smaller than -0.092, b) We have very strong evidence against the null hypothesis c) -0.092/0.025=3.68. The observed statistic is 3.68 standard deviations from the mean of the simulated null distribution d) Yes, both give very strong evidence against the null hypothesis e) -0.092 $\pm$ 0.05 = (-0.142, -0.042). We are 95% confident that the difference in the proportions of people who know the census is required by law when comparing people with a college a degree and people without is -0.142 and -0.042; people with a college degree are between 4.2% and 14.2% more likely to know this. f)

Yes, 0 is not in the interval. g) We have very strong evidence of an association between education level and knowledge that the census is required, with between 4.2% and 14.2% more people with a college education knowing the census is required when compared to people without a college education. This result generalizes to the population of all US adults, but does not suggest a cause-and-effect relationship between having a college degree and knowledge of the census law.

5.2.26 a) A, b) B

5.2.29 a) Experiment, because some households were assigned to receive incentive and others not b)

		Incentive- Yes	Incentive – No	Total
Participated?	Yes	286	245	531
	No	82	122	204
Total		368	367	735

c) $286/368 - 245/367 = 0.11$, d) There is no effect of the incentive, and the difference we saw was due to random chance or there is an effect of the incentive

5.2.30 a) Null: There is no association between whether or not the household received the incentive and participation, Alt: There is an association between whether or not the household received the incentive and participation b) Null: $\pi_{incentive} = \pi_{no\text{-}incentive}$, Alt: $\pi_{incentive} \neq \pi_{no\text{-}incentive}$ where π is the long-run proportion of households that will participate c) Use 531 red cards to represent the households that participated and 204 black cards to represent the households that didn't. Shuffle the 735 cards thoroughly. Deal out 368 cards to represent the households receiving the incentive and 367 to represent the households that didn't. Count how many red cards are in each group. Calculate the proportion of participators in each group, and subtract those proportions. Repeat this process a large number of times, say 1000. The p-value will be approximated by the proportion of repetitions in which this difference is 0.11 or larger or -0.11 or smaller.

5.2.31 a) 0, none of the simulated statistics were 0.11 or larger, or -0.11 or smaller b) We have very strong evidence against the null hypothesis c) $0.11/0.033 = 3.33$, the observed statistic is 3.33 standard deviations from the mean of the null distribution d) Yes, both give very strong evidence against the null hypothesis e) $0.11 \pm 2 \times 0.033 = (0.044, 0.176)$. We are 95% confident that the proportion of homes that will participate when given a financial incentive is between 4.4% and 17.6% higher than when no incentive is given. f) Yes, 0 is not in the interval g) We have very strong evidence that the proportion of participating households is higher when a financial incentive is used than when no incentive is used (4.4% to 17.6% more households participate). This is a cause-and-effect relationship that generalizes to all households in the nation.

SECTION 5.3

5.3.1 The appropriate conclusion is:

    A. The sample data provide strong evidence that a higher proportion of American teenagers had hearing loss in 2005-06 than in 1988-94.

because we can be 95% confident that $\pi_{05\text{-}06} - \pi_{88\text{-}94}$ is between 0.0015 and 0.0467, so we can be 95% confident that $\pi_{05\text{-}06}$ is greater than $\pi_{88\text{-}94}$ by between 0.0015 and 0.0467.

5.3.2

Notice that the parameter here subtracts in the opposite order, so a 95% CI would turn out to be

    B. Negative of above: (-0.0467, -0.0015)

5.3.3

A. A larger confidence level will increase the length of the confidence interval.

5.3.4

The midpoint for the 99% confidence interval will be the same as the midpoint for the 95% confidence interval. The midpoint of the 95% confidence interval is $(0.0015 + .0467)/2 = 0.0241$. Therefore, we know that the 2005-6 sample proportion with hearing loss was 0.0241 larger than the 1988-94 sample proportion.

5.3.5

We cannot find the width exactly without knowing the sample sizes involved.

5.3.7

a) The parameters of interest are the population proportion of all American households that would respond to the survey when offered a monetary incentive ($\pi_{incentive}$), and the population proportion of all American households that would respond to the survey when not offered a monetary incentive (π_{non}).

b) The null hypothesis is that the population proportion of all American households that would respond to the survey is the same whether offered a monetary incentive or not. The alternative hypothesis is that the population proportion of all American households that would respond to the survey is greater when offered a monetary incentive than when not offered a monetary incentive.

c) $H_0: \pi_{incentive} = \pi_{none}$, $H_a: \pi_{incentive} > \pi_{none}$

d) All four values in the two-way table (286, 245, 82, and 122) are larger than 10, so the theory-based approach should be valid.

e) Using the Theory-Based Inference applet, we find the following output:

The following applet simulation reveals an approximate p-value of 0.000.

The standardized statistic is $z = 3.32$.

The theory-based one-sided p-value is 0.0005.

The 95% confidence interval is (0.0453, 0.1739).

g) The p-value says that if there were really no difference in the population proportions who would respond to the survey between those offered a monetary incentive and those not offered such an incentive, then there's only a 0.0005 probability of obtaining sample proportions as far apart, in the direction of favoring the incentive group, as was observed in this study (namely, a difference of 0.1096).

h) We are 95% confident that the proportion who will respond to a telephone survey after receiving an incentive is between 4.5% and 17.4% higher after receiving an incentive.

i) The p-value is small enough to indicate that the sample data provide very strong evidence that offering a monetary incentive does help to increase the response proportion to the survey. This is a randomized

1205

experiment, so we can conclude that the monetary incentive caused the higher response proportion. Furthermore, the households were randomly selected, so we can generalize this conclusion to all households in the U.S. The confidence interval reveals that the magnitude of the increase in response proportion is between 4.5 and 17.4 percentage points higher when the household is offered a monetary incentive than when no such incentive is offered.

5.3.11

a) These data rise from sampling from two processes.

b) Let π_A represent the long-run probability of author A winning and π_B the long-run probability of author B winning.

c) The null hypothesis would be that the authors have the same probability of winning. The (one-sided) alternative hypothesis is that author B has a higher probability of winning.

d) $H_0: \pi_A = \pi_B$
$H_a: \pi_A < \pi_B$

e) Using the Two Proportions applet, we find a p-value of approximately 0.11.

f) It would be ok to use a theory-based approach here because all four cell counts (25, 74, 192, 444) are larger than 10.

g) Using the Theory-Based Inference applet, we find a standardized statistic of $z = -1.69$, a p-value of 0.0917 and a confidence interval of (-0.1042, 0.005)

h) The p-value tells us that approximately 4% of random samples from processes with the same probability of success, would yield a difference in sample proportions as large as 0.050 or larger by chance alone (assuming the null hypothesis is true).

i) There is moderate evidence against the null hypothesis; At the 5% significance level there is not enough evidence to reject the null hypothesis

j) A 95% confidence interval for the difference in their winning probabilities ($\pi_A - \pi_B$) is (-0.1042, 0.005), indicating that B's win probability could be us much as 0.104 higher than A's or as much as 0.005 smaller than A's. This analysis depends on assuming that the sample of games is essentially a random sample for each player.

k) Zero is inside the 95% confidence interval, meaning that it's plausible there is no difference in the proportions.

l) The data provide only moderate evidence of a difference in the performance of Author A and Author B (p=0.09; 95% CI: -0.10 to 0.005). These results cannot be generalized to the typical performances of Author's A and B, unless the games in the sample are representative of their typical performance. There is no potential for a cause-and-effect conclusion because this is not a randomized experiment.

5.3.13

a) The parameters are the long term probabilities of men that have localized prostate cancer dying of the disease if they have surgery (call this proportion π_s) and if they are just observed (call this proportion π_o).

b) Null: The probability for men with localized prostate cancer and have surgery dying of the disease is equal to the probability for men with localized prostate cancer and are just observed dying of the disease. Alternative: The probability for men with localized prostate cancer and have surgery dying of the disease is different than the probability for men with localized prostate cancer and are just observed dying of the disease.

c) $H_0: \pi_s = \pi_o$ $H_a: \pi_s \neq \pi_o$

d) We can use the theory-based approach to test the hypotheses because at least 10 men died and 10 men survived in each group.

e) Our p-value is 0.1587 and the standardized statistic is -1.41.

f) We can be 99% confident that the probability of dying of prostate cancer for those that get surgery is between -0.0757 lower to 0.0221 higher than those that are just observed for 10 years.

g) Since our p-value is large we don't have strong evidence that the probability for men with localized prostate cancer and have surgery dying of the disease different than the probability for men with localized prostate cancer and are just observed dying of the disease (99%CI: -0.0757, 0.0221). If we did have significance, we could determine causation since it was an experiment. The ability to generalize this conclusion to a broader population is suspect since this study is not based on a random sample.

5.3.17 No, it is not valid to use a theory-based approach here because there are cells in the table with only 3 and 4. They all should be at least 10.

5.3.19

a) The parameters are the proportion of all U.S. adults with some college or less that knew responding to the Census is required by law (π_{SC}) and the proportion of all U.S. adults with a college degree or more that knew responding to the Census is required by law (π_{CD}).

b) Null: The proportion of all U.S. adults with some college or less that knew responding to the Census is required by law is the same as the proportion of all U.S. adults with a college degree or more that knew responding to the Census is required by law. Alternative: The proportion of all U.S. adults with some college or less that knew responding to the Census is required by law is different than the proportion of all U.S. adults with a college degree or more that knew responding to the Census is required by law.

c) H_0: $\pi_{SC} = \pi_{CD}$ H_a: $\pi_{SC} \neq \pi_{CD}$

d) Theory-based methods are valid since we have at least 10 people that knew and 10 people that didn't know responding to the Census is required by law for both groups.

e) The p-value is 0.0002 and the standardized statistic is -3.68.

f) We can be 95% confident that the proportion of U.S. adults with a college degree or more that knows that responding to the Census is required by law is 0.0422 to 0.1422 higher than for those with some college or less.

g) We have strong evidence that the proportion of all U.S. adults with some college or less that knew responding to the Census is required by law is different than the proportion of all U.S. adults with a college degree or more that knew responding to the Census is required by law (0.0422 to 0.1422 higher among those with a college degree). We can generalize this to the population of all U.S. adults since it was a random sample of all U.S. adults, but this does not provide cause-and-effect between education level and Census knowledge because it is not a randomized experiment.

5.3.21

a) The parameters are the probabilities of someone with cardiac disease dying of that disease for those with depression (π_D) and for those without depression (π_N).

b) Null: The probability of a depressed person with cardiac disease dying of that disease is the same as the probability of a non-depressed person with cardiac disease dying of that disease. Alternate: The probability of a depressed person with cardiac disease dying of that disease is different than the probability of a non-depressed person with cardiac disease dying of that disease.

c) H_0: $\pi_D = \pi_N$ H_a: $\pi_D \neq \pi_N$

d) Theory-based methods are valid since we have at least 10 people that died and 10 people that didn't in both groups.

e) The p-value is 0.0263 and the standardized statistic is 2.22.

f) We can be 95% confident that the probability of dying for those with cardiac disease and depression is between 0.0039 and 0.2091 higher than those with cardiac disease and no depression.

g) Since we have a small p-value, we have strong evidence that the probability of a depressed person with cardiac disease dying of that disease is different than the probability of a non-depressed person with cardiac disease dying of that disease, with depressed individuals risk of death between 0.0284 Since it is an observational study we can't conclude depression caused the deaths. We can probably generalize this to people similar to those that were in the study.

5.3.23

a) This is an experiment since the physicians were randomly assigned to treatment groups.

b) The explanatory variable is taking aspirin or placebo. The response variables is whether they suffered a heart attack or not.

c) Double blind means neither the tester nor the subject knows which treatment is being received. This helps control for any bias that could develop affecting the behavior of either the tester or the subject.

d)

		Placebo	Aspirin	Total
Heart Attack?	Yes	189	104	293
	No	10,845	10,933	21,778
	Total	11,034	11,037	22,071

e) -0.0077

f) Null: The probability of a heart attack for those taking aspirin is the same as for those taking the placebo. Alternative: The probability of a heart attack for those taking aspirin is different than that for those taking the placebo.

g) We get a p-value of 0.

h) Since our p-value is very small, we have very strong evidence that the probability of a heart attack for those taking aspirin is different (and less) than that for those taking the placebo.

i) We can use theory-based since there are more than 10 deaths and non-deaths in each group.

j) The p-value is 0 and the standardized statistic is 5.00.

k) Simulation-based and theory-based p-values are the same.

5.3.24

a) The parameters are the probability of a heart attack for those taking aspirin (π_A) and the probability of a heart attack for those taking not taking aspirin (π_p).

b) $H_0: \pi_A = \pi_P$ $H_a: \pi_A \neq \pi_P$

c) We can use theory-based since there are more than 10 deaths and non-deaths in each group.

d) We can be 99% confident that the probability of having a heart attack is between 0.0037 to 0.0117 higher for those not taking aspirin compared to those that do take aspirin.

e) We get a p-value of 0.

f) There is very strong evidence of a difference in the probability of heart attacks between aspirin takers and non-aspirin-takers since our p-value is so small.

g) We have strong evidence that the probability of having a heart attack is higher for those not taking aspirin compared to those that do (between 0.0037 and 0.0117 higher). We can conclude aspirin is causing this difference since this was a randomized experiment. We can probably infer these results to middle-aged to older men similar to those in the study.

h) Relatively speaking, the 99% confidence interval is narrow because we had such large sample sizes.

i) It all depends on your perspective if you think this is a very large difference or not. If you just look at the difference in sample proportions, it doesn't look that much different. However, if you looked at relative risk the difference in probabilities is quite large. If you think you are likely to have a heart attack, no matter what you look at, this difference might seem huge.

5.3.25

a) Rounding your final answer to two decimal places, you should get 5.00, the same standardized statistic as that in the applet.

b) Rounding your final answer to four decimal places, you should get 0.0047 to 0.0107, the same confidence interval as that in the applet.

5.3.26

a) The relative risk is 1.82.

b) Since our p-value was very small, we should not expect 1 to be contained in the confidence interval.

End of Chapter Exercises

5.CE.1a) Yes, the sample sizes in the two groups can be different. While this makes it unreasonable to compare the number of successes in each group, computing the conditional proportions between the two groups allows for meaningful comparisons to be made. b) Yes, small sample sizes alone are not enough to prevent comparisons to be made. The challenge, of course, will be that you will need to have a very large difference in the groups to be convinced. Consider a sample size of 5 in each group, where all the observations in Group A are success, and all in Group B are failure. This is fairly compelling evidence of a difference in the groups (p=0.008).

5.CE.3 For the first research question, the response variable (number of Harry Potter books) is quantitative, not categorical, so comparing proportions (as we are doing in this chapter) doesn't make sense. Similar to the first research question, the response variable for the second research question is also quantitative (milk production).

5.CE.5

a) Observational study
b) Explanatory variable: Minority group, Response: Type of coach
c)

		Minority	Non-minority	Total
First base	Yes	20	10	30
	No	7	23	30
	Total	27	33	60

d)   p-value is 0.001
e)   We have strong evidence that minority coaches are more likely to be first base coaches than non-minorities
f)   No, cause-and-effect conclusions are not possible. This is an observational study, not a randomized experiments.

5.CE.10 a) Random sampling only, random assignment of gender is not possible b)

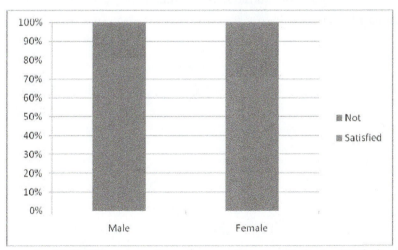

a) Null: The proportion of men who are satisfied with their attractiveness is the same as the proportion of women Alt: The proportions are different
b) The sample size in each group

5.CE.11 a) z=1.66, p=0.0978, b) z=3.7, p=0.0002, c) As sample size increases and the difference in proportions remains the same, the strength of evidence against the null hypothesis increases

5.CE.16

The proportion of males who consider life to be exciting in the sample is 52.8%; for women the proportion is only 48.6%, as shown in the bar chart below.

$$\hat{p}_1 - \hat{p}_2 = 0.042$$

This difference (4.2%) is not statistically significant (p=0.13; 95% CI on difference: -1.3%, 9.8%), meaning that we don't have evidence of a difference in the proportion of men and women who think that life is exciting.

5.CE.17

The proportion of males who consider life to be exciting or routine in the sample is 94%; for women the proportion is 95%%, as shown in the bar chart below.

$$\hat{p}_1 - \hat{p}_2 = -0.013$$

This difference (1.3%) is not statistically significant (p=0.30; 95% CI on difference: -3.8%, 1.2%), meaning that we don't have evidence of a difference in the proportion of men and women who think that life is exciting or routine.

Chapter 6

6.1.1. B

6.1.3. B

6.1.5. IQR

6.1.7. B

6.1.9. a) All 5-numbers will increase by 5; the IQR will not change because it is a measure of variability—the difference in the 3rd and 1st quartiles will not change b) The maximum will change, there will be little to no impact on the other four values in the five number summary or on the IQR c) The minimum will change, there will be little to no impact on the other four values in the five number summary or on the IQR

6.1.11. a) 25th value is 6 and 26th value is 7, so median is 6.5 b) lower quartile is 6, upper quartile is 9, c) Larger since the data is right-skewed

6.1.12.

Mean:	SMALLER	SAME	LARGER
Median:	SMALLER	SAME	LARGER
Standard deviation:	SMALLER	SAME	LARGER
Inter-quartile range:	SMALLER	SAME	LARGER

6.1.15 The shape of the female study hours distribution is right skewed, while the distribution for males is more symmetric. In particular, no males report more than 20 hours of studying per week, while a number of females report more than 20 (as many as 45 hours) of studying per week. Thus, the spread of the females distribution is quite a bit larger than for males. Finally, the center of the females study hour distribution is closer to 15-20 hours p/week, compared to only approximately 10 hours p/week for males.

6.1.16 a) Males: 10, Females: 15. Females tend to study more than males b) Males: 11, Females: 14. The variability in study hours p/week is larger for females

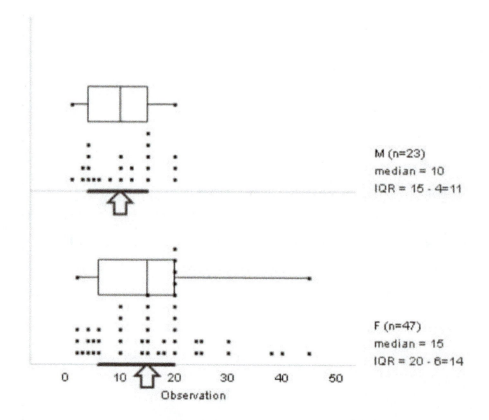

M (n=23)
median = 10
IQR = 15 - 4=11

F (n=47)
median = 15
IQR = 20 - 6=14

6.1.18 a) Min=-0.5, first quartile = 0.5, Median= 1.25, third quartile = 2, Max=4, b) 2-0.5=1.5 IQR c) No change to IQR, change to standard deviation

6.1.21 a) 85 seconds is Wendy's, 173 seconds is Hot N' Now, b) The IQR for Hot N' Now is 116.5 seconds, for Wendy's is 75 seconds c) For both restaurants the mean will be larger than the median since the data is right-skewed

6.1.22 a) Hot N' Now mean is 203 seconds, Wendy's is 93.7 seconds b) SD of Hot N' Now is 89.6 seconds, for Wendy's it is 46.7 seconds

6.1.24 a) Right b) Mean larger c) No because the distributions look fairly similar

6.1.27 a) Explanatory: Shelf level, Response: Sugar content, b) The distributions suggest less sugar content on high shelves (an association) c) Mean, Low: 11.925, Mean, High: 9.625—supports an association since they are different d) Median, low: 13, Median, high: 9—supports an association since they are different e) The medians are farther apart than the means are

Solutions to Section 6.2

6.2.1 a, b, c and d are all appropriate

6.2.2 a

6.2.3 b,c

6.2.5 a) Observational study b) Each student c) Explanatory: Gender (Cat), Response: Study hours (Quant) d) B e) F

6.2.11 a) Observational study b) Statistics students c) Explanatory: Gender, Response: Number of flip flops d) B, e) A f) B, C g) C

h)

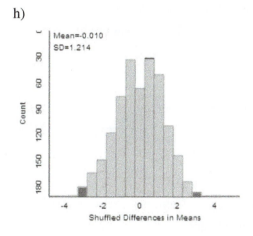

6.2.12 a) 3.68/1.214 = 3.03. b) We have strong evidence against the null hypothesis

6.2.13 a) In many simulations of the null hypothesis we got values of 3.68 or larger, or -3.68 or smaller 0.1% of the time. b) We have strong evidence against the null hypothesis

6.2.14 a) 3.68 ± 2×1.214 = 3.68 ± 2.43 = (1.25, 6.11) b) We are 95% sure that females own between 1.25 and 6.11 more flip-flops on average than males c) Yes, the confidence interval does not include 0

6.2.15 a)

Mean=0.014
SD=2.348

Count

Shuffled Differences in Medians

b) $5/2.348 = 2.13$

6.2.20 a) Obituaries=obs unit, had children yes/no=explanatory, age of death = response b) Observational study because the researcher did not assign variable values to the observational units c)

	Sample size	Sample mean	Sample SD
Had children	70	78.43	14.36
No children	20	63.9	25.81

d) $78.43 - 63.9 = 14.53$

e) There is a difference in the population mean lifespans of the two groups; There isn't a difference in the population mean lifespans, and the observed difference happened by chance

6.2.21 a) The population average lifespan of men with children (μ_{Had}); The population average lifespan of men without children (μ_{None}) b) Null: The population average lifespan with children is the same as without Alt: The population average with children is longer than without c) Null: $\mu_{Had} = \mu_{None}$ Alt: $\mu_{Had} > \mu_{None}$ d) Write out all 90 lifespans on 90 separate slips of paper. Shuffle the papers and deal into two stacks (70 and 20, respectively). Find the difference in the averages of the two stacks. Repeat many times, and count what proportion of the time values greater than 14.53 occur in the simulated data.

6.2.22 a) The p-value is approximately 0, meaning that we never got values of 14.529 or larger in the simulation b) We have very strong evidence against the null hypothesis c) $14.529/4.987 = 2.91$ d) $14.529 \pm 9.974 = (4.56, 24.503)$. We are 95% sure that the men with children lived between 4.56 and 24.503 years longer than those without children e) Yes, 0 is not in the interval f) We have strong evidence that the average life span of men with children is longer than those without; between 4.56 and 24.503 years longer on average. This is not a cause-and-effect conclusion (necessarily), but can be generalized to men with obituaries in the San Luis Obispo Tribune in 2012.

6.2.23 a) The p-value is approximately 0.009, meaning that we rarely got values of 16.5 or larger in the simulation b) We have very strong evidence against the null hypothesis (that the medians are the same in the two groups) c) $16.5/6.5 = 2.54$, d) $16.5 \pm 2*6.5 = (3.5, 29.5)$. We are 95% confident that the difference in population medians is between 3.5 and 29.5. e) Yes, 0 is not in the interval f) We have very strong evidence of a difference in the medians, in fact, men with children have a median lifespan between 3.5 and 29.5 years longer than men without children. While not necessarily a cause-and-effect relationship, the result can be generalized to men with obituaries in the San Luis Tribune in 2012.

6.2.26 a) Yes, the SD for carbon is higher b) Any value between 0 and infinity c) $6.25/4.89 = 1.28$ d) Write the 56 commute times on 56 slips of paper. Shuffle and deal into two piles of 26 and 30 each. Compute the SD of each pile, and then compute the ratio of the two SDs (SD of pile of 26 over SD of pile of 30). Repeat many times and see how often a ratio of 1.28 or larger is obtained in the simulations. e) 1 since having the SDs be equal is what happens if the null hypothesis is true f)

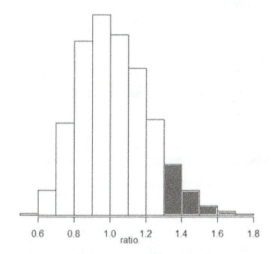

g) B
h) D

Solutions to 6.3 exercises

6.3.1. a) The difference in the population mean number of hours spent online daily, in particular how many more hours are spent by females then males b) (i)-no change (ii)- no change (iii) Will change in sign (negative to positive or vice versa) (iv)-no change (v) Will change in sign (vi)- no change (vii) Will change in sign (viii) Both will change in sign (ix) No change c) Yes, the exact same conclusion will be reached about strength of evidence against the null hypothesis. The CI will be different because it is estimating a different parameter, but will mean the same thing in terms of whether boys or girls are, on average, spending more hours online each day

6.3.4. Null: The population average VDAS score for early birds is the same as the population average VDAS score for night owls Alt: The population averages are different for the two groups. Since the sample sizes in each group are over 20 and the data within each group is not strongly skewed a theory-based approach is appropriate here. The p-value for this test is <0.0001, with a t-statistic of -5.06, meaning that we have very strong evidence that the population averages are different. In particular, the Early Bird group has between 0.5298 and 1.2102 lower VDAS scores on average (95% confidence) than the Night Owl group. We cannot infer cause-and-effect from this study because students were not randomly assigned to a sleep group. Furthermore, we cannot generalize to a broader population because this is not a random sample—more information about the characteristics of the students in the sample is needed to infer more broadly.

6.3.5. The test statistic (t) would have been even farther from 0, the p-value would be even smaller (stronger evidence against null) and the confidence interval would be centered on the same value, but narrower.

6.3.6. The test statistic (t) would have been closer to 0, the p-value would be larger (weaker evidence against the null) and the confidence interval would be centered at a different value, but would be the same width.

6.3.7. The test statistic (t) would have been closer to 0, the p-value would be larger (weaker evidence against the null) and the confidence interval would be centered at the same value, but would be wider.

6.3.8. Null: The population average fear score is the same for early birds as it is for night owls, Alt: The population average fear scores are different in the two groups Since the sample sizes are both over 20 and the data within each group is not strongly skewed a theory-based approach is appropriate here. The p-value for this test is 0.1722, with a t-statistic of -1.37, meaning that we have little to no evidence that the population averages are different. A 95% confidence intervals indicates that the Early Bird group average fear scores are between 0.6594 lower and 0.1194 higher than the Night Owl group. If the result was statistically significant, we could not infer cause-and-effect from this study because students were not randomly assigned to a sleep group. Furthermore, we cannot generalize to a broader population because this is not a random sample—more information about the characteristics of the students in the sample is needed to infer more broadly.

6.3.10 Null: The long-run average MRT score for men is the same as for women Alt: The long-run average MRT scores are different between the two groups. The validity conditions are met because the sample sizes are over 20 in each group and there is not strong skewness. The p-value (p<0.0001; t=5.57), means that there is very strong evidence that the long-run averages are different. A 95% confidence interval indicates that men are have an average MRT score between 2.92 and 6.16 higher than women. This does not suggest cause-and-effect because it is not a randomized experiment, and the results should be generalized to other men and women with caution because this was not a random sample and little information is given about the characteristics of the men and women in the study.

6.3.12 a) To see if the price of the pill was associated with it's perceived effectiveness (measured via pain tolerance) b) Null: The average maximum tolerance when taking the regular price pill is the same as when taking the discount price pill Alt: The average maximum tolerance for the regular price pill is different than for the discount price pill c) two-sample t test d) At least 20 people are in each group, and the data is not strongly skewed in either group and so the validity conditions are met e) p-value=0.51, t=-0.66. f) The probability of getting a t-statistic of -0.66 or smaller or 0.66 or larger by chance if the null hypothesis is true is quite likely (51% of the time) g) (-12.39, 6.19) h) We have little to no evidence of a difference in the average maximum tolerance between the regular-price and discount price pill (95% CI: -12.39, 6.19; p=0.51). This study gives the potential for cause-and-effect conclusion due to the use of random assignment, but not the ability to generalize to a larger population without some caution since a random sample was not used to obtain the studies participants.

6.3.14 a) Null: The average spatial score for men is the same as for women Alt: The average spatial score for men is different than for women b) The sample sizes are at least 20 in both groups, and the data is not strongly skewed within either group c) p-value=0.004; The probability of obtaining a t-statistic of 3.07 or larger, or -3.07 or smaller is 0.004 if the null hypothesis is true d) 95% CI: 0.81, 3.99. Men have an average spatial score between 0.81 and 3.99 higher than women e) We have strong evidence that the average spatial score is significantly higher for men (95% confident between 0.81 and 3.99 higher) than for women. This is not necessarily a cause-and-effect conclusion, nor can the result be confidently generalized to a broader population since the sample was not obtaining randomly from a larger population. f) Null: The average verbal score for men is the same as for women Alt: The average verbal score for men is different than for women. The sample sizes are at least 20 in both groups, and the data is not strongly skewed within either group for the theory-based test on this data is valid. p-value=0.0002; The probability of obtaining a t-statistic of 4.24 or larger, or -4.24 or smaller is 0.0002 if the null hypothesis is true d) 95% CI: -6.22, -2.18. Men have an average verbal scores between 2.18 and 6.22 lower than women e) We have strong evidence that the average verbal score is significantly higher for women (95% confident between 2.18 and 6.22 higher) than for men. This is not necessarily a cause-and-effect conclusion, nor can the result be confidently generalized to a broader population since the sample was not obtaining randomly from a larger population.

6.3.15 a) The difference in the average spatial test scores between men who breathe through their left nostril as compared to their right nostril b) Null: The average spatial score for men who breathe through their left nostril is the same as for men who breathe through their right nostril, Alt: The average spatial scores are different for the two groups, c) two-sample t-test d) Since the sample size is less than 20 in both groups, the data should be distributed symmetrically in both groups, and it is e) p-value is 0.035 (t = -2.30). f) We have strong evidence that the average spatial score for men is different based on whether or not they breathe through their left or right nostril (0.2 to 4.8 higher on average when comparing the two groups). This suggests a cause-and-effect relationship between spatial score and nostril due to the random-assignment, however this result should be generalized with caution since the sample was not obtained randomly g) p-value will get smaller h) No change i) p-value would get smaller

6.3.16 a) We don't have enough evidence of a difference at the 1% significance level since 0.0162 is not less than 0.01 b) There would be enough evidence since 0.0162<0.05 c) ii) and iii) are both true, i) and iv) are both false.

6.3.19 The samples in each group are above 20 (40 in each group) and the data is not strongly skewed, so a two-sample t-test is appropriate here. Null: The population average time in the bathroom for men is the same for women, Alt: The population average time in the bathroom is different. t=1.68, p=0.097. We have moderate evidence that the population average wait times are different between the two restaurants.

6.3.20 We are 95% confident that the average time in the bathroom is 5 seconds less to 58.7 seconds longer for women than men. The interval does include 0 which is consistent with the test of significance in the previous question..

6.3.22 a) Randomized experiment because randomness was used to determine whether or not the waitress gave her name b) Explanatory: gave name yes/no, Response: Amount of tip c) Null: The average tip amount when the waitress gives her name is the same as when she doesn't give her name, Alt: The average tip amount is larger when she gives her name d) $n_{Name} = 20, \bar{x}_{Name} = \$5.44, s_{Name} = \$1.75, n_{NoName} = 20, \bar{x}_{NoName} = \$3.49, s_{NoName} = \$1.13$. e) Yes, sample size is 20 in both groups, without strong skewness f) t=4.19, p=0.0001, g) We have strong evidence that the average tip amount is higher when she states her name. This means the waitress will probably start stating her name every time to boost tips amounts on average.

6.3.23 a) 95% CI: 1.00 to 2.90. b) Tip amounts are between $1 and $2.90 larger on average when she states her name c) Yes, 0 is not in the interval d) We have very strong evidence that the waitress receives larger tip amounts, on average, when she states her name (between $1.00 and $2.90 more). This is cause-

and-effect relationship due to the random assignment, but the waitress should be cautious in generalizing since this was only with the 40 customers she had on a particular Sunday morning for brunch.

6.3.24 a) Midpoint same, wider interval b) (0.81, 3.09). As predicted.

6.3.25 a) t-statistic farther from 0, p-value smaller and narrower confidence interval b) t=5.92, p<0.0001, 95% CI: 1.29, 2.61. As predicted.

6.3.26 a) t-statistic closer to 0, p-value larger and wider confidence interval b) t=3.08, p=0.0019, and 95% CI: 0.67 to 3.23. As predicted.

Solutions to end of chapter problems

6.CE.1. a) The difference in the population mean number of Harry Potter books read by girls vs. boys, in particular how many more books are read by girls compared to boys b) (i)-no change (ii)- no change (iii) Will change in sign (negative to positive or vice versa) (iv)-no change (v) Will change in sign (vi)- no change (vii) Will change in sign (viii) Both will change in sign (ix) No change c) Yes, the exact same conclusion will be reached about strength of evidence against the null hypothesis. The CI will be different because it is estimating a different parameter, but will mean the same thing in terms of whether boys or girls are, on average, reading more Harry Potter books.

6.CE.3 a) Each game is an observational unit. The variables are explanatory (categorical: Regular vs. Replacement), and response (duration, quantitative) b) Neither c) The dotplots suggest that games with replacements tended to be longer d) Null: The long-run average game duration is the same for replacement referees and regular referees, Alt: The long-run average game durations are different e) p-value is 0.009, f) The probability of obtaining a difference of 8.035 or larger (or -8.035 or smaller) in the sample means if there is no difference in the long-run average durations only occurs about 0.9% of the time g) $8.035 \pm 3.05 \times 2$, (1.94, 14.14) h) Games with replacement referees take between 1.94 and 14.14 minutes longer on average than games with regular referees. i) We have strong evidence of a difference in the long-run average length of games officiated by replacement referees, with such games taking between 1.94 and 14.14 minutes longer on average than games officiated by regular referees. This is not a cause-and-effect relationship, nor can this result, necessarily, but generalized to a broader set of NFL games.

6.CE.8 a) Confession: Min=2, Q1=9, Median=13, Q3=17, Max=39, Boone: Min=3, Q1=11, Median=14, Q3=20, Max=47 b)

Boone (n=29)
median = 14
IQR = 20 - 11=9

Confession (n=55)
median = 13
IQR = 17 - 9=8

Score

c) Based on the five number summaries and boxplots, there does not appear to be much of a difference in the sentence lengths between the two books

6.CE.9 a) Null: The average sentence length in Confession is the same as in Boone, Alt: The average sentence lengths are longer in confession b) Boone, mean=15.931, SD=9.075, Confession, mean=13.982, SD=8.027. c) The sentence lengths are longer in the sample from Boone, so there will not be evidence they are longer in Confession!

6.CE.11 a) Null hypothesis: Average SAT score improvement with coaching is the same as without coaching, Alt hypothesis: Average SAT score improvement with coaching is more than without coaching. t=4.28, p<0.0001. We have strong evidence that SAT scores improve more with coaching than without coaching b) 99% CI: (0.72, 2.88) Average SAT improvement scores are between 0.72 and 2.88 higher in the with coaching than the without coaching group. c) Yes, the CI does not include 0 d) Yes, the p-value is very small e) No, the CI only indicates a small (0.7 to 2.9) improvement on average f) Just because something is statistically significant it does mean that it is, necessarily, practically important

Chapter 7

7.1.1a) False b) True c) True d) True

7.1.3a) True b) False c) True

7.1.5a) Appropriate b) Not appropriate

7.1.7 When using repeated measures, the same observational units are measured twice. When using matching similar (but not the same) observational units are measured twice.

7.1.9d)

7.1.11 a) Have 15 people play with Brand A and 15 people play with Brand B. Compare the average distances between the two groups b) Have each player play with each ball, with the order determined randomly; find the difference in distance hit for each player c) b) is better because it uses pairing. It is better because some players will tend to hit the ballfarther than others and the paired design accounts for that.

7.1.13 a) Experiment, paired b) Experiment, not paired c) Observational study, not paired

7.1.17 a) Randomly assign people to sit or exercise first. Have each person do the memorization activity while doing each activity. Find the difference in performances for each person. B) Yes, because random assignment is used to determine which activity people did when

7.1.18 a) Randomly assign each of the 20 people to two groups of 10, with one group exercising and the other sitting. Compare average performance between the two groups b) Paired with repeated measures, because there is probably quite a bit of person-to-person variability in memorization ability

7.1.19 a) Have each person play the game once to get a baseline memorization score. Pair up people who have similar scores. Randomly assign one member of each pair to exercise and the other to sit. Find difference in scores within each pair of people. b) Less variation within pairs when using the paired with repeated measures design, so that one is more appropriate

7.1.23 a) Randomly assign people to use caffeine supplement or not. Have each person run the 5K twice: once with supplement and once without. Find the difference in performances for each person. B) Yes, because random assignment is used to determine which activity people did when

1223

7.1.24 a) Randomly assign each of the 30 people to two groups of 15, with one group getting caffeine supplement and the other group not. Compare average performance between the two groups b) Paired with repeated measures, because there is probably quite a bit of person-to-person variability in memorization ability

7.1.25 a) Have each person run the 5K once to get a baseline time. Pair up people who have similar times. Randomly assign one member of each pair to run the 5K having taken the supplement, and the other member not. Find difference in scores within each pair of people. b) Less variation within pairs when using the paired with repeated measures design, so that one is more appropriate

7.2.1 B

7.2.3 A

7.2.5 a) 1 b) 3 c) 4 d) 5

7.2.7 a) Null: There is no association between gender and body flexibility Alt: There is an association between gender and body flexibility such that females are more flexible than males b) Explanatory: Gender, Response: Flexibility c) Independent groups

7.2.9 Differences since that is what is actually being tested.

7.2.11 The standard deviation of the difference is quite a bit smaller than the standard deviation of the response variable (weight) within each group (Fresh/Soph)

7.2.14 a) Music (yes/no; categorical), b) Number of words memorized (quantitative) c) 2.3 d) Yes, 2.3 is in the tail of the distribution e) 2.3/0.974 = 2.4 f) Yes, the standardized statistic is more than 2

7.2.15 a) $2.3 \pm 2 \times 0.974 = 2.3 \pm 1.948$, (0.352, 4.278) b) Yes, because the confidence interval does not include 0

7.2.16 a) -2.3, b) $\mu_d < 0$ c) No Change d) No Change

7.2.20 a) Explanatory: Exercise or sit down, Response: Number of words memorized b) Null: The long-run average difference in words memorized is 0 Alt: The long-run average difference in words memorized is not 0 c) -0.07 is the average number of words memorized d) p-value is approximately 0.78 e) We do not have strong evidence that exercising while trying to memorize words helps or hinders the process

7.2.24a) Explanatory: Dominant vs. Non-dominant hand, Response: Reaction time b) Null: The long-run average difference in reaction times between hands (Dominant – Non-dominant) is 0, Alt: The long-run average difference is less than 0, c) Yes, the average difference is -0.026 and most people's average differences are less than d) -0.026, the observed average difference in reaction times e) p-value= 0.011 f) Yes, there is strong evidence that reaction times are slower when people use their non-dominant hand

7.2.25 a) -0.026/0.011 = -2.36 b) Yes, the standardized statistic is less than 2

7.2.26 a) -0.026 ± 0.022 = (-0.048, -0.04) b) We are 95% confident that the long-run average difference in reaction times is between 0.04 and 0.048 seconds faster with the dominant hand c) Yes, 0 is not in the interval

7.3.2 a) The average difference in husband and wife marriage ages in Cumberland county; μ_d b) Null: The average difference in husband and wife marriage ages in Cumberland county is 0, Alt: The average difference in husband and wife marriage ages in Cumberland county (male-female) is not0, c) Null: μ_d=0, Alt: $\mu_d \neq 0$ d) 24 pairs, and distribution of paired differences in ages is not strongly skewed

e) t=1.91, p=0.0688

f) We will obtain differences as extreme as the one we actually observed (>1.875 or <-1.875) approximately 6.88% of the time when the null hypothesis is true. We have moderate evidence that the average difference in ages is different than zero.

g) (-0.157, 3.91). We are 95% confident that, on average, males are between 0.157 less and 3.91 years older than females when they get married

h) We have moderate evidence that the average difference in ages is different than zero, being 95% confident that males are between 0.157 less and 3.91 years older than females when they get married. This is not a randomized experiment so no cause-and-effect conclusion is possible. Furthermore, the sample is only representative of Cumblerand County, PA and should not be generalized further.

7.3.6a) Each infant is paired with itselfb) Whether the helper or hinderer is being watched c)Time looking at the approach d) Null: The average time looking at the approach with the helper is the same as with the hinderer Alt: The average time looking at the approach with the helper is longer than the same with the hinderer e) You need the data in each condition and it is not provided

7.3.7 a) The long-run average difference in times looking at the helper and the hinderer condition (μ_d) b) Null: $\mu_d=0$ and Alt: $\mu_d=0$ c) Because the sample size is less than 20, the distribution of differences should be fairly symmetric—and it is (stated as such in the problem). d) t=2.60, p=0.01 e) The probability of obtaining an average difference of 1.14 seconds or larger if the average difference is 0 is 1% f) The long-run average difference in looking times is between 0.21 and 2.07 seconds longer when infants are looking at the hinderer as opposed to the helper g) We have strong evidence that the average difference looking times is larger than 0 (infants, on average, look at the hinderer toy between 0.21 and 2.07 seconds longer than the helper toy). This suggests a cause-and-effect between condition and looking time because random assignment was used, but the result should be generalized to other infants with caution because this was not a random sample.

7.3.8a) Smaller b) Larger c) Same d) Smaller

7.3.9a) Smaller b) Larger c) Larger d) Same

7.3.10 a) Smaller b) Larger c) Same d) Smaller

7.3.14 a) Overestimate, the sample average is greater than zero b) Because each person has two measurements of their kcal burning and we want to know if average difference is different than 0 c) Null: The average difference in estimated and actual calories burned is 0, Alt: The average difference in estimated and actual calories burned is not 0, d) Data is not strongly skewed e) t=2.93, p=0.0044, f) 41.4, 216.6 g) Confidence interval does not include 0 and p-value is less than 0.05 h) We have very strong evidence that the average difference in estimated and actual kcal burned is different than 0, with men tending to overestimate calories burned by between 41.4 and 216.6 on average. This is not a cause-and-effect because no random assignment was used, and this result should be generalized with caution because these men are not necessarily representative of all men with regards to their ability to estimate calories burned.

7.3.15a)
$$t = \frac{\bar{x}_d - 0}{\frac{s_d}{\sqrt{n}}} = \frac{129}{\frac{393.5}{\sqrt{80}}} = 2.93$$
b) $\bar{x}_d \pm multiplier \times \frac{s_d}{\sqrt{n}} = (41.0, 216.9)$

7.3.16 a) t=2.27, p=0.03 b) 16.05, 241.95, c) p-value larger now, t-statistic smaller now (with wrong analysis) d) Suggests pairing was effective at reducing variability and improving strength of evidence

7.3.17 a) Larger, b) Smaller c) Same d) Larger

7.3.18 a) Larger, b) Smaller c) Smaller d) Same

7.3.19 a) Larger, b) Smaller c) Same d) Larger

7.3.22 a) H_0: $\mu_{new} = 0$ versus H_a: $\mu_{new} > 0$ b) p=0.1172. No change in p-value from changing the order of subtraction c) (-7.78, 29.54). The signs of both endpoints of the interval will change due to changing the order of subtraction.

7.3.24 a) The long-run average difference in RMET scores between oxytocin and placebo (μ_d) b) Null: μ_d Alt: $\mu_d > 0$ c) The sample size is more than 20 and the difference scores are not strongly skewed d) t=2.18, p=0.0188 e) The probability of obtaining a difference of 3 or larger if the long-run average difference is 0 is 1.88% f) 0.1844, 5.8156 g) We have strong evidence that the long-run average difference is greater than zero (between 0.18 and 5.82 higher scores, on average, after oxytocin). This result suggests a cause-and-effect relationship between oxytocin and placebo, but the result should be generalized with caution since little is known about how representative the volunteers in the sample are of any populations of interest

7.3.25 a)
$$t = \frac{\bar{x}_d - 0}{\frac{s_d}{\sqrt{n}}} = \frac{3}{\frac{7.54}{\sqrt{30}}} = 2.18$$

b) $\bar{x}_d \pm multiplier \times \frac{s_d}{\sqrt{n}} = (0.25, 5.75)$

Solutions to End of Chapter Exercises

7.CE.1 a) A, b) D, c) C, d) B

7.CE.3 a) Incorrect because the hypotheses are about statistics instead of parameters b) Incorrect, the value in the null hypothesis (e.g., 40) should match the value in the alt hypothesis c) Incorrect, the null and alternative hypothesis here both use 'equal to'—only one should

7.CE.5 a) Have children, separately, use both smoke detectors, record time needed to leave house both times (randomly decide which detector to use first) b) Some children (e.g., older) will be able to leave much faster than others (e.g., younger) c) The first time, children may learn what they need to do and be faster the second time around, regardless of smoke detector or not.

7.CE.7a) The husband's ages in the sample are not connected to the wives ages in the sample b) Collect both husband and wife ages from each marriage license c) Because there is likely quite a bit of variation in ages at marriage, but less variation in difference in husband and wife ages

7.CE.9 a) Null: Average pretest flexibility same for males and females, Alt: Average posttest flexibility is different for males and females. Sample sizes are 98 (F) and 81 (M), without strong skewness in either pretest distribution meaning that at two-sample t test is valid. T=3.73, p=0.0003. We have very strong evidence of a difference in the average flexibility of males and females (Females are between 0.747 and 2.43 inches more flexible than males; 95% CI). b) Matched pairs test to evaluate if there is a significant change in flexibility, this is different than an independent samples t-test to compare male and female flexibility c) Null: Average difference in flexibility pre vs. post is 0, Alt: Average difference is less than zero (pre minus post).

The distribution of flexibility change scores is not strongly skewed and there are more than 20 (actually 179) scores, so one-sample t-test (paired t-test) is appropriate. We have very strong evidence (t=-4.76, p<0.0001) that the average difference (pre minus post) is less than 0. A 95% confidence interval finds an average increase in flexibility between 0.27 and 0.65 inches. This result generalizes to all students taking the class around this time, but does not demonstrate cause-and-effect because random assignment was not used. d) Explanatory: Gender, Response: Change score. Use a two-sample t-test e) First, create the difference scores for each person, then use Multiple means applet to look at distributions of change scores for males and females.

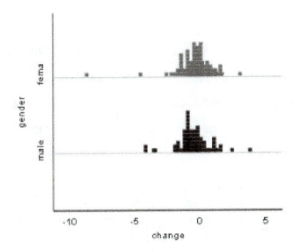

Summary Statistics:

	n	Mean	SD
fema	98	-0.36	1.33
male	81	-0.59	1.25
pooled	179	-0.46	1.30

The sample sizes are above 20 in each group, but there is one potential outlier for females. Using a simulation-based approach yields a p-value of 0.21 (two-sided), compared a two-sample t-test (theory-based) yields a p-value of 0.24 (t=1.19), so the outlier isn't making much of an impact. Null: The average change in flexibility is the same for men and women, Alt: The average change is different for men and women. Conclusion: We do not have evidence of a difference in the average change in flexibility for men and women. A 95% confidence interval finds that men, on average, change in flexibility between 0.15 less and 0.61 more than women over the course of the semester. This result can generalize to all students taking the course around that time, but does not suggest cause-and-effect between gender and change in flexibility since gender was not randomly assigned to the participants.

7.CE.11a) Null: The average difference in the percentage of times chimps open the gate is 0 (percentage of times open gate when food platform width is wide minus narrow) Alt: The average difference is greater than 0 b) The fact that the data is paired. This may yield a p-value larger than you could get if you analyzed the data (correctly) as paired c) The p-value from a simulation using Matched Pairs applet gives a very small p-value (<0.001). We have strong evidence that chimps are more likely to open the gate in the collaborative condition than the solo condition.

7.CE.15 a) Each cat had negative interactions recorded before and after exposure to catnip b) Null: There is no association between exposure to catnip and negative interactions, Alt: There is an association between exposure to catnip and negative interactions (more neg. interactions after catnip) c) p-value = 0.004, The probability of obtaining a difference in the number of negative interactions of -1 or less is 0.004. d) We have strong evidence against the null hypothesis

7.CE.16 a) While there are only 15 pairs (cats) in the study, the distribution of their negative interaction scores is not strongly skewed b) t=-3.24, p=0.003 c) We have strong evidence against the null hypothesis

7.CE.17a) -1.0 ± 0.396×2 = -1.0 ± 0.792, (-1.79, -0.21). We are 95% confident that, on average, cats have between 0.21 and 1.79 more negative interactions after catnip than before b) (-1.66, -0.34) We are 95% confident that, on average, cats have between 0.34 and 1.66 more negative interactions after catnip than before

7.CE.18 a) Null: The average number of negative interactions before catnip is the same as after, Alt: The average number of neg. interactions before catnip is less than after, t=-1.34, p-value=0.096. We have moderate evidence that the number of negative interactions before catnip is less than after catnip b) The p-value has gotten quite a bit larger (0.004 to 0.096) c) Yes, the p-value is much smaller indicating that, because of the pairing, the study yields a statistically significant result, compared to the same study without pairing which would not have yielded a statistically significant result

Chapter 8

8.1.1.a, b, e

8.1.3. C

8.1.5.
$$\frac{|0.25-0.30|+|0.25-0.35|+|0.30-0.35|}{3} = \frac{0.05+0.10+0.05}{3} = \frac{0.20}{3} = 0.067$$

8.1.7. If you don't take absolute values there are two potential problems, automatic zeros and ambiguity.

(1) Automatic zeros. Recall Example 8.1 Coming to a Stop. The differences were: $\hat{p}_S - \hat{p}_L = -0.047$, $\hat{p}_F - \hat{p}_S = -0.062$, and $\hat{p}_L - \hat{p}_F = 0.12$. If you add these without taking absolute values first, you get zero. This is not a coincidence, but an algebraic fact. Check that $(\hat{p}_S - \hat{p}_L) + (\hat{p}_F - \hat{p}_S) + (\hat{p}_L - \hat{p}_F) = 0$.

(2) Ambiguity. If you take absolute values, order doesn't matter: $|\hat{p}_S - \hat{p}_L| = |\hat{p}_L - \hat{p}_S|$. If you don't take absolute values, order does matter. There's a difference between $(\hat{p}_S - \hat{p}_L) + (\hat{p}_F - \hat{p}_S) + (\hat{p}_L - \hat{p}_F) = 0$ and $(\hat{p}_L - \hat{p}_S) + (\hat{p}_F - \hat{p}_S) + (\hat{p}_L - \hat{p}_F) = 2(\hat{p}_L - \hat{p}_S)$.

8.1.9. a) Random sampling. Random assignment is not possible – you can't use random numbers to tell a person which party to belong to. Random sampling is both possible and important to avoid bias. (Even if you can't get access to voter registration lists, you can take a random sample of individuals and ask their party affiliation.)

b) You would need to know the sample sizes – the number of Democrats, Republicans, and Independents in the sample.

8.1.11.a) Males: 0.50, Females: 0.60 b) 0.10 c) In this case the MAD statistic is just the absolute value of the observed statistic in this example. This will only be the case when both variables have two categories d)

Difference in proportions:

Total Shuffles = 1000

MAD:

Total Shuffles = 1000

The distribution of the MAD statistic looks like the distribution of the difference in proportions, except all of the negative values have been made positive so it's right skewed and always positive, instead of bell-shaped and centered at 0.

e) Because the MAD statistic uses an absolute value.

8.1.15 a) Control group: (i) Little or no baldness $(331+221)/771 = 71.50\%$, (ii) Some or much baldness 100% - 71.50% = 28.50% b) Those with heart disease (i) Little or no baldness $(251+165)/661 = 62.93\%$ (ii) Some or much baldness 100% - 62.93% = 37.07%

At this stage, there does seem to be an association between baldness and heart attacks. Among those with heart disease, baldness is more common (37%) than among those without heart disease (29%). Moreover, the samples are large.

c) In words, the null hypothesis is that there is no association between baldness and heart disease. More specifically, the probability of "some, much or extreme" baldness is the same for both the control group and those with heart disease. The alternative is that the two probabilities are different. In symbols, $H_0: \pi_C = \pi_H$, where π_C is the probability that a man in the control group has at least some baldness, and π_H is the probability that a man in the group with heart disease has at least some baldness.

d) The value of the MAD is 0.146, with a null distribution that is unimodal and somewhat skewed to the right. The simulated p-value based on 10,000 repetitions is 0.0044. The value of the chi-square statistic is 14.06, with a null distribution that is highly skewed to the right. The p-value, based on 10,000 repetitions, is 0.0033. Both tests give very strong evidence against the null hypothesis, which should be rejected. The conclusion is that there is in fact an association between degree of baldness and presence of heart disease. Degree of baldness could not be randomly assigned, so inference about cause and effect is not supported. The subjects were a sample chosen from the 22,000 doctors in the Physicians Health Study. If the sample was chosen by random sampling, then the results of the study generalize to all the physicians in the population.

8.1.17 a)

	Leno	Letterman	Stewart
Political joke	315	136	83
Non-political joke	998	512	169
Total	1313	648	252

b) Comedian (categorical), c) Type of joke (categorical) d) Null: There is no association between comedian and whether or not the joke is political e) Leno: 315/1313=0.24, Letterman: 136/648=0.21, Stewart: 83/252=0.329, f) MAD=0.08 g) It is very small h) We have very strong evidence of an association between comedian and political nature of jokes used

8.1.20 a) 18-34: 68/232=0.293, 35-54: 128/332=0.386, 55+: 173/338=0.512 b) Yes, the proportions are different between the groups. Older individuals are more likely to approve of the practice than younger individuals c) Null: There is no association between age and whether or not someone approves of Stop and Frisk, Alt: There is an association between age and whether or not someone approves of Stop and Frisk d) 0.146, e) p<0.001 f) We have strong evidence of an association between age and whether or not someone approves of stop and frisk

8.1.21 Null: There is no difference between 18-34 year olds and 55+ in their likelihood of saying "approve" about the police practice, Alt: There is a difference (proportions are different). P-value is <0.001. We have very strong evidence of a difference in the approval rates of the Stop and Frisk practice when comparing 18-34 year olds and 55+.

8.1.23 a) White: 3801/4223=0.900, Black: 611/664=0.920, Hispanic: 614/682=0.900, b) Not much of an association—the proportions are quite close together c) Null: There is no association between race and owning a cell phone (the population proportions are the same in each group), Alt: There is an association between race and owning a cell phone (at least one of the population proportions is different) d) MAD=0.013, e) p-value=0.37, f) We don't have strong evidence of an association between race and whether or not someone owns a cellphone

8.1.24 Null: The population proportion of White's that own a cell phone is the same as for Black's, Alt: The population proportions of cell phone owners are different for while's and blacks. The p-value is 0.35. We don't have evidence of a difference in the proportion of Black's and White's that own a cell phone. Note: We used white's, but Hispanics also fine since white and Hispanic proportions are the same in this sample.

8.1.25 a) White: 2238/4223=0.592, Black: 372/664=0.609, Hispanic: 382/682=0.622 b) Not much of an association—the proportions are quite close together c) Null: There is no association between race and owning a smartphone (the population proportions are the same in each group), Alt: There is an association between race and owning a smartphone (at least one of the population proportions is different) d) MAD=0.020, e) p-value=0.40, f) We don't have strong evidence of an association between race and whether or not someone owns a smartphone

8.1.26

a) Smartphone ownership among cell phone owners

		White	Black	Hispanic	Total
Own a	Yes	2238	372	382	2992
smartphone?	No	1563	239	232	2034
	Total	3801	611	614	5026

b) White: 2238/3801=0.589, Black: 372/611=0.609, Hispanic: 382/614=0.622

8.1.28 a) Null: The proportion of cars that stopped when lead is the same as when following, Alt: The proportions are different. The max-min statistic is 0.905-0.776=0.129. The p-value from simulation is approximately 0.089. We have moderate evidence that the proportion of cars that come to a complete stop when leading is different than when following b) The null distributions are both right skewed and always positive, but the null distribution of the MAD statistic is less variable (SD=0.026) than the null distribution of the max-min statistic (SD=0.039) c) The strength of evidence is similar between the two approaches

8.2.1 a) The statistic can never be negative. The statistic is a sum of terms of the form $n_i(\hat{p} - \hat{p}_i)^2$, all divided by $\hat{p}(1 - \hat{p})$. All of the individual terms are non-negative, and the overall divisor is positive. b) The chi-square can be zero only if all the individual conditional probabilities $\hat{p}_i$ are equal to each other (and thus equal to $\hat{p}$). c) (B) The distribution is skewed toward high values, with a long right tail. (More detail: For degrees of freedom greater than 2, the curve starts at (0,0), increases steadily to a unique peak, and then decreases steadily toward the x-axis. For degrees of freedom of 1 or 2, the curve has a vertical asymptote at $x=0$, and decreases steadily for x>0.) d) The long right tail comes from the fact that there is a "wall" at zero. The statistic can take on any positive value, but can never be negative. (More detail: In the formula for chi-square, the values of the n_i and $\hat{p}$ are fixed by the marginal totals; only the values of the squared terms $(\hat{p}_i - \hat{p})^2$ vary as you re-randomize. For some randomizations, just by chance, one of these terms will be unusually large.)

8.2.6 a) Null: There is no association between generation and marriage views, Alt: There is an association b) All cell counts are larger than 10 c) chi-square=22.26, p=0.0001 d) We have very strong evidence of an association between generation and marriage views e) Cause-and-effect is not possible here because this is not a randomized experiment; generalizing these findings to all adult Americans is possible because a random sample was used. f) Mill – GenX (-0.0444, 0.0662), Mill-Boom (0.0401, 0.1407)*, Mill – 65 (0.0422, 0.1938)*, GenX-Boom (0.0341, 0.1248)*, GenX-65 (0.0345, 0.1797)*, Boom-65 = (-0.0412, 0.0965). The confidence intervals show that younger generations tend to more often report that marriage is obsolete. Millenials and GenX are both significantly more likely than Boomers and 65+ to report marriage being obsolete.

8.2.11 a) Each person in the study b) The population proportion of men with heart disease in each baldness group (π) c) Null: There is no association between baldness and heart disease, Alt: There is an association between baldness and heart disease, d) Null: $\pi_{None} = \pi_{Little} = \pi_{Some} = \pi_{Much}$, Alt: At least one population proportion is different than the others e) chi-square: 14.51, p=0.0023. We have strong evidence of an association between heart disease and baldness, but this is not cause-and-effect because it is an observational study. Furthermore, the results should be generalized with caution since the sample was not obtained by taking a random sample.

8.2.12 a) None: 0.431, Little: 0.427, Some: 0.513, Much: 0.598 b) Two largest: Much vs. None and Much vs. Little, two smallest: None vs. Little and None vs. Some, c) Much vs. None: (-0.277, -0.056)*, Much vs. Little (-0.285, -0.056)*, None vs. Little: (-0.06, 0.07), None vs. Some: (-0.146, -0.018)* d) Yes, there are least 10 with heart disease and 10 without heart disease in each group.

8.2.17 a) Observational study, individuals were not assigned values of either variable b) Each participant in the survey (1883) c) Explanatory: Political leaning, Response: Favor tax? Both are categorical d) The population proportion that favor tax on unhealthy food/soda within each political group (π) e) (Null: $\pi_{Rep} = \pi_{Demo} = \pi_{Indep,.}$ Alt: At least oneof the $\pi$is different f) Null: There is no association between Political leaning and whether or not someone favors a tax on food/soda, Alt: There is an association, g) There are at least 10 individuals in each cell of the table h) chi-square=56.41, p<0.001 i) The likelihood of getting a chi-squared statistic of 56.41 or larger if the null hypothesis is true is <0.0001 j) We have very strong evidence of an association between political party and opinion on the tax; This does not, however, suggest a cause-and-effect relationship between political leaning and the tax, but the result does generalize to all US adults. K) Rep vs. Dem (-0.26, -0.16)*, Rep vs. Ind (-0.14, -0.04)*, Dem vs. Ind (0.07, 0.17)*. All groups are significantly different from each other with Democrats most in favor, followed by Independents, and Republicans showing the lowest support for the tax.

8.2.19 a) Observational study. Students were not randomly assigned values of either variable. b) Each of the 50 students. c) Explanatory: Gender (male/female), Response: Eat breakfast frequency, both are categorical d) Null: There is no association between gender and breakfast eating frequency, Alt: There is an association e) Some of the cell counts are less than 10 f) Get 50 cards, have 26 blue, 11 red, 10 green and 3 black. Shuffle and deal into two stacks of 37 and 13, respectively. Compute the chi-square statistic. Repeat many times. The p-value is the proportion of times the observed chi-square statistic or larger occurred in the simulations. g) chi-square=1.4, p=0.79. h) We obtain a chi-square value of 1.4 or larger approximately 0.79 of the time i) We have little to no evidence of an association between gender and whether or not someone eats breakfast. This result should be generalized with caution and, if statistical significance was established a cause-and-effect conclusion would not be possible.

8.2.21

a)

		DesipramineHCl	Lithium Carbonate	Placebo	Total
Relapse?	**Yes**	10	18	20	**48**
	No	14	6	4	**24**
	Total	**24**	**24**	**24**	**72**

b) Experiment. Subjects were assigned values of one of the variables c) Each of the 72 individuals in the study d) Explanatory: Type of medicine, Response: Did the patient relapse again? d) Explanatory: Type of medicine, Response: Relapse? Both variables are categorical. e) Null: There is no association between type of medicine and relapse, Alt: There is an association f) Two of the cell counts are less than 10 g) Get 72 playing cars: 48 black, 24 red. Shuffle and deal into three stacks of 24 each. Compute the chi-square statistic and repeat. H) chi-square: 10.50, p=0.0052 i) The chances of getting a chi-square statistic of 10.50 or larger if the null hypothesis is true is 0.0052 j) We have strong evidence of an association between treatment group and chance of relapse. This is a cause-and-effect conclusion because random assignment was used. The results cannot be generalized easily, since little is known about the participants.

8.2.23 a) Experiment, people are randomly assigned to a group b) Null: There is no association between treatment group and race, Alt: There is an association, c) Some of the cell counts are less than 10 d) chi-square=3.97, p-value=0.46 e) Shows that random assignment 'worked' in the sense that it kept the two treatment groups roughly the same in terms of race distribution (this is what the p-value tells you.

8.2.26

a) $\pi = P(One) = \frac{1}{6}$. Expected number of ones $= n\pi = 600(1/6) = 100$.

b) $\pi = P(Even) = \frac{3}{6}$. Expected number of even numbers $= n\pi = 600(3/6) = 300$.

c) $\pi = P(> 2) = \frac{4}{6}$. Expected number of values greater than two $= n\pi = 600(4/6) = 400$.

d) $\pi = P(> 2) = \frac{4}{6}$. Expected number of values greater than one $= n\pi = 600(5/6) = 500$.

8.2.28 a) Marine. The sample size of n=33 comes from Marine's 33 tries. There are 5 choices for each try, so if Marine is just guessing, p = P(Correct) = 1/5. The expected number correct is np = 33/5 = 6.6. The expected number wrong is $n(1-p)$ = 33(4/5) = 26.4.

b) Nurse Gilbert. The overall probability of a death on a shift is p = 74/1641 = 0.045. For the shifts with Gilbert present, n = 257, and the expected number of shifts with a death is np = (257)(0.045) = 11.57.

c) Organ Donors. The overall proportion of donors is 109/161 = 0.677. Applying this proportion to each of the group sizes gives the expected number of donors in each group. Here are the expected values:

	Opt In	Opt Out	Neutral
Yes	36.56	36.56	35.88
No	17.44	17.44	17.12

8.CE.1 a) Sex and coffee consumption, both variables are categorical (sex=2 categories, coffee consumption = 3 categories) b) It could involve random sampling (give the survey to a random sample), but not random assignment unless you told people how much coffee they have to drink c) No, the data is not cross-tabulated (e.g., number of men, every day)

8.CE.2 a)

	Men	Women	Total
Every day	12	20	22
Sometimes	12	20	22
Almost never	14	12	26
Total	38	32	70

The graph reveals that there is little difference in the men/women proportion comparing the three bars.

b)

	Men	Women	Total
Every day	22	0	22
Sometimes	16	8	22
Almost never	0	26	26
Total	38	32	70

8.CE.3 a)

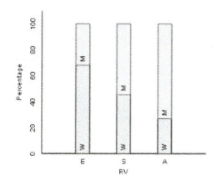

The graph reveals a strong association between gender and coffee consumption. b)

Count Samples Greater Than ▼ 8.18   Count
Count = 20/1000 (0.0200)
☐ Overlay Chi-squared distribution

Approximate p-value = 0.02

c) We have strong evidence of an association between sex and coffee d) While we have strong evidence of an association (with women drinking coffee more than men), this result cannot generalize because a random sample wasn't taken, nor does this suggest a cause-and-effect relationship between random assignment was not used.

8.CE.4 No, two counts are less than 10. Chi-square statistic is 8.18. The theory-based p-value is 0.0168, compared to 0.02 by simulation. We have strong evidence of an association between coffee consumption and gender

8.CE.5 a) None b) Smaller c) Stronger

8.CE.9 a) Explanatory: Educational degree received, Response: Generally trust people , b) Less than HS: 0.138, HS: 0.272, JC: 0.308, Bach: 0.494, Grad: 0.634

Sample Data

The proportions and the segmented bar graph both indicate that as educational level increases people are generally more trusting.

c) Null: There is no association between educational degree and trusting others, Alt: There is an association between educational degree and trusting others d) MAD=0.243, e) The approximate p-value is 0.

Total Shuffles = 2000

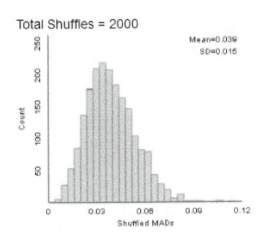

f) The probability of obtaining a MAD statistic of 0.243 or larger if the null hypothesis is true is approximately 0. g) Yes h) We have strong evidence of an association between educational degree attained and trust

8.CE.10 Chi-square: 138.26, p-value<0.0001. We have strong evidence of an association between educational degree attained and trust. Individuals with less than a HS education are significantly less trusting than individuals with more education (95%CIs: 0.076 to 0.191 less than HS, 0.275 to 0.064 less than JC, 0.444 to 0.278 less than Bach, 0.585 to 0.406 less than Grad). HS is significantly less than Bach (0.293 to 0.151 less) and Grad (0.446 to 0.279 less), JC is significantly less than Bach (0.300 to 0.073 less) and Grad (0.448 to 0.205 less), and Bach is significantly less than Grad (0.239 to 0.041). The only pair of groups that was not significantly different was High School and JC (-0.136 to 0.065).

8.CE.11 chi-square: 156.67, p<0.0001. We have strong evidence of an association between educational level attained and trust level, with trust increasing with educational level. There is little difference between this analysis and the one with the two-category response variable in the previous exercises.

8.CE.13 To answer these questions, it helps to have the row and column totals:

	Boy	Girl	Total
Grades	117	130	247
Popular	50	91	141
Sports	60	30	90
Total	227	251	478

To get the conditional and marginal proportions, divide a cell entry by the corresponding column total.

	Boy	Girl	Total
Grades	0.515	0.518	0.517
Popular	0.220	0.363	0.295
Sports	0.264	0.120	0.188
Total	1.000	1.000	1.000

    a) 117/227 = 0.515 of boys think getting good grades is most important

b) $130/251 = 0.518$ of girls think getting good grades is most important

c) $50/227 = 0.220$ of boys think being popular is most important

d) $91/251 = 0.363$ of girls think being popular is most important

e) $60/227 = 0.264$ of boys think excelling in sports is most important

f) $30/251 = 0.188$ of girls think excelling in sports is most important

g) Overall, $247/478 = 0.517$ of boys and girls think getting good grades is most important

h) Overall, $141/478 = 0.295$ of boys and girls think being popular is most important

i) Overall, $90/478 = 0.188$ of boys and girls think excelling in sports is most important

j) On balance, it looks like boys and girls do have differing goals. Boys and girls seem to be the same in terms of the proportion who think getting good grades is most important, but a higher proportion of girls think being popular is most important, and a higher proportion of boys think excelling in sports is more important.

8.CE.14 The value of the chi-square statistic is 21.46 on 2 df, with a theory-based p-value of 0.0000. (A simulation-based p-value was 0.0002 based on 10,000 repetitions.) The null hypothesis is that the conditional proportion of students choosing each goal are the same for boys and girls. The alternative is that for at least goal, the two conditional proportions in a pair are different. There is very, very strong evidence that the null hypothesis should be rejected.

The conditional proportions, difference in proportions, p-value for testing the difference, and the confidence interval for the difference for each goal given the sex of the students are:

	Boy	Girl	Difference	z	p	Lower	Upper
Grades	0.515	0.518	-0.003	-0.05	0.9564	-0.092	0.087
Popular	0.220	0.363	-0.142	-3.04	0.0007	-0.222	-0.062
Sports	0.264	0.120	0.145	4.04	0.0000	0.075	0.215

Thus we find that there is no evidence that the proportion choosing good grades is different for boys and girls, but that boys rank doing well in sports more highly than girls, while girls rank being popular higher than do boys. Inference about cause and effect is not supported because the values of the explanatory variable (boy and girl) could not be randomly assigned. Generalization to a larger population is not justified on the basis of sampling, because it was not random. However, it is reasonable to think that the results here are typical of other US students in grades 4 – 6.

8.CE.18

a) The observational units are the 47 Harvard undergraduates.

b) The (categorical) explanatory variable is whether the student received a bonus or a rebate.

c) The (categorical) response variable is whether the student spent or saved the money.

d) Abstractly, the null hypothesis is that there is no association between the explanatory and response variables. More concretely, in the context of the study, the null hypothesis is that the conditional probability of saving the money does not depend on whether it was called a rebate or a bonus. The alternative hypothesis is that the conditional probabilities differ.

8.CE.19

a) The number in the upper left cell is 9. The difference in conditional probabilities is $9/25 – 16/22 = -0.3673$. Both statistics lead to the same two-sided p-value of 0.019.

b) For 2x2 tables, the MAD and chi-square statistics are equivalent. They both lead to the same simulated p-value. For this data set, the MAD is 0.184, and the chi-square statistic is 6.34. Based on 10,000 repetitions, both statistics lead to a p-value of 0.0082.

c) The two p-values are quite close, and lead to the same conclusion. Mathematically, it can be proved that for 2x2 tables of counts, you always get the same p-value from a given set of shuffles, regardless of whether you use the count in the upper left cell, the difference in conditional proportions, the MAD, or the chi-square statistic.

Chapter 9

9.1.1 D

9.1.3 D

9.1.5 That the means are the same in each class level

9.1.7 If you don't take absolute values there are two potential problems, automatic zeros and ambiguity. (1) Automatic zeros: Finding the sum of the differences could always result in zero if you subtracted in a certain direction. (2) Ambiguity: If you take absolute values, the order you subtract doesn't matter, but if you don't take absolute values, order does matter.

9.1.9 $((5-4)+(10-5)+(10-4))/3 = 12/3=4$

9.1.11 a)The observational units are the teams, the response is the total payroll, and the explanatory variable is the league and division. b) For plots like the one above:
i.Each point represents a unit.ii.Each horizontal cluster corresponds to a value of the explanatory variable.iii.Values along the horizontal axis represent values of the response. c) B. The overall distribution of the response is skewed right. d) D. Group means show substantial differences, and overall variability within divisions is small enough by comparison to differences in group means that a p-value will show evidence, possibly strong evidence, of differences.

9.1.13 a) Explanatory: Class level, Response: Monthly hours in extracurricular activities b) Null: There is no association between class level and hours in extracurricular activities, Alt: There is an association between class level and hours in extracurricular activities, c) $((43.54-33.83)+(43.54-28.63)+(43.54-26.24)+(33.83-28.63)+(33.83-26.24)+(28.63-26.24))/6 = (9.71+14.91+17.3+5.2+7.59+2.39)/6 = 57.1/6= 9.52$ d) The p-value will be larger since 9.52 is near the middle of the simulated null distribution of MAD statistics e) We do not have strong evidence of an association between hours in extracurricular activities per month and class level when using the MAD statistic.

9.1.15 a) Explanatory: Video watched, Response: Emotional state rating, b) Null: The average emotional rating is the same for all three videos, Alt: At least one of the average emotional ratings is different for one of the videos c) The emotional state ratings seem lower for the sad group than for the other two groups, d) MAD=2.467, the average difference in mean emotional states is 2.467 when comparing the three groups, e) The p-value is <0.001. f) Because the p-value is so small, we have very strong evidence of an association between video watched and emotional state. This suggests a cause-and-effect relationship between video watched and emotional state rating since this was a randomized experiment.

However, the result cannot, necessarily, be generalized to a larger population since random sampling was not used to obtain the sample.

9.1.16 a) Explanatory: Video watched, Response: Mood rating, b) Null: The average mood rating is the same for all three videos, Alt: At least one of the average mood ratings is different for one of the videos c) The mood ratings seem lower for the sad group than for the other two groups, d) MAD=2.489, the average difference in mean mood states is 2.489 when comparing the three groups, e) The p-value is <0.001. f) Because the p-value is so small, we have very strong evidence of an association between video watched and mood. This suggests a cause-and-effect relationship between video watched and mood since this was a randomized experiment. However, the result cannot, necessarily, be generalized to a larger population since random sampling was not used to obtain the sample.

9.1.17 a) Explanatory: Video watched, Response: Stress level, b) Null: The average stress level is the same for all three videos, Alt: At least one of the average stress levels is different for one of the videos c) The stress levels seem lower for the sad group than for the no video group, and higher for the happy video group, d) MAD=2.000, the average difference in stress is 2.000 when comparing the three groups, e) The p-value is <0.001. f) Because the p-value is so small, we have very strong evidence of an association between video watched and stress level. This suggests a cause-and-effect relationship between video watched and stress since this was a randomized experiment. However, the result cannot, necessarily, be generalized to a larger population since random sampling was not used to obtain the sample.

9.1.19 a) The null hypothesis is that there is no association between the number of kills (response) and the level of the game (explanatory variable). More specifically, the mean number of kills (the long-run expected value) is the same for all three levels. The alternative hypothesis is that at least one mean differs from the others. b) Let $\Box_i$ be the mean number of kills for level I, with I = 1, 2, or 3 c) The null hypothesis is H_0: $\mu_1 = \mu_2 = \mu_3$. The alternative is that at least one long term mean is different. d) The simulation is based on shuffling, which in turn requires that all 90 observed values be interchangeable. In more detail, suppose that there were 3 different players, and that each played 10 games at each level. Then all the games played by a given person would be interchangeable and could be shuffled, but games played by two different people would not be. e) The MAD has value 0.306. f) A set of 10,000 shuffles of the response resulted in 9551 with a MAD of 0.306 or more, for an estimated p-value of 0.9551. g) The p-value is very far from being significant. It is extremely likely to get a MAD of .306 just by random chance. We have no evidence that number of kills is associated with level of the game. The exercise tells nothing about how the levels were assigned, so inference about cause is not supported. The game players were not a random sample, and generalization to some larger population is not supported.

9.1.20 Deaths: MAD = 3.054, p=0.016. We have strong evidence of an association between deaths and level played.

9.1.21 Assists: MAD=1.785, p=0.13. We do not have evidence of an association between assists and level played

9.1.22 Medals: MAD=1.652, p=0.15. We do not have evidence of an association between medals and level played

9.1.26 a) Means: A=5, B=6, C=7, SD=2.74 for all 3 groups. MAD=1.33 b) Mean is 1.05, SD of null distribution is 0.544, p-value is 0.31 c) Means: A=5, B=6, C=7, SD=1.22 for all 3 groups. MAD=1.33. d) p-value is 0.005; Mean is 0.55/SD=0.28. e) The means of the 3 groups are the same f) In Study 2 the SDs within each group are smaller, and so the null distribution has a lower mean and lower SD meaning that the same MAD statistic is unlikely in one case, but not the other g) No, MAD statistics are not 'standardized' they are not comparable across studies.

9.2.1 An ANOVA test is used to test a null hypothesis that all group means are equal. In more detail: We have response values on observational units. With each unit belonging to exactly one of two or more groups. We assume as part of our model that each group has a mean response value. ANOVA is used to test the null hypothesis that all group means are the same.

9.2.3

a) If the differences in the population means increase, you will be **more** likely to reject the null hypothesis.

a) If the population standard deviations increase, you will be **less** likely to reject.

b) If your significance level increases, you will be **more** likely to reject.

c) If your sample size increases, you will be **more** likely to reject.

9.2.5

a) C. The mean absolute pairwise difference will be the same size for both.

b) A. The p-value will be larger for A than for B

c) C. The mean square for groups will be the same size for both.

d) A. The mean square error (within groups): will be larger for A than for B

e) A. The p-value will be larger for A than for B.

9.2.7

a) Abstractly, the null hypothesis is that there is no association between response and explanatory variables. In the context of the study, the null hypothesis is that the mean heart rates are the same for men and women.

b) The value of the t-statistic is 0.63, with an associated p-value of 0.7338, a value that is not even close to significant. There is no evidence that average heart rates differ for men and women.

Sex could not be randomly assigned, so there could well be hidden factors that have influenced the heart rates in the study. Subjects were not chosen by random sampling from a larger population, so generalization beyond the observed sample is not automatic.

c) The value of the F statistic is 0.40 on 1 and 128 df, with an associated p-value of 0.5286. The p-value is far from being significant. The conclusion and its scope are the same as in (b) above.

d) The conclusions and their scope are the same.

e) ANOVA requires that the group standard deviations be equal. This is not a requirement for the t-test.

9.2.9a) F=0.05, p-value=0.9558, b) We do not have evidence of an association between kills and level

9.2.10 a) F=4.18, p=0.0185, b) We have evidence of an association between deaths and level

9.2.11 a) F=1.88, p=0.1591, b) We do not have evidence of an association assists and level

9.2.12 a) F=2.35, p=0.101, b) We do not have evidence of an association between medals and level

9.2.13 a) chi-square=4.79, p=0.0914, b) We have moderate evidence of an association between winning and losing and level played.

9.2.17 a) Each student b) Explanatory: Group (CA, KS or ME; categorical), Response: $ willing to donate, c) the long-run average amount willing to donate in each of the three groups (μ_{CA}, μ_{KS}, μ_{ME}) d) Null: $\mu_{CA} = \mu_{KS} = \mu_{ME}$, Alt: At least one of the three means is different than the others, e) MAD=3.988, p-value=0.84, f) F=0.19, p-value=0.83, g) We do not have evidence of a difference in the long-run average amount willing to donate between the three groups. Cause-and-effect would be possible due to the random assignment, but generalizing should be done with caution due to the fact that the sample was not obtained by taking a random sample h) The ANOVA/F-test may not be valid due to the small sample size combined with strong skewness/outliers in the data i) No follow-up analysis is needed since there is not evidence of an association j) Not only the states were different, but the type of disaster was also different, so if there had been evidence of an association we couldn't attribute it necessarily to state, could have been type of disaster

9.2.18 a) Bigger, b) Smaller, c) Stronger d) F-statistics increase as the sample size increase, all else being equal. This will make the p-value smaller and increase the strength of evidence

9.2.19 No, there would be two categorical variables and a chi-square test would be more appropriate.

9.2.22 a) Null hypothesis: The long-run average distance reached is the same for each of the three groups, Alt: At least one of the long-run averages is different than the others, b) Null: $\mu_{yoga} = \mu_{walking} = \mu_{control}$, Alt: At least one of the three means is different than the others, μ=long-run average distance reached c) We have strong evidence (based on the small p-value) that at least one of the three means is different than the others. This suggests a cause-and-effect relationship between group and distance reached. Generalizing this result to a larger group should be done with caution since the sample was not obtained randomly. d) The sample sizes are above 20 in each group, so we are assuming that the SDs are within a factor of 2 of each other, and that there is not strong skewness/outliers in the data.

9.2.23a) Null hypothesis: The long-run average reaction time is the same for each of the three groups, Alt: At least one of the long-run averages is different than the others, b) Null: $\mu_{yoga}=\mu_{walking}=\mu_{control,}$ Alt: At least one of the three means is different than the others, $\mu=$long run average reaction time c) We have strong evidence (based on the small p-value) that at least one of the three means is different than the others. This suggests a cause-and-effect relationship between group and reaction time. Generalizing this result to a larger group should be done with caution since the sample was not obtained randomly. d) The sample sizes are above 20 in each group, so we are assuming that the SDs are within a factor of 2 of each other, and that there is not strong skewness/outliers in the data.

9.2.24a) Null hypothesis: The long-run average words recalled is the same for each of the three groups, Alt: At least one of the long-run averages is different than the others, b) Null: $\mu_{yoga}=\mu_{walking}=\mu_{control,}$ Alt: At least one of the three means is different than the others, $\mu=$long run average words recalled c) We have strong evidence (based on the small p-value) that at least one of the three means is different than the others. This suggests a cause-and-effect relationship between group and words recalled. Generalizing this result to a larger group should be done with caution since the sample was not obtained randomly. d) The sample sizes are above 20 in each group, so we are assuming that the SDs are within a factor of 2 of each other, and that there is not strong skewness/outliers in the data.

9.2.25a) Null hypothesis: The long-run average perception of health is the same for each of the three groups, Alt: At least one of the long-run averages is different than the others, b) Null: $\mu_{yoga}=\mu_{walking}=\mu_{control,}$ Alt: At least one of the three means is different than the others, $\mu=$long-run average perception of health c) We have strong evidence (based on the small p-value) that at least one of the three means is different than the others. This suggests a cause-and-effect relationship between group and perception of health. Generalizing this result to a larger group should be done with caution since the sample was not obtained randomly. d) The sample sizes are above 20 in each group, so we are assuming that the SDs are within a factor of 2 of each other, and that there is not strong skewness/outliers in the data.

9.CE.1a) Smaller b) Larger c) Smaller

9.CE.3μ is the population (or long-run) average-it is not observed, but is what we are trying to learn about. $\bar{x}$ is the sample mean, it is observed.

9.CE.5 The overall average will equal the average of the group averages when the size of each group is the same. For example, if each section had 25 students, then the average of each group of 25 students, and the overall average over 100 students would be the same. At the other extreme, if one section had 97 students and the other three 'sections' each had one student each then, averaging the four section averages won't necessarily be the same as the overall average.

9.CE.7 a) Study A: $(0+20+20+20+20+0)/6=80/6=13.3$, Study B: $(10+10+20+0+10+10)/6=60/6=10$, b) Study A, c) The denominators of the F statistic will the same for both studies. However, the group means in Study A are more spread out than in Study B. So, just like Study A had a larger MAD statistic, Study A will also have a larger F statistic

9.CE.9

a)

b) Sample size is 50 for each of the 5 groups. Doc: M=97.4, SD=40.5, Mas: M=66.0, SD=38.4, Bac: M=55.2, SD=32.2, Assoc: M=36.8, SD=28.5, Some: M=32.51, SD=20.80

c) The distribution of incomes are right skewed within each group, with a few extreme values in each group. The means and medians, however, follow a trend so that each additional 'higher' level degree is associated with higher earnings.

d) The MAD statistic is 31.799, which yields a p-value of <0.001. There is very strong evidence of an association between degree received and individual yearly earnings.

e) F=31.53, p<0.0001. There is very strong evidence of an association between degree received and individual yearly earnings.

f)

☑ Compute 95% confidence interval(s)
1:Some - 2:Associate: (-17.23, 8.65)
1:Some - 3:Bachelor: (-35.64, -9.75)*
1:Some - 4:Master: (-46.44, -20.55)*
1:Some - 5:Doctorate: (-77.84, -51.95)*
2:Associate - 3:Bachelor: (-31.35, -5.46)*
2:Associate - 4:Master: (-42.15, -16.26)*
2:Associate - 5:Doctorate: (-73.55, -47.66)*
3:Bachelor - 4:Master: (-23.74, 2.15)
3:Bachelor - 5:Doctorate: (-55.14, -29.26)*
4:Master - 5:Doctorate: (-44.35, -18.46)* ·

g) Higher education is strongly associated with higher income. Average yearly income ranges from $32.5K for individuals with only some higher education (but no degree), up to $97.4K for individuals with a doctorate. The doctorate degree increases average yearly income over $30K p/year over a Masters' degree.

9.CE.11 a) There is a right skewed distribution of the extent head injury on the dummy's head within each of the three groups, with at least 1 outlier in the other group. The average head injury score in the other group is 1138, compared to only 865.5 and 811.5 in the four and two-door groups, respectively. b) No, there is an outlier and severe skew---especially in the 'other' group. c) The MAD statistic is 217.82. The p-value is <0.0001. There is very strong evidence of an association between number of doors and extent of head injury

9.CE.12 a) The distribution of the head injury measurement is fairly symmetric within each of the three groups, with the highest mean (3.01) for the other group. The technical conditions now look to be met since the data is not strongly skewed within each group and the SDs are close (they are all approximately 0.20). b) $F=11.96$, $p<0.0001$. Yes, there is strong evidence that head injury measurements differ based on the number of doors of the vehicle c) The vehicles in the 'other' group have significantly higher average head injury scores than the other two groups (95% CIs: Two vs. other (0.09, 0.21), Four vs. other (0.06, 0.18)). There is not a significant difference in the average head injury scores between the two door and four door groups.

9.CE.14.

  a) The observational unit is a location. The response is the noise level in decibels. The explanatory variable is the area of New York city.
  b) B. Within-group variability is roughly constant (roughly the same in in each group) but at least one group mean differs substantially from the others. A theory-based test is appropriate, and the p-value may be significant.

c) The scope of inference is extremely limited, because it is impossible to separate the effect on noise level from three different influences: location within New York, type of location (store, restaurant, etc.) and time of day.

9.CE.16.

a)

Music Type	Mean	SD
None	21.69	3.38
Pop	21.91	2.73
Classical	24.13	

b) The response is the amount spent. The explanatory variable is the type of music. The experimental units are the customers.

c)

d) The null hypothesis is that there is no association between the type of music and the mean amount spent by customers. The alternative hypothesis is that the mean amount spend is associated with the type of music. More formally, define parameters

μ_N = mean amount spent by a customer when there is no music playing

μ_P = mean amount spent by a customer when there is pop music playing

μ_C = mean amount spent by a customer when there is classical music playing

Then the null hypothesis is H_0: $\square_N = \square_P = \square_C$

and the alternative is that at least one of the $\square$s is different from the rest.

A simulation-based test using the observed value of the MAD (1.623) has an estimated p-value of 0.0000 using 10,000 repetitions. This is highly significant, and so we reject the null hypothesis and conclude that there is in fact an association between type of music and size of the check. Because the type of music was randomly assigned, we can conclude that the type of music is the cause of the difference. However, because customers were not chosen using a random sample, we cannot automatically generalize to a larger population.

e) The distributions are roughly symmetric, with no outliers. The group SDs are about the same, and the sample sizes are substantial. In short, a theory-based test is justified.

f) The theory-based test uses the observed value of the F-statistic, equal to 27.82 on 2 and 390 df. The p-value is 0.0000, which agrees with the simulation-based test. The conclusion is the same as the one stated in (d) above.

g) Here are confidence intervals from the applet Theory-based Inference:

> Classical – Pop from 1.52 to 2.91
>
> Classical – None   from 1.73 to 3.14
>
> Pop – None   from -0.46 to 0.90

The intervals tell us that there is no detectable difference in average amount paid whether there is pop music or no music, but that when classical music is playing, the average check amount is distinctly higher.

h) The results in (d) and (f) are essentially the same.

i) See answer to (d) above.

Chapter 10

10.1.1: The correlation will be negative: larger distances go with lower exam scores and vice-versa.

10.1.3: The correlation will be exactly 1.0, because there is a perfect linear relationship with positive slope: Exam2 = Exam1 − 5.

10.1.5: In a scatterplot, the points are observational units; the x-axis represents the explanatory variable; and the y-axis represents the response.

10.1.7: D. There may be some nonlinear relationship.

10.1.9: A

10.1.11: (a) Observational study. (b) The 20 houses. (c) The explanatory variable is square footage and the response is the price. Both of these are quantitative. (d) There is a fairly strong positive linear association beween square footage and price with no unusal observations. (e) The correlation coefficient is 0.780 and this backs up the description given in part (d) since it is a number fairly close to 1.

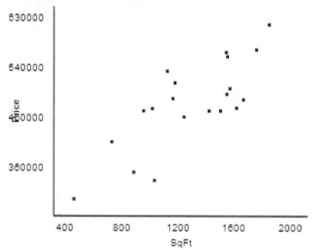

10.1.13: (a) Observational. (b) The 34 people in the study. (c) The explanatory variable is finger length and the response is height. Both of these are quantitative. (d) There is a moderate positive linear association beween finger length and height with no unusual observations. (e) The correlation coefficient is 0.474 and this backs up the description given in part (d) since it is just a little bit below 0.5.

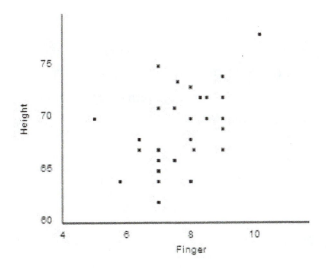

10.1.14: (a) The point (5,70) is circled in the scatterplot. It is a bit unusual since it has the shortest finger length, but is average height. (b) r = 0.474. (c) r = 0.566. (d) The correlation increased. This makes sense because the point removed did not nicely fall into the positive linear pattern of the data, so removing it would make the data better fit that pattern.

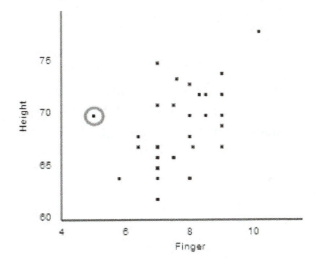

10.1.15: (a) The point (10.16,78) is circled in the scatterplot. It is a bit unusual since it represents the longest finger and the greatest height. (b) r = 0.474. (c) r = 0.353 (d) The correlation decreased. This makes sense because the point removed very nicely fell into the positive linear pattern of the data (and was extreme on the x-axis), so removing it would decrease the correlation.

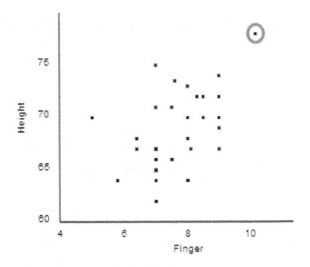

10.1.19: (a) Observational. (b) The 16 students. (c) The explanatory variable is name length and response is Scrabble score. Both of these are quantitative. (d) There is a moderate positive association beween name length and Scrabble socre. There are a couple of scores that have fairly few number of letters but give high scrabble scores. (e) The correlation coefficient is 0.476 and this backs up the description given in part (d) since it is around 0.5.

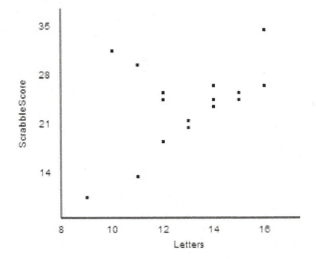

10.1.20: (a) There is a fairly strong positive linear (perhaps a bit concave downward) association between the two variables. (b) The correlation coefficient is 0.782. This supports the answer in part (a) since it is fairly close to 1.

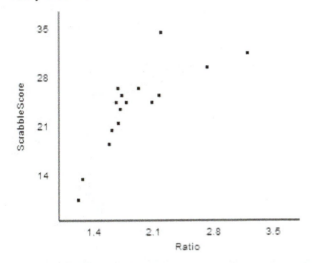

10.1.21: (a) The linear association is stronger between the Scrabble points and the ratio since the correlation is closer to 1. (b) This makes sense because the scrabble ratio will take into account the score of each letter, while the name length doesn't. These individual letter scores can be quite high or low so they have a great impact on the total word score.

10.2.1: (a) **Invalid**: The p-value is computed *assuming* there is no association. It does not tell the chance of no association. (b) **Invalid**, for the same reason as in (a) above. (c) **Invalid**, but close: See the answer to (d) below. (d) **Valid**: If there were no association between height and hand span, the p-value is probability of observing the association observed in the sample data or an even stronger association in a sample of 10 students is 0.022. (e) **Invalid**, for the same reason as in (a) above.

10.2.3: (a) B. (b) With Lisa's larger sample we have more information and thus the strength of evidence is stronger.

10.2.7: (a) Observational. (b) The 90 Honda Civics. (c) The explanatory variable is age and the response is height. They are both quantitative. (d) There is no association between age and price of used Honda Civics. (e) There is an association between age and price of used Honda Civics. (f) There is a fairly strong negative association between age and price of the cars. Not much unusual except perhaps the oldest and cheapest car. (g) The correlation of -0.820 backs up the answer to part (f) since it is close to -1. (h) Put the 90 ages on 90 index cards and the 90 prices on 90 other index cards. Shuffle the price index cards and randomly match them up with the 90 age cards. From this data set, compute the correlation.

That simulated correlation will be one dot in the null distribution. Repeat this many, many times to create a null distribution. Find the proportion of simulated correlations that are -0.820 and below along with the simulated correlations that are 0.820 and above. This proportion will be the p-value.

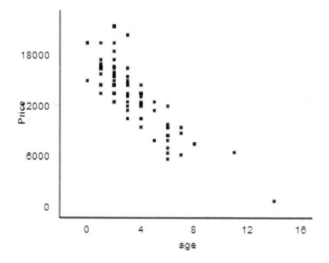

10.2.8: (a) The p-value is about 0. (b) If there is no association between age and price of used Honda Civics, the probability we would get a sample correlation as extreme as -0.820 is about 0. (c) We have extremely strong evidence that there is an association between a car's price and its age in the population of all used Honda Civics advertised online. Since our sample correlation is negative, we can go further and say the association is also negative in the population.

10.2.11: (a) Null: There is no association between maximum speed and maximum height of roller coasters. Alternative: There is an association between maximum speed and maximum height of roller coasters. (b) Put the 129 heights on 129 index cards and the 129 speeds on 129 other index cards. Shuffle the speed index cards and randomly match them up with the 129 height cards. From this data set, compute the correlation. That simulated correlation will be one dot in the null distribution. Repeat this many, many times to create a null distribution. Find the proportion of simulated correlations that are as extreme as the observed correlation from the sample data.

10.2.12: (a) The p-value is about 0. (b) If there is no association between maximum height and maximum speed for roller coasters, the probability we would get a sample correlation as extreme as 0.895 is about 0. (c) We have extremely strong evidence that there is an association between maximum height and maximum speed for roller coasters around the world. Since our sample correlation is positive, we can go further and say the association is also positive in the population.

10.2.17: (a) Null: There is no association between length of names and their Scrabble score. Alternative: There is an association between length of names and their Scrabble word scores. (b) Put the 16 name lengths on 16 index cards and the 16 word scores on 16 other index cards. Shuffle the word score cards and randomly match them up with the 16 name length cards. From this data set, compute the correlation. That simulated correlation will be one dot in the null distribution. Repeat this many, many times to create a null distribution. Find the proportion of simulated correlations that are as extreme as the observed correlation from the sample data.

10.2.18: (a) The p-value is about 0.06. (d) If there is no association between name lengths and their scores, the probability we would get a sample correlation as extreme as 0.476 is about 0.06. (e) We have do not have strong evidence that there is a association between name length and its Scrabble word score, but we do have moderate evidence of such an association and it is positive.

10.2.19: (a) Null: There is no association between Scrabble points per letter of names and their scrabble score. Alternative: There is an association between Scrabble points per letter of names and their scrabble score. (b) Put the 16 points per name on 16 index cards and the 16 word scores on 16 other index cards. Shuffle the word score cards and randomly match them up with the 16 points per name cards. From this data set, compute the correlation. That simulated correlation will be one dot in the null distribution. Repeat this many, many times to create a null distribution. Find the proportion of simulated correlations that are as extreme as the observed correlation from the sample data.

10.2.20: (a) The p-value is about 0.0006. (b) If there is no association between name lengths and their scores, the probability we would get a sample correlation as extreme as 0.782 is about 0.0006. (c) We have strong evidence that there is an association between Scrabble points per letter of names and their scrabble score and that relationship is positive.

10.3.1 B

10.3.3 a) Yes, the mean of the residuals is calculated as the sum of the residuals divided by the number of residuals. If the sum is zero, then zero divided by a positive number is also zero. b) No, suppose that the least squares regression line does not go through any of the points on the scatterplot. Then none of the residuals would be zero, and if the 50th percentile fell on a residual it would necessarily be either negative or positive.

10.3.5

a) negative slope coefficient

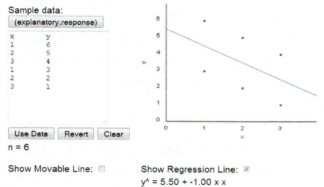

Sample data:

(explanatory,response)

x	y
1	6
2	5
3	4
1	3
2	2
3	1

Use Data   Revert   Clear

n = 6

Show Movable Line: ☐   Show Regression Line: ☑

$\hat{y} = 5.50 + -1.00 \times x$

b) slope coefficient greater than 1

Sample data:

(explanatory,response)

x	y
1	2
2	4
3	6
1	8
2	10
3	12

Use Data   Revert   Clear

n = 6

Show Movable Line: ☐   Show Regression Line: ☑

$\hat{y} = 3.00 + 2.00 \times x$

c) slope coefficient of zero

Sample data:

(explanatory,response)

x	y
1	5
2	5
3	5
1	10
2	10
3	10

Use Data   Revert   Clear

n = 6

Show Movable Line: ☐   Show Regression Line: ☑

$\hat{y} = 7.50 + 00 \times x$

10.3.7: a) explanatory: number of pieces, response: price b) The scatterplot shows a strong positive linear association with no unusual observations c) r=0.974 d) r^2=0.949 this means that 94.9% of the variation in price is explained by the linear association with number of pieces.

10.3.8: a) predicted price = 4.86 + 0.1050(pieces) b) for each additional lego piece, the price increases 0.105 cents. c) If there are no pieces in the lego set, the price is $4.86 d) the y-intercept is an extrapolation beyond the data and it does not make sense that one would pay almost $5 for a lego set with no pieces in it.

10.3.9: a) correct but incomplete b) incorrect c) correct d) correct but incomplete e) correct f) correct

10.3.10: a) incorrect b) correct c) incorrect d) incorrect

10.3.11: a) $57.36 b) $162.36 c) $105; 1500 is 1000 more pieces than 500, each additional piece increases the predicted price by .105 .105 (and 1000 times slope coefficient) is $105 d) $529.86 e) The data ranges from 34 pieces to 3803 pieces. 5000 pieces is beyond the range of the data and it is not clear that the linear relationship will hold beyond the range of the data, so there is not a lot of confidence in the predicted price for a 5000 piece set.

10.3.12: a) $48.435 b) 49.99-48.44 = 1.55 c) the observed price is $1.55 above the predicted price for the let d) above, the residual is positive

10.3.13: a) It depends on which Lego set is offered for $0. If it is a Lego set that already doesn't cost very much, then no the least squares line would not be impacted very much. If it is a Lego set that originally was very expensive, then yes it would affect the least squares regression line b) Either of the sets that are $399.99, Choose the one with 3803 pieces since this is the furthest away from the range of values for pieces c) predicted price = 15.03 + 0.0756(pieces) r^2 = .636 yes, they have changed quite a bit.

10.3.15: a) moderately strong negative linear association b) predicted takeoff velocity = 394.47 - 0.0122(bodymass) c) for each additional gram of body mass the cat's predicted takeoff velocity decreases 0.0122 cm/second d) For a cat whose body mass is zero their takeoff velocity is 394.47 cm/second this doesn't make sense in context of the study because a cat wouldn't have body mass of zero e) 24.6% of the variability seen in takeoff velocity is explained by the linear association with body mass of the cat

10.3.16: a)333.47 cm/sec b) 272.47 cm/sec a 10,000 gm cat is beyond the range of the data and the linear relationship may not hold outside of this range c) 7930 gm cat has observed takeoff velocity of 286.3 predicted takeoff velocity of 297.72, residual is -11.4. The predicted takeoff velocity is 11.4 cm/sec lower than the predicted takeoff velocity

10.3.17: a) the cat with body mass 2660 b) the cat with body mass 7930 c) the cat with body mass 5600 d) the cat with body mass 3550

10.3.18: a) percent body fat has the strongest association with takeoff velocity. It is a strong negative linear relationship b) predicted takeoff velocity = 397.65 -1.95(bodyfat) c) for each 1% increase in body fat the predicted takeoff velocity decreases 1.95 cm/sec d) r2 = 0.424 42.4% of the variability in takeoff velocities can be explained by the linear relationship with percent body fat

10.3.20   a)The explanatory variable is the number of pages in the textbook. The response variable is the price of the textbook b) The equation of the least squares line is: predicted price = -3.42 + 0.1473×pages c) The predicted price of a 500-page textbook is: -3.42 + 0.1473×500 ≈ 70.23 dollars.The predicted price of a 1500-page textbook is: -3.42 + 0.1473×1500 ≈ 217.53 dollars.The first prediction is more believable, because 500 pages is within the range of the sample data, but no textbooks in the sample had close to 1500 pages d) The slope coefficient of 0.1473 means that for each additional page in a textbook, the predicted price increases by about 0.1473 dollars, which is almost 15 cents e) The proportion of variability in textbook prices that is explained by knowing the number of pages is the square of the correlation coefficient, which is $r^2 = 0.677$, or 67.7%.

10.3.24: a) The answer to both questions is Yes. The least squares line always goes through the "point of averages." If the fitted slope is positive, then values of x above the mean correspond to fitted values for y above the mean, and vice-versa b) If someone scores a 79.8 on test 2, their predicted score for test 3 is 84.8. If they actually get a 90 on test 3, their residual is 90 − 84.8 = 5.2 c) If someone scores 10 points *above* the mean on test 2, their expected score on test 3 is (0.3708)(10) = 3.708 points above the mean d) If someone scores 10 points *below* the mean on test 2, their expected score on test 3 is also 3.708 points below the mean.

10.3.25: a) 75.594 or 76 b) 90.426 or 90 c) The student who scored a 55 on test 2 is predicted to achieve a higher score on test 3, the student who scored a 95 on test 2 is predicted to achieve a lower score on test 3

10.4.1. A

10.4.3 Write the heights on 10 different slips of paper and foot lengths on 10 different slips of paper. Lay the 10 slips of paper with the heights written on them in a line on a flat surface. Shuffle the 10 slips of paper with the foot lengths on them and deal one out to each of the 10 slips of paper with the heights on them. Calculate the least squares regression equation for these 10 pairs of data and record the slope. Repeat this procedure 1000 times to get 1000 simulated slopes which make up the null distribution

10.4.5: Lay the 20 slips of paper with the heights written on them in a line on a flat surface. Shuffle the 20 slips of paper with the vertical leap distances on them and deal one out to each of the 20 slips of paper with the heights on them. Calculate the least squares regression equation for these 20 pairs of data and record the slope. Repeat this procedure 1000 times to get 1000 simulated slopes which make up the null distribution

10.4.9: a) hours sleep the previous night b) seconds needed to complete a paper and pencil maze c)i 1 ii) No, strong evidence is found when the observed statistic is 2 or more SDs away from the mean under the null, 7.76 is only 1 SD away

10.4.10: a) For each additional hour of sleep per night the predicted number of seconds needed to complete the maze decreases by 7.76 b) 19.33 is the predicted amount of time for someone who didn't sleep the previous night. This is extrapolation because no one in the sample pulled an all-nighter.

10.4.13: a) H_0: No association between year of manufacture and price and H_a: Association between year of manufacture and price b) Count the number of simulated slopes that are as extreme as the observed slope and divide by 1000. P-value is 0 c) With a p-value of 0 we have very strong evidence against the null and in favor of the alternative that the slope describing the association between the number of years before 2006 a Honda Civic was manufactured and the sale price is different from zero. Based on our simulation, which assumed there was no association between number of years before 2006 a Honda Civic was manufactured and its sale price, not once did we find a simulated slope that was as steep as the one from the original data.

10.4.14: a) Predicted price = 18785.31 - 1397.47(age) b) Slope: For every additional year old the car is the average price drop $1397.47, Intercept: A car from 2006 (current model year) has a predicted price of $18,785.31

10.4.17: a) Null: There is no association between height and haircut price in the population of students, Alt: There is an association between height and haircut price in the population of students b) p-value is 0.007 c) With a p-value of 0.007 we have very strong evidence against the null and in favor of the alternative that the slope describing the association between height and amount of last haircut is different from zero. d) No, this is an observational study and there are many confounding variables that may explain the variability in haircut costs

10.4.18 a) Predict haircut price = 160.66-1.92(height) b) Slope: For every additional inch taller someone is their predicted haircut price drops $1.92, Intercept: for someone who is 0 inches tall, the predicted price of their last haircut is $160.66

10.4.21: a) H_0: There is no association between missing class and GPA in the population students and H_a: There is an association between missing class and GPA in the population studentsb) p-value is 0 c) With a p-value of 0 we have very strong evidence against the null and in favor of the alternative that the slope describing the association between number of classes skipped in the past three weeks and GPA is different from zero. There is a negative association; that is as the number of classes skipped increases, the predicted GPA decreases d) No, this is an observational study and there are many confounding variables that may explain the variability in GPA

10.4.23: a) H_0: There is no association between age and the time it takes to read a list of 20 color words. H_a: There is an association between age and the time it takes to read a list of 20 color words b) b=0.2383 for each additional year it takes 0.2383 seconds longer to read the 20 color words c) p-value = 0.003 d) A p-value of 0.003 provides very strong evidence against the null and in favor of there being an association between age and the time in seconds it takes for someone to read 20 color words. The association is positive, as your age increases, it takes longer to read the 20 color words

10.4.24: a) p-value = 0.188, results are no longer significant b) p-value = 0, yes is a significant result c) For ages 18 to 80, there is very strong evidence of an association between age and time to read 20 color points. As one moves beyond the range of the data, the linear model may not be the best model to describe the data and predictions using the linear model beyond the 18 to 80 age range may not be good predictions.

10.5.1: A

10.5.3: A

10.5.5: To see whether or not you get a different conclusion

10.5.7: a) predicted time in seconds = 198.33 – 7.76(hours sleep) b) 0.3052 c) 0.1526

10.5.10: a) In the population, $\square$ is the amount the guessed weight changes when the guessed height increases by 1 inch b) $H_0: \beta = 0$ and $H_a: \beta \neq 0$ c) Yes, the three validity conditions (linear trend, similar distributions and equal spread around the line) are all reasonably well met for this dataset d) p-value = 0.0159 provides strong evidence against the null and in support of there being an association between the guessed height and guessed weight for the professor

10.5.13: a)In the population of all Honda Civics, slope tells how much the price changes for each year older the car is, $\square$ is the symbol for the population slope b) $H_0: \beta = 0$ and $H_a: \beta \neq 0$ c) c) Yes, the three validity conditions (linear trend, similar distributions and equal spread around the line) are all reasonably well met for this dataset d) p-value = 0 e) With a p-value of 0 we have very strong evidence against the null and in favor of the alternative that the slope describing the association between the number of years before 2006 a Honda Civic was manufactured and the sale price is different from zero

10.5.14: a) (-1603.94, -1191.01) We are 95% confident that in the populatino of all Honda Civics for each year older the gets the sale price of the car decreases between $1191.01 an $1603.94 b) Yes, there is a negative association because as one variable increases (age of car) the other decreases (sale price)

10.5.16: a) In the population, slope tells us how much BMI changes for each additional inch in height, $\square$ is the symbol for the population slope b) H_0: $\beta = 0$ and H_a: $\beta \neq 0$ c) c) Yes, the three validity conditions (linear trend, similar distributions and equal spread around the line) are all reasonably well met for this dataset d) p-value = 0.0417 e) The p-value provides strong evidence against the null and in support of there being an association between BMI and height

10.5.17: a) (-0.3033, -0.0059) We are 95 confident that in the population for each additional inch in height , the BMI decreases by between 0.0059 and .03033 b) Yes, zero is not in the confidence interval and thus is not a plausible value for the population slope

10.5.19: a) In the population the slope, $\square$, is how much height in inches changes for each additional hour of sleep b) H_0: $\beta = 0$ and H_a: $\beta < 0$ c) p-value = 0.0328 provides us with strong evidence against the null and in support of there being a negative association between amount of sleep per night in hours and height in inches d) new p-value = 0.3622 now we do not have strong evidence of a negative association between hours slept and height in inches

Chapter 10 End-of-Chapter Exercises

10.CE.1: a) r=1 as midterm scores increase so do final exam scores and every final exam score is the same amount higher than the midterm score b) r=1 as midterm scores increase so do final exam scores and every final exam score is the same amount lower than the midterm score c) r=1 as midterm scores increase so do final exam scores and every final exam score is the same factor higher than the midterm score

10.CE.3: age is in years and distance is in minutes from the school

a)

b)

c)

d)

e)

Test-Taking Time

10.CE.5: a) Weak negative linear association between the time to take the test and the score on the test b) r = -0.035 c) p-value = 0.8420 d) t = -0.19, p-value = 0.8481 e) Both simulation-based and theory-based tests of significance offer weak evidence against the null and so it is plausible that there is no association between time to take a test and score on the test

Count Samples Beyond   ▾ -.035   Count
Count = 842/1000 (0.8420)

10.CE.6: a) predicted score = 73.74 – 0.0604(time) b) For each additional minute needed to take the test, the predicted test score drops 0.0604 points c) If you take 0 minutes to take the exam, your predicted score is 73.74. This doesn't make sense in the context of the problem. d) $r^2 = 0.001$ so 0.1% of the variation in test scores can be explained by the linear association with time to take the test

10.CE.7: a) t = -0.19 p-value = 0.8481 b) (-0.6978, 0.5771) c) We are 95% confident that in the population for each additional minute taken on the test the score decreases by 0.6978 up to increases by 0.5771 d) This is consistent with the test result since 0 is a plausible value for the population slope coefficient and we found weak evidence against the null of no association between time to take test and score on test

10.CE.10: a) H_0: $\beta = 0$ and H_a: $\beta \neq 0$, t = 19.05 , p-value = 0 We have very strong evidence against the null and in support of an association between distance in miles of a hike and time in minutes the hike takes, specifically the more miles in a hike, the longer it will take b) (28.18 , 34.77) we are 95% confident that in the long run each additional mile in a hike increases the time it takes to make the hike by between 28.18 minutes to 34.77 minutes

10.CE.11: a) H_0: $\beta = 0$ and H_a: $\beta \neq 0$, t = 3.07 , p-value = 0.003 We have very strong evidence against the null and in support of an association between elevation gain in feet of a hike and time in minutes the hike takes, specifically the higher the elevation gain of a hike, the longer it will take b) (0.0217, 0.1020) we are 95% confident that in the long run each additional foot gain in elevation in a hike increases the time it takes to make the hike by between 0.0217 minutes to 0.102 minutes

10.CE.13: a) $H_0: \beta = 0$ b) $H_a: \beta > 0$ c) Professor's arrival time in minutes early (or late) and students' average arrival time in minutes early (or late) d) $0.0107 < $ p-value $ < 0.0155$ e) We have strong evidence against the null and in support of the alternative that there is a positive correlation between the time a Professor arrives to class and the average time the students arrive to class. f) $H_a: \beta \neq 0$ and $0.190 < $ p-value $ < 0.304$ g) No, this is an observational study and there are many confounding variables present so cause/effect conclusions can't be made.

10.CE.15 Correlation and slope: (a) F (b) C (c) D (d) A (e) E (f) B

10.CE.16: Outliers and influential points.

A (All locations): a) No big outliers, no influential points; strong linear relationship.

B (All restaurants): c) No outliers, but two influential points. Removing these points will decrease the correlation and increase the fitted slope.

C (Stores): b) No big outliers, no influential points; no evidence of a linear relationship.

D (Other locations): b) No big outliers, no influential points; no evidence of a linear relationship.

E (Restaurants, evening): a) No big outliers, no influential points; strong linear relationship.

F (Other, no gym): b) No big outliers, no influential points; no evidence of a linear relationship.

10.CE.17: Interpretation.

a) For restaurants, the correlation between average noise level and time becomes weaker if you exclude the two lunch-time measurements.

b) For stores, the noise level is not related to time of day.

c) For the "other" locations, the suggestion of a possible relationship between noise level and time of day becomes stronger if you exclude the gym measured at 6:30 pm.

d) On average, restaurants are more noisy than stores.

e) On average, decibel levels measured later in the day tend to be higher than those measured earlier in the day.

f) Time of day is confounded with type of location.

g) For restaurants the fitted noise level increases by more than 10 decibels during the three hours from 7 pm to 10 pm.

h) If you go by the fitted line for all restaurants, the predicted decibel level at 4 pm is closest to _____ (round to the nearest 10 dB).

10.CE. 18: Study Design.

a) The obstacle to reaching a conclusion about the effect of time is that location and time of day are confounded. For example, most restaurants were observed late in the day and were loud. Most stores were observed early in the day and were not loud. There is no way – based on this badly planned data set – to disentangle the effects of location from effects of time of day.

b) A better design would measure noise levels at different times for each location.

10.CE. 19: Quieter than a cemetery.

a) At the y-intercept the decibel level is about 55. (Between 12 and midnight the fitted level goes up 20, so from 12 back to zero the fitted level goes down 20, from 75 to 55.) b) The fitted level of 55 cannot be right, because 0 corresponds to midnight (24), for which the fitted level is 95.

10.CE. 20: Breaking the association.

Histogram A is for Plot A, all locations. Histogram B is for Plot B, restaurants. Histogram C is for Plot C, stores. All histograms are centered at 0 (the null hypothesis of no association), but the histograms differ in the size of their variability. The upper left histogram is least spread out, with almost all outcomes between -0.5 and 0.5. The bottom histogram is the most spread out, with the smallest percentage of outcomes between -0.5 and 0.5. The size of variability is negatively correlated with sample size: The larger the sample, the smaller the variability. Thus Histogram A is for the largest of the three samples, and C is for the smallest of the three.

10.CE.22:

a) 2012 popular vote for Romney and 2012 popular vote for Obama: Correlation should be **strong and positive**: For states with large populations, both vote totals will be large. For states with small populations, both vote totals will be small.

b) Total 2012 popular vote for president and 2012 number of electoral votes: Correlation will be **positive and very strong**. A state's electoral vote total is roughly proportional to its population, so states with large populations will have large numbers of electoral votes, and vice-versa. (In more detail, the number of electoral votes is equal to its number of Senators plus number of Representatives. Each state has 2 Senators and its number of Representatives is proportional to its population, so (Electoral vote) $\approx$ 2 + k*(Population size) where k is a positive constant.)

c) Number of Macdonald's restaurants and number of attorneys: Correlation will be **strong and positive**, because both numbers are related to population size. States with larger populations will have more attorneys and more MacDonald's restaurants. States with smaller populations will have fewer of both.

d) High school graduation rate and percentage of people over 25 with a four-year college degree: Correlation will be **strong and positive** because as a rule, graduation from high school is a requirement for applying to college. States with high rates of graduation from high school will tend to have higher percentages of college graduates.

e) Year of statehood and number of years as a state: Correlation will be **strong and negative**. In fact the correlation will equal -1.0, because there is a perfect linear relationship with slope -1: (Year of statehood) + (Years as a state) = (Current Year).

f) Year of statehood and longitude (east/west): Correlations will be **positive and moderate**. The US was settled from east to west, which means that as a rule the year of statehood is earlier (lower) for eastern states, and later (higher) for western states. Longitude also increases as you go from east to west in the US. Thus both quantitative variables tend to increase together as you go from east to west.

g) Longitude and latitude (north/south): Correlation will be **near zero**, because there is no systematic relationship between a state's north-south location and its east-west location.

h) Latitude and average annual temperature: Correlation will be **negative and moderate**. As a rule, southern states (lower latitudes) tend to have warmer (larger) average annual temperatures, whereas northern states (higher latitudes) tend to have colder (lower) annual temperatures.